Stimuli-Responsive Interfaces

Takeshi Kawai · Mineo Hashizume
Editors

Stimuli-Responsive Interfaces

Fabrication and Application

Springer

Editors
Takeshi Kawai
Tokyo University of Science
Tokyo
Japan

Mineo Hashizume
Tokyo University of Science
Tokyo
Japan

ISBN 978-981-10-9622-8 ISBN 978-981-10-2463-4 (eBook)
DOI 10.1007/978-981-10-2463-4

Softcover reprint of the hardcover 1st edition 2016

Printed on acid-free paper

This Springer imprint is published by Springer Nature
The registered company is Springer Nature Singapore Pte Ltd.
The registered company address is: 152 Beach Road, #22-06/08 Gateway East, Singapore 189721, Singapore

Preface

When there is a material, there is absolutely an interface between the material and the surrounding environment. Therefore, it is important to look at the interface for the design and fabrication of materials. It is also a lesson for us in fundamental sciences lying there. This book introduces recent progress in stimuli-responsive interfaces constructed on colloidal materials such as micelles and vesicles and on solid material surfaces.

The publication of this book is a project of the Division of Colloid and Interface Science, Research Institute for Science and Technology (RIST), Organization for Research Advancement, Tokyo University of Science (TUS), Japan. The division was established in 1981 and has been playing a leading role in research related to colloid and interface science and technology both in Japan and internationally.

This book was designed and edited in order to:

- Include not only schematic illustrations, but also visual demonstrations of many actual samples' morphological or functional changes
- Cover interdisciplinary fields from fundamental sciences to practical applications
- Be useful for graduate students and young researchers in academia and industry

The book is not an introduction of the division; contributions are from various universities' or institutes' scientists who are actively involved in the field of stimuli-responsive interfaces. About one-third of the chapters are written by division members. We hope this volume serves as an excellent guide to designing and fabricating novel, functional, eco-friendly, stimuli-responsive interfaces and related materials.

Finally, we acknowledge Dr. Shin'ichi Koizumi and Ms. Taeko Sato of Springer for their help in the publication of the book. We also thank RIST, TUS for the financial support needed for the editing processes.

Tokyo, Japan

Takeshi Kawai
Mineo Hashizume

Contents

Chapter 1
Introduction

Takeshi Kawai and Mineo Hashizume

In our daily life, the stability of the material property is one important point needed for actual materials. For example, structural materials used for buildings require high mechanical stability, in other words, durability, against changes of weather conditions including natural disasters. The stabilities against sunlight and water are required for most of the actual materials. The body frame materials used for vehicles require high mechanical strength against external mechanical shocks. As for those used for airplanes or spaceships, high thermal stability is additionally required. Many of the articles for daily use are also required morphological and mechanical stabilities to use them for a longer time.

On the other hand, there are also many kinds of materials that exhibit specific functions respond to certain kinds of outer stimuli, for example, disposable diapers that adsorb water by gelation, air bags that expand against mechanical shock, and thermostats control the temperature of the devises by responding to the temperature changes. Thermoplastics, photoresists, and various kinds of sensors can also be classified into the materials that respond to the outer stimuli. Recently, self-healing materials that mend mechanical scratches are used in some commercial products. These materials can be called as "stimuli-responsive materials."

Recently, stimuli-responsive materials (or systems) have also been received much attention in the field of chemistry-based researches [1, 2]. Here, the term "stimuli-responsive material (or system)" is used a little bit more specifically compared to the case used for general industrial materials. In stimuli-responsive materials (systems) treated from the viewpoint of materials chemistry, researchers surely pay attention to the unit components, that is, molecules, molecular assemblies, and nanomaterials, whereas the development of actual materials having stimuli-responsibility sometimes tends to be focused on only the input (stimuli) and

T. Kawai (✉) · M. Hashizume
Department of Industrial Chemistry, Faculty of Engineering,
Tokyo University of Science, Tokyo, Japan
e-mail: kawai@ci.kagu.tus.ac.jp

T. Kawai and M. Hashizume (eds.), *Stimuli-Responsive Interfaces*,
DOI 10.1007/978-981-10-2463-4_1

the output (the reaction of the system) but not on the behavior of the unit components of the system.

This book mainly focuses on the design of stimuli-responsive systems based on molecular or nanolevel of novel concepts and their realization. Therefore, most of actual materials as described above are not the main targets of this book. Most of the systems described in this book mainly focus on much fundamentals: evaluation of property of the systems, clarification of the mechanism, explorer the feasibility, evolvability, and applicability limit of the concepts, and so on. Of course, it is definitely true that the success of these systems leads to the development of actual stimuli-responsive materials having high quality.

The development of stimuli-responsive systems is the challenges from the viewpoint of conversion of the changes, in other words, actions, at molecular or nanolevel of the unit components by the stimuli, to the huge output (changes) of the whole systems, in addition to the fabrication of the unit components to realize efficient conversion of the stimuli to the changes (actions).

When constructing stimuli-responsive systems, generally we need to consider several factors. They are listed below with the examples relating to each factor.

(a) Substance (element): the materials for constructing the system
Organic, inorganic, metallic, and hybrid; molecule, ion, molecular assembly, and nanomaterial; dimensions (1D, 2D, 3D); and sizes;
(b) Stimulus: the matter that affect to the target molecule or nanomaterial
Light, heat, pH, magnetism, electricity, pressure, mechanical stress, molecule, and ion;
(c) Action: what the stimuli cause to the target (at molecular or nanolevel)
Change of structure, conformation, and electronic state;
(d) Output: reaction of the system as the result of the action
Change of morphology, viscosity, wettability, electronic state, and mechanical strength; release of material; emit; and catalyze reactions;
(e) Reversibility
Efficiency, repeatability, and hysteresis; and
(f) Kinetics
Response speed and efficiency.

Factors from (a) to (d) are related to the object of the system, and (e) and (f) are temporal factors. The stimulus (b) is not limited to physical ones such as light and heat: Chemical stimuli such as molecules and ions are also useful. Here, specific molecular or ion recognitions are involved. Such systems are sometimes inspired by the molecular recognitions in biological systems. It should be noted that the action (c) that the stimuli cause to the target (molecules or nanomaterials), in other words, unit components, and the output (d) are basically different. The conversion from (c) to (d) is one of the key points to construct the system. In addition to stimulus (b) and action (c), the examples of the output (d) also contain the changes of both physical and chemical properties. The present classification might be not perfect but contains sufficient factors to design, fabrication, and evaluation of the stimuli-responsive systems.

One particular point of this book is that it looks at various stimuli-responsive systems accompanying with the viewpoint of "interface." For many systems described in this book, the stimuli caused specific action of the target (unit components: molecules, molecular assemblies, and nanomaterials) to change the properties of the interfaces between the object and the medium, among unit components, and so on, which results in the change of the physical properties or functions of the whole systems. We learn the importance of the design and construction of the interfaces in the stimuli-responsive systems by reviewing recent progresses in these research fields. The chapters of this book are written by the specialists engaging in various kinds of stimuli-responsive systems. We hope the readers realize the scientific significance, feasibility, and potential application of the "stimuli-responsive interfaces" created using molecular and nanolevel designs.

References

1. Theato P, Sumerlin BS, O'Reilly RK, Epps III TH (eds) (2013) Stimuli responsive themed issue. Chem Soc Rev 42(17):7045–7486
2. Stuart MAC, Huck WTS, Genzer J, Muller M, Ober C, Stamm M, Sukhorukov GB, Szleifer I, Tsukruk VV, Urban M, Winnik F, Zauscher S, Luzinov I, Minko S (2010) Emerging applications of stimuli-responsive polymer materials. Nat Mater 9:101–113

References

Chapter 2
Photo-Induced Demulsification

Yutaka Takahashi and Yukishige Kondo

Abstract Demulsification is an aggressive phase-separation phenomenon of stable emulsions. Several studies have examined demulsification by chemical reactions using external stimuli, such as temperature variation, pH variation, and light irradiation. This study focused on the interfacial and emulsification properties of light-responsive surfactants because the use of light as an external stimulus is relatively simple. This chapter presents an overview of the photo-induced demulsification of emulsions prepared using light-responsive surfactants containing an azobenzene skeleton. Stable emulsions are obtained when mixtures of *n*-octane and aqueous solutions of azobenzene-containing surfactants are homogenized. Ultraviolet irradiation of the stable emulsions induces complete phase separation of the octane and aqueous surfactant solution phases. The phase separation, i.e., demulsification, results from a change in the interfacial properties, such as occupied areas per molecule and interfacial tension, of the light-responsive surfactants at the oil/water interface.

Keywords Photoresponsive surfactant · Azobenzene · Demulsification

2.1 Introduction

Photoresponsive surfactants exhibit interfacial properties that can be controlled by light irradiation at specific wavelengths [1–3]. Azobenzene, diarylethene, and spiropyran are well-known photoresponsive compounds that are converted reversibly between two isomers by a particular wavelength of light [4–7]. Several studies have examined ways of controlling the interfacial properties of photoresponsive surfactants containing an azobenzene [8–12]. The *trans* and *cis* isomers of azobenzene derivatives can be switched reversibly using a particular wavelength of

Y. Takahashi · Y. Kondo (✉)
Department of Industrial Chemistry, Tokyo University of Science, Tokyo, Japan
e-mail: ykondo@rs.tus.ac.jp

T. Kawai and M. Hashizume (eds.), *Stimuli-Responsive Interfaces*,
DOI 10.1007/978-981-10-2463-4_2

light; the *trans* isomer is converted to the *cis* by ultraviolet (UV) light, and visible light irradiation converts the *cis* isomer to the *trans* isomer [7, 13].

Emulsions are metastable systems consisting of two immiscible liquids, such as water and oil. Surfactants have generally been used to stabilize emulsions. Therefore, the interfacial properties of the surfactants at the oil/water interface are important for the stability of emulsions. On the other hand, demulsification involves the breaking of stable emulsions into separate oil and water phases. The phenomenon is used as "emulsion liquid membrane extraction" [14], which is a technique to extract harmful substances from aqueous waste fluids. Thus far, demulsification has been carried out by adding demulsifiers to stable emulsions and applying some form of physical energy [15–18]. Recent studies have reported that the properties of emulsions containing stimuli-responsive materials can be controlled by external stimuli, such as pH variation and light irradiation. The stability of emulsions prepared using pH-responsive polymers changed drastically on varying the pH using acids or bases [19–26]. In some cases, however, the addition of acids or bases to emulsions is undesirable because the resulting emulsions are contaminated. In contrast, Eastoe et al. showed that UV irradiation induced the destabilization of the microemulsions containing a photodestructible surfactant and sodium bis(2-ethylhexyl)sulfosuccinate (AOT) and reported their studies of photosensitive microemulsion [27–29]. They also demonstrated the photoinduced phase separation of single-phase microemulsions consisting of decane and aqueous solution of a photoresponsive surfactant and AOT mixtures by UV light irradiation [30]. In addition, Perrin et al. reported control of the emulsion types by UV and visible light irradiation [31]. Emulsions consisting of *n*-dodecane, water, sodium nitrate, nonionic surfactant, and amphiphilic photoresponsive polymers containing azobenzene groups were converted from the oil-in-water (O/W) type to the water-in-oil (W/O) type by UV light irradiation [32, 33].

This chapter provides an overview of two systems of photo-induced demulsification that were reported recently. One is the photo-induced demulsification of stable O/W-type emulsions prepared using a photoresponsive gemini surfactant with an azobenzene as a spacer [34], while the other system involves the photo-induced demulsification of stable emulsions prepared using mixtures of anionic and cationic surfactants [35].

2.2 Demulsification of Photoresponsive Emulsions Using a Photoresponsive Gemini Surfactant

A few studies of gemini surfactants containing photoresponsive groups as a spacer have been reported. Shimomura et al. succeeded in inducing an orientational change in a gemini surfactant with a diarylethene group at the air/water interface by UV irradiation [36]. Oda et al. reported the formation and disruption of vesicles formed from photoresponsive gemini surfactants containing an azobenzene group by irradiation with UV and visible light [37, 38]. This chapter focuses on the photo-induced changes in the interfacial and emulsion properties of a photoresponsive gemini surfactant (C_7-azo-C_7; Fig. 2.1) at the oil/water interface [34].

Stable emulsions were obtained from mixtures of *n*-octane and aqueous *trans* C_7-azo-C_7 solutions at specific weight fraction and surfactant concentration. Emulsions consisting mixtures of *n*-octane and 10 mM of aqueous *trans* C_7-azo-C_7 solution (weight fraction of aqueous *trans* C_7-azo-C_7 solution (f_w): 0.3) exhibited no phase separation over 1 week after preparation (Fig. 2.2b). When the fluorescent probes were added to the stable emulsions, fluorescence microscopy showed that the emulsions were the O/W type. As shown in Fig. 2.2c, the UV irradiation of stable O/W-type emulsions consisting of octane/10 mM of aqueous *trans* C_7-azo-C_7 solution (f_w: 0.3) for 6 h led to complete phase separation, i.e., demulsification. The *trans* C_7-azo-C_7 molecules in the emulsions were converted to the *cis* isomer by UV light irradiation. No phase separation was observed when stable O/W-type emulsions were stored for 6 h in the absence of UV irradiation in a dark room. Optical microscopy under UV light irradiation revealed the coalescence of octane droplets in the stable O/W-type emulsion. In contrast, 10 mM of aqueous *cis* C_7-azo-C_7 solutions were prepared by UV irradiation of the aqueous solutions of the *trans* isomer before homogenization. The emulsions showed no phase separation by the homogenization of mixtures consisting of *n*-octane and the aqueous *cis* C_7-azo-C_7 solution (Fig. 2.2d). No change in the stability of the emulsion was observed after irradiation with visible light. These results suggest that photo-induced demulsification occurs via a photoisomerization process from the *trans* to *cis* isomers in emulsions.

The photo-induced demulsification of stable O/W-type emulsions consisting mixtures of *n*-octane and 10 mM of aqueous *trans* C_7-azo-C_7 solution is discussed

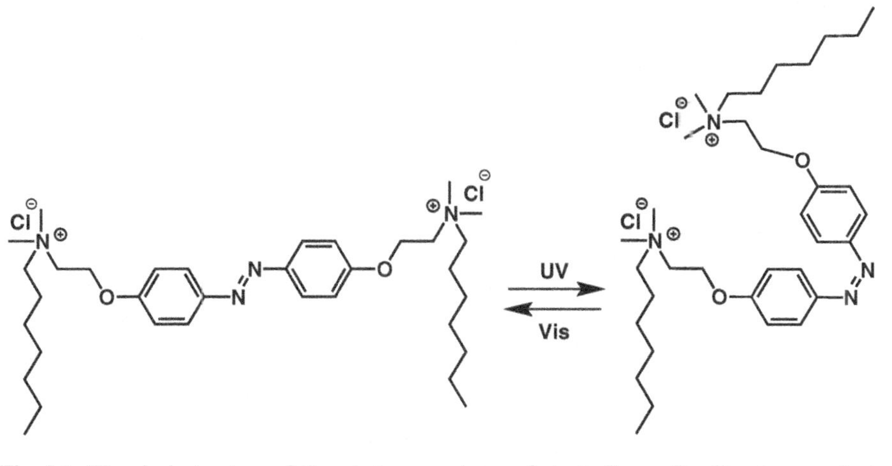

Fig. 2.1 Chemical structure of the photoresponsive surfactant, C_7-azo-C_7 The *trans* and *cis* isomers are shown. Reprinted with the permission from Ref. [34]. Copyright 2013 American Chemical Society

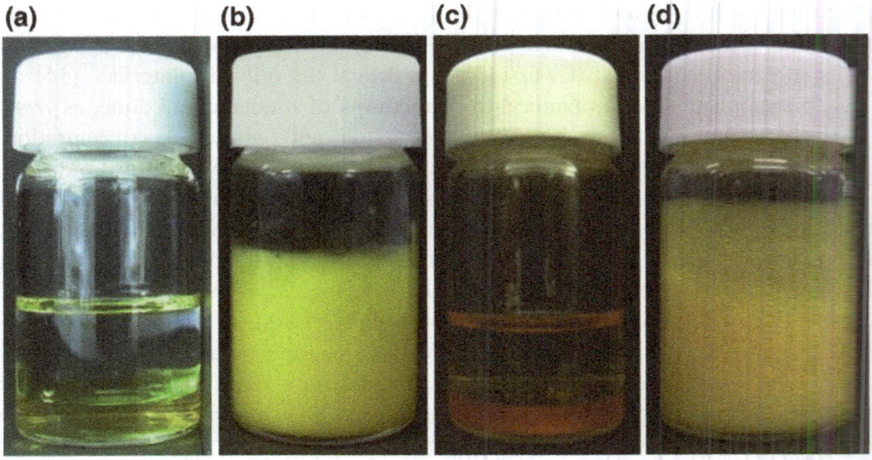

Fig. 2.2 Photographs of **a** a mixture of *n*-octane and 10 mM of aqueous *trans* C_7-azo-C_7 solution (f_w: 0.3) before homogenization and **b** a stable emulsion obtained by homogenization. **c** Photo-induced demulsification of the emulsion shown in (**b**) by UV light irradiation on macroscopic scale. **d** An emulsion consisting of *n*-octane and the aqueous *cis* isomer of C_7-azo-C_7 solution (no phase separation). Reprinted with the permission from Ref. [34] Copyright 2013 American Chemical Society

from the results of the interfacial tensions measurements. An *n*-octane/aqueous C_7-azo-C_7 solution of *trans* or *cis* isomers showed similar interfacial tension at a surfactant concentration of 10 mM. The occupied areas per molecule for *trans* and *cis* C_7-azo-C_7 at the octane/water interface were estimated to be 3.2 and 1.6 nm^2, respectively, from the Gibbs's adsorption isotherm and the slopes of the interfacial tension versus the surfactant concentration. Therefore, photoisomerization from the *trans* to *cis* isomer of C_7-azo-C_7 molecule at the octane/water interface induced a decrease in the occupied areas per molecule at the interface. In addition, dynamic interfacial tension between octane and 10 mM of aqueous C_7-azo-C_7 solutions with UV light irradiation was measured. When the octane/water interface-adsorbed *trans* C_7-azo-C_7 molecules were exposed to UV light, the interfacial tension increased abruptly and then decreased gradually with increasing UV irradiation time. The value reached an equilibrated interfacial tension of *cis* C_7-azo-C_7. The abrupt increase in the dynamic interfacial tension with UV irradiation indicated an increase in the Gibbs free energy at the interface. This resulted from a reduction of the occupied areas per C_7-azo-C_7 molecule at the octane/water interface with *trans*–*cis* photoisomerization, followed by the temporal and partial exposure (direct contact) of octane and water phases at the interface, contributing to an increase in the interfacial tension. From these results of the equilibrated and dynamic interfacial tension measurements, the authors suggested the following mechanism for the photo-induced demulsification of stable O/W-type emulsions prepared using C_7-azo-C_7. As shown in Fig. 2.3a, the C_7-azo-C_7 molecules of the *trans* isomer were

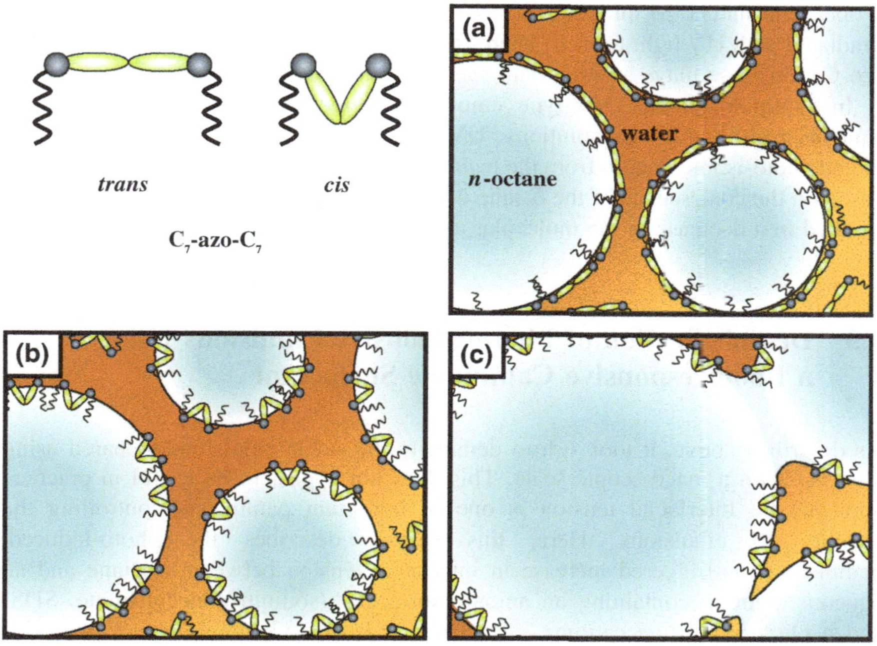

Fig. 2.3 Possible mechanism of the photo-induced demulsification of stable O/W-type emulsions prepared using C_7-azo-C_7. **a** *trans* C_7-azo-C_7 molecules were adsorbed on the entire interface of *n*-octane and aqueous surfactant solution in the stable emulsions. **b** UV irradiation led to partial direct contact between the water and *n*-octane phases. **c** Exposure of the *n*-octane/water interface induced the coalescence of *n*-octane droplets in the emulsions. Finally, the *n*-octane and water phases were fully separated on a macroscopic scale. Reprinted with the permission from Ref. [34]. Copyright 2013 American Chemical Society

adsorbed on the entire interface between octane and the aqueous surfactant solution in the stable emulsions. UV irradiation induced a decrease in molecular area with photoisomerization from the *trans* to *cis* isomer and a temporal increase in interfacial tension between octane and the aqueous surfactant solution. This suggests that octane and water phases were contacted directly by the reduction of the azobenzene group in the C_7-azo-C_7 molecule at the octane/water interface (Fig. 2.3b). Exposure of octane/water interface led to the coalescence of octane droplets in the O/W-type emulsions to reduce the interfacial Gibbs free energy, followed by complete phase separation, i.e., demulsification (Fig. 2.3c).

Additional experiments were performed to confirm the mechanism of photo-induced demulsification described above. No phase separation was observed when a stable emulsion consisting of octane and a 10 mM aqueous solution of a gemini surfactant containing no azobenzene (1,2-bis(dodecyldimethylammonium) ethane dibromide) was exposed to UV light for 6 h. In addition, the emulsion remained the same after a stable emulsion consisting of octane containing a 10 mM azobenzene and a 20 mM aqueous sodium dodecyl sulfate (SDS) solution was irradiated with UV light for 6 h. Therefore, the azobenzene group of a spacer in C_7-azo-C_7 molecule plays an important role in photo-induced demulsification.

In summary, stable O/W-type emulsions were obtained from *n*-octane and aqueous *trans* C_7-azo-C_7 solutions. UV irradiation of the stable emulsions promoted photo-isomerization from the *trans* to *cis* isomer and induced demulsification based on the coalescence of the octane droplets. The photo-induced demulsification resulted in a decrease in the molecular area of C_7-azo-C_7 at the oil/water interface.

2.3 Demulsification of Photoresponsive Emulsions Using a Photoresponsive Catanionic Surfactant

As described above, it took 6 h to demulsify the stable emulsions prepared using C_7-azo-C_7 on a macroscopic scale. This does not appear to be useful in practical applications. Interfacial tension is one of important parameters controlling the stability of emulsions. Here, this section describes the photo-induced, demulsification-triggered increase in interfacial tension between *n*-octane and an aqueous solution containing an anionic surfactant (sodium dodecylsulfate, SDS) and a photoresponsive cationic surfactant (2-(4-(4-butylphenyl)diazenylphenoxy) ethyltrimethylammonium bromide, C4AzoTAB; Fig. 2.4) with UV irradiation [35]. In addition, the photo-induced demulsification rates were also investigated.

The emulsions were prepared by homogenizing mixtures of *n*-octane and an aqueous SDS/*trans*-C4AzoTAB solution (weight fraction of aqueous mixed-surfactant solution (f_w): 0.2) at 25 °C. Stable O/W-type emulsions that exhibited no phase separation over 1 week after the preparation were obtained in the regions of the specific total surfactant concentrations (>5 mM) and molar ratios (X_{SDS}) of SDS in an aqueous SDS/C4AzoTAB solution ($0 \leq X_{SDS} \leq 0.31$, $0.7 \leq X_{SDS} \leq 1$). When stable O/W-type emulsions consisting of octane and 10 mM of aqueous SDS/C4AzoTAB solution with $X_{SDS} = 0.1$ (f_w: 0.2) were irradiated with UV light, these octane and aqueous surfactant solution phases were separated gradually from the emulsions. As shown in Fig. 2.5c, the stable emulsions were demulsified by UV irradiation for 90 min, which was four times more rapid than the photo-induced demulsification of the stable emulsions prepared using C_7-azo-C_7 (6 h). UV irradiation caused the *cis* isomerization of *trans*-C4AzoTAB in the emulsions because the color of the aqueous SDS/C4AzoTAB solutions after UV irradiation for 90 min changed from yellow to reddish, corresponding to the *cis* isomer (Fig. 2.5). In addition, emulsions were obtained when mixtures of octane and a 10 mM aqueous SDS/*cis*-C4AzoTAB solution $X_{SDS} = 0.1$ and $f_w = 0.2$ in the photostationary state of the *cis* isomer were homogenized (Fig. 2.5d). Therefore, the UV irradiation of stable O/W-type emulsions induced the *cis* photo-isomerization of *trans*-C4AzoTAB molecules in the emulsions leading to photo-induced demulsification. No phase separation was observed when stable

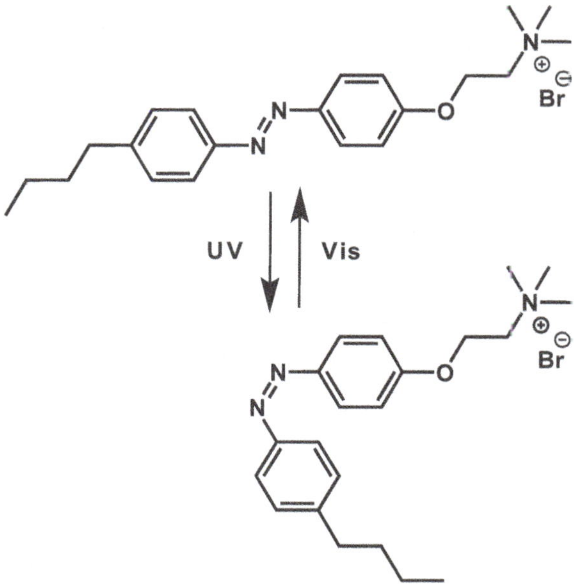

Fig. 2.4 Chemical structure of the *trans* and *cis* isomers of a photoresponsive surfactant, 2-(4-(4-butylphenyl)diazenylphenoxy)ethyltrimethylammonium bromide (C4AzoTAB). Reprinted with the permission from Ref. [35]. Copyright 2016 American Chemical Society

Fig. 2.5 Photographs of a mixture consisting of *n*-octane and a 10 mM aqueous SDS/*trans*-C4AzoTAB solution with X_{SDS} = 0.1 (f_w: 0.2) at 25 °C **a** before homogenization and **b** a stable emulsion obtained by homogenization. **c** Photo-induced demulsification of the emulsion shown in (**b**) by UV irradiation on the macroscopic scale. **d** The mixture of octane and aqueous SDS/*cis*-C4AzoTAB solution in the photostationary state of the *cis* isomer was homogenized. No emulsion was obtained. Reprinted with the permission from Ref. [35]. Copyright 2016 American Chemical Society

emulsions obtained from octane and aqueous C4AzoTAB solution without SDS were irradiated with UV light. SDS in aqueous SDS/C4AzoTAB solutions plays an important role in photo-induced demulsification.

The authors examined possible mechanisms for the photo-induced demulsification of stable O/W-type emulsions prepared using SDS/C4AzoTAB mixtures. The number distributions of the diameter of octane droplets in the O/W-type emulsions consisting of octane and a 10 mM aqueous SDS/C4AzoTAB solution with $X_{SDS} = 0.1$ (f_w: 0.2) exposed to UV light irradiation were estimated by observing the oil droplets through optical microscopy. The number-averaged diameter before UV light irradiation was 14.6 ± 9.2 μm. A change in the number distributions of the diameter of the octane droplets in the O/W-type emulsions with UV irradiation and optical microscopy observation under UV irradiation indicated that UV irradiation led to the coalescence of small octane droplets to larger ones in the emulsions, i.e., demulsification. Therefore, the photo-induced demulsification of stable emulsions prepared using SDS/C4AzoTAB mixtures resulted from Ostwald ripening. Equilibrated interfacial tension measurements indicated that the interfacial tension between *n*-octane and aqueous SDS/C4AzoTAB solutions with $X_{SDS} = 0.1$ for both *trans* and *cis* isomers decreased with increasing surfactant concentration and reached constant values at specific concentrations (Fig. 2.6). However, the interfacial tension of the *cis* isomer was higher than that of the *trans* isomer over the range, 0.1–15 mM. In other words, UV irradiation induced increases in the interfacial tensions at specific surfactant concentrations. The authors also suggested the following mechanism for the photo-induced demulsification of stable O/W-type emulsions prepared from SDS/C4AzoTAB mixtures. SDS and *trans*-C4AzoTAB molecules were absorbed onto the entire octane/aqueous surfactant solution

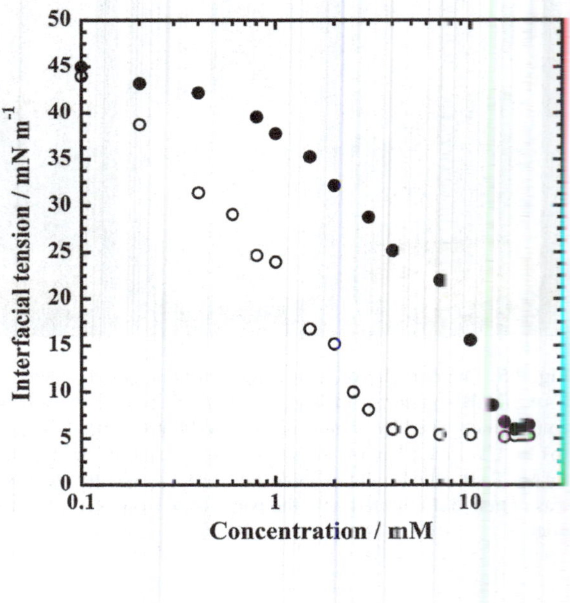

Fig. 2.6 Relationship between the interfacial tensions between *n*-octane and aqueous SDS/C4AzoTAB solution and the mixed-surfactant concentrations. *Open circles trans* isomer. *Filled circles cis* isomer. Reprinted with the permission from Ref. [35]. Copyright 2016 American Chemical Society

interfaces in the stable emulsions. UV irradiation led to *trans–cis* photoisomerization and an increase in interfacial tension at the octane/aqueous surfactant solution interface. The increase in interfacial tension with *trans–cis* photoisomerization caused the coalescence of smaller octane droplets to larger ones (Ostwald ripening) in O/W-type emulsions to reduce the interfacial area of the emulsions. Moreover, Ostwald ripening progresses with *trans–cis* photoisomerization until the octane and water phases were separated completely.

The photo-induced demulsification rate of stable emulsions prepared from SDS/C4AzoTAB mixtures was investigated. Ostwald ripening has been described theoretically by Lifshitz, Slezov, and Wagner (LSW theory) [39, 40]. According to LSW theory, the rate of demulsification based on Ostwald ripening accelerates with increasing interfacial tension between the oil and water phases. Stable O/W-type emulsions consisting of *n*-octane and aqueous SDS/*trans*-C4AzoTAB solutions with $X_{SDS} = 0.1$ (f_w: 0.2) were obtained with a total surfactant concentration greater than 6 mM. The octane and aqueous surfactant solution phases were fully separated from the stable emulsions at total surfactant concentrations of 7 and 10 mM, whereas the emulsions at a total surfactant concentration of 20 mM were not demulsified completely by UV irradiation for 6 h. UV irradiation caused complete phase separation within 60 min with a total surfactant concentration of 7 mM and in 90 min with a total surfactant concentration of 10 mM. The time required for demulsification at 7 mM was shorter than that at 10 mM. The differences in interfacial tension between the *trans* and *cis* isomers at 7 and 10 mM were 16.6 and 10.2 mN m^{-1}, respectively, whereas there was almost no difference in interfacial tension between the *trans* and *cis* isomers at 20 mM. Therefore, a change in interfacial tension with UV light irradiation was correlated with the photoinduced demulsification rate. On the macroscale, it takes 90 min for stable emulsions consisting of octane and 10 mM of aqueous SDS/*trans*-C4AzoTAB solution with $X_{SDS} = 0.1$ (f_w: 0.2) to separate completely under UV light irradiation. When the emulsions are placed in vials, UV light illuminates only the surface of the vial, barely penetrating the emulsions because the emulsions are turbid, as shown in Fig. 2.5b. Therefore, UV irradiation will accelerate demulsification if the path length of the emulsions can be reduced. Here, a microreactor was used to reduce the path length of the emulsions. The stable emulsions were pumped through the microreactor using a syringe pump under UV irradiation (Fig. 2.7). When the emulsions flowed through the microreactor at 70 mL h^{-1} with UV irradiation, the octane and aqueous surfactant solution phases were separated completely within 3.5 min. The time required for photoinduced demulsification in the microreactor was 26 times shorter than that in a vial. In addition, no phase separation was observed when the microreactor was used in the absence of UV irradiation. Thus, the demulsification in the microreactor was triggered only by UV light.

The homogenization and photoinduced demulsification of mixtures consisting of *n*-octane and 10 mM of an aqueous SDS/C4AzoTAB solution with $X_{SDS} = 0.1$ (f_w: 0.2) were repeated. The emulsions prepared, using a mixture of SDS/*trans*-C4AzoTAB,

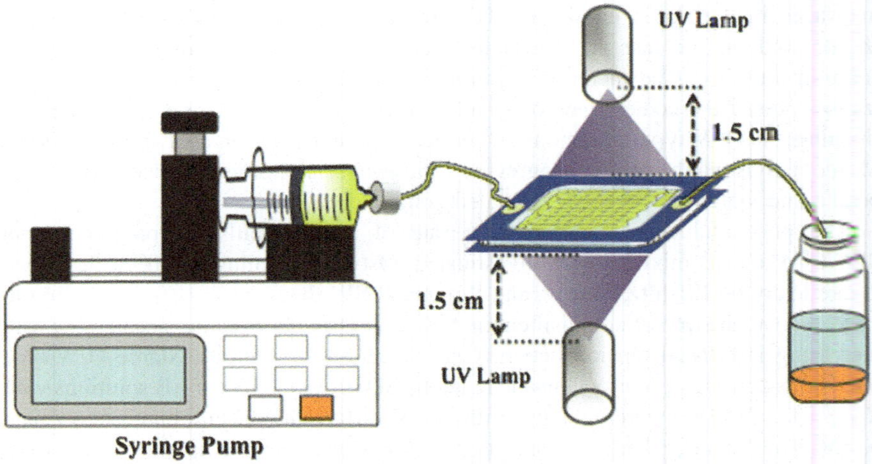

Fig. 2.7 Geometry of the demulsification of stable emulsions in a microreactor. An emulsion consisting of *n*-octane and an aqueous SDS/*trans*-C4AzoTAB solution was pumped through the microreactor using a syringe pump with UV light irradiation. Reprinted with the permission from Ref. [35]. Copyright 2016 American Chemical Society

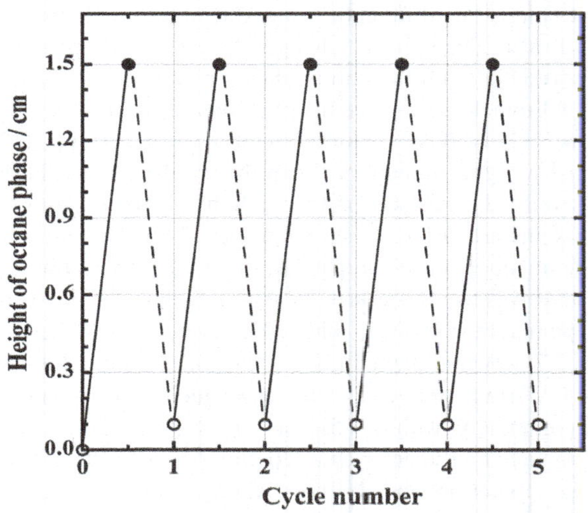

Fig. 2.8 Change in the height of the *n*-octane phase separated from stable emulsions consisting of octane and a 10 mM aqueous SDS/*trans*-C4AzoTAB solution with X_{SDS} = 0.1 (f_c: 0.2) with repetitive *trans–cis* photoisomerization and homogenization. The height of 1.5 cm corresponds to complete phase separation. *Solid lines* photoisomerization with UV light irradiation; *dashed lines* visible light irradiation and homogenization. *Open circles* SDS/*trans*-C4AzoTAB; *filled circles* SDS/*cis*-C4AzoTAB. Reprinted with the permission from Ref. [35]. Copyright 2016 American Chemical Society

were irradiated with UV light. Subsequently, visible light irradiation of the aqueous surfactant solution phase, which had been separated from the emulsions, induced photoisomerization from the *cis* to *trans* isomer. The separated octane and aqueous surfactant solution phases were then homogenized to prepare the emulsions. Thus, the reversible process of homogenization and photoinduced demulsification was observed (Fig. 2.8). Repetitive *trans–cis* photoisomerization led to complete phase separation. Therefore, the SDS/C4AzoTAB mixtures can be applied repetitively to the photoinduced demulsification.

In conclusion, stable O/W-type emulsions could be prepared from *n*-octane and aqueous SDS/*trans*-C4AzoTAB solutions, whereas no emulsions were obtained using SDS/*cis*-C4AzoTAB mixtures. UV irradiation led to a change in the emulsification abilities of the SDS/C4AzoTAB mixtures and an increase in interfacial tension between octane and aqueous mixed-surfactant solution. The stable emulsions were demulsified fully by UV irradiation for 90 min, which was shorter than the photo-induced demulsification of stable emulsions prepared using C_7-azo-C_7 (6 h). The use of a microreactor reduced the time required for the photoinduced demulsification to 3.5 min. This photo-induced demulsification was caused by the Ostwald ripening of octane droplets. Repetitive *trans–cis* photoisomerization of the aqueous SDS/C4AzoTAB solutions in the emulsions resulted in photo-induced demulsification. Overall, the photo-induced demulsification demonstrated in this chapter will be useful for industrial applications.

References

1. Liu X, Abbott NL (2009) Spatial and temporal control of surfactant systems. J Colloid Interface Sci 339(1):1–18
2. Eastoe J, Vesperinas A (2005) Self-assembly of light-sensitive surfactants. Soft Matter 1 (5):338–347
3. Brown P, Butts CP, Eastoe J (2013) Stimuli-responsive surfactants. Soft Matter 9(8):2365–2374
4. Irie M (2000) Diarylethenes for memories and switches. Chem Rev 100(5):1685–1716
5. Tian H, Yang S (2004) Recent progresses on diarylethene based photochromic switches. Chem Soc Rev 33(2):85–97
6. Berkovic G, Krongauz V, Weiss V (2000) Spiropyrans and Spirooxazines for memories and switches. Chem Rev 100(5):1741–1754
7. Tamai N, Miyasaka H (2000) Ultrafast dynamics of photochromic systems. Chem Rev 100 (5):1875–1890
8. Cicciarelli BA, Hatton TA, Smith KA (2007) Dynamic surface tension behavior in a photoresponsive surfactant system. Langmuir 23(9):4753–4764
9. Shin JY, Abbott NL (1999) Using light to control dynamic surface tensions of aqueous solutions of water soluble surfactants. Langmuir 15(13):4404–4410
10. Chevallier E, Mamane A, Stone HA, Tribet C, Lequeux F, Monteux C (2011) Pumping-out photo-surfactants from an air-water interface using light. Soft Matter 7(17):7866–7874
11. Chevallier E, Monteux C, Lequeux F, Tribet C (2012) Photofoams: remote control of foam destabilization by exposure to light using an azobenzene surfactant. Langmuir 28(5):2308–2312

12. Shang T, Smith KA, Hatton TA (2003) Photoresponsive surfactants exhibiting unusually large, reversible surface tension changes under varying illumination conditions. Langmuir 19 (26):10764–10773
13. Kumar GS, Neckers DC (1989) Photochemistry of azobenzene-containing polymers. Chem Rev 89(8):1915–1925
14. Sastre AM, Kumar A, Shukla JP, Singh RK (1998) Improved techniques in liquid membrane separations: an overview. Sep Purif Rev 27(2):213–298
15. Allende D, Cambiella Á, Benito JM, Pazos C, Coca J (2008) Destabilization-enhanced centrifugation of metalworking oil-in-water emulsions: effect of demulsifying agents. Chem Eng Technol 31(7):1007–1014
16. Ferreira BMS, Ramalho JBVS, Lucas EF (2013) Demulsification of water-in-crude oil emulsions by microwave radiation: effect of aging, demulsifier addition, and selective heating. Energy Fuels 27(2):615–621
17. Kundu P, Agrawal A, Mateen H, Mishra IM (2013) Stability of oil-in-water macro-emulsion with anionic surfactant: effect of electrolytes and temperature. Chem Eng Sci 102:176–185
18. Silva EB, Santos D, Alves DRM, Barbosa MS, Guimarães RCL, Ferreira BMS, Guarnieri RA, Franceschi E, Dariva C, Santos AF, Fortuny M (2013) Demulsification of heavy crude oil emulsions using ionic liquids. Energy Fuels 27(10):6311–6315
19. Liu H, Wang C, Zou S, Wei Z, Tong Z (2012) Simple, reversible emulsion system switched by pH on the basis of chitosan without any hydrophobic modification. Langmuir 28 (30):11017–11024
20. Morse AJ, Dupin D, Thompson KL, Armes SP, Ouzineb K, Mills P, Swart R (2012) Novel pickering emulsifiers based on pH-responsive poly(tert-butylaminoethyl methacrylate) latexes. Langmuir 28(32):11733–11744
21. Binks BP, Murakami R, Armes SP, Fujii S (2006) Effects of pH and salt concentration on oil-in-water emulsions stabilized solely by nanocomposite microgel particles. Langmuir 22 (5):2050–2057
22. Read ES, Fujii S, Amalvy JI, Randall DP, Armes SP (2005) Effect of varying the oil phase on the behavior of pH-responsive latex-based emulsifiers: demulsification versus transitional phase inversion. Langmuir 21(4):1662–1662
23. Fujii S, Cai Y, Weaver JVM, Armes SP (2005) Syntheses of shell cross-linked micelles using acidic ABC triblock copolymers and their application as pH-responsive particulate emulsifiers. J Am Chem Soc 127(20):7304–7305
24. Yi C, Liu N, Zheng J, Jiang J, Liu X (2012) Dual-responsive poly(styrene-alt-maleic acid)-graft-poly(N-isopropyl acrylamide) micelles as switchable emulsifiers. J Colloid Interface Sci 380(1):90–98
25. Chen Q, Cao X, Liu H, Zhou W, Qin L, An Z (2013) pH-responsive high internal phase emulsions stabilized by core cross-linked star (CCS) polymers. Polym Chem 4(15):4092–4102
26. Morse AJ, Armes SP, Thompson KL, Dupin D, Fielding LA, Mills P, Swart R (2013) Novel Pickering emulsifiers based on pH-responsive poly(2-(diethylamino)ethyl methacrylate) latexes. Langmuir 29(18):5466–5475
27. Eastoe J, Sanchez-Dominguez M, Cumber H, Burnett G, Wyatt P, Heenan RK (2003) Photoresponsive microemulsions. Langmuir 19(17):6579–6581
28. Eastoe J, Sanchez Dominguez M, Cumber H, Wyatt P, Heenan RK (2004) Light-sensitive microemulsions. Langmuir 20(4):1120–1125
29. Eastoe J, Wyatt P, Sanchez-Dominguez M, Vesperinas A, Paul A, Heenan RK, Grillo I (2005) Photo-stabilised microemulsions. Chem Commun 22:2785–2786
30. Tabor RF, Oakley RJ, Eastoe J, Faul CFJ, Grillo I, Heenan RK (2009) Reversible light-induced critical separation. Soft Matter 5(1):78–80
31. Porcar I, Perrin P, Tribet C (2001) UV-visible light: a novel route to tune the type of an emulsion. Langmuir 17(22):6905–6909
32. Khoukh S, Perrin P, Bes de Berc F, Tribet C (2005) Reversible light-triggered control of emulsion type and stability. ChemPhysChem 6(10):2009–2012

33. Khoukh S, Tribet C, Perrin P (2006) Screening physicochemical parameters to tuning the reversible light-triggered control of emulsion type. Colloids Surf A 288(1–3):121–130
34. Takahashi Y, Fukuyasu K, Horiuchi T, Kondo Y, Stroeve P (2014) Photoinduced demulsification of emulsions using a photoresponsive gemini surfactant. Langmuir 30 (1):41–47
35. Takahashi Y, Koizumi N, Kondo Y (2016) Active demulsification of photoresponsive emulsions using cationic-anionic surfactant mixtures. Langmuir. doi:10.1021/acs.langmuir.5b03912
36. Karthaus O, Shimomura M, Hioki M, Tahara R, Nakamura H (1996) Reversible photomorphism in surface monolayers. J Am Chem Soc 118(38):9174–9175
37. Faure D, Gravier J, Labrot T, Desbat B, Oda R, Bassani DM (2005) Photoinduced morphism of gemini surfactant aggregates. Chem Commun 9:1167–1169
38. Ahmad RK, Faure D, Goddard P, Oda R, Bassani DM (2009) Photosensitive vesicles from a cis-azobenzene gemini surfactant show high photoresponse. Org Biomol Chem 7(15):3173–3178
39. Lifshitz IM, Slyozov VV (1961) The kinetics of precipitation from supersaturated solid solutions. J Phys Chem Solids 19(1–2):35–50
40. Wagner C (1961) Theorie der Alterung von Niederschlägen durch Umlösen (Ostwald-Reifung). Elektrochem 65(7–8):581–591

Chapter 3
Stimuli-Responsible Viscoelastic Surfactant Solutions

Hideki Sakai, Koji Tsuchiya and Kenichi Sakai

Abstract In this chapter, viscoelasticity control of aqueous solutions using novel stimuli-responsive surfactants is described. The formation of viscoelastic wormlike micellar solutions by these surfactants is switchable by external stimuli such as light irradiation, redox reaction, and pH change. This chapter consists of following three sections: (1) photochemical control of viscoelasticity using (a) photoresponsive azobenzene-modified surfactants, (b) a photoresponsive counter ion, and (c) a novel photocleavable surfactant. (2) Redox-responsive wormlike micellar solution using a ferrocenyl surfactant. (3) pH-responsive wormlike micellar solution by acylglutamic acid-alkylamine complex. Formation-disintegration control of wormlike micelles, and thus, the solution viscosity change, using these stimuli-responsive surfactants, may lead to applications for controlled release of fragrances, flavors, and drugs incorporated into the micelles, and for prevention of printer ink bleeding.

Keywords Stimulus-responsive surfactant · Wormlike micelle · Viscoelasticity · Azobenzene · Cinnamic acid · Ferrocene · Acylglutamic acid · Alkylamine

3.1 Introduction

Amphiphilic substances (surfactants) that have both hydrophilic and hydrophobic groups in one molecule form molecular assemblies such as micelles, wormlike micelles, and vesicles in aqueous solutions. Formation of these molecular assemblies gives various functions to solvents (aqueous solutions and organic media). For example, formation of wormlike micelles causes drastic increase in the viscosity of

H. Sakai (✉) · K. Sakai
Department of Pure and Applied Chemistry, Faculty of Science and Technology, Tokyo University of Science, Chiba, Japan
e-mail: hisakai@rs.noda.tus.ac.jp

H. Sakai · K. Tsuchiya · K. Sakai
Research Institute for Science and Technology, Tokyo University of Science, Chiba, Japan

T. Kawai and M. Hashizume (eds.), *Stimuli-Responsive Interfaces*,
DOI 10.1007/978-981-10-2463-4_3

the solutions due to the three-dimensional entanglement of the wormlike structure. Wormlike micellar solutions with high viscoelasticity have been applied to various industrial fields including cosmetics, foods, and detergents.

Attempts have attracted attention to control the formation and disruption of these surfactant molecular assemblies and the properties of the solutions by means of external stimuli such as pH [1, 2], temperature [3, 4], redox reaction [5, 6], and light irradiation [7, 8], because of the important role of the assemblies in such applications as release rate control of perfumes and drugs held in their interior and targeting of drug delivery systems. The authors have also reported the reversible formation control of micelles [9–12] and vesicles [13–17] using redox-active and photoresponsive surfactants. By controlling the formation of the micelles and vesicles, release rate controls of oil-soluble and water-soluble drugs incorporated in the assemblies were made possible, respectively.

Aqueous solutions containing wormlike micelles formed by surfactants exhibit high viscoelasticity, due to their three-dimensional entanglement. Typical examples are mixed systems of cetyltrimethylammonium bromide (CTAB)/sodium salicylate (NaSal) [18] and mixed systems of nonionic surfactants [19]. If the formation and disintegration of the wormlike micelles, and thus the solution viscosity change, can be effectively controlled by external stimulations, it may lead to applications for controlled release of fragrances, flavors, and drugs incorporated into the micelles, and for prevention of printer ink bleeding (Fig. 3.1).

In this chapter, the authors review the studies regarding the wormlike micellar systems whose viscoelasticity can be controlled by (1) light irradiation, (2) redox reaction, and (3) pH change, using novel surfactants responsive to these external stimuli.

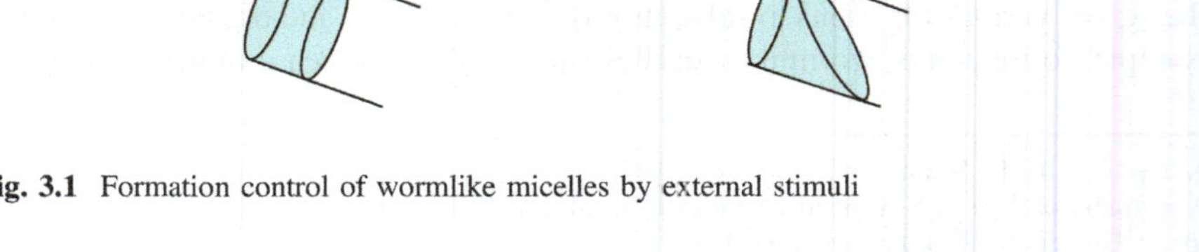

Fig. 3.1 Formation control of wormlike micelles by external stimuli

3.2 Photoresponsive Wormlike Micellar Solutions

3.2.1 *Photochemical Control of Viscoelasticity Using "Photoswitchable" Azobenzene-Modified Surfactant [20]*

Reversible control of interfacial properties and the formation and disruption of molecular assemblies such as micelles and vesicles in solution by light irradiation [7, 8, 13–17, 20–23] has attracted considerable attention since the response is relatively fast and the addition of the additional component to the system is not necessary. Our past studies also demonstrated the surfactant systems accompanying a dynamic aggregated state change induced by light irradiation, for example, active control of micellar solubilization [13, 17] and the amount of model drug entrapped in vesicles [14, 15].

On the other hand, if the formation and disruption of the wormlike micelles are able to be effectively controlled by light irradiation, it can be applied for controlled release of fragrances and drugs solubilized in the micelles, and for prevention of printer ink bleeding. In this section, "photoresponsive" viscoelastic solutions using novel "photoswitchable" surfactants are described. The present systems are fundamentally different in utilizing a dynamic change in the aggregated state of the photoresponsive (low molecular weight) surfactants from previously reported polymer systems [24] and organogels [25] that undergo a photoresponsive gel-to-sol transition.

We used an azobenzene-modified cationic surfactant (4-butylazobenzene-4′-(oxyethyl)trimethylammonium bromide, AZTMA; Scheme 3.1), which exhibits reversible *trans-cis* photoisomerization, as a "photoswitchable" agent.

CTAB is known to form wormlike micelles in water even at low concentrations when NaSal is added to the solution [18]. In the present system, a small amount of

Scheme 3.1 *trans-cis* Photoisomerization of AZTMA (reprinted with permission from Ref. [20]. Copyright (2005) American Chemical Society)

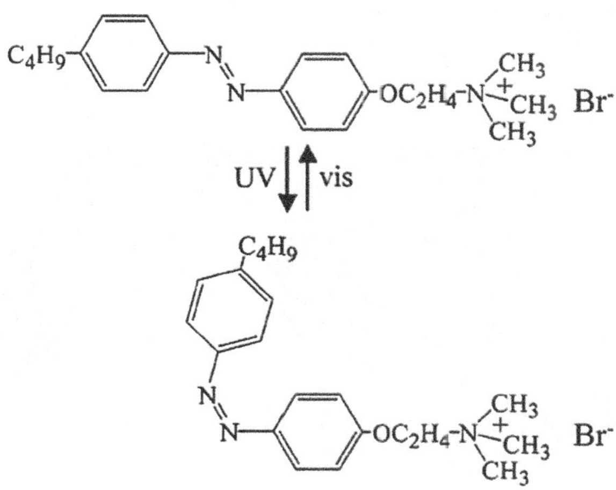

trans-AZTMA (10 mM) was mixed with aqueous CTAB solution (50 mM), and then NaSal (50 mM) was added to the solution to give viscoelastic wormlike micellar solution. AZTMA alone formed no wormlike micelles in water even in the presence of NaSal.

The CTAB/NaSal/AZTMA solution was placed in a quartz cuvette and then irradiated with UV light (260–390 nm) using a Hg-Xe lamp while the solution was stirred. UV–Vis spectra for the solution showed that UV light irradiation decreased the absorption band at 344 nm assigned to the *trans*-isomer and caused absorption due to the *cis*-isomer to appear at 440 nm. Visible light irradiation (>410 nm) after the UV light irradiation changed *cis*-AZTMA back to its *trans*-isomer, but not completely. However, repeated UV and visible light irradiation resulted in reversible *trans*-*cis* photoisomerization of AZTMA.

Panels a and b of Fig. 3.2 show the photographs of an aqueous CTAB/NaSal/AZTMA mixture before and after UV light irradiation, respectively. The viscosity of the mixture irradiated with UV light (Fig. 3.2b) was obviously lower than that of the nonirradiated one (Fig. 3.2a).

Figure 3.3 shows the dependence of storage (G') and loss (G'') moduli on angular frequency (ω) measured using a stress-controlled cone-plate rheometer (CSL 100, Carri-Med, Ltd.). Addition of *trans*-AZTMA to wormlike micellar solution of CTAB and NaSal shifted the G' and G'' curves to the low-frequency side, which indicates an enhancement in the degree of entanglement of wormlike micelles. For the nonirradiated solution, a plateau appeared on the G' curve and a maximum on the G'' curve. This viscoelastic behavior corresponds to a Maxwell model behavior as reported for typical wormlike micellar solutions [18, 25]. On the other hand, the UV-irradiated solution had no plateau on the G' curve in the measured frequency range. This rheological behavior is similar to that observed for unentangled chain polymer solutions.

We estimated the solution viscosity in terms of the zero-shear viscosity (η_0). Figure 3.4 shows the η_0 values of the mixture measured during the repeated cycles between *trans*-AZTMA and *cis*-AZTMA caused by UV and visible light irradiation. Addition of AZTMA (*trans*-form) to aqueous CTAB/NaSal solution increased

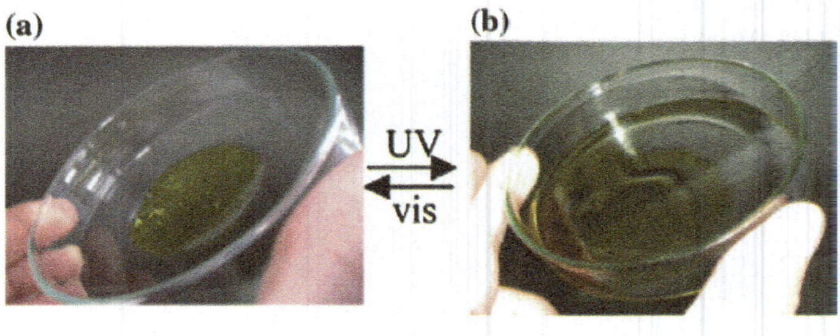

Fig. 3.2 Photographs showing the appearance of aqueous CTAB/NaSal/AZTMA mixture **a** before and **b** after UV irradiation (reprinted with permission from Ref. [20] Copyright (2005) American Chemical Society)

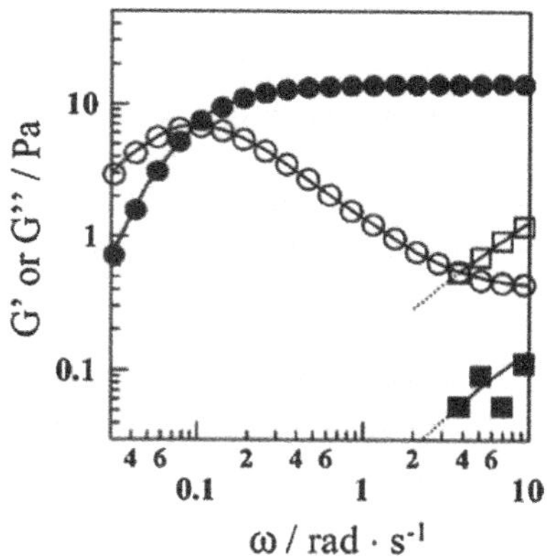

Fig. 3.3 Storage modulus G' and loss modulus G'' as a function of angular frequency ω for aqueous 50 mM CTAB/50 mM NaSal/10 mM AZTMA mixture before (*filled circle G'*, *open circle G''*) and after (*filled square G'*, *open square G''*) UV light irradiation (reprinted with permission from Ref. [20]. Copyright (2005) American Chemical Society)

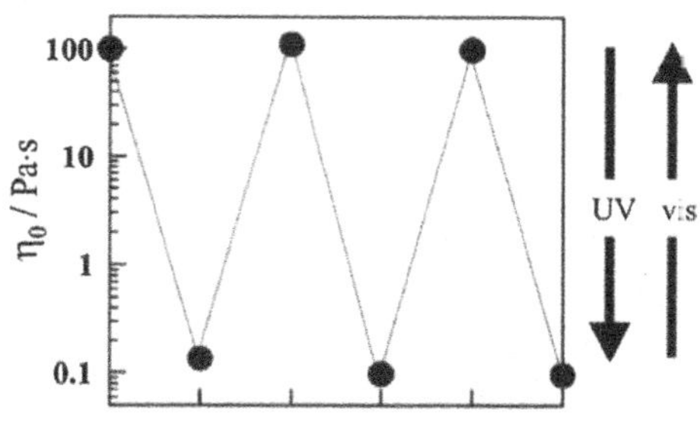

Fig. 3.4 Zero-shear viscosities (η_0) of aqueous solution of CTAB (50 mM)/NaSal (50 mM)/ AZTMA (10 mM) measured during the repeated cycles of AZTMA between the *trans*-form and *cis*-form. The high values of η_0 correspond to *trans*-AZTMA solution, and the low values correspond to *cis*-AZTMA solution (reprinted with permission from Ref. [20]. Copyright (2005) American Chemical Society)

its η_0 value from 60 to 100 Pa s, which is 5 orders of magnitude higher than the viscosity of water. When UV light was irradiated on the CTAB/NaSal/AZTMA (*trans*-form) solution, the η_0 value decreased 4 orders of magnitude (0.1 Pa s) over that of the *trans*-AZTMA solution. The following visible light irradiation made the viscosity return to the value of the *trans*-AZTMA solution. This η_0 change was completely reversible between the *trans*- and *cis*-AZTMA solutions.

The drastic viscosity change reported here is likely to be attributed to a change in the aggregation state of surfactant molecules. The reversible *trans-cis* photoisomerization of AZTMA (Scheme 3.1) causes changes in the geometrical structure and interfacial properties of the surfactant, thus leading to the reversible formation and disruption of molecular assemblies. *trans*-AZTMA is easily incorporated into wormlike micelles of CTAB and NaSal due to the linear structure of its hydrophobic tail. On the other hand, the bulky structure of *cis*-AZTMA is likely to destroy the network structure of wormlike micelles, and smaller aggregates (probably short rods) are formed in the solution. AZTMA incorporated in wormlike micelles, even though it is a minor component, permits a drastic change in solution viscosity through the *trans-cis* photoisomerization. The principle we report here thus can be generalized to other wormlike micellar solutions.

Recently, we have also studied photochemical viscosity control of organic solvents by using reverse wormlike micellar system containing cinnamic acid [27] and azobenzene derivatives. These results will be reported elsewhere.

3.2.2 Photochemical Control of Viscoelasticity Using "Photoswitchable" Counter Ion [28]

As described in the previous section, studies at our group have successfully led to light-controlled reversible viscosity changes in mixed aqueous solutions known to form wormlike micelles, consisting of the cationic surfactant, CTAB, and NaSal together with a photoresponsive azobenzene-modified surfactant, AZTMA [20]. Further advantages may be obtained, however, if the organic salt added to induce micelle formation is itself photoresponsive. This would result in a simpler system that might exhibit higher efficiency for light-induced viscosity changes. Among the various known photoresponsive molecules, cinnamic acid is similar in structure to NaSal and, unlike to AZTMA, requires no complex process for its synthesis. Its derivatives are already widely used as UVB region sunscreen agents. In this section, we therefore investigated the formation of high-viscosity micelle solutions, consisting of CTAB aqueous solutions containing sodium cinnamate (NaCin), a cinnamic acid salt, and the control of their viscosity by light.

Figure 3.5 shows appearance of 50 mM CTAB aqueous solutions in the presence of NaCin with various concentrations. It was found that the consistency of the aqueous CTAB/NaCin solution proceeded toward a gel state with increasing addition of NaCin to a fixed concentration of 50 mM CTAB until the NaCin concentration reached approximately 50 mM, after which no readily observable change in the solution viscosity occurred. We therefore performed the dynamic viscoelasticity measurements using the equimolar CTAB (50 mM)/NaCin (50 mM) aqueous solution. From the frequency dependence of G' (storage modulus) and G'' (loss modulus), it was found that this solution exhibits viscous-body behavior ($G' < G''$) at lower frequencies and elastic-body behavior ($G' > G''$) at higher frequencies and that G' reaches and maintains a plateau in the high-frequency region whereas G'' declines from a maximum. This behavior resembles that of the Maxwell model observed for diluted polymer solutions. In addition, we have observed molecular assemblies with

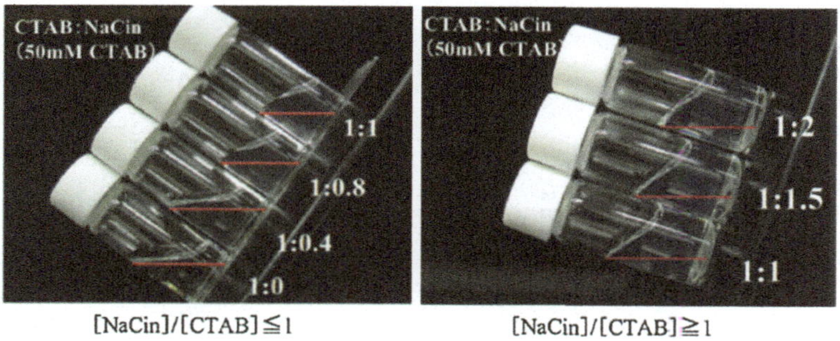

Fig. 3.5 Visual observation of CTAB/NaCin mixed solutions (reprinted with permission from Ref. [28]. Copyright (2012) Chemical Society of Japan)

wormlike morphology by cryo-TEM observation of the CTAB/NaCin solution. The above results confirm that wormlike micelles form in the CTAB/NaCin aqueous solution.

We performed UV irradiation of the CTAB/NaCin aqueous solution (50 mM/50 mM) and observed it for changes in viscosity. As shown in Fig. 3.6, it was found that the gel-form solution samples changed to liquid-form under the irradiation, and it was thus found possible to control the viscosity of this system by UV irradiation. In measurements of its UV/vis absorption before and after the photoirradiation, the *trans*-form solution before UV irradiation exhibited a maximum absorption wavelength λ_{max} of 269 nm, but the absorbance was reduced by UV irradiation and the position of maximum absorption shifted to shorter wavelength (255 nm). This finding indicates that the NaCin molecules underwent photoisomerization during the irradiation, thus changing from the *trans*-form to the *cis*-form.

We investigated the rheological behavior of the solution following UV irradiation. It was difficult to measure the dynamic viscoelasticity of the solution in the photostationary state, but in its rheological behavior, the solution exhibited Newtonian flow, as determined by flow-curve measurements, and a solution viscosity of 2.1×10^{-3} Pa s. This extremely small value was approximately 1/30,000

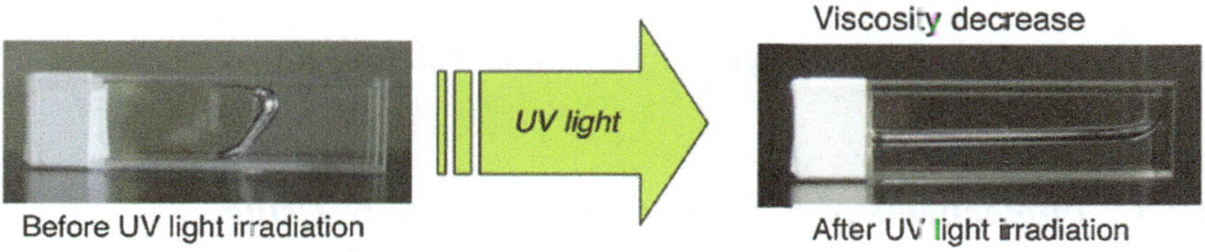

Fig. 3.6 Viscosity decrease observed by UV light irradiation to the CTAB (50 mM)/NaCin (50 mM) solution (reprinted with permission from Ref. [28]. Copyright (2012) Chemical Society of Japan)

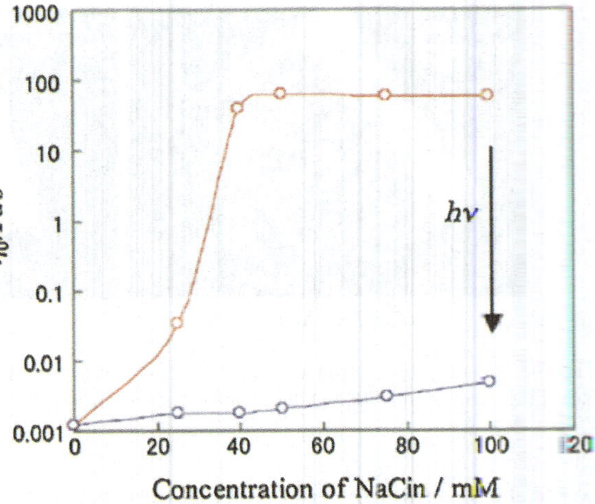

Fig. 3.7 Zero-shear viscosity of CTAB (50 mM)/NaCin solutions as a function of NaCin concentration before and after UV light irradiation (reprinted with permission from Ref. [28]. Copyright (2012) Chemical Society of Japan)

of the η_0 before UV irradiation and represents a reduction far larger than that (approximately 1/1000) of the CTAB/NaSal/AZTMA mixed aqueous solutions previously reported by us [20]. Figure 3.7 shows the results of our investigation of the change in viscosity under UV irradiation for systems containing NaCin at different concentrations with a fixed CTAB concentration. As shown, we observed a remarkable decrease in viscosity by UV irradiation in the aqueous solutions containing NaCin at concentrations of 40–100 mM.

The results of the rheological measurements suggest that the wormlike micelles were gradually changed to small molecular aggregates without entanglement by the irradiation, resulting in decreased viscosity. Moreover, when we investigated the environment of the NaCin molecules by ^{1}H NMR measurements before and after UV irradiation, the results suggested that in its transformation to the *cis*-form, the carboxylic acid moiety of the NaCin becomes positioned in a more hydrophilic environment. From the results of these investigations, we surmise that the change in viscosity under irradiation involves two mechanisms: (1) photoisomerization of the NaCin molecules solubilized within the CTAB micelles induces the formation of the bulky *cis*-form and a corresponding increase in the distance between the polar CTA^+ groups; and (2) some of the *cis*-Cin^- ions which form under the UV irradiation separate from the micelle surfaces, resulting in a transformation of the wormlike micelles to nonentangling spherical or small rod-form micelles.

3.2.3 Photochemical Control of Viscoelasticity Using a Novel "Photocleavable" Surfactant [29]

Recently, we have reported a new photocleavable surfactant, C4-C-N-PEG9 (Scheme 3.2b) that undergoes photocleavage through UV-induced cyclization in aqueous solutions, yielding two functional materials including a coumarin derivative and an aminated polyoxyethylene compound (Scheme 3.2c) [30]. It is expected that the cleaved compounds may act as a perfume and a moisturizing agent, respectively. To the best of our knowledge, the photocleavable mechanism has not been applied earlier in the wormlike micelles studies. In this section, we study the effect of photocleavable surfactant, C4-C-N-PEG9, on the rheological behavior of environmentally friendly viscoelastic wormlike micelles formed by aqueous mixture of polyoxyethylene phytosteryl ether (PhyEO20, Scheme 3.2a), and tetraoxyethylene dodecyl ether (C12EO4) [19].

The 4.9 wt% PhyEO20/H_2O + 2.4 wt% C12EO4 solution forms wormlike micelles, and its viscosity is $\sim$10 Pa s. We have found that the addition of C4-C-N-PEG9 into this viscous, non-Newtonian fluid system decreases the viscosity. Viscosity decreases in parallel to the C4-C-N-PEG9 concentration reaching $\sim$0.003 Pa s at 2.5 wt% of C4-C-N-PEG9. However, viscosity of the C4-C-N-PEG9 incorporated system increases significantly ($\sim$200 times at 1.5 wt% of C4-C-N-PEG9 system) upon UV irradiation as shown in Fig. 3.8.

Small-angle X-ray scattering studies have shown that addition of C4-C-N-PEG9 favors wormlike-to-sphere-type transition in the micellar structure. However, UV irradiation in the C4-C-N-PEG9 incorporated system causes one-dimensional micellar growth. Since C4-C-N-PEG9 has relatively bigger headgroup size compared to the C12EO4, addition of C4-C-N-PEG9 into wormlike micelles reduces the critical packing parameter, resulting in the formation of spherical aggregates. UV irradiation-induced one-dimensional micellar growth is caused due to photocleavage of the C4-C-N-PEG9 into a less surface-active coumarin derivative and an aminated polyoxyethylene compound. The hydrophobic coumarin derivative

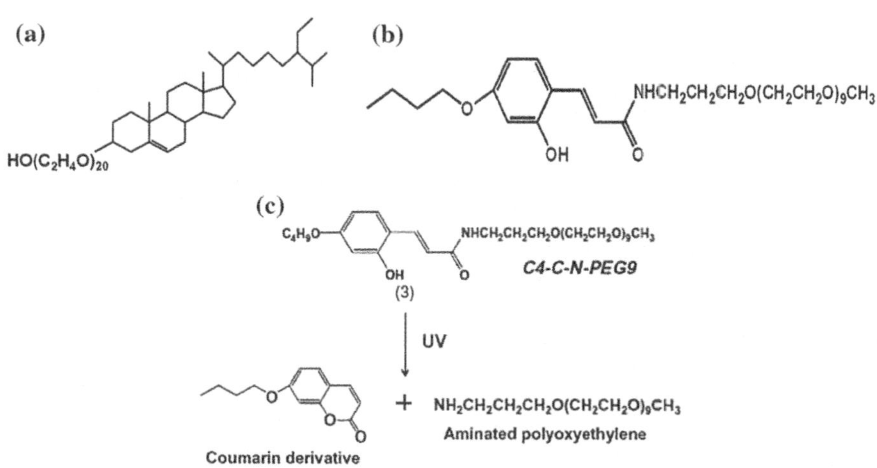

Scheme 3.2 Molecular structures of polyoxyethylene phytosteryl ether **a** PhyEO20, photocleavable surfactant **b** C4-C-N-PEG9, and **c** photocleavage reaction of C4-C-N-PEG9 (reprinted with permission from Ref. [29]. Copyright (2013) American Chemical Society)

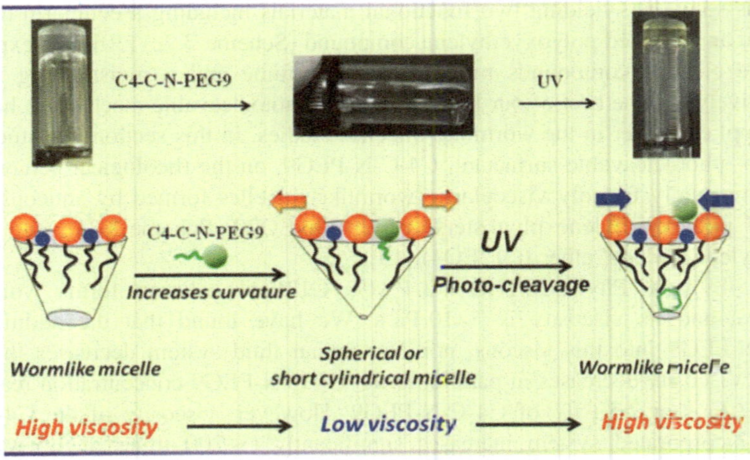

Fig. 3.8 Wormlike micellar structure formed by PhyEO20/C12EO4 system, and the effect of C4-C-N-PEG9 and UV irradiation. The PhyEO20 molecule is shown as an *orange head* and *black tail*, the C12EO4 molecule is shown with a *blue head* and *black tail*, while the C4-C-N-PEG9 is shown with a *green head* and *green tail* (reprinted with permission from Ref. [29] Copyright (2013) American Chemical Society)

formed after cleavage of C4-C-N-PEG9 goes to the micellar core and is responsible for decreasing the viscosity. However, the hydrophilic aminated polyoxyethylene prefers to reside at the vicinity of headgroup of PhyEO20 reducing the interhead repulsion, increasing the critical packing parameter and the viscosity as well.

3.3 Redox-Responsive Wormlike Micellar Solution Using a Ferrocenyl Surfactant

Electrorheological (ER) fluids [31, 32] are colloidal dispersions whose viscosity can be controlled by an electric field. When an alternating potential is applied to an ER fluid, colloidal particles dispersed in the fluid align to form strings parallel to the electric field because of the attraction of induced dipoles between the dispersed particles, which causes a viscosity rise in the direction perpendicular to the field. These smart fluids can be applied to clutches, valves, dampers, and precise control of mechanical positions. However, the conventional ER fluids have some disadvantages in terms of their industrial applications because of their poor dispersion stability and the need for high voltage and high particle concentration.

To improve the disadvantages in the conventional ER fluids, we have developed a novel class of ER fluids based on the redox reaction of a redox-active ferrocenyl surfactant, (11-ferrocenylundecyl)trimethylammonium bromide (FTMA, Scheme 3.3). Ferrocenyl surfactants have been used as a redox-active surfactant in

Scheme 3.3 Redox reaction of FTMA

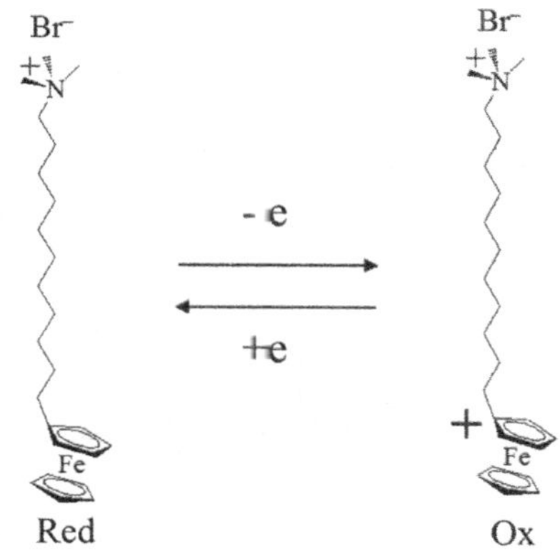

many studies because ferrocene is chemically stable, and its redox potential does not lead to electrolysis of water [33]. The redox reaction of the ferrocenyl surfactants permits control of formation and disruption of aggregates such as micelles [9, 34, 35] and vesicles [10–12]. FTMA has a redox-active ferrocenyl group positioned at the end of its hydrophobic tail. The ferrocenyl group in the reduced form is electrically neutral and shows poor water solubility. However, the ferrocenium cation in the oxidized form behaves as a hydrophilic group. This remarkable difference in the hydrophilic-lipophilic balance (HLB) between the reduced form and the oxidized form induces a change in aggregation state through the redox reaction of FTMA. In the present ER fluid, FTMA self-assembles into wormlike micelles in the presence of NaSal, and the viscoelasticity can be controlled by changing in the degree of entanglement of wormlike micelles through the redox reaction of FTMA [36].

Figure 3.9a shows a photograph of the appearance of aqueous FTMA (50 mM)/ NaSal mixture at molar ratio of [NaSal]/[FTMA] = 0.4.

The FTMA/NaSal mixture is viscous and spinnable. The solution does not flow even though it is tiled by 90°. This suggests the formation of wormlike micelles in the solution.

Figure 3.10a shows G', G'', and the real part of the complex viscosity ($|\eta^*|$) as a function of angular frequency (ω). The G' curve shows a plateau and is higher than the G'' curve at high frequencies. In addition, G'' is higher than G at low frequencies. This viscoelastic behavior is similar to that of concentrated colloidal polymer solutions with sufficient entanglement. Figure 3.10b shows G'' as a function of G', a Cole-Cole plot, for the mixture. The semicircular-shaped curve indicates a behavior of a Maxwell model with a single relaxation [18, 37]. This viscoelastic behavior is similar to that of typical wormlike micellar solutions. The η_0 is given by $\lim_{\omega\to 0}|\eta^*|$. The η_0 value obtained for the mixture is about 15 Pa s, which is four orders of magnitude higher than the viscosity of water (about 1×10^{-3} Pa s). The viscoelastic

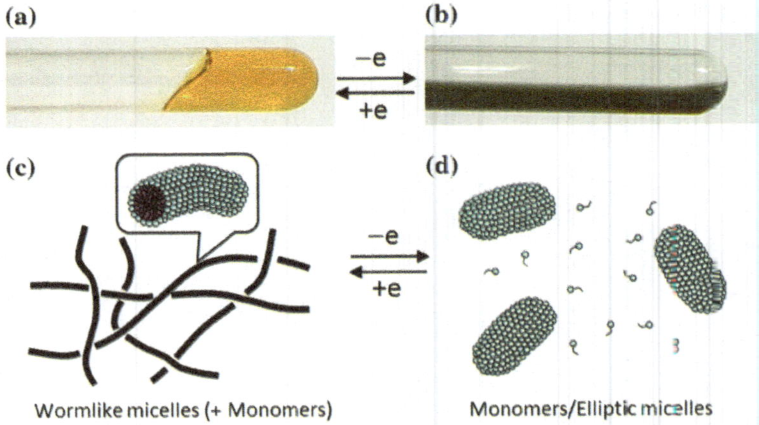

Fig. 3.9 Photographs of the appearance of aqueous FTMA/NaSal solution ([FTMA] = 30 mM, [NaSal]/[FTMA] = 0.4) **a** before and **b** after electrolytic oxidation. Schematic illustrations of aggregation states for **c** reduced and **d** oxidized solutions (reprinted with permission from Ref. [36]. Copyright (2004) American Chemical Society)

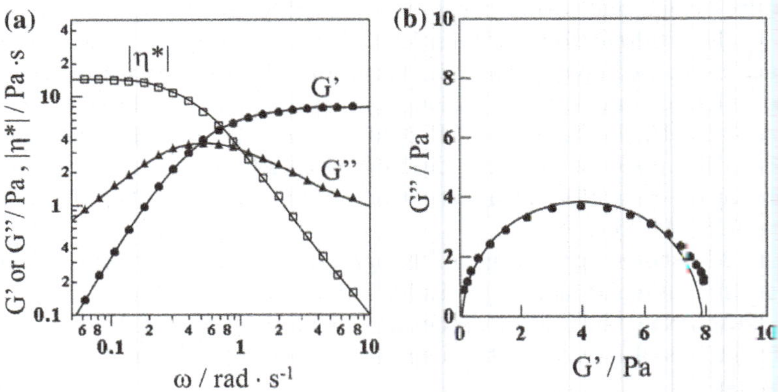

Fig. 3.10 **a** Storage modulus (G'), loss modulus (G'') and the real part of the complex viscosity ($|\eta^*|$) as a function of angular frequency ω for aqueous FTMA (reduced form)/NaSal solution. **b** Cole-Cole plot for aqueous FTMA (reduced form)/NaSal solution (reprinted with permission from Ref. [36]. Copyright (2004) American Chemical Society)

behavior of the aqueous FTMA/NaSal mixture is quite different from that of conventional ER fluids, which show Bingham or Newtonian properties.

Figure 3.11 shows a freeze-fracture transmission electron microscopic (FF-TEM) image for the FTMA (reduced form)/NaSal mixture. The extremely elongated structure with the diameter of 8 nm and lengths in the micrometer range observed in the micrograph clearly indicates the formation of wormlike micelles in the solution.

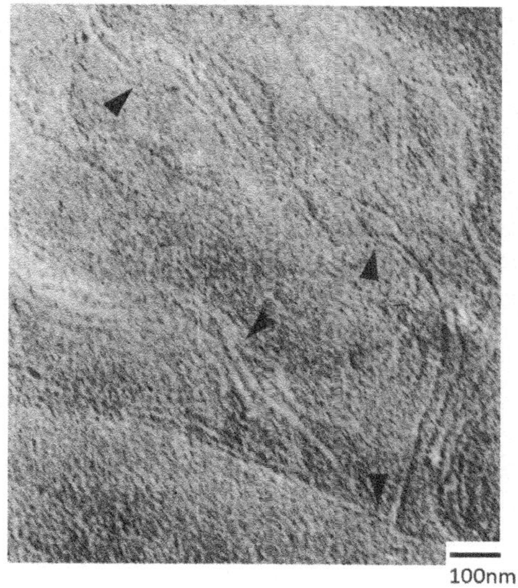

Fig. 3.11 FF-TEM micrograph of aqueous FTMA (reduced form)/NaSal mixture (reprinted with permission from Ref. [36]. Copyright (2004) American Chemical Society)

Figure 3.9b shows a photograph of the appearance of the mixture after electrolytic oxidation at +0.5 V vs. SCE, a potential much more positive than the equilibrium potential (+0.15 V vs. SCE) of FTMA [35]. The electrolytic oxidation induces a pronounced decrease in viscosity. When the oxidized solution is tilted by 90°, it flows immediately. The oxidized mixture behaves as a Newtonian behavior. The shear stress is proportional to the shear rate, and there is no hysteresis between the flow curves when the shear rate is increased and then decreased (Fig. 3.12). The viscosity of the oxidized solution is 2.5×10^{-3} Pa s, which is about 1/6000 of the η_0 value of the reduced solution. This remarkable viscosity decrease is probably responsible for a considerable change in aggregation state from wormlike micelles

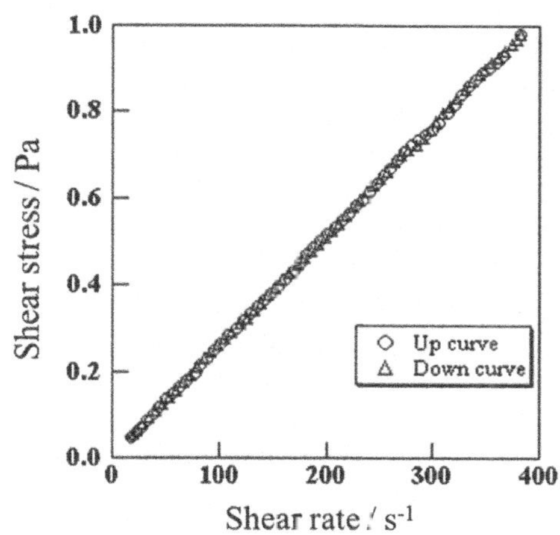

Fig. 3.12 Flow curve for aqueous FTMA/NaSal solution after electrolytic oxidation (reprinted with permission from Ref. [36]. Copyright (2004) American Chemical Society)

to monomers or small elliptic micelles (Fig. 3.9c, d) through a simple redox reaction of FTMA.

The present viscoelasticity-controlling system has several advantages over conventional ER fluids. We can achieve a remarkable change in viscoelasticity for *dilute* and thermodynamically *stable* wormlike micellar solution by applying a significantly *low* DC voltage. Under electrolytic conditions, the time to gain a useful change in viscosity is rather long compared with that for conventional ER fluids. However, the time can be reduced by decreasing the total sample volume and increasing the surface area of the electrode. Hence, thin-layer electrodes are useful for our system to control the viscoelasticity during short times.

The findings obtained in our study give us the key with regard to a method of efficiently controlling the viscoelasticity of other systems. The present system would be a useful candidate for novel ER fluid which will be applied to inkjet printers, flow rate controlling systems, and the controlled release systems.

3.4 pH-Responsive Wormlike Micellar Solution by Acylglutamic Acid-Alkylamine Complex

pH-responsive viscoelastic fluids have also been developed on the basis of self-assembly of surfactants [38–40]. In these systems, a pH change induces a protonation (or deprotonation) of pH-responsive functional groups, leading to a change in packing geometry of surfactant molecules in micelles. Hence, the micelle structure is changed in response to the solution pH. We have developed a pH-responsive wormlike micelle system by means of acylglutamic acid (alkyl dicarboxylic acid)-alkylamine complex [41]. Acylglutamic acid is a kind of amino acid-based surfactants and typically forms spherical or ellipsoidal micelles in aqueous solutions. The addition of cosurfactants or organic counterions is usually required when preparing viscoelastic wormlike micellar solutions in such systems [42–47]. In our case, we employed alkylamine as an organic counterion of dodecanoylglutamic acid (C12Glu) to neutralize the carboxylic acid headgroups as well as an agent to control the packing geometry being suitable for wormlike micelles. The complex formation occurs stoichiometrically as a result of a proton transfer from the acid to the amine; that is, a tertiary alkylamine (dodecyldimethylamine, C12DMA) or a secondary alkylamine (dodecylmethylamine, C12MA) yields a 1:1 complex with C12Glu, whereas a primary alkylamine (dodecylamine, C12A) forms a water-insoluble 1:2 complex.

Figure 3.13 shows η_0 of aqueous C12Glu-C12DMA complex solutions as a function of pH, assessed by steady-shear viscosity measurements. Clearly, the rheological property of the aqueous C12Glu-C12DMA solutions is strongly dependent on the pH; a maximum η_0 value is seen at pH 5.7. We also performed dynamic rheological measurements and confirmed the formation of wormlike micelles in the pH range 5.5–6.0.

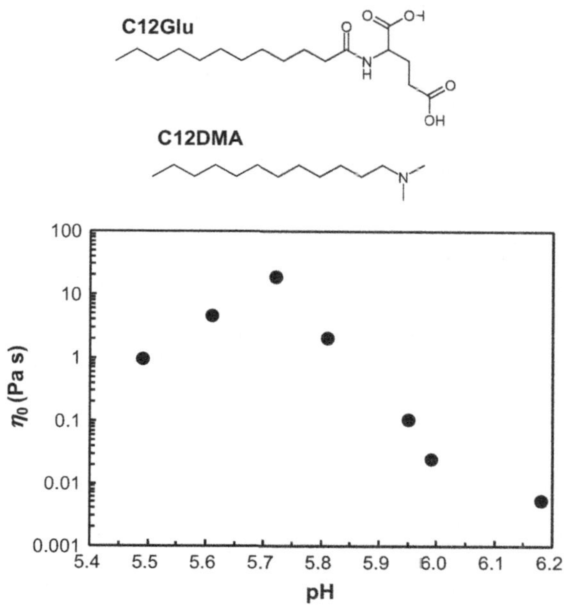

Fig. 3.13 Zero-shear viscosity (η_0) of aqueous C12Glu-C12DMA complex solutions as a function of pH. The complex concentration is fixed at 3 wt%. Chemical structures of C12Glu and C12DMA are also shown in this figure (reprinted with permission from Ref. [41]. Copyright (2012) American Chemical Society)

Newtonian fluids suggesting the formation of spherical or rodlike micelles are obtained above the solution pH of 6.2. Then, the decreased pH induces a micellar growth of the spherical or rodlike micelles into wormlike micelles, and hence, viscoelastic non-Newtonian fluids are obtained in the narrow pH range 5.7–6.0. Around the pH achieving the maximum η_0 value, the wormlike micelles are highly entangled with each other. The further decrease in the solution pH to 5.5 results in the decreased η_0. This results from branching or interconnection of the wormlike micelles, accompanying with an increased number of entanglements between wormlike micelles. These structural transitions occur toward lesser positive curved aggregates with decreasing pH, as schematically shown in Fig. 3.14. The decreased pH results in the protonation of a carboxylic acid headgroup nonreacted with the

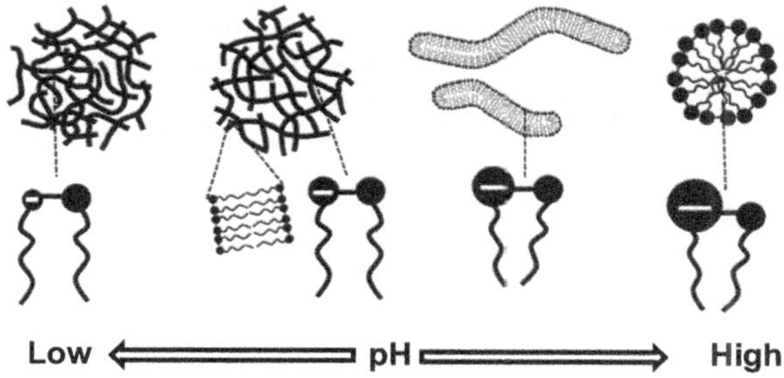

Fig. 3.14 Schematic illustration of pH-dependent wormlike micelle structures (reprinted with permission from Ref. [41]. Copyright (2012) American Chemical Society)

alkylamine compound, leading to the decreased head-to-head repulsion, and hence, the positive curvature of the micellar aggregates tends to be decreased.

In this example, the alkyl dicarboxylic acid-alkylamine complex can be obtained through a very simple and handy reaction; that is, mixing at room temperature in water or organic solvents such as ethanol. We believe, therefore, that this example offers a unique and simple way to yield high-performance materials including various stimulus-responsive amphiphiles.

Acknowledgments Studies reported by the authors were partially supported by Grant-in-Aid for Scientific Research (KAKENHI, No. 22107003) on Innovative Areas of "Fusion Materials" (No. 2206) from the Ministry of Education, Culture, Sports, Science and Technology (MEXT). The authors are also thankful to Prof. Masahiko Abe, Prof. Tetsuo Saji, Prof. Yukishige Kondo, Dr. Yasushi Kakizawa, and Dr. Atsutoshi Matsumura for their continuous support and fruitful discussion.

References

1. Ali M, Jha M, Das SK, Saha SK (2009) Hydrogen-bond-induced microstructural transition of ionic micelles in the presence of neutral naphthols: pH dependent morphology and location of surface activity. J Phys Chem B 113:15563–15571
2. Scarzello M, Klijn JE, Wagenaar A, Stuart MCA, Hulst R, Engberts JBFN (2006) pH-dependent aggregation properties of mixtures of sugar-based gemini surfactants with phospholipids and single-tailed surfactants. Langmuir 22:2558–2568
3. Tobita K, Sakai H, Kondo Y, Yoshino N, Iwahashi M, Momozawa N, Nishiyama K, Abe M (1997) Thermoreponsive viscoelasticity of sodium 1-oxo-1-[4-(tridecafluorohexyl) phenyl]-2-hexanesulfonate aqueous solutions. Langmuir 13:5054–5055
4. Abe M, Tobita K, Sakai H, Kamogawa K, Momozawa N, Kondo Y, Yoshino N (2000) Thermoresponsive viscoelasticity of concentrated solutions with a fluorinated hybrid surfactant. Colloids Surf A 167:47–60
5. Aydogan N, Abbott NL (2001) Comparison of the surface activity and bulk aggregation of ferrocenyl surfactants with cationic and anionic headgroups. Langmuir 17 5703–5706
6. Anton P, Heinze J, Laschewsky A (1993) Redox-active monomeric and polymeric surfactants. Langmuir 9:77–85
7. Eastoe J, Vesperinas A (2005) Self-assembly of light-sensitive surfactants. Soft Matter 1:338–347
8. Zhao Y (2012) Light-responsive block copolymer micelles. Macromolecules 45:3647–3657
9. Takei T, Sakai H, Kondo Y, Yoshino N, Abe M (2001) Electrochemical control of solubilization using a ferrocene-modified nonionic surfactant. Colloids Surf, A 183–185:757–765
10. Kakizawa Y, Sakai H, Yamaguchi A, Kondo Y, Yoshino N, Abe M (2001) Electrochemical control of vesicle formation with a double-tailed cationic surfactant bearing ferrocenyl moieties. Langmuir 17:8044–8048
11. Kakizawa Y, Sakai H, Nishiyama K, Abe M, Shouji H, Kondo Y, Yoshino N (1996) Solution properties of double-tailed cationic surfactants having ferrocenyl groups in their hydrophobic moieties. Langmuir 12:921–924
12. Sakai H, Imamura H, Kondo Y, Yoshino N, Abe M (2004) Reversible control of vesicle formation using electrochemical reaction. Colloids Surf, A 232:221–228

13. Orihara Y, Matsumura A, Saito Y, Ogawa N, Saji T, Yamaguchi A, Sakai H, Abe M (2001) Reversible release control of an oily substance using photoresponsive micelles. Langmuir 17:6072–6076
14. Sakai H, Matsumura A, Saji T, Abe M (1999) Photochemical switching of vesicle formation using an azobenzene-modified surfactant. J Phys Chem 103:10737–10740
15. Matsumura A, Tsuchiya K, Torigoe K, Sakai K, Sakai H, Abe M (2011) Photochemical control of molecular assembly formation in a catanionic surfactant system. Langmuir 27:1610–1617
16. Sakai H, Ebana H, Sakai K, Tsuchiya K, Ohkubo T, Abe M (2007) Photo-isomerization of spiropyran-modified cationic surfactants. J Colloid Interface Sci 316:1027–1030
17. Akamatsu M, FitzGerald P, Shiina M, Misono T, Tsuchiya K, Sakai K, Abe M, Warr GG, Sakai H (2015) Micelle structure in a photo-responsive surfactant with and without solubilized ethylbenzene from small-angle neutron scattering. J Phys Chem C 119:5904–5910
18. Shikata T, Hirata H, Kotaka T (1987) Micelle formation of detergent molecules in aqueous media: viscoelastic properties of aqueous cetyltrimethylammonium bromide solutions. Langmuir 3:1081–1086
19. Naito N, Acharya DP, Tanimura J, Kunieda H (2005) Phase behavior of polyoxyethylene phytosterol/polyoxyethylene dodecylether/water systems. J Oleo Sci 54:7–13
20. Sakai H, Orihara Y, Kodashima H, Matsumura A, Ohkubo T, Tsuchiya K, Abe M (2005) Photoinduced reversible change of fluid viscosity. J Am Chem Soc 127:13454–13455
21. Takahashi Y, Fukuyasu K, Horiuchi T, Kondo Y, Stroeve P (2014) Photoinduced demulsification of emulsions using a photoresponsive gemini surfactant. Langmuir 30:41–47
22. Shin JY, Abbott NL (1999) Using light to control dynamic surface tensions of aqueous solutions of water soluble surfactants. Langmuir 15:4404–4410
23. Kumar R, Raghavan SR (2009) Photogelling fluids based on light-activated growth of zwitterionic wormlike micelles. Soft Matter 5:797–803
24. Irie M, Hirano Y, Hashimoto S, Hayashi K (1981) Photoresponsive polymers. 2. Reversible solution viscosity change of polyamides having azobenzene residues in the main chain. Macromolecules 14:262–267
25. Koumura N, Kudo M, Tamaoki N (2004) Photocontrolled gel-to-sol-to-gel phase transitioning of meta-substituted azobenzene bisurethanes through the breaking and reforming of hydrogen bonds. Langmuir 20:9897–9900
26. Shikata T, Hirata H, Kotaka T (1988) Micelle formation of detergent molecules in aqueous media. 2. Role of free salicylate ions on viscoelastic properties of aqueous cetyltrimethylammonium bromide-sodium salicylate solutions. Langmuir 4:354–359
27. Shrestha RG, Agari N, Tsuchiya K, Sakamoto K, Sakai K, Abe M, Sakai H (2014) Phosphatidylcholine-based nonaqueous photorheological fluids: effect of geometry and solvent. Colloid Polym Sci 292:1599–1609
28. Sakai H, Taki S, Tsuchiya K, Matsumura A, Sakai K, Abe M (2012) Photochemical control of viscosity using sodium cinnamate as a photoswitchable molecule. Chem Lett 41:247–248
29. Aikawa S, Shrestha RG, Ohmori T, Fukukita Y, Tezuka T, Endo T, Torigoe K, Tsuchiya K, Sakamoto K, Sakai K, Abe M, Sakai H (2013) Photorheological response of aqueous wormlike micelles with photocleavable surfactant. Langmuir 29:5668–5676
30. Sakai H, Aikawa S, Matsuda W, Ohmori T, Fukukita Y, Tezuka Y, Matsumura A, Torigoe K, Tsuchiya K, Arimitsu K, Sakamoto K, Sakai K, Abe M (2012) A cinnamic acid-type photo-cleavable surfactant. J Colloid Interface Sci 376:160–164
31. Winslow WM (1949) Induced fibration of suspensions. J Appl Phys 20:1137–1140
32. Halsey TC (1992) Electrorheological fluids. Science 258:761–766
33. Anton P, Heinze J, Laschewsky A (1993) Redox-active monomeric and polymeric surfactants. Langmuir 9:77–85
34. Hoshino K, Saji T (1987) Electrochemical formation of an organic thin film by disruption of micelles. J Am Chem Soc 109:5881–5883
35. Saji T, Hoshino K, Ishii Y, Goto M (1991) Formation of organic thin films by electrolysis of surfactants with the ferrocenyl moiety. J Am Chem Soc 113:450–456

36. Tsuchiya K, Orihara Y, Kondo Y, Yoshino N, Ohkubo T, Sakai H, Abe M (2004) Control of vscoelasticity using redox reaction. J Am Chem Soc 126:12282–12283
37. Soltero JFA, Puig JE, Manero O (1996) Rheology of the cetyltrimethylammonium tosilate-water system. 2. Linear viscoelastic regime. Langmuir 12:2654–2662
38. Verma G, Aswal VK, Hassan P (2009) pH-responsive self-assembly in an aqueous mixture of surfactant and hydrophobic amino acid mimic. Soft Matter 5:2919–2927
39. Chu Z, Feng Y (2010) pH-switchable wormlike micelles. Chem Commun 46:9028–9030
40. Lu H, Shi Q, Huang Z (2014) pH-responsive anionic wormlike micelle based on sodium oleate induced by NaCl. J Phys Chem B 118:12511–12517
41. Sakai K, Nomura K, Shrestha RG, Endo T, Sakamoto K, Sakai H, Abe M (2012) Wormlike micelle formation by acylglutamic acid with alkylamines. Langmuir 28:17617–17622
42. Shrestha RG, Shrestha LK, Aramaki K (2007) Formation of wormlike micelle in a mixed amino-acid based anionic surfactant and cationic surfactant systems. J Colloid Interface Sci 311:276–284
43. Shrestha RG, Shrestha LK, Aramaki K (2008) Wormlike micelles in mixed amino acid-based anionic/nonionic surfactant systems. J Colloid Interface Sci 322:596–604
44. Shrestha RG, Shrestha LK, Aramaki K (2009) Rheology of wormlike micelles in aqueous systems of a mixed amino acid-based anionic surfactant and cationic surfactant. Colloid Polym Sci 287:1305–1315
45. Aramaki K, Iemoto S, Ikeda N, Saito K (2010) Composition-insensitive highly viscous wormlike micellar solutions formed in anionic and cationic surfactant systems. J Oleo Sci 59:203–212
46. Shrestha RG, Shrestha LK, Matsunaga T, Shibayama M, Aramaki K (2011) Lipophilic tail architecture and molecular structure of neutralizing agent for the controlled rheology of viscoelastic fluid in amino acid-based anionic surfactant system. Langmuir 27:2229–2236
47. Shrestha RG, Nomura K, Yamamoto M, Yamawaki Y, Tamura Y, Sakai K, Sakamoto K, Sakai H, Abe M (2012) Peptide-based gemini amphiphiles: phase behavior and rheology of wormlike micelles. Langmuir 28:15472–15481

Chapter 4
Stimuli-Responsive Charge-Free Reverse Micelles in Non-Aqueous Media

Lok Kumar Shrestha and Katsuhiko Ariga

Abstract In this chapter, formulation of charge-free nonionic surfactant reverse micelles in non-aqueous media and their structural transformations upon external stimuli are demonstrated. Contrary to the general postulation that reverse micelles usually assembled into spheroid shape with small aggregation number, formation of reverse rodlike micelles and evidence of sphere-to-wormlike reverse micelle transition are demonstrated in surfactant/oil system without the addition of water.

Keywords Nonionic surfactants · Reverse micelles · Stimuli responsive · SAXS

4.1 Introduction

Self-assembly is the independent organization of atoms, molecules, or functional units into the well-defined patterns or structures without external interference. This process is common in nature and highly been utilized in technology. Self-assembly is one of the few practical strategies for creation of ensembles of nanostructures in different length scales. Therefore, it has become an essential part of current advanced nanotechnology. It has been proposed that self-assembly occurs at all length scales from microscopic molecular level to macroscopic objects of gigantic size [1, 2]. Inter-object interactions lead to their assembly with the resulting structure depending on shape, surface–surface interactions, electrostatics, and homogeneity of the assembling building blocks [3, 4]. At the molecular level, self-assembly may be also determined by morphology or geometry of the assembling units [5, 6]. Self-assembly can obviously be extended to the crystal state where the building blocks are arranged in a three-dimensional (3D) pattern [7].

Amphiphilic molecules that possess both hydrophilic and hydrophobic (lipophilic) moieties are architecturally simple molecules that can spontaneously

L.K. Shrestha (✉) · K. Ariga
World Premier International Center for Materials Nanoarchitectonics (WPI-MANA),
National Institute for Materials Science (NIMS), Tsukuba, Japan
e-mail: SHRESTHA.Lokkumar@nims.go.jp

T. Kawai and M. Hashizume (eds.), *Stimuli-Responsive Interfaces*,
DOI 10.1007/978-981-10-2463-4_4

self-assemble into complex hierarchical structures from lower to higher dimensions in the bulk phase or at an interface [8–12]. Recent developments in multifunctional nanostructure design using the advanced concept of nanoarchitectonics utilize this simple process of self-assembly [13–18]. Amphiphilic self-assemblies involving lipids or proteins mimic the structure of biological systems, thus highlighting the necessity of a fundamental physical understanding of amphiphilic self-assembly toward a realization of the complex mechanisms operating in nature.

It is well known that amphiphilicity is one of the main driving forces for the self-assembly of surfactants. The thermodynamic properties of surfactants in solution are controlled by the strong tendency of hydrophobic tails to avoid a direct contact with water (i.e., strong hydrophobicity). This unfavorable situation or interaction can be minimized by aggregation of surfactant molecules into aggregate structure called micelles in which the hydrophilic headgroup of surfactant molecules is exposed to water phase and the hydrophobic groups are shielded at the interior of aggregates. This overall process of aggregation of self-assembling is referred to as micellization [19]. Another aspect of micellization is the entropy effect [20, 21]. The entropic contribution arises from the local structure of water due to hydrogen bonding. The segregated hydrocarbon chains of surfactant molecules interrupt the hydrogen bonding between water molecules causing entropically unfavorable locally ordered structure. To avoid the disruption of the water structure and entropically unfavorable situation, entropically more favorable aggregated micellar structures are formed. Micellization has also been considered as an alternative mechanism to adsorption at interfaces for avoiding direct contact of hydrophobic groups with water, thereby reducing the free energy of the system. It should be noted that micellization is the prototype for self-assembly of lipid amphiphiles or surfactants and indicates the importance of structural segregation based on the differing regions of hydrophobicity/hydrophilicity of a molecule.

Amphiphilic molecules such as surfactants or block copolymers usually show remarkable phase behavior in aqueous system forming various phases depending on composition and temperature [22–25]. Just above critical micelle concentration (*cmc*), surfactants generally tend to form spheroid type micelles, whose morphology can be modulated into other geometries such as prolate, rods or disk or ordered liquid crystalline phases, bilayer vesicles, and also reverse micelles with the change of control parameters such as molecular structure of surfactant or solvent, concentration, temperature, salinity, and pH [26–28]. Scheme 4.1 shows a schematic representation of a typical amphiphilic molecule and the different structures available both in normal and reverse (inverted) forms. Other more exotic forms are also available including nanotubes, oblate micelles, and even toroidal structures (not shown). These self-assembled structures have extensively been utilized in various realistic applications including food, cosmetic and pharmaceutical formulations, and in technology. However, despite the enormous applications in industry and technology, self-assembly of surfactants in non-aqueous media and their fundamental physicochemical properties, structure, and dynamic behavior of reverse micelles have not been deeply investigated compared to the aqueous system [29].

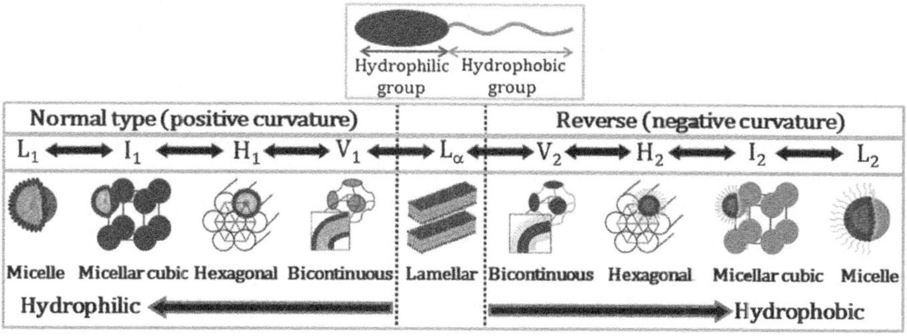

Scheme 4.1 Schematic representation of an amphiphilic surfactant molecule with normal and reverse types of self-assembled structures. L_1, I_1, H_1, V_1, and L_α represents normal micelle, micellar cubic phase, hexagonal phase, bicontinuous cubic phase, and lamellar phase, respectively. V_2, I_2, H_2, and L_2, respectively, represent reverse phases. Adopted with permission from Ref. [10]

In this chapter, we will describe the self-assembly of various nonionic surfactants by referring to several examples in non-aqueous media, i.e., charge-free reverse micelles and its structural transformations upon various external stimuli. Initially, we will describe the foundations of self-assembly from a molecular perspective and reverse micelle microstructure transformation based on surfactant molecular architecture. This will involve the free structure control of reverse micelles tuning surfactant molecular size and amphiphilicity in non-aqueous solvent systems. We will then discuss how molecular size, molecular structure, and solvent polarity determine the shape, size, and internal structure of reverse micelles. Surfactant composition followed by thermo-responsive behavior of reverse micelles will be our next topics of discussion. We will discuss examples of temperature-induced rod-to-sphere type of transition in the nonionic surfactant reverse micellar structure. We will also focus on water-induced sphere-to-rod transition in the reverse micellar structure. Finally, we will discuss sphere-to-wormlike reverse micelles microstructure transformation in water-free systems.

4.2 Self-Assembly in Non-Aqueous Media

Micelles are important structures formed by the association of most of all types of surfactants in solution. Conventional normal micelles consist of an inner hydrophobic core (oily environment) which is shielded from water by the surrounding corona formed by the hydrophilic headgroups of the surfactant. On the other hand, reverse micelles have an inverted structure, i.e., the core is hydrophilic (watery or polar environment) and the shell is hydrophobic [29, 30]. Therefore, they are often known as inverse or inverted micelles. Similar to the normal micelles in aqueous system, reverse micelles also grow into rodlike or cylindrical, and flexible wormlike micelles [31]. However, solution behavior of surfactant in non-aqueous media markedly differs from the aqueous one. It is well known in aqueous systems that the physicochemical, colligative, and spectral properties of the system undergo an abrupt change over a narrow region of concentration (near *cmc*) upon increasing surfactant concentration at a constant temperature. However, in non-aqueous media, these properties change gradually without any sharp transition indicating that the existence of well-defined *cmc* is questionable in surfactant/oil systems. Nevertheless, series of previous investigations based on advanced characterization techniques have provided enough evidence of aggregate structures formation in non-aqueous solvent systems even at lower surfactant concentrations. Dipole–dipole interaction between the hydrophilic headgroups has been regarded as one of the driving forces for the formation of reverse micelles in non-aqueous media.

Shape-controlled reverse micelles offer wide range of practical applications. They serve as a model system for size-controlling microreactor for different aqueous chemical reactions. Reverse micelles also potentially stabilize reactive species that are insoluble in nonpolar organic solvent [32]. Furthermore, they have been found to be an excellent model for biological membranes [33]. One of the most intriguing features of reverse micelle is its water solubilization capacity. Similarly, it can potentially solubilize other polar solvents. Water is often solubilized in the core of the reverse micelles, causing parent reverse micelle to swell several time sometimes a few orders of magnitude [34, 35]. In material nanoarchitectonics, reverse micelles have also been used as soft template for the synthesis of nanomaterials and also structure guiding agent of nanomaterials [36–40]. Investigations have shown that shape and size of the template reverse micelles mimic the morphology of the derived nanomaterials [41–43]. Owing to these various technological application, fundamental studies of the formulations of reverse micelles and their free structure control continue to grow not only in colloid and interface society but also in material research society.

In the following section, we will discuss some of recent results on the formulation of charge-free reverse micelles in surfactant/oil binary systems and their microstructure transformation upon external stimuli, molecular structure of surfactant and solvent, concentration, temperature, water, and co-surfactant. We describe the shape, size, and internal cross-sectional structure characterizations of these charge-free reverse micelle using small-angle X-ray scattering (SAXS) technique and also bulk rheology. We will particularly focus on glycerol- and sucrose-based fatty acid esters' nonionic surfactants. These surfactants have drawn considerable attention in food, cosmetic and pharmaceutical industries due to their biocompatibility, biodegradability, and mildness to the skin. Polyglycerol fatty acid esters are basically edible nonionic surfactants having polyglycerol chain as a hydrophilic moiety, and depending on the surfactant molecular structure, solvent nature, composition, and temperature can potentially form various self-assembled structure including α-gel phase, lyotropic lamellar liquid crystal (L_α), vesicles, and reverse micelles [29, 30, 44, 45]. The α-gel and L_α phase of these surfactants show outstanding non-aqueous foam stabilization property [46–49].

4.2.1 Effect of Surfactant Molecular Structure on Reverse Micellar Structure

Here, we discuss effect of surfactant molecular structures (both headgroup size and length of hydrophobic chain) on the reverse micellar structure in non-aqueous media without the addition of water, i.e., surfactant/oil binary mixtures. First, we discuss the effect of headgroup size. Polyglycerol monooleate (abbreviated as $C_{18:1}G_n$, $n = 1$, 2, and 4) and polyglycerol monolaurate (abbreviated as $C_{12}G_n$, $n = 1$, and 2) nonionic surfactants are considered to demonstrate a simple route to the growth control of charge-free reverse micelles by headgroup size. SAXS technique is used for the structural characterizations. The scattering data are evaluated by *generalized indirect Fourier transformation* (GIFT) method [50, 51].

Figure 4.1 shows the normalized X-ray scattered intensities, $I(q)$, and the corresponding pair-distance distribution functions, $p(r)$, obtained by the GIFT

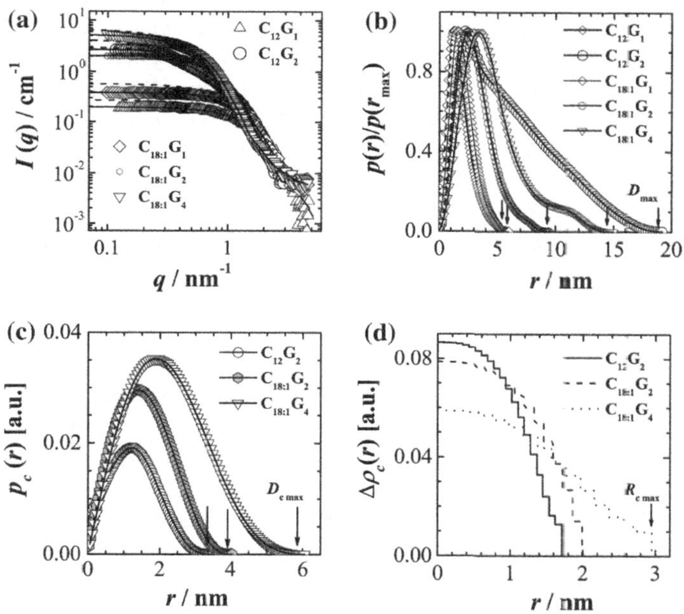

Fig. 4.1 **a** The normalized X-ray scattering intensities, $I(q)$, of 5 % $C_{18:1}G_n$ ($n = 1$, 2, and 4)/decane at 25 °C and 5 % $C_{12}G_n$ ($n = 1$, and 2)/decane in absolute units at 60 °C, **b** the corresponding pair-distance distribution functions, $p(r)$, **c** the cross-sectional pair-distance distribution function, $p_c(r)$, and **d** the corresponding electron density profile, $\Delta\rho_c(r)$, calculated via deconvolution from $p_c(r)$ for the selected systems. The *solid and broken lines* in panel **a** and in rest of all $I(q)$ curves in this chapter represent the GIFT fit and the calculated form factor for n particles existing in unit volume, $nP(q)$, respectively. The *arrows* in panel **b** and rest of the $p(r)$ curves indicate the maximum dimensions of the micellar core, D_{max}. Data for 5 % $C_{12}G_n$ ($n = 1$, and 2)/decane are adopted from Ref. [53] with permission from the PCCP Owner Societies. Data for $C_{18:1}G_n$ ($n = 1$, 2, and 4)/decane are adopted with permission from Ref. [52] © 2010 American Chemical Society

evaluation of SAXS data for 5 % glycerol monolaurate ($C_{12}G_1$) and 5 % diglycerol monolaurate ($C_{12}G_2$) in decane at 60 °C, and 5 % polyglycerol monooleate ($C_{18:1}G_n$, $n = 1$, 2, and 4) in decane at 25 °C. It should be noted that short-chain mono- and diglycerol fatty acid esters form reverse micelles only at elevated temperatures after melting of solid or lamellar liquid crystal phase [45]. However, polyglycerol monooleate surfactants self-assemble into reverse micelles at 25 °C [52]. As suggested by the fundamental critical packing parameter (*cpp*) concept, bigger reverse micelles are expected with surfactants having larger headgroups. However, here we note that the extent of micellar growth or the types of microstructure transitions caused by the headgroup size also depend on the lipophilic chain of the surfactants, i.e., in overall surfactant molecular structure. As shown in Fig. 4.1, scattering functions largely depend on the headgroup size of surfactant. The forward scattering intensity, $I(q = 0)$, of $C_{12}G_2$ is much higher (~one order of magnitude) compared to $C_{12}G_1$ demonstrating bigger reverse micelles in $C_{12}G_2$ system (Fig. 4.1a). Furthermore, the low q scattering intensity of $C_{12}G_1$/decane follows $I(q) \sim q^0$ behavior (typical of spheroid type micelles), whereas the low q scattering intensity of $C_{12}G_2$/decane decays following $I(q) \sim q^{-1}$ behavior indicating rodlike structure in $C_{12}G_2$ system [51, 53]. This is a clear illustration of sphere-to-rod transition in reverse micellar structure. A similar trend in the scattering function is observed with $C_{18:1}G_n$ ($n = 1$, 2, and 4)/decane systems. The $I(q = 0)$ increases with increase in the headgroup size from mono- to tetraglycerol. However, note that the low q scattering intensities of $C_{18:1}G_2$ and $C_{18:1}G_4$ do not follow $I(q) \sim q^{-1}$ behavior. The low q slope is less than 1. Thus, these observations confirm that reverse micelle microstructure transitions caused by headgroup size of surfactant do not always favor sphere-to-rod transition. The microstructure transition rather depends on the overall hydrophilic–lipophilic balance of the surfactant.

The real-space structural information presented in the form of $p(r)$ curves gives clear picture of microstructure and its transition with headgroup size of surfactant. For better visibility, normalized $p(r)$ curves ($p(r)/p(r_{max})$) are presented in Fig. 4.1b. Note that all the $p(r)$ curves described in this chapter should be recognized as a measure of the core structure of reverse micelles. This is because in non-aqueous media the contrast (electron density difference) between solvent and lipophilic part of the surfactant is negligibly small; as a result, SAXS selectively detects only the hydrophilic micellar core. A pronounced peak in the low r region with an extended linear tail in the high r side of the $p(r)$ curve indicates the rodlike structure of micelles in the $C_{12}G_2$ system [51, 54]. An asymmetry in the $p(r)$ curve with a pronounced peak in the low r side with a long tail has a downward convex shape in an indication of elongated ellipsoid prolate type of reverse micelles [51, 54, 55]. $C_{18:1}G_2$ and $C_{18:1}G_4$ surfactants showed such behavior in the $p(r)$ curves. On the other hand, symmetric bell-shaped $p(r)$ curves of $C_{12}G_1$ and $C_{18:1}G_1$ confirm the globular micelles.

For the scattering particles with elongated geometries such as rods or ellipsoidal prolates, a direct cross-sectional analysis based on *indirect Fourier transformation* (IFT) technique is available under assumption that the scattering length density profile of the cross section is simply a function of the radial position [56, 57]. However, for such analysis, the axial length of the elongated particles should be at least three times longer than the cross-sectional diameter. Figure 4.1c, d shows the results from direct cross-sectional analysis, i.e., deconvolution of the cross-sectional pair-distance distribution function, $p_c(r)$, into the radial difference electron density distribution profile $\Delta\rho_c(r)$. Note that the $p_c(r)$ and $\Delta\rho_c(r)$ which reach to zero at higher r side refer to the quantitative measure of maximum cross-sectional diameter (D_{cmax}) and maximum cross-sectional radius (R_{cmax}) of the micellar core, respectively. The cross-sectional diameter is ca. 3.4 nm from the $p_c(r)$ and R_{cmax} is ca. 1.7 nm from the $\Delta\rho_c(r)$ for the $C_{12}G_2$ system. Similarly, the cross-sectional diameters of 4.0 and 6.0 nm are calculated for $C_{18:1}G_2$, and $C_{18:1}G_4$ systems, respectively. In principle, the micellar core should consist of only glycerol molecules. However, experimental core radius (R_{cmax} = 1.7 nm for $C_{12}G_2$ micelles) is higher than twice of the extended length of a glycerol moiety: one glycerol moiety of surfactant accounts ~0.4–0.5 nm. This demonstrates that the reverse micellar core also contains some impure water present in the $C_{12}G_2$ (contains ~0.8 % water) and decane (contains ~0.007 % water).

Next, we discuss the effect of surfactant chain length in diglycerol fatty acid ester nonionic surfactants [58]. Diglycerol monocaprate ($C_{10}G_2$), diglycerol monolaurate ($C_{12}G_2$), and diglycerol monomyristate ($C_{14}G_2$) surfactants having exactly the same hydrophilic unit are considered. Figure 4.2 shows SAXS patterns and the calculated $p(r)$ functions for $C_{10}G_2$, $C_{12}G_2$, and $C_{14}G_2$ (5 %) in styrene at 25 °C. Decreasing forward scattering intensities ($I(q = 0)$) without affecting the scattering behavior at the higher q region upon increasing hydrocarbon chain length of surfactant from C_{10} to C_{14} (Fig. 4.2a) indicates a decrease in the micellar size. As shown in the $p(r)$ curves (Fig. 4.2b), D_{max} estimated from the $p(r)$ curves decreases from ca. 6.0 to 3.0 nm upon the increase in surfactant chain from C_{10} to C_{14} keeping the position of maximum in $p(r)$ apparently unchanged. This trend corresponds to the ellipsoidal prolate-to-sphere type transition. Figure 4.2c, d shows optimum fit curves to the experimental $I(q)$ and the resulting $p(r)$ curves. Note that an ellipsoidal prolate model is able to explain the shape and size of the $C_{10}G_2$, $C_{12}G_2$, and $C_{14}G_2$ reverse micelles obtained by the GIFT method. The geometrical parameters such as short (a) and long (b) axes of the ellipsoidal prolate, the radius of the gyration (R_g), and the micellar aggregation number (N_{agg}) obtained from the model fittings are given in Table 4.1.

Table 4.1 shows that the value of a remains apparently the same, whereas b decreases upon changing surfactant from the $C_{10}G_2$ to $C_{14}G_2$ indicating an ellipsoidal prolate-to-sphere type transition in the reverse micellar structure. This transition is anticipated by the packing parameter. Micellar shrinkage may also be caused due to repulsive excluded volume interactions, which increases in magnitude with the increase in the volume of the hydrophobic part of surfactant. As a result, the micelle interface tends to become more curved and as a consequence micelles shrink.

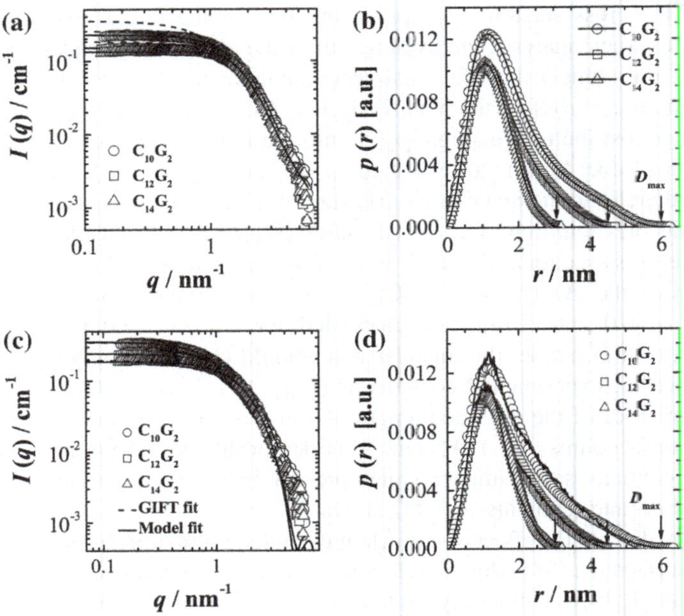

Fig. 4.2 **a** SAXS intensities, $I(q)$, of 5 % $C_{10}G_2$, 5 % $C_{12}G_2$, and 5 % $C_{14}G_2$ in styrene at 25 °C in absolute scales, **b** the corresponding $p(r)$ curves, **c** model fitting (*solid lines*) to the experimental scattering $I(q)$ (*symbols*) with GIFT fit (*dashed lines*), and **d** comparison of $p(r)$ curves obtained from the GIFT method (*symbols*) and model fittings (*solid lines*). The data were fitted by considering a homogeneous ellipsoidal prolate model. Adopted with permission from Ref. [58] © 2011 American Chemical Society

Table 4.1 Effect of surfactant chain length on the geometrical parameters such as short axis (a), and long axis (b) of ellipsoid prolate, radius of gyration (R_g), and aggregation number (N_{agg})

Surfactant	Short axis (a)/nm	Long axis (b)/nm	Radius of gyration (R_g)/nm	Aggregation number (N_{agg})
$C_{10}G_2$	1.40	3.1	1.5	49
$C_{12}G_2$	1.40	2.2	1.2	36
$C_{14}G_2$	1.39	1.5	0.9	24

Reprinted with permission from Ref. [58] © 2011 American Chemical Society

4.2.2 Solvent Effect on the Reverse Micellar Structure

Here, we discuss solvent molecular structure-induced structural evolution in the structure of $C_{14}G_2$ and $C_{10}G_2$ nonionic surfactant reverse micelles. First, we discuss the structure of $C_{14}G_2$ reverse micelles in cyclic and straight chain alkanes having different chain lengths. Since $C_{14}G_2$ forms self-assembled reverse micelles only at higher temperatures, micellar structure at 50 °C is compared. Figure 4.3 shows $I(q)$

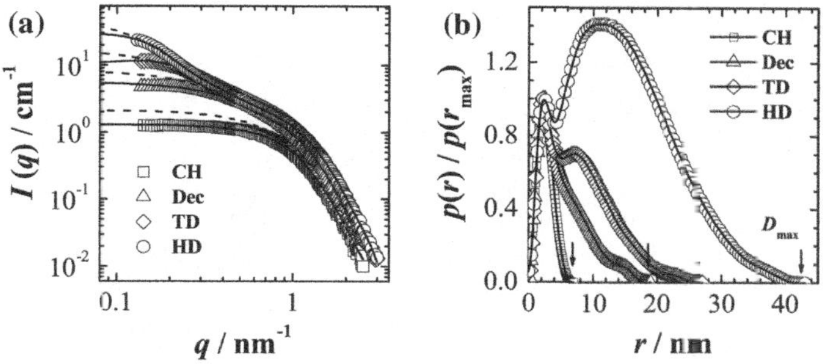

Fig. 4.3 **a** Normalized SAXS intensities, $I(q)$s, of 5 % $C_{14}G_2$ in CH, Dec, TD, and HD at 50 °C in absolute scales, **b** the corresponding real-space structural information, $p(r)$ curves. Adopted with permission from Ref. [59] © 2007 American Chemical Society

s and the corresponding $p(r)$ curves. Slightly asymmetric shape of the $p(r)$ curve in cyclohexane (CH) confirms an ellipsoid prolate type of micelle with a nearly homogenous scattering length density distribution inside the core. The micellar size is ca. 6.5 nm. When CH is replaced with straight chain alkane oils, micellar structure becomes highly sensitive to the chain length. Increase in hydrocarbon chain of alkane oil favored 1D micellar growth [59]. As shown in Fig. 4.3a, the $I(q = 0)$ and the low q slope continuously increase upon changing oil from *n*-decane (Dec) to *n*-hexadecane (HD) indicating the micellar growth. The $p(r)$ curve in Dec exhibits a typical signature of ellipsoid prolate. Whereas a clear feature of rodlike micelle is observed with *n*-tetradecane (TD). The low q slope becomes noticeably steeper and reaches $\sim q^{-2}$ in HD system. Additionally, a small bump seen in $p(r)$ of TD system grew into a pronounced broad maximum accompanying a drastic increase of D_{max} (Fig. 4.3b). Although this is a typical feature of planar or disklike 2D aggregate, the presence of a strong attractive interactions cannot be avoided as the system is very close to the phase separation line and the obtained $p(r)$ curves may simply be due to branched cylinders [60].

Next, we discuss the $C_{10}G_2$ nonionic surfactant reverse micelles in aromatic solvents such as alkylated benzenes; ethylbenzene (EB), hexylbenzene (HB), octylbenzene (OB), decylbenzene (DB), dodecylbenzene (DDB), and tetradecylbenzene (TDB). Note that alkylbenzenes are less nonpolar than alkane oils. Therefore, the Krafft temperature of $C_{10}G_2$ decreases in these solvents and thus forms self-assembled reverse micelles at 25 °C [61]. Figure 4.4a, b shows $I(q)$s and $p(r)$ curves for the 5 % $C_{10}G_2$/alkylbenzenes. Figure 4.4c, d shows the results obtained from geometrical model fittings for the selected systems as typical example. As shown in Fig. 4.4a, the scattering curve highly depends on the alkyl chain length of solvent. Forward scattering intensity increases with the alkyl chain length while the scattering in the higher q region ($q > 1.5$ nm^{-1}) remains essentially unchanged demonstrating micellar growth. The low q scattering behavior of EB system follows approximately q^0 behavior (typical of spheroid micelle). The low q slope increases in parallel to the alkyl chain length and follows q^{-1} behavior in the

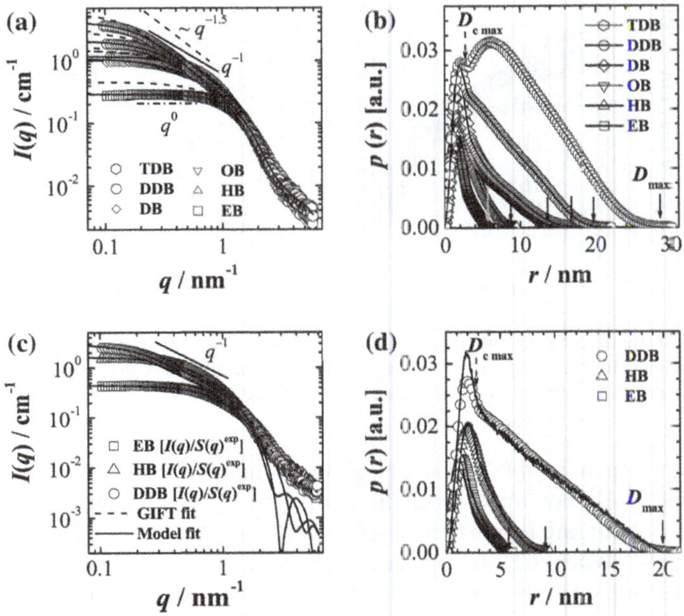

Fig. 4.4 **a** X-ray scattered intensities, $I(q)$, of the 5 % $C_{10}G_2$ in alkylbenzene at 25 °C in absolute scales, **b** the $p(r)$ functions, **c** model fitting (*solid lines*) to the experimental scattering intensities (*symbols*) with GIFT fit (*dashed lines*), and **d** comparison of $p(r)$ curves obtained from GIFT method (*symbols*) and model fittings (*solid lines*). The data were fitted considering a homogeneous ellipsoid prolate (for EB and HB) and homogeneous cylinder (DDB) models. Adopted with permission from Ref. [61] © 2013 American Chemical Society

DDB system. Growth to rodlike micelles leads to the q^{-1} behavior in the scattering curve, at least for systems with negligible interaction effects. With further increase in alkyl chain, the slope further increases and reaches $q^{-1.5}$ in TDB system, which is again the signature of the attractive interactions. The shape of the $p(r)$ curve for EB system corresponds to typical shape of globular micelle with maximum core diameter ca. 5.7 nm. These globular micelles transform into rodlike micelles with a cross-sectional diameter approximately 3.0 nm (confirmed from the inflection point after the $p(r)$ peak, which semiquantitatively indicates cross-sectional diameter of rodlike micelle) and maximum length of about 9.5 nm in HB. Here again, the transition is sphere-to-rod type. The axial growth of $C_{10}G_2$ micelles with successive increase in the alkyl chain length of alkylbenzene is shown in Fig. 4.4b. Length of rodlike micelles is ca. 13.8 nm (OB), 17.0 nm (DB), 20.9 nm (DDB), and 28.5 nm (TDB). The additional broad peak after the typical cross-sectional peak in TDB is the signature of the attractive interaction.

The optimum model fits to the experimental $I(q)$, and the derived $p(r)$ curves are presented in Fig. 4.4c, d. Homogeneous ellipsoidal prolate and cylinder models well explain the shape and size of the micelles derived from the GIFT method with a notable divergence in the higher q region of the experimental $I(q)$ curves. The absence of the theoretically predicted minima in the high q part of the experimental $I(q)$ is an indication of polydispersity in size and/or small electron density fluctuations inside the micellar core.

The solvent-induced 1D micellar growth can be explained in terms of polarity of the solvent. With increase in chain length, the solvent polarity decreases which perturbs the interpenetration of hydrophobic chains of surfactant and solvent. As a result, the effective cross-sectional area of hydrophilic moiety of the surfactant and consequently critical packing at the micellar interface decreases. In non-aqueous system, decrease in critical packing causes micellar growth. Moreover, transfer free energy of hydrophilic part of the surfactant from hydrophilic to oily hydrophobic environment with different hydrocarbon chain length may be crucial for the determination of the self-assembly of amphiphiles in non-aqueous media. It can be expected that the oil-to-oil transfer free energy of hydrophobic moiety of $C_{10}G_2$ is negligibly small for the different chain lengths of hydrocarbon oils. However, considering highly hydrophilic nature of diglycerol moiety, the transfer free energy from hydrophilic to oily environment will be considerably different depending on the chain length of solvent. This unfavorable free energy of diglycerol moiety with the longer alkyl chain alkylbenzene will decrease the cross-sectional area occupied by the $C_{10}G_2$ molecule on the hydrophilic/hydrophobic interface to optimize the contact with solvent and promotes the formation of longer cylindrical micelles.

4.2.3 *Effect of Concentration on the Reverse Micellar Structure*

In this section, we discuss the effect of surfactant concentration on the reverse micellar structure. We will discuss two examples; in first case, reverse micelles grow with concentration, and in second case, shape and size of reverse micelles remain apparently unaffected by the change in surfactant concentration.

First, we discuss the concentration-induced 1D micellar growth in $C_{10}G_2$/styrene system at 25 °C [58]. Figure 4.5 shows the experimental SAXS intensities, $I(q)$s, and the GIFT derived $p(r)$ curves in the dilute regions of $C_{10}G_2$/styrene system at 25 °C at different surfactant concentrations (5 % $\leq W_s \leq$ 25 %). In this system, the D_{max} of the ellipsoidal prolate micelles increases with the surfactant concentration and eventually transforms into rodlike micelles at $W_s \geq$ 15 %. As it can be seen in the $I(q)$ curves increasing surfactant concentration from 5 to 10 %, the $I(q)$ increase throughout the q range, which is more than expected due to an increase in the number density of micelles in unit volume. Careful observation of $I(q)$ function for 15 % system reveals that although the scattering intensity is higher throughout the q range, low q scattering intensity ($I(q = 0)$) is smaller than expected. Moreover, a weak interaction peak that appears at $q_{max} \sim 1.2\ nm^{-1}$ grows with concentration and also shifts toward the higher q side ($q_{max} \sim 1.52\ nm^{-1}$ for

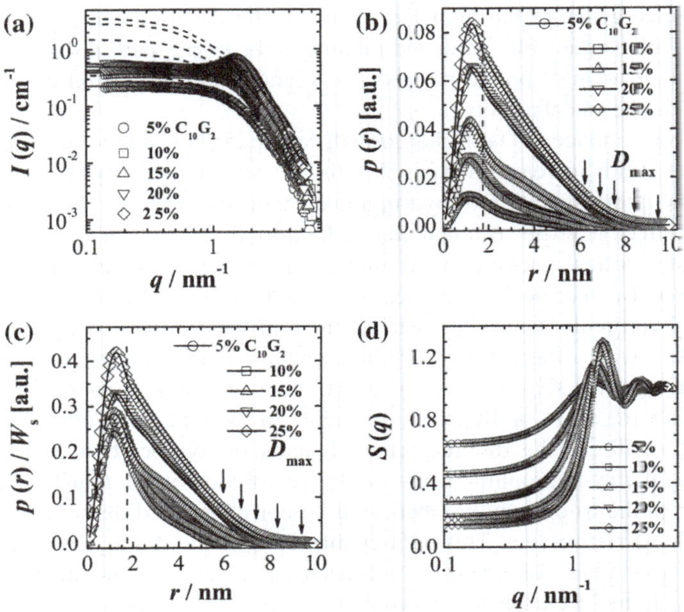

Fig. 4.5 **a** The $I(q)$ curves of the $C_{10}G_2$/styrene system at different surfactant concentrations ($5\ \% \le W_s \le 25\ \%$) on absolute scale at 25 °C, **b** the corresponding $p(r)$ functions, **c** the normalized $p(r)$ by surfactant weight fraction, $p(r)/W_s$, and **d** the static structure factor curves, $S(q)$. The *broken lines* in panel **b** and **c** highlight the cross-sectional diameter, D_c. Adopted with permission from Ref. [58] © 2011 American Chemical Society

$W_s = 25\ \%$ system), i.e., the intermicellar distance between the neighboring micelles decreases.

D_{max} increases from ca. 6.0 to 9.3 nm upon increasing $C_{10}G_2$ concentration from $W_s = 5$ to 25 %. However, the inflection point after the maximum of $p(r)$ curves remains essentially constant at ca. 1.75 nm suggesting that the micellar growth is apparently 1D. Monotonous increase of curve height with surfactant concentration in the normalized $p(r)$ curves by surfactant weight fraction (W_s), ($p(r)/W_s$), further confirms such growth (Fig. 4.5c). Careful observations reveal that the $p(r)$ curves at concentrations $W_s \ge 15\ \%$ decay linearly at higher r side showing the signature of ellipsoidal prolate-to-rodlike transition.

The static structure factor, $S(q)$, calculated by the GIFT evaluation of the SAXS data considering hard sphere interaction potential model reveals the presence of considerable intermicellar interactions mainly at the higher concentrations $W_s \ge 15\ \%$. The $S(q)$ peak position, which measures the mean distance between the neighboring micelles, shifts slightly toward higher q regions with increasing concentration inferring that the nearest neighboring micellar distance decreases. Note that the nature of actual structure factor curves may differ from what is predicted for mono-dispersed hard spheres, for which the extrapolated $S(q)$ to zero

Table 4.2 Effect of surfactant concentration on the geometrical parameters of $C_{10}G_2$ micelles in styrene such as short axis (a), and long axis (b) of ellipsoid prolate, cross-sectional radius (R_c), length of cylinder (L_{max}), radius of gyration (R_g), and aggregation number (N_{agg})

Surfactant weight fraction (W_s)	Short axis (a)/nm	Long axis (b)/nm	Cross-sectional radius (R_c)/nm	Maximum length of cylinder (L_{max})/nm	Radius of gyration (R_g)/nm	Aggregation number (N_{agg})
0.05	1.40	3.10	–	–	1.56	49
0.10	1.44	3.35	–	–	1.78	53
0.15	–	–	0.82	6.95	2.07	162
0.20	–	–	0.87	8.05	2.38	264
0.25	–	–	0.87	8.41	2.50	276

q, $S(q = 0)$, value reflects the osmotic compressibility of the system and is determined by the packing fraction of hard spheres.

Geometrical models of homogeneous ellipsoid prolate (5 % $\leq W_s \leq$ 10 %) and homogeneous cylinder (15 % $\leq W_s \leq$ 25 %) well fit the SAXS data. The calculated geometrical parameters obtained from model fittings are summarized in Table 4.2.

Next, we discuss the effect of surfactant concentration in $C_{18:1}G_2$/decane system at 25 °C [53]. Figure 4.6 shows the $I(q)$ and the corresponding $p(r)$ curves for the $C_{18:1}G_2$/decane system at different concentrations (5–25 %) at 25 °C. From the $p(r)$ curves, it is quite obvious that reverse micellar size does not change despite a wide variation in the surfactant concentrations. Figure 4.6a shows that with increasing surfactant concentration from 5 to 15 %, the scattering intensities increas throughout the q range. This could be attributed to the increase in the number density of reverse micelles in unit volume. However, with further increasing concentration say above 15 %, the scattering intensity in the low q region decreases and again a weak but growing interaction peak appears at intermediate q range ($q \sim 0.9\ \text{nm}^{-1}$) due to repulsive intermicellar interactions.

Depending on the systems, reverse micelles may or may not grow with concentration. Usually, elongated micelles show growth with concentration as was seen in $C_{10}G_2$/styrene system. On the other hand, spheroid micelles, which are commonly observed in lipophilic surfactants in non-aqueous systems, do not grow with composition, for instance, in $C_{18:1}G_2$/decane system. In case where micelles do not grow with concentration, the number density of micelles increases with increase in surfactant concentration, and a repulsive intermicellar force, which works perpendicular to the interface, comes into play. The repulsive intermicellar interactions can induce micelle ordering with a symmetry depending on the micellar shape. As we have discussed, $C_{18:1}G_2$/decane system contains slightly elongated prolate type reverse micelles at 25 °C; however, shape and size of the micelles remain unchanged upon increasing surfactant concentration from 5 to 25 %. The apparently unchanged position of the maximum of $p(r)$ in the normalized $p(r)$ function

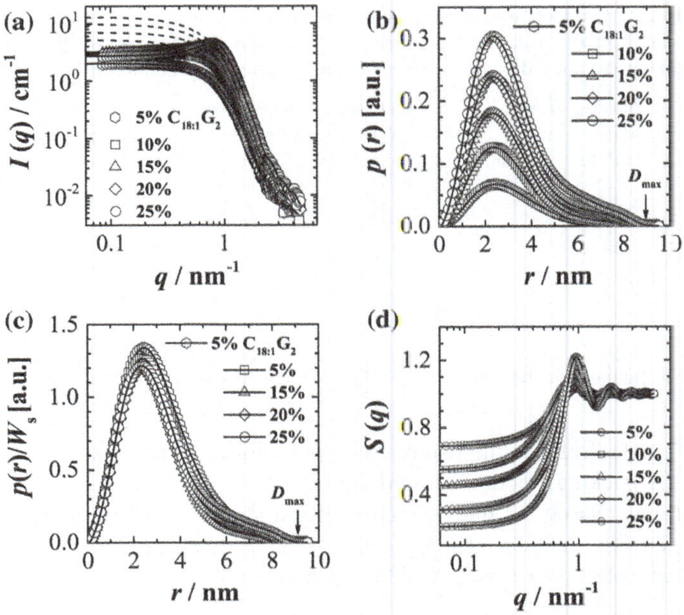

Fig. 4.6 **a** The scattering curves $I(q)$ of the $C_{18:1}G_2$/decane systems at different surfactant concentrations (5–25 %) obtained on absolute scale at 25 °C, **b** the corresponding real-space $p(r)$ functions, **c** the $p(r)$ functions normalized by weight fraction of the surfactant (W_s), and **d** the structure factor curves obtained from the GIFT evaluation of the SAXS data. Adopted with permission from Ref. [52] © 2010 American Chemical Society

(Fig. 4.6c) highlights the fact that the radius of micellar core also remains unchanged despite a wide variation in the surfactant concentration. All the curves lie almost on top of each other. In case of micellar growth, the curves height would increase with the volume fraction of the surfactant. Thus, in this system, the only change occurred with increase in surfactant concentration is the increased number density of micelles at higher surfactant concentrations.

Static structure factor reveals the presence of considerable intermicellar interactions particularly at higher concentrations (>15 %), and the $S(q)$ peak position shifts slightly toward higher q regions with increasing concentration (Fig. 4.6d). The $S(q = 0)$ value decreases monotonically with surfactant concentration due to the reduced osmotic compressibility of micelles. From the virtually concentration independent structure factor peak at $q \sim 0.9$ nm^{-1} in the $S(q)$ curves, the mean distance between the micelles is ca. ~7.0 nm, which is slightly smaller than the average maximum diameter of the micelles (~9.0 nm). This highlights the fact that the micelles are not approaching laterally but from other possible orientations with a possibility of interpenetration of hydrophobic chain of surfactant and decane, mainly at higher volume fraction of the surfactant.

4.2.4 Thermo-Responsive Behavior of Reverse Micelles

Like normal micelles in aqueous systems, nonionic surfactant reverse micelles in non-aqueous media also show thermo-responsive behavior [58–61]. However, contrary to the normal micelles, charge-free reverse micelles show rod-to-sphere type transition with increase in temperature [58–61]. Herein, we will discuss two examples. In the first example, we discuss temperature-induced shrinkage of ellipsoid prolate type reverse micelles observed in $C_{12}G_2$/EB system [62]. In second example, rod-to-sphere type transition in $C_{10}G_2$/TDB system will be discussed.

Figure 4.7 shows $I(q)$ and $p(r)$ functions for 5 % $C_{12}G_2$/EB system at different temperatures (25, 40, 50, 60, and 75 °C). One can clearly see in Fig. 4.3a that successive increase in temperature from 25 to 75 °C decreases the forward scattering intensity and the calculated form factor $P(q)$ in the low q regime of the SAXS data, providing evidence of decrease in micelle size. D_{max} decreases from 5.0 to 3.8 nm with increase in temperature from 25 to 75 °C, i.e., the micellar size decreases by ~30 %.

Next, we discuss the effect of temperature on the micellar structure of $C_{10}G_2$/TDB system [61]. The forward scattering decreases monotonically with the rise of temperature while keeping the scattering in the higher q region ($q > 1.5$ nm^{-1}) essentially unchanged (Fig. 4.8a). Note that this scattering behavior is exactly opposite to the increase in alkyl chain length of alkylbenzene (Fig. 4.4a) inferring that an increase in temperature decreases the micellar size. Furthermore, the low q slope of the scattering curve which follows approximately $q^{-1.5}$ behavior at 25 °C decreased and follows q^{-1} behavior at 85 °C (broken line in Fig. 4.8a). These observations confirmed that increase in temperature decreases the micellar size and also reduces the attractive interactions that are present at lower temperature (25 °C), which can be best seen in $p(r)$ functions (Fig. 4.8b). A broad peak after the typical cross-sectional peak (attractive interaction) decreases, and the maximum axial length of micelles as indicated by arrows also decreases with the rise in temperature. At highest SAXS measurement temperature (85 °C), the $p(r)$ apparently

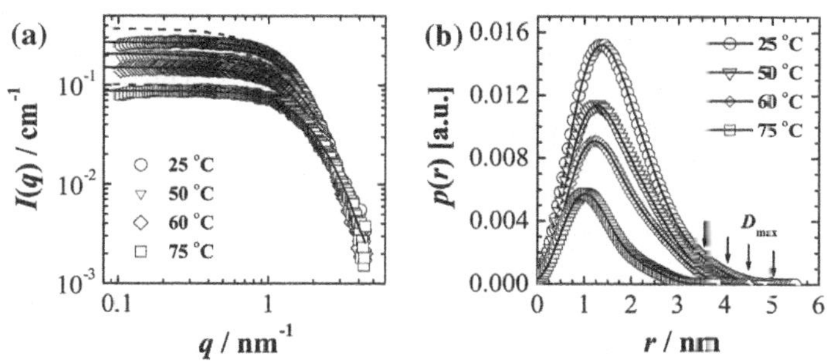

Fig. 4.7 **a** X-ray scattering intensity, $I(q)$, for 5 % $C_{12}G_2$/EB system in absolute scales at different temperatures (25–75 °C), and **b** the corresponding real-space $p(r)$ functions. Adopted with permission from Ref. [62] © Elsevier

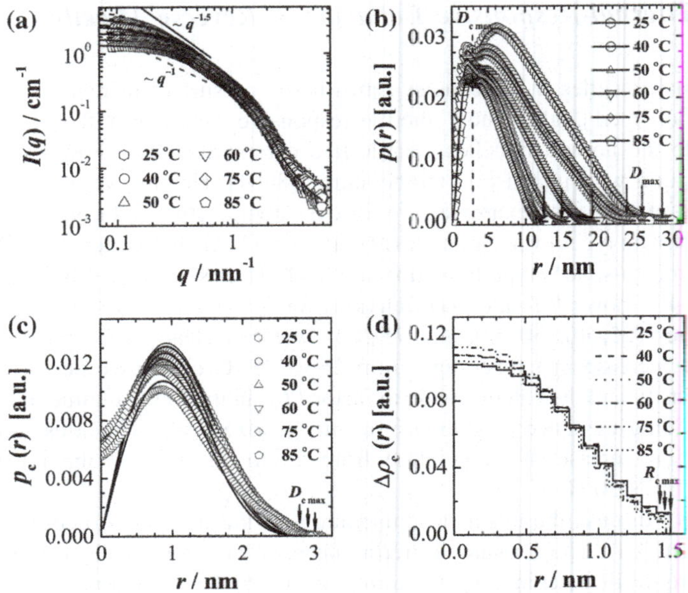

Fig. 4.8 **a** X-ray scattered intensities, $I(q)$, of the 5 % $C_{10}G_2$/TDB at different temperatures (25, 40, 50,60, 75, and 85 °C) in absolute scales, **b** the corresponding $p(r)$ functions, **c** the $p_c(r)$ curves, and **d** the corresponding $\Delta\rho_c(r)$ profiles, calculated with the deconvolution procedure of the $p_c(r)$ using DECON program. Adopted with permission from Ref. [61] © 2013 American Chemical Society

corresponds to the form factor and has the typical shape for a rod with maximum length ca. 11.5 nm and cross-sectional diameter approximately 3.0 nm (indicated by a broken line in Fig. 4.8b). Although micelles still exhibit rodlike structure at 85 °C, globular type micelles are expected to be formed at higher temperature say above 100 °C, i.e., rod-to-sphere transition is expected. Figure 4.8c, d shows the results of direct cross-sectional analysis for the quantitative estimation of cross-sectional internal structure. From the $p_c(r)$ functions, the maximum cross-sectional diameter, D_{cmax}, is ca. 3.09 nm and it remains unchanged until 60 °C and then decreases slightly at higher temperatures, e.g., D_{cmax} is ca. 2.99 and 2.95 nm for 75 and 85 °C, respectively. A similar trend is observed in core cross-sectional radius estimated from the $\Delta\rho_c(r)$ profiles (Fig. 4.8d); R_{cmax} is ca. 1.50 until 60 °C and then decreases to 1.46 (75 °C) and 1.44 (85 °C).

Temperature-induced decrease in reverse micellar size in non-aqueous media is anticipated to be caused by (a) an increase in temperature that enhances the penetration tendency of solvent molecules into the lipophilic chin of surfactant, and micellar curvature tends to be more curved at higher temperature; in other words, interpenetration reduces the effective cross-sectional area of the surfactant head group and increases the *cpp*, (b) temperature increasing the effective hydrophobic volume of the surfactant chain by increasing the kink states in the chain, and (c) the temperature effect on the aggregate size could also be related to dehydration observed for ethoxylated surfactants upon heating.

4.2.5 Water-Induced Reverse Micellar Growth

Reverse micelles are very well known for their ability to encapsulate water or other polar solvent molecules [59, 62, 63]. Usually, water or other polar solvent molecules tend to be soluble in the reverse micellar core causing microstructure transformations of the parent micelles.

Investigations have shown that reverse micelles swell with water and also induced 1D micellar growth producing bigger or elongated micelles [63]. Sometimes, addition of small amount of water causes sphere-to-wormlike micelle transition [35, 63]. Here, we discuss the microstructure transformations induced by the addition of small amount water in the reverse micelles system of 5 % $C_{10}G_2$/EB. This system potentially solubilizes 2.0 % water at 25 °C and caused a significant micellar growth [62]. Both the maximum dimension and cross-sectional radius of the micellar core increase with water. As expected, water causes significant modification in the scattering function of 5 % $C_{10}G_2$/EB system (Fig. 4.9).

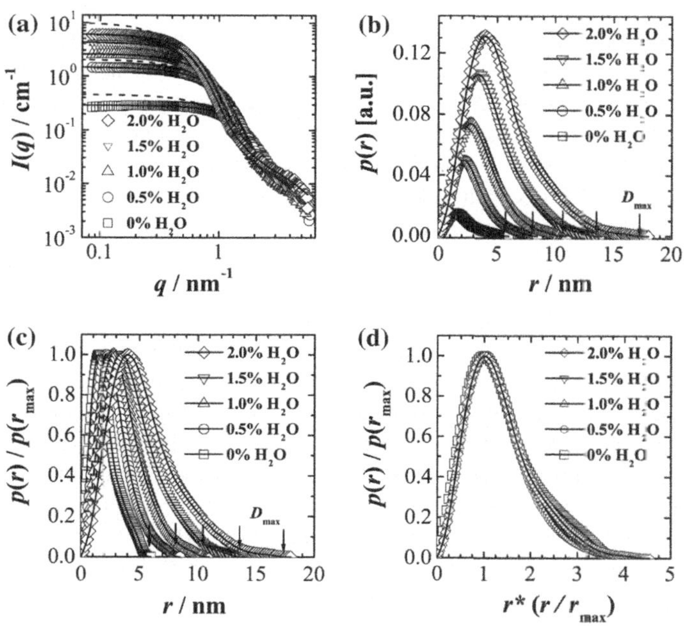

Fig. 4.9 **a** The $I(q)$ curves of 5 % $C_{10}G_2$/EB system at different water concentrations (0, 0.5, 1.0, 1.5, and 2.0 %) on absolute scale at 25 °C, **b** the corresponding $p(r)$ functions, **c** the normalized $p(r)$ function, $p(r)/p(r_{max})$, and **d** master curves, plot of $p(r)/p(r_{max})$ versus r^* (r/r_{max}). Adopted with permission from Ref. [62] © Elsevier

The low q scattering intensity, $I(q = 0)$, increases more than one order of magnitude, and the $I(q)$ curve in the cross-sectional region ($1.8 \leq q \leq 3.0\ nm^{-1}$) shifts toward the lower q side upon the addition of water (Fig. 4.9a). This is a clear indication of water-induced simultaneous changes in maximum dimension and the cross-sectional structure of $C_{10}G_2$ reverse micelles. In the normalized $p(r)$ curves (Fig. 4.9c), both the positions of maximum of $p(r)$, r_{max}, and D_{max} shift toward higher r values providing evidence of micellar growth. A monotonous increase in D_{max} with water is shown in $p(r)$ curves (Fig. 4.9b); D_{max} are ca. ~5.7 nm (0 % H_2O), 8.0 nm (0.5 % H_2O), 10.7 nm (1.0 % H_2O), 13.6 nm (1.5 % H_2O), and 17.2 nm (2.0 % H_2O). Note that size of the 2.0 % H_2O incorporated micelles is approximately 3 times bigger than the parent micelles without water. The master plot (plot of normalized $p(r)$ vs. r^* (r/r_{max})) shows that all the curves lie almost on top of each other indicating that shape of the micelles remains virtually the same even after water addition. In this case, there is a water pool at the micellar core.

4.2.6 Charge-Free Reverse Wormlike Micelles

Wormlike micelles are elongated flexible self-assembled structures formed by the aggregation of surfactants and polymers [64]. Elongated micelles above a certain threshold concentration entangle into a dynamic network structure and display remarkable viscoelastic properties [65, 66], which have been exploited in numerous industrial and technological fields. Like the conventional normal micelles in aqueous system, reverse spheroid micelles can grow into wormlike micelle network structure and show viscoelastic properties upon applying external stimuli such as water and co-surfactants. Ternary mixtures of lecithin/water/oil are one such example [63, 67, 68]. Lecithin; a zwitterionic phospholipid with two alkyl chains, usually forms globular reverse micelles in a variety of organic solvents, which can grow axially into flexible cylinders upon the addition of trace amount of water. The growth continues with water and after a certain concentration the sufficiently elongated cylindrical micelles form a transient network of reverse wormlike micelles with viscosity several orders of magnitude higher than those of organic solvent or water, and the system exhibits viscoelastic properties [69–71]. Thus, water has been regarded as an essential component for the formation of reverse wormlike micelles in lecithin-based aggregates, and it is the molar ratio of water to lecithin, which controls the reverse micellar growth. Some advancement in lecithin-based reverse wormlike micelles turned out where water could be replaced with polar additives such as formamide glycerol [72–74].

Here, we discuss charge-free reverse wormlike micelles formulation in a non-aqueous system without the addition of water. Sucrose-based lipophilic nonionic surfactant, sucrose trioleate (abbreviated as STO), spontaneously self-assembled into small globular reverse micelles in n-hexadecane (HD) without water under ambient conditions [75]. Addition of relatively less lipophilic sucrose dioleate (SDO) into the semidilute solution of STO micelles causes the micellar growth into long cylinders, which eventually form entangled transient network of viscoelastic wormlike micelles [75]. Evolution in the micellar structure can be judged from the SAXS (Fig. 4.10) and rheology data (Fig. 4.11). Figure 4.10 shows that the low q scattering intensity reaches to zero in parallel to the abscissa, i.e., the low q scattering intensity follows q^0 behavior in the forward direction, which is an indication of the formation of globular micelles in the SDO-free system. The scattering behavior of STO micelles modifies significantly in the SDO incorporated system. The forward scattering intensity, $I(q = 0)$, and also the low q slope increase with SDO weight fraction (W_1 = SDO/(STO + SDO)), and eventually at $W_1 = 0.90$, the $I(q)$ decays following q^{-1} behavior in the low q region, which infers the formation of cylindrical micelles, i.e., SDO-induced sphere-to-rod type transition.

The real-space $p(r)$ function for the systems with $W_1 = 0$ to 0.30 represents globular type of micelles with D_{max} below 7.0 nm. An asymmetry in the shape of the $p(r)$ curves appears with further increases in W_1 showing a shape transition from globular to elongated ellipsoidal prolate or short rodlike micelles for W_1 = 0.50–0.70 systems. Micellar size increases with W_1; D_{max} is ca. 10 nm ($W_1 = 0.50$) and

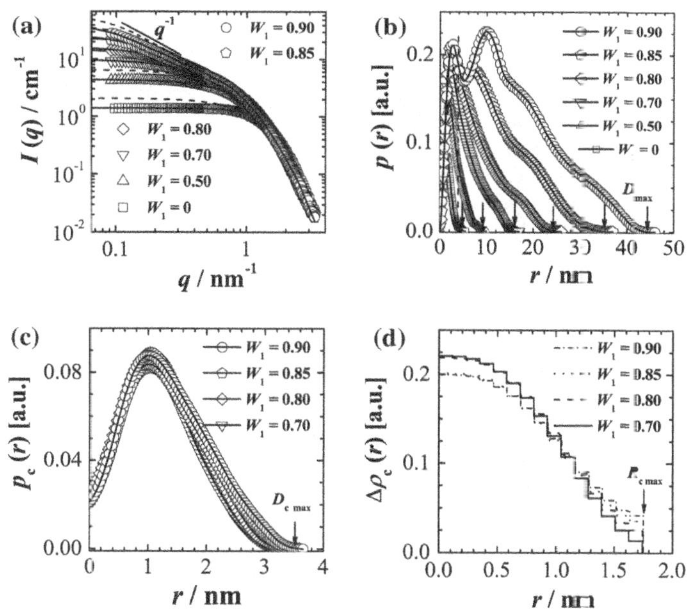

Fig. 4.10 **a** X-ray scattered intensities, $I(q)$, of the 10 % STO/hexadecane/SDO systems at different mixing fraction of SDO (W_1 = SDO/(STO + SDO)) at 25 °C in absolute scales, and **b** the corresponding $p(r)$ curves, **c** $p_c(r)$, and **d** $\Delta\rho_c(r)$ curves of selected systems. Adopted with permission from Ref. [75] © 2011 American Chemical Society

$\sim$16.5 nm (W_1 = 0.70). The $p(r)$ curves at $W_1 \geq 0.80$ represent typical of long cylinders with D_{max} ca. 35.5 nm (W_1 = 0.70) and 45 nm (W_1 = 0.90). Note that the inflection point after the pronounced maximum in the lower r regime, as indicated by broken line (Fig. 4.10b), remains essentially unchanged. This shows that the growth caused by SDO is purely one-dimensional. Appearance of a deviation in the intermediate r region, mainly at W_1 = 0.85 and W_1 = 0.90, more significant in the latter case, can be attributed to the strong attractive interactions as the system is approaching the phase separation line (above W_1 = 0.90, the system phase separates out into a turbid solution of vesicular dispersions).

Quantitative information of internal cross-sectional structure can be seen in Fig. 4.10c, d, which shows the $p_c(r)$ and $\Delta\rho_c(r)$ for the elongated micellar systems (W_1 = 0.70–0.90). The positive values of $\Delta\rho_c(r)$ in the entire r region are a clear confirmation of the electron-rich hydrophilic reverse micellar core. The calculated cross-sectional diameters from the position of D_{cmax} in $p_c(r)$ are almost identical, D_{cmax} ca. 3.6 nm, which is close to those roughly estimated from the inflection point of the total $p(r)$ curves. The core radius, R_{core} ca. $\sim$1.75 nm, from the $\Delta\rho_c(r)$ is nearly half of the cross-sectional diameter and remains unchanged at all W_1.

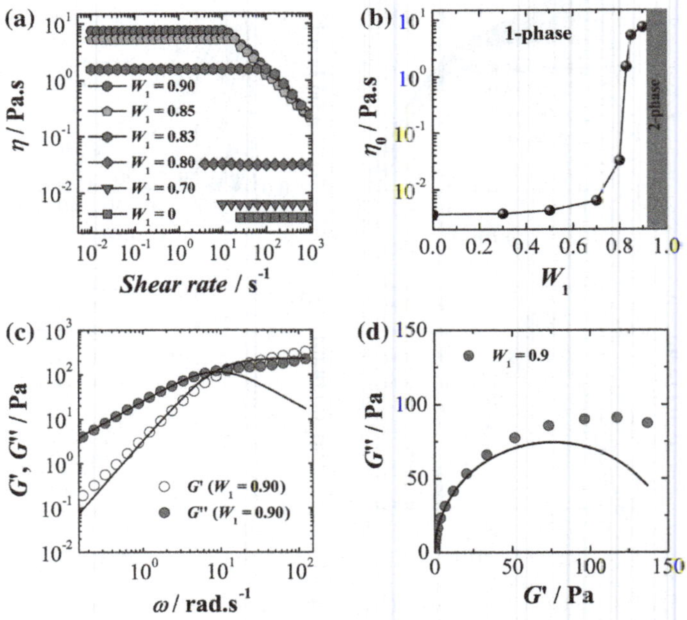

Fig. 4.11 **a** Steady-shear viscosity at different mixing fraction of SDO, **b** the corresponding zero-shear viscosity, **c** dynamic rheological behavior, and **d** Cole–Cole plot (variation of G'' vs. G'). In panel **a,** symbols (for G' and G'') represent the experimental data points and solid lines are the Maxwell model fit. Adopted with permission from Ref. [75] © 2011 American Chemical Society

Next, we discuss the modification in bulk rheological properties of the system induced by the microstructure transitions from sphere-to-cylinder or wormlike micelles. Steady-shear rheological results presented in Fig. 4.11a show that the bulk viscosity of STO globular micelles is equivalent to that of pure hexadecane. Viscosity is independent to the applied shear (Newtonian fluid). Addition of SDO (up to $W_1 = 0.5$) increases viscosity only slightly, which is an indication of a micellar growth. However, as confirmed by SAXS, micelles are only slightly elongated; network structure is yet to be formed. Viscosity increases significantly until $W_1 \leq 0.8$, but the solution still exhibits a Newtonian fluidlike behavior. A shear-thinning, an indication of formation of entangled network structure, behavior is observed only at $W_1 \geq 0.83$ (Fig. 4.11a). The samples with compositions $W_1 > 0.83$ show a shear birefringent behavior (birefringent upon shaking or when a small force is applied but isotropic at rest). Further, increase in W_1 not only increases the viscosity but also shifts the critical shear rate (shear rate at which shear-thinning occurs) toward the low shear rate showing that system is getting more structured. Zero-shear viscosity (η_0) calculated by extrapolation of the plateau viscosity to zero shear rates increases only slightly until W_1 reaches $W_1 = 0.70$ and then increases significantly (Fig. 4.11b). Above $W_1 > 0.80$, the micelles are long enough to form a transient network of wormlike micelles. The η_0 increases by ~4th orders of magnitude as W_1 increases from 0 to 0.90. Above $W_1 = 0.90$, there is a phase separation to a turbid solution.

A representative plot of dynamic oscillatory shear rheology, elastic modulus (G') and viscous modulus (G''), as a function of frequency (ω), presented in Fig. 4.11c clearly show that the system display viscoelastic properties; elastic behavior ($G' > G''$) at higher ω or short time scales and viscous behavior ($G'' > G'$) at low ω or long time scales [76]. Maxwell model with a single stress relaxation mode could explain the viscoelasticity of the system. The semicircular Cole–Cole plot (Fig. 4.11d: variation of G'' vs. G') further verifies the viscoelastic behavior of wormlike micelles in this system [64].

SDO-induced sphere-to-wormlike micelle microstructure transformation can be understood in terms of critical packing. Contrary to the normal micelles in aqueous system, for a spherical micelle to be formed in non-aqueous media, the *cpp* must have a value well above 1, i.e., the molecular geometry of surfactant should have an inverted cone shape, which is possible in lipophilic surfactants. Furthermore, unlike in aqueous system for a reverse sphere to grow into a cylinder, the *cpp* must decrease. Here, we anticipate that the SDO, which itself is insoluble in hexadecane, is soluble at the surfactant palisade layer of STO micelles. Consequently, the spontaneous curvature of the system tends to be less negative. On the other hand, under this condition, the *cpp* tends to decrease favoring transition from spheres to

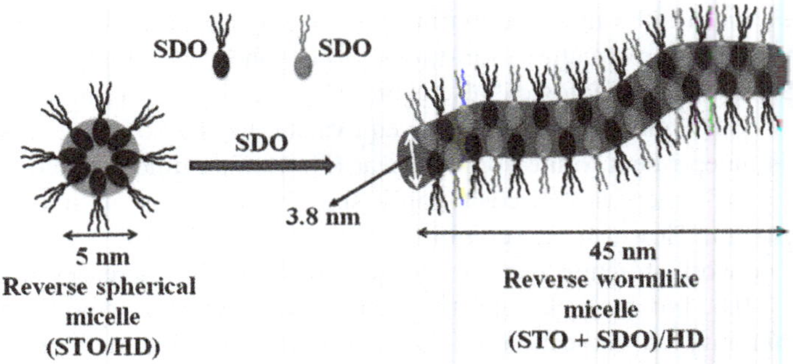

Fig. 4.12 Schematic representation of charge-free reverse micelle formation in non-aqueous media

long cylinders or wormlike micelles. It is also possible that the headgroups of SDO and STO surfactants are fastened together by hydrogen bonding and, as a result, the overall headgroup size of the surfactant increases so that *cpp* decreases and favors micellar growth. The possible mechanism of micellar growth is shown in Fig. 4.12.

References

1. Whitesides GM, Grzybowski B (2002) Self-assembly at all scales. Science 295:2418–2421
2. Cademartiri R, Stan CA, Tran VM, Wu E, Friar L, Vulis D, Clark LW, Tricard S, Whitesides GM (2012) A simple two-dimensional model system to study electrostatic self-assembly. Soft Matter 8:9771–9791
3. Lopes WA, Jaeger HM (2001) Hierarchical self-assembly of metal nanostructures on diblock copolymer scaffolds. Nature 414:735–738
4. Stupp SI, LeBonheur VV, Walker K, Li LS, Huggins KE, Keser M, Amstutz A (1997) Supramolecular materials: self-organized nanostructure. Science 276:384–389
5. Rodríguez-Abreu C, Shrestha RG, Shrestha LK, Harush E, Regev O (2013) Wormlike soft nanostructures in nonionic systems: principles, properties and application as templates. J Nanosci Nanotechnol 13:4497–4520
6. Mucic RC, Storhoff JJ, Mirkin CA, Letsinger RL (1998) DNA-directed synthesis of binary nanoparticle network materials. J Am Chem Soc 120:12674–12675
7. Thota BNS, Urner LH, Haag R (2016) Supramolecular architectures of dendritic amphiphiles in water. Chem Rev 116:2079–2102
8. Yu G, Jie K, Huang F (2015) Supramolecular amphiphiles based on host-guest molecular recognition motifs. Chem Rev 115:7240–7303
9. Barclay TG, Constantopoulos K, Matisons J (2014) Nanotubes self-assembled from amphiphilic molecules via helical intermediates. Chem Rev 114:10217–10291
10. Hill JP, Shrestha LK, Ishihara S, Ji Q, Ariga K (2014) Self-assembly: from amphiphiles to chromophores and beyond. Molecules 19:8589–8609
11. Ramanathan M, Shrestha LK, Mori T, Ji Q, Hill JP, Ariga K (2013) Amphiphile nanoarchitectonics: from basic physical chemistry to advanced applications. Phys Chem Chem Phys 15:10580–10611

12. Shrestha LK, Strzelczyk KM, Shrestha RG, Ichikawa K, Aramaki K, Hill JP, Ariga K (2015) Nonionic amphiphile nanoarchitectonics: self-assembly into micelles and lyotropic liquid crystals. Nanotechnology 26:204002–204011
13. Ariga K, Li J, Fei J, Ji Q, Hill JP (2016) Nanoarchitectonics for dynamics functional materials from atomic-/molecular-level manipulation to macroscopic action. Adv Mater 28:1251–1286
14. Ariga K, Li J (2016) Nanoarchitectonics for advanced materials: strategy beyond nanotechnology. Adv Mater 28:987–988
15. Aono M, Ariga K (2016) The way to nanoarchitectonics and the way of nanoarchitectonics. Adv Mater 28:989–992
16. Ariga K, Ji Q, Nakanishi W, Hill JP, Aono M (2015) Nanoarchitectonics: a new materials horizon for nanotechnology. Mater Horiz 2:406–413
17. Nakanishi W, Minami K, Shrestha LK, Ji Q, Hill JP, Ariga K (2014) Bioactive nanocarbon assemblies: nanoarchitectonics and application. Nano Today 9:378–394
18. Shrestha LK, Ji Q, Mori T, Miyazawa K, Yamauchi Y, Hill JP, Ariga K (2013) Fullerene nanoarchitectonics: from zero to higher dimensions. Chem Asian J 8:1662–1679
19. Tanford C (1980) *The hydrophobic effect*: formation of micelles and biological membrances. Wiley, New York
20. Nagarajan R, Ruckenstein E (1991) Theory of surfactant self-assembly: a predictive molecular thermodynamic approach. Langmuir 7:2934–2969
21. Chandler D (2005) Interfaces and the driving force of hydrophobic assembly. Nature 437:640–647
22. Dong R, Hao J (2010) Complex fluids of poly(oxyethylene) monoalkyl ether nonionic surfactants. Chem Rev 110:4978–5022
23. Kekicheff P, Grabielle-Madelmont C, Ollivon M (1989) Phase diagram of sodium dodecyl sulfate-water system: 1. A calorimetric study. J Colloid Interface Sci 131:112–132
24. Auvray X, Petipas C, Anthore R, Rico I, Lattes A (1989) X-ray diffraction study of mesophases of cetylmethylammonium bromide in water, formamide, and glycerol. J Phys Chem 93:7458–7464
25. Pieranski P, Sittler L, Sotta P, Imperor-Clerc M (2001) Growth and shapes of a cubic lyotropic liquid crystal. Eur Phys J E 5:317–328
26. Shrestha RG, Shrestha LK, Shrestha SC, Aramaki K (2008) Phase behavior and microstructure of nonionic fluorocarbon surfactant in aqueous system. J Phys Chem B 112:10520–10527
27. Shrestha LK, Shrestha RG, Iwanaga T, Aramaki K (2007) Aqueous phase behavior of diglycerol fatty acid esters. J Dispers Sci Technol 28:883–891
28. Shrestha RG, Rodriguez-Abreu C, Aramaki K (2009) Wormlike micelles in mixed amino acid surfactant/nonionic surfactant aqueous systems and the effect of added electrolyte. J Oleo Sci 58:243–254
29. Shrestha LK, Kaneko M, Sato T, Acharya DP, Iwanaga T, Kunieda H (2006) Phase behavior of diglycerol fatty acid esters-nonpolar oil systems. Langmuir 22:1449–1454
30. Shrestha LK, Sato T, Acharya DP, Iwanaga T, Aramaki K, Kunieda H (2006) Phase behavior of monoglycerol fatty acid ester in nonpolar oils: reverse rodlike micelles at elevated temperatures. J Phys Chem B 110:12266–12273
31. Palazzo G (2013) Wormlike reverse micelles. Soft Matter 9:10668–10677
32. Pileni MP (1989) Structure and reactivity in reverse micelles, vol 65. Pileni MP Ed., Elsevier, Amsterdam
33. Luisi PL, Strab BE (ed) (1987) Reverse micelles: biological and technological relevance of amphiphilic structures in apolar media. Plenum Press, New York
34. Shrestha LK, Shrestha RG, Aramaki K (2011) Growth control of nonionic reverse micelles by surfactant and solvent molecular architecture and water addition. J Nanosci Nanotechnol 11:4863–4873
35. Scartazzini R, Luisi PL (1988) Organogels from lecithins. J Phys Chem 92:829–833
36. Boutonnet M, Kizling J, Stenius P, Maire G (1982) The preparation of monodisperse colloidal metal particles from microemulsions. Colloids Surf 5:209–225

37. Lisiecki I, Pileni MP (1993) Synthesis of copper metallic clusters using reverse micelles as microreactors. J Am Chem Soc 115:3887–3896
38. Pileni MP (1997) Nanosized particles made in colloidal assemblies. Langmuir 13 3266–3276
39. López-Quintela MA (2003) Synthesis of nanomaterials in microemulsions: formation mechanism and growth control. Curr Opin Colloid Interface Sci 8:137–144
40. Pileni MP (2003) The role of soft colloidal templates in controlling sthe size and shape of inorganic nanocrystals. Nat Mater 2:145–150
41. Pileni MP (1993) Reverse micelles: a microreactor. J Phys Chem 97:9661–9668
42. Cushing BL, Kolesnichenko VL, O'Connor CJ (2004) Recent advances in the liquid-phase syntheses of inorganic nanoparticles. Chem Rev 104:3889–3946
43. Shrestha LK, Shrestha RG, Neus U, Rodriguez-Abreu C, Ariga K (2014) In-situ formation of silver nanoparticles using nonionic surfactant reverse micelles as nanoreactors. J Nanosci Nanotechnol 14:2238–2244
44. Shrestha LK, Sato T, Aramaki K (2007) Phase behavior and self-organized structure of diglycerol monolaurate in different nonpolar organic solvents. Langmuir 23:6606–6613
45. Shrestha LK, Aramaki K (2007) Phase behavior of diglycerol monomyristate in different nonpolar organic solvent systems. J Dispers Sci Technol 28:1236–1241
46. Shrestha LK, Aramaki K, Kato H, Takase Y, Kunieda H (2006) Foaming properties of monoglycerol fatty acid ester in nonpolar oil system. Langmuir 22:8337–8345
47. Kunieda H, Shrestha LK, Acharya DP, Kato H, Takase Y, Gutiérrez JM (2007) Super-stable nonaqueous foams in diglycerol fatty acid esters-nonpolar oil systems. J Dispers Sci Technol 28:133–142
48. Shrestha LK, Shrestha RG, Solans C, Aramaki K (2007) Effect of added water on the foaming properties of diglycerol fatty acid ester-oil systems. Langmuir 23:6918–6926
49. Shrestha RG, Shrestha LK, Solans C, Gonzalez C, Aramaki K (2010) Nonaqueous foam with outstanding stability in diglycerol monolaurate/olive oil system. Colloid Surf A 353:157–165
50. Brunner-Popela J, Glatter O (1997) Small-angle scattering of interacting particles. I. Basic principles of a global evaluation technique. J Appl Crystallogr 30:431–442
51. Weyerich B, Brunner-Popela J, Glatter O (1999) Small-angle scattering of interacting particles. II. Generalized indirect fourier transformation under consideration of the effective structure factor for polydisperse systems. J Appl Crystallogr 32:197–209
52. Shrestha LK, Dulle M, Glatter O, Aramaki K (2010) Structure of polyglycerol oleic acid ester nonionic surfactant reverse micelles in decane: Growth control by headgroup size. Langmuir 26:7015–7024
53. Shrestha LK, Sato T, Aramaki K (2009) Intrinsic parameters for structural variation of reverse micelles in nonionic surfactant (glycerol α-monolaurate)/oil systems: a SAXS study. Phys Chem Chem Phys 11:4251–4259
54. Shrestha LK, Sato T, Dulle M, Glatter O, Aramaki K (2010) Effect of lipophilic tail architecture and solvent engineering on the structure of trehalose-based nonionic surfactant reverse micelles. J Phys Chem B 114:12008–12017
55. Stradner A, Glatter O, Schurtenberger P (2000) A hexanol-induced sphere-to-flexible cylinder transition in aqueous alkyl polyglucoside solutions. Langmuir 16:5354–5364
56. Glatter O (1980) Determination of particle-size distribution functions from small-angle scattering data by means of the Indirect Transformation method. J Appl Crystallogr 13:7–11
57. Glatter O (1979) The interpretation of real-space information from small-angle scattering experiments. J Appl Crystallogr 12:166–175
58. Shrestha LK, Shrestha RG, Aramaki K (2011) Intrinsic parameters for the structure of nonionic reverse micelles in styrene: SAXS and rheometry studies. Langmuir 27:5862–5873
59. Shrestha LK, Sato T, Aramaki K (2007) Shape, size and structure control of reverse micelles in diglycerol monomyristate nonionic surfactant system. J Phys Chem B 111:1664–1671
60. Shrestha LK, Shrestha RG (2013) Noionic reverse micelles near the critical point. J Oleo Sci 62:1073–1081

61. Shrestha LK, Shrestha RG, Aramaki K, Yoshikawa G, Ariga K (2013) Demonstration of solvent-induced one-dimensional nonionic reverse micell growth. J Phys Chem Lett 4:2585–2590
62. Shrestha LK, Shrestha RG, Aramaki K, Hill JP, Ariga K (2012) Nonionic reverse micelle formulation and their microstructure transformations in an aromatic solvent ethylbenzene. Colloids Surf A 414:140–150
63. Schurtenberger P, Scartazzini R, Magid LJ, Leser ME, Luis PL (1990) Structural and dynamic properties of polymer-like reverse micelles. J Phys Chem 94:3695–3701
64. Dreiss CA (2007) Wormlike micelles: where do we stand? Recent developments, linear rheology and scattering techniques. Soft Matter 3:956–970
65. Shrestha RG, Shrestha LK, Aramaki K (2007) Formation of wormlike micelles in a mixed amino-acid based anionic surfactant and cationic surfactant systems. J Colloid Interface Sci 311:276–284
66. Shrestha RG, Abezgauz L, Danino D, Sakai K, Sakai H, Abe M (2011) Structure and dynamics of poly(oxyethylene) cholesteryl ether wormlike micelles: rheometry, SAXS, and Cryo-TEM studies. Langmuir 27:12877–12883
67. Yu Z-J, Neuman RD (1994) Giant rodlike reverse micelles formed by sodium bis (2-ethylhexyl) phosphate in n-heptane. Langmuir 10:2553–2558
68. Tung S-H, Huang Y-E, Raghavan SR (2006) A new reverse wormlike micellar system: mixture of bile salt and lecithin in organic liquids. J Am Chem Soc 128:5751–5756
69. Kumar R, Ketner AM, Raghavan SR (2010) Nonaqueous photorheological fluids based on light-responsive reverse wormlike micelles. Langmuir 26:5405–5411
70. Tung S-H, Huang Y-E, Raghavan SR (2007) Constrasting effects of temperature on the rheology of normal and reverse wormlike micelles. Langmuir 23:372–376
71. Shrestha RG, Agari N, Tsuchiya K, Sakamoto K, Saakai K, Abe M, Sakai H (2014) Phosphatidylcholine-based nonaqueous photorheological fluids: effect of geometry and solvent. Colloid Polym Sci 292:1599–1609
72. Hashizaki K, Chiba T, Taguchi H, Saito Y (2009) Highly viscoelastic reverse worm-like micelles formed in a lecithin/urea/oil system. Colloid Polym Sci 287 927–932
73. Hashizaki K, Taguchi H, Saito Y (2009) A novel reverse worm-like micelle from a lecithin/sucrose fatty acid ester/oil system. Colloid Polym Sci 287:1099–1105
74. Hashizaki K, Sakanishi Y, Yako S, Tsusaka H, Imai M, Taguchi H, Saito Y (2012) New lecithin organogels from lecithin/polyglycerol/oil systems. J Oleo Sci 61:267–275
75. Shrestha LK, Yamamoto Y, Arima S, Aramaki K (2011) Charge-free reverse wormlike micelles in nonaqueous media. Langmuir 27:2340–2348
76. Rehage H, Hoffmann H (1988) Rheological properties of viscoelastic surfactant system. J Phys Chem 92:4712–4719

Chapter 5
Stimuli-Responsive Self-Healing Viscoelastic Gels

Rekha Goswami Shrestha and Kenji Aramaki

Abstract A viscoelastic gel (VEG) can exhibit both viscous and elastic properties. Such gels can respond dramatically to the external stimulus like temperature, pH, and CO_2 showing change in macroscopic properties with minor variation in the environment. Some smart viscoelastic gels show switchable self-healing properties on switching ON–OFF of stimuli. They revert to their original form on removing stimuli imposed. They have been studied extensively by the theoreticians and experimentalists—owing to their unique rheological properties and prospective applications and great potential in various industrial applications ranging from microfluidics, oil production, drug delivery, to drag reduction. Recently, smart viscoelastic gels (SVEGs) have attracted considerable interest due to the tunability of their viscoelasticity with imposed stimuli, such as electric currents, UV–Vis, temperature, redox reaction, and pH.

Keywords Viscoelastic gel · Photoresponsive · pH responsive · CO_2 responsive

5.1 Introduction

A gel is a solid, jelly-like material that can have properties ranging from soft and weak to hard and tough. Gels are defined as a substantially dilute cross-linked system, which exhibits no flow when in the steady state. By weight, gels are mostly liquid, yet they behave like solids due to a three-dimensional cross-linked network within the liquid. It is the crosslinking within the fluid that gives a gel its structure (hardness) and contributes to the adhesive stick. The gel itself has the mechanical

R.G. Shrestha (✉)
Nano Funtionality Integration Group, National Institute of Material Science, Sengen, Tsukuba 305-0047, Japan
e-mail: rekhashrestha3@hotmail.com

K. Aramaki
Graduate School of Environment and Information Sciences, Yokohama National University, Tokiwadai, 79-7, Hodogaya-ku, Yokohama 240-8501, Japan

T. Kawai and M. Hashizume (eds.), *Stimuli-Responsive Interfaces*,
DOI 10.1007/978-981-10-2463-4_5

properties of a solid, i.e., it can maintain its form under the stress of its own weight, and under any mechanical stress, it shows the phenomenon of strain, referring structural aspects of gels and their viscoelastic properties [1]. Gel is coherent colloid disperse systems of at least two components that exhibit mechanical properties characteristic of the solid state and both the dispersed component and the dispersion medium extend themselves continuously throughout the whole system.

We encounter different forms of gels in everyday life. Their applications range from personal care products, foodstuffs to electronic devices to drug delivery and tissue engineering [2]. They are used in the art restoration, pumping of gels into oil wells to increase the efficiency of crude oil recovery [3], in the gelation of gasoline to make napalm [4]. The gels having the mechanical properties [5] can help protect the skyscrapers from crumbling during earthquakes.

Nature is abundant in gel. They impart unique properties to many macrobiological systems. Instances are fibrous actin [6] and clathrin [7], tubulin [8] keratin [9], insulin [10], collagen gels [11], silk processing by insects and spiders [12], amyloid fibrils that have been implicated in Alzheimer's and other neurodegenerative diseases [13], sickle cell anemia hemoglobin in its deoxy state [14], and fibrinogen, the blood clotting factor [15]. Similarly, some species secrete gels that are effective in parasite control. For example, the long-finned pilot whale secretes an enzymatic gel that rests on the outer surface of this animal and helps prevent other organisms from establishing colonies on the surface of these whales' bodies [16]. Gels existing naturally in the body include mucus, the vitreous humor of the eye, cartilage, tendons and blood clots. Their viscoelastic nature results in the soft tissue component of the body, disparate from the mineral-based hard tissue of the skeletal system. Researchers are actively developing synthetically derived tissue-replacement technologies derived from hydrogels, for both temporary implants (degradable) and permanent implants (nondegradable).

A huge number of research works are ongoing in this field, and the publications and citations per year, Fig. 5.1, itself highlight the importance of these materials. We, therefore, believe that the review chapter on stimuli-responsive viscoelastic gels (SVEGs) should be documented.

This chapter emphasizes on the responsive behavior of viscoelastic gels (VEGs), especially in surfactants: the gels that can dynamically alter their structures and properties in response to environmental changes mainly redox potential, ultraviolet (UV)/visible (Vis) light, pH and CO_2 (more emphasis as it is now more hot topic) [17] and leaving temperature effect. The activation of the trigger changes the micellar structures and consequently alters the related rheological response. These functionalities and responsive alterations in the aggregates provide a basis to design smart materials such as diagnostics, biosensors [18], drug delivery [19], tissue engineering [20], and smart optical systems [21], to a range of potential applications, like clutches [22], microfluidics, valves [23], vibration control [24] clean fracturing fluids [25], tissue engineering [26], human muscle stimulators [27].

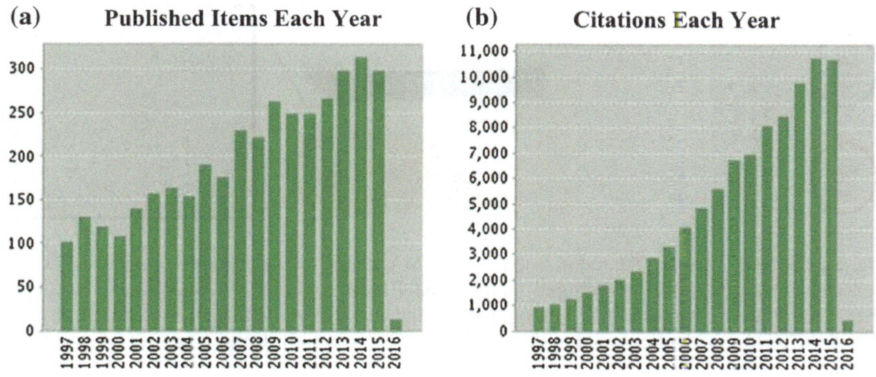

Fig. 5.1 The number of publications and citations on "viscoelastic gels" on recent 20 years. Data compiled from ISI web of Science as of March 1, 2016

5.2 Redox-Responsive Viscoelastic Gels

The rheology of the redox-responsive VEGs or viscoelastic electrorheological (ER) gels can be controlled by applying redox potential [28]. This particular class of gel finds applications in optical switches, biosensors, detectors, and electrochromic devices.

First, Saji et al. reported a low molecular weight surfactants with a ferrocene-based redox moiety for which the cmc and micellization behavior changed with the stimulus of electron transfer. Through spectroscopic studies, they observed that micelles could be broken up into monomers by oxidation and re-formed [29, 30]. Abbott et al. demonstrated that redox-active surfactants containing ferrocene could reversibly and dramatically control the surface tension of solutions [31]. Changes in oxidation state can also lead to changes in the microstructure of aggregates [32–34].

Abe et al. used the concept of mixing a ferrocenyl-based surfactant, 11-ferrocenylundecyltrimethylammonium bromide (FTMA), with sodium salicylate to control form viscoelastic gel, the viscoelasticity of which can be controlled using a "Faradaic (redox) reaction" [35]. FTMA, a cationic "redox-switchable" ferrocenyl surfactant, forms VEGs in the presence of NaSal. The authors have characterized solutions containing 50 mM FTMA in the presence of various concentrations of NaSal. Figure 5.1 shows appearance of FTMA system in oxidized and reduced states. In the absence of or with 10 mM NaSal, the solutions of FTMA flow like water. On increasing the concentration of NaSal to 20 mM, a highly viscous solution is formed. Electrochemical oxidation of FTMA to $FTMA^+$ (see Scheme in Fig. 5.2) in the presence of 20 mM NaSal resulted in the reappearance of flow behaviors comparable to those seen in the absence of NaSal. The qualitative observations described above were accompanied by rheological measurements, as a function of concentration of NaSal and oxidation state of FTMA. The viscosity of a

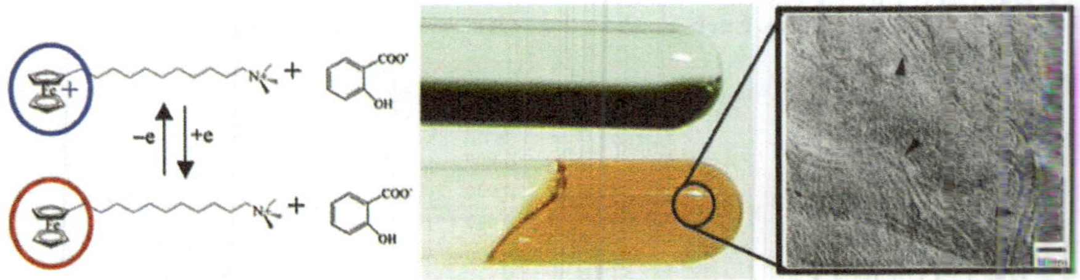

Fig. 5.2 Scheme showing the redox reaction of FTMA. The visual appearance of aqueous FTMA (reduced form, down) (50 mM)/NaSal mixture and FTMA (oxidized form, up)/NaSal mixtures after electrolytic oxidation FFEM micrograph of aqueous FTMA (reduced form)/NaSal mixture (*C*NaSal/*C*FTMA) 0.4). Reprinted with permission from Ref. [35]. Copyright © 2004 American Chemical Society

solution containing reduced FTMA and 20 mM NaSal is 15 Pa. s. Electrochemical oxidation of FTMA to $FTMA^+$ results in a decrease in viscosity (four orders of magnitude) to 0.0025 Pa s. The long, flexible, cylindrical micelles are present in solutions containing 50 mM reduced FTMA and 20 mM NaSal (as confirmed by FFEM). Attempts to recover the initial viscosity by reducing a solution electrochemically containing $FTMA^+$ back to FTMA were hindered by gel formation near the electrode surface, which dramatically reduced the diffusion rate of surfactant to the electrode. Ferrocenyl surfactant, FTMA, has a redox-active ferrocenyl group positioned at the end of its alkyl chain. The ferrocenyl group in the reduced form is hydrophobic, while the ferricinium cation (oxidized form) acts as a hydrophilic group. This causes a remarkable change in the hydrophilic–lipophilic balance of FTMA, thereby altering its aggregation state.

5.3 Photoresponsive VEGs

Light is a noninvasive trigger that can avoid changes in the composition or thermodynamic conditions. It can be directed at a precise spatial location, particularly important for medical applications. Light-responsive solution can be easily patterned and modulated. The benefits of using the light as a trigger include miniaturization and limited chemical contamination of the solution, promoting the development of green sustainable systems [36]. Light-responsive VEGs involve light-induced cis–trans isomerization or dimerization of light-sensitive surfactants or additives containing a suitable chromophore, which alters the packing of surfactant molecules in the aggregates and drives micellar transitions between VEGs and other structures, thus tuning the rheological properties. Basically, two types of strategies are followed to formulate light-sensitive VEGs:

(1) incorporate light-sensitive additives into existing VEGs solutions,
(2) introduce light-sensitive functional moieties on the surfactant molecules.

In both cases, the micelles self-assemble to form transient network or disassemble upon alternate UV/Vis light irradiation. The report on light-responsive VEGs dates back to the early 1980s by Wolff and co-workers. They introduced substituted polyaromatic group like anthracene [37], acridizinium salts [38], coumarins [39] to the micellar solutions of CTAB, and they form light-responsive highly VEGs. Upon UV irradiation, photodimerization of polyaromatic compound occurs, which changes the rheological properties by several orders. The authors further developed systems with photoswitchable viscosity by introducing coumarin derivatives into the aqueous solutions of Triton X-100 and CTAB. Similarly, incorporation of photoresponsive molecules like azobenzene [40], stilbene [41], spiropyran [42], and cinnamic acid and its derivatives [23, 43–49] can also form light-responsive VEGs. Reports on reversible *trans–cis* photoisomerization of azobenzene-modified cationic surfactants [40b, 50–53] and effect of light on self-assembly of aqueous mixtures of sodium dodecyl sulfate and a cationic, bolaform surfactant containing azobenzene [54] are also available. The change in viscoelastic properties of gels has been confirmed as a result of the *trans–cis* photoisomerization of photoresponsive molecule present in the solution.

Aikawa et al. [55] have reported photosensitive VEGs by adding a novel photocleavable surfactant, C4-C-N-PEG9, to nonphotosensitive VEGs formed in an aqueous mixtures of 2.4 wt% tetraoxyethylene dodecyl ether, $C_{12}EO_4$, and 4.9 wt% poly(oxyethylene) phytosterol ether, $PhyEO_{20}$. Upon addition of C4-C-N-PEG9, decrease in viscosity is observed, while on UV irradiation it reverts back to the original value, as shown in Fig. 5.3. Small-angle X-ray scattering studies have shown that addition of C4-C-N-PEG9 favors long cylinders-to-sphere-type micellar

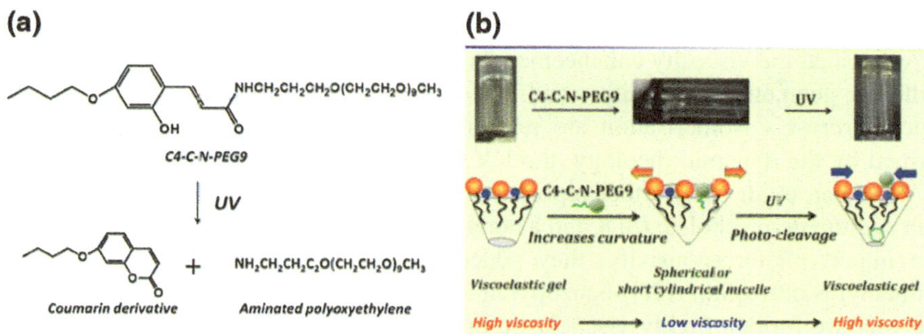

Fig. 5.3 VEGs formed by 4.9 wt% $PhyEO_{20}/H_2O$ + 2.4 wt% $C_{12}EO_4$ and the effect of C4-C-N-PEG9 and UV irradiation are schematically represented in Fig. 5.2. **a** Schematic representation of C-N-PEG9 before and after UV irradiation. **b** $PhyEO_{20}$ molecule is shown as bigger spherical head and black tail, $C_{12}EO_4$ molecule with a bigger spherical head and black tail, while the C4-C-N-PEG9 star-shaped head and black tail. Reprinted with permission from Ref. [55]. Copyright © 2013 American Chemical Society

transition. As confirmed by the UV–Vis spectrometry and HPLC measurements, C4-C-N-PEG9 experiences a photocleavage through UV-induced cyclization in aqueous solution, yielding a less surface-active coumarin derivative (7-butoxy-2H-chromen-2-one) and an aminated poly(oxyethylene) compound, and it is due to this transformation that UV irradiation induces 1D micellar growth.

Phospholipid-based surfactants are well-known candidate for forming nonaqueous photoresponsive VEGs instances includes a ternary mixture of phospholipids/oil/H_2O [56] or phospholipids/oil/and substituent for water like urea, sucrose fatty acid esters [57], sugars of RNA and DNA, polyglycerol [58], bile salt, and sodium deoxycholate [59]. Phospholipids having two alkyl chains form globular or ellipsoidal micelles in a variety of organic solvents. A trace amount of water or hydrophilic additives can form hydrogen bonds with the phosphate group of phospholipids reducing the interfacial curvature of molecular assembly inducing an axial growth of micelles and leading to the formation of VEGs [60]. The η_0 increases by several orders of magnitude. The increase depends upon the molar ratio of water to phospholipids. Raghvan et al. reported the formation of UV-responsive VEGs in a ternary mixture of lecithin/*p*-coumaric acid/oils [42a, 61].

Shrestha et al. have reported photoinduced rheological changes in nonaqueous photoresponsive gels obtained with 1,2-diacyl-sn-glycero-3-phosphocholine-based nonaqueous VEGs, 1-palmitoyl-2-oleophosphatidylcholine (POPC)/cyclohexane/H_2O, POPC/isooctane/H_2O, and L-α-dioleoylphosphatidylcholine (DOPC)/isooctane/H_2O systems [62]. Initially, the ternary mixtures form highly viscoelastic fluids. On adding *trans*-CA the viscoelasticity increases, while it decreases upon UV irradiation, as shown in Fig. 5.4a–c. The effect of nature of the substituent on the benzene ring of *trans*-CA on rheology has been effectively coined. When hydroxycinnamic acid (HCA) (hydrophilic, –OH group attached to the benzene ring of CA) is added to DOPC/isooctane/H_2O phase, separation occurs, while with the methoxycinnamic acid (MOCA) and methylcinnamic acid (MCA) (hydrophobic groups, $-OCH_3$, and $-CH_3$ attached to the benzene ring of CA, respectively) leads to higher viscoelasticity. The study on the effects of position of the substitution on CA revealed the viscosity enhancement is in the order of *p*- > *m*- > *o*-isomers. The different geometries of additive and that obtained upon substitution and photoinduced *trans-cis* isomerisation are responsible for the different rheology as confirmed by the dynamic rheology, the UV absorption, and the ^1H NMR spectrums.

Raghvan et al. [63] further reported VEGs combining the lecithin and the bile salt sodium deoxycholate (SDC) in a nonpolar organic solvent such as cyclohexane. To impart photoresponsivity, they added spiropyran (SP) derivative 1′, 3′, 3′-trimethyl-6-nitrospiro[1(2H)-benzopyran-2,2′-indoline]. Initially, the lecithin/SDC/SP mixture formed highly viscoelastic fluids. Under UV irradiation, the colorless SP was isomerized to the open colored merocyanine (MC) form, causing the fluid viscosity to decrease by 10-fold. The closed SP form is hydrophobic and nonionic, whereas the open MC form is zwitterionic and hydrophilic. These two photoisomers are known to interact differently with the headgroups of lecithin. On switching off the trigger, the MC reverts to the SP form, and the viscosity recovers its initial value. This cycle could be repeated several times without loss of response.

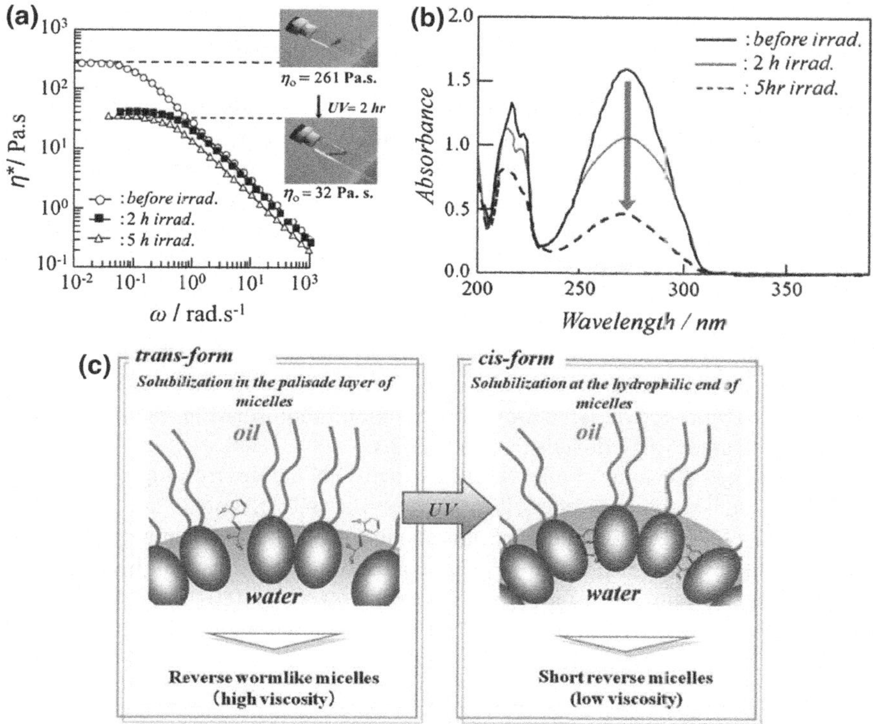

Fig. 5.4 **a** Frequency-dependent complex viscosity $|\eta^*|$, **b** the corresponding UV–Vis absorption spectra for DOPC(50 mM)/Isooctane/H_2O(150 mM)/CA(10 mM) solution before and after UV irradiation at 25 °C, and **c** view of change in solubilization sites of *trans* and *cis* isomers in micellar solution responsible for the different rheology on UV irradiation. Reprinted with permission from Ref. [62]. Copyright © 2014 Springer

An interesting system has been reported by Scheven and co-workers [22] where a multistate photoactive compound, 2,4,4′-trihydroxychalcone (Ct), is added to impart photorheological properties to aqueous solutions of weakly entangled VEGs formed by CTAB in the presence of salicylic acid (HSal). Significantly, a very small amount of Ct ([Ct]/[CTAB] = 1 %) could give rise to a reversible photochromic and photorheological response of CTAB/HSal solutions, in the poorly entangled regime where viscoelastic properties depend strongly on the concentration of the intercalant. The presence of Ct leads to a multiequilibrium system, involving several chemical species interconvertible by light and pH, and its UV-induced *trans–cis* isomerization indirectly caused changes in micellar morphology, which consequently affected the viscoelastic response of the micellar fluid. The observed rheological phenomenon is reversible.

Meanwhile, Zhao and co-workers [23] have reported a single-component light-responsive VEG in azobenzene introduced gemini surfactant, sodium-2,2′-(diazene-1,2-diylbis(4,1- phenylene))-didodecanoate (C12-azo-C12). A 60 mM aqueous solution of this surfactant produced a very viscous transparent fluid of orange color, with η_0 as high as 734 Pa s. Upon UV irradiation for 3 h, η_0 drops to 0.007 Pa s. The recovery of viscosity is attained to its original upon irradiation with light for 2 h and for several cycles without any decay. Authors attribute high viscosity to the presence of the rigid azospacer, which keep the two hydrophobic tails apart, resulting in a "pseudo-volume" between them favoring the formation of VEGs. Light-induced *trans–cis* photoisomerization of azobenzene group in C12-azo-C12 reduces the distance between the two ionic headgroups, breaking down the entangled network into small and discrete micelles. The worms are constructed and destructed upon alternate UV/Vis light irradiation, which take advantage of dimerization, *cis–trans* isomerization, photoscission, polymerization, or polarity changes of light-sensitive molecules.

A novel pyrrolidinium ionic liquid N-methyl-N-cetylpyrrolidinium bromide (C_{16}MPBr) and sodium (4-phenylazo-phenoxy)-acetate (AzoNa) are formed by facile mixing of C_{16}MPBr and a photosensitive azobenzene-containing molecule *trans*-AzoNa in water [53]. Under ambient conditions, it exhibits a high viscosity of about 10 Pa s. Upon UV irradiation, viscosity lowers to 0.01 Pa s. Due to the *trans–cis* photoisomerization of the AzoNa incorporated in the system, the transient network morphology transforms into spherical micelles.

5.4 pH-Responsive Viscoelastic Gels

Among the stimuli, pH has attracted particular interest as pH variation is relatively easy to introduce and control. It has been found that the pH-switchable systems are easier to prepare in both laboratory and industrial scales compared with other stimuli-responsive viscoelastic gels. Thus far, the synthesis of peculiar molecules with functional groups and the introduction of a pH-sensitive monomer molecular into the system are the most widely used methods to prepare the pH-switchable viscoelastic gels. Most of the pH-responsive viscoelastic gels so far reported involve mainly zwitterionic alkyldimethylamine oxide, C_nDMAO [64], 2-6 cationic CTAB in the presence of hydrotropes [65], and through the design of surfactant mixture composition [66] and surfactant architecture [67].

A more common route is directly use pH-sensitive surfactants to form viscoelastic gels [5, 20–22]. Maeda et al. [68] demonstrated reversible conversion of high-viscoelastic solutions into low-viscous vesicles when increasing the ionization degree of oleyldimethylamine oxide by the addition of HCl/NaOH. Engberts group [21, 22, 69] examined pH-dependent behaviors of a series of carbohydrate-based gemini surfactants with two amine functionalities for vesicles—VEGs—spherical micelles transitions. A control on the micellar aggregation and thereby the bulk properties of viscoelastic gels can be introduced by altering the pH of the solution [70].

The alkyldimethylamine oxides, C_nDMAO exists as either a nonionic or a cationic (protonated form) species depending on pH of their aqueous solution. At neutral pH, they carry no net charge and spontaneously self-organize into a transient network of long, flexible micelles without any additives because of weak repulsion between the head groups [68, 71, 72]. It has been reported that near the point of half-ionization ($\alpha = 0.5$, where α is the degree of ionization) alkyldimethylamine oxides exhibit a characteristic maximum in micellar size and weight, bulk viscosity (in $C_{14}DMAO$), and a minimum in cmc and surface tension. There are several reports on pH-sensitive VEGs with CnDMAO taking into account its pH-dependent micellar aggregations [73–76]. Besides, the pH-responsive VEGs formed by commercially available alkyldimethylamine oxide, tailor-made p-dodecyloxybenzyldimethylamine oxide (pDoAO) [77], showed pH responsiveness in terms of micellar length and hence viscoelasticity. Similarly, Ghosh et al. developed [66b] pH-responsive micellar aggregates by using the amino acid-based zwitterionic surfactant N-(n-dodecyl-2-aminoethanoyl)-glycine ($C_{12}Gly$) in the presence of the anionic surfactant, SDS. Fluorescence depolarization in combination with DLS, TEM, and fluorescence showed a transition between vesicles, spheres, and branched worms, as the pH or the surfactant composition was altered.

There are reports on pH-responsive viscoelastic gels in cationic surfactants solutions by incorporating hydrotopes [78] and other surfactants [66, 67b, c, d, e]. Huang and co-workers [7] introduced the pH-sensitive potassium phthalic acid into cetyltrimethylammonium bromide (CTAB) solutions and found reversible switchability between VEGs with long cylindrical micelles and short cylindrical micelles on altering the pH of the solutions. Hassan and co-workers [78c] designed pH-responsive viscoelastic gels based also on CTAB, replacing the potassium phthalic acid with anthranilic acid.

Some pH-responsive VEGs can also be formulated by varying surfactant composition in appropriate mixtures [11, 12, 66b]. Lin et al. [11] developed pH-regulated surfactant self-assemblies by mixing CTAB and decylphosphoric acid (DPA) at a 1:1 molar ratio. Upon increasing the pH, the 1:1 cationic/anionic pair turned into a 1:2 cationic/anionic pair, which showed a weaker aggregating ability due to the steric effect and the larger hydration effect. Consequently, the self-assembled organization could be effectively tailored between spheres, worms, vesicles, and lamellar structures, thereby in turn controlling the viscoelasticity of the solution.

Zhao et al. [79] have reported a pH-sensitive viscoelastic gels, based on a simple ammonium surfactant, a weak acid, N-cetyl-N,N-dihydroxyethylammonium bromide (CDHEAB). When NaOH is added to CDHEAB solution, the relative ratio of CDHEAB and its counterpart, CDHEA, varies with the concentration of NaOH, i.e., the pH. An aqueous solution of CDHEAB is clear at pH range = 3.31-4.97. On start from 4.97 till 5.53, the viscosity increases sharply, up to 10^6-fold. The rheological measurements confirm the presence of transient network of cylindrical micelles. The viscosity, however, decreases at pH 5.63. The variation of rheological behavior has been attributed to the molecular structure that allows the combination of strong hydrophobic interactions and effective hydrogen bonding upon protonation at higher pH.

In yet another unique report, Sakai et al. [80] have reported a pH-sensitive viscoelastic gels in a complex formed by an amino acid-based surfactant (dodecanoylglutamic acid, C_{12}Glu) alkyl dicarboxylic acid—and a secondary alkylamine (dodecylmethylamine, C_{12}MA) and a tertiary alkylamine (dodecyldimethylamine, C_{12}DMA). C_{12}Glu forms a stoichiometric 1:1 complex with C_{12}DMA and C_{12}MA. Rheological measurements have suggested that the complex systems yield viscoelastic gels and the rheological behavior is strongly dependent on the pH of the solution. This pH-dependent behavior results from the structural transformation of the viscoelastic micelles to occur in the narrow pH range 5.5–6.2 (in the case of C_{12}Glu-C_{12}DMA system), i.e., positive curved aggregates such as spherical or rodlike micelles tend to be formed at high pH values, Fig. 5.5.

Chu and Feng [1, 13, 67a, 70] developed a pH-switchable VEC by using the neutralized surfactant from the insoluble long-chain tertiary amine N-erucamidopropyl-N,N-dimethylamine (UC22AMPM) with maleic acid in 2:1 molar ratio. A reversible Newtonian-to-viscoelastic transition with an increase in η_0 of five orders of magnitude is observed upon altering the pH. Varying the pH from 6.20 to 7.29, partial deprotonation of the quaternary ammonium headgroup occurs; correspondingly, the micellar aggregate showed a structural transformation from wormlike micelles to spheres. Upon further increasing the pH above ~9.80, UC22AMPM precipitated from the bulk solution. A relatively higher viscosity can be obtained when maleic acid is replaced with hydrochloric acid in this mixture.

The sugar-based (reduced glucose) gemini surfactant GS3 shows pH responsiveness [67b, c]. A vesicle-to-worm-to-sphere transition can be observed within a narrow pH region from 6.0 to 5.6 and lower pH. The nature of the sugar and spacer was found to have little influence on this process. Yang et al. [67f] synthesized a C22-tailed sarcosinate anionic surfactant, 2-(N-erucacyl-N-methyl amido) acetate (EMAA), and observed that its solutions display a pH-controllable transition from micelles to vesicles. Jaeger et al. [67d] synthesized and investigated a bolaform surfactant, ω-carboxyl hexadecyltrimethylammonium bromide (HOOC-CTAB), and found that it formed ribbons at pH 6.8, rods at pH 2.2, and

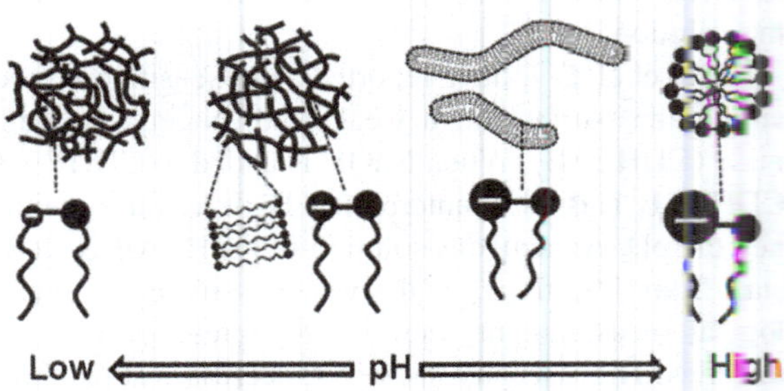

Fig. 5.5 Schematic figure of micellar structures as a function of pH. Reprinted with permission from Ref. [80]. Copyright © 2014 Springer

worms at pH 11.5. Recently, the pH-responsive behavior of another bola amphiphile, dotriacontane-1,32-diyl-bis[2-(dimethylammonio)ethylphosphate] (Me2PE-C32-Me2PE), was reported by Graf et al. [67e].

5.5 CO_2-Responsive Viscoelastic Gels

A weak acid, biocompatible and renewable, inexpensive gas, CO_2 represents interesting alternative stimuli. Basically, CO_2 functions as a pH trigger; however, using CO_2 is more advantageous over pH; in particular, uptaken CO_2 can be completely removed by streaming an inert gas into the solution and/or mild heating, which allows the CO_2-reactive functional groups to resort to their initial forms from both the products and the media. CO_2 can switch the properties of some solvents [81], emulsifiers [82], polymers [83, 84], and initiators [85], which bear CO_2-sensitive moieties, including amidine, guanidine, and amine, that allows these molecule to form cationic adducts—reversible reaction with CO2 in water can form cationic adducts.

Following on from the pioneering work of Jessop team on CO_2-switchable solvents in 2005 [86], many studies have been reported on CO_2-responsive VEGs at ambient temperature and pressure. Feng and co-workers [87] have reported CO_2-switchable VEGs, in a mixture of anionic surfactant, SDS and N,N,N′,N′-tetramethyl-1,3- propanediamine in 2:1 molar ratio. A reversible sphere-to-worm transition could be obtained by bubbling CO_2 and N_2 into an aqueous mixture solution alternately and repeated for several cycles, without loss of response to CO_2. The TMPDA molecules are protonated into quaternary ammonium species that serves both as a spacer to link two C12 hydrophobic tails and two positive counterions neutralizing the negative sulfate moieties, behaving like a 'pseudo'-gemini surfactant, and forming VEGs as verified by cryo-TEM. Upon removing CO_2 by N_2, the spacers get deprotonated back into tertiary amines, breaking the 'pseudo'-geminis back to individual molecules that form low-viscosity spherical micelles. The team further developed CO_2-switchable VEGs with 2wt% aqueous solution of octadecyl dipropylene triamine (ODPTA). On bubbling CO_2 for $\sim$2 min, low-viscosity emulsion-like ODPTA solution immediately transforms into a transparent viscoelastic gel with η_0 as high as 2 × 104 mPas. 5000 times higher than the initial solution. When CO_2 is displaced with N_2 at 75 °C, the gel recovers its initial low-viscosity cloudy state. Alternative bubbling of CO_2 and N_2 gives reversible cycles. Cryo-TEM observation before and after CO_2 trigger reveals a microstructural transition from spherical micelles and vesicles to transient networks, which exactly explains the observed rheological response. In order to attain 'symmetrical' reversibility, bubbling of inert gas accompanied by heating is necessary for complete displacement of the uptaken CO_2.

Further, the team has developed VEGs composed of C22-tailed tertiary amine, N-erucamidopropyl-N,N-dimethylamine (UC22AMPM) [88]. Upon bubbling CO_2 for $\sim$10 min at room temperature, a 100 mM UC22AMPM solution immediately becomes homogeneous, transparent, and highly viscoelastic and can capture bubbles over a long timescale. The gel-like fluid returned to a low-viscosity emulsion-like solution by bubbling air for 30 min or exposing to air for ~4 days at ambient temperature. The η_0 increases by more than 5 orders of magnitude with CO_2 trigger. Cryo-TEM observation shows elongated, flexible entangle cylindrical micelles present in "UC22AMPM—CO_2" solution, and only small spherical micelles are evidenced in "UC22AMPM–air" solution. Besides, this process can be repeated over several cycles without any alteration of the response.

The same group reports a CO_2-induced anionic wormlike micellar fluid by introducing triethylamine (TEA) into anionic surfactant sodium erucate (NaOEr) at a molar ratio (CNaOEr:CTEA) of 3:10. CO_2-induced smart viscoelastic fluids based on mixtures of sodium erucate and trimethylamine [89]. Upon bubbling of CO_2, pH of the solution decreases from 12.3 to 10.0, protonating TEA into a quaternary ammonium salt. It promotes the growth of micelles as a hydrotrope by screening electrostatic repulsion between the anionic headgroups in NaOEr molecules, resulting in the formation of transient network of micelles and viscoelasticity buildup; upon removal of CO_2, the quaternized TEA is deprotonated back into a nonionic tertiary amine, and electrostatic repulsion strengthens causing the viscoelastic fluid–low-viscosity spherical micellar solution transition. Reversible sphere-to-worm transition could be repeated in the pH range of 12.3–10.0 by introducing and displacing CO_2 alternately; however, excess with CO_2 pH < 9.50, the solution transforms into a cloudy dispersion with low viscosity due to the de-neutralization of NaOEr, and this process is irreversible.

A sodium erucate (NaOEr) solution is a homogeneous viscoelastic fluid at pH 11.34 with η_0 as high as 60,000 mPa s [90]. Upon bubbling CO_2 into the NaOEr aqueous solution at 60 °C forms a low-viscosity opaque solution within 1 min. At pH 11.34, sodium erucate behaving like ultra-long-chain anionic surfactants can self-assemble into long wormlike micelles at a low concentration and then entangle into transient networks responsible for the viscoelasticity. When CO_2 is bubbled η_0 drops dramatically. The pH value of the solution decreases due to the formation of H_2CO_3, resulting the conversion of erucate ions into nonionic erucic acid molecules, and breaking away from the surfactant aggregates. Finally, the original viscoelastic wormlike micelle solutions are converted into low-viscosity spherical micelles or free monomer solutions. On bubbling N_2 to displace CO_2 at 80 °C for one day, the pH value of the sample only increases back to ~9.22, which is insufficient to drive wormlike micelle formation, and the viscoelastic fluid does not recover its original state by depleting CO_2 with N_2.

Su et al. [90] also developed a CO_2-thinning solution system but showing viscosity switchability by bubbling/removing CO_2 into/from the aqueous mixture of sodium stearate ($C_{18}CNa$) and $NaNO_3$ at 80 °C. Before introducing CO_2, the 200 mM solution shows highest viscosity η_0 of 12,400 mPa. s. at 80 °C; upon CO_2 bubbling for 10 min at the same temperature, the mixture gets milky and η_0 only 2.0 mPa s. While displacing CO_2 with N_2 for about 40 min at 80 °C, the viscosity reverts back to 12,360 mPa s. By adding tetrabutylammonium bromide (TBAB) to an aqueous solution of $C_{18}CNa$ [the Krafft point of $C_{18}CNa$ is reduced] these authors also obtain switchable viscosity at 60 °C. They demonstrated the reversible conversion of highly viscoelastic wormlike micellar solutions into low-viscosity spherical micelles when bubbling CO_2 in a mixture of sodium octadecyl sulfate ($C_{18}SNa$) and 2-(dimethylamino) ethanol (DMAE). The mixture of 200 mM $C_{18}SNa$ and 200 mM DMAE was switched between low and high zero-shear viscosity states by alternating CO_2 and N_2 treatment over three cycles, combined with high temperature (60 °C). Another common hydrotrope, N,N,N′,N′-tetramethyl-1,4-butanediamine (TMDAB), was also successfully used for switching the viscosity of the aqueous solution with C18SNa. The viscosity can be reversibly controlled by the addition and removal of CO_2, and the viscosity reached a maximum for a molar ratio of TMDAB: $C_{18}SNa$ of 1:2.

In yet another report [91], there was a facile approach to confer CO_2-switchability to classic cetyltrimethylammonium bromide (CTAB)-sodium salicylate (NaSal) system. An introduction of CO_2-responsive trimethylamine (TEA) leads to a decrease in the solution viscosity; upon bubbling CO_2, TEA transforms into protonated TEA+, regaining the viscoelasticity. TEA can break up the binding capacity of CTAB-NaSal, which is unfavorable for the growth of micelles, while the negative effect disappears after bubbling CO_2, and the binding capacity of CTAB and NaSal was recovered again, which favors the growth of micelles. The rheological behavior can be reversibly switched via a "green" way of bubbling or removing CO_2.

5.6 Applications

The field of stimuli-responsive gels has grown considerably over the past decade [92]. An abundance of 'intelligent' nanostructured systems has emerged and been proposed for a wide range of applications, for instance in electronic devices, as 'smart' optical systems [93], micro-electromechanical systems, in coatings, and quite significantly in the biomedical field for drug delivery, diagnostics, as biosensors, for bioseparation and artificial organs and tissues. Traditionally, these materials have predominantly been based on polymers; however, VEGs displaying substantial viscoelasticity [94] provides an opportunity for capture and release, relevant to the areas of drug delivery [95], the release of perfumes. It is still a very young field. Responsive viscoelastic gel could provide an alternative to conventional polymer-based materials for biomedical applications; they are easier to degrade and therefore likely to be more biocompatible; they are more amenable to being responsive to a range of various triggers particularly relevant to the biomedical field, since they often appear as the result of a pathology; these are: pH, temperature and redox potential [96]. They naturally incorporate a hydrophobic core, which could serve as a solubilization locus for pharmaceutical oils and drugs [97]. In tissue engineering, scaffolds which gel upon exposure to body temperature offer the significant advantage of avoiding surgical procedures and complications associated with implants. The incorporation of stimuli responsiveness in 'intelligent' materials promises to address some of the systemic and intracellular delivery barriers. In addition, external stimuli can also be envisaged to target delivery to specific sites, in particular via the application of (localized) light, ultrasound or magnetic field. They have potential as vehicles for controlled and targeted delivery. One could also envisage imparting responsiveness to chemical and biological stimuli, such as glucose concentration for the treatment of diabetes [98] or the presence of metabolites [99]. Light-responsive gels provide the additional benefit of precise localization, making it useful in numerous nanotechnological applications, sensor systems, nanoelectronics, microfluidics, molecular devices, logic gates, and information storage devices [100]. Applications of these gels are anticipated in bioseparation, capillary eletrophoresis as they are likely to tolerate the addition of electrolytes, macromolecules, or nanoparticles [101]. The pH stimulus is of especially pharmaceutical, biomedical and related fields interest. It is often triggered in the controlled release of encapsulated drugs. The VEGs whose viscosity rheology is pH-sensitive can be useful in the extraction process of $CaCO_3$ in calcite or $MgCO_3$ in dolomite.

In oil industry, controllable rheology of VEGs can be extensively used in different stages of oil production process, including drilling, gravel packing, fracturing fluid and self-diverting acidizing [102], to tertiary oil recovery [103]. They have long been used as cleaning and personal care products [104]. Recently, Zakin et al. demonstrated a very promising approach to developing effective drag reducing systems for use in cooling and heating installations [105]. Several researchers have exploited the ‘sol–gel’ transition and network structure of VEGs to build-up nanofibers [106], silica nanoparticles [107], as building blocks for the fabrication of smart surface coatings [108], porous nanostructured materials [109]. It could be envisioned as optical sensors, by mixing with solvatochromic species or aggregachromic dyes. In order to implement VEG into the actual applications studying the effects of various stimuli on rheology is must.

5.7 Conclusion

Aqueous and nonaqueous responsive viscoelastic gels are possible to formulate. The molecular packing in micellar aggregates is affected by adding cosurfactant, altering the temperature, pH, and electrical stimulant. The basic principle relies on an introduction of a responsive surfactant or additive unit, either chemically to the gelling system or, simply, by adding responder into the surfactant solutions. An appropriate stimulus can modify critical packing parameter, therefore inducing a spontaneous reorganization of the surfactants from highly VEGs to spheres, short rods or vesicles, and vice versa leading to the modification of the rheological properties. The responsive VEGs expand the field of applications and provide new insight on the study of living polymers. These gels can find real applications in very diverse range of applications, as sensors, in microfluidics, drug delivery, biomedical applications, nanostructured materials, and in the oil industry. Future trends in this field may see an expansion toward multistimuli-responsive VEGs, as triggers to ultrasounds, microwaves, and magnetism. The hybridization of stimuli-responsive gels with polymers, colloids, nanoparticles, biomaterials, and any other functional materials could also open a facile way to fabricate stimuli-responsive hybrid materials. Responsive VEGs on aqueous or nonaqueous media have been largely unexplored and may become the focus of further interest.

References

1. Kruyt HR (ed) (1949) Colloid Science; vol II, Amsterdam: Elsevier, p. 484 (Chapter XII)
2. Wallace DG, Rosenblatt (2003) Collagen gel systems for sustained delivery and tissue engineering. J Adv Drug Deliv Revs 55:1631–1649
3. Clark JB (1952) Treatment of wells; U.S. Patent 2,596,844; May 13
4. Fieser LP, Harris GC, Hershberg EB, Morgana M, Novello FC, Putnam ST (1946) Napalm. Ind Eng Chem 38:768–773
5. See for instance the architecture and sculpture art of Buckminster Fuller and Kenneth Snelson for applications of the concept of "tensegrity" [a]. [a] Ingber DE (1998) Sci Amer 278:48–57
6. Greer SC (2002) Reversible polymerization and aggregations. Ann Rev Phys Chem 53:173–200
7. Kirchhausen T (2000) Clathrin. Ann Rev Biochem 69:699–727
8. (a) Tuszynski JA, Brown JA, Sept D (2003) Models of the collective behavior of proteins in cells: Tubulin, actin and motor proteins. J Biol Phys 29:401–428. (b) Oakley BR, Akkari YN (1999) γ-Tubulin at ten: progress and prospects. Cell Struct Funct 24:365–372
9. (a) Fuchs E (1995) Keratins and the skin. Ann Rev Cell Dev Biol 11:123–153. (b) Smack DP, Korge BP, James WD (1994) Keratin and keratinization. J Am Acad Dermatol 30:85–102
10. Waugh DF (1946) A fibrous modification of insulin. I. The heat precipitate of insulin. J Am Chem Soc 68:247–250
11. Caria A, Bixio L, Kostyuk O, Ruggiero C (2004) Elastic scattering and light transport in three-dimensional collagen gel constructs: a mathematical model and computer Simulation approach. IEEE Trans Nanobiosci 3:85–89
12. (a) Jin HJ, Kaplan DL (2003) Mechanism of silk processing in insects and spiders Nature 424:1057–1061. (b) Valluzzi R, Jin HJ (2004) Park templated native silk smectic gels. J PCT Int Appl WO 41,845 (Cl. C07 K), 21 May 2004
13. (a) Jimenez JL, Nettleton EJ, Bouchard, M, Robinson CV, Dobson, CM, Saibil HR (2002) The protofilament structure of insulin amyloid fibrils. Proc Natl Acad Sci USA 99:9196–9201. (b) Liu W, Prausnitz JM, Blanch HW (2004) Amyloid fibril formation by peptide LYS (11–36) in aqueous trifluoroethanol. Biomacromolecules 5:1818–1823. (c) Nilson MR (2004) Techniques to study amyloid fibril formation in vitro methods 34:151–160. (d) Waterhouse SH, Gerrard JA (2004) Amyloid fibrils in bionanotechnology. Aust J Chem 57:519–523
14. Galkin O, Vekilov PG (2004) Mechanisms of homogeneous nucleation of polymers of sickle cell anemia hemoglobin in deoxy state. J Mol Biol 336:43–59
15. (a) Drukman S, Kavallaris M (2002) Microtubule alterations and resistance to tubulin-binding agents Int J Oncol 21:621–628. (b) Reinhart WH (2001) Molecular biology and self-regulatory mechanisms of blood viscosity: a review. J Biorheol 38:203–212
16. Dee EM, McGinley M, Michael Hogan C (2010) Long-finned pilot whale. In: Saundry P, Cleveland C (eds) Encyclopedia of earth. National Council for Science and the Environment, Washington DC
17. Chu Z, Dreiss CA, Feng Y (2013) Smart wormlike micelles. Chem Soc Rev 42:7174–7203

18. Tokarev I, Minko S (2009) Multiresponsive, hierarchically structured membranes: new, challenging, biomimetic materials for biosensors, controlled release, biochemical gates, and nanoreactors. Adv Mater 21:241–247
19. Bajpai AK, Shukla SK, Bhanu S, Kankane S (2008) Responsive polymers in controlled drug delivery. Prog Polym Sci 33:1088–1118
20. Lee KY, Mooney DJ (2001) Hydrogels for tissue engineering. Chem Rev 101:1869–1879
21. Pucci A, Bizzarri R, Ruggeri G (2011) Polymer composites with smart optical properties. Soft Matter 7:3689–3700
22. Hartsock DL, Novak RF, Chaundy GJ (1991) ER fluid requirements for automotive devices. J Rheol 35:1305–1326
23. Ketner AM, Kumar R, Davies TS, Elder PW, Raghavan SR (2007) A simple class of photorheological fluids: surfactant solutions with viscosity tunable by light. J Am Chem Soc 129:1553–1559
24. Stanway R, Sproston JL, El-Wahed AK (1996) Applications of electro-rheological fluids in vibration control: a survey. Smart Mater Struct 5:464–482
25. Boek ES, Jusufi A, LoÅNwen H, Maitland GC (2002) Molecular design of responsive fluids: molecular dynamics studies of viscoelastic surfactant solutions. J Phys Condens Matter 14:9413–9430
26. Chu Z, Feng Y (2011) Thermo-switchable surfactant gel. Chem Commun 47:7191–7193
27. Bohon K, Krause S (1998) An electrorheological fluid and siloxane gel based electromechanical actuator: working toward an artificial muscle. J Polym Sci Part B Polym Phys 36:1091–1094
28. Stanway R, Sproston JL, El-Wahed AK (1996) Applications of electro-rheological fluids in vibration control: a survey. Smart Mater Struct 5:464–482
29. Saji T, Hoshino K, Aoyagui S (1985) Reversible formation and disruption of micelles by control of the redox state of the surfactant tail group. J Chem Soc Chem Commun 865–866
30. Saji T, Hoshino K, Aoyagui S (1985) Reversible formation and disruption of micelles by control of the redox state of the head group. J Am Chem Soc 107:6865–6868
31. Aydogan N, Gallardo BS, Abbott NL (1999) A molecular-thermodynamic model for gibbs monolayers formed from redox-active surfactants at the surfaces of aqueous solutions: redox-induced changes in surface tension. Langmuir 15:722–730
32. Aydogan N, Abbott NL (2001) Comparison of the surface activity and bulk aggregation of ferrocenyl surfactants with cationic and anionic headgroups. Langmuir 17:5703–5706
33. Sakai H, Imamura H, Kondo Y, Yoshino N, Abe M (2004) Reversible control of vesicle formation using electrochemical reaction. Colloids Surf A 232:221–228
34. Tsuchiya K, Sakai H, Saji T, Abe M (2003) Electrochemical reaction in an aqueous solution of a ferrocene-modified cationic surfactant mixed with an anionic surfactant. Langmuir 19:9343–9350
35. Tsuchiya K, Orihara Y, Kondo Y, Yoshino N, Ohkubo T, Sakai H, Abe M (2004) Control of viscoelasticity using redox reaction. J Am Chem Soc 126:12282–12283
36. Yang D, Piech M, Bell NS, Gust D, Vail S, Garcia AA, Schneider J, Park CD, Hayes MA, Picraux ST (2007) Photon control of liquid motion on reversibly photoresponsive surfaces. Langmuir 23:10864–10872
37. Wolff T, Kerperin K (1993) Influence of solubilized 2,2,2-Trifluoro-1-(9-anthryl)-ethanol and its photodimerization on viscoelasticity in dilute aqueous cetyltrimethylammonium bromide solutions. J Colloid Interface Sci 157:185–195
38. Lehnberger C, Wolff T (1999) Photorheological effects in aqueous micellar tetramethylammoniumhydrogen-2-dodecyl malonate via photodimerization of acricizinium bromide. J Colloid Interface Sci 213:187–192
39. Yu XL, Wolff T (2003) Rheological and photorheological effects of 6-alkyl coumarins in aqueous micellar solutions. Langmuir 19:9672–9679
40. (a) Shang T, Smith KA, Hatton, TA (2006) Self-Assembly of a nonionic photoresponsive surfactant under varying irradiation conditions: a small-angle neutron scattering and cryo-tem study. Langmuir 22:1436–1442. (b) Sakai H, Orihara Y, Kodashima H,

Matsumura A, Ohkubo T, Tsuchiya K, Abe M (2005) Photoinduced reversible change of fluid viscosity. J Am Chem Soc 127:13454–13455. (c) Sakai H. Matsumura A, Yokoyama S, Saji T, Abe M (1999) Photochemical switching of vesicle formation using an azobenzene-modified surfactant. J Phys Chem B 103:10737-10740. (d) Hubbard Jr FP, Santonicola G, Kaler EW, Abbott NL (2005) Small-angle neutron scattering from mixtures of sodium Dodecyl sulfate and a cationic, bolaform surfactant containing azobenzene. Langmuir 21:6131–6136. (e) Lee CT, Smith KA, Hatton TA (2005) Photocontrol of protein folding: the interaction of photosensitive surfactants with bovine serum albumin. Biochemistry 44:524–536. (f) Orihara Y, Matsumura A, Saito Y, Ogawa, N, Saji T, Yamaguchi A, Sakai H, Abe M (2001) Reversible release control of an oily substance using photoresponsive micelles. Langmuir 17:6072–6076. (g) Shin JY, Abbott N L (1999) Using light to control dynamic surface tensions of aqueous solutions of water soluble surfactants. Langmuir 15:4404–4410. (h) Bradley M, Vincent B, Warren N, Eastoe J, Vesperinas A (2006) Photoresponsive surfactants in microgel dispersions. Langmuir 22:101–105. (i) Bi Y, Wei H, Hu Q, Xu W, Gong Y, Yu L (2015) Wormlike micelles with photoresponsive viscoelastic behavior formed by surface active ionic liquid/azobenzene derivative mixed solution. Langmuir 31:3789–3798

41. (a) Eastoe J, Dominguez MS, Wyatt P, Beeby A, Heenan, RK (2002) Properties of a stilbene-containing gemini photosurfactant: light-triggered changes in surface tension and aggregation. Langmuir, 18:7837–7844. (b) Eastoe J, Dominguez MS, Wyatt P, Heenan RK (2004) UV causes dramatic changes in aggregation with mixtures of photoactive and inert surfactants. Langmuir, 20:6120–6126. (c) Kozlecki T, Wilk KA (1996) Photochemical behavior of micellized 4-(4′-alkylstyryl) pyridinium salts. J Phys Org Chem 9:645–651. (d) Kozlecki T, Wilk KA Syper L (1998) Photoisomerizable cationic surfactants as microviscosity probes. Prog Colloid Poly Sci 110:193–198
42. (a) Sun C, Arimitsu K, Abe K, Ohkubo T, Yamashita T, Sakai H, Abe M (2004) synthesis and photochemical properties of a cationic surfactant having a spiropyran group M. Mater Technol 22:229–232. (b) Liu S, Fujihira M, Saji T (1994) Formation of an organic thin film by photochemical isomerization of a surfactant with a spiropyran moiety J Chem Soc Chem Comm 16:1855–1856
43. (a) Kumar R, Ketner AM, Raghavan SR (2010) Nonaqueous photorheological fluids based on light-responsive reverse wormlike micelles. Langmuir 26:5405–5411. (b) Kumar R, Raghavan SR (2009) Photogelling fluids based on light-activated growth of zwitterionic wormlike micelles. Soft Matter 5:797–803
44. Sakai H, Taki S, Tsuchiya K, Matsumura A, Sakai K, Abe M (2012) Photochemical control of viscosity using sodium cinnamate as a photoswitchable molecule. Chem Lett 41:247–248
45. Kumar R, Raghavan SR (2009) Photogelling fluids based on light-activated growth of zwitterionic wormlike micelles. Soft Matter 5:797–803
46. Baglioni P, Braccalenti E, Carretti E, Germani R, Goracci L, Savelli G, Tiecco M (2009) Surfactant-based photorheological fluids: effect of the surfactant structure. Langmuir 25:5467–5475
47. Li J, Zhao M, Zhou H, Gao H, Zheng L (2012) Photo-induced transformation of wormlike micelles to spherical micelles in aqueous solution. Soft Matter 8:7858–7864
48. Du M, Dai C, Chen A, Wu X, Li Y, Liu Y, Li W, Zhao M (2015) Investigation on the aggregation behavior of photo-responsive system composed of 1-hexadecyl-3-methylimidazolium bromide and 2-methoxycinnamic acid. RSC Adv 5:68369–68377
49. Oh H, Ketner AM, Heymann R, Kesselman E, Danino D, Falvey DE, Raghavan SR (2013) A simple route to fluids with photo-switchable viscosities based on a reversible transition between vesicles and wormlike micelles. Soft Matter 9:5025–5033
50. Matsumura A, Sakai K, Sakai H, Abe M (2011) Photoinduced increase in surfactant solution viscosity using azobenzene dicarboxylate for molecular switching. J Oleo Sci 60:203–207
51. Yan H, Long Y, Song K, Tung C-H, Zheng L (2014) Photo-induced transformation from wormlike to spherical micelles based on pyrrolidinium ionic liquids. Soft Matter 10:115–121

52. Lu Y, Zhou T, Fan Q, Dong J, Li X (2013) Light-responsive viscoelastic fluids based on anionic wormlike micelles. J Colloid Interface Sci 412:107–111
53. Takahashi Y, Yamamoto Y, Hata S, Kondo Y (2013) Unusual viscoelasticity behaviour in aqueous solutions containing a photoresponsive amphiphile. J Colloid Interface Sci 407:370–374
54. Jr Hubbard FP, Abbott NL (2007) Effect of light on self-assembly of aqueous mixtures of sodium dodecyl sulfate and a cationic bolaform surfactant containing azobenzene. Langmuir 23:4819–4829
55. Aikawa S, Shrestha RG, Ohmori T, Fukukita Y, Tezuka Y, Endo T, Torigoe K, Tsuchiya K, Sakamoto K, Sakai K, Abe M, Sakai H (2013) Photorheological response of aqueous wormlike micelles with photocleavable surfactant. Langmuir 29:5668–5676
56. (a) Shchipunov YA (2001) Colloid Surf A 183–185:541–554. (b) Willard DM, Riter RE, Levinger NE (1998) Dynamics of polar solvation in lecithin/water/cyclohexane reverse micelles. J Am Chem Soc 120:4151–4160. (c) Scartazzini R, Luisi PL (1988) Organogels from lecithins. J Phys Chem 92:829–833
57. (a) Hashizaki K, Taguchi H, Saito Y (2009) A novel reverse worm-like micelle from a lecithin/sucrose fatty acid ester/oil system. Colloid Polym Sci 287:1099–1105. (b) Hashizaki K, Chiba T, Taguchi H, Saito Y (2009) Highly viscoelastic reverse worm-like micelles formed in a lecithin/urea/oil system. Colloid Polym Sci 287:927–932
58. (a) Hashizaki K, Taguchi H, Saito Y (2009) New Lecithin organogels with sugars of RNA and DNA. Chem Lett 38:1036–1037. (b) Hashizaki K, Sakanishi Y, Yako S, Tsusaka H, Imai M, Taguchi H, Saito, Y (2012) New lecithin organogels from lecithin/polyglycerol/oil systems. J Oleo Sci 61:267–275
59. (a) Tung SH, Huang YE, Raghavan SR (2006) A new reverse wormlike micellar system: mixtures of bile salt and lecithin in organic liquids. J Am Chem Soc 128:5751–5756. (b) Tung SH, Huang, YE, Raghavan SR (2007) Contrasting effects of temperature on the rheology of normal and reverse wormlike micelles. Langmuir 23:372–376
60. (a) Schurtenberger P, Scartazzini R, Magid, LJ, Leser ME, Luisi PL (1990) Structural and dynamic properties of polymer-like reverse micelles. J Phys Chem 94:3695-3701. (b) Schurtenberger, P, Magid LJ, King SM, Lindner, P (1991) Cylindrical structure and flexibility of polymerlike lecithin reverse micelles J Phys Chem 95:4173–4176
61. Kumar R, Ketner MA, Raghavan SR (2010) Nonaqueous photorheological fluids based on light-responsive reverse wormlike micelles. Langmuir 26(8):5405–5411
62. Shrestha RG, Agari N, Tsuchiya K, Sakamoto K, Sakai K, Abe M, Sakai H (2014) Phosphatidylcholine-based nonaqueous photorheological fluids: effect of geometry and solvent. Colloid Polym Sci 292:1599–1609
63. Lee HY, Diehn KK, Sun K, Chen T, Raghavan SR (2011) Reversible photorheological fluids based on spiropyran-doped reverse micelles. J Am Chem Soc 133:8461
64. (a) Kawasaki H, Souda M, Tanaka S, Nemoto N, Karlsson G, Almgren M, Maeda H (2002) Reversible vesicle formation by changing pH. J Phys Chem B 106:1524–1527. (b) Maeda H, Yamamoto A, Souda M, Kawasaki H, Hossain KS, Nemoto N, Almgren M (2001) Effects of protonation on the viscoelastic properties of tetradecyldimethylamine oxide micelles J Phys Chem B 105:5411–5418. (c) Maeda H, Tanaka S, Ono Y, Miyahara M, Kawasaki H, Nemoto N, Almgren M (2006) Reversible micelle-vesicle conversion of oleyldimethylamine oxide by pH changes. J Phys Chem B 110:12451–12458. (d) Majhi PR, Dubin PL, Feng X, Guo X (2004) Coexistence of spheres and rods in micellar solution of dodecyldimethylamine oxide. J Phys Chem B 108:5980–5988
65. (a) Lin Y, Han X, Huang J, Fu H, Yu C (2009) A facile route to design pH-responsive viscoelastic wormlike micelles: smart use of hydrotropes. J Colloid Interface Sci 330:449–455. (b) Yan H, Zhao M, Zheng L (2011) A hydrogel formed by cetylpyrrolidinium bromide and sodium salicylate. Colloid Surf A Physicochem Eng Asp 392:205–212. (c) Verma G, Aswal VK, Hassan P (2009) pH-responsive self-assembly in an aqueous mixture of surfactant and hydrophobic amino acid mimic. Soft Matter 5:2919–2927. (d) Ali M, Jha M, Das SK, Saha SK (2009) Hydrogen-bond-induced microstructural transition of ionic

micelles in the presence of neutral naphthols: pH dependent morphology and location of surface activity. J Phys Chem B 113:15563–15571

66. (a) Lin Y, Han X, Cheng X, Huang J, Liang D, Yu C (2008) pH-regulated molecular self-assemblies in a cationic–anionic surfactant system: from a "1–2" surfactant pair to a "1-1" surfactant pair. Langmuir 24:13918–13924 12. (b) Ghosh S, Khatua D, Dey J (2011) Interaction between zwitterionic and anionic surfactants: spontaneous formation of zwitanionic vesicles. Langmuir 27:5184–5192
67. (a) Smart fluids: switchable viscosity. NPG Asia Mater Research Highlight (2011). doi:10.1038/asiamat.2011.29. (b) Johnsson M, Wagenaar A, Engberts JBFN (2003) Sugar-based gemini surfactant with a vesicle-to-micelle transition at acidic pH and a reversible vesicle flocculation near neutral pH. J Am Chem Soc 125:757–760. (c) Johnsson M, Wagenaar A, Stuart MCA, Engberts JBFN (2003) Sugar-based gemini surfactants with pH-dependent aggregation behavior: vesicle-to-micelle transition, critical micelle concentration, and vesicle surface charge reversal. Langmuir 19:4609–4618. (d) Jaeger DA, Li G, Subotkowski W, Carron KT (1997) Fibers and other aggregates of omega-substituted surfactants. Langmuir 13:5563–5569. (e) Graf G, Drescher S, Meister A, Dobner B, Blume A (2011) Self-assembled bolaamphiphile fibers have intermediate properties between crystalline nanofibers and wormlike micelles: formation of viscoelastic hydrogels switchable by changes in pH and salinity. J Phys Chem B 115:10478–10487. (f) Yao R Qian J. Li H. Yasin A. Xie Y. Yang H (2014) Synthesis and high performance of a new sarcosinate anionic surfactant with a long unsaturated tail. RSC Adv 4:2865–2872
68. Ono Y, Shikata T (2005) Dielectric behavior of aqueous micellar solutions of betaine-type surfactants. J Phys Chem B 109:7412–7419
69. (a) Ikeda S, Tsunoda M, Maeda H (1979) Effects of ionization on micelles size of dimethyldodecylamine oxide. J Colloid Interface Sci 70:448–455. (b) Herrmann KW (1962) Nonionic–cationic micellar properties of dimethyldodecylamine oxide. J Phys Chem 66:295–300
70. Chu Z, Feng Y (2010) 'Smart fluids: switchable viscosity', NPG Asia Mater, featured highlight. Chem Commun 46:9028–9030
71. Hoffmann H (1994) Viscoelastic surfactant solutions. In: Herb CA, Prud'homme RK (eds) Structure and flow of surfactant solution. ACS Symp Ser vol 578. American Chemical Society, Washington, DC, pp 2–31
72. Hashimoto K, Imae T (1991) Rheological properties of aqueous solutions of alkyldimethylamine and oleyldimethylamine oxides—spinnability and viscoelasticity. Langmuir 7:1734–1741
73. Rathman JF, Christian SD (1990) Determination of surfactant activities in micellar solutions of dimethyldodecylamine oxide. Langmuir 6:391–395
74. Maeda H (1996) Dodecyldimethylamine oxide micelles: stability, aggregation number and titration properties. Colloids Surf A 109:263–271
75. Zhang H, Dubin PL, Kaplan JI (1991) Potentiometric and dynamic light scattering studies of micelles of dimethyldodecylamine oxide. Langmuir 7:2103–2107
76. Mille M (1981) Effect of nearest-neighbor interactions on surface titrations. J Colloid Interface Sci 81:169–179
77. Brinchi L, Germani R, Profio PD, Marte L, Savelli G, Oda R, Berti D (2010) Viscoelastic solutions formed by worm-like micelles of amine oxide surfactant. J Colloid Interface Sci 346:100–106
78. (a) Lin Y, Han X, Huang J, Fu H, Yu C (2009) A facile route to design pH-responsive viscoelastic wormlike micelles: smart use of hydrotropes. J Colloid Interface Sci 330:449–455. (b) Yan H, Zhao M, Zheng L (2011) A hydrogel formed by cetylpyrrolidinium bromide and sodium salicylate. Colloid Surf A Physicochem Eng Asp 392:205–212. (c) Verma G, Aswal VK, Hassan P (2009) pH-responsive self-assembly in an aqueous mixture of surfactant and hydrophobic amino acid mimic. Soft Matter 5:2919–2927. (c) Ali M, Jha M, Das SK, Saha SK (2009) Hydrogen-bond-induced microstructural transition of ionic

micelles in the presence of neutral naphthols: pH dependent morphology and location of surface activity. J Phys Chem B 113:15563–15571
79. Zhao L, Wang K, Xu L, Liu Y, Zhag S, Li Z, Yan Y, Huang J (2012) Extremely pH-sensitive fluids based on a rationally designed simple amphiphile. Soft Matter 8:9079
80. Sakai K, Nomura K, Shrestha RG, Endo T, Sakamoto K, Sakai H, Abe M (2012) Wormlike micelle formation by acylglutamic acid with alkylamines. Langmuir 28(51):17617–17622
81. Jessop PG, Heldebrant DJ, Li XW, Eckert CA, Liotta CL (2005) Green chemistry reversible nonpolar-to-polar solvent. Nature 436:1102
82. Liu YX, Jessop PG, Cunningham M, Eckert CA, Liotta CL (2006) Switchable surfactants. Science 313:958–960
83. Guo Z, Feng Y, Wang Y, Wang J, Wu Y, Zhang Y (2011) A novel smart polymer responsive to CO_2. Chem Commun 47:9348–9350
84. Guo Z, Feng Y, He S, Qu M, Chen H, Liu H, Wu Y, Wang Y (2013) CO_2-responsive "smart" single-walled carbon nanotubes. Adv Mater 25:584–590
85. Su X, Jessop PG, Cunningham MF (2012) Surfactant-free polymerization forming switchable latexes that can be aggregated and re-dispersed by CO_2 removal and then re-addition. Macromolecules 45:666–670
86. Zhang Y, Feng Y, Wang Y, Li X (2013) CO_2 switchable viscoelastic fluids based on a pseudogemini surfactant. Langmuir 29(13):4187–4192
87. Zhang Y, Chu Z, Dreiss CA, Wang Y, Fei C, Feng Y (2013) Smart wormlike micelles switched by CO2 and air†. Soft Matter 9:6217–6221
88. Zhang Y, Feng Y (2015) CO_2-induced smart viscoelastic fluids based on mixtures of sodium erucate and trimethylamine. J Colloid Interface Sci 447:173–181
89. Zhang Y, Yin H, Feng Y (2014) CO_2-responsive anionic wormlike micelles based on natural erucic acid. Green Mater 2:95–103
90. Su X, Cunningham MF, Jessop PG (2013) Switchable viscosity triggered by CO_2 using smart worm-like micelles, Chem Commun 49:2655–2657
91. Zhang Y, An P, Liu X, Fang Y, Hu X (2015) Smart use of tertiary amine to design CO_2-triggered viscoelastic fluids Colloid Polym Sci 293:357–367
92. Peppas NA, Hiltz JZ, Khademhosseini A, Langer R (2006) Hydrogels in biology and medicine: from molecular principles to bionanotechnology. Adv Mater 18:1345–1360
93. Lo¨wik DWPM, Leunissen EHP, van den Heuvel M, Hansen MB, van Hest JCM (2010) Stimulus responsive peptide based materials. Chem Soc Rev 39:3394–3412
94. (a) Raghavan SR, Kaler EW (2001) Highly viscoelastic wormlike micellar solutions formed by cationic surfactants with long unsaturated tails. Langmuir 17:300–306. (b) Chu Z, Feng Y (2010) Soft Matter 6:6065–6067. (c) Kumar R, Kalur, GC, Ziserman, L, Danino D, Raghavan SR (2007) Wormlike micelles of a C22-tailed zwitterionic betaine surfactant: from viscelastic solutions to elastic gels. Langmuir 23:12849–12856
95. Meng F, Zhong Z, Feijen J (2009) Stimuli-responsive polymersomes for programmed drug delivery. Biomacromolecules, 10:197–209. (b) Xiong W, Wang W, Wang Y, Zhao Y, Chen H, Xu H, Yang X (2011) Dual temperature/pH-sensitive drug delivery of poly (N-isopropylacrylamide-co-acrylic acid) nanogels conjugated with doxorubicin for potential application in tumor hyperthermia therapy. Colloids Surf B 84:447-453. (c) Su J, Chen F, Cryns VL, Messersmith PB (2011) Catechol polymers for pH-responsive, targeted drug delivery to cancer cells. J Am Chem Soc 133:11850–11853
96. (a) Ganta S, Devalapally H, Shahiwala A, Amiji M (2008) A review of stimuli-responsive nanocarriers for drug and gene delivery. J Controlled Release 126:187–204. (b) Hatefi... Kretlow JD, Klouda L, Mikos AG (2007) Injectable matrices and scaffolds for drug delivery in tissue engineering. Adv Drug Delivery Rev 59:263–273. (c) You YZ, Zhou QH, Manickam DS, Wan L, Mao GZ, Oupicky D (2007) Dually responsive multiblock copolymers via reversible addition—fragmentation chain transfer polymerization: synthesis of temperature- and redox-responsive copolymers of poly(N-isopropylacrylamide) and poly(2-(dimethylamino) ethyl methacrylate). Macromolecules 40:8617–8624

97. (a) Afifi H, Karlsson G, Heenan R.K, Dreiss CA (2011) Solubilization of oils or addition of monoglycerides drives the formation of wormlike micelles with an elliptical cross-section in cholesterol-based surfactants: a study by rheology, SANS, and cryo-TEM. Langmuir 27:7480–7492. (b) Afifi H, Karlsson G, Heenan RK, Dreiss CA (2012) Structural transitions in cholesterol-based wormlike micelles induced by encapsulating alkyl ester oils with varying architecture. J Colloid Interface Sci 378:125–134
98. Wu Q, Wang L, Yu H, Wang J, Chen Z (2011) Organization of glucose-responsive systems and their properties. Chem Rev 111:7855–7875
99. Dave El-Hamed F, Liu NJ (2011) Stimuli-responsive releasing of gold nanoparticles and liposomes from aptamer-functionalized hydrogels. Nanotechnology 22:494011
100. Ferri V, Elbing M, Pace G, Dickey MD, Zharnikov M, Samori P, Mayor M, Rampi MA (2008) Light-powered electrical switch based on cargo-lifting azobenzene monolayers. Angew Chem Int Ed 47:3407–3409
101. (a) Baglioni P, Braccalenti E, Carretti E, Germani R, Goracci L, Savelli G, Tiecco M (2009) Langmuir 25:5467–5475. (b) Pereira M, Leal CR, Parola AJ, Scheven UM (2010) Reversible photorheology in solutions of cetyltrimethylammonium bromide, salicylic acid, and *trans*-2,4,4′-trihydroxychalcone. Langmuir 26:16715–16721
102. Kefi S, Lee J, Pope TL, Sullivan P, Nelson E, Hernandez AN, Olsen T, Parlar M, Powers B, Roy A, Wilson A, Twynam A (2004) Expanding applications for viscoelastic surfactants. Oilfield Rev 16(4):10–23
103. Sullivan P, Nelson EB, Anderson V, Hughes T (2007) Oilfield applications of giant micelles. In: Zana R, Kaler EW (ed) Giant micelles: properties and applications. CRC Press, Boca Raton, pp 453–472
104. Dreiss CA Wormlike micelles: where do we stand? recent developments, linear rheology and scattering techniques. Soft Matter 3:956–970
105. Shi HF, Wang Y, Fang B, Talmon Y, Ge W, Raghavan SR, Zakin JL (2011) Light-responsive threadlike micelles as drag reducing fluids with enhanced heat-transfer capabilities. Langmuir 27:5806–5813
106. Li G, Zhang Z (2004) Synthesis of dendritic polyaniline nanofibers in a surfactant gel. Macromolecules 37:2683–2685
107. (a) Tan B, Dozier A, Lehmler HJ, Knutson BL, Rankin SE (2004) Elongated silica nanoparticles with a mesh phase mesopore structure by fluorosurfactant templating. Langmuir 20:6981–6984. (b) Nagamine S, Kurumada KI, Tanigaki M (2001) Growth of silica particles in surfactant gel. Adv. Powder Technol 12:145–156. (c) Broz P, Driamov S, Ziegler J, Ben-Haim N, Marsch S, Meier W, Hunziker P (2006) Toward intelligent nanosize bioreactors: a pH-switchable, channel-equipped, functional polymer nanocontainer. Nano Lett 6:2349–2353
108. Sakai K, Smith EG, Webber GB, Baker M, Wanless EJ, Bütün V, Armes SP, Biggs S (2006) Comparison of the adsorption of cationic diblock copolymer micelles from aqueous solution onto mica and silica. Langmuir 22:8435–8442
109. Choi D, Kumta PN (2007) Surfactant based sol–gel approach to nanostructured $LiFePO_4$ for high rate Li-ion batteries. J Power Sources 163:1064–1069

Chapter 6
Stimuli-Responsive and Soft-Template Functions of Novel Amphiphiles Having Amidoamine Groups

Takeshi Kawai and Yoshiro Imura

Abstract In this chapter, we demonstrate that practical applications of a long-chain amidoamine derivative with a simple molecular structure (C18AA) are equivalent to or higher than those of supramolecules with complex molecular structures. C18AA acts as a thermal and pH-responsive soft material. Namely, C18AA can form normal organogels in apolar solvents, which show a phase transition from gel to sol upon heating, whereas novel worm-like micelles of C18AA formed in aqueous solution show a heat-induced viscosity transition from sol to gel upon heating. Interestingly, O/W emulsions made by the addition of an aqueous solution of HCl into a C18AA organogel have a double stimuli-responsive heat-induced gel property, and the sol–gel transition can be regulated by both temperature and pH. Moreover, we show that C18AA emulsions containing a quaternary ammonium salt develop an iridescent color in a specific temperature range, and the color and the coloring temperature can be independently controlled by adjusting the C18AA and salt concentrations, respectively. We also present the soft-template function of C18AA for preparing shape-controlled Au nanocrystals, in addition to soft materials' functions. C18AA has the ability to synthesize ultrathin Au nanowires in organogels, two-dimensional dendritic Au nanowires, and novel neuron-shaped Au nanocrystals in aqueous solutions.

Keywords Heat-induced gel · Emulsions · Viscosity control · Thermal and pH response · Coloring emulsions · Soft template · Au nanowires

---

T. Kawai (✉) · Y. Imura
Department of Industrial Chemistry, Faculty of Engineering, Tokyo University of Science, Shinjuku, Japan
e-mail: kawai@ci.kagu.tus.ac.jp

T. Kawai and M. Hashizume (eds.), *Stimuli-Responsive Interfaces*,
DOI 10.1007/978-981-10-2463-4_6

6.1 Introduction

Materials with stimuli-responsive viscosity are of interest due to their potential uses in pharmacy, micromachinery, and pollutant removal [1–3]. In recent years, self-assemblies of low molecular weight amphiphiles, e.g., supramolecular fibrils or micellar aggregates, have received great attention as high-performance materials with stimuli-responsive viscosity [4, 5]. In particular, long and highly flexible aggregates, called worm-like micelles, are promising viscoelastic materials because of their unique mechanical and rheological properties, which are analogous to flexible polymer solutions [6]. Formation of worm-like micelles has been reported in various amphiphilic compounds, and their distinguishing feature is the entanglement of micellar segments into a transient network. The viscoelasticity of worm-like micellar solutions is heat dependent, with viscosity generally decreasing with increasing temperature due to decay in the micellar length [7, 8]. In contrast, some worm-like micelles show an increase in viscosity upon heating due to the growth of worm-like micelles [9, 10] or the transition of vesicles to worm-like micelles [11, 12]. Unfortunately, these structural changes occur gradually and a broad temperature range is needed to complete the thermal viscosity transitions.

Recently, we have demonstrated that an amphiphilic long-chain amidoamine derivative (N-(2-amino-ethyl)-3-{[2-(2-amino-ethylcarbamoyl)-ethyl]-octadecyl-amino}-propioamide (C18AA) (Fig. 6.1) has multifunctional capabilities, besides acting as an organogelator in some apolar solvents, despite its quite simple molecular structure [13–16]. The multifunctional capability originates from the following: (i) it is pH responsive due to the protonation ability of the terminal amine and tertiary amine groups [17–19] and (ii) the amide groups and the terminal amine groups of the hydrophilic part can form a hydrogen-bonded network structure in water and organic solvents [6, 14–16]. The hydrogen bond strength depends strongly on temperature; consequently, the molecular assemblies formed in water or organic solvents should change with temperature. Therefore, C18AA can be used in thermal- and pH-responsive soft materials [6, 14–16, 20]. In this review, we demonstrate that C18AA solubilized in water forms unique straight elongated micelles, which are more similar to fibrils made by self-assembled supramolecular amphiphiles or organogelators than to conventional worm-like micelles, and that the aqueous solution of C18AA acts as a heat-induced gel, which undergoes a phase transition from solution to gel upon heating [6]. Further, we show that heat-induced emulsion gels of C18AA have a highly sensitive response to temperature and pH.

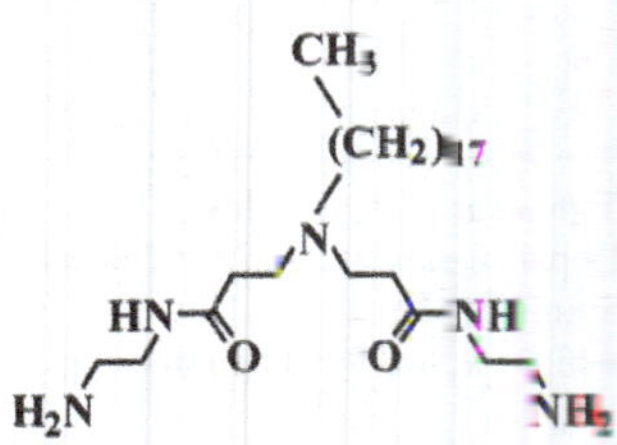

Fig. 6.1 Molecular structure of C18AA

and the sol to gel transition can be independently regulated by pH and C18AA concentration [15, 16].

Additionally, morphological control of metal nanomaterials is important in the development of modern material chemistry, because the physical and chemical properties can be easily and widely tuned by tailoring the particle size and shape [21, 22]. Recently, there have been many reports on the fabrication of various shaped metal nanomaterials, such as cubes, plates, rods, and wires [22–29]. The general approach to the shape control of metal nanomaterials is the conventional seeded growth method: Seed nanoparticles (NPs) with specific shapes and/or particular crystal facets are grown to obtain the desired nanostructure shape [22, 29–31]. C18AA has a selective affinity to specific crystal facets of Au, namely the terminal amine groups of C18AA adsorb selectively on (100) and (110) crystal facets of Au, but barely adsorb on the (111) crystal facet [17, 32]. This adsorption property provides an opportunity to prepare shape-controlled Au nanocrystals by controlling the growth direction of the nanocrystals. The combination of self-assembled aggregates of C18AA and the selective adsorption property provide a good soft template for Au nanocrystals [13, 14]. In the latter part of this review, we present the soft-template function for Au nanocrystals and demonstrate that C18AA has the ability to prepare ultrathin straight Au nanowires, two-dimensional dendritic Au nanowires, and neuron-shaped Au nanowires [13–15, 32–34].

6.2 Heat-Induced Gel in Water

In this section, we describe that C18AA containing two amide moieties forms uniaxial elongated aggregates in water, and the elongated micelles possess thermo-sensitive behavior different from conventional worm-like micelles. Figure 6.2 shows a TEM image of the aggregate formed in water. The aggregates

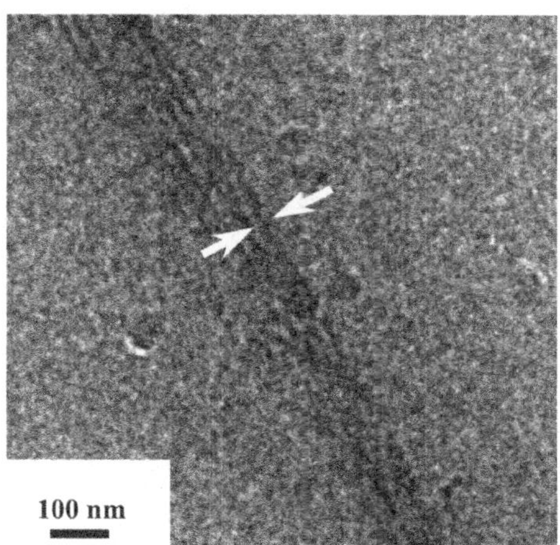

Fig. 6.2 Cryo-TEM image of C18AA aggregate and the bundle in a 17 wt% C18AA aqueous solution at room temperature

have a straight elongated fibrous structure with a width of ~10 nm. This characteristic is more similar to those of conventional fibrils made by self-assembled supramolecular amphiphiles or organogelators than to those of conventional worm-like micelles. In general, the straightness of conventional fibrils is derived from intermolecular interactions, such as hydrogen bonding. The molecular interactions between C18AA in the elongated aggregates can be monitored by FT-IR spectroscopic measurements of the aqueous C18AA solution [6]. The amide groups being in a strong hydrogen bonding state was evidenced by the peak position of the amide I band in the FT-IR spectrum. Thus, it was proven that the straightness of the C18AA aggregates is caused by a strong intermolecular hydrogen bonding network between the amide groups.

Steady shear rheological measurements on aqueous C18AA solutions were made at 25 °C as a function of C18AA concentration. An abrupt increase in the zero-shear viscosity (η_0) occurred at a C18AA concentration of ~17 wt%. This abrupt increase is probably caused by overlap of the straight aggregates. However, the η_0 value of ~1 Pa·s in the concentration range of [C18AA] = 17–33 wt% is less than that of conventional worm-like micelle systems. Furthermore, the dynamic viscoelasticity curve below 40 °C and a high concentration of C18AA do not indicate Maxwell behavior, although the curves of conventional worm-like micelles should be described by the Maxwell model. This indicates that C18AA aggregates rarely become entangled at ~25 °C. The straightness of the C18AA worm-like micelles physically prevents micellar entanglement, resulting in low viscosity. Interestingly, the dynamic viscoelasticity shows a typical Maxwell fluid-like response at a high temperature of 52 °C, similar to that of typical worm-like micelles. Thus, elevated temperatures bring about a transformation from straight elongated micelles to flexible worm-like micelles. This transformation is induced by a defect of intermolecular hydrogen bonding between the amide groups upon heating.

Figure 6.3 shows the temperature responses of storage (G′) and loss (G″) moduli at a C18AA concentration of 33 wt% and a constant oscillatory frequency of 1 Hz. At room temperature, aqueous C18AA solutions show a lower dynamic modulus value and G′ is less than G″, while the modulus value drastically increases at 48 °C

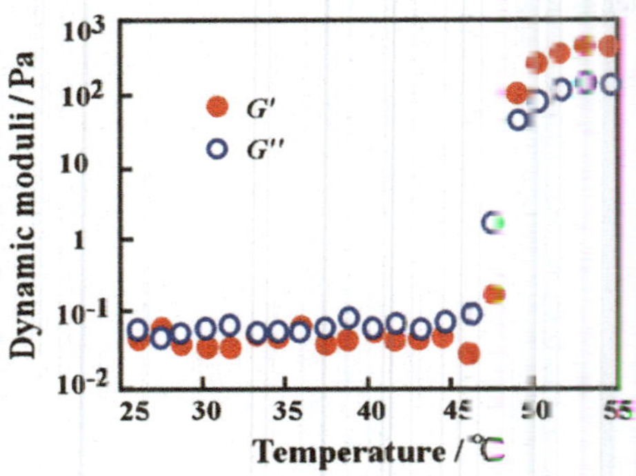

Fig. 6.3 Temperature response of dynamic modulus of a 33 wt% C18AA aqueous solution at oscillatory frequency of 1 Hz. 3 °C/min

and G′ eventually exceeds G″. The viscosity profiles indicate that aqueous solutions of C18AA have a heat-induced gel property and undergo a thermally reversible phase transition from a low viscosity solution to a rigid gel upon heating. Moreover, the shape response of the viscosity to temperature changes is very different from that of conventional worm-like micelles, which show a broad change in viscosity with temperature changes.

6.3 Heat-Induced Emulsion Gel

In the previous section, we described the heat-induced gel property of C18AA micelles formed in water. Interestingly, addition of LiCl, a hydrogen bond breaker, into the aqueous solution of C18AA promotes the formation of many defects in the hydrogen-bonding network of the elongated micelles, and, consequently, results in a lowering of the transition temperature. However, the controllable temperature is in a narrow temperature range of 41–48 °C. In this section, we demonstrate that the addition of aqueous LiCl into the toluene gel converts it from the normal organogel to a heat-induced emulsion gel. Furthermore, the transition temperatures from sol to gel states can be controlled over a wide temperature range of 7–52 °C simply by changing the concentration of C18AA.

C18AA can form an organogel in some apolar solvents such as toluene, benzene, and cyclohexane, as mentioned in the introduction. In the organogel, C18AA assembles into an ordered lamellar structure, where the hydrocarbon chains of C18AA are interdigitated, and the amide groups are linked through hydrogen bonding to the adjacent amide group [16]. The complex viscosity η* profiles for the organogels (Fig. 6.4c) show that the phase transition from gel to sol occurs at 45 °C

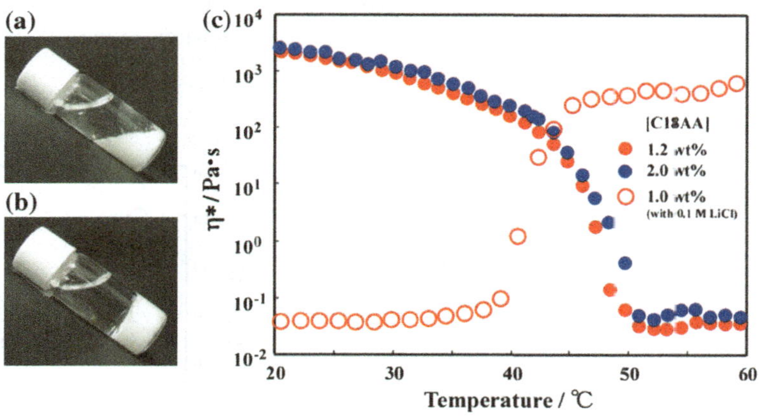

Fig. 6.4 Photographic images of heat-induced gel at **a** 25 °C and **b** 60 °C. **c** Temperature dependence of η* of C18AA/toluene gel (*closed circles*) and heat-induced gel (*open circles*) after addition of 0.1 M LiCl

upon heating, and that the transition temperatures remain fairly constant independent of C18AA concentration. The rigidity of the toluene gel is almost equal to those observed for typical low molecular weight gelators.

Addition of small amounts of 0.1 M LiCl aqueous solution into the C18AA toluene gel produces opaque emulsions. Interestingly, the emulsion phase shows a heat-induced gel property, namely it becomes a rigid gel state at higher temperatures (Fig. 6.4a, b). In viscoelastic measurements of the opaque emulsion phase, the η* value increases drastically from 10^{-1} to 10^{3} Pa·s upon heating (Fig. 6.4c). This remarkable change in fluidity upon heating and cooling is completely reversible. Thus, by simple addition of LiCl, the toluene gel of C18AA can be converted into a heat-induced gel with the opposite thermal property as that of the normal gel.

The amount of LiCl plays an important role in the rigidity of the heat-induced gel, but has little influence on the sol–gel transition temperatures. On the other hand, the transition temperatures largely depend on the concentration of C18AA, as shown in Fig. 6.5. The transition temperatures decrease linearly with increasing C18AA concentration, but the η* values in the gel state are independent of [C18AA], i.e., the C18AA concentration affects the transition temperature, but not the rigidity of the gel. Thus, the sol–gel transition temperature can be controlled simply by changing the C18AA concentration. Furthermore, the linear region in the transition temperature at 7–52 °C provides opportunities for potential application in useful new stimuli-responsive materials.

A vitally important question is the difference in the molecular assembly of C18AA between the sol and gel states. The optical and fluorescence microscopic observations of the opaque emulsions demonstrated that both the sol and gel states consist of O/W emulsions, and that the emulsion structure does not differ much between the sol and gel states (Fig. 6.6a, b). On the other hand, the measurements of C18AA distributions among water, toluene, and the water/toluene interface revealed that C18AA molecules transfer from the O/W interface to the continuous water phase upon heating. Since the increase in C18AA concentration in water brings about the formation of larger aggregates, O/W emulsions in the gel state are

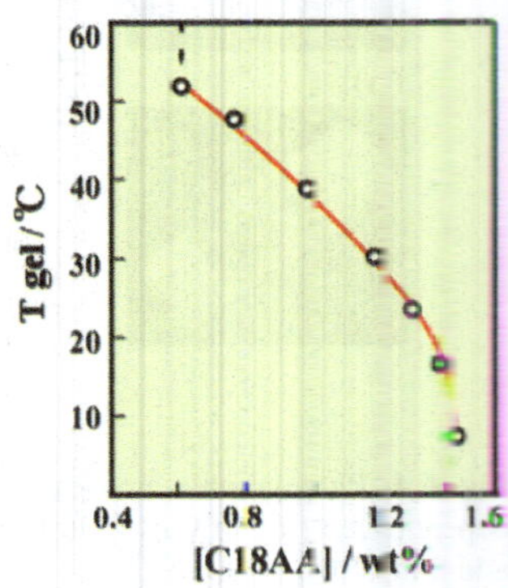

Fig. 6.5 Sol–gel transition temperature (T*gel*) with C18AA concentration

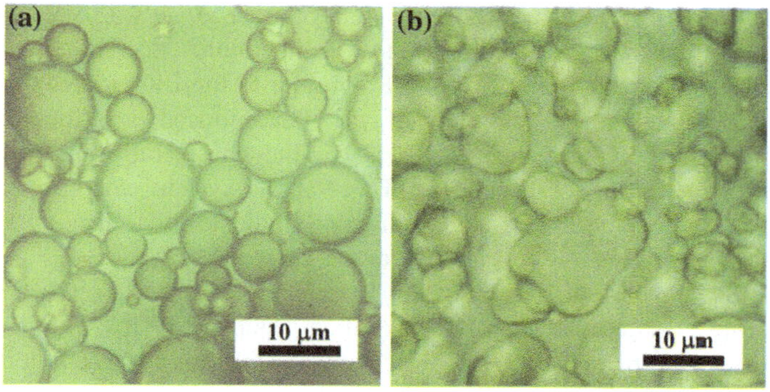

Fig. 6.6 Optical micrographs of 1.2 wt% C18AA/toluene/0.1 M LiCl system. **a** Sol state at 25 °C and **b** Gel state at 40 °C

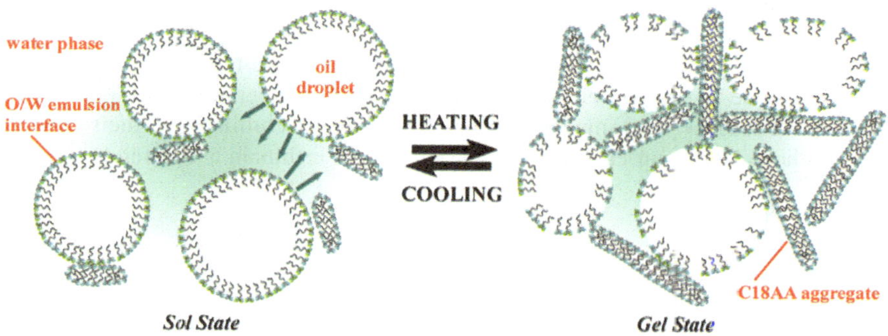

Fig. 6.7 Schematic illustration of proposed mechanism for heat-induced gel formation

trapped and pinned by the network of C18AA aggregates formed in the continuous water phase, leading to a high viscosity for the system (Fig. 6.7).

6.4 pH- and Thermo-Responsive Emulsion Gel

LiCl was absolutely essential for the formation of heat-induced emulsion gels, but it remains an unsettled question whether compounds apart from LiCl are effective in forming O/W emulsions of C18AA that can act as heat-induced gels. Moreover, the reversible sol–gel transition of heat-induced gels produced by a LiCl electrolyte cannot be regulated by electrolyte concentration, because there is no way to achieve a decrease in concentration. If we can control the sol–gel transition by pH, it would provide a double stimuli-responsive gel, namely the sol–gel phase transition could be reversibly regulated by both pH and temperature.

We confirmed that heat-induced emulsion gels based on C18AA can also be prepared using HCl. The sol–gel transition temperatures decrease linearly with increasing C18AA concentration (Fig. 6.8), similar to the LiCl system. Figure 6.9 shows the resultant phase diagram of the heat-induced gel for a 1.0 wt% C18AA/toluene/aq. HCl system. At pH > 9.5, the phase transition temperatures increase with increasing pH, while at pH < 9.3 the system exists in a gel state regardless of temperature. The phase diagram indicates that the sol–gel phase transition is highly sensitive to pH and can be controlled by pH.

The pH-responsive sol–gel transition is related to protonation of the amine groups, because the terminal and tertiary amine groups of C18AA act as protonation sites at lower pH. The ^{1}H NMR chemical shifts of the methylene groups adjacent to the protonation sites revealed that at pH > 9.5 there is no protonation of either amine group, while at pH < 9.5 the terminal amine groups start to protonate [19]. Thus, the gradual decrease in sol–gel phase transition temperature with decreasing pH above pH ~9.5 in Fig. 6.9 is not caused by protonation of C18AA, but by a salting-in effect of H^{+}, which acts as a chaotropic cation. On the other hand, the abrupt decrease at pH ~9.5 is directly related to protonation of the terminal amine groups. The mechanism of the sol–gel phase transition regulated by pH is the same as that which occurs by a change in temperature. The protonated C18AA molecules are more soluble in water according to their higher hydrophilicity. As a consequence, decreasing the pH results in a transfer of C18AA

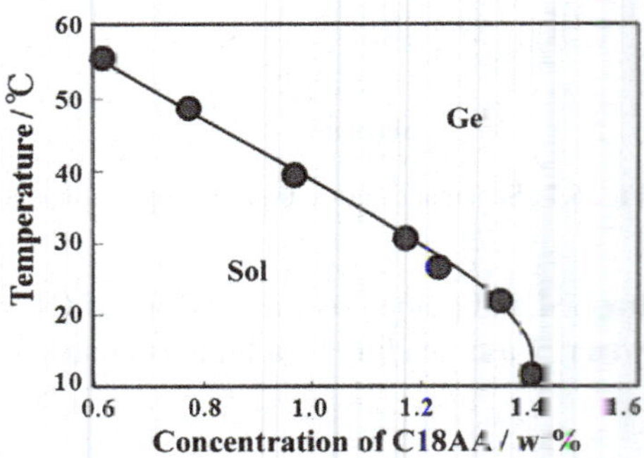

Fig. 6.8 Schematic illustration of proposed mechanism for heat-induced gel formation

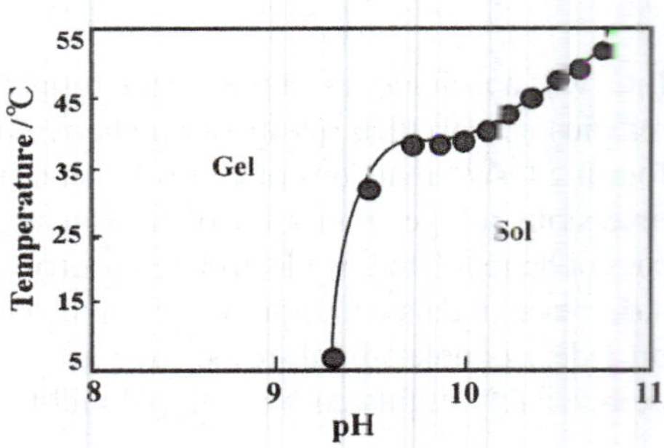

Fig. 6.9 Sol–gel phase diagram of 1.0 wt%-C18AA/toluene/HCl

molecules from the O/W interface to the continuous water phase and subsequent formation of larger aggregates of C18AA in the water phase. The network of the C18AA aggregates formed in the continuous water phase traps and pins the O/W emulsions, leading to gel formation.

6.5 Thermo-Responsive Coloring Emulsions

Thermo-responsive coloring materials, which are capable of switching their color reversibly in response to temperature, have been extensively studied, because they can be used as imaging and sensing materials. The coloring materials are classified into two main groups according to the origin of their color. Namely, one is derived from the light absorption phenomena of organic or inorganic compounds such as dyes and quantum dots, and the other is derived from interference phenomena of periodically layered structures such as colloidal crystals and laminate films. The interference coloring materials have the potential for possessing highly sensitive thermo-responsive color switching when the materials are functionalized with or embedded in thermo-responsive compounds. There are two important features, color and coloring temperature, but it is generally difficult to fabricate coloring materials with independently tunable colors and coloring temperatures. Here, we review coloring emulsions comprising C18AA and tetraoctylammonium bromide (TOAB), which develop an iridescent color derived from the interference of multilayered liquid crystalline-like structures. The coloring temperature and the color can be controlled completely independently by the concentrations of C18AA and TOAB, respectively.

A water/C18AA + TOAB/toluene system at [C18AA] = 30 mM and [TOAB] = 6 mM is an opaque emulsion at room temperature, whereas with increasing temperature the emulsion develops an iridescent color (Fig. 6.10). A subsequent increase in temperature brings about the formation of an opaque emulsion again. As a result, the iridescent color disappears above ~47 °C. The iridescent color is observed in a narrow temperature range from 38 to 47 °C, and the coloring phenomena are completely reversible for many heating and cooling cycles [20]. The reflection spectra of the iridescent emulsion show a sharp peak, and

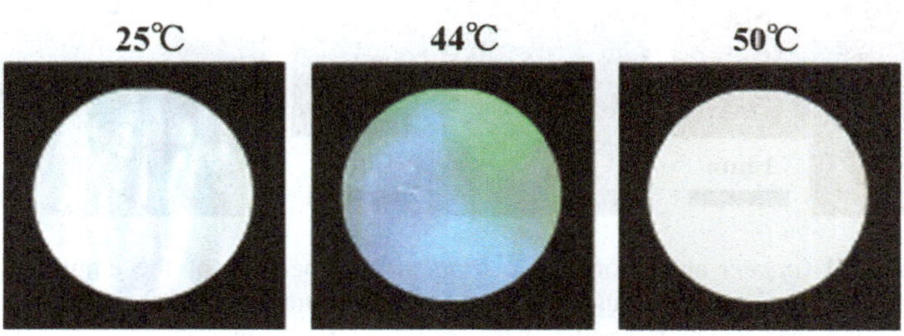

Fig. 6.10 Photographs of the iridescent emulsion at [C18AA] = 30 mM and [TOAB] = 6 mM

the peak position λ is shifted to a shorter wavelength with increasing incident angle θ [20]. The plot of $\sin^2\theta$ against λ^2 shows a linear relationship, which is the typical characteristic feature of interference color. The lattice spacing d calculated on the basis of the Bragg equation is $\sim$180 nm.

The water/C18AA + TOAB/toluene system passes through at least three states upon heating: (i) an opaque emulsion phase at a lower temperature, (ii) an iridescent emulsion phase in the middle temperature range, and (iii) an opaque emulsion phase at a higher temperature. Electrical conductivity measurements and the optical and fluorescence microscopic observations of the system (Fig. 6.11) reveal that the emulsions show a phase inversion from O/W to W/O through a swollen lamellar phase, which develop an iridescent color derived from the interference of a multilayered liquid crystalline-like structure.

For the thermo-responsive coloring materials, the coloring temperature and the color, i.e., the temperature range and the spectral range of the coloring materials, are the essential features. The spectral range of interference color can be tuned by controlling the lattice spacing d of the lamellar structure of the iridescent emulsions, whereas the coloring temperature is subjected to the phase inversion temperatures of the emulsions.

Figure 6.12 shows the plots of the lattice spacing d against the reciprocal concentration of C18AA at a constant value of [TOAB]/[C18AA] = 0.2 and those against the surfactant ratio ([TOAB]/[C18AA]) at a constant concentration of [C18AA] = 30 mM. The ratio of [TOAB] to C18AA has no effect on the lamellar spacing d, while the values of d are inversely proportional to the concentration of C18AA. If the lamellar structure is homogeneous in all the iridescent emulsions and C18AA molecules are totally consumed by the formation of the lamellar layers at the oil–water interface, the number of lamellar layers should increase linearly with

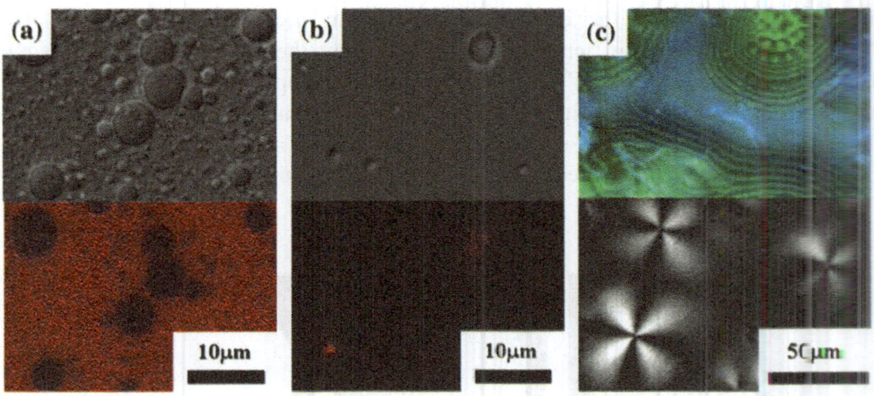

Fig. 6.11 Optical (*upper*) and fluorescence (*bottom*) micrographs of the emulsion in opaque emulsion phase at **a** lower temperature, and **b** higher temperature. **c** Optical reflection (*upper*) and crossed polarizing (*bottom*) micrographs of the iridescent emulsion

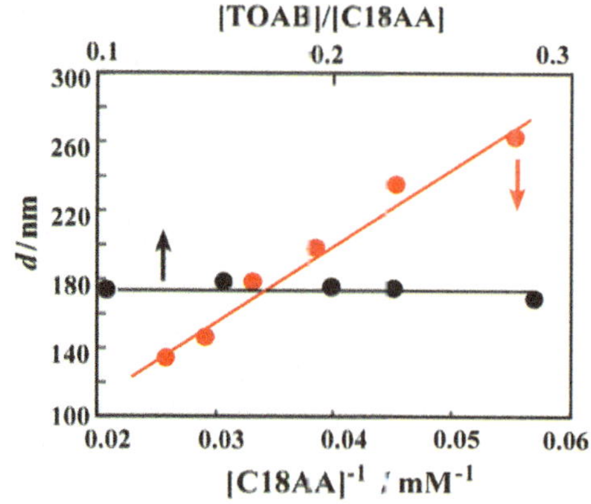

Fig. 6.12 Plots of the lattice spacing d as a function of $[C18AA]^{-1}$ at [TOAB]/[C18AA] = 0.2 (*red*) and of the [TOAB]/[C18AA] ratio at [C18AA] = 30 mM (*black*)

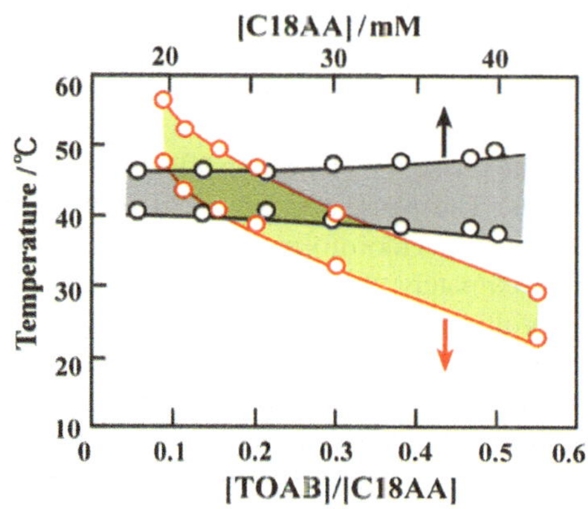

Fig. 6.13 The variation in the coloring temperature range with [C18AA] at [TOAB]/[C18AA] = 0.2 (*black*) and with the [C18AA]/[TOAB] ratio at [C18AA] = 30 mM (*red*). Shadowed areas indicate the coloring temperature regions

increasing C18AA concentration for constant volumes of toluene and water. Thus, the linear relationship between d and $[C18AA]^{-1}$ in Fig. 6.12 is a reasonable result.

On the other hand, the coloring temperature is shifted to a considerably lower temperature with increasing [TOAB]/[C18AA], but is not affected by [C18AA], as shown in Fig. 6.13. Namely, the coloring temperature can be controlled by the ratio of [TOAB]/[C18AA], i.e., TOAB concentration. The lowering of the phase inversion temperature at a high ratio of [TOAB]/[C18AA] is probably due to a lowering of the hydrophilicity of the system, which is caused by the addition of the high lipophilic TOAB surfactant into the C18AA surfactant system.

The coloring temperature and the color can be controlled completely independently by the concentrations of C18AA and TOAB, respectively. Significantly, noble metal nanoparticles (NPs) or semiconductor quantum dots having optical functions can be incorporated into the current iridescent emulsion system without influencing the interference color and the coloring temperature of the emulsions, because the NPs are dispersed within the bulk organic phase and do not affect the interfacial property of the emulsions. Therefore, the iridescent emulsion of the present system has a flexible thermo-responsibility and great availability for coloring materials, thermometers, and visual sensing.

6.6 pH-Responsive Phase Transfer of Gold Nanoparticles

In this section, we review that C18AA can be used as a useful phase transfer reagent for Au NPs, and the phase transfer of Au NPs between water and an organic solvent is performed by a pH change. We also demonstrate that the phase transfer process has the potential to fractionate Au NPs according to crystal facet distinction.

Au NPs (15.7 nm diameter) dispersed in water were poured onto a chloroform solution of C18AA. When the pH value of the water phase was adjusted to 13.0 and the mixture was allowed to stand for a long time, Au NPs were transferred from the water to the chloroform phase (Fig. 6.14). Adjustment of the pH to 8.0 resulted in a phase transfer of some of the Au NPs from the chloroform to the water phase (Fig. 6.14). The reversible phase transfer between water and chloroform phases is promoted by protonation–deprotonation of the terminal amine groups of C18AA.

The phase transfer of Au NPs by alternating pH change is reversible for at least 6 cycles. The phase transfer yield from water to chloroform was 100 %, while the yield from chloroform to water was ~17 %. The yields were reproducible to approximately plus or minus 10 % for repeated phase transfer processes. When the chloroform and water phases were divided into glass bottles, it was found that the yield of Au NPs collected from water was up to 92 %, while that from chloroform was nearly 0 %. This indicates that there are two types of Au NPs: One that is easily transferred to the water phase and another that is not.

Au NPs in chloroform had a larger proportion of (111) crystal facets than those in the water phase, although there was no significant size difference between the two samples. Therefore, the phase transfer method is very effective for the fractionation of Au NPs according to their crystal facet distinction. The driving force of

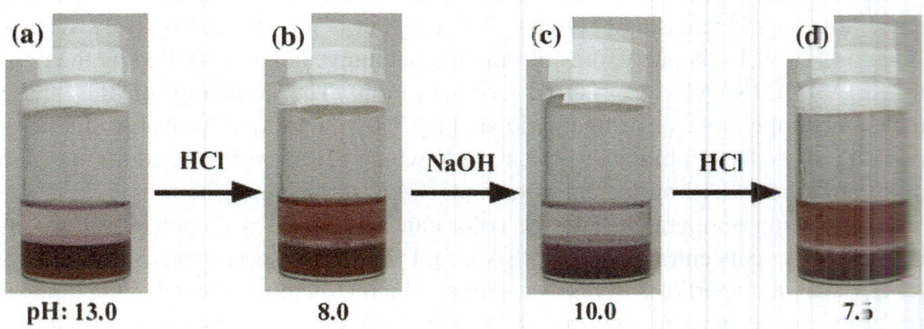

Fig. 6.14 Photographs for the reversible phase transfer of Au NPs between water and chloroform by changing pH

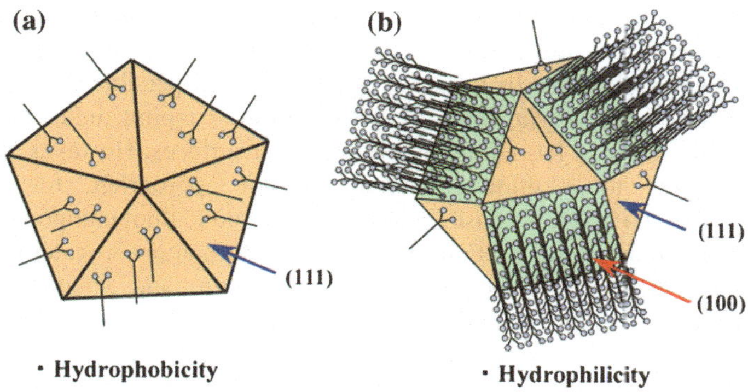

Fig. 6.15 Proposed adsorption structures of C18AA on Au NPs having **a** a large content and **b** a small content of Au (111) crystals facets

crystal facet fractionation can be explained by selective affinity to specific crystal facets of Au [17]. Namely, C18AA molecules adsorb on the (100) and (110) crystal facets of Au, but barely adsorb on the (111) crystal facet [17]. This indicates that the adsorption density of C18AA on Au NPs remaining in the chloroform phase should be low, due to the larger content of (111) crystal facets. Accordingly, Au NPs consisting primarily of (111) crystal facets are difficult to cover with a bilayer of C18AA and thus possess insufficient hydrophilicity to disperse in water (Fig. 6.15a). In contrast, Au NPs bearing a larger proportion of (100) and (110) crystal facets achieve sufficient hydrophilicity by the formation of a C18AA bilayer on the Au NPs (Fig. 6.15b). Au NPs with a larger proportion of (100) and (110) facets can transfer reversibly between water and chloroform by the formation and destruction of C18AA bilayer structures induced by pH, while Au NPs with a larger portion of (111) facets remain in the chloroform phase, regardless of a change in the pH.

6.7 Water-Dispersible Au Nanowires Synthesized in C18AA Organogel

One-dimensional noble metal nanowires (NWs) have attracted attention due to their unique electrical, optical, and magnetic properties and potential applications in nanodevices and catalysis [25, 26]. Chemical synthesis of noble metal NWs in the liquid phase is generally based on template assistance, nanoparticle assembly, and polyol reduction [35–37]. However, the noble metal NWs resulting from conventional chemical synthesis are polycrystalline with relatively large diameters (>5 nm) and rough surfaces [23–26]. Recently, ultrathin and straight single-crystal Au NWs with diameters of ca. 2 nm have been fabricated by the soft-template method. In all these reports, oleylamine has been used as the key material for the soft template, and the dispersion medium is an organic solvent [23–26]. The formation of water-dispersible ultrathin Au NWs is desirable from both a fundamental point of view and for practical applications in hybridization with biomolecules such as peptides or biocopolymers. In order to obtain water-dispersible Au NWs, it is necessary to exchange oleylamine with a hydrophilic capping molecule, such as hydrophilic thiols, which has strong affinities for Au surfaces. However, the ligand exchange process to a thiol compound is not appropriate for obtaining water-dispersible Au NWs, because ligand exchange to the thiol compound leads to shape transformation. In this section, we review the formation of water-dispersible ultrathin Au NWs with diameters of less than 2 nm and lengths of a few micrometers in the organogel of C18AA. Here, the lamellar structure of C18AA acts as a useful template, and moreover, the phase transfer ability of C18AA effectively forms water-dispersible Au NWs without ligand exchange.

An oil-soluble reducing agent is added into the C18AA toluene gel containing $HAuCl_4$, and the organogel is left without stirring for 8 h at 55 °C. Au NWs are formed in the organogel of C18AA (Fig. 6.16). The average diameter and length of the Au NWs are 1.8 nm and a few micrometers, respectively. A periodic fringe corresponding to the (111) lattice spacing is observed along the long axis of the Au NWs, indicating that ultrathin Au NWs are grown in the (111) direction [13].

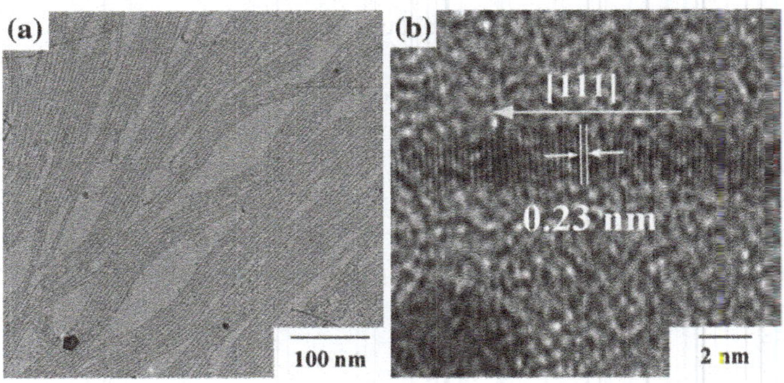

Fig. 6.16 **a** TEM and **b** HR-TEM images of ultrathin Au NWs

Fig. 6.17 Selective adsorption of C18AA on Au NWs and growth of a (111) crystal facet in toluene

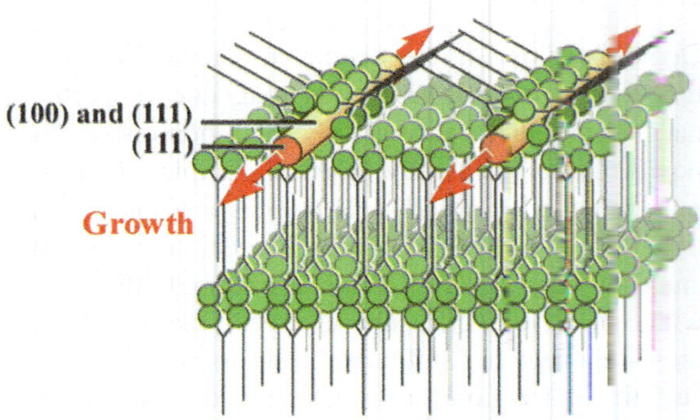

The presence of the C18AA lamellar structure is crucial to obtain ultrathin Au NWs, because the reduction of $HAuCl_4$ in the non-gelation solvents results in the formation of only Au NPs without Au NWs. The importance of the lamellar structure is also confirmed by the lack of Au NW formation at a lower concentration of 0.5 wt% C18AA toluene solution at 55 °C or at a 2 wt% C18AA solution in toluene at 85 °C, at which all the C18AA molecules are dissolved in toluene. The selective adsorption property of the terminal amine groups onto the (100) or (110) crystal facets of Au is also essential to fabricate the ultrathin Au NWs. Figure 6.17 shows a schematic illustration of the formation mechanism of Au NWs in the C18AA organogel.

We demonstrated in the previous section that C18AA is a useful phase transfer agent for Au NPs between aqueous and organic solvent phases. When water is added to the C18AA toluene gel containing the as-prepared Au NWs, many Au NWs are obtained in the aqueous phase. This indicates that C18AA forms a bilayer structure on the sides of Au NWs consisting of (100) and/or (110) facets, and thus water-dispersible Au NWs can be simply prepared by the phase transfer process.

6.8 Au Nanowires Synthesized at Room Temperature

C18AA is superior to oleylamine for the synthesis of water-dispersible ultrathin Au NWs, but the technique still has temperature limitations, such that the nanowires must be prepared at elevated temperatures in the range of 55 °C rather than at room temperature. C18AA derivatives with different alkyl chain lengths, C*n*AA (*n*; 12, 14, 16, 18), were synthesized to overcome this limitation. C*n*AA compounds exhibit gelation ability in a variety of apolar solvents, including toluene, benzene, and cyclohexane, and the minimum gelation concentration decreases with increasing alkyl chain length [14]. The sol–gel transition temperatures (T*gel*) of 5 wt% C12AA, C14AA, C16AA, and C18AA toluene gel are 25, 31, 39, and 45 °C, respectively. C*n*AA in the organogels forms a lamellar structure and has the ability to function as a soft template for Au NWs. In the case of C16AA, ultrathin Au NWs are produced at 40, 55, and 75 °C, with the optimal preparation temperature for the NWs being 55 °C. There are no Au NWs at the lower temperature of 25 °C or at the higher temperature of 85 °C. In contrast, employing C12AA results in the formation of ultrathin Au NWs even at 25 °C. Table 6.1 summarizes the products obtained from C*n*AA systems at a variety of temperatures. The NWs dimensions are independent of the alkyl chain length, although the optimal preparation temperature increases with increasing alkyl chain length.

The optimal temperatures are related to the state of the C*n*AA in toluene. Table 6.2 summarizes the solubility states of C*n*AA in toluene at various temperatures. Here, a turbid sol consists of small crystals, which are the same as the lamellar aggregates present in the toluene gel. The particular solubility state is found to correlate precisely with the form of Au nanocrystals summarized in Table 6.1. Interestingly, in the case of C12AA and C14AA, it is possible to

Table 6.1 Types of Au nanocrystals prepared from C*n*AA at various temperatures

	25 °C	40 °C	55 °C	70 °C	85 °C
C12AA	***NWs***	***NWs***	*NWs*	*NPs*	*NPs*
C14AA	*NWs*	***NWs***	***NWs***	*NWs*	*NPs*
C16AA	*NPs*	*NWs*	***NWs***	*NWs*	*NPs*
C18AA	*NPs*	*NWs*	***NWs***	***NWs***	*NPs*

NWs Ultrathin Au NWs with high yield (>10 %). *NWs* Ultrathin Au NWs with low yield (<10 %). *NPs* Only NPs

Table 6.2 Temperature dependence of the state of C*n*AA/toluene mixture

	25 °C	40 °C	55 °C	70 °C	85 °C
C12AA	***SC***	***SC***	***SC***	*Sol*	*Sol*
C14AA	***SC***	***SC***	***SC***	***SC***	*Sol*
C16AA	*Gel*	***SC***	***SC***	***SC***	*Sol*
C18AA	*Gel*	***SC***	***SC***	***SC***	*Sol*

Gel Rigid gel. ***SC*** Turbid sol containing small crystals. *Sol* Transparent solution

fabricate ultrathin Au NWs at room temperature. A comparison between Table 6.1 and Table 6.2 demonstrates that Au NWs are prepared in the turbid sols containing small C*n*AA crystals, while either small or large Au NPs (but no NWs) are obtained from the rigid organogels and clear solutions. It is evident that the absence of NWs in the transparent solutions is due to the absence of lamellar aggregates, which serve as the soft template. In the case of the rigid gels, in which C*n*AA lamellar aggregates surely exist, the reasons for the lack of NW formation are less obvious. One plausible explanation is inhomogeneous dispersion of the reducing agent due to the greater viscosity of these systems.

Another interesting feature of Au nanowires is the two-dimensional self-assembly of Au nanowires in a closely packed parallel array, as shown in Fig. 6.18. The average gap distances between NWs increase in an almost linear fashion with increasing hydrocarbon chain length (Fig. 6.18), demonstrating that it is possible to precisely control the gap distance between Au NWs by changing the hydrocarbon chain length of C*n*AA. Furthermore, the hydrocarbon chains of C*n*AA adsorbed onto Au NWs are aligned in an interdigitated manner, because the observed gap distances were longer than the corresponding molecular lengths, but shorter than double the molecular lengths.

6.9 SiO_2-Coated Au Nanowires

Ultrathin Au NWs are generally easily broken by external stimuli such as electron beam irradiation [24]. When Au nanowires were exposed to a TEM electron beam with accelerating voltage of 100 kV for 10 s, all ultrathin Au NWs were broken down into fragments (10–100 nm in length) due to the relatively high localized

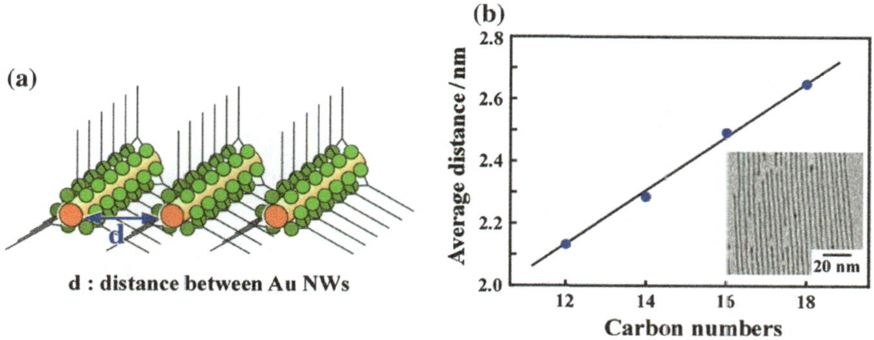

Fig. 6.18 **a** Schematic illustration of ordered ultrathin Au NWs and **b** Au NWs spacing as a function of C*n*AA alkyl chain length. TEM image in the inset is of two-dimensional ultrathin Au NWs

temperature on the NWs [38, 39]. Silica coating is a very effective method to improve the thermal stability of ultrathin Au NWs. The conventional preparation method for SiO_2 coating of metal nanocrystals is performed in an ethanol solution of tetraethoxysilane (TEOS) under basic conditions of ammonia. Thus, an ethanol solution of TEOS and an ammonia solution were added to the dispersion of Au NWs capped with 3-mercaptopropanoic acid, and the mixture was left for 24 h without stirring at room temperature. Figure 6.19 shows the representative as-prepared silica-coated Au NWs. The thickness of the silica-coating layer is controllable from 6 to 18 nm by adjusting the TEOS concentration. The silica-coated Au NWs have high dispersibility in water and can be redispersed in water even after drying.

The silica-coated ultrathin Au NWs maintain their original morphology after 10 min of exposure to a TEM electron beam with accelerating voltage of 100 kV. Even a longer exposure time of ~10 min does not affect the morphology, proving that silica-coating improves the resistance of Au NWs to a high-energy electron beam [33]. Furthermore, the silica-coated Au NWs have high thermal and pH resistances. When Au NWs on a TEM copper grid are heated at 150 °C for 3 h, the bare Au NWs turn into NPs and completely disappear, whereas the silica-coated Au NWs retain their original shape [33]. In addition, the bare Au NWs are easily transformed under acidic conditions of pH 3, whereas the silica-coated Au NWs retain their morphology between pH 3 and pH 10. Therefore, silica coating is a very valuable method for imparting environmental resistance to ultrathin Au NWs. Providing high thermal stability would be effective for new nanocatalysts since they can then function under high-performance conditions at higher temperature.

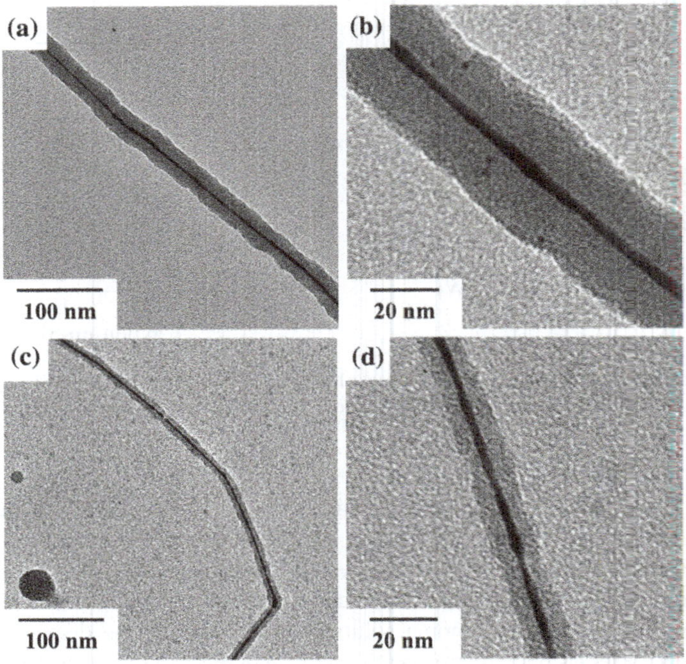

Fig. 6.19 TEM images of silica-coated ultrathin Au NWs synthesized at TEOS concentrations of **a**, **b** 1.1 mM and **c**, **d** 0.5 mM

6.10 Neuron-Shaped Au NWs Synthesized in Water

In the previous sections, we showed that toluene gels of C18AA are useful soft templates for preparing ultrathin Au NWs. Here, we review that aqueous solutions of C18AA can be used for the preparation of two-dimensional (2D) dendritic Au nanowires (DNWs) with diameters of 100–200 nm. We also show that the seeded growth method in the presence of C18AA enables the synthesis of larger 2D Au DNWs with diameters of 400–700 nm or neuron-shaped Au NWs consisting of two DNWs dangling from both ends of ultrathin Au NWs.

A typical synthetic protocol of 2D Au DNWs is to add $HAuCl_4$ to a given concentration of C18AA aqueous solution and then leave the mixture for 8 h without stirring. In the case of [C18AA]/[$HAuCl_4$] = 6.67, Au DNWs comprising 4–7 nm branches are successfully obtained as shown in Fig. 6.20a. However, spherical and 2-D plate-like Au nanocrystals are obtained at lower and higher [C18AA]/[$HAuCl_4$] ratios of 1.67 and 33.5, respectively [32]. The formation of Au DNWs is affected mainly by the [C18AA]/[$HAuCl_4$] ratios, but not by the reaction temperature. Indeed, 2D Au DNWs are produced in the temperature range of 25–70 °C. It is noted that the formation of 2D Au DNWs is a rare case, because it is common for 3D DNWs to be obtained during the preparation of dendritic structures without using two-dimensional templates or reaction spaces. Interestingly, the formation of Au NPs as a by-product is negligible, according to transmission electron microscopy (TEM) observations. This result is consistent with the absence of the surface plasmon band of Au NPs at ~530 nm in the UV-vis spectra [32].

The HR-TEM image of 2D Au DNWs has fringes with a periodicity of 0.235 nm, which corresponds to the (111) lattice spacing observed along the long axis of the branch in the Au DNWs (Fig. 6.20b). Further, HR-TEM observations of various branches revealed that Au DNWs are grown in the (111) direction. This indicates that the sides of DNW branches consist of (100) and (110) crystal facets covered with a high-density monolayer of C18AA, whereas the ends consist of (111) crystal facets covered with a low-density monolayer. This provides the opportunity to prepare larger DNWs by the seeded growth method (Fig. 6.21a). An aqueous solution of C18AA and $HAuCl_4$ is added to a seed solution of the original Au DNWs, and the mixture is left for 24 h at room temperature under the condition of $[C18AA]/[HAuCl_4] = 6.67$. The morphology of larger 2D Au DNWs with

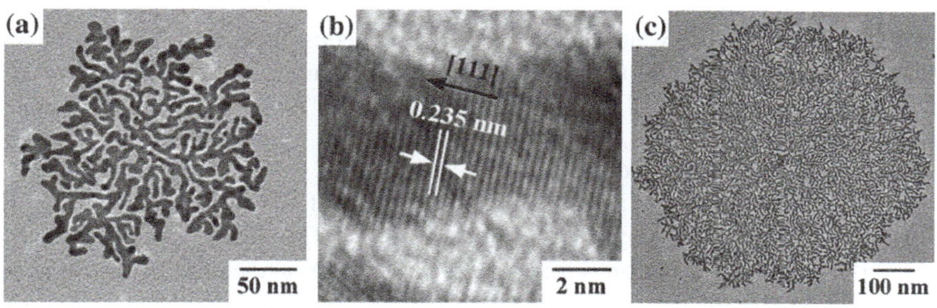

Fig. 6.20 **a** TEM and **b** HR-TEM images of two-dimensional Au nanowires. **c** TEM image of larger two-dimensional Au nanowires prepared by seeded growth method

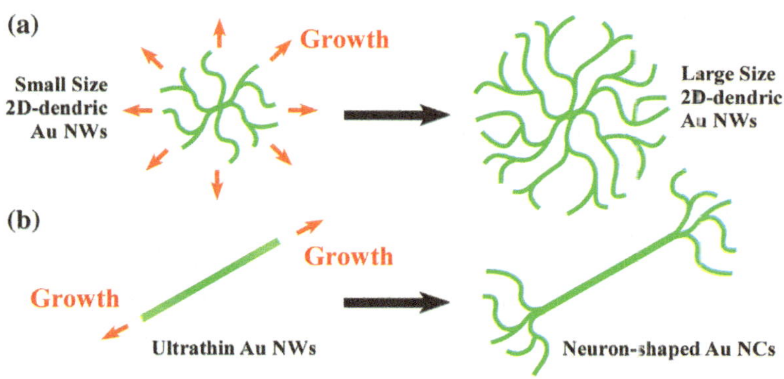

Fig. 6.21 Schematic illustration of **a** large two-dimensional Au nanowires and **b** neuron-shaped Au nanocrystals

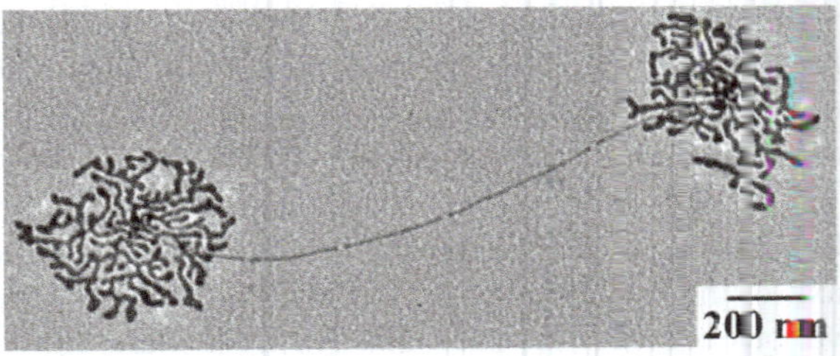

Fig. 6.22 TEM image of neuron-shaped Au nanocrystals

diameters of 400–700 nm (Fig. 6.20c) is very similar to that of the original seed, and it is difficult to discern the connection between the seed and growth parts.

The seeded growth method using the selective adsorption property of C18AA proved to be very useful for the preferential growth of Au from the (111) crystal facet. Considering that both ends of the ultrathin Au NWs prepared in C18AA-toluene gel have (111) crystal facets (Fig. 6.20b), it is expected that application of the seeded growth method to ultrathin Au NWs would be useful to obtain a hybrid nanostructure of straight NWs and DNWs (Fig. 6.21b). Thus, C18AA and $HAuCl_4$ in water were added to an aqueous dispersion of ultrathin Au NWs, and the mixed solution was left at room temperature for a couple of days at [C18AA]/[$HAuCl_4$] = 6.67. Novel neuron-shaped Au nanocrystals consisting of two DNWs dangling from both ends of an ultrathin Au NW were successfully obtained, as shown in Fig. 6.22.

These results indicate that the successive growth of Au from open surfaces is possible with a lower adsorption density of C18AA. This feature provides a possible application for a bottom-up technique to achieve nanoconnections from Au nanomaterials to another nanomaterial or substrate, such as electrodes. Therefore, the seeded growth method using C18AA could be used in a wide range of applications to produce complicated nanostructures.

6.11 Conclusion

We demonstrated that C18AA with two amide moieties acts as a high-performance soft material with thermal and pH-responsive viscosity. C18AA in water forms novel straight elongated micelles through the intermolecular hydrogen bonding of the amide groups, and the elongated micelles transform from straight to flexible worm-like structures upon heating, together with the viscosity phase transition from sol to gel states. We also reported that O/W emulsions of C18AA/toluene/LiCl system show a heat-induced gel property with a reversible sol–gel transition, and that the transition temperature can be precisely tuned over a wide temperature range. Furthermore, O/W emulsions of the C18AA/toluene/HCl system can be induced to undergo a sol–gel phase transition by a change in temperature or pH, suggesting their potential for highly thermo-responsive or pH-responsive materials. Another characteristic feature of the C18AA emulsions was that the water/C18AA + TOAB/toluene system develops an iridescent color derived from interference of the multilayered liquid crystalline structure, and the coloring temperature and the color can be controlled completely independently by the concentration of C18AA and TOAB, respectively. The iridescent emulsions offer great potential in coloring materials, and thermometric and visual-sensing applications.

We also demonstrated that the selective adsorption property of C18AA gives us an opportunity to practice the nanoengineering of Au nanocrystals. Namely, the presence of C18AA enables the transfer of Au NPs reversibly between water and chloroform phases via pH changes. In addition, the phase transfer method was found to be very effective for the fractionation of Au NPs according to their crystal facet distinction. We expect that this method has the potential to be expanded to other NPs and substantially contributes to progress in the NPs catalyst field. Ultrathin Au NWs with average diameters of 1.8 nm and lengths of a few micrometers were successfully prepared in the lamellar structure of the C18AA organogel. We also proved that the seeded growth method is an effective technique to fabricate larger 2D Au DNWs and neuron-shaped DNWs. This seeded growth method from a specific crystal facet of Au could lead to various applications, such as nanoconnectors and the fabrication of complex Au nanostructures.

References

1. Beebe DJ, Moore JS, Bauer JM, Yu Q, Liu RH, Devadoss C, Jo BH (2000) Functional hydrogel structures for autonomous flow control inside microfluidic channels. Nature 404:588–590
2. Vasudevan M, Buse E, Lu D, Krishna H, Kalyanaraman R, Shen AQ, Khomami B, Sureshkumar R (2010) Irreversible nanogel formation in surfactant solutions by microporous flow. Nat Mater 9:436–441
3. Kiyonaka S, Sugiyasu K, Shinkai S, Hamachi I (2002) First thermally responsive supramolecular polymer based on glycosylated amino acid. J Am Chem Soc 124:10954–10955
4. Matsumoto S, Yamaguchi S, Ueno S, Komatsu H, Ikeda M, Ishizuka K, Iko Y, Tabata KV, Aoki H, Ito S, Noji H, Hamachi I (2008) Photo gel–sol/sol–gel transition and its patterning of a supramolecular hydrogel as stimuli-responsive biomaterials. Chem Eur J 14:3977–3986
5. Bouteiller L, Qiao Y, Lin Y, Wang Y, Li Z, Huang J (2011) Metal-driven viscoelastic wormlike micelle in anionic/zwitterionic surfactant systems and template-directed synthesis of dendritic silver nanostructures. Langmuir 27:1718–1723
6. Morita C, Imura Y, Ogawa T, Kurata H, Kawai T (2013) Thermal-sensitive viscosity transition of elongated micelles induced by breaking intermolecular hydrogen bonding of amide groups. Langmuir 29:5450–5456
7. Kumar R, Kalur GC, Ziserman L, Danino D, Raghavan SR (2007) Wormlike micelles of a C22-tailed zwitterionic betaine surfactant: from viscoelastic solutions to elastic gels. Langmuir 23:12849–12856
8. Varade D, Rodríguez-Abreu C, Shrestha LK, Aramaki K (2007) Wormlike micelles in mixed surfactant systems: effect of cosolvents. J Phys Chem B 111:10438–10447
9. Chu Z, Feng Y (2011) Thermo-switchable surfactant gel. Chem Commun 47:7191–7193

10. Kalur GC, Frounfelker BD, Cipriano BH, Norman AI, Raghavan SR (2005) Viscosity increase with temperature in cationic surfactant solutions due to the growth of wormlike micelles. Langmuir 21:10998–11004
11. Davies TS, Ketner AM, Raghavan SR (2006) Self-assembly of surfactant vesicles that transform into viscoelastic wormlike micelles upon heating. J Am Chem Soc 128:6669–6675
12. Hassan PA, Valaulikar BS, Manohar C, Kern F, Bourdieu L, Candau SJ (1996) Vesicle to micelle transition: rheological investigations. Langmuir 12:4350–4357
13. Imura Y, Tanuma H, Sugimoto H, Ito R, Hojo S, Endo H, Morita C, Kawai T (2011) Water-dispersible ultrathin Au nanowires prepared using a lamellar template of a long-chain amidoamine derivative. Chem Commun 47:6380–6382
14. Morita C, Tanuma H, Kawai C, Ito Y, Imura Y, Kawai T (2013) Room-temperature synthesis of two-dimensional ultrathin gold nanowire parallel array with tunable spacing. Langmuir 29:1669–1675
15. Morita C, Kawai C, Tsujimoto K, Kasai K, Ogue Y, Imura Y, Kawai T (2013) High organogelation ability and soft-templateing for ultrathin Au nanowires of long-chain amidoamine derivatives. J Oleo Sci 62:81–87
16. Morita C, Sugimoto H, Matsue K, Kondo T, Imura Y, Kawai T (2010) Changes in viscosity behavior from a normal organogelator to a heat-induced gelator for a long-chain amidoamine derivative. Chem Commun 46:7969–7971
17. Imura Y, Morita C, Endo H, Kondo T, Kawai T (2010) Reversible phase transfer and fractionation of Au nanoparticles by pH change. Chem Commun 46:9206–9208
18. Imura Y, Morita C, Kawai T (2013) Reversible dispersion–precipitation of single-walled carbon nanotubes by pH change and addition of organic components. New J Chem 37:3607–3611
19. Morita C, Sugimoto H, Imura Y, Kawai T (2011) Double-stimuli responsive O/W emulsion gel based on a novel amidoamine surfactant. J Oleo Sci 60:557–562
20. Morita C, Aoyama T, Imura Y, Kawai T (2011) Novel thermo-responsive coloring phenomena in water/surfactant/oil emulsions. Chem Commun 47:11760–11762
21. Lim B, Xia Y (2011) Metal nanocrystals with highly branched morphologies. Angew Chem Int Ed 50:76–85
22. Xia Y, Xiong Y, Lim B, Skrabalak SE (2009) Shape-controlled synthesis of metal nanocrystals: Simple chemistry meets complex physics. Angew Chem Int Ed 48:60–103
23. Halder A, Ravishankar N (2007) Ultrafine single-crystalline gold nanowire arrays by oriented attachment. Adv Mater 19:1854–1858
24. Lu X, Yavuz MS, Tuan HY, Korgel BA, Xia Y (2008) Ultrathin gold nanowires can be obtained by reducing polymeric strands of oleylamine—AuCl complexes formed via aurophilic interaction. J Am Chem Soc 130:8900–8901
25. Huo Z, Ck Thung, Huang W, Zhang X, Yang P (2008) Sub-two nanometer single crystal Au nanowires. Nano Lett 8:2041–2044
26. Feng H, Yang Y, You Y, Li G, Guo J, Yu T, Shen Z, Wu T, Xing B (2009) Simple and rapid synthesis of ultrathin gold nanowires, their self-assembly and application in surface-enhanced Raman scattering. Chem Commun. 1984–1986
27. Kim F, Connor S, Song H, Kuykendall T, Yang P (2004) Platonic gold nanocrystals. Angew Chem Int Ed 43:3673–3677
28. Imura Y, Tsujimoto K, Morita C, Kawai T (2014) Preparation and catalytic activity of Pd and bimetallic Pd–Ni nanowires. Langmuir 30:5026–5030
29. Murphy CJ, Jana NR (2002) Controlling the aspect ratio of inorganic nanorods and nanowires. Adv Mater 14:80–82
30. Tsuji M, Miyamae N, Lim S, Kimura K, Zhang X, Hikino S, Nishio M (2006) Crystal structures and growth mechanisms of Au@Ag core-shell nanoparticles prepared by the microwave-polyol method. Cryst Growth Des 6:1801–1807
31. Tsuji M, Matsumoto K, Miyamae N, Tsuji T, Zhang X (2007) Rapid preparation of silver nanorods and nanowires by a microwavepolyol method in the presence of Pt catalyst and polyvinylpyrrolidone. Cryst Growth Des 7:311–320

32. Imura Y, Maezawa A, Morita C, Kawai T (2012) Neuron-shaped gold nanocrystals and two-dimensional dendritic gold nanowires fabricated by use of a long-chain amidoamine derivative. Langmuir 28:14998–15004
33. Imura Y, Hojo S, Morita C, Kawai T (2014) Preparation of silica-coated ultrathin gold nanowires with high morphological stability. Langmuir 30:1888–1892
34. Imura Y, Morita C, Kawai T (2012) Fractionation of Au nanomaterials using selective adsorption of a long-chain amidoamine derivative. Chem Lett 41:603–605
35. Bai H, Xu K, Xu Y, Matsui H (2007) Fabrication of Au nanowires of uniform length and diameter using a monodisperse and rigid biomolecular template: collagen-like triple helix. Angew Chem Int Ed 46:3319–3322
36. Pei LH, Mori K, Adachi M (2007) Formation process of two-dimensional networked gold nanowires by citrate reduction of $AuCl_4^-$ and the shape stabilization. Langmuir 20:7837–7843
37. Sun Y, Xia Y (2002) Large-scale synthesis of uniform silver nanowires through a soft, self-seeding, polyol process. Adv Mater 14:833–837
38. Pazos-Pérez N, Baranov D, Irsen S, Hilgendorff M, Liz-Marzán LM, Giersig M (2008) Synthesis of flexible, ultrathin gold nanowires in organic media. Langmuir 24:9855–9860
39. Wang C, Sun S (200) Facile synthesis of ultrathin and single-crystalline Au Nanowires. Chem Asian J 4:1028–1034

Chapter 7
Photoresponsive Gold Clusters

Sachil Sharma, Yoshiki Niihori, Wataru Kurashige and Yuichi Negishi

Abstract Advances in nanotechnology have encouraged the fabrication of stable and highly functional nanomaterials. Thiolate-protected gold clusters ($Au_n(SR)_m$) have attracted much attention as new functional nanomaterials because they exhibit size-specific properties and functions. Particularly, $Au_n(SR)_m$ clusters that can display high thermodynamic and chemical stabilities have great potential application as a nanomaterial. Our studies have focused on developing effective approaches to improve the functionality of such stable $Au_n(SR)_m$ clusters toward the fabrication of stable and highly functional nanomaterials. Specifically, herein, we present our findings regarding the fabrication of stable $Au_n(SR)_m$ clusters with photoresponsive properties.

Keywords Gold clusters · High stability · Photoresponsive property · Nanomaterials

7.1 Introduction

The preparation of nanomaterials that can exhibit both stability and functionality causes progress in the nanotechnology field. Metal clusters that consist of several to hundreds of metal atoms exhibit size-specific physical and chemical properties, which are not observed in bulk metals. Such properties vary considerably

S. Sharma · Y. Niihori · W. Kurashige · Y. Negishi
Department of Applied Chemistry, Faculty of Science, Tokyo University of Science, 1-3 Kagurazaka, Shinjuku-ku, Tokyo 162-8601, Japan

Y. Negishi
Photocatalysis International Research Center, Tokyo University of Science, 2641 Yamazaki, Noda, Chiba 278-8510, Japan

Y. Negishi (✉)
Department of Materials Molecular Science, Institute for Molecular Science, Myodaiji, Okazaki, Aichi 444-8585, Japan
e-mail: negishi@rs.kagu.tus.ac.jp

T. Kawai and M. Hashizume (eds.), *Stimuli-Responsive Interfaces*,
DOI 10.1007/978-981-10-2463-4_7

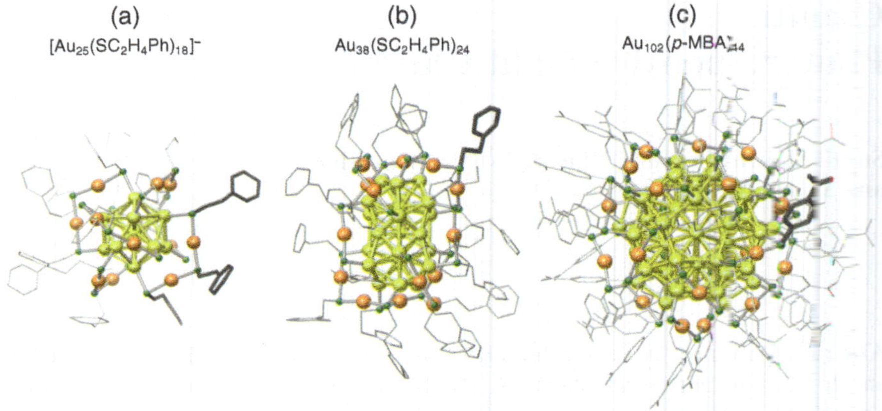

Fig. 7.1 Geometrical structures of **a** $[Au_{25}(SC_2H_4Ph)_{18}]^-$, **b** $Au_{38}(SC_2H_4Ph)_{24}$, and **c** $Au_{102}(p\text{-MBA})_{44}$ (*p*-MBA = *p*-mercaptobenzoic acid). The *yellow*, *orange*, and *green* spheres represent Au, Au, and S atoms, respectively, whereas the *gray* structures represent the ligands (reprinted with permission from Refs. [20, 23] and reprinted from Ref. [24] with permission from AAAS, Copyright 2008 American Chemical Society, Copyright 2010 American Chemical Society)

depending on the number of constituent atoms [1]. Accordingly, metal clusters have generated significant interest as potentially useful nanomaterials.

Numerous studies regarding the synthesis of metal clusters, particularly phosphine-protected metal clusters, have been reported since the 1960s [2–5]. However, phosphine-protected metal clusters are susceptible to degradation in solution and in the solid state, and such a poor stability restricts their application [6]. In contrast, thiolate-protected gold clusters ($Au_n(SR)_m$; Fig. 7.1) exhibit higher stability owing to strong bonds formed between the gold and thiolate ligands. These clusters are stable both in solution and in the solid state. Hence, since the first report in 1994 [7], $Au_n(SR)_m$ clusters have attracted much attention as new functional nanomaterials in diverse fields from basic science to applied chemistry [8–19].

7.2 Stable $Au_n(SR)_m$ Clusters

Within the $Au_n(SR)_m$ series, some clusters with a specific chemical composition, e.g., $Au_{25}(SR)_{18}$ (Fig. 7.1), exhibit higher stability in solution than those with a different chemical composition [10, 20–26]. Specifically, reacting $Au_n(SG)_m$ clusters ($n < 25$; SG = glutathionate) with an excess of glutathione enables size reduction via etching to generate Au(I)-thiolate oligomers (Fig. 7.2). Although a similar etching reaction occurs in larger $Au_n(SG)_m$ clusters ($25 \leq n \leq 39$), size reduction stops at the $Au_{25}(SG)_{18}$ stage rather than proceeding to the formation of oligomers (Fig. 7.2) [27]. This phenomenon indicates that $Au_{25}(SG)_{18}$ exhibits particularly high stability among members of the series of $Au_n(SG)_m$ clusters shown in Fig. 7.2.

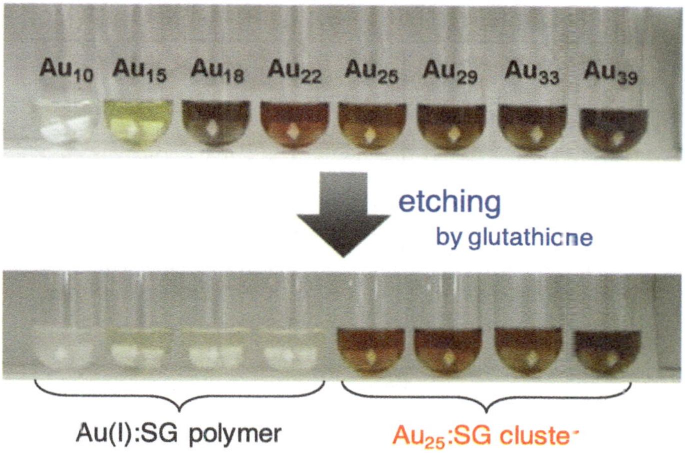

Fig. 7.2 Comparison of the stability of a series of $Au_n(SG)_m$ clusters (n = 10–39) against etching (reproduced from Ref. [27] with permission from John Wiley & Sons Ltd., Copyright 2007)

Such stable clusters can be synthesized size-selectively. $Au_n(SR)_m$ clusters are generally prepared by adding a reducing agent to a solution containing thiol and gold ions. The $Au_n(SR)_m$ clusters thus prepared typically have a cluster-size distribution. If this mixture is exposed to severe conditions, less stable clusters are converted into stable clusters as described above. For example, when exposing $Au_n(SR)_m$ clusters (SR = SG [28] or SC_2H_4Ph [29]; $25 \leq n < 38$) to severe conditions, unstable $Au_n(SR)_m$ clusters are converted into a stable $Au_{25}(SR)_{18}$ cluster. Stable clusters can be precisely synthesized on a large scale using this size-focusing method.

Numerous studies have been conducted on the physical and chemical properties of such stable clusters. The findings have shown that small $Au_n(SR)_m$ clusters exhibit size-specific properties such as photoluminescence [30], optical activity [31], and catalytic activity [32] (Fig. 7.3). Furthermore, the surface ligands of these stable clusters can be replaced with other ligands through ligand-exchange reactions [33] while maintaining the number of constituting atoms of the metal core. Thus, the physical and chemical properties of stable $Au_n(SR)_m$ clusters can be modified by ligand-exchange reactions.

In summary, stable $Au_n(SR)_m$ clusters can be size-selectively synthesized and exhibit size-specific physical and chemical properties, thereby endowing them with great potential as new nanomaterials.

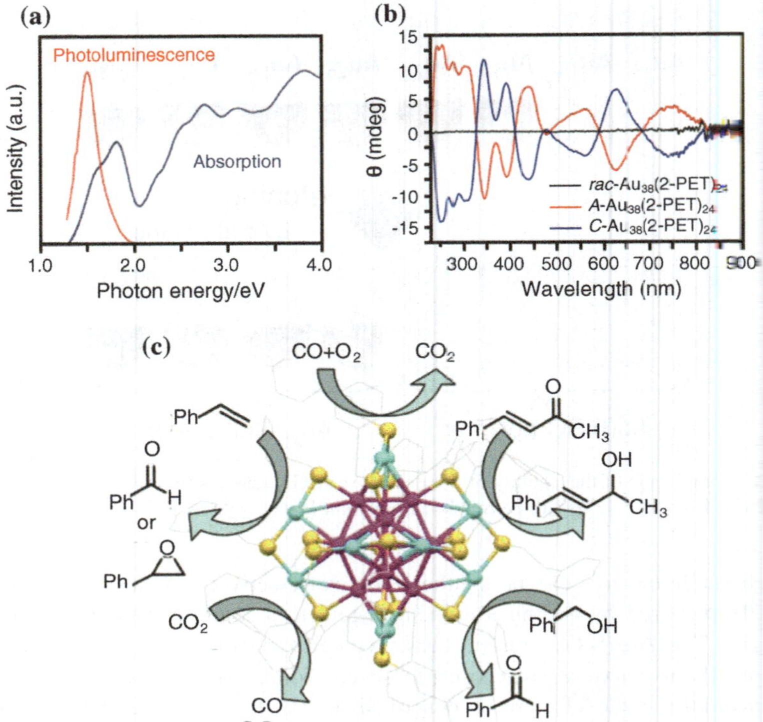

Fig. 7.3 Size-specific physical and chemical properties of small $Au_n(SR)_m$ clusters: **a** photoluminescence of $Au_{25}(SG)_{18}$, **b** optical activity of two optical isomers of $Au_{38}(SC_2H_4Ph)_{24}$, and **c** reactions catalyzed by $Au_{25}(SR)_{18}$ (reprinted with permission from Refs. [10, 31, 32]. Copyright 2012 Chemical Society of Japan, Copyright 2009 American Chemical Society, Copyright 2014 American Chemical Society)

7.3 Synthesis of Stable, Functionalized $Au_n(SR)_m$ Clusters

Our research efforts are aimed to synthesize stable and highly functional nano-materials by imparting high functionality to stable $Au_n(SR)_m$ clusters [34, 35]. As demonstrated in many previous gas-phase studies, doping small metal clusters with heteroatoms has a significant effect on the stability as well as the physical and chemical properties [36] of the resulting clusters. Accordingly, it is expected that the properties of such stable $Au_n(SR)_m$ clusters can also be significantly modified by doping with heteroatoms [37]. Additionally, protection by other chalcogenates or functionalized ligands is considered to be useful for functionalization [37]. Therefore, the effects of heteroatom doping and protection by functionalized ligands on the stability and physical and chemical properties of these clusters were investigated [34, 35, 38]. The primary aim of such investigations is to establish

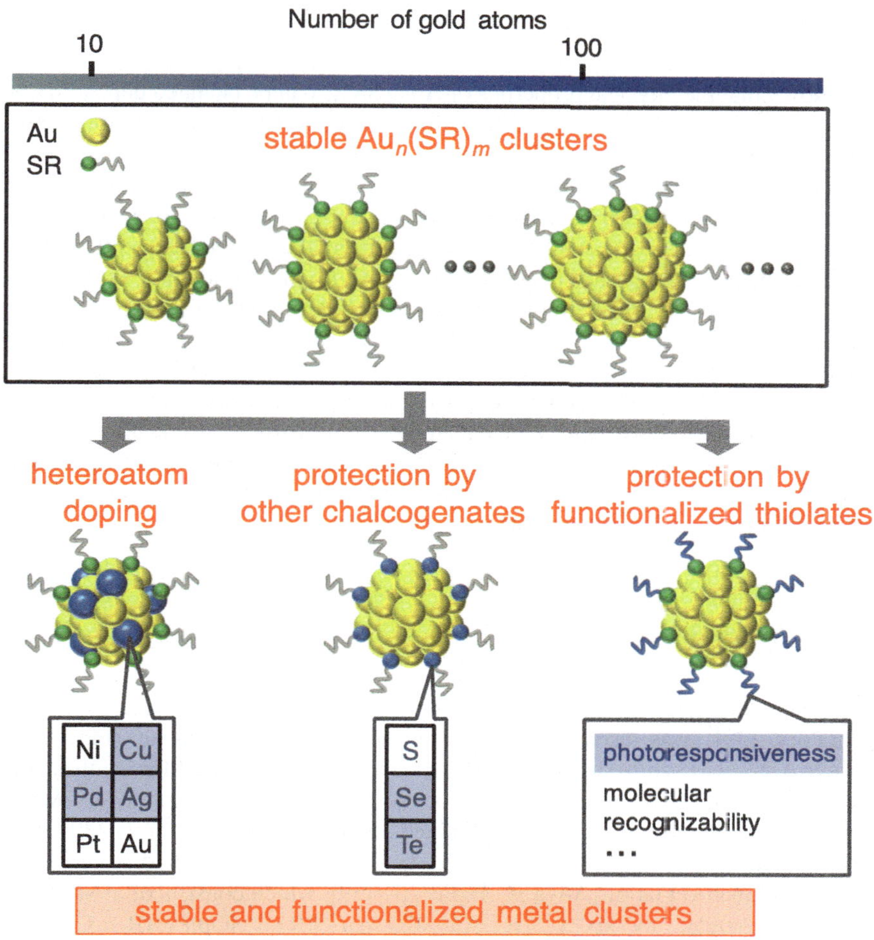

Fig. 7.4 Strategies employed toward the fabrication of stable and functionalized metal clusters

means to impart high functionality to stable $Au_n(SR)_m$ clusters for the subsequent fabrication of stable and highly functionalized nanomaterials (Fig. 7.4) [34, 35, 38]. Particularly, in this chapter, we discuss research related to the preparation of photoresponsive Au_{25} clusters [39].

7.4 Synthesis of Photoresponsive Au_{25} Clusters

It is known that the physical properties of $Au_{25}(SR)_{18}$ can be changed depending on the ligand structure. For example, the redox potential [40–42] and photoluminescence quantum yield [30] of $Au_{25}(SR)_{18}$ can be varied according to the ligand structure. Thus, it can be expected that the combination of Au_{25} clusters with thiolates that can undergo reversible structural changes in response to photoirradiation would result in the synthesis of Au_{25} clusters that can subsequently exhibit reversible physical properties changes in response to photoirradiation exposure.

Azobenzene is one of the most well-studied photoresponsive molecules. It can undergo isomerization from the *trans* to *cis* configuration upon exposure to UV irradiation and from the *cis* to *trans* configuration upon exposure to visible light irradiation (Fig. 7.5a). Thus, combining Au_{25} clusters with azobenzene compounds is expected to afford clusters, of which the physical properties can change reversibly in response to photoirradiation exposure. Therefore, Au_{25} clusters protected with azobenzene derivative thiolates were synthesized (Fig. 7.5b) with the aim of generating photoresponsive Au_{25} clusters, as discussed in Sect. 7.4.1.

7.4.1 Synthesis

4-(4-(Phenylazo)phenoxy)butyl-1-thiol (Az-SH), which was synthesized in our laboratory [39], was used as the azobenzene derivative thiolate (Fig. 7.5b). $Au_{25}(S\text{-}Az)_{18}$ was synthesized using a process similar to that used for synthesizing $Au_{25}(SC_2H_4Ph)_{18}$ [29] (Fig. 7.6). Mass spectrometry of the product ($Au_{25}(S\text{-}Az)_{18}$) revealed that $[Au_{25}(S\text{-}Az)_{18}]^-$ was obtained in high purity (Fig. 7.7) [39]. The various characterization studies indicated that $[Au_{25}(S\text{-}Az)_{18}]^-$ had a geometric structure similar to that of $[Au_{25}(SC_2H_4Ph)_{18}]^-$ [39], which consists of a Au_{13} core that is surrounded by six [–S(R)–Au–S(R)–Au–S(R)–] staples (R = C_2H_4Ph) (Fig. 7.1a) [20, 21].

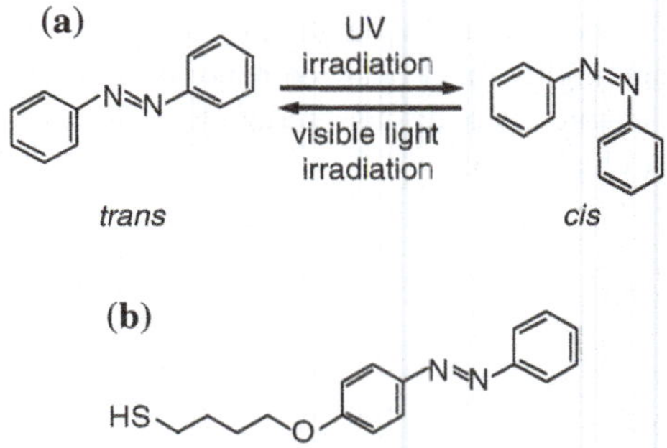

Fig. 7.5 **a** Photoisomerization of azobenzene. **b** Molecular structure of the azobenzene derivative thiol used in our study

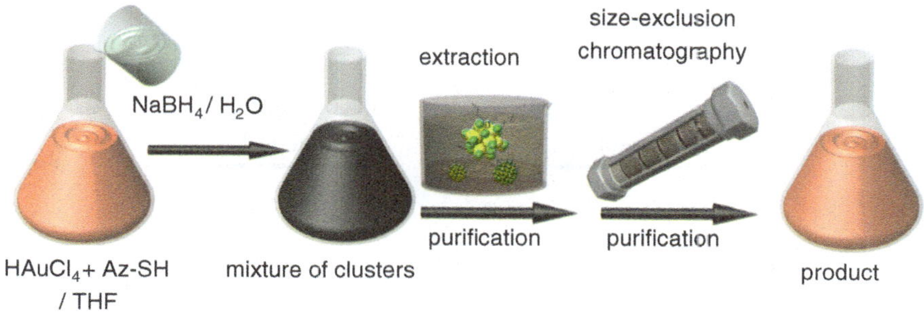

Fig. 7.6 Synthesis procedure of $[Au_{25}(S\text{-}Az)_{18}]^-$

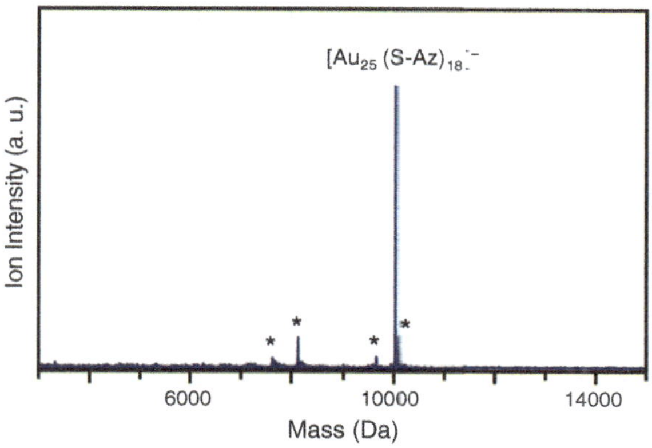

Fig. 7.7 Negative ion matrix-assisted laser desorption/ionization mass spectrum of $[Au_{25}(S\text{-}Az)_{18}]^-$. Peaks denoted by * are attributed to fragment ions or phenyl radical adducts [39] (reproduced from Ref. [39] by permission of The Royal Society of Chemistry)

7.4.2 Photoisomerization

The azobenzene ligands in $Au_{25}(S\text{-}Az)_{18}$ were isomerized as follows. First, $Au_{25}(S\text{-}Az)_{18}$ was dissolved in toluene. Then, this solution was placed in a quartz cell (10 mm × 10 mm) with a lid and vigorously stirred using a cell stirrer while irradiated with UV light (λ = 360 nm) for 30 s. Consequently, azobenzene ligands isomerized from the *trans* to *cis* configuration (Fig. 7.8). The cluster solution was then irradiated with visible light (λ = 420 nm) for 30 s, and the azobenzene ligands isomerized from the *cis* to *trans* configuration (Fig. 7.8). The distance between the cell and light source was fixed at 100 mm. At this distance, the temperature of the solution did not increase upon exposure to light irradiation.

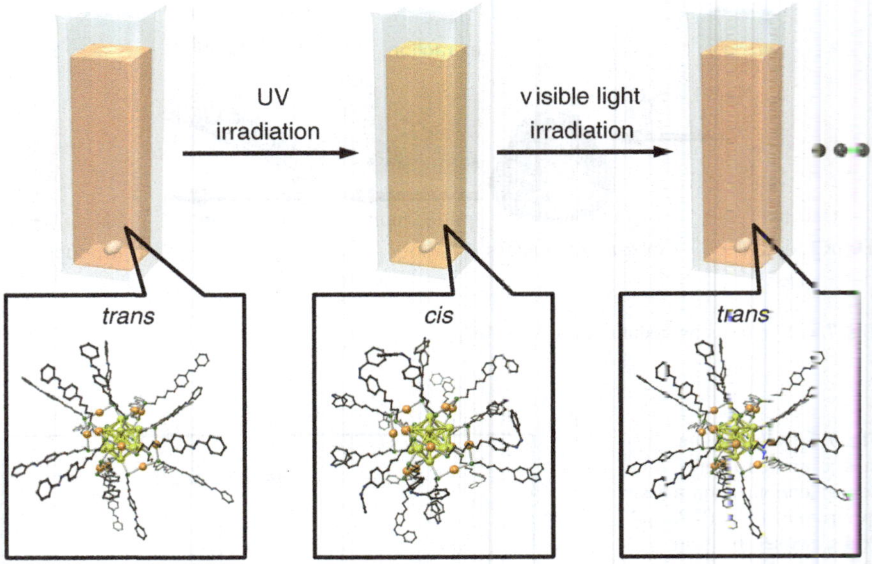

Fig. 7.8 Schematic of photoisomerization

7.4.3 Isomerization of Ligands

Figure 7.9a shows the optical absorption spectra of toluene solutions of Az-SH. Upon synthesis, Az-SH featured a *trans* configuration, which displayed a strong absorption peak at 3.54 eV (350 nm), originating from π–π^* absorption. In contrast, when Az-SH was isomerized to the *cis* configuration, the peak at 3.54 eV became weaker, and the absorption peak at 2.75 eV (450 nm), which could be ascribed to n–π^* absorption of the *cis* configuration, became stronger. The spectrum reverted back to its original features when Az-SH adopted the original *trans* configuration (Fig. 7.9a).

Similar spectral changes were observed for isolated $[Au_{25}(S\text{-}Az)_{18}]^-$ [39]. Figure 7.9b shows the photoresponsive behavior of $[Au_{25}(S\text{-}Az)_{18}]^-$ in toluene solution. Strong absorption was originally observed at 3.54 eV. Upon irradiation of the solution with UV, the absorption at 3.54 eV decreased, whereas that at 2.75 eV increased. In contrast, when the solution was irradiated with visible light, absorption at 3.54 eV increased, whereas that at 2.75 eV decreased. These results demonstrate that azobenzene can reversibly isomerize upon photoirradiation even when present in $[Au_{25}(S\text{-}Az)_{18}]^-$ clusters.

The (*trans*-to-*cis* or *cis*-to-*trans*) isomerization efficiency of S-Az, estimated by changes in the absorption intensity at 3.54 eV, was nearly 100 % [39]. These results considerably differed from those obtained for two-dimensional self-assembled monolayers (SAMs). Specifically, isomerization does not occur in two-dimensional SAMs consisting of azobenzene derivative thiolates featuring a

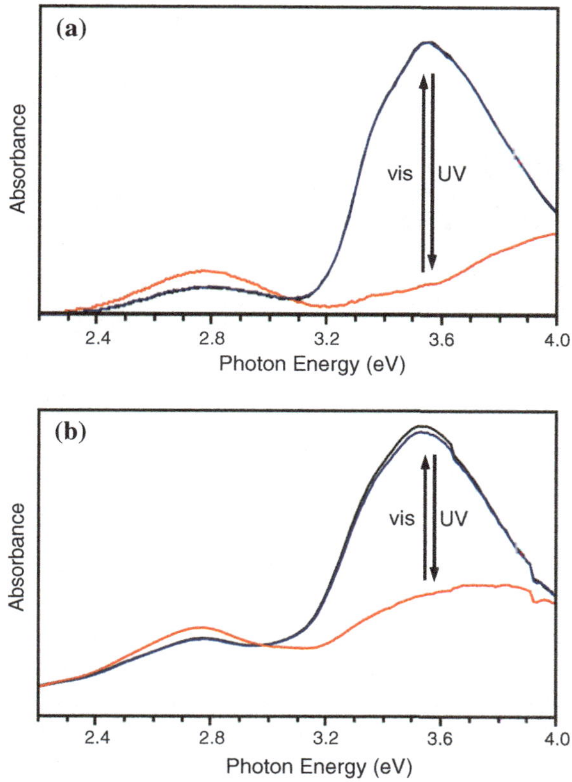

Fig. 7.9 Optical absorption spectra showing the photoresponsive behavior of **a** Az-SH in toluene solution (2.0×10^{-5} M) and **b** $[Au_{25}(S\text{-}Az)_{18}]^-$ in toluene solution (5.0×10^{-7} M) in the energy range of 2.2–4.0 eV. In both **a** and **b**, the visible photostationary state (*blue line*) overlaps with the initial state (*black line*) (reproduced from Ref. [39] by permission of The Royal Society of Chemistry)

structure similar to that of S-Az without spacer molecules (Fig. 7.5b) [43]. This is because the neighboring molecules in two-dimensional SAMs are very close, and consequently there is insufficient space for isomerization to occur (Fig. 7.10a). In contrast, the azobenzene-derived thiolates in $[Au_{25}(S\text{-}Az)_{18}]^-$ are radially distributed at the Au_{13} core, and thus there is sufficient space for isomerization to occur between the ligands (Fig. 7.10b). Such a configuration is considered to be responsible for the superior photoisomerization efficiency of azobenzene in $[Au_{25}(S\text{-}Az)_{18}]^-$.

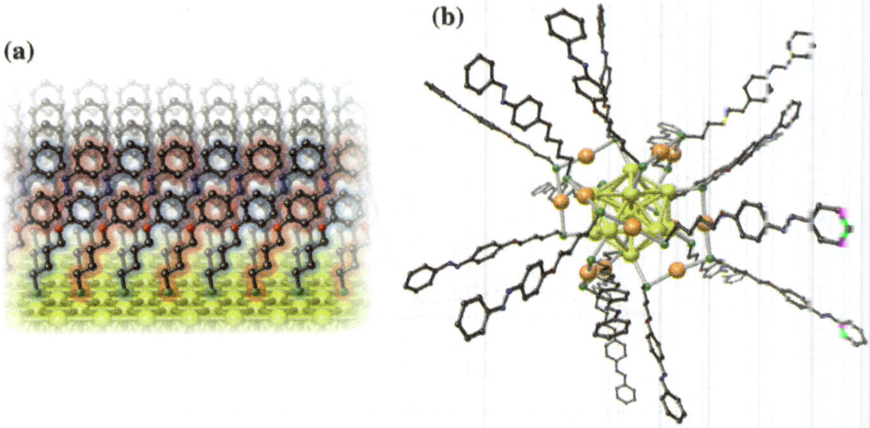

Fig. 7.10 Structural images of **a** two-dimensional SAMs consisting of azobenzene derivative thiolates and **b** $[Au_{25}(S\text{-}Az)_{18}]^-$

7.4.4 *Photoresponsive Behavior of Redox Potential*

Variations in the redox potential of $[Au_{25}(S\text{-}Az)_{18}]^-$ owing to the photoisomerization of azobenzene were investigated by differential pulse voltammetry (DPV). The results clearly demonstrated that the redox potential of $Au_{25}(S\text{-}Az)_{18}$ changed reversibly owing to the photoisomerization of azobenzene [39]. Figure 7.11 shows the DPV curves of $[Au_{25}(S\text{-}Az)_{18}]^-$ before and after being subjected to photoirradiation. The voltammogram obtained prior to photoirradiation displayed peaks at −206 and 126 mV. The voltammogram was similar to that of $[Au_{25}(SC_2H_4Ph)_{18}]^-$ [30, 40]; however, the peak positions differed because the ligands were different. The peaks at −206 and 126 mV were attributed to the redox potentials of $[Au_{25}(S\text{-}Az)_{18}]^{-/0}$ and $[Au_{25}(S\text{-}Az)_{18}]^{0/+}$, respectively, by comparison with the voltammogram of $[Au_{25}(SC_2H_4Ph)_{18}]^-$. In contrast, in the voltammogram obtained after UV irradiation, the peaks shifted to higher energies (−169 and 139 mV) (Table 7.1). However, the peak positions in the voltammogram obtained after visible light irradiation were similar to those in the voltammogram obtained prior to irradiation (−206 and 121 mV) (Table 7.1). These results indicate that the redox potentials of $[Au_{25}(S\text{-}Az)_{18}]^{-/0}$ and $[Au_{25}(S\text{-}Az)_{18}]^{0/+}$ vary reversibly with photoirradiation. $[Au_{25}(SC_2H_4Ph)_{18}]^-$ was subjected to the same analysis; however, peak shifts were not observed upon photoirradiation [39]. These results indicate that the peak shifts were caused by the isomerization of azobenzene.

This variation in the redox potential was attributed to variations in the intramolecular dipole moment of S-Az owing to the isomerization of azobenzene [44, 45]. Variations in the intramolecular dipole moment of ligands were considered to alter the orbital energies of the cluster, thereby causing the variation in the redox potential of $[Au_{25}(S\text{-}Az)_{18}]^-$. In contrast, the isomerization of azobenzene is known

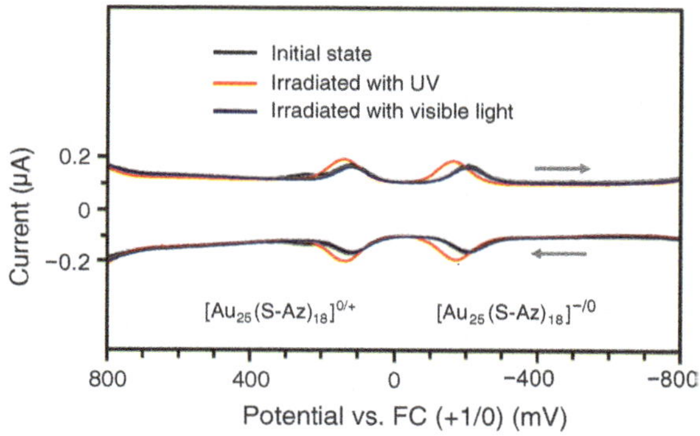

Fig. 7.11 DPV curves showing the photoresponse behavior of $[Au_{25}(S\text{-}Az)_{18}]^-$. In this experiment, $[Au_{25}(S\text{-}Az)_{18}]^-$ in toluene solution (1.0×10^{-5} M) was irradiated with UV and subsequently with visible light. After photoirradiation, the resulting suspensions were evaporated to dryness, and the residual solids were dissolved in 0.1 M $(C_4H_9)_4NPF_6$ CH_2Cl_2 solution. DPV analysis was performed at room temperature (reproduced from Ref. [39] by permission of The Royal Society of Chemistry)

Table 7.1 Formal potentials of $[Au_{25}(S\text{-}Az)_{18}]^-$ in 0.1 M $(C_4H_9)_4NPF_6$ CH_2Cl_2 solution

State	Redox potential (mV)	
	$[Au_{25}(S\text{-}Az)_{18}]^{-1/0}$	$[Au_{25}(S\text{-}Az)_{18}]^{0/+1}$
Initial state	−206	126
UV photostationary state	−169	139
Visible photostationary state	−206	121

Formal potentials are averages of the reduction and oxidation peak potentials determined in the DPV potential scans

to alter the dispersion of gold nanoparticles (>1.5 nm) in solution [46–49]. Accordingly, the dispersion of $[Au_{25}(S\text{-}Az)_{18}]^-$ was investigated to assess the effect of variations in interparticle interactions (differences in the interparticle distance), owing to changes in dispersion, on the redox potential. The results revealed that $[Au_{25}(S\text{-}Az)_{18}]^-$ could be dispersed in toluene solution regardless of the (*trans*/*cis*) configuration of the clusters [39]. Based on these results, we concluded that changes in the redox potential in Fig. 7.11 are not caused by differences in the dispersion of $[Au_{25}(S\text{-}Az)_{18}]^-$.

7.4.5 Photoresponsive Behavior of Optical Absorbance

Figure 7.12a shows the optical absorption photoresponsive behavior of $[Au_{25}(S\text{-}Az)_{18}]^-$ in the energy range of 1.1–2.2 eV. The absorbance of the peaks at 1.60 and 1.82 eV changed upon photoirradiation; the absorbance decreased upon

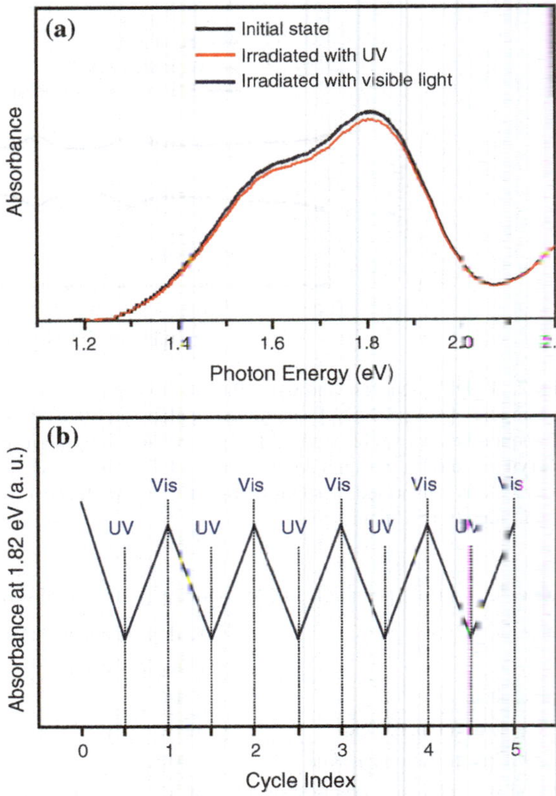

Fig. 7.12 **a** Photoresponsive behavior of the optical absorption spectrum of a toluene solution of $[Au_{25}(S\text{-}Az)_{18}]^{-}$ (1.0×10^{-5} M) in the energy range of 1.1–2.2 eV. In **a**, the visible photostationary state (*blue line*) overlaps with the initial state (*black line*). **b** Changes in the absorbance of $[Au_{25}(S\text{-}Az)_{18}]^{-}$ toluene solution at 1.82 eV owing to photoirradiation with UV and visible light (reproduced from Ref. [39] by permission of The Royal Society of Chemistry)

exposure of $[Au_{25}(S\text{-}Az)_{18}]^{-}$ to UV irradiation and increased upon exposure to visible light irradiation. This reversible photoresponsive behavior was observed repeatedly (Fig. 7.12b) in UV–visible light irradiation exposure cycles. These results indicate that changes in the ligand structure (or the dipole moment of the ligands) also vary the absorbance characteristics of these peaks. The vibrational excited states in the electronic ground state (or the electronic excited state) may depend on the ligand structure (or the dipole moment of the ligands) [50]. Theoretical calculations are required to determine the exact origin of this phenomenon.

In contrast, negligible changes were observed in the peak energies in the absorption spectra even after isomerization of azobenzene (Fig. 7.12a). In the absorption spectra of $[Au_{25}(SR)_{18}]^{-}$, the peaks observed in the range of 1.4–2.0 eV are related to transition from the highest occupied molecular orbital (HOMO) to the lowest unoccupied molecular orbital (LUMO) [51]. As stated above, the isomerization of azobenzene can change the redox potential of $[Au_{25}(S\text{-}Az)_{18}]^{-/0}$ (i.e., the HOMO energy). However, changes in the dipole moment of the ligand exert a nearly identical electronic effect on the LUMO [51, 52]. Such phenomena are considered to be responsible for the non-responsiveness of the peak positions in the optical absorption spectrum of $[Au_{25}(S\text{-}Az)_{18}]^-$.

7.5 Summary and Future Directions

The functionalization of stable $Au_n(SR)_m$ clusters plays an important role in the fabrication of stable and highly functional nanomaterials. Particularly, in this chapter, we discussed our research strategies on the conjugation of $Au_{25}(SR)_{18}$ with azobenzene. Such conjugation afforded the successful synthesis of $Au_{25}(SR)_{18}$ that exhibit flexible redox potentials in response to light stimulation. To our knowledge, this is the first example of the preparation of $Au_{25}(SR)_{18}$ that can display such variable redox potential responses upon light stimulation.

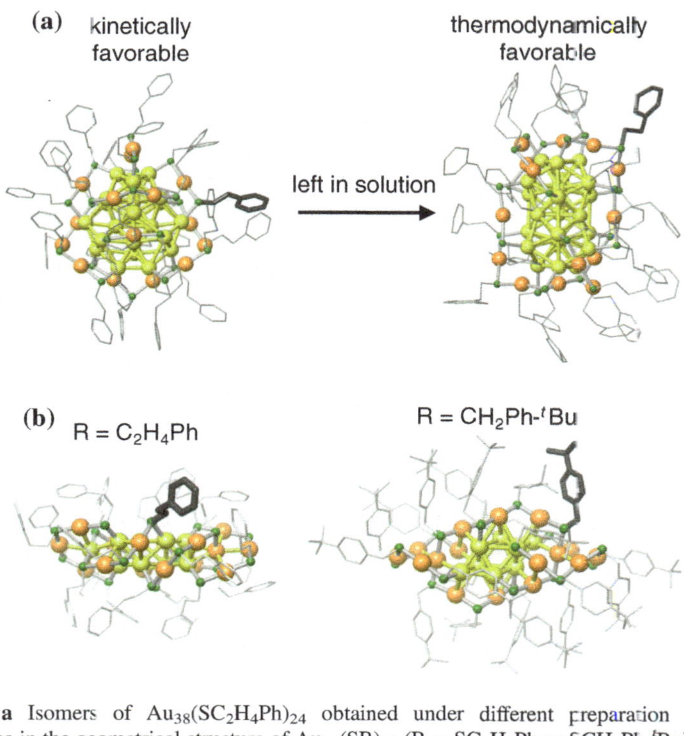

Fig. 7.13 **a** Isomers of $Au_{38}(SC_2H_4Ph)_{24}$ obtained under different preparation conditions. **b** Variations in the geometrical structure of $Au_{24}(SR)_{20}$ (R = SC_2H_4Ph or $SCH_2Ph\text{-}^t Bu$) according to ligand structure. The *yellow*, *orange*, and *green* spheres represent Au, Au, and S atoms, respectively, whereas the *gray* structures represent ligands (reprinted from Ref. [53] with permission from Macmillan Publishers Ltd., Copyright 2015 and reproduced from Ref. [56] with permission from The Royal Society of Chemistry)

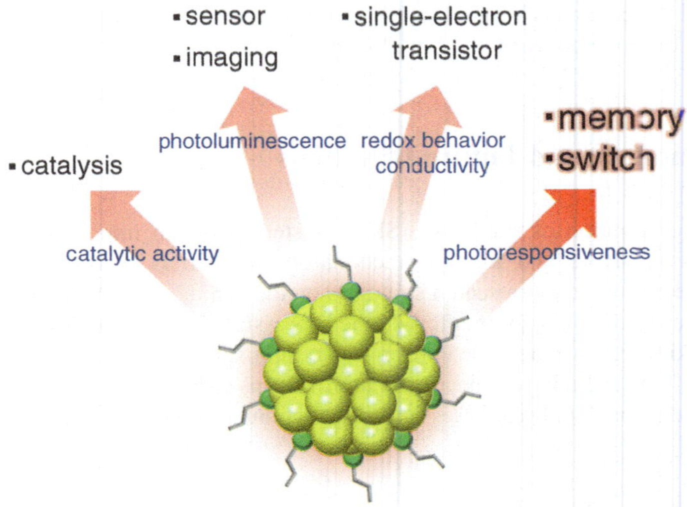

Fig. 7.14 Potential applications of $Au_n(SR)_m$ clusters

Additionally, the possibility of modulating other physical and chemical properties, i.e., photoluminescence and catalytic activity, that are specific to small $Au_n(SR)_m$ clusters (Fig. 7.3) is of interest. However, tailoring such physical and chemical properties in $[Au_{25}(S\text{-}Az)_{18}]^-$ is currently more challenging as structural deformation does not occur in the metal core upon light stimulation. With continuing research development, we expect that $Au_n(SR)_m$ clusters, of which the geometrical structure of the metal core itself can be modified by light stimulation, will be synthesized in the future. For instance, such $Au_n(SR)_m$ clusters could potentially be synthesized by using $Au_n(SR)_m$ clusters with a soft metal core. A recent study has revealed that $Au_{38}(SR)_{24}$ have isomerizable structural isomers (Fig. 7.13a) [53], indicating that $Au_{38}(SR)_{24}$ has a soft metal core. Furthermore, as reported, $Au_{24}(SR)_{20}$ and $Au_{24}(SeR)_{20}$ have different metal core structures despite having the same number of metal atoms and ligands (Fig. 7.13b) [54, 55]. This implies that $Au_{24}(SR)_{20}$ also has a soft metal core, which can undergo structural changes depending on the structure of the ligand [56]. The conjugation of such $Au_n(SR)_m$ clusters with a photoresponsive ligand is expected to enable the creation of $Au_n(SR)_m$ clusters with a variable metal core structure, thereby accordingly displaying variable photoluminescence property and/or catalytic activity upon light stimulation. The creation of such clusters would expand the range of applications of $Au_n(SR)_m$ clusters (Fig. 7.14).

Acknowledgments We thank Mr. Ukyo Kamimura and Ms. Mao Ide for technical assistance. This work was supported by Grants-in-Aid for Scientific Research (Nos. 25288009 and 25102539) and by the Canon Foundation, Kajima Foundation, Nippon Sheet Foundation for Materials Science and Engineering, Sumitomo Foundation, and Suzuki Foundation.

References

1. Castleman AW Jr, Bowen KH Jr (1996) Clusters: structure, energetics, and dynamics of intermediate states of matter. J Phys Chem 100:12911–12944. doi:10.1021/jp961030k
2. McPartlin M, Mason R, Malatesta L (1969) Novel cluster complexes of gold(0)-gold(I). J Chem Soc D 334. doi:10.1039/C29690000334
3. Briant CE, Theobald BRC, White JW, Bell LK, Mingos DMP, Welch AJ (1981) Synthesis and X-ray structural characterization of the centred icosahedral gold cluster compound $[Au_{13}(PMe_2Ph)_{10}Cl_2](PF_6)_3$; the realization of a theoretical prediction. J Chem Soc Chem Commun 5:201–202. doi:10.1039/C39810000201
4. Teo BK, Shi X, Zhang H (1992) Pure gold cluster of 1:9:9:1:9:9:1 layered structure: a novel 39-metal-atom cluster $[(Ph_3P)_{14}Au_{39}Cl_6]Cl_2$ with an interstitial gold atom in a hexagonal antiprismatic cage. J Am Chem Soc 114:2743–2745. doi:10.1021/ja00033a073
5. Kamei Y, Shichibu Y, Konishi K (2011) Generation of small gold clusters with unique geometries through cluster-to-cluster transformations: octanuclear clusters with edge-sharing gold tetrahedron motifs. Angew Chem Int Ed 50:7442–7445. doi:10.1002/anie.201102901
6. Kurashige W, Negishi Y (2012) Synthesis, stability, and photoluminescence properties of $PdAu_{10}(PPh_3)_8Cl_2$ clusters. J Clust Sci 23:365–374. doi:10.1007/s10876-011-0437-8
7. Brust M, Walker M, Bethell D, Schiffrin DJ, Whyman R (1994) Synthesis of thiol-derivatised gold nanoparticles in a two-phase liquid-liquid system. J Chem Soc Chem Commun 801–802. doi:10.1039/C39940000801
8. Whetten RL, Shafigullin MN, Khoury JT, Schaaff TG, Vezmar I, Alvarez MM, Wilkinson A (1999) Crystal structures of molecular gold nanocrystal arrays. Acc Chem Res 32:397–406. doi:10.1021/ar970239t
9. Parker JF, Fields-Zinna CA, Murray RW (2010) The story of a monodisperse gold nanoparticle: $Au_{25}L_{18}$. Acc Chem Res 43:1289–1296. doi:10.1021/ar100048c
10. Tsukuda T (2012) Toward an atomic-level understanding of size-specific properties of protected and stabilized gold clusters. Bull Chem Soc Jpn 85:151–168. doi:10.1246/bcsj.20110227
11. Jin R (2010) Quantum sized, thiolate-protected gold nanoclusters. Nanoscale 2:343–362. doi:10.1039/B9NR00160C
12. Dass A (2012) Nano-scaling law: geometric foundation of thiolated gold nanomolecules. Nanoscale 4:2260–2263. doi:10.1039/C2NR11749E
13. Häkkinen H (2012) The gold-sulfur interface at the nanoscale. Nat Chem 4:443–455. doi:10.1038/nchem.1352
14. Aikens CM (2011) Electronic structure of ligand-passivated gold and silver nanoclusters. J Phys Chem Lett 2:99–104. doi:10.1021/jz101499g
15. Jiang D-e, Nobusada K, Luo W, Whetten RL (2009) Thiolated gold nanowires: metallic *versus* semiconducting. ACS Nano 3:2351–2357. doi:10.1021/nn900498c
16. Heinecke CL, Ni TW, Malola S, Mäkinen V, Wong OA, Häkkinen H, Ackerson CJ (2012) Structural and theoretical basis for ligand exchange on thiolate monolayer protected gold nanoclusters. J Am Chem Soc 134:13316–13322. doi:10.1021/ja3032339
17. Dolamic I, Knoppe S, Dass A, Bürgi T (2012) First enantioseparation and circular dichroism spectra of Au_{38} clusters protected by achiral ligands. Nat Commun 3:798. doi:10.1038/ncomms1802
18. Udayabhaskararao T, Pradeep T (2013) New protocols for the synthesis of stable Ag and Au nanocluster molecules. J Phys Chem Lett 4:1553–1564. doi:10.1021/jz400332g
19. Yu Y, Luo Z, Yu Y, Lee JY, Xie J (2012) Observation of cluster size growth in CO-directed synthesis of $Au_{25}(SR)_{18}$ nanoclusters. ACS Nano 6:7920–7927. doi:10.1021/nn3023206
20. Heaven MW, Dass A, White PS, Holt KM, Murray RW (2008) Crystal structure of the gold nanoparticle $[N(C_8H_{17})_4][Au_{25}(SCH_2CH_2Ph)_{18}]$. J Am Chem Soc 130:3754–3755. doi:10.1021/ja800561b

21. Zhu M, Aikens CM, Hollander FJ, Schatz GC, Jin R (2008) Correlating the crystal structure of a thiol-protected Au_{25} cluster and optical properties. J Am Chem Soc 130:5883–5885. doi:10.1021/ja801173r
22. Zhu M, Eckenhoff WT, Pintauer T, Jin R (2008) Conversion of anionic $[Au_{25}(SCH_2CH_2Ph)_{18}]^-$ cluster to charge neutral cluster via air oxidation. J Phys Chem C 112:14221–14224. doi:10.1021/jp805786p
23. Qian H, Eckenhoff WT, Zhu Y, Pintauer T, Jin R (2010) Total structure determination of thiolate-protected Au_{38} nanoparticles. J Am Chem Soc 132:8280–8281. doi:10.1021/ja103592z
24. Jadzinsky PD, Calero G, Ackerson CJ, Bushnell DA, Kornberg RD (2007) Structure of a thiol monolayer-protected gold nanoparticle at 1.1 Å resolution. Science 318:430–433. doi:10.1126/science.1148624
25. Negishi Y, Nobusada K, Tsukuda T (2005) Glutathione-protected gold clusters revisited: bridging the gap between gold(I)-thiolate complexes and thiolate-protected gold nanocrystals. J Am Chem Soc 127:5261–5270. doi:10.1021/ja042218h
26. Negishi Y, Chaki NK, Shichibu Y, Whetten RL, Tsukuda T (2007) Origin of magic stability of thiolated gold clusters: a case study on $Au_{25}(SC_6H_{13})_{18}$. J Am Chem Soc 129:11322–11323. doi:10.1021/ja073580+
27. Shichibu Y, Negishi Y, Tsunoyama H, Kanehara M, Teranishi T, Tsukuda T (2007) Extremely high stability of glutathionate-protected Au_{25} clusters against core etching. Small 3:835–839. doi:10.1002/smll.200600611
28. Shichibu Y, Negishi Y, Tsukuda T, Teranishi T (2005) Large-scale synthesis of thiolated Au_{25} clusters via ligand exchange reactions of phosphine-stabilized Au_{11} clusters. J Am Chem Soc 127:13464–13465. doi:10.1021/ja053915s
29. Dharmaratne AC, Krick T, Dass A (2009) Nanocluster size evolution studied by mass spectrometry in room temperature $Au_{25}(SR)_{18}$ synthesis. J Am Chem Soc 131:13604–13605. doi:10.1021/ja906087a
30. Wu Z, Jin R (2010) On the ligand's role in the fluorescence of gold nanoclusters. Nano Lett 10:2568–2573. doi:10.1021/nl101225f
31. Knoppe S, Bürgi T (2014) Chirality in thiolate-protected gold clusters. Acc Chem Res 47:1318–1326. doi:10.1021/ar400295d
32. Li G, Jin R (2013) Atomically precise gold nanoclusters as new model catalysts. Acc Chem Res 46:1749–1758. doi:10.1021/ar300213z
33. Niihori Y, Kurashige W, Matsuzaki M, Negishi Y (2013) Remarkable enhancement in ligand-exchange reactivity of thiolate-protected Au_{25} nanoclusters by single Pd atom doping. Nanoscale 5:508–512. doi:10.1039/c2nr32948d
34. Negishi Y, Kurashige W, Niihori Y, Nobusada K (2013) Toward the creation of stable, functionalized metal clusters. Phys Chem Chem Phys 15:18736–18751. doi:10.1039/C3CP52837E
35. Negishi Y (2014) Toward the creation of functionalized metal nanoclusters and highly active photocatalytic materials using thiolate-protected magic gold clusters. Bull Chem Soc Jpn 87:375–389. doi:10.1246/bcsj.20130288
36. Akutsu M, Koyasu K, Atobe J, Hosoya N, Miyajima K, Mitsui M, Nakajima A (2006) Experimental and theoretical characterization of aluminum-based binary superatoms of $Al_{12}X$ and their cluster salts. J Phys Chem A 110:12073–12076. doi:10.1021/jp065161p
37. Kurashige W, Niihori Y, Sharma S, Negishi Y (2014) Recent progress in the functionalization methods of thiolate-protected gold clusters. J Phys Chem Lett 5:4134–4142. doi:10.1021/jz501941p
38. Kurashige W, Yamazoe S, Kanehira K, Tsukuda T, Negishi Y (2013) Selenolate-protected Au_{38} nanoclusters: isolation and structural characterization. J Phys Chem Lett 4:3181–3185. doi:10.1021/jz401770y
39. Negishi Y, Kamimura U, Ide M, Hirayama M (2012) A photoresponsive Au_{25} nanocluster protected by azobenzene derivative thiolates. Nanoscale 4:4263–4268. doi:10.1039/C2NR30830D

40. Lee D, Donkers RL, Wang G, Harper AS, Murray RW (2004) Electrochemistry and optical absorbance and luminescence of molecule-like Au_{38} nanoparticles. J Am Chem Soc 126:6193–6199. doi:10.1021/ja049605b
41. Holm AH, Ceccato M, Donkers RL, Fabris L, Pace G, Maran F (2006) Effect of peptide ligand dipole moments on the redox potentials of Au_{38} and Au_{140} nanoparticles. Langmuir 22:10584–10589. doi:10.1021/la061553b
42. Devadas MS, Kwak K, Park J-W, Choi J-H, Jun C-H, Sinn E, Ramakrishna G, Lee D (2010) Directional electron transfer in chromophore-labeled quantum-sized Au_{25} clusters: Au_{25} as an electron donor. J Phys Chem Lett 1:1497–1503. doi:10.1021/jz100395p
43. Evans SD, Johnson SR, Ringsdorf H, Williams LM, Wolf H (1998) Photoswitching of azobenzene derivatives formed on planar and colloidal gold surfaces. Langmuir 14:6436–6440. doi:10.1021/la980450t
44. Suda M, Kameyama N, Suzuki M, Kawamura N, Einaga Y (2008) Reversible phototuning of ferromagnetism at Au-S interfaces at room temperature. Angew Chem Int Ed 47:160–163. doi:10.1002/anie.200704027
45. Suda M, Kameyama N, Ikegami A, Einaga Y (2009) Reversible phototuning of the large anisotropic magnetization at the interface between a self-assembled photochromic monolayer and gold. J Am Chem Soc 131:865–870. doi:10.1021/ja808231c
46. Manna A, Chen P-L, Akiyama H, Wei T-X, Tamada K, Knoll W (2003) Optimized photoisomerization on gold nanoparticles capped by unsymmetrical azobenzene disulfides. Chem Mater 15:20–28. doi:10.1021/cm0207696
47. Klajn R, Bishop KJM, Grzybowski BA (2007) Light-controlled self-assembly of reversible and irreversible nanoparticle suprastructures. Proc Natl Acad Sci USA 104:10305–10309. doi:10.1073/pnas.0611371104
48. Kawai T, Sumi A, Morita C, Kondo T (2008) Preparation and photocoagulation in chloroform of Au nanoparticles capped with azobenzene-derivatized alkanesulfides. Colloids Surf A 321:308–312. doi:10.1016/j.colsurfa.2008.01.029
49. Raimondo C, Reinders F, Soydaner U, Mayor M, Samorì P (2010) Light-responsive reversible solvation and precipitation of gold nanoparticles. Chem Commun 46:1147–1149. doi:10.1039/B915491D
50. Devadas MS, Bairu S, Qian H, Sinn E, Jin R, Ramakrishna G (2011) Temperature-dependent optical absorption properties of monolayer-protected Au_{25} and Au_{38} clusters. J Phys Chem Lett 2:2752–2758. doi:10.1021/jz2012897
51. Aikens CM (2010) Geometric and electronic structure of $Au_{25}(SPhX)_{18}^-$ (X = H, F, Cl, Br, CH_3, and OCH_3). J Phys Chem Lett 1:2594–2599. doi:10.1021/jz1009828
52. Guo R, Murray RW (2005) Substituent effects on redox potentials and optical gap energies of molecule-like $Au_{38}(SPhX)_{24}$ nanoparticles. J Am Chem Soc 127:12140–12143. doi:10.1021/ja053119m
53. Tian S, Li Y-Z, Li M-B, Yuan J, Yang J, Wu Z, Jin R (2015) Structural isomerism in gold nanoparticles revealed by X-ray crystallography. Nat Commun 6:8667. doi:10.1038/ncomms9667
54. Das A, Li T, Li G, Nobusada K, Zeng C, Rosi NL, Jin R (2014) Crystal structure and electronic properties of a thiolate-protected Au_{24} nanocluster. Nanoscale 6:6458–6462. doi:10.1039/C4NR01350F
55. Song Y, Wang S, Zhang J, Kang X, Chen S, Li P, Sheng H, Zhu M (2014) Crystal structure of selenolate-protected $Au_{24}(SeR)_{20}$ nanocluster. J Am Chem Soc 136:2963–2965. doi:10.1021/ja4131142
56. Tang Q, Ouyang R, Tian Z, Jiang D-e (2015) The ligand effect on the isomer stability of $Au_{24}(SR)_{20}$ clusters. Nanoscale 7:2225–2229. doi:10.1039/C4NR05826G

Chapter 8
Stimuli-Responsive Structure Control of Self-Assembled Gold Nanoparticles

Hideyuki Mitomo, Kenichi Niikura and Kuniharu Ijiro

Abstract Metal nanostructures have attracted a great deal of attention as components of functional materials, and there is a great demand for the development of functional devices composed of these metal nanostructures. In particular, the fabrication of metal nanostructures with dynamic structure control by external stimuli is a major focus. In this chapter, several studies related to "the stimuli-responsive structure control of metal nanostructures" will be introduced, with a particular focus on our own research. Our fabrication strategy is based on self-organization by controlling the surface properties of nanoparticles and tuning their self-assembly through the design and synthesis of surface-modifying ligands. When gold nanoparticles (AuNPs) were modified with fluorinated tetra(ethylene glycol) derivatives, the AuNPs could spontaneously form a well-packed thin film during the drying process or hollow capsules in THF solutions. The AuNP thin film could be transferred onto hydrogel and the intervals in the AuNP array tuned by changes in the size of the gel. One of the potential applications of this tunable plasmonic structure is in sensing devices using surface-enhanced Raman scattering (SERS). This control of the gap distance in the AuNP assembly could effectively work in the SERS detection of proteins, the signal of which was enhanced more than 10-fold in comparison with that of a conventional system. AuNP capsules were cross-linked with PEG to improve the stability of the capsules in water and to give thermo-responsiveness. Cross-linked AuNP capsules showed a rapid response upon light irradiation, suggesting that they have potential applications as a drug delivery carrier with a controlled release function triggered by light.

Keywords Gold nanoparticles · Self-assembly · Active control · Plasmon · Surface-enhanced Raman scattering · Stimuli-responsive · Drug delivery system · Smart nanodevice

---

H. Mitomo · K. Niikura (✉) · K. Ijiro
Research Institute for Electronic Science, Hokkaido University, Sapporo, Japan
e-mail: kniikura@poly.es.hokudai.ac.jp

T. Kawai and M. Hashizume (eds.), *Stimuli-Responsive Interfaces*,
DOI 10.1007/978-981-10-2463-4_8

8.1 Introduction

Living creatures, which are basically composed of organic molecules, show excellent functionality, such as molecular transport or homeostasis. These functions emerged from sophisticated and elegant molecular assembly systems. In fact, self-assembled molecular structures are widely found in living bodies. For example, a lipid membrane, such as a cell membrane or the endoplasmic reticulum (ER), is composed of lipid molecules and proteins and is well known as a fluid mosaic model explained by Singer and Nicolson [1]. It works dynamically to provide a vesicular transportation via vesicle fusion and division in living cells (Fig. 8.1a) [2]. Actin fibers and tubulin, as components of the cytoskeleton, are formed through the self-assembly of proteins (actin and microtubule, respectively) and are also deformed to dispersed proteins at the same time (Fig. 8.1b) [3, 4]. These kinds of dynamic structure changes in living things provide a role model of functional materials or systems.

On the other hand, artificial objects prepared from inorganic materials have shown little evidence of emergent functions from sophisticated structures. Thanks to the development of nanotechnology, it is well known that novel properties can appear due to a quantum effect when the size of materials, such as metals or semiconductors, is reduced to a nanoscale [5]. In particular, nanoparticles made from noble metals, such as gold or silver, show photocatalytic activity and plasmonic properties, providing enhanced electric fields around nanoparticles on irradiation by specific light frequencies that correspond to the plasmon peak. Further, the close proximity of metal nanostructures induces plasmon coupling in the gaps between them, so that the optical, electrical, and magnetic properties of the assemblies differ from those of monodispersed nanostructures due to the collective plasmon characteristic. Therefore, a great deal of effort has been spent on the study of the fabrication of sophisticated metal nanostructures to develop plasmon-originated enhanced electric fields and other practical applications such as enhanced fluorescence or Raman scattering. Recently, there have been several reports on the fabrication of one-, two-, and three-dimensional structures composed of metal nanoparticles through self-assembly [6]. Several kinds

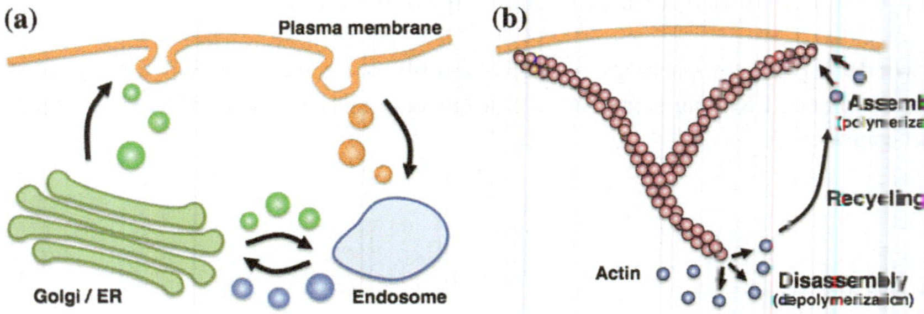

Fig. 8.1 Schematic illustration of lipid membranes dynamics in a cell showing vesicular transportation (**a**) and actin dynamics in a cell showing the assembly/disassembly of cytoskeletal architecture (**b**)

of applications, such as in energy-capturing devices [7] and highly sensitive sensing devices [8] or pressure sensors based on color changes [9], using the plasmon coupling phenomenon of self-assembled nanoparticles have already been reported.

External stimuli, such as temperature, pH, light, and metal ions, have been examined with the aim of developing advanced functions through the controlled assembly/disassembly of AuNPs [10]. A direct and promising approach to realizing stimuli-responsive colloidal assembly has been the use of stimuli-responsive molecules for the surface modification of nanoparticles. However, it is still difficult to control the shape or size of the assembly to form fibril structures, bilayer membranes, or hollow capsules, using these approaches, even though biomolecules can form. In order to achieve better control of self-assembly, the preparation and self-assembly of Janus-like gold nanoparticles, which have two faces consisting of a hydrophobic and hydrophilic surface, is now in progress through the application of a biomimetic approach [11]. In this context, it currently appears that the best way to fabricate dynamically controllable metal nanostructures is the initial preparation of a nanoparticle assembly with a defined structure by self-assembly followed by its combination with stimuli-responsive polymers.

To date, we have succeeded in fabricating gold nanoparticle self-assembled thin films with a well-packed structure through an easy drying process [12] as well as the gold nanoparticle self-assembled vesicles with a hollow structure [13] by focusing on the surface ligands, which form protective layers on the nanoparticles of a few nanometer in thickness, and adapting their chemical structures to control the self-assembled structure by tuning surface properties such as a frictionality or the balance of dispersiveness and cohesiveness in solution. Further, we have recently achieved the dynamic control of these self-assembled structures through stimuli or external environmental changes via their combination with stimuli-responsive polymers [14, 15]. The resultant structures showed great potential for practical applications. In this chapter, we will introduce several studies related to "the stimuli-responsive structure control of gold nanoparticle assemblies" with a particular focus on our recent research (Fig. 8.2).

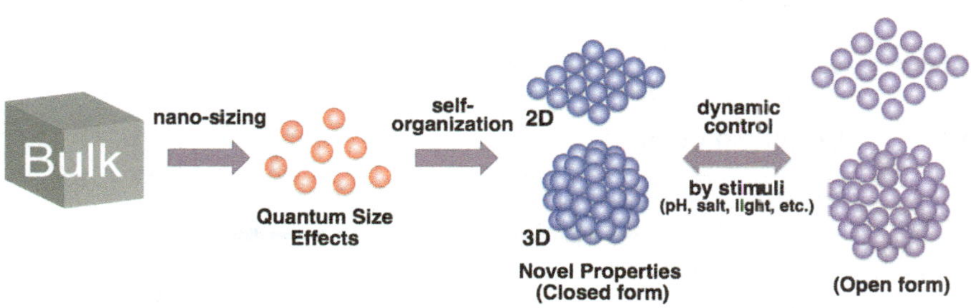

Fig. 8.2 Schematic illustration showing the concept of the emergence of size effects through the down-sizing of materials to a nanoscale, the emergence of novel properties through the assembly or organization of these nanoparticles, and the dynamic control of their structures by the use of external stimuli to switch or tune their properties or functions

8.2 Two-Dimensional Self-Assembly and Stimuli-Responsive Control

In this section, we introduce the dynamic control of the nanoparticle gaps in two-dimensional self-assembled AuNP thin films embedded on a hydrogel and their application to biomolecular sensing based on Raman scattering (Fig. 8.3).

8.2.1 Formation of Two-Dimensional Gold Nanoparticle Thin Films Through a Self-Assembly Process

Superlattice structures composed of nanoparticles have been variously prepared using interactions between DNA molecules [16, 17], the Langmuir–Blodgett technique [18, 19], layer-by-layer deposition [20], and self-assembly during solvent evaporation (self-assembly) [21, 22]. In particular, the self-assembly method has attracted much attention because of its simplicity, as it is based on the simple process of casting on a solid substrate followed by drying. However, to obtain a well-ordered array structure, the general approach involves the slow and regulated evaporation using high-boiling point solvents which take a long time to prepare. To improve this time-consuming and unstable system, we focused on the fluorination of the nanoparticle surface ligands. Previous reports based on theoretical analysis implied that weaker interparticle interactions lead to the formation of an ordered structure by allowing the rearrangement of nanoparticles to a thermodynamically stable and close-packed structure after initial amorphous aggregation [23, 24]. The surface of nanoparticles modified with fluorinated ligands is expected to provide a

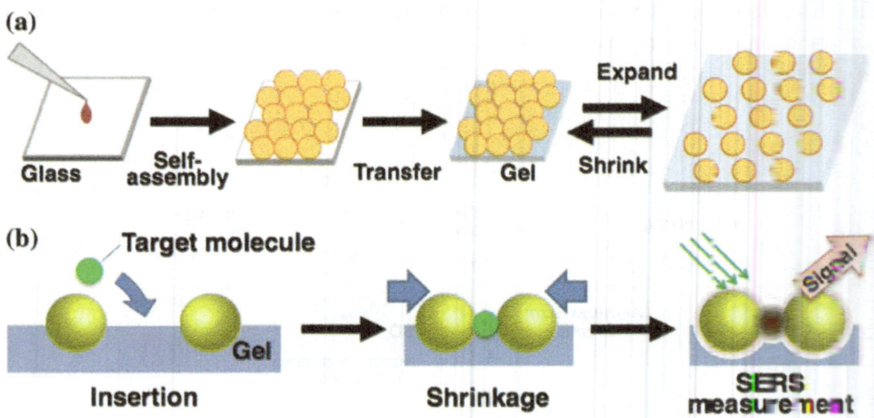

Fig. 8.3 Schematic illustration of the process described in this section. **a** Preparation of self-assembled AuNP films on a gel for the stimuli-responsive control of structure and **b** their application to sensitive SERS detection

low cohesion force, so that well-ordered self-assembled structures can be formed even though highly volatile solvents are used.

The simple preparation of closely packed AuNP films has already been reported by Kimizuka et al., through perfluoroalkanethiol modification in fluorinated solvents [25]. However, due to the poor solubility of ligands, the applications of these films are limited. In order to improve solubility while preserving the merits of fluorination, we have designed and synthesized a tetra(ethylene glycol) (TEG)-capped undecanethiol derivative surface ligand, in which the hydrogens of TEG are substituted for fluorine, as TEG has good solubility in several polar organic solvents. TEG- and fluorinated TEG (FTEG)-modified AuNPs were then prepared through a ligand exchange reaction with citrate-protected AuNPs (20 nm in diameter). The FTEG-modified AuNPs were found to be well dispersed in commonly used organic solvents, such as alcohols and tetrahydrofuran (THF).

TEG- or FTEG-modified AuNPs dispersed in methanol, as a general organic solvent with a low boiling point, were cast on carbon-coated TEM grids and dried. Scanning transmission electron microscopy (STEM) images and the diffraction pattern obtained from the Fourier transformation of STEM images showed that a hexagonal-packed layered structure (3D superlattice) was formed over a wide area (~ 10 μm), in the case of FTEG-AuNPs (Fig. 8.4a) [12]. On the other hand, assembled films composed of TEG-modified AuNPs did not show an ordered structure (Fig. 8.4b). These results indicated that the fluorination of TEG could work to prevent the formation of amorphous aggregations and enhance the rearrangement of the assembled structures until they reached a stable state or final stage of solvent drying by reducing interparticle interactions as a lubricator (Fig. 8.4c). This FTEG modification could efficiently provide a highly ordered two-dimensional self-assembled AuNP films by a simple evaporation method from a general organic solvent.

8.2.2 *Stimuli-Responsive Two-Dimensional Metal Patterns on Hydrogels*

Fabrication techniques of metal nanopatterns on solid substrates have been well established using both top–down and bottom–up approaches. However, the dynamic or active control of these fabricated patterns remains limited, even though nanotechnological techniques allow a single atom to be arranged on a silicon substrate by scanning probe microscope (SPM) [26]. We focused on a hydrogel, which can change its volume greatly in response to stimuli or external environmental changes. Hydrogel is a 3D polymer network swollen with water and a flexible material. It is well known that a hydrogel composed of a thermo-responsive polymer, known as poly(N-isopropylacrylamide) (pNIPAm), shows phase transition at around 32 °C. Below this temperature (LCST: lower critical solution temperature), the pNIPAm gel can be expanded in a swollen state; however, it shrinks by deswelling at temperature over the LCST [27]. In addition, poly(acrylic acid)

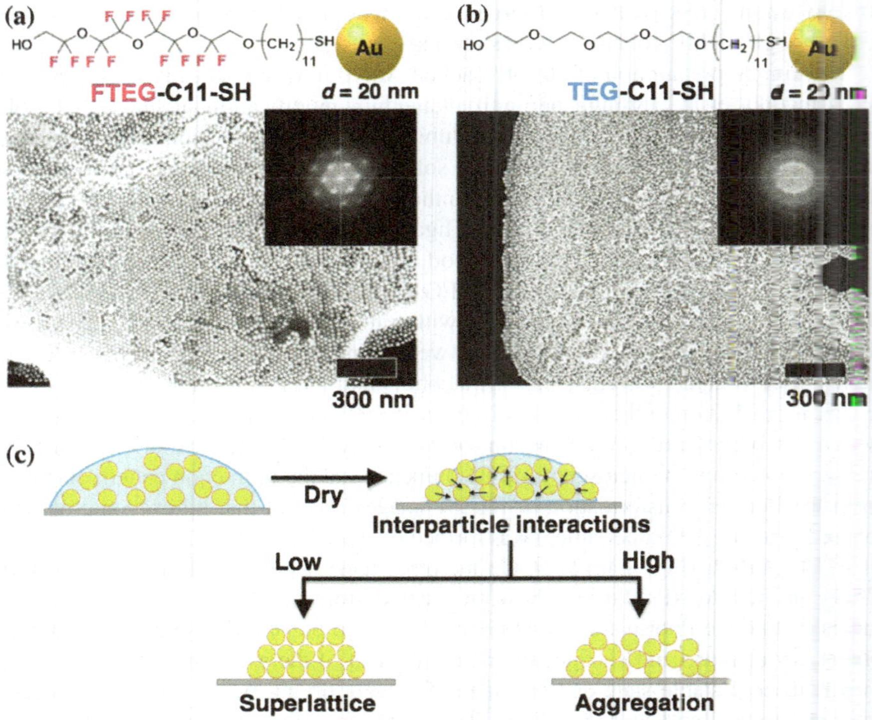

Fig. 8.4 Self-organization of gold nanoparticles (AuNPs) through the fast drying process. **a**, **b** STEM images of cast AuNP films on a TEM grid formed from a methanolic solution; FTEG-AuNPs (**a**), and TEG-AuNPs (**b**). The insets show the diffraction patterns obtained by Fourier transformations of the STEM images. **c** Hypothetical mechanism for the formation of the close-packed structure of FTEG-AuNPs during the fast drying process. The *arrows* indicate the motion (rearrangement) of nanoparticles in a concentrated state due to the strong lubricating property of the FTEG-modified AuNPs [12]. Copyright (2010) the Royal Society of Chemistry

(pAAc) gel, which has a carboxylic acid on a side chain, shows large volume change in response to changes in pH or salt concentration. If a metal pattern can be fabricated on these kinds of hydrogels, the periodicity or intervals in those patterns can be tuned by volume changes in the gel. Due to the characteristics of hydrogels, however, the fabrication of fine structures on them has been quite difficult.

Therefore, we developed a fabrication method for metal patterns on a hydrogel based on the transfer of metal patterns prepared on a solid substrate onto the water-swollen hydrogels [28]. First, micron-scale gold patterns were fabricated via the conventional photolithographic method on a silicon substrate. The transfer of the metal patterns was then performed by the preparation of the hydrogels on this substrate through the polymerization of a monomer and their subsequent gentle release from the substrate (Fig. 8.5). When the interaction between Au and a gel is stronger than that between Au and a solid substrate, the Au pattern can be

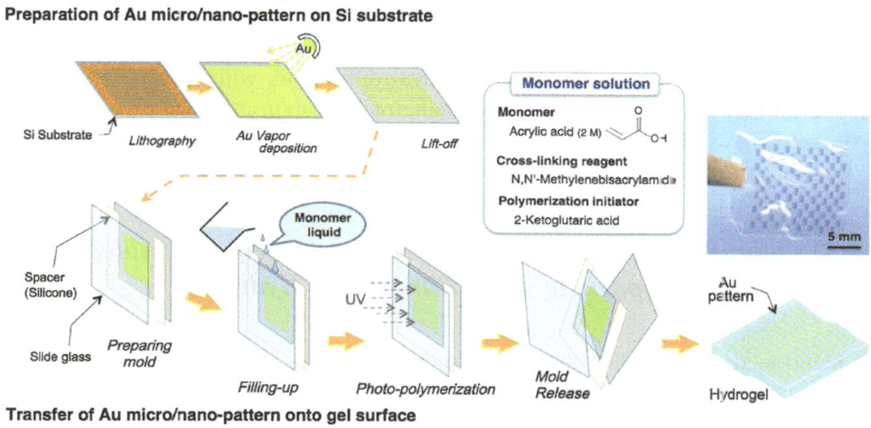

Fig. 8.5 Schematic illustration of the fabrication of Au patterns on poly(acrylic acid) gel. Adapted and reproduced with permission from [28]. Copyright © (2012) Advanced Materials

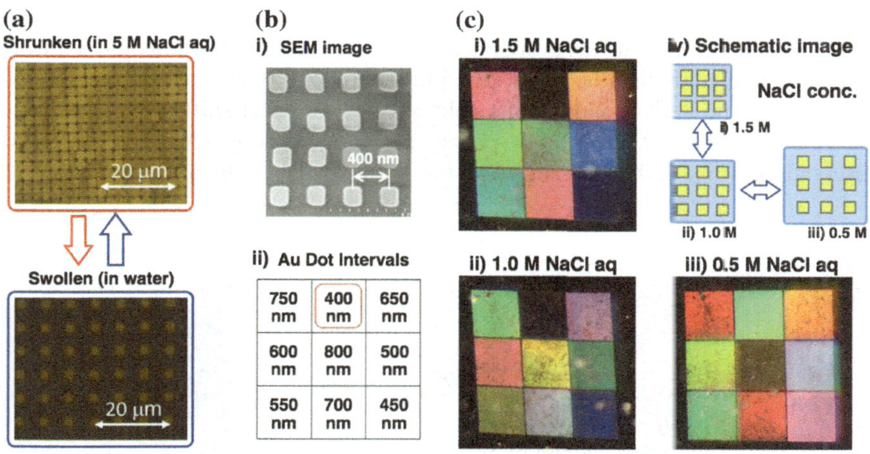

Fig. 8.6 Micron or submicron size Au patterns on PAA gel. **a** Optical microscopic images of micron patterned Au on PAA gel in a shrunken state in 5 M NaCl solution and swollen state in MilliQ water. **b** SEM image of submicron patterned Au film with 400-nm intervals (*i*) and intervals at each position (400, 450, 500, 550, 600, 650, 700, 750, and 800 nm, respectively) (*ii*). **c** Color changes in submicron Au patterns on PAA gel immersed in different NaCl concentrations: (*i*) 1.5 M, (*ii*) 1.0 M, and (*iii*) 0.5 M NaCl, and (*iv*) schematic illustration of changes in Au Dot pattern by position. Adapted and reproduced with permission from [28] Copyright © (2012) Advanced Materials

successfully transferred. Optical microscopic observations of micron-scale Au patterns transferred onto pAAc gel showed a clear change in periodic distance in response to changes in the salt concentration between 0 and 5 M NaCl (Fig. 8.6a).

Further, in the case of submicron-scale Au patterns, which were similar in size to the wavelength of visible light and fabricated by electron beam lithography, structural colors were observed due to optical diffraction, with their colors, which depend on their periodic distance, also changing in response to size changes in the hydrogel (Fig. 8.6b, c).

8.2.3 *Stimuli-Responsive Two-Dimensional Self-Assembled Gold Nanoparticle Thin Films on a Hydrogel*

As mentioned above, when the size of metal structures is reduced to a nanoscale, which is smaller than submicron size, various plasmonic phenomena are observed, such as the specific absorption of light and enhancement of the electric field, on irradiation with light. These plasmonic phenomena could provide useful properties and lead to the fabrication of functional materials. Therefore, we prepared self-assembled AuNPs films on a solid substrate as shown above (8.2.1) and transferred the films onto pAAc gel in the same manner as shown in 8.2.2.

FTEG ligand-covered AuNPs were cast on glass substrates and subjected to drying. When the solvent was evaporated, a thin blue-colored film was formed from a red colloidal solution (Fig. 8.7a). The plasmon coupling effects due to the close proximity of the AuNPs induce this color change (spectral shift). The assembled structure was then visualized by atomic force microscopy (AFM) and STEM observations (Fig. 8.7b). By adjusting the volume of a drop of AuNP colloidal

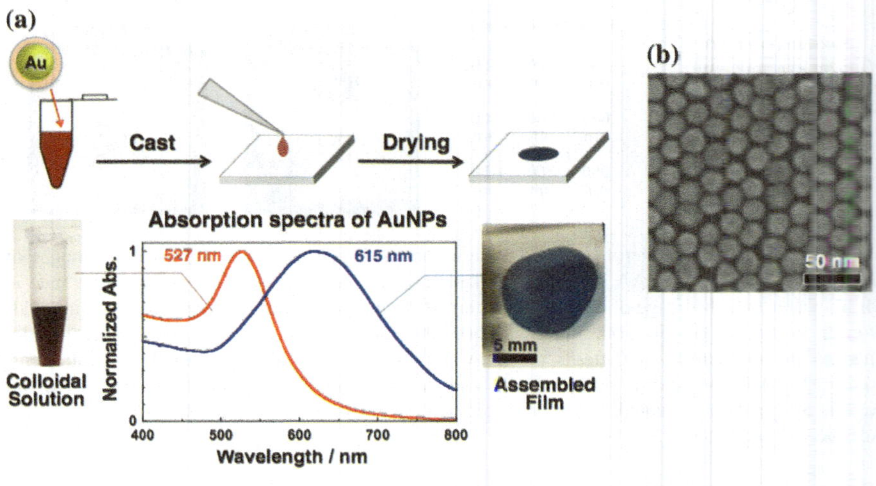

Fig. 8.7 Schematic illustration of the AuNP film preparation, images of a gold nanoparticle colloidal solution and subsequent thin film, and their extinction spectra (**a**) and scanning electron microscopic (SEM) images of the thin film on a supported carbon membrane (**b**). Adapted and reproduced with permission from [14]. Copyright © (2016) Advanced Optical Materials

Fig. 8.8 Photographic images of the process (transfer of a AuNP thin film from a glass substrate onto pAAc gel). Scale bars represent 5 mm. Adapted and reproduced with permission from [14]. Copyright © (2016) Advanced Optical Materials

solution, a well-packed almost monolayered AuNP film with minimum defects could be prepared. The complete transference of the AuNP thin film was performed by in situ polymerization on the substrates and confirmed by eye as well as absorption spectra of the film on a glass substrate and on a gel (Fig. 8.8).

The pAAc gel undergoes volume change in response to external stimuli such as changes in salt concentration. First, the gel was immersed in pure water and demonstrated a swollen state. This swollen gel was moved into an aqueous solution with a high salt concentration (1 M NaCl). The contraction of the gel occurred quickly (mostly within 5 min) and was almost completed within 30 min (Fig. 8.9a). This size change in the gel also caused spectral changes in the AuNP thin film. When the salt concentration of outer solution increased, the intensity of the extinction spectra increased and the extinction peak was red-shifted (Fig. 8.9b, c). The reversibility of this spectral change was confirmed through several cycles of change. A change in extinction intensity is mainly caused by a change in the AuNP density in the measurement area. Direct observation of the AuNPs on the hydrogel is quite difficult using conventional surface analytical techniques as the gel is soft and swollen with water. However, these results strongly indicate that the gap distances in the AuNP self-assembled thin film could be actively controlled by salt concentration providing a tunable plasmonic substrate.

8.2.4 An Application of Active Control of Two-Dimensional Self-Assembled Gold Nanoparticles in Bio-Sensing

The plasmonic phenomena provide a wide range of potential applications, such as use in optical or photonic devices and sensors [29–32]. In particular, surface-enhanced Raman scattering (SERS) is a powerful tool for the detection of materials based on intrinsic chemical fingerprint information and affords a highly sensitive sensing method. Raman spectroscopy is a useful experimental approach to

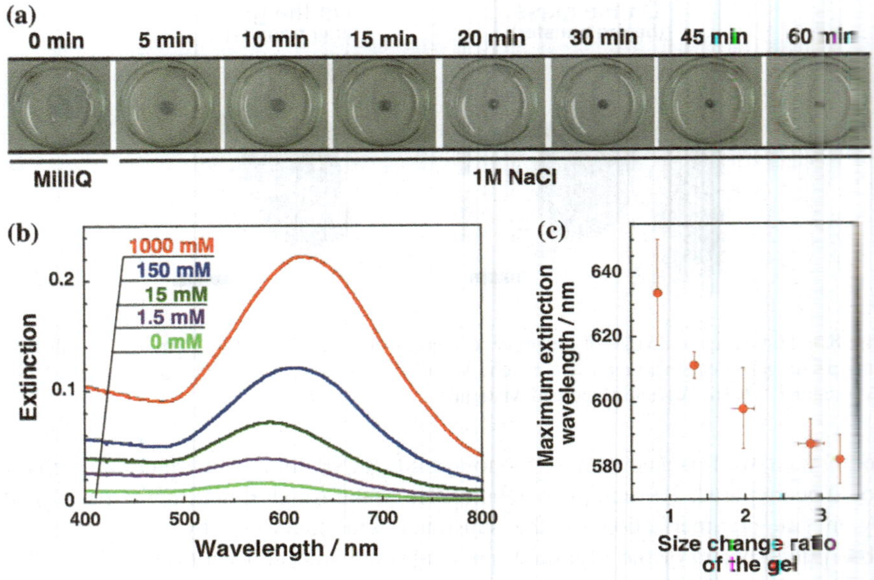

Fig. 8.9 Spectral changes resulting from NaCl concentration-dependent gel size change. **a** Time course of the gel shrinkage. The expanded gel (in MilliQ water) shrunk within 30 min after immersion into the 1 M NaCl solution. **b** Extinction spectra of a AuNP thin film-attached gel under various salt conditions. (c) Plasmon shifts against gel size change. Adapted and reproduced with permission from [14]. Copyright © (2016) Advanced Optical Materials

obtaining structural information on molecules; however, its sensitivity is quite low. Enhanced electric fields resulting from localized surface plasmon could greatly enhance Raman signals and improve its sensitivity. The morphology of metal nanostructures is a primary factor in determining the magnitude of signal enhancement in Raman scattering [33–35]. The gap structure is a leading candidate for improving the signal-enhancing effect as a significantly enhanced electric field can be formed at the gap, which is known as a hot spot. In general, although a narrower gap between metal nanostructures can provide a stronger electromagnetic field and, in turn enhance Raman scattering, the insertion of analytes into the gap becomes more difficult. In brief, this is a trade-off against the use of conventional SERS substrates. In the case of the detection of macromolecules, such as DNA or proteins, steric hindrance poses a significant problem as the analytes approach the narrow gaps. At this point, a tunable plasmonic device, in which nanoparticle distance can be dynamically changed in response to stimuli, could address this issue by allowing the efficient insertion of analytes into a widened gap and thus improving the enhancement of Raman signals at the narrowed gap on a single substrate (Fig. 8.3b).

To investigate this hypothesis, AuNP self-assembled thin films on pAAc gel, as tunable plasmonic substrates, were applied to the detection of proteins by SERS.

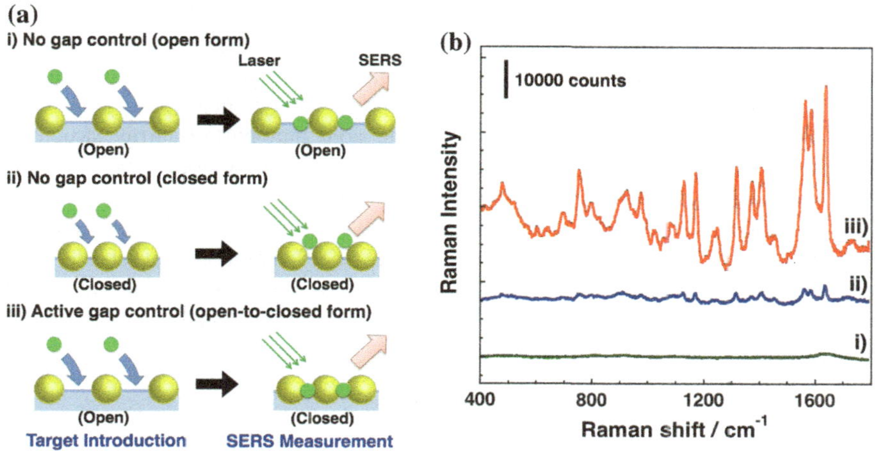

Fig. 8.10 SERS measurements of cytochrome *c* using tunable plasmonic substrates. **a** Schematic illustrations of three approaches; (*i*) No gap control in "open form": Analytes were injected onto a gel is an expanded state (in MilliQ water) and analyzed as is, (*ii*) No gap control in "closed form": Analytes were injected onto a gel is a contracted state (in 1 M NaCl solution) and analyzed as is, and (*iii*) Active gap control as "open to closed form": Target molecules were injected onto a gel is an expanded state (in MilliQ water) and analyzed after contraction (in 1 M NaCl solution). **b** SERS spectra of cytochrome *c* for each approach. Adapted and reproduced with permission from [14]. Copyright © (2016) Advanced Optical Materials

Analyses were performed through three approaches to verify the effect of active gap control during SERS measurement (Fig. 8.10a). The first approach was "no gap control in open form"; that is, the gap distance was widened during the insertion of analytes and detection of SERS signals. The second approach was "no gap control in closed form"; that is, the gap distance was reduced during insertion and detection. The third approach was "active gap control"; that is, the gap distance was widened at insertion and then reduced for detection. In the case of cytochrome *c* (Cyto *c*), which is a hemeprotein with a molecular weight of 12 kDa, no signals were detected using approach (i) (no gap control in open form), but weak signals belonging to Cyto *c* were detected using approach (ii) (no gap control in closed form) (Fig. 8.10b (i) and (ii)). This difference comes from the enhancing effect on the plasmonic field. On the other hand, when a Cyto *c* solution was put on the open-formed AuNP thin film immobilized on the gel and SERS was measured after closing the gap (approach (iii)), significantly stronger SERS signals were obtained. The SERS intensities for approach (iii) were ca. 10 times stronger than those for approach (ii), even when the SERS detection conditions were the same (Fig. 8.10b (ii) and (iii)). This indicates that the enhancement of SERS was not caused by the effects of salt, but by the sandwiching system via the active gap control. Binding analysis of Cyto *c* onto the Au plate covered with the same surface ligand (FTEG-alkane thiol) also supported the notion that the effect of salt on the binding to the surface was not significant. The analyte proteins bound to the surface of the

AuNP by hydrophobic interactions with the hydrophobic part of FTEG, which was displayed on the surface. A larger surface area was exposed on the open-form structure than on the closed-form structure, particularly around the gaps in the AuNPs. This appears to be the main reason that more analyte protein, Cyto *c*, could exist at a hot spot, which is a nanogap in the gold nanoparticle array structure. When hemoglobin (MW: 65 kDa) was used as an analyte hemoprotein, a similar effect with regard to SERS measurement was observed.

This demonstrates that the active gap control system using volume change in the hydrogel is an effective method for enhancing the SERS signals of macromolecules. As the insertion of macromolecules into the narrow hot spots of the array is a critical problem, this approach offers a solution for large-sized analytes, particularly biomolecules.

8.3 Three-Dimensional Self-Assembly and Stimuli-Responsive Control

This section introduces the formation of hollow capsules made of gold nanoparticles by self-assembly in solution and their application to drug delivery systems (DDSs) with the potential for controlled release with good responsiveness (Fig. 8.11).

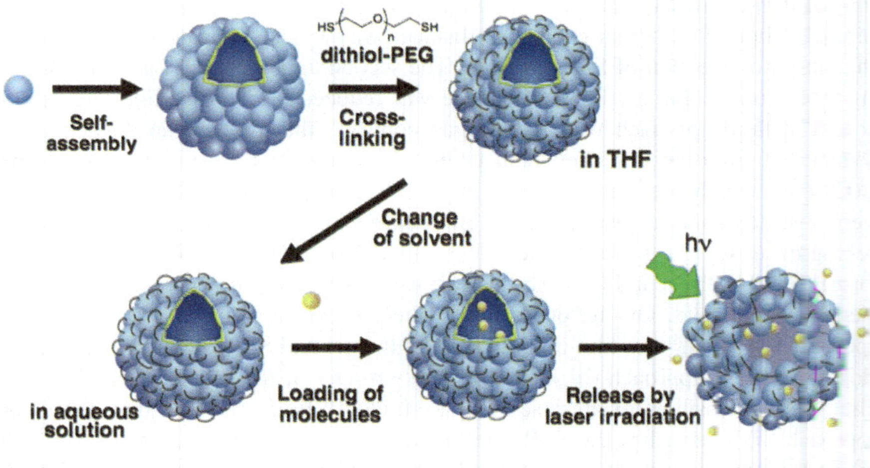

Fig. 8.11 Schematic illustration of the process described in this section. Fabrication of gold nanoparticle vesicles (AuNVs) through self-assembly and the encapsulation of dye or drug molecules followed by their light-triggered release through changes in the conformation of the AuNVs

8.3.1 Formation of Three-Dimensional Gold Nanoparticle Capsules Through a Self-Assembly Process

Living organisms contain various kinds of capsule structures. For example, lipid capsules, such as exosomes or lysosomes, contribute to vesicle transportation to carry proteins and RNAs. In viruses, capsule structures made of proteins encapsulate DNAs or RNAs for their protection. These examples suggest that capsule structures possess merits as carriers; therefore, liposomes and virus capsules have been studied toward the development of novel drug delivery systems [36, 37]. Moreover, the existence of these capsules in nature indicates that the self-assembly process is particularly suited to the preparation of capsule structures. Recently, it has been reported that surface modification with several kinds of polymers can provide capsule structures composed of metal nanoparticles by self-assembly [38–40]. For the application of these structures to drug delivery systems, the size of the carrier is of vital importance. A size of around 100 nm is thought to be suitable due to its high retentivity in blood and effective accumulation into tumor tissue more than into normal tissues, known as the enhanced permeation and retention (EPR) effect [41, 42]. However, only a few papers have reported the construction of vesicles of 100 nm in size by the self-assembly of NPs. So the development of a method for the preparation of metal nanoparticle vesicles is in high demand. Therefore, we have focused on the surface ligands on nanoparticles and prepared AuNP vesicles of sub-100 nm in size through self-assembly by surface modification with semi-fluorinated oligo(ethylene glycol), referred to as SFL, as a small molecular ligand in THF [13].

SFL-coated AuNPs were synthesized through a ligand exchange reaction from citrate-protected AuNPs (5, 10, and 20 nm in diameter). SFLs in THF were added into a citrate-protected AuNP solution. Soon after mixing the color of the solution quickly changed from red to purple due to plasmonic coupling (Fig. 8.12a). This color change indicated that AuNP assemblies were formed in the solution. The SEM and TEM images supported the formation of the assemblies, the interparticle distance of which was small (estimated to be 2.5 nm). SEM also showed the efficient formation of spherical assemblies of NPs (Fig. 8.12b). The mean diameter of the vesicles consisting of 10 nm AuNPs was ca. 70 nm, as measured by the SEM images, and the size distribution was relatively narrow. Dynamic light scattering (DLS) measurements in THF also confirmed this. The hollow structure was confirmed by transmission electron microtomography (TEMT).

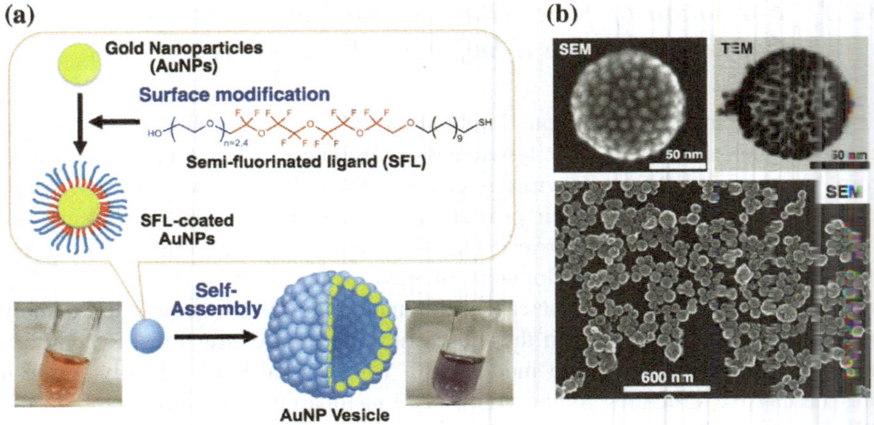

Fig. 8.12 **a** Schematic illustration of the chemical structure of the SFL and self-assembly of SFL-AuNPs to generate AuNP vesicles. **b** SEM and TEM images of AuNP vesicles made up of 10 nm SFL-AuNPs. Adapted and reproduced with permission from [13]

8.3.2 *Stimuli-Responsive Three-Dimensional Self-Assembled Gold Nanoparticle Vesicles and Their Application as a Drug Delivery System (DDS)*

The size of our AuNVs (sub-100 nm) is potentially applicable to in vivo use. However, their stability in water was so poor that the vesicle structures collapsed. To improve the stability in water, cross-linkage between AuNPs in a single vesicle with polyethylene glycol (PEG) was performed [15]. For cross-linking, after the removal of excess SFL, dithiol-PEG was mixed with the AuNV solution in THF and this mixture was incubated. Then, AuNVs cross-linked with PEG showed good dispersibility in water and retained their hollow structure.

As PEG is known to be a thermo-responsive polymer, AuNVs cross-linked with PEG should possess thermo-responsiveness. To investigate the thermo-responsiveness of the interparticle distance between each AuNP in the AuNVs, absorption spectra were measured at various temperatures (Fig. 8.13). AuNVs cross-linked with dithiol-PEG gave a plasmon peak at 541 nm. This peak was red-shifted by about 20 nm compared to that of the monodispersed Au nanoparticles. When the temperature was increased above 60 °C, the plasmon peak was drastically blue-shifted by about 10 nm. This result indicates that the inter-particle distance of the cross-linked AuNVs was increased. Conformational change in the PEG is speculated to assist in the widening of the nanogaps [43].

To demonstrate the utility of AuNVs as a drug carrier, we examined dye encapsulation into and release from cross-linked AuNVs as a model for drugs based on the temperature-dependent change in structure from a closed form to an open form (Fig. 8.14a, b). When the mixture of the dye and cross-linked AuNVs was

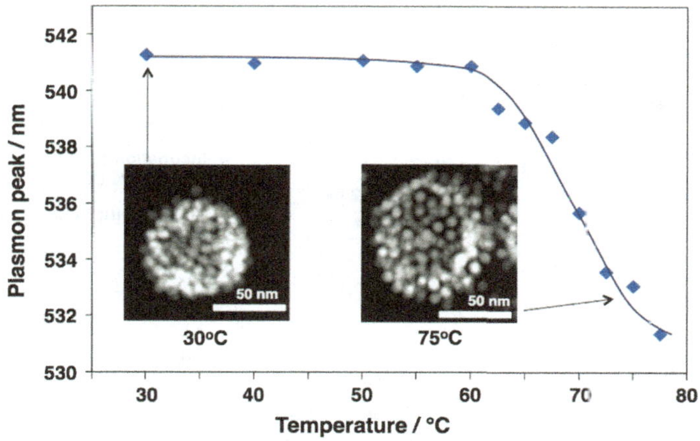

Fig. 8.13 Plasmon peak shift of AuNP vesicles modified with dithiol-PEG in water as a function of temperature. The insets show STEM images of AuNP vesicles at 30 and 75 °C. The scale bars represent 50 nm. Adapted and reproduced with permission from [15]

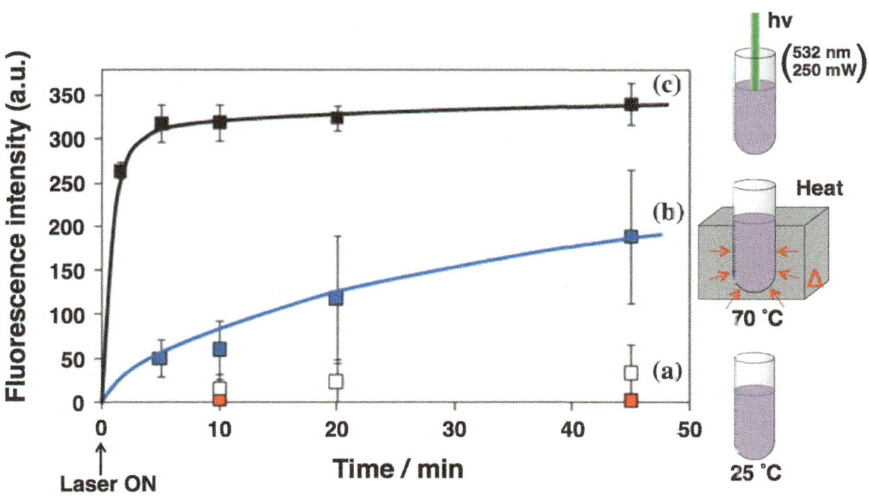

Fig. 8.14 Fluorescence intensity of rhodamine B released from cross-linked AuNP vesicles at 25 °C (**a**), at 70 °C (**b**), and upon laser irradiation (**c**). Adapted and reproduced with permission from [15]

heated to 62.5 °C, the interparticle gap was opened and dye entered the vesicles. Free dye molecules were removed by dialysis at room temperature. Cross-linked AuNVs held rhodamine dyes at 25 °C for at least 150 min. Then, when the temperature was increased to 70 °C, dye molecules were gradually released. These results indicated that cross-linked AuNVs could encapsulate dyes without leakage

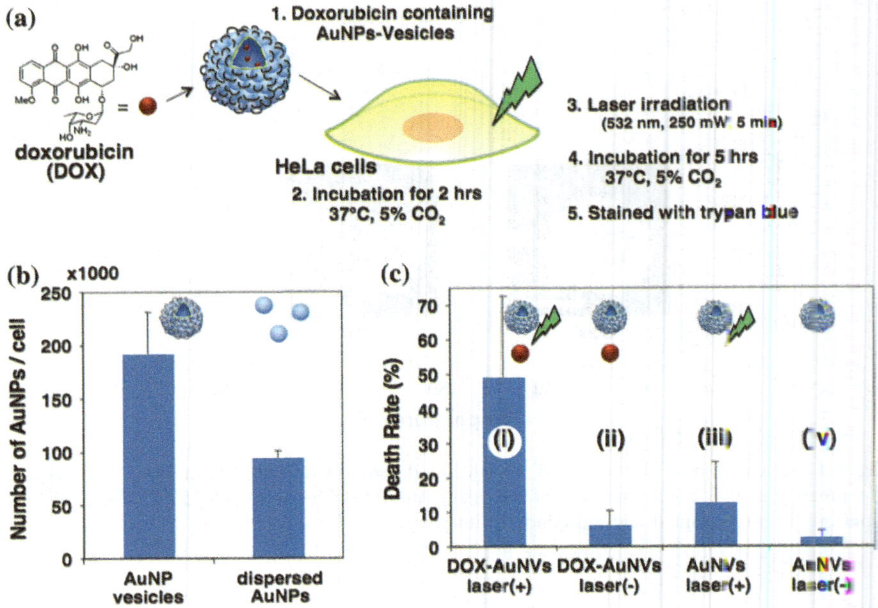

Fig. 8.15 **a** Schematic illustration of experimental procedures. **b** Cellular uptake of cross-linked AuNP vesicles (*left*) and dispersed AuNPs (*right*) per cell as determined by ICP-ES. **c** Death rate of HeLa cells treated with DOX-containing AuNP vesicles and laser irradiation (*i*), treated with DOX-containing AuNP vesicles and no laser irradiation (*ii*), treated with DOX-free AuNP vesicles and laser irradiation (*iii*), and treated with DOX-free AuNP vesicles and no laser irradiation (*iv*). The rates were obtained by counting the cells stained with trypan *blue*. Adapted and reproduced with permission from [15]

at around room temperature and release them in response to heating. The AuNVs were irradiated with a diode laser (532 nm, 250 mW), and dye release as a function of irradiation time was measured (Fig. 8.14c). The release of the dye from AuNVs triggered by laser irradiation was extremely fast and was completed within 5 min. Even using a lower power laser (25 mW), a similar release profile was observed. This result supports the notion that AuNVs are highly responsive to light.

Finally, the controlled release of drugs by light irradiation in a living cell was examined (Fig. 8.15). In this experiment, doxorubicin (DOX), which is known as an anticancer drug, and HeLa cells, which are widely used cultured cells, were used. The AuNVs were incubated with HeLa cells attached on a tissue plate for 2 h in a culture medium. The number of AuNPs incorporated into a cell was calculated from the result of inductively coupled plasma emission spectrometry (ICP-ES). Cross-linked AuNVs showed twice the level of cellular uptake compared to that of dispersed AuNPs with the same surface chemistry (Fig. 8.15b). In the same way, DOX-encapsulated AuNVs were incubated with HeLa cells and washed with FBS. The cells were irradiated with a diode laser for 5 min in the cultured medium, and the cell death ratio was analyzed by counting the number of dead cells. The ratio of dead cells after adding DOX-AuNVs was largely increased by laser irradiation, while the viability of cells without laser irradiation was not affected (Fig. 8.15c). Further, treatment with AuNVs themselves or laser irradiation alone without AuNVs did not induce significant cell death.

Cross-linking of AuNVs with PEG provided better stability in water, thermo-responsiveness, and a possible improvement in retentivity in blood. As a result, AuNVs cross-linked with dithiol-PEG could be dispersed in water and show potential for application as drug delivery carriers with controlled release triggered by light irradiation.

8.4 Conclusion

Developments in nanoscience and nanotechnologies have opened up new possibilities in terms of material fabrications. In particular, plasmonic materials are one of the most attractive materials due to their practical applications in energy-capturing devices or sensitive sensor devices. To exploit their plasmonic phenomena, structure control is a fundamental issue. The recently developed top–down process could provide precise structures on a sub-100 nm scale on a substrate. Such structures have been used to provide advanced microprocessors in computers. However, it is still difficult to fabricate 3D nanostructures or control the fabricated structures dynamically in response to external stimuli, which may be referred to as 4D control. This next-generation technology has the potential to open new doors. In this chapter, we introduced the fabrication of AuNP assemblies, such as 2D films and 3D capsules, by using fluorinated surface ligands to confer thermo-responsiveness through their combination with thermo-responsive polymers, and discussed their potential applications to bio-sensing and drug delivery systems by making the best use of dynamic control of their assembled structures. In the near future, the controlled assembly/disassembly of nanoparticles of a desired size and shape in a stimuli-responsive manner, as observed in lipids or proteins in nature, could be realized. There is the potential to be successfully applied as artificial nanomachines or smart nanodevices in a wide range of fields.

References

1. Singer SJ, Nicolson GL (1972) The fluid mosaic model of the structure of cell membranes. Science 175(80):720–731
2. Bonifacino JS, Glick BS (2004) The mechanisms of vesicle budding and fusion. Cell 116:153–166
3. Fletcher DA, Mullins RD (2010) Cell mechanics and the cytoskeleton. Nature 463:485–492
4. Lee SH, Dominguez R (2010) Regulation of actin cytoskeleton dynamics in cells. Mol Cells 29:311–325
5. Roduner E (2006) Size matters: why nanomaterials are different. Chem Soc Rev 35:583–592

6. Nie Z, Petukhova A, Kumacheva E (2010) Properties and emerging applications of self-assembled structures made from inorganic nanoparticles. Nat Nanotechnol 5:15–25
7. Mubeen S, Lee J, Singh N, Krämer S, Stucky GD, Moskovits M (2013) An autonomous photosynthetic device in which all charge carriers derive from surface plasmons. Nat Nanotechnol 8:247–251
8. Alvarez-Puebla RA, Agarwal A, Manna P, Khanal BP, Aldeanueva-Potel P, Carbó-Argibay E, Pazos-Pérez N, Vigderman L, Zubarev ER, Kotov NA, Liz-Marzán LM (2011) Gold nanorods 3D-supercrystals as surface enhanced Raman scattering spectroscopy substrates for the rapid detection of scrambled prions. Proc Natl Acad Sci USA 108:8157–8161
9. Han X, Liu Y, Yin Y (2014) Colorimetric stress memory sensor based on disassembly of gold nanoparticle chains. Nano Lett 14:2466–2470
10. Grzelczak M, Vermant J, Furst EM, Liz-Marzán LM (2010) Directed self-assembly of nanoparticles. ACS Nano 4:3591–3605
11. Iida R, Kawamura H, Niikura K, Kimura T, Sekiguchi S, Joti Y, Bessho Y, Mitomo H, Nishino Y, Ijiro K (2015) Synthesis of Janus-like gold nanoparticles with hydrophilic/hydrophobic faces by surface ligand exchange and their self-assemblies in water. Langmuir 31:4054–4062
12. Nishio T, Niikura K, Matsuo Y, Ijiro K (2010) Self-lubricating nanoparticles: self-organization into 3D-superlattices during a fast drying process. Chem Commun (Camb) 46:8977–8979
13. Niikura K, Iyo N, Higuchi T, Nishio T, Jinnai H, Fujitani N, Ijiro K (2012) Gold nanoparticles coated with semi-fluorinated oligo(ethylene glycol) produce sub-100 nm nanoparticle vesicles without templates. J Am Chem Soc 134:7632–7635
14. Mitomo H, Horie K, Matsuo Y, Niikura K, Tani T, Naya M, Ijiro K (2016) Active gap SERS for the sensitive detection of biomacromolecules with plasmonic nanostructures on hydrogels. Adv Opt Mater 4:259–263
15. Niikura K, Iyo N, Matsuo Y, Mitomo H, Ijiro K (2013) Sub-100 nm gold nanoparticle vesicles as a drug delivery carrier enabling rapid drug release upon light irradiation. ACS Appl Mater Interfaces 5:3900–3907
16. Park SY, Lytton-Jean AKR, Lee B, Weigand S, Schatz GC, Mirkin CA (2008) DNA-programmable nanoparticle crystallization. Nature 451:553–556
17. Nykypanchuk D, Maye MM, van der Lelie D, Gang O (2008) DNA-guided crystallization of colloidal nanoparticles. Nature 451:549–552
18. Achermann M, Petruska MA, Crooker SA, Klimov VI (2003) Picosecond energy transfer in quantum dot Langmuir–Blodgett nanoassemblies. J Phys Chem B 107:13782–13787
19. Kanehara M, Kodzuka E, Teranishi T (2006) Self-assembly of small gold nanoparticles through interligand interaction. J Am Chem Soc 128:13084–13094
20. Lin MH, Chen HY, Gwo S (2010) Layer-by-layer assembly of three-dimensional colloidal supercrystals with tunable plasmonic properties. J Am Chem Soc 132:11259–11263
21. Rabani E, Reichman DR, Geissler PL, Brus LE (2003) Drying-mediated self-assembly of nanoparticles. Nature 426:271–274
22. Dong A, Chen J, Vora PM, Kikkawa JM, Murray CB (2010) Binary nanocrystal superlattice membranes self-assembled at the liquid–air interface. Nature 466:474–477
23. Fujita M, Nishikawa H, Okubo T, Yamaguchi Y (2004) Multiscale simulation of two-dimensional self-organization of nanoparticles in liquid film. Japanese J Appl Phys Part 1 Regul Pap Short Notes Rev Pap 43:4434–4442
24. Hu M, Chujo S, Nishikawa H, Yamaguchi Y, Okubo T (2004) Spontaneous formation of large-area monolayers of well-ordered nanoparticles via a wet-coating process. J Nanopart Res 6:479–487
25. Yonezawa T, Onoue S, Kimizuka N (2001) Self-organized superstructures of fluorocarbon-stabilized silver nanoparticles. Adv Mater 13:140–142
26. Schweizer DK, Eigler EK (1990) Positioning single atoms with a scanning tunneling microscop. Nature 344:524–525

27. Hirokawa Y, Tanaka T (1984) Volume phase transition in a nonionic gel. J Chem Phys 81:6379
28. Shimamoto N, Tanaka Y, Mitomo H, Kawamura R, Ijiro K, Sasaki K, Osada Y (2012) Nanopattern fabrication of gold on hydrogels and application to tunable photonic crystal. Adv Mater 24:5243–5248
29. Chen H, Ming T, Zhao L, Wang F, Sun L-D, Wang J, Yan C-H (2010) Plasmon–molecule interactions. Nano Today 5:494–505
30. Willets KA, Van Duyne RP (2007) Localized surface plasmon resonance spectroscopy and sensing. Annu Rev Phys Chem 58:267–297
31. Stewart ME, Anderton CR, Thompson LB, Maria J, Gray SK, Rogers JA, Nuzzo RG (2008) Nanostruct Plasmonic Sens 494–521
32. Wang G, Mitomo H, Matsuo Y, Shimamoto N, Niikura K, Ijiro K (2013) DNA-templated plasmonic Ag/AgCl nanostructures for molecular selective photocatalysis and photocatalytic inactivation of cancer cells. J Mater Chem B 1:5899
33. Tian Z, Ren B, Wu D (2002) Surface-enhanced Raman scattering: from noble to transition metals and from rough. J Phys Chem 106:9463–9483
34. Yokota Y, Ueno K, Misawa H (2011) Essential nanogap effects on surface-enhanced Raman scattering signals from closely spaced gold nanoparticles. Chem Commun (Camb) 47:3505–3507
35. Banholzer MJ, Millstone JE, Qin L, Mirkin CA (2008) Rationally designed nanostructures for surface-enhanced Raman spectroscopy. Chem Soc Rev 37:885
36. Allen TM, Cullis PR (2013) Liposomal drug delivery systems: from concept to clinical applications. Adv Drug Deliv Rev 65:36–48
37. Yoo J-W, Irvine DJ, Discher DE, Mitragotri S (2011) Bio-inspired, bioengineered and biomimetic drug delivery carriers. Nat Rev Drug Discov 10:521–535
38. Song J, Cheng L, Liu A, Yin J, Kuang M, Duan H (2011) Plasmonic vesicles of amphiphilic gold nanocrystals: self-assembly and external-stimuli-triggered destruction. J Am Chem Soc 133:10760–10763
39. Nikolic MS, Olsson C, Saldier A, Kornowski A, Rank A, Schubert R, Frömsdorf A, Weller H, Förster S (2009) Micelle and vesicle formation of amphiphilic nanoparticles. Angew Chem Int Ed 48:2752–2754
40. Nie Z, Fava D, Kumacheva E, Zou S, Walker GC, Rubinstein M (2007) Self-assembly of metal–polymer analogues of amphiphilic triblock copolymers. Nat Mater 6:609–614
41. Moghimi SM, Hunter AC, Murray JC (2001) Long-circulating and target-specific nanoparticles: theory to practice. Pharmacol Rev 53:283–318
42. Cabral H, Matsumoto Y, Mizuno K, Chen Q, Murakami M, Kimura M, Terada Y, Kano MR, Miyazono K, Uesaka M, Nishiyama N, Kataoka K (2011) Accumulation of sub-100 nm polymeric micelles in poorly permeable tumours depends on size. Nat Nanotechnol 6:815–823
43. Bjorling M, Karlstrom G, Linse P (1991) Conformatlonal adaptlon of Poly(ethylene oxide). J Phys Chem 95:6706–6709

Chapter 9
Supramolecular Metal Complex Nanoarchitectures via Various Amphiphiles

Keita Kuroiwa

Abstract This chapter discusses the supramolecular control of metal complexes using lipid amphiphiles, oligomeric amphiphiles, and diblock copolypeptide amphiphiles. In the studies reported herein, various lipid amphiphiles with glutamic acid connectors (**L1** and **L2**), amphiphilic *N*-isopropylacrylamide (NIPAAm) oligomers and polymers (**2**–**10**), and diblock copolypeptide amphiphiles with lysine (**11**–**14**) and glutamate (**15**–**17**) were prepared. Lipid-packaged dinuclear ruthenium (II, III) complexes in the class III mixed-valency state were found to exhibit a reversible hypochromic effect and morphological evolution from microtapes to microtubes, giving rise to changes in absorption spectral intensities. Moreover, these morphological and spectral changes could be reversed by simple standing and shaking. In addition, aqueous solutions of **2**–**10** were found to form stable rods and vesicles in aqueous solution. Nanorods and nanosquares were also obtained from various metal ions or metal cyanide complexes. The electrostatic interactions between the oligomeric amphiphiles and the metal complexes, as well as the degree of polymerization (in the vicinity of 100), were both found to play significant roles in the morphology and thermoresponsiveness of these hybrids. Composites composed of metal cyanide complexes or a cobalt terpyridine complex with diblock copolypeptide amphiphiles were also prepared, and supramolecular structures were obtained. Photoluminescence resulting from metal–metal interactions and reversible reverse spin transitions was successfully obtained. The concept of composites based on amphiphiles could be expanded to allow the synthesis of other useful coordination compounds and should allow increased development of the nanochemistry of element-block composites.

Keywords Self-assembly · Supramolecular control · Functional metal complex · Electronic state · Spin crossover · Luminescence

K. Kuroiwa (✉)
Department of Nanoscience, Faculty of Engineering, Sojo University, Kumamoto, Japan
e-mail: keitak@nano.sojo-u.ac.jp

T. Kawai and M. Hashizume (eds.), *Stimuli-Responsive Interfaces*,
DOI 10.1007/978-981-10-2463-4_9

9.1 Introduction

The spatial self-assembly and integration of functional metal complexes have attracted significant attention due to the potential for the development of fundamental molecular systems [1]. Naturally occurring examples of such systems include the metal complexes that play an important role in various biological functions, such as the transportation of oxygen, gene activation, and catalytic reactions of enzymes [2, 3]. Thus, biomimetic chemistry has been employed to both understand and develop the functional reactions and properties of metal complexes, including studies of intermolecular and intramolecular interactions among natural metal complexes designed for specific functions. However, to date, our knowledge of self-assembly on the molecular level is still minimal and is largely limited to chemical structures composed of only one type of molecule. The more complex self-assembly of both metal complexes and organic compounds, resulting in supramolecular control of various functions, is controlled by strong interactions such as covalent bonding and coordination and also by weak interactions, including hydrogen bonding, hydrophobic interactions and van der Waals interactions [4–6]. Of these, non-covalent interactions are expected to play an important role in the supramolecular control of coordination systems.

Metal complexes have been investigated in the fields of coordination chemistry and supramolecular chemistry, as well as in other fields of chemistry and physics. Of particular interest are macrocyclic metal complexes (or "molecular boxes") [7], in which metals are bridged by linear bidentate ligands, and permanent porous coordination polymers (metal–organic frameworks) [8, 9] with large channels or cavities capable of hosting various molecules. These both represent high priorities with regard to the investigation of supramolecular coordination compounds. In addition, low-dimensional coordination polymers and coordination systems have been examined as means of generating functional nanomaterials. Kimizuka et al. first reported the one-dimensional supramolecular assembly of metal complexes and demonstrated color changes resulting from aggregation-induced metal-to-ligand charge transfer, showing that such assemblies possess different functions from the metal complex in the bulk state [10–12]. Following the development of such functional self-assembling systems, many chemists and physicists began to examine low-dimensional self-assembly and low-dimensional coordination systems [13–16]. However, to date, almost all supramolecular systems based on low-dimensional compounds have relied on strong interactions such as covalent bonding and coordination, and the functions of the resulting supramolecules have been similar to those of the original metal complexes in the solid state.

Recently, our group developed a flexible supramolecular system composed of metal complexes, using lipophilic and amphiphilic compounds [17–20]. This system exhibited flexible, metastable properties such as solvent-induced metal-to-ligand charge transfer [17], heat-set gel-like network formation [18, 19], and supramolecular spin crossover via adaptive molecular clefts [20], properties that are not observed in the solid state. The design of supramolecular systems such as these should allow the fabrication of flexible, stimuli-responsive supramolecules with unique and specific functions, and is also expected to improve our understanding of multifunctional biomimetic systems.

In this chapter, we briefly describe the chemical structures and properties that result in a flexible supramolecular system, focusing on the use of lipid amphiphiles [27, 28], oligomeric amphiphiles [29, 30], and diblock copolypeptide amphiphiles [31–33]. The important roles that flexibility and weak interactions play in the supramolecular control of nanostructure morphologies and in the generation of dynamic, metastable functions involving electronic, magnetic, and optical properties are also discussed based on the most recent findings of our research group.

9.2 Dynamic Self-Assembly of Mixed-Valence Metal Complexes Exhibiting Reversible Transformations in Response to External Stimuli Using Lipid Amphiphiles

9.2.1 Self-Assembly of Metal Complexes Using Lipid Amphiphiles

When Kunitake set out to develop new bilayer-forming compounds in 1977 [34], his group assumed that the unique structures of the polar head groups of biolipid molecules were determined by the biosynthetic and physiological requirements of membrane formation, rather than by physical chemistry [35]. The first candidates were, therefore, simple double-chain synthetic lipids that formed clear aqueous dispersions. When these dispersions were observed using electron microscopy, the formation of single- and multiwalled vesicles and multilayers was established. The aggregation characteristics of these compounds, such as the critical aggregate concentrations and molecular weights, were also consistent with the generation of bilayer vesicles [34, 35].

It is well known that the use of lipid amphiphiles makes possible the aggregation and self-assembly of functional coordination polymers [10–13, 17, 22, 25]. In particular, the controlled self-assembly of metal complexes via lipid amphiphiles is of significant scientific and technological importance with regard to the development of multifunctional materials and devices. Mixed-valence complexes have attracted much attention, since these compounds exhibit unique physical and chemical properties resulting from charge transfer interactions between metal ions linked via bridging ligands [36–38]. As an example, the low-dimensional assembly of such complexes gives rise to specific electronic, magnetic, and optical properties. Moreover, the assembly of discrete binuclear mixed-valence complexes has been suggested as a basis for forming molecular communication systems, including quantum cellular automata [39, 40]. Ideally, the characteristics of such systems would be tunable by controlling the spatial arrangement of the mixed-valence complexes, resulting in electronic interactions among the metal complexes without covalent or coordinative linkages.

In this context, supramolecular strategies have been developed to construct nanoassemblies of coordination compounds, such as one-dimensional (1D), two-dimensional (2D), and three-dimensional (3D) metal complexes [13–15]. However, studies to date have focused on the conversion of crystalline coordination polymers to nanowires, nanosheets, and nanoparticles, which can be regarded as the isolation of low-dimensional structures from 3D solids (Fig. 9.1a). In contrast, there have been no reports concerning the reversible and hierarchical self-assembly of discrete mixed-valence metal complexes (which normally do not interact with one another) into 1D nanowires, 2D nanosheets, or 3D nanoarchitectures, a concept that represents the very essence of bottom-up nanotechnology.

Kimizuka et al. have fabricated self-assembled nanowires by modifying such mixed-valence coordination compounds using amphiphiles (Fig. 9.1b) [10–12].

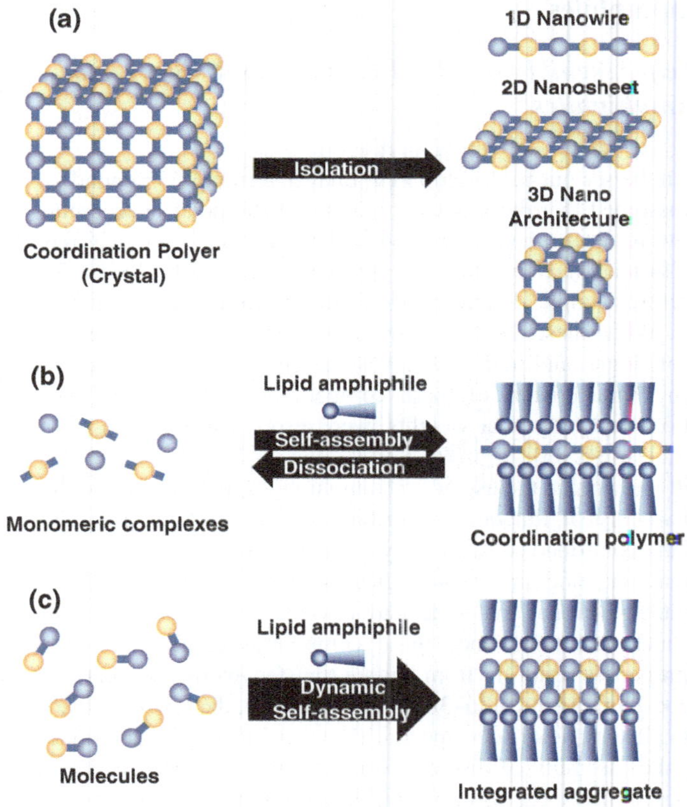

Fig. 9.1 Schematic illustrations of **a** isolation from a crystal, **b** self-assembly of a coordination polymer, and **c** integration of molecules to construct a nanoassembly of coordination compounds

The introduction of lipid counter anions to cationic mixed-valence coordination polymers has been shown to lead to the formation of nanowires in organic media, resulting in the emergence of various functionalities that do not appear in the bulk solid state, such as reversible dissociation [10], solvatochromism [41], helical ribbon morphologies [42], and honeycomb structures [43]. Nanocomposites of triazole complexes [22] and metal clusters [25] with lipid amphiphiles have also been developed based on supramolecular strategies.

This section addresses the dynamic, flexible self-assembly of discrete mixed-valence complexes with lipid amphiphiles (Fig. 9.1c) [27, 28]. Studies along these lines have shown that such assemblies generate morphological evolution that gives rise to hypochromic effects as well as the reversible stimuli-responsive transformation of the self-assembled structures. Moreover, these assemblies represent the first-ever observations of a reversible, hypochromic effect exhibited by discrete mixed-valence complexes in organic media.

9.2.2 Self-Assembly of Discrete Mixed-Valence Metal Complexes [27, 28]

Mixed-valence ruthenium complexes with lipid amphiphiles were prepared using the two amphiphiles **L1** and **L2** (Fig. 9.2). Composite **1** was obtained by replacing the counteranion of $[Ru_2(\mu\text{-}Cl)_3(tacn)_2](PF_6)_2$ (tacn = 1,4,7-triazacyclononane) with **L1**. The product had an indigo color, demonstrating that it included the mixed-valence complex. In addition, peaks ascribable to P-F stretching (556 and 824 cm^{-1}) that were present in the infrared (IR) spectrum of the initial complex were not observable in the IR spectrum of **1**, indicating that the counter anions of $[Ru_2(\mu\text{-}Cl)_3(tacn)_2](PF_6)_2$ had been completely replaced by **L1**. Interestingly, no other combinations of the three complexes and the two lipids (Fig. 9.2) were found to form complexes and to be somehow unstable or impure and thus not worth studying. This result indicated that the unique geometries of $[Ru_2(\mu\text{-}Cl)_3(tacn)_2]^{2+}$ and **L1** were well suited to the formation of a composite.

The UV–vis absorption spectrum of an indigo-colored dispersion of **1** in dichloromethane showed two strong bands, centered at 306 and 602 nm. The band at 602 nm was similar to that generated by pure $[Ru_2(\mu\text{-}Cl)_3(tacn)_2](PF_6)_2$ [12a], indicating the formation of a halogen-bridged mixed-valence structure with $[Ru_2^{II,III}(\mu\text{-}Cl)_3(tacn)_2]^{2+}$ in the class III state was present in the composite. The other strong band (λ_{max} = 306 nm) corresponded to the absorption of hydrolyzed $[Ru_2(\mu\text{-}Cl)_3(tacn)_2]^{2+}$ ($[Ru_2^{III,III}Cl_2(\mu\text{-}OH)_2(tacn)_2]^{2+}$, Fig. 9.3) [12]. This result suggested that the cationic moiety of **1** consisted primarily of a 1:1 mixture of $[Ru_2(\mu\text{-}Cl)_3(tacn)_2]^{2+}$ and $[Ru_2Cl_2(\mu\text{-}OH)_2(tacn)_2]^{2+}$ (Fig. 9.3). Elemental analysis revealed that composite **1** was in fact $[Ru_2(\mu\text{-}Cl)_3(tacn)_2][Ru_2Cl_2(\mu\text{-}OH)_2\text{-}(tacn)_2]$ $(\mathbf{L1})_4\cdot 4Na(\mathbf{L1})\cdot 5H_2O$, which was consistent with the formula estimated from UV–vis spectroscopy.

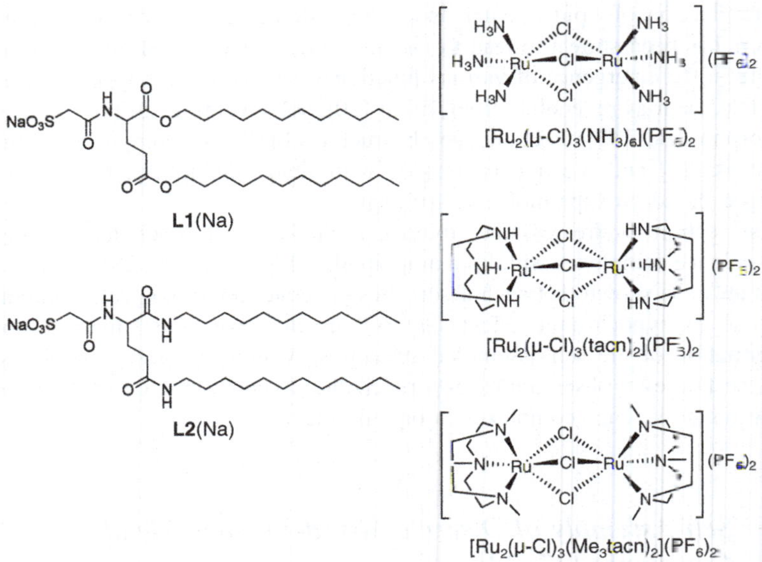

Fig. 9.2 Chemical structures of lipids (**L1** and **L2**) and ruthenium complexes ($[Ru_2(\mu\text{-}Cl)_3(NH_3)_6]^{2+}$, $[Ru_2(\mu\text{-}Cl)_3(tacn)_2]^{2+}$, and $[Ru_2(\mu\text{-}Cl)_3(Me_3tacn)_2]^{2+}$)

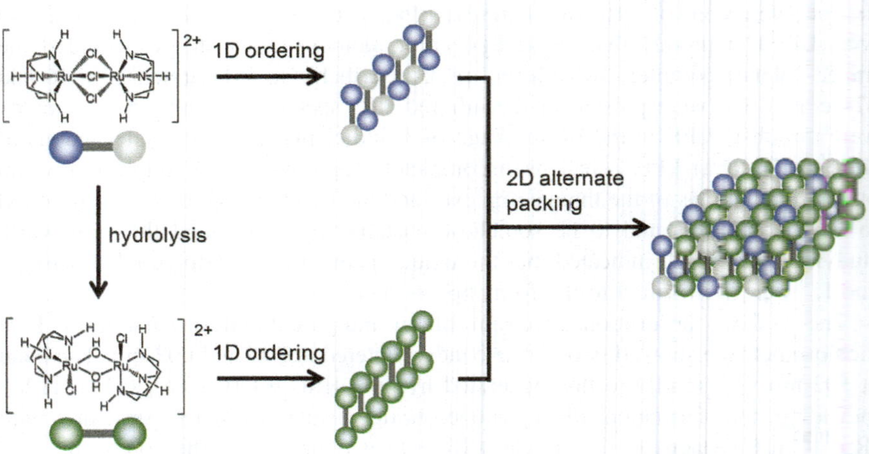

Fig. 9.3 Schematic illustration of the self-assembly of discrete dinuclear ruthenium complexes into a lamellar structure

Thus, a nanocomposite was formed, consisting of a mixed-valence ruthenium complex with a lipid, via a unique process including partial oxidation to $[Ru_2Cl_2(\mu\text{-}OH)_2\text{-}(tacn)_2]$. The nanocomposite was stable in air for a period of several months, indicating that the 1:1 combination of $[Ru_2(\mu\text{-}Cl)_3(tacn)_2]^{2+}$ and $[Ru_2Cl_2(\mu\text{-}OH)_2(tacn)_2]^{2+}$ was also stable. Therefore, such lipid-packaged nanocomposites can evidently generate geometries suitable for the formation of stable structures through self-assembly.

9.2.3 Hypochromic Behavior of Nanocomposites of Discrete Mixed-Valence Metal Complexes with Lipid Amphiphiles [27]

As noted, the UV–vis spectrum of **1** in dichloromethane showed two strong bands centered at 306 and 602 nm (plot A in Fig. 9.4a). Unexpectedly, following the dissolution of **1** in the solvent, these absorption bands gradually decreased, displaying a so-called hypochromic effect (Fig. 9.4a, c). This phenomenon suggested that the transition dipole moments of the dinuclear ruthenium complexes were in a parallel arrangement as a result of the supramolecular assembly of the dinuclear ruthenium complexes and lipid amphiphiles, as discussed below (see Fig. 9.3; Sect. 9.2.4). The initial spectrum (Fig. 9.4a, plot A) includes the contributions of light scattering from irregular, submicron aggregates. The subsequent absorbance changes, which proceeded for more than 3 h after dissolution, are attributed solely to the hypochromic effect since baseline changes arising from the scattering absorption (at ~500 nm) were not observed. It is also noteworthy that the baseline of the absorption spectra remained largely unchanged, even during the growth of the significant supramolecular structures.

Even more surprisingly, the initial absorption bands were recovered by simply agitating the solution (compare plots C and D in Fig. 9.4a, b). This absorbance change was found to be reversible; the absorbance could be repeatedly decreased and recovered through allowing the solution to stand and then applying agitation (Fig. 9.4c, d).

The absorption spectra of **1** before and after shaking at the 11.5 h mark were equivalent to those before and after shaking for 24 h, respectively. It is also noteworthy that the $[Ru_2(\mu\text{-}Cl)_3(tacn)_2]^{2+}$ in **1** remained intact in contact with air, while in aqueous solution it was completely hydrolyzed to $[Ru_2^{III}Cl_2(\mu\text{-}OH)_2(tacn)_2]^{2+}$. The enhanced stability of this complex in air may be due to isolation of the $[Ru_2(\mu\text{-}Cl)_3(tacn)_2]^{2+}$ from water and oxygen as a result of the lipid-bilayer packaging. In addition, the absorption spectra after the second and third shakings (plots D and F in Fig. 9.3c) were at almost the intensities as on plot B in Fig. 9.3a. Therefore, the decrease in the overall absorbance soon after dissolution may be attributed not only to the hypochromic effect but also to changes in the degree of light scattering based on variations in the dispersion state of **1**.

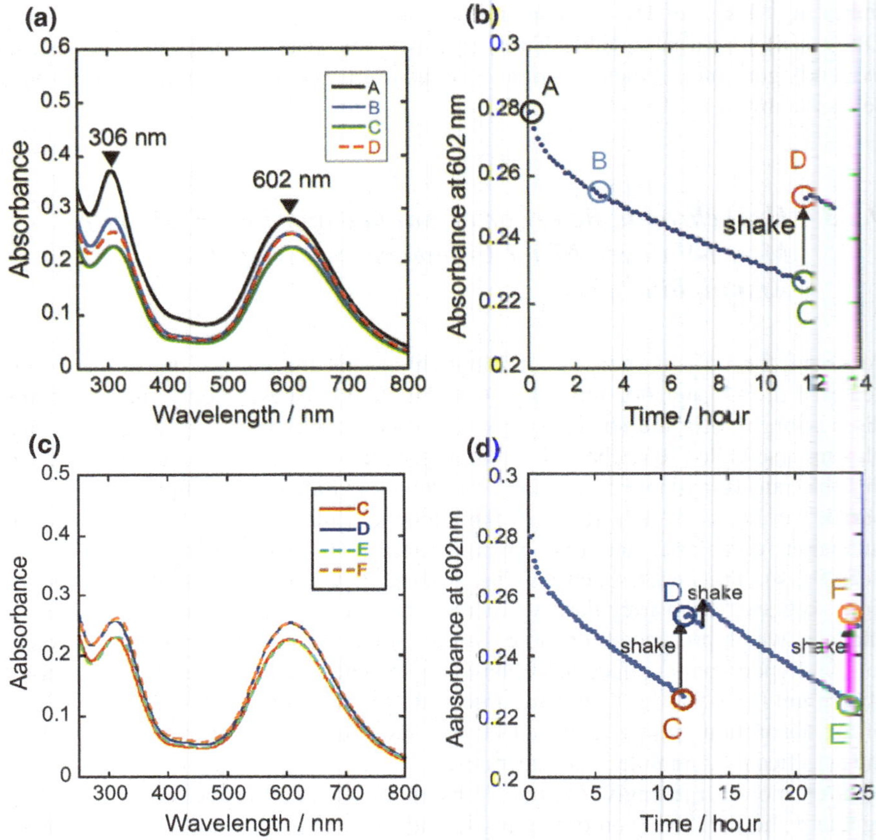

Fig. 9.4 **a** and **c** UV–vis absorption spectra of composite **1** (0.021 mM) in dichloromethane (*A*) immediately, (*B*) 3 h and (*C*) 11.5 h after dissolution, and (*D*) immediately after agitation, (*E*) 24 h later and (*F*) after being agitated three times. **b** and **d** Time dependence of the absorption intensity at 602 nm after dissolution of **1** in dichloromethane. Points *C* and *D* (before and after shaking, $t = 11.5$ h) have the same intensities as points *E* and *F* (before and after shaking, $t = 24$ h), respectively

9.2.4 Highly Ordered Alignment of Mixed-Valence Metal Complexes and Their Responsiveness to External Stimuli [27]

To investigate the cause of the hypochromic effect and its reversible behavior, the self-assembly structures were examined by transmission electron microscopy (TEM). TEM images of **1** in dichloromethane after aging for various intervals are shown in Fig. 9.5a–c. The images immediately after dissolution exhibit irregular, submicrometer-sized aggregates. After aging for 6 h, the blue dispersion evidently

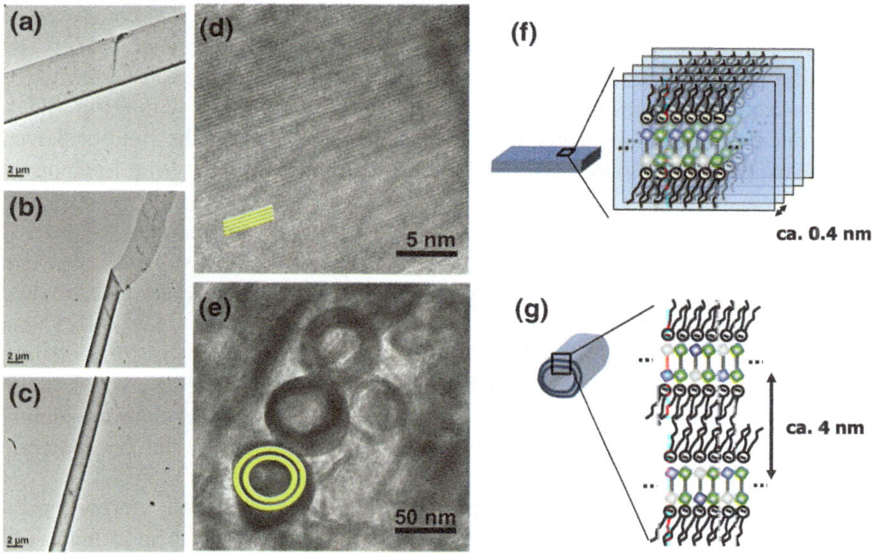

Fig. 9.5 TEM images of composite **1** upon dissolution in dichloromethane after **a** 6 h and **b** and **c** 12 h at room temperature. HR-TEM images of **1** in dichloromethane after **d** 6 h and **e** 12 h. The added *lines* and *circles* indicate the apparent positions of layers and microtubes of **1**. Samples were observed without negative staining. Schematic views of **f** the 2D stacking of a lamellar structure of **1** after 6 h and **g** the double-layered microtube structure of **1** after 12 h, as estimated from the HR-TEM images

generated microtapes with widths of 6–7 μm and lengths of 100–500 μm (Fig. 9.5a). Surprisingly, upon further aging of the microtapes in dichloromethane for 12 h, tubular aggregates with diameters of 2 μm were obtained (Fig. 9.5b, c). The image in Fig. 9.5b clearly indicates the structural transformation from microtapes to microtubes. These aggregates remained dispersed without forming precipitates. Interestingly, upon shaking the dispersion, the microtube structures were transformed to microtape structures, similar to those observed at the aging period of 6 h. This aggregation behavior indicated that the hypochromic effect observed in the UV–vis absorption spectra was related to the supramolecular self-assembly of the dinuclear ruthenium complexes. It is noteworthy that these structures could be controlled by external stimuli such as shaking in the case of composite **1**. These results demonstrate that a combination of discrete metal complexes and lipid amphiphiles can enable certain delicate transformations between dynamic structures.

High-resolution TEM (HR-TEM) images of **1** were subsequently obtained to confirm the detailed supramolecular assembly structures. After 6 h, lamellar patterns were observed in the microtapes with an average spacing of 0.4 nm

(Fig. 9.5d). This spacing indicated that the microtape structure was composed of lamellar structures formed from lipid-packaged complexes (Fig. 9.5f). The double-layered microtubes were observed in abundance after 12 h, as determined by cross-sectional images that showed a spacing of approximately 4 nm (Fig. 9.5e). These microtubes contained no more than two layers, and so the lipid-packaged complexes produced double-layered lamellar microtubes with 4-nm spacing in dichloromethane (Fig. 9.5g).

Wide-angle X-ray diffraction (WAXD) analysis of **1** demonstrated an ordered lamellar structure with a long period of 36.2 Å. Importantly, this value is consistent with a double-layered structure having approximately 4 nm spacing, as estimated on the basis of the HR-TEM images (Fig. 9.5f, g). The period value is less than twice the molecular length of **L1** (ca. 24 Å, as estimated using the CPK model), suggesting that the lipid compounds were tilted with respect to the layers making up the lipid-packaged dinuclear ruthenium complexes. Thus, the alkyl chains evidently adopted a more tilted orientation as a means of adapting to the coordination compound layers.

The lipid-packaged mixed-valence complex **1** also displayed morphological changes with aging of the solution in dichloromethane (Fig. 9.6). The formation of a bilayer structure led to morphological evolution from microtapes to microtubes, giving rise to changes in the absorption spectral intensities. Moreover, these morphological and spectral changes could be reversed by allowing the solution to stand or, conversely, by agitation. The technique of combining lipid molecules and discrete coordination compounds evidently makes it possible to design flexible, reversible, and signal-responsive supramolecular coordination systems.

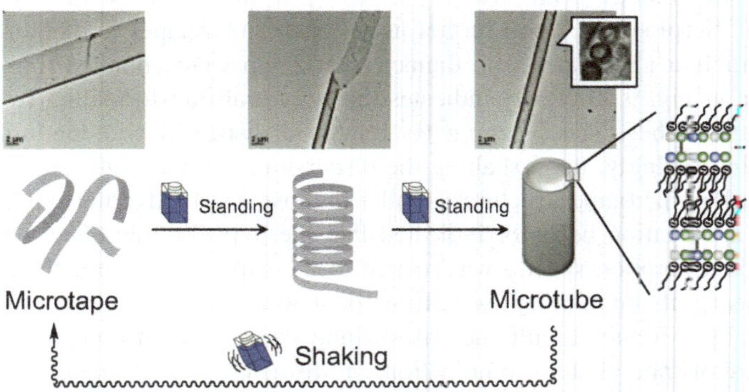

Fig. 9.6 Schematic view of the dynamic self-assembly of composite **1**. The formation of a bilayer structure causes a morphological evolution from microtapes to microtubes, giving rise to changes in absorption spectral intensities. Moreover, morphological and spectral changes could be reversed by standing and shaking

9.3 Composites of Amphiphilic *N*-isopropylacrylamide Oligomers and Polymers with Inorganic Compounds

9.3.1 Morphological Control of Amphiphilic NIPAAm

It is well known that poly(*N*-isopropylacrylamide) (PNIPAAm) exhibits a significant and abrupt thermal transition at its lower critical solution temperature (LCST) [44, 45]. For this reason, PNIPAAm synthesized by living polymerization has been used to develop highly ordered functional materials showing thermoresponsiveness. As an example, thermoresponsive hybrids of PNIPAAm with inorganic compounds, such as gold nanoparticles [46–54], gold nanocages [55], silver nanoparticles [56], magnetic nanoparticles [57–60], silica nanoparticles [61–63], and silsesquioxanes [64–67], have been developed with the aim of producing functional organic/inorganic nanomaterials. In addition, copolymers of PNIPAAm with acrylic acid [68], carboxylates [69], crown ethers [70], aza crowns [71], polythioethers [72], phenanthroline [73], terpyridine [74], and iminodiacetic acid derivatives [75] have been synthesized in order to tune the LCST transition and to achieve the extraction and recognition of metal ions. However, there have been no reports concerning control over the alignment of the metal complexes in these materials. Ideally, the characteristics of such metal complexes would be tunable by controlling their spatial arrangement within the composite, resulting in supramolecular nanocomposites without covalent or coordinative linkages.

This section focuses on the self-assembly of nanocomposites composed of amphiphilic NIPAAm oligomers/polymers with inorganic compounds, such as metal ions and metal complexes, as a means of obtaining hierarchical control over morphology [29, 30]. It is increasingly clear that an amphiphilic NIPAAm oligomer having an LCST can form a molecular hybrid with various inorganic species. Our own work has concentrated on the use of macromolecules at the transition between oligomers and polymers (with a degree of polymerization of approximately 100) in conjunction with inorganic compounds that undergo aggregation, based on coordination and metal–metal interactions, to form nanoarchitectures. Thus, oligomeric/polymeric NIPAAm amphiphiles have been integrated with metal complexes, and the morphological regulation and transformation as well as the thermodynamic control of the resulting composites have been assessed.

9.3.2 Composites of Amphiphilic NIPAAm Oligomers and Polymers with Metal Ions [29]

Amphiphilic NIPAAm oligomers and polymers were employed to obtain an understanding of the effects of a degree of polymerization in the vicinity of 100. Amphiphilic poly(*N*-isopropylacrylamide) (DTC-PNIPAAm$_n$-COOH, where DTC is 2,2-dimethyltrimethylene carbonate) was obtained by combining NIPAAm with

DTC-COOH through reversible addition–fragmentation chain transfer (RAFT) polymerization to synthesize DTC-NIPAAm$_{35}$-COOH (**2**), DTC-NIPAAm$_{52}$-COOH (**3**), DTC-NIPAAm$_{74}$-COOH (**4**), and DTC-NIPAAm$_{271}$-COOH (**5**) (Fig. 9.7). Gel permeation chromatography (GPC) and ^{1}H NMR data indicated that each of the oligomers and polymers had a low polydispersity index, ranging from 1.05 to 1.32. When these amphiphilic polymers were dissolved in water at a concentration of 5 mM, transparent solutions were obtained. To study the thermal phase transitions at the LCST, the LCST values of 5 mM aqueous solution of these amphiphilic PNIPAAms were determined by turbidimetry at different pH values. These data indicated that **2** had an LCST of 16.4 °C at pH 4.0, whereas no LCST was observed above pH 5.0. From these results, it was evident that aggregation and/or settling of the oligomers occurred due to low solubility resulting from deprotonation of the amine group. In contrast, **3** and **4** exhibited higher LCST values below pH 5.0–6.0 and an increase in M_n appeared to elevate the LCST at pH 4.0 (**2**: 16.4 °C; **3**: 22.1 °C; **4**: 25.5 °C). In addition, **4** showed a constant LCST in the vicinity of 30 °C despite changing the pH from 4.0 to 8.0. These results showed that the end group effect of the carboxylate moiety was more pronounced in the case of shorter chains. Moreover, the PNIPAAm chain effect was the most significant for the high M_n sample **4**, whereas the end group effect was negligible in this sample. Therefore, the balance between the effect of the hydrophilic carboxyl group, the thermal transition of the PNIPAAm group, and the effect of the hydrophobic alkyl chain is the most important factor affecting the LCST.

The PNIPAAms are useful since they may integrate with and align metal ions. The associated aggregation and self-assembly of metal ions were investigated by TEM. Interestingly, **2**/Ag^+ (Fig. 9.8a) exhibited fibrous nanostructures with a weave pattern of several nm. In addition, **2**/Co^{2+}, **2**/Ni^{2+}, **2**/Cu^{2+}, and **2**/Fe^{3+} adopted supramolecular nanostructures consisting of assembled nanorods (**2**/Co^{2+}, **2**/Ni^{2+}), nanonetworks (**2**/Cu^{2+}, Fig. 9.8b), and nanoparticles (**2**/Fe^{2+}). These nanostructures presumably formed as the result of the self-assembly of metal ion/amphiphile composites. Diblock copolymers containing hydrophilic and hydrophobic parts tend to form one-directionally stacked lamellar structures due to the nucleation of lamellar layers, and these lamellar layers, when grown anisotropically, generate nanowires (Fig. 9.8c). These amphiphilic PNIPAAms in combination with the metal complexes were therefore capable of interacting not only at the molecular level but also on the nanometer scale to form hierarchical

Fig. 9.7 Chemical structures of amphiphilic NIPAAm oligomers and polymers synthesized from DTC

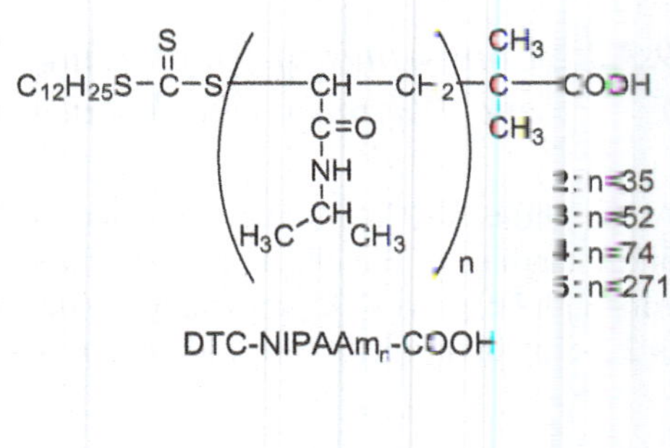

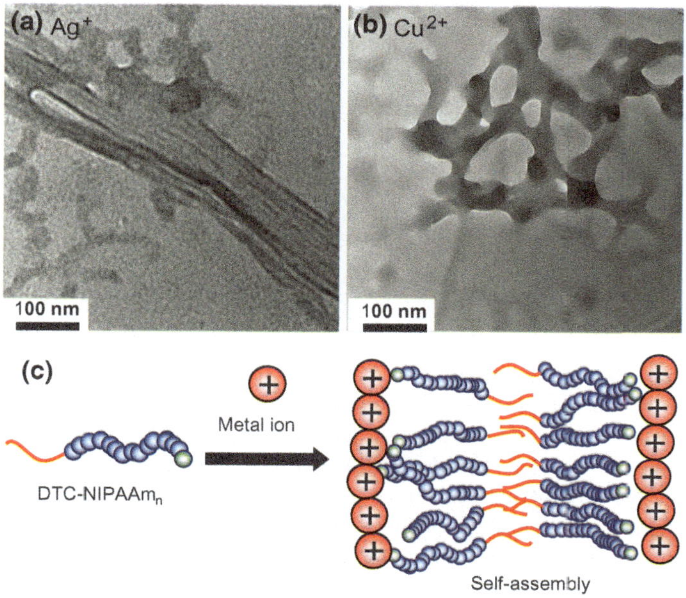

Fig. 9.8 TEM images of samples prepared from DTC-NIPAAm$_{35}$-COOH (**2**)/metal ions in aqueous solution (pH 8.0, 5 mM): **a** Ag^+ and **b** Cu^{2+}. **c** Schematic view of the 2D stacking of lamellar structures of **2**/Ag^+ as estimated based on TEM images

structures, a phenomenon similar to the manner in which the quaternary structures of proteins are formed.

Composites of **3** and **4** with metal ions (Ag^+, Co^{2+}, Ni^{2+}, Cu^{2+}, and Fe^{3+}) were found to generate nanostructures, whereas composites of **5** with metal ions displayed either undefined or networked structures ranging from several hundred nanometers to several micrometers in length. It is likely that these nanostructures were randomly generated as a consequence of the assembly of partially amorphous polymers into larger structures in association with random-coil copolymers with higher degrees of polymerization.

Confirming the relationship between thermoresponsiveness and coordination, Fig. 9.9 presents representative photographs of samples prepared in solution at different temperatures. At 15 °C, a **2**/metal ion mixture formed a transparent solution (Fig. 9.9, left). In contrast, at 50 °C, an aggregated solution with a definite coloration (owing to the Cu^{2+}) appeared (Fig. 9.9, right), resulting from the aggregation of the nanocomposite above the LCST. Nanocomposites of these amphiphilic oligomers/polymers with Cu^{2+} were also assessed by UV–vis spectroscopy. The addition of 5 mM aqueous solutions of Cu^{2+} to **2** resulted in the appearance of a new absorption shoulder and/or peaks in the region around 700 nm at temperatures above the LCST. This phenomenon suggests the presence of a *d-d* transition ($^2E_g \rightarrow {}^2A_{1g}$ and $^2E_g \rightarrow {}^2B_{1g}$) based on distorted octahedral

Fig. 9.9 Photographs of a DTC-NIPAAm$_{35}$-COOH (**2**)/Cu^{2+} composite (*left*) at 15 °C and (*right*) 50 °C in aqueous solution (pH 8.0, 5 mM)

coordination [76, 77]. It appears that, based on solvation dynamics, composites of Cu^{2+} with the amphiphilic oligomers/polymers underwent integration in the nanocomposites associated with LCST transformation. PNIPAAm itself is hydrophilic below the LCST but becomes hydrophobic above that temperature. Amphiphilic NIPAAm/Cu^{2+} thus undergoes a phase transition at the LCST because of the cooperative dehydration of the PNIPAAm chains and the concomitant collapse of individual chains from hydrated coils to hydrophobic globules. As a result of these two factors, these nanocomposites showed remarkable color-changing sensitivity to temperature.

The LCST-induced coordination between the carboxylic acid segments and the metal ions, as well as the degree of polymerization of approximately 100, both played a significant role in this morphological control. It is therefore evident that the characteristics of the amphiphile and the LCST value were both involved in inducing interesting integrations and nanostructures in aqueous solutions of the metal ions.

9.3.3 Composites of Amphiphilic NIPAAm Oligomers and Polymers with Metal Cyanide Complexes [30]

Our group has previously investigated combinations of macromolecules at the transition between oligomers and polymers with inorganic compounds that tend to self-aggregate through metal–metal interactions to form 1D and/or 2D assemblies.

In this work, K[Au(CN)$_2$] or K$_2$[Pt(CN)$_4$] were added to aqueous solutions of various poly(*N*-isopropylacrylamide) amphiphiles (DTC-PNIPAAm$_n$-NH$_3$Cl), DTC-NIPAAm$_{13}$-NH$_3$Cl (**6**), DTC-NIPAAm$_{61}$-NH$_3$Cl (**7**), DTC-NIPAAm$_{78}$-NH$_3$Cl (**8**), DTC-NIPAAm$_{119}$-NH$_3$Cl (**9**), and DTC-NIPAAm$_{274}$-NH$_3$Cl (**10**) (Fig. 9.10) at 1:1 molar ratios (based on $[Au(CN)_2]^-$) or at 2:1 molar ratios (based on $[Pt(CN)_4]^{2-}$) relative to the amine moiety.

The associated aggregation and self-assembly of the metal cyanide complexes were investigated by TEM. Interestingly, **6**/[Au(CN)$_2$] (Fig. 9.11a) exhibited square nanostructures with widths in the range of 100–200 nm. In addition, **6**/[Pt(CN)$_4$]

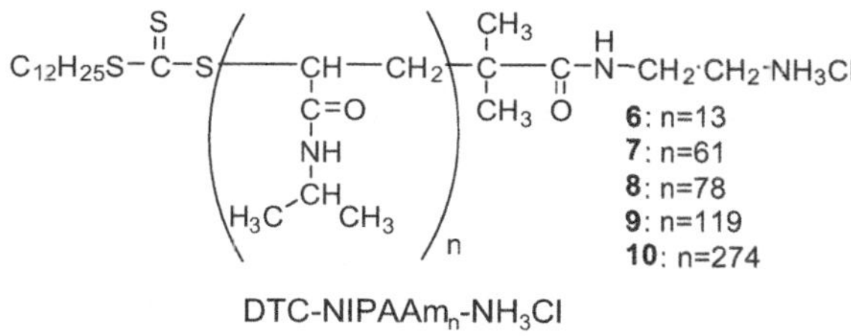

Fig. 9.10 Chemical structures of amphiphilic NIPAAm oligomers and polymers synthesized from an amine derivative of DTC

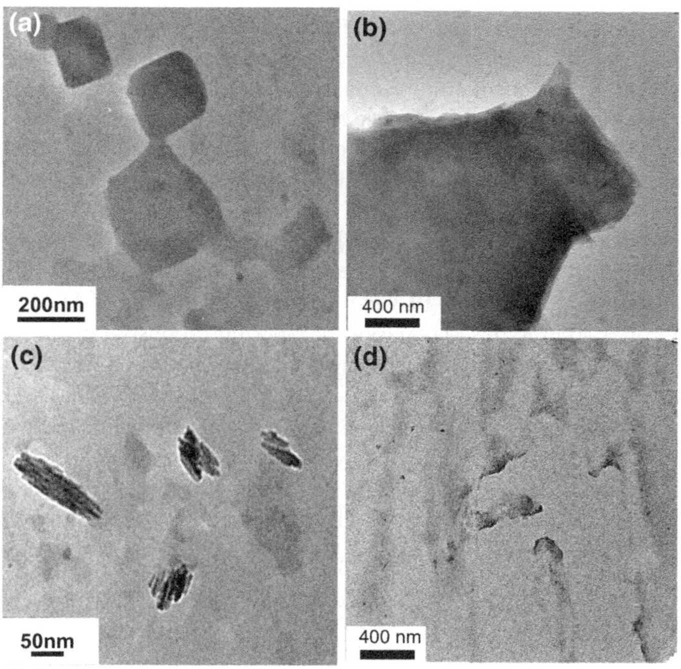

Fig. 9.11 TEM images of DTC-NIPAAm$_{13}$-NH$_3$ (**6**)/[Au(CN)$_2$] samples at **a** 21 °C and **b** 45 °C, and DTC-NIPAAm$_{13}$-NH3 (**6**)/[Pt(CN)$_4$] samples at **c** 21 °C and **d** 45 °C in water at pH 7.0 ([6] = [Au(CN)$_2$] = 1/2[Pt(CN)$_4$] = 0.5 mM)

(Fig. 9.11b), **7**/[Au(CN)$_2$], and **7**/[Pt(CN)$_4$] adopted supramolecular nanostructures consisting of assembled nanorods (**6**/[Pt(CN)$_4$]: length 40–50 nm, width 10–20 nm; **7**/[Au(CN)$_2$]: length 40–150 nm, width 15–30 nm; **7**/[Pt(CN)$_4$]: length 50–150 nm, width 15–20 nm). These square- and rod-like nanostructures presumably formed as the result of the self-assembly of metal complex/amphiphile composites. In contrast, composites of **8–10** with $[Au(CN)_2]^-$ or $[Pt(CN)_4]^{2-}$ displayed either undefined or networked structures ranging from several hundred nanometers to several micrometers in length. It is more likely that these nanostructures were randomly generated as a consequence of the assembly of partially amorphous polymers into larger structures in association with random-coil copolymers with higher degrees of polymerization. In addition, based on the results of analyses by TEM, dynamic light scattering (DLS), and small-angle X-ray scattering (SAXS), it appears that hybrids of PNIPAAms and metal complexes in water can generate both nanostructures and cylindrical aggregates with radii in the range of 20–70 nm (or spherical aggregations several nm in size) in rod, square, and sheet structures with widths of several hundred nanometers (Fig. 9.12).

To illustrate the relationship between thermoresponsiveness and morphology, Fig. 9.11 displays a set of representative TEM images of samples prepared in solution at different temperatures. As described above, at 21 °C, **6**/[Au(CN)$_2$] generated square structures (Fig. 9.11a). In contrast, at 45 °C, sheet-like structures appeared (Fig. 9.11b), resulting from the aggregation of the nanocomposite due to passing the LCST. The thermoresponsive nanocomposite **6**/[Au(CN)$_2$] thus undergoes assembly due to the extension of the PNIPAAm chain and the close metal–metal distance. Below the LCST of the nanocomposite (30 °C), the interaction among PNIPAAm chains was weak. It is noteworthy that, above the LCST, two-dimensional sheets with widths of several μm were self-assembled with increasing temperature. In the case of **6**/[Pt(CN)$_4$], thermoresponsive interactions were also found (Fig. 9.11c, d), while the self-assembling structures of the nanocomposites composed of **7–10** were fully agglomerated and TEM images did not show clear morphologies. Therefore, the thermoresponsive phenomenon may be attributed to delicate interactions among the metal complexes via van der Waals forces between the chains of **6**, whereas **7–10**, having longer PNIPAAm chains than **6**, undergo interactions that are too strong to allow them to maintain the assembled structures in water.

Nanocomposites of these amphiphilic oligomers/polymers with $[Au(CN)_2]^-$ or $[Pt(CN)_4]^{2-}$ were also assessed by UV–vis spectroscopy. The addition of 0.5 mM

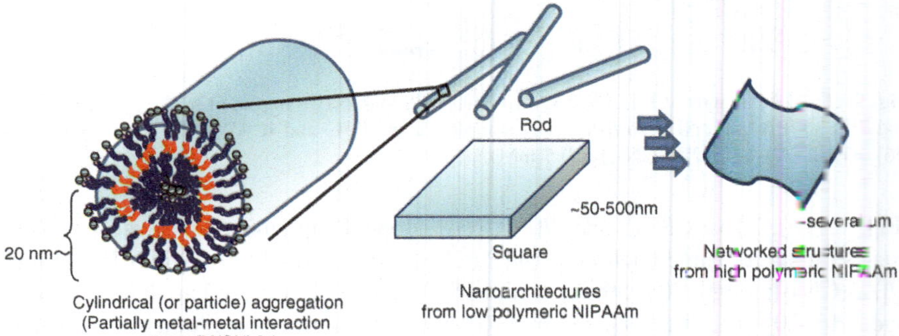

Fig. 9.12 Illustration of the self-assembly of DTC-NIPAAm$_n$-NH$_3$/[Au(CN)$_2$] and DTC-NIPAAm$_n$-NH$_3$/[Pt(CN)$_4$] as estimated from TEM, DLS, and SAXS data

aqueous solutions of K[$Au(CN)_2$] to **6–10** resulted in the appearance of a new absorption shoulder and/or peaks in the region of 250–350 nm. These observations suggest the presence of $[Au(CN)_2]^-$ and $[Pt(CN)_4]^{2-}$, based on both electrostatic interactions with the positively charged amine moieties of the NIPAM and metallophilicity. We also investigated the aggregation of $[Au(CN)_2]^-$ or $[Pt(CN)_4]^{2-}$ and their metallophilic interactions by obtaining luminescence spectra of the composites. In these trials, 0.5 mM solutions of $[Au(CN)_2]^-$ and $[Pt(CN)_2]^{2-}$ exhibited weak luminescence in the range of 340–450 nm. Based on solvation dynamics, it appears that composites of $[Au(CN)_2]^-$ or $[Pt(CN)_4]^{2-}$ with the amphiphilic oligomers/polymers underwent alignment such that there was sufficient distance between the metal complexes even in the nanocomposites.

As noted, we have demonstrated the formation of amphiphilic PNIPAAm/metal cyanide complex composites, with significant variations in nanostructure depending on the structure of the PNIPAAm. The formation of composite materials produced by combining these amphiphiles with metal complexes demonstrates that it is possible to promote the formation of one-dimensional structures such as rods, as well as more complex architectures including square structures. The technique of combining amphiphilic molecules with discrete coordination compounds thus makes it possible to design flexible supramolecular coordination systems. This general concept of an amphiphilic polymer composite could be expanded to include other useful compounds and should provide valuable information leading to further advances in the fields of coordination materials and synthetic polymer chemistry.

9.4 Self-Assembly of Discrete Metal Complexes in Aqueous Solution via Diblock Copolypeptide Amphiphiles

9.4.1 Self-Assembly of Diblock Copolypeptide Amphiphiles

Diblock copolypeptide amphiphiles are synthetic materials with many features that make them of interest to those working in the field of protein engineering, and which have various applications such as in drug delivery systems and tissue engineering [75–84]. Their unique properties are due to the propensity of these amphiphiles to form double-walled vesicles or biocompatible fibrillar nanostructures based on the self-assembly of their hydrophilic and hydrophobic blocks. Moreover, the resulting hierarchical microstructures are suitable for the integration with organic molecules, tissue cells, and proteins. In addition, these copolypeptide structures can be used to tune various inorganic materials. They may, for example, be employed to modify the molecular structure, porosity, and morphology of silica [85, 86]. The inherent functional self-assembly abilities of these copolypeptide amphiphiles could potentially lead to their application not only as structural templates for inorganic compounds but also for the intelligent transformation of inorganic materials via dynamic tuning of the electronic state of the materials.

At present, the main focus is on the use of amphiphilic compounds as a means of developing supramolecular composites with metal complexes that will function as flexible nanostructural materials. In particular, metalloproteins and enzymes containing flexible internal portions based on apoproteins may be obtained from the complex amino acids and metal centers found in biological systems [2]. Therefore, the presence of a flexible region is a key factor in the operation of functional metal complexes.

In this section, we focus on the dynamic structural transformations of metal complexes, achieved through the use of diblock copolypeptide amphiphiles having the general structural formula poly-L-Lysine-*block*-L-Leucine ($(Lys)_m$-*block*-$(Leu)_n$) [31, 32] and poly-L-Glutamate-*block*-L-Leucine ($(Glu)_m$-*block*-$(Leu)_n$) [33]. These amphiphiles are known to assemble into fibrillary structures, resulting in the formation of hydrogels and vesicles [79–84]. Our work investigated not only the morphological evolution associated with the intermolecular interactions of the metal complexes but also the hierarchical transformations of composite materials composed of combinations of the copolypeptides with metal complexes.

In particular, we developed an alternative approach to the design of metal complexes by compounding these materials with diblock copolypeptide amphiphiles, resulting in intermolecular interactions that generated photoluminescence in water or reverse spin transitions. Various diblock copolypeptide amphiphiles with metal complexes were found to exhibit unique morphologies, based on the alignment of the metal complexes. This technique of combining amphiphilic molecules with discrete coordination compounds makes it possible to design flexible, reversible, and signal-responsive supramolecular coordination systems.

9.4.2 Self-Assembly of Discrete Metal Cyanide Complexes in Aqueous Solution via Diblock Copolypeptide Amphiphiles [31, 32]

A number of diblock copolypeptide amphiphiles were employed in this work, including $(Lys)_{96}$-*block*-$(Leu)_1$ (**11**), $(Lys)_{183}$-*block*-$(Leu)_{19}$ (**12**), $(Lys)_{299}$-*block*-$(Leu)_{385}$ (**13**), and $(Lys)_{989}$-*block*-$(Leu)_{137}$ (**14**) (Fig. 9.13). All these exhibited low polydispersity values, ranging from 1.13 to 1.19, as determined by GPC and ^1H NMR integration data regarding the lysine (Lys) moieties. When $K[Au(CN)_2]$ was added to aqueous solutions of these amphiphiles at 1:1 molar ratios ($[Au(CN)_2]^-$ to lysine units), dispersion solutions were obtained. The associated aggregation and self-assembly of $[Au(CN)_2]^-$ were investigated by UV–vis spectroscopy. The addition of 0.2 mM aqueous solutions of $K[Au(CN)_2]$ to **11–14** resulted in the appearance of a new absorption shoulder in the region of 250–280 nm. Interestingly, the absorbance in this region was found to increase in the following

Fig. 9.13 Chemical structures of diblock copolypeptide amphiphiles **11–14**

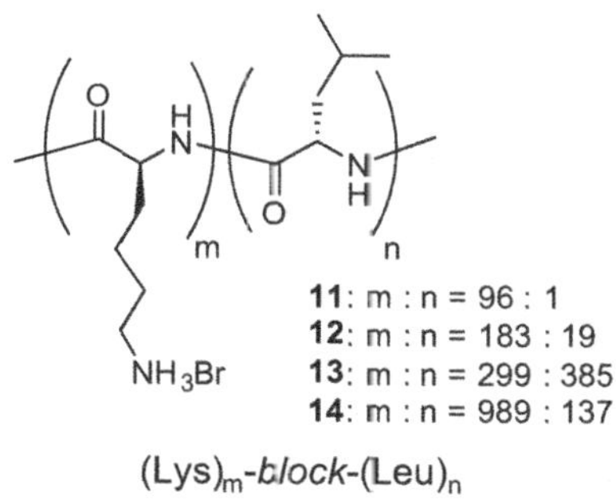

order: **12**/[Au(CN)$_2$] > **13**/[Au(CN)$_2$] > **14**/[Au(CN)$_2$] > **11**/[Au(CN)$_2$]. These observations suggest the oligomerization and/or polymerization of [Au(CN)$_2$]$^-$ based on both electrostatic interactions with the positively charged side groups of the amphiphile's Lys moieties and aurophilicity. It is therefore noteworthy that the copolypeptide amphiphiles in this work were synthesized to a degree of polymerization such that they were well suited to self-assembly and also were able to form structures that promoted aurophilic interactions. In the composite mixtures, induced circular dichroism (ICD) appeared in the absorbance region of 250–300 nm associated with [Au(CN)$_2$]$^-$, while signals corresponding to α-helix or β-sheet conformations were not observed. These results suggested the assembly of [Au(CN)$_2$]$^-$ anions around those segments of the amphiphile backbone containing the cationic Lys units and a concurrent random-coil arrangement of the polymer. Therefore, both aurophilic and electrostatic interactions between the Lys segments and the [Au(CN)$_2$]$^-$, as well as hydrophobic interactions between the Leu segments, appear to have induced a random backbone conformation. In the case of this study, interactions of the anionic metal complex with the Lys segments of the copolypeptide amphiphiles induced conformational changes that transformed the polymeric backbone to a random coil, which in turn allowed aurophilic interactions.

We also investigated the aggregation and aurophilic interactions of [Au(CN)$_2$]$^-$ by obtaining luminescence spectra of amphiphile/[Au(CN)$_2$]$^-$ composites. The excitation and emission spectra of composites **11–14**/[Au(CN)$_2$]$^-$ (Lys units : K[Au(CN)$_2$] = 1 : 1) are shown in Fig. 9.14. A 0.2 mM solution of [Au(CN)$_2$]$^-$ in the absence of the copolypeptide did not exhibit luminescence, while the luminescent intensity increased following the addition of **11–14**. Amphiphile **12** in particular resulted in a dramatic increase in intensity (λ_{max} = 465 nm), suggesting that the Au–Au bonds may result from polynuclear [Au(CN)$_2$$^-$]$_n$ excimers and exciplexes. The addition of amphiphile **13** also led to an increase in the emission intensity, although the emission band shifted to 365 nm as a result of the formation of the trimer [Au(CN)$_2$$^-$]$_3$. In contrast, the use of **11** and **14** only generated slight increases in luminescent intensity. Together, these results demonstrate that electrostatic and van der Waals interactions of the complex with the amphiphiles both played a significant role in generating luminescence from Au–Au interactions.

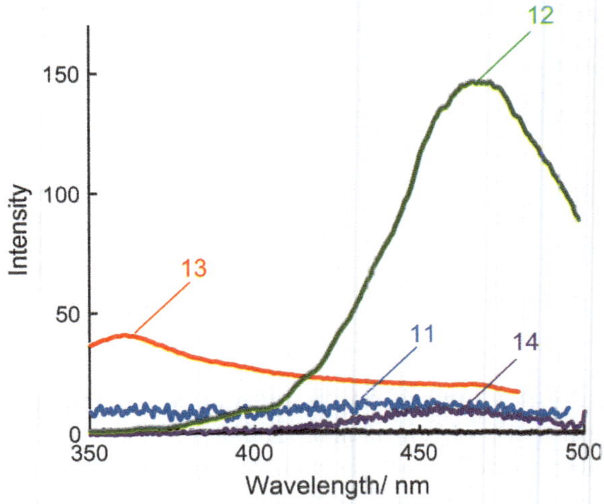

Fig. 9.14 Photoluminescence spectra of **11–14**/[Au(CN)$_2$] in deionized water ([Au(CN)$_2$] = 0.2 mM, Lys units : K[Au(CN)$_2$] = 1 : 1)

The detailed nanostructures of the amphiphile/[Au(CN)$_2$]$^-$ complexes were examined by acquiring TEM images of DI water solutions of polypeptide/[Au(CN)$_2$]$^-$ mixtures (Fig. 9.15). These samples were not stained, and so the dark regions are ascribed to the presence of Au. Both **11**/[Au(CN)$_2$]$^-$ and **14**/[Au(CN)$_2$]$^-$ were found to generate indefinite or sheet structures several hundred nanometers long. It is likely that these nanostructures were randomly generated as a consequence of partial dissociation of the polynuclear $[Au(CN)_2{}^-]_n$ complexes into shorter structures in association with the random-coiled copolypeptides.

Surprisingly, **12**/[Au(CN)$_2$]$^-$ (Fig. 9.15a) and **13**/[Au(CN)$_2$]$^-$ (Fig. 9.15b) exhibited numerous nanorods (length: 90–150 nm, width: 15–30 nm) and nanowires (width: 20–50 nm). These nanostructures were more developed than those of **11** and **14,** and so it was clear that the regular supramolecular structures resulting from **12** to **13** were due to the self-assembly of [Au(CN)$_2$]$^-$ in conjunction with the copolypeptide amphiphiles. These results were consistent with the oligonuclear and polynuclear behaviors deduced from UV–vis, CD, and luminescence spectral data. When one molar equivalent of a 5 mM aqueous solution of K[Au(CN)$_2$] was added to 5 mM (in terms of Lys units) aqueous solutions of **12** or **13**, colloidal dispersions was obtained. Sample prepared using the **12** solution displayed a nanostructure vaguely similar to the weave pattern of traditional Japanese *waraji* (straw sandals, see Fig. 9.15e), with regular sections consisting of assembled nanorods 15–30 nm in length (Fig. 9.15c). This *waraji*-like nanostructure presumably formed as a result of self-assembly of nanorods composed of the metal complex/amphiphile composite. In addition, **13**/[Au(CN)$_2$] showed a fractal pattern consisting of nanowires. It is known that diblock copolymers with hydrophilic and hydrophobic parts form one-directionally stacked lamellar structures due to the nucleation of lamellar

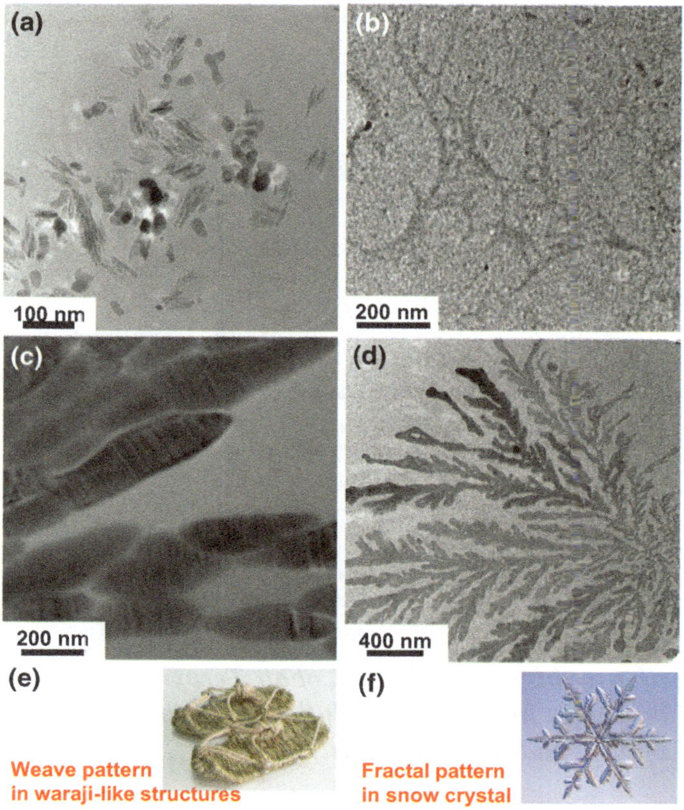

Fig. 9.15 TEM images of samples prepared from **12**/$[Au(CN)_2]^-$ **a** and **13**/$[Au(CN)_2]^-$ **b** in aqueous solution {$[Au(CN)_2]$ = 0.2 mM (Lys units : $[Au(CN)_2]$ = 1 : 1)}. TEM images of samples prepared from **12**/$[Au(CN)_2]^-$ **c** and **13**/$[Au(CN)_2]^-$ **d** in aqueous solution {$[Au(CN)_2]$ = 5 mM (Lys units : $[Au(CN)_2]$ = 1 : 1)}. **e** and **f** Schematic view of the Japanese *waraji*-(straw sandal) photographs and snow crystal photographs compared with the TEM images

layers, and that these lamellar layers grow anisotropically in the nanostructures [87, 88]. Diblock copolypeptide amphiphiles and metal complexes are therefore capable of interacting not only at the molecular level but also at the subnanometer scale to form hierarchical structures, a phenomenon similar to the manner in which the quaternary structures of proteins are formed, as illustrated in Fig. 9.16. The data from UV–vis absorption and luminescence intensity analyses showed that these amphiphile/$[Au(CN)_2]^-$ complexes included polynuclear $[Au(CN)_2]_n$ species that undergo Au–Au bonding interactions. In particular, the observations concerning aurophilic interactions indicated that $[Au(CN)_2]^-$ forms ordered arrays together with the copolypeptide amphiphiles. It is therefore evident that the amphiphile was capable of inducing interesting alignment structures in aqueous solutions of the metal complex.

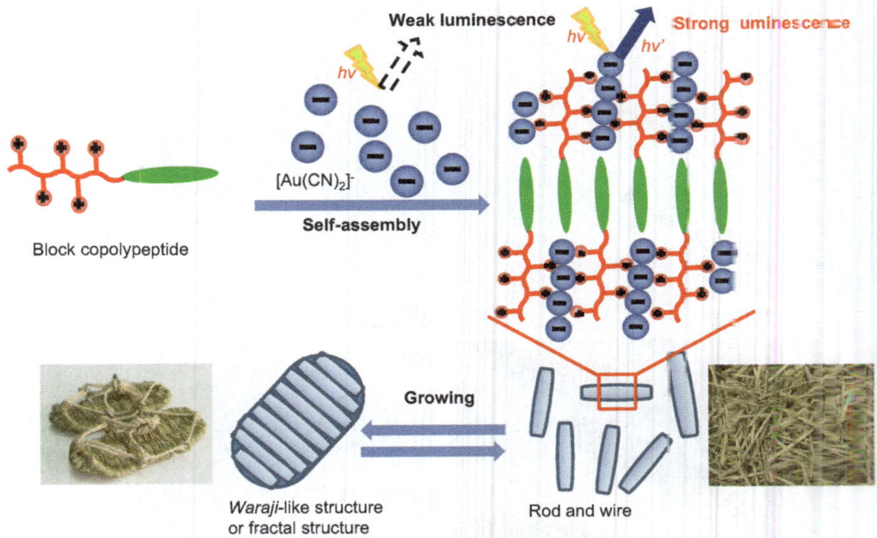

Fig. 9.16 Schematic illustration of the self-assembly of copolypeptide amphiphile/Au complexes to develop functional nanoarchitectures such as eclipse-like (*waraji*-like) and fractal structures

9.4.3 Self-Assembly of Discrete Cobalt Terpyridine Complexes in Aqueous Solution via Diblock Copolypeptide Amphiphiles [33]

Our research group has also focused on combining cobalt(II) terpyridine complexes with diblock copolypeptide amphiphiles and examining the spin crossover (SCO) characteristics of the resulting complexes in water. The diblock copolypeptide amphiphiles **15** and **16** and the polypeptide **17** (Fig. 9.17) were synthesized so as to have a suitable degree of polymerization, since this is known to enhance the solubility of these compounds in water and also improves the packing of the polypeptides in supramolecular assemblies, resulting in the formation of hydrogels [84]. In contrast, polymer **17** was synthesized to a higher degree of polymerization since it was not intended to produce a supramolecular effect. Composites of polypeptides **15–17** and a cobalt(II) terpyridine complex ($[Co^{II}(MeO\text{-}tery)_2]^{2+}$) were obtained by mixing solutions of the respective compounds, followed by precipitation of the resulting composites in which the counter anion of the $[Co^{II}(MeO\text{-}tery)_2]^{2+}$ was replaced. Elemental analysis indicated the presence of water molecules in the final products: **15**/$[Co^{II}(MeO\text{-}terpy)_2]\cdot H_2O$, **16**/$[Co^{II}(MeO\text{-}terpy)_2]\cdot 3H_2O$ and **17**/$[Co^{II}(MeO\text{-}terpy)_2]\cdot 4H_2O$.

The morphologies of the supramolecular structures generated by dispersions of these complexes were subsequently assessed by TEM. The TEM image of **15**/$[Co^{II}(MeO\text{-}terpy)_2]$ following transfer to a carbon-coated Cu grid exhibited

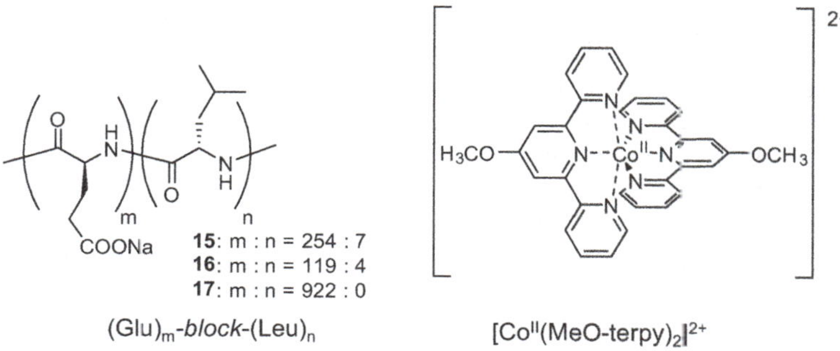

Fig. 9.17 Chemical structures of diblock copolypeptide amphiphiles **15** and **16**, the polypeptide **17** and the cobalt(II) terpyridine complex $[Co^{II}(MeO\text{-}terpy)_2]^{2+}$

rectangular structures with widths of 700 nm–6 μm, while **16**/$[Co^{II}(MeO\text{-}terpy)_2]$ showed nanostructures with widths ranging from 500 nm to 2 μm.

High-resolution scanning transmission electron microscopy coupled with energy-dispersive X-ray spectroscopy (HR-STEM EDX) also confirmed that the composites consisted of cobalt(II) complexes and polymers. Figure 9.18 presents STEM EDX mapping results (Fig. 9.18b, Co; Fig. 9.18c, O) for a nanocomposite of **15** within the boxed area indicated in Fig. 9.18a. These data confirm the formation of nanostructures in which the cobalt complex and the amphiphile are evenly matched. Composite **17**/$[Co^{II}(MeO\text{-}terpy)_2]$ did not show a specific structure in TEM observations, indicating that combinations of **17** and metal complexes did not form a supramolecular structure. Samples of the composites in water were freeze-dried and wide-angle X-ray scattering (WAXS) analysis of the powdered composites was performed, during which no crystalline peaks were observed. This lack of peaks indicated the presence of an amorphous phase in the rectangular supramolecular structures. In addition, SAXS analysis showed the formation of nanostructures of various sizes, which was consistent with the nanostructures having widths of several hundred nm and several mm observed via TEM and DLS.

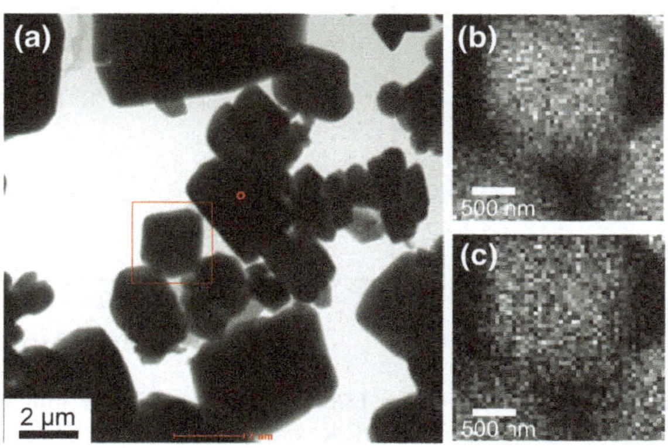

Fig. 9.18 **a** STEM image and **b** Co and **c** O STEM-EDX maps of **15**/$[Co^{II}(MeO\text{-}terpy)_2]$ within the *red square* shown in (**a**)

The variations in the magnetic susceptibilities of **15**/[Co^{II}(MeO-terpy)$_2$], **16**/[Co^{II}(MeO-terpy)$_2$], and **17**/[Co^{II}(MeO-terpy)$_2$] with temperature were examined. Composite **15**/[Co^{II}(MeO-terpy)$_2$] was found to exist in the high spin (HS) state at all temperatures and exhibited a $\chi_m T$ value within the range of 1.69–2.25 cm^3 K mol^{-1} over the temperature range of 5–300 K (the blue plot in Fig. 9.19). On further heating to 400 K, the $\chi_m T$ value was found to decrease at 337 K, consistent with the loss of water molecules (the red plot in Fig. 9.19). However, after annealing, the pre-heated compound displayed markedly different behavior. The $\chi_m T$ value gradually decreased from 1.60 cm^3 K mol^{-1} at 400 K to 1.05 cm^3 K mol^{-1} at 274 K, representing normal thermal SCO behavior (the red plot in Fig. 9.19). Upon further cooling, the $\chi_m T$ value increased abruptly below the $T_{1/2}\downarrow$ value of 260 K, to 1.96 at 222 K. On further cooling, the $\chi_m T$ varied between 1.70 and 2.20 cm^3 K mol^{-1} in the temperature range between 5 and 220 K. On further heating, the $\chi_m T$ value abruptly dropped ($T_{1/2}\uparrow$ = 345 K), showing the transition from HS to low spin (LS). Finally, the $\chi_m T$ value gradually increased between 361 and 400 K. The wide thermal hysteresis loop (ΔT = 85 K) near room temperature was maintained through successive thermal cycles. Thus, reverse spin transition was obtained from the composite in its solvated state based on intermolecular interactions among the metal complexes [89–93]. Composites **16**/[Co^{II}(MeO-terpy)$_2$] and **17**/[Co^{II}(MeOterpy)$_2$] also showed abnormal reverse spin transitions during heating and cooling cycles (5–300–5–400–5 K). Reversibility between the HS and LS states in the reverse spin transition, however, was dependent on the polymer employed. In particular, the diblock copolypeptide amphiphile **15** evidently possessed a suitable degree of polymerization and a balance between hydrophilic and hydrophobic portions, leading to perfect reversibility

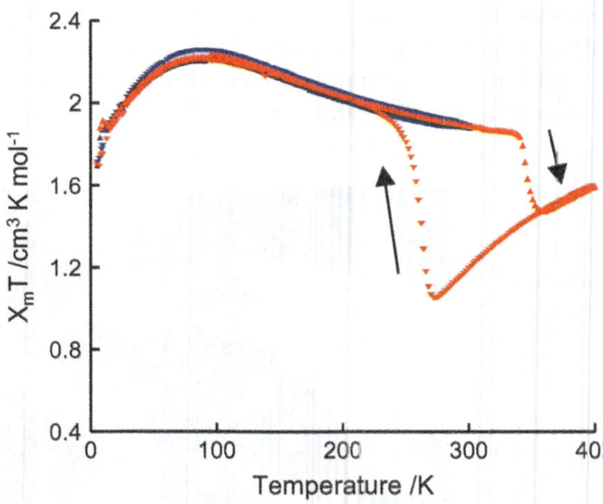

Fig. 9.19 $\chi_m T$ versus T for **15**/[Co^{II}(MeO-terpy)$_2$] on warming (↑) and cooling (↓). The *blue* and *red plots* indicate the first and second cycles, from 5 to 300 K and 400 K, respectively

between the HS and LS states to generate reverse spin transition. The $[Co^{II}(MeO\text{-}terpy)_2]^{2+}$ complex with BF_4^- anions is typically observed to undergo a gradual SCO transition centered around $T_{1/2}$ = 100–300 K [94]. During this process, water molecules in the solvent have been found to play an important role in the SCO behavior, due to either two-step SCO (involving an H_2O-solvated complex) or one-step SCO (non-solvated complex). In addition, reverse spin transition can be achieved using an organic-solvated metal complex in solvents such as acetone in association with a structural phase transition [94]. These prior results suggest that the SCO of the aqueous dispersions of solid composites of polymers and a cobalt(II) terpyridine complex should be accompanied by a transition between solvated and non-solvated phases, since the metal complex is dispersed in the amphiphilic polypeptide-induced nanostructure. Therefore, reversible reverse spin transition is thought to result from the balance between the amorphous diblock copolypeptide amphiphile and the loose packing of the cobalt(II) terpyridine complexes (Fig. 9.20).

This work demonstrated that composites consisting of a cobalt(II) terpyridine complex with diblock copolypeptide amphiphiles generate supramolecular structures in water. The formation of these nanostructures resulted in the evolution of morphologies ranging in size from subnanometer to several micrometers. Moreover, these supramolecular composites displayed reverse spin transition depending on the polypeptide structures. Changes in both the morphology and magnetic properties of these materials can be induced by variations in the temperature and solvated phase.

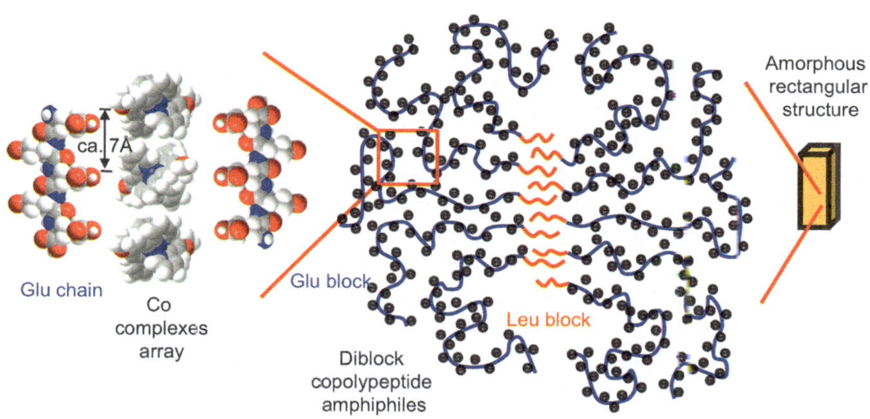

Fig. 9.20 Schematic illustration of the self-assembly of diblock copolypeptide amphiphile/$[Co^{II}(MeO\text{-}terpy)_2]$, showing how a rectangular structure is formed from a diblock copolypeptide amphiphile with an amorphous structure

9.5 Conclusion and Outlook

The integration of amphiphile-directed self-assembly and the chemistry of metal complexes has led to the development of dynamic supramolecular assemblies. Imparting an amphiphilic nature to supermolecules has been found to drive their hierarchical self-assembly. Weak interactions, such as hydrophobic interactions, hydrogen bonding, van der Waals interactions, and intermolecular interactions, play a pivotal role in the determination of the supramolecular architecture, distinct from observations reported during earlier supramolecular chemistry studies. The combinatorial supramolecular approach is also an effective means of developing functional nanomaterials. The combination of supramolecular nanoarchitectures and metastable engineering should allow new approaches to the design of nanomaterials. In addition to these nanocomposite approaches, synthetic biomolecules also provide powerful scaffolds to construct nanoarchitectures. Therefore, well-designed nanoarchitectures may be formed from natural lipids, peptides, sugars, and similar compounds. We have also recently reported that combinations of graphene and metal complexes can lead to specially tuned physical properties [95, 96]. These materials hierarchically self-assemble into mesoscopic architectures with special morphologies on the graphene. This approach is simple and much easier than those devised for previous supermolecule construction. Supramolecular assemblies having such functional components are anticipated to find increased applications in future.

Finally, it appears that the generation of metastable electronic structures based on self-assembly will be an important issue in nanochemistry. Studies of amphiphilic metal complexes have shown that discrete inorganic complexes form self-assembling nanoarchitectures in solution, the electronic structures of which are tunable with the help of amphiphilic molecular assemblies. The dynamic formation of various nanocomposite architectures and networks has been an unexpected result of some of this work. These unique self-assembling properties are not accessible from isolated coordination polymers or thermodynamically stable supramolecular chemistry. The growth of metastable networks in this manner is reminiscent of many central processes of life. Thus, we envisage that self-assembling nanoarchitecture will be applied to the design of biomimetic networks that display growth and self-organization in response to external stimuli. These self-assembling systems would contribute to the development of chemical learning systems. They may also provide an opportunity to design molecular (or supramolecular) machines, a new field of molecularly organized chemistry that integrates element blocks [97] to generate transduction, translation, amplification, chemical or physical outputs, dynamic control, and self-propagation.

Acknowledgments This work was financially supported in part by a grant-in-aid for Young Scientists (A) (No. 24685019) and a grant-in-aid for Scientific Research on Innovative Areas (new polymeric materials based on element blocks, #2401) (Nos. 25102547 and 15H00770).

References

1. Gispert JR (2008) Coordination chemistry. Wiley-VCH, Weinheim
2. Lippard SJ, Berg JM (1994) Principles of bioinorganic chemistry. University of Science Books
3. Anastassopoulou J, Theophanaides T (1995) The role of metal ions in biological systems and medicine. In: Kessissoglou DP (ed) Bioinorganic chemistry: an inorganic perspective of life. Nato science series C 459, pp 209–218
4. Lehn J-M (1995) Supramolecular chemistry-concept and perspectives. Wiley-VCH, Germany
5. Awod JL, Davies JED, MacNicol DM, Vögtle F, Lehn J-M (eds) (1996) Comprehensive supramolecular chemistry, vol. 9. Pergamon, Oxford
6. Whitesides GM, Mathias JP, Seto CT (1991) Molecular self-assembly and nanochemistry: a chemical strategy for the synthesis of nanostructures. Science 254(5036):1312–1319
7. Fujita M, Yazaki J, Ogura K (1990) Preparation of a macrocyclic polynuclear complex, [(en)Pd(4,4′-bpy)]$_4$(NO3)$_8$ (en = ethylenediamine, bpy = bipyridine), which recognizes an organic molecule in aqueous media. J Am Chem Soc 112(14):5645–5647
8. Kondo M, Yoshitomi T, Seki K, Matsuzaka H, Kitagawa S (1997) Three-dimensional framework with channeling cavities for small molecules: $\{[M_2(4,4'\text{-bpy})_3(NO_3)_4]\cdot xH_2O\}_n$ (M = Co, Ni, Zn). Angew Chem Int Ed 36(16):1725–1727
9. Li H, Eddaoudi M, Groy TL, Yaghi OM (1998) Establishing microporosity in open metal–organic frameworks: gas sorption isotherms for Zn(BDC) (BDC = 1,4-benzenedi carboxylate). J Am Chem Soc 120(33):8571–8572
10. Kimizuka N, Oda N, Kunitake T (1998) Supramolecular assemblies comprised of one-dimensional mixed valence platinum complex and anionic amphiphiles in organic media. Chem Lett 27(7):695–696
11. Kimizuka N (2000) Towards self-assembling inorganic molecular wires. Adv Mater 12 (19):1461–1463
12. Kimizuka N (2008) Self-assembly of supramolecular nanofibers, Adv Polym Sci 219:1–26 (and references therein)
13. Kurth DG, Lehmann P, Schütte M (2000) A route to hierarchical materials based on complexes of metallosupramolecular polyelectrolytes and amphiphiles. Proc Natl Acad Sci USA 97(11):5704–5707
14. Caseri WR, Chanzy HD, Feldman K, Fontana M, Smith P, Tervoort TA, Goossens JGP, Meijer EW, Schenning APHJ, Dolbnya IP, Debije MG, de Haas MP, Warman JM, van de Craats AM, Friend RH, Sirringhaus H, Stutzmann N (2003) "(Hot-)Water-Proof", semiconducting, platinum-based chain structures: processing, products, and properties. Adv Mater 15 (2):125–129
15. Niel V, Martinez-Agudo JM, Muñoz MC, Gaspar AB, Real JA (2001) Cooperative spin crossover behavior in cyanide-bridged Fe(II)-M(II) bimetallic 3D Hofmann-like networks (M=Ni, Pd, and Pt). Inorg Chem 40(16):3838–3839
16. Gaspar AB, Ksenofontov V, Seredyuk M, Gütlich P (2005) Multifunctionality in spin crossover materials. Coord Chem Rev 249(23):2661–2676
17. Kuroiwa K, Oda N, Kimizuka N (2006) Supramolecular solvatochromism. Effect of solvents on the self-assembly and charge transfer absorption characteristics of lipid-packaged, linear mixed valence platinum complexes. Sci Tech Adv Mater 7(7):629–634
18. Kuroiwa K, Shibata T, Takada A, Nemoto N, Kimizuka N (2004) Heat-set gel-like networks of lipophilic Co(II) triazole complexes in organic media and their thermochromic structural transitions. J Am Chem Soc 126(7):2016–2021
19. Kuroiwa K, Kimizuka N (2010) Electrochemically controlled self-assembly of lipophilic Fe^{II} 1,2,4-triazole complexes in organic media. Chem Lett 39(7):790–791
20. Kuroiwa K, Shibata T, Sasaki S, Ohba M, Takahara A, Kunitake T, Kimizuka N (2006) Supramolecular control of spin crossover phenomena in lipophilic Fe(II) 1,2,4-triazole complexes. J Polym Sci A: Polym Chem 44(17):5192–5202

21. Kume S, Kuroiwa K, Kimizuka N (2006) Photo-responsive molecular wires of Fe(II) triazole complexes in organic media and light-induced morphological transformations. Chem Commun 42(23):2442–2444
22. Matsukizono H, Kuroiwa K, Kimizuka N (2008) lipid-packaged linear iron(II) triazole complexes in solution: controlled spin conversion via solvophobic self-assembly. J Am Chem Soc 130(17):5622–5623
23. Matsukizono H, Kuroiwa K, Kimizuka N (2008) Self-assembly-directed spin conversion of iron(II) 1,2,4-triazole complexes in solution and their effect on photorelaxation processes of fluorescent counter ions. Chem Lett 37(4):446–447
24. Kuroiwa K, Kikuchi H, Kimizuka N (2010) Spin crossover characteristics of nanofibrous Fe^{II}-1,2,4-triazole complexes in liquid crystals. Chem Commun 46(8):1229–1231
25. Noguchi T, Chikara C, Kuroiwa K, Kaneko K, Kimizuka N (2011) Controlled morphology and photoreduction characteristics of polyoxometalate(POM)/lipid complexes and the effect of hydrogen bonding at molecular interfaces. Chem Commun 47(22):6455–6457
26. Kuroiwa K, Kimizuka N (2013) Self-assembly and functionalization of lipophilic metal-triazole complexes in various media. Polymer J 45(4):384–390
27. Kuroiwa K, Yoshida M, Masaoka S, Kaneko K, Sakai K, Kimizuka N (2012) Self-assembly of tubular microstructures from mixed-valence metal complexes and their reversible transformation via external stimuli. Angew Chem Int Ed 51(3):656–659
28. Kuroiwa K (2012) Dynamic self-assembly from mixed-valence metal complexes and their reversible transformations by external stimuli. Kobunshi Ronbunshu 69(8):485–492
29. Kuroiwa K, Higuma C, Shimogawa Y, Hachisako H, Sakurai S (2014) Development of amphiphilic *N*-isopropylacrylamide oligomers and polymers, and their composites with metal ions. Kobunshi Ronbunshu 71(10):457–466
30. Kuroiwa K, Koga Y, Ishimaru Y, Nakashima T, Hachisako H, Sakurai S (2016) Morphological control of hybrid amphiphilic poly(*N*-isopropylacrylamide)/metal cyanide complexes. Polymer J 48(6):729–739
31. Kuroiwa K, Masaki Y, Koga Y, Deming TJ (2013) Self-assembly of discrete metal complexes in aqueous solution via block copolypeptide amphiphiles. Int J Mol Sci 14(1):2022–2035
32. Arie T, Masaki Y, Kuroiwa K (2014) Synthesis of amphiphilic polypeptide and their composites of luminescent metal complexes. Polym Prep Jpn 3Pc005 at the 53rd SPSJ Symposium on Macromoeules, Nagasaki University, Nagasaki, 24–26 Sept 2014
33. Kuroiwa K, Arie T, Sakurai S, Hayami S, Deming TJ (2015) Supramolecular control of reverse spin transitions in cobalt(II) terpyridine complexes with diblock copolypeptide amphiphiles. J Mater Chem C 3(30):7779–7783
34. Kunitake T, Okahata Y (1977) A totally synthetic bilayer membrane. J Am Chem Soc 99 (11):3860–3861
35. Kunitake T (1992) Synthetic bilayer membranes: molecular design, self-organization, and application. Angew Chem Int Ed 31(16):709–726 (and references therein)
36. Okamoto H, Yamashita M (1998) Solitons, polarons, and excitons in quasi-one-dimensional halogen-bridged transition metal compounds. Bull Chem Soc Jpn 71(9):2023–2039
37. Wada Y, Mitani T, Yamashita M, Koda T (1985) Charge transfer exciton in halogen-bridged mixed-valent Pt and Pd complexes: analysis based on the Peierls-Hubbard model. J Phys Soc Jpn 54(8):3143–3153
38. Toriumi K, Wada Y, Mitani T, Bandow S, Yamashita M, Fujii Y (1989) Synthesis and crystal structure of a novel one-dimensional Halogen-Bridged Ni^{III}-X-Ni^{III} compound, $\{[Ni(R,R\text{-}chxn)_2Br]Br_2\}_\infty$. J Am Chem Soc 111(6):2341–2342
39. Lent CS (2000) Bypassing the transistor paradigm. Science 288(5471):1597–1599
40. Qi H, Sharma S, Li Z, Snider GL, Orlov AO, Lent CS, Fehlner TP (2003) Molecular quantum cellular automata cells. Electric field driven switching of a silicon surface bound array of vertically oriented two-dot molecular quantum cellular automata. J Am Chem Soc 125 (49):15250–15259

41. Lee CS, Kimizuka N (2002) Solvatochromic nanowires self-assembled from cationic, chloro-bridged linear platinum complexes and anionic amphiphiles. Chem Lett 31(12):1252–1253
42. Lee CS, Hatanaka Y, Kimizuka N (2002) Effect of lipid-packaging on the charge transfer characteristics of one-dimensional. Mixed-valence platinum complexes. Int J Nanosci 1(5–6):391–395
43. Lee CS, Kimizuka N (2002) Pillared honeycomb nano-architectures formed on solid surfaces by the self-assembly of lipid-packaged one-dimensional Pt complexes. Proc Natl Acad Sci 99 (8):4922–4926
44. Heskins M, Guillet JE, James E (1968) Solution properties of poly(*N*-isopropylacrylamide). J Macromol Sci Chem 2(8):1441–1455
45. Schild HG (1992) Poly(*N*-isopropylacrylamide): experiment, theory and application. Prog Polym Sci 17(2):163–249
46. Yusa S-i, Yamago S, Sugahara M, Morikawa S, Yamamoto T, Morishima Y (2007) Thermo-responsive diblock copolymers of poly(*N*-isopropylacrylamide) and poly(*N*-vinyl-2-pyrroridone) synthesized via organotellurium-mediated controlled radical polymerization (TERP). Macromolecules 40(16):5907–5915
47. Shan J, Zhao Y, Granqvist N, Tenhu H (2009) Thermoresponsive properties of *N*-isopropylacrylamide oligomer brushes grafted to gold nanoparticles: effects of molar mass and gold core size. Macromolecules 42(7):2696–2701
48. Nagai A, Yoshii R, Otsuka T, Kokado K, Chujo Y (2010) BODIPY-based chain transfer agent: reversibly thermoswitchable luminescent gold nanoparticle stabilized by BODIPY-terminated water-soluble polymer. Langmuir 26(19):15644–15649
49. Smith AE, Xu X, Savin DA, McCormick CL (2010) Reversible gold "locked" synthetic vesicles derived from stimuli-responsive diblock copolymers. Polym Chem 1(5):628–630
50. Lian X, Jin J, Tian J, Zhao H (2010) Thermoresponsive nanohydrogels cross-linked by gold nanoparticles. ACS Appl Mater Interfaces 2(8):2261–2268
51. Liang M, Lin I-C, Whittaker MR, Minchin RF, Monteiro MJ, Toth I (2010) Cellular uptake of densely packed polymer coatings on gold nanoparticles. ACS Nano 4(1):403–413
52. Liu Y, Tu W, Cao D (2010) Synthesis of gold nanoparticles coated with polystyrene-*block*-poly(*N*-isopropylacrylamide) and their thermoresponsive ultraviolet-visible absorbance. Ind Eng Chem Res 49(6):2707–2715
53. Sistach S, Beija M, Rahal V, Brulet A, Marty J-D, Destarac M, Mingotaud C (2010) Thermoresponsive amphiphilic diblock copolymers synthesized by MADIX/RAFT: properties in aqueous solutions and use for the preparation and stabilization of gold nanoparticles. Chem Mater 22(12):3712–3724
54. Ebeling B, Vana P (2013) RAFT-polymers with single and multiple trithiocarbonate groups as uniform gold-nanoparticle coatings. Macromolecules 46(12):4862–4871
55. Li W, Cai X, Kim C, Sun G, Zhang Y, Deng R, Yang M, Chen J, Achilefu S, Wang LV, Xia Y (2011) Gold nanocages covered with thermally-responsive polymers for controlled release by high-intensity focused ultrasound. Nanoscale 3(4):1724–1730
56. Sun X–L, He W–D, Li J, He N, Han S–C, Li L–Y (2008) Preparation of polypyrrole-graft-poly(*N*-isopropylacrylamide)/silver nanocomposites from pyrrolyl-capped macromonomer by $AgNO_3$ and their stimuli responsibility of light emission. J Polym Sci A: Polym Chem 46(20):6950–6960
57. Li Q, Zhang L, Bai L, Zhang Z, Zhu J, Zhou N, Cheng Z, Zhu X (2011) Multistimuli-responsive hybrid nanoparticles with magnetic core and thermoresponsive fluorescence-labeled shell via surface-initiated RAFT polymerization. Soft Matter 7 (15):6958–6966
58. Wu X, He X, Zhong L, Lin S, Wang D, Zhu X, Yan D (2011) Water-soluble dendritic-linear triblock copolymer-modified magnetic nanoparticles: preparation, characterization and drug release properties. J Mater Chem 21(35):13611–13620

59. Wang H, Luo W, Chen J (2012) Fabrication and characterization of thermoresponsive Fe_3O_4@PNIPAM hybrid nanomaterials by surface-initiated RAFT polymerization. J Mater Sci 47(16):5918–5925
60. Budgin AM, Kabachii YA, Shifrina ZB, Valetsky PM, Kochev SS, Stein BD, Malyutin A, Bronstein LM (2012) Functionalization of magnetic nanoparticles with amphiphilic block copolymers: self-assembled thermoresponsive submicrometer particles. Langmuir 28 (9):4142–4151
61. You Y-Z, Kalebaila KK, Brock SL, Oupický D (2008) Temperature-controlled uptake and release in PNIPAM-modified porous silica nanoparticles. Chem Mater 20(10):3354–3359
62. Chen J, Liu M, Chen C, Gong H, Gao C (2011) Synthesis and characterization of silica nanoparticles with well-defined thermoresponsive PNIPAM via a combination of RAFT and click chemistry. ACS Appl Mater Interfaces 3(8):3215–3223
63. Xu L, Pan J, Xia Q, Shi F, Ji Dai, Wei X, Yan Y (2012) Composites of silica and molecularly imprinted polymers for degradation of sulfadiazine. J Phys Chem C 116(48):25309–25318
64. Masuda T, Yamamoto S-i, Moriya O, Kashio M, Sugizaki T (2007) Preparation of amphiphilic polysilsesquioxane by grafting of block copolymer of acrylamide monomers. Polymer J 39(3):220–229
65. Zhang W, Liu L, Zhuang X, Li X, Bai J, Chen Y (2008) Synthesis and self-assembly of tadpole-shaped organic/inorganic hybrid poly(*N*-isopropylacrylamide) containing polyhedral oligomeric silsesquioxane via RAFT polymerization. J Polym Sci A: Polym Chem 46 (21):7049–7061
66. Wang L, Zeng K, Zheng S (2011) Hepta(3,3,3-trifluoropropyl) polyhedral oligomeric silsesquioxane-capped poly(*N*-isopropylacrylamide) telechelics: synthesis and behavior of physical hydrogels. ACS Appl Mater Interfaces 3(3):898–909
67. Zhang W, Wang S, Li X, Yuan J, Wang S (2012) Organic/inorganic hybrid star-shaped block copolymers of poly(l-lactide) and poly(*N*-isopropylacrylamide) with a polyhedral oligomeric silsesquioxane core: synthesis and self-assembly. Eur Polym J 48(4):720–729
68. Lehto J, Vaaramaa K, Vesterinen E, Tenhu H (1998) Uptake of zinc, nickel, and chromium by *N*-isopropyl acrylamide polymer gels. J Appl Polym Sci 68(3):355–362
69. Zhang L-J, Cao L–Q, Wang X–H, Wang J–D (2012) Preparation of the thermosensitive metal ion imprinted polymer in supercritical carbon dioxide: application in the selective recognition of copper (II). Polym Adv Tech 23(8):1174–1180
70. Irie M, Misumi Y, Tanaka T (1993) Stimuli-responsive polymers: chemical induced reversible phase separation of an aqueous solution of poly(*N*-isopropylacrylamide) with pendent crown ether groups. Polymer 34(21):4531–4535
71. Kuckling D, Pareek P (2003) Synthesis of transition-metal-ion-selective poly(*N*-isopropylacrylamide) hydrogels by the incorporation of an aza crown ether. J Polym Sci A: Polym Chem 41(11):1594–1602
72. Chayama K, Morita Y, Iwatsuki S (2010) Thermosensitive Gels incorporating polythioether units for the selective extraction of class b metal ions. J Chromatogr A 1217(43):6785–6790
73. Liu T, Hu J, Yin J, Zhang Y, Li C, Liu S (2009) Enhancing detection sensitivity of responsive microgel-based Cu(II) chemosensors via thermo-induced volume phase transitions. Chem Mater 21(14):3439–3446
74. Chiper M, Fournier D, Hoogenboom R, Schubert US (2008) Thermosensitive and switchable terpyridine-functionalized metallo-supramolecular poly(*N*-isopropylacrylamide). Macromol Rapid Commun 29(20):1640–1647
75. Qin L, He X–W, Zhang W, Li W–Y, Zhang Y–K (2009) Macroporous thermosensitive imprinted hydrogel for recognition of protein by metal coordinate interaction. Anal Chem 81 (17):7206–7216
76. Lever ABP (1984) Inorganic electronic spectroscopy. Elsevier, Amsterdam
77. Nag JK, Pal S, Sinha C (2005) Synthesis and characterization of cobalt(II), nickel(II), copper (II), palladium(II) and dioxouranium(VI) complexes of the antipyrine Schiff base of 3-formylsalicylic acid. Transit Metal Chem 30(5):523–526

78. Pochan DJ, Pakstis L, Ozbas B, Nowak AP, Deming TJ (2002) SANS and cryo-TEM study of self-assembled diblock copolypeptide hydrogels with rich nano-through microscale morphology. Macromolecules 35(14):5358–5360
79. Novak AP, Breedveld V, Pakstis L, Ozbas B, Pine DJ, Pochan D, Deming TJ (2002) Rapidly recovering hydrogel scaffolds from self-assembling diblock copolypeptide amphiphiles. Nature 417(6887):424–428
80. Deming TJ (2002) Methodologies for preparation of synthetic block copolypeptides: materials with future promise in drug deliverly. Adv Drug Deliv Rev 54(8):1145–1155
81. Breedveld V, Nowak AP, Sato J, Deming TJ, Pine DJ (2004) Rheology of block copolypeptide solutions: hydrogels with tunable properties. Macromolecules 37(10):3943–3953
82. Deming TJ (2005) Polypeptide hydrogels via a unique assembly mechanism. Soft Matter 1 (1):28–35
83. Holowka EP, Sun VZ, Kamei DT, Deming TJ (2007) Polyarginine segments in block copolypeptides drive both vesicular assembly and intracellular delivery. Nat Mater 6(1):52–57
84. Yang C-Y, Song B, Ao Y, Nowak AP, Abelowitz RB, Korsak RA, Havton LA, Deming TJ, Sofroniew MV (2009) Biocompatibility of amphiphilic diblock copolypeptide hydrogels in the central nervous system. Biomaterials 30(15):2881–2898
85. Tomczak MM, Glawe DD, Drummy LF, Lawrence CG, Stone MO, Perry CC, Pochan DJ, Deming TJ, Naik RR (2005) Polypeptide-templated synthesis of hexagonal silica platelets. J Am Chem Soc 127(36):12577–12582
86. Bellomo EG, Deming TJ (2006) Monoliths of aligned silica-polypeptide hexagonal platelets. J Am Chem Soc 128(7):2276–2279
87. Higuchi T, Tajima A, Yabu H, Shimomura M (2008) Spontaneous formation of polymer nanoparticles with inner micro-phase separation structures. Soft Matter 4(6):1302–1305
88. Yabu H, Koike K, Motoyoshi K, Higuchi T, Shimomura M (2010) A novel route for fabricating metal-polymer composite nanoparticles with phase-separated structures. Macromol Rapid Commun 31(14):1267–1271
89. Hayami S, Shigeyoshi Y, Akita M, Inoue K, Kato K, Osaka K, Takata M, Kawajiri R, Mitani R, Maeda Y (2005) Reverse spin transition triggered by a structural phase transition. Angew Chem Int Ed 44(31):4899–4903
90. Hayami S, Moriyama R, Shuto A, Maeda Y, Ohta K, Inoue K (2007) Spin transition at the mesophase transition temperature in a cobalt(II) compound with branched alkyl chains. Inorg Chem 46(19):7692–7694
91. Hayami S, Murata K, Urakami D, Kojima Y, Akita M, Inoue K (2008) Dynamic structural conversion in a spin-crossover cobalt(II) compound with long alkyl chains. Chem Commun 48:6510–6512
92. Hayami S, Kato K, Komatsu Y, Fuyuhiro A, Ohba M (2011) Unique spin transition and wide thermal hysteresis loop for a cobalt(II) compound with long alkyl chain. Dalton Trans 40 (10):2167–2169
93. Komatsu Y, Kato K, Yamamoto Y, Kamihata H, Lee YH, Akita F, Kawata S, Hayami S (2012) Spin-crossover behaviors based on intermolecular interactions for cobalt(II) complexes with long alkyl chains. Eur J Inorg Chem 16:2769–2775
94. Hayami S, Nakaya M, Ohmagari H, Alao AS, Nakamura M, Ohtani R, Yamaguchi R, Kuroda-Sowa T, Clegg JK (2015) Spin crossover behaviors in solvated cobalt(II) compounds. Dalton Trans 44(20):9345–9348
95. Karim MR, Shinoda H, Nakai M, Hatakeyama K, Kamihata H, Matsui T, Taniguchi T, Koinuma M, Kuroiwa K, Kurmoo M, Matsumoto Y, Hayami S (2013) Electrical conductivity and ferromagnetism in a reduced graphene-metal oxide hybrid. Adv Funct Mater 23(3): 323–332
96. Murashima Y, Karim MR, Saigo N, Takehira H, Ohtani R, Nakamura M, Koinuma M, Lindoy LF, Kuroiwa K, Hayami S (2015) Graphene oxide and reduced graphene oxide hybrids with spin crossover iron(III) complexes. Inorg Chem Front 10(2):886–892
97. Chujo Y, Tanaka K (2015) New polymeric materials based on element-blocks. Bull Chem Soc Jpn 88(5):633–643

Chapter 10
Langmuir and Langmuir–Blodgett Monolayers Having Photo-Responsibilities

Ken-ichi Iimura

Abstract In this chapter, after a brief introduction of the advantage in Langmuir–Blodgett (LB) technique and its applications, recent development of our research on photocurrent generation with LB films of amphiphilic helicene derivatives is addressed. The controllability of film structures using a Langmuir trough, in addition to the variety of available compounds, has been utilized to induce and/or maximize photo-response functions in the ultrathin film systems. Among them, the photo-induced electron transfer has attracted continuous interest. Helicenes are highly π-conjugated, polyaromatic helical compounds receiving much attention for applications especially in the optoelectrical field. Amphiphilic oxa[9]helicene derivatives were newly synthesized and used for preparation of monolayers by the LB technique. The molecules formed very homogeneous monolayers in which the hydrophobic aromatic planes of helicene moiety oriented vertically to the monolayer plane and the hydrophilic poly(ethylene oxide) (EO) chain was compactly packed with bended/twisted conformation. The photocurrent generation was confirmed for monolayers deposited on FTO substrates, and the current density was dependent on the number of EO unit.

Keywords Langmuir monolayer · Langmuir–Blodgett (LB) films · Photocurrent · Helicene · Organic semiconductor

10.1 Introduction

Langmuir–Blodgett (LB) technique is a very unique and useful method to prepare organized thin films of amphiphilic compounds on solid surfaces [1, 2]. In this technique, amphiphilic molecules are spread from a dilute solution with a volatile solvent on the water surface, and a resultant monolayer of amphiphile, referred to as

---

K. Iimura (✉)
Department of Material and Environmental Chemistry, Graduate School of Engineering, Utsunomiya University, Utsunomiya, Japan
e-mail: emlak@cc.utsunomiya-u.ac.jp

T. Kawai and M. Hashizume (eds.), *Stimuli-Responsive Interfaces*,
DOI 10.1007/978-981-10-2463-4_10

spreading monolayer or Langmuir monolayer, is laterally compressed by barriers with monitoring surface pressure. A measurement of surface pressure (π) against molecular area (A) at a constant temperature gives a so-called π − A isotherm, and from it one can know fundamental properties of the monolayer including a two-dimensional phase and its transition. The monolayer is deposited onto a solid substrate at a desired surface pressure (or molecular area) by a vertical dipping method to prepare a LB film. Other deposition techniques such as Langmuir–Schaefer (LS) and horizontal scooping-up methods are also utilizable for preparation of multilayers and monolayers on solid supports.

Photo-responsibility in organic thin films has attracted considerable interest for both fundamental researches and potential applications for decades. In this trend, Langmuir monolayers and LB films have played an important role as ideal organized ultrathin film systems with defined molecular orientation, density, arrangement, thickness, and so on. The controllability of such film structures by the lateral compression of monolayer with barriers and/or preparation conditions (e.g., subphase temperature, dissolved ions, deposition rate and method) is also very advantageous and favorable for studying photo-induced functions of thin film systems.

There are many achievements in previous researches on photo-induced functions with the LB film systems. A representative example is photoisomerization of azopolymers in LB films. The azobenzene side chains undergo a *cis-trans* isomerization due to irradiation of light of appropriate wavelength, and this change enabled to control liquid-crystal alignment [3]. The corresponding phenomenon to chromophores induces dichroism and birefringence in films, which can be applied to optical storage [4, 5]. The surface morphological change due to the photoisomerization in LB films is also of considerable interest [6, 7].

10.2 Photocurrent Generation in LB Films

Photocurrent generation from LB films of organic dyes or semiconductors is also a subject of continuous attention. Researches have been motivated by fundamental interest for understanding and/or mimicking the biological photo-synthesis system and technological applications for development of photoelectric devices and photovoltaic cells [8].

Since pioneering works by Kuhn with amphiphilic dyes [9, 10], a lot of researches with a variety of compounds have reported photocurrent generation with LB films [11, 12]. Fullerenes have excellent electron-acceptor properties and thus frequently used in charge separation systems. However, it is difficult to prepare high-quality LB films with pristine fullerene because of intrinsic hydrophobicity. The introduction of hydrophilic groups to the fullerene molecule solved the problem: Some fullerene derivatives with hydrophilic groups formed stable and high-quality monolayers [13–15]. Amphiphilic donor–acceptor dyads are also unique and useful compounds, in which the porphyrin donor and the fullerene acceptor are covalently double-bridged to fix their spatial arrangement [16, 17].

10.3 LB Films of Oxa[9]Helicenes

Helicenes are polycyclic aromatic semiconductor compounds with *ortho*-fused benzene rings. The molecules are helical in shape owing to the steric hindrance between the rings, and this structural feature with highly conjugated π-electron system causes unique optical and electronic properties, and self-assembly behavior. Because of them, there has been a continuous interest in potential applications of helicenes in the fields of asymmetric molecular recognition [18–21], nonlinear optics [22, 23], circular polarized luminescence in display devices [24–26], organic thin-film transistors [27, 28], and so on. In particular, a recent interest of helicene-based science focuses on the use in optical and photoelectrochemical applications. For this purpose, it is important to find a suitable approach which enables fabrication of homogeneous organized thin films and possibly fine tuning of film structures such as thickness, molecular density, orientation, and arrangement.

A helicene backbone is intrinsically hydrophobic and less soluble in organic solvents, and furthermore, the π-conjugated polyaromatic rings have a tendency to induce co-facial stacking between the molecules in both bulk and film leading to formation of macroscopic fibers. Hence, it is not easy to prepare homogeneous thin films with helicene compounds through wet processes. The introduction of an appropriate polar substituent to the helicene backbone is expected to be a valid way to increase solubility and to weaken the tendency of self-aggregation, if molecular functionality is not impaired due to the introduction. In our study, a hydrophilic poly(ethylene oxide) (EO) chain was introduced to oxa[9]helicene backbone (Fig. 10.1), and resultant derivatives were used for preparation of monomolecular films by the LB method [29, 30]. The chemical modification successfully provided high solubility in polar organic solvents including chloroform, and additionally amphiphilic characteristics to the resultant derivatives. The formation of

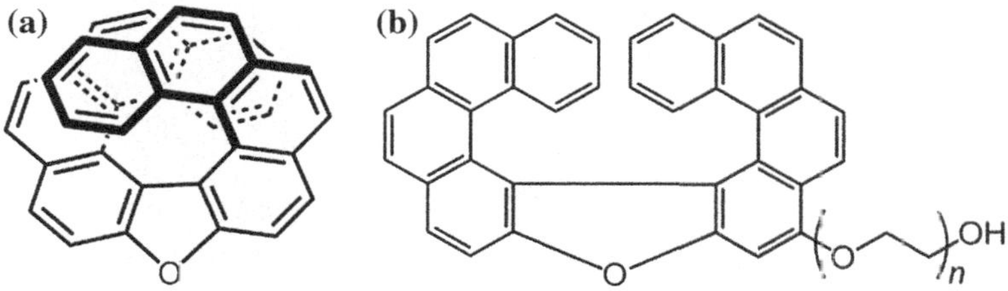

Fig. 10.1 Molecular structures of **a** oxa[9]helicene (O9H) in a stereoscopic display and **b** its derivative with poly(ethylene oxide) chain (*n* = 2; 2EO9H, *n* = 4; 4EO9H) used for preparation of Langmuir monolayers

three-dimensional aggregation was also prevented to form very homogeneous monolayers with photocurrent generation ability.

Surface pressure (π)—molecular area (A) isotherms for the oxa[9]helicene compounds are shown in Fig. 10.2, together with Brewster angle microscopic (BAM) images for O9H film and 2EO9H monolayers. The isotherm for O9H shows steep increase in surface pressure at around 20 Å^2/molec., and the BAM image taken before the surface pressure rise reveals a considerable number of bright spots coexistent with homogeneous islands. The molecular area of 20 Å^2/molec. is much smaller than that expected from a size of helicene ring, and the very high brightness of the spots indicates they are much thicker than a monolayer. On the other hand, for 2EO9H, the surface pressure monotonously increases upon compression up to a kink at about 25 mN/m. BAM observation confirms large two-dimensional islands before surface pressure rise (Fig. 10.2b) and their fusion during compression to give a homogeneous monolayer. For 4EO9H, the surface pressure behavior and surface morphological features are very similar to those for 2EO9H. These observations indicate that O9H molecules spontaneously form three-dimensional aggregates on the water surface after the spreading because of less amphiphilicity, whereas 2EO9H and 4EO9H work as water-insoluble amphiphiles with ability to form homogeneous monolayers.

The monolayers of 2EO9H and 4EO9H were deposited on silicon wafer, and molecular orientation and arrangement in the monolayers were evaluated by using X-ray reflectometry (XR) (Fig. 10.3). The model analysis was performed under assumption that the monolayers consist of two layers: a hydrophobic helicene layer exposed to air and a hydrophilic EO chain layer on SiO_2 of silicon wafer. As seen

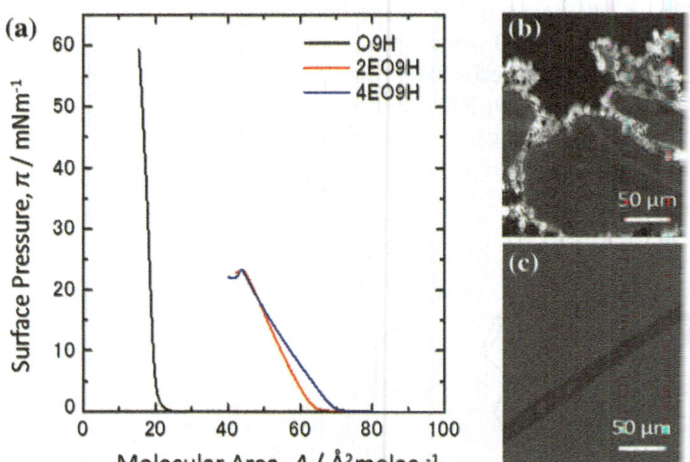

Fig. 10.2 **a** π-A isotherms for Langmuir monolayers of oxa[9]helicenes on the ultrapure surface at 24 °C, and **b**, **c** BAM images observed for the monolayers of **b** O9H and **c** 2EO9H at 28 and 75 Å^2/molec., respectively

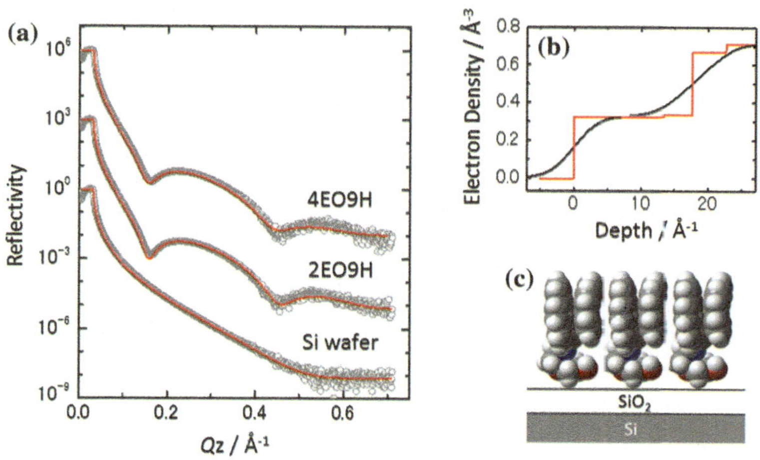

Fig. 10.3 **a** XR data (*circles*) and fitted curves (*solid lines*) for a bare silicon wafer, and 2EO9H and 4EO9H monolayers deposited on silicon wafers at 10 mN/m. The reflectivity scales of XR curves for the monolayers are shifted for clarity. **b** An electron density profile for the 2EO9H monolayer obtained from the fitting analysis of XR data. A *red line* is drawn by setting values of roughness between layers to zero, for display purpose. The monolayer/air interface is set depth of zero. **c** Schematic illustration of orientation/arrangement of 2EO9H molecules in the monolayer

from Fig. 10.3a, this assumption finally brought good agreement of calculated curves and XR data, and vertical electron density profiles were obtained (for example, the profile for the 2EO9H monolayer is shown in Fig. 10.3b). The estimated thicknesses of helicene layer are 13.5 and 13.4 Å for the 2EO9H and 4EO9H monolayers, respectively. These values are close to a diameter of disk-like helicene part rather than its thickness, indicating the hydrophobic aromatic planes of helicene moiety orient perpendicularly to the monolayer plane, as drawn schematically in Fig. 10.3c. On the other hand, the estimated thicknesses of hydrophilic EO layers are 4.2 and 4.9 Å for the 2EO9H and 4EO9H monolayers, respectively, which are much less than those expected for fully stretched chains and thus suggest that the EO chains are bended/twisted. It also should be noted that absorption peaks in UV/visible spectra for 2EO9H and 4EO9H in monolayers on quartz substrates are red-shifted by about 3–5 nm compared to those in chloroform solutions. The shift implies the π-stacking of the molecules in the monolayers [31–33]. However, considering the degree of shift, it is expected that the stacking interaction is not so strong compared to typical planar aromatic molecules, because of the helical structure of helicene molecules.

Helicene monolayers deposited on FTO-coated glass substrates showed photocurrent response (Fig. 10.4). The measurements were performed in a TEMPAXFloat® vessel by using a three-electrode system composed of a monolayer-deposited FTO substrate as working electrode, an aqueous reference electrode of Ag/Ag^+ (3.0 M NaClaq), and a platinum counter electrode, in a mixed

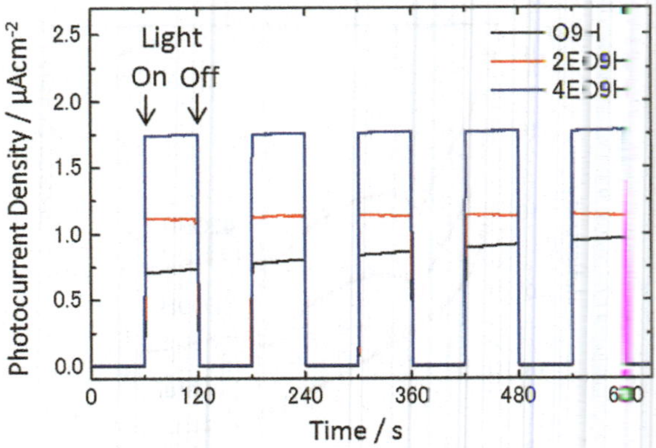

Fig. 10.4 Photocurrent response of the O9H film and the monolayers of 2EO9H and 4EO9H transferred at 10 mN/m on FTO substrates. The process of light irradiation and non-irradiation was repeated several times with 60-s intervals

electrolyte solution of 0.1 M methyl viologen, 0.1 M Na_2SO_4, and 0.1 M KH_2PO_4. The pH of the solution was maintained at 5.9. A xenon lamp was used to illuminate the samples under a constant distance from the lamp to the sample. All measurements were scanned at an applied voltage of zero.

The photocurrent was generated from all the samples when the light was irradiated. The photocurrent density observed for the 4EO9H monolayer was larger than that for the 2EO9H although the molecular density and orientation in both monolayers are almost the same according to the isotherm data and the XR analysis mentioned above. In contrast, the O9H film showed the lowest photocurrent density in spite of the largest molecular density among three samples, as known from the comparison of molecular areas at the deposition pressure in the isotherms. These results indicate that the photocurrent density is not simply proportional to the molecular density on the substrates. One plausible explanation for the difference in the photocurrent density is that the electron-donating nature of EO group increases electron density in the π-cloud of helicene moiety [34, 35], resulting in an increase of photocurrent. For the O9H film, in addition to this effect, a higher probability of excited-state quenching in the three-dimensional aggregates may also contribute to lowering the photocurrent density.

10.4 Future Perspective

As mentioned above, one of the advantages in the LB technique is capability of constructing highly ordered thin films with a defined molecular arrangement, orientation, and/or thickness. On the other hand, helicenes possess axial chirality due to their helical molecular structure despite the lack of chiral centers. Therefore, although we have used racemic compounds, the use of their enantiomers in the LB technique will enable to form films with selective response to the circularly polarized light, as a combined function of the chirality, the electrochemical property, and the orientation/arrangement of the molecules in the films.

References

1. Petty MC (1996) Langmuir-Blodgett films: an introduction. Cambridge University Press Cambridge
2. Iimura K, Kato T (2012) Langmuir-Blodgett (LB) film, in organized organic ultrathin films: fundamentals and applications. In: Ariga K. Wiley-VCH, pp 43–105
3. Seki T, Sakuragi M, Kawanishi Y, Suzuki Y, Tamaki T, Fukuda R, Ichimura K (1993) "Command surfaces" of Langmuir-Blodgett films. Photoregulations liquid crystal alignment by molecularly tailored surface azobenzene layers. Langmuir 9:211–218
4. Oliveira ON Jr, dos Santos Jr DS, Balogh DT, Zucolotto V, Mendonça CR (2005) Optical storage and surface-relief gratings in azobenzene-containing nanostructured films. Adv Colloid Interface 116:179–192
5. Ceridório LF, Balogh DT, Caseli L, Cardoso MR, Viitala T, Mendonça CR, Oliveira ON Jr (2010) Molecular-level interactions of an azopolymer and poly(dodecylmethacrylate) in mixed Langmuir and Langmuir–Blodgett films for optical storage. J Colloid Interface Sci 346:87–95
6. Matsumoto M, Miyazaki D, Tanaka M, Azumi R, Manda E, Kondo Y, Yoshino N, Tachibana H (1998) Reversible light-induced morphological change in Langmuir-Blodgett films. J Am Chem Soc 120:1479–1484
7. Haro M, Giner B, Gascón I, Royo FM, López MC (2007) Isomerization behavior of an azopolymer in terms of the Langmuir-Blodgett film thickness and the transference surface pressure. Macromolecules 40:2058–2069
8. Choi H-G, Oh B-K, Lee WH, Choi J-W (2001) Deposition behavior and photoelectrochemical characteristics of chlorophyll *a* Langmuir-Blodgett films. Biotechnol Bioprocess Eng 6:183–188
9. Kuhn H (1979) Synthetic molecular organizates. J Photochem 10:111–132
10. Kuhn H (1979) Electron transfer in monolayer assmblies. Pure Appl Chem 51:341–352
11. Zheng J, Li F, Huang C-H, T-t Liu, Zhao X-s YuX, Wu N (2001) Enhancement of second-harmonic response and photocurrent generation from a benzothiazolium styryl dye LB film through an interfacial self-assembly reaction. J Phys Chem B 105:3229–3234
12. Sakomura M, Fujihira M (2007) Langmuir-Blodgett (LB) films on electrodes, in encyclopedia of electrochemistry. In: Bard AJ, Stratmann M, Fujihira M, Rubinstein I, Rusling JF (eds) Modified electrodes. Wiley-VCH, pp 171–202
13. Maliszewskyj NC, Heiney PA, Jones DR, Strongin RM, Cichy MA (1993) Smith AB III (1993) Langmuir films of fullerene C_{60}, fullerene epoxide $C_{60}O$, and dihydrofulleroid $C_{61}H_2$. Langmuir 9:1439–1441
14. Ravaine S, Le Pecq F, Mingotaud C, Delhaes P, Wudl F, Patterson LK (1995) Synthesis, physical characterizations, and langmuir films of new methanofullerenes. J Phys Chem 99:9551–9557
15. Zhang W, Shi Y, Gan L, Wu N, Huang C, Wu D (1999) Formation and photoelectric properties of C_{60}EDTA-M^{n+} monolayer films. Langmuir 15:6921–6924
16. Vuorinen T, Kaunisto K, Tkachenko NV, Efimov A, Lemmetyinen H, Alekseev AS, Hosomizu K, Imahori H (2005) Photoinduced electron transfer in Langmuir-Blodgett monolayers of porphyrin-fullerene dyads. Langmuir 21:5383–5390
17. Vivo P, Vuorinen T, Chukharev V, Tolkki A, Kaunisto K, Ihalainen P, Peltonen J, Lemmetyinen H (2010) Multicomponent molecularly controlled Langmuir-Blodgett systems for organic photovoltaic applications. J Phys Chem C 114:8559–8567
18. Laarhoven WH, Prinsen WJC (1984) Carbohelicenes and heterohelicenes. Top Curr Chem 125:63–130
19. Meurer KP, Vögtle F (1985) Helical molecules in organic chemistry. Top Curr Chem 127:1–76
20. Rowan AE, Nolte RJM (1998) Helical molecular programming. Angew Chem Int Ed 37:63–68
21. Katz TJ (2000) Syntheses of functionalized and aggregating helical conjugated molecules. Angew Chem Int Ed 39:1921–1923

22. Murguly E, McDonald R, Branda NR (2000) Chiral discrimination in hydrogen-bonded[7] helicenes. Org Lett 2:3169–3172
23. Bossi A, Licandro E, Maiorana S, Rigamonti C, Righetto S, Stephenson GR, Spassova M, Botek E, Champagne B (2008) Theoretical and experimental investigation of electric field induced second harmonic generation in tetrathia[7]helicenes. J Phys Chem C 112:7900–7907
24. Weix DJ, Dreher SD, Katz TJ (2000) [5]HELOL phosphite: a helically grooved sensor of remote chirality. J Am Chem Soc 122:10027–10032
25. Phillips KES, Katz TJ, Jockusch S, Lovinger AJ, Turro NJ (2001) Synthesis and properties of an aggregating heterocyclic helicene. J Am Chem Soc 123:11899–11907
26. Field JE, Muller G, Riehl JP, Venkataraman D (2003) Circularly polarized luminescence from bridged triarylamine helicenes. J Am Chem Soc 125:11808–11809
27. Kim C, Marks TJ, Facchetti A, Schiavo M, Bossi A, Maiorana S, Licandro E, Todescato F, Toffanin S, Muccini M, Graiff C, Tiripicchio A (2009) Synthesis, characterization, and transistor response of tetrathia-[7]-helicene precursors and derivatives. Org Electron 10:1511–1520
28. Yang Y, Correa da Costa R, Fuchter MJ, Campbell AJ (2013) Circularly polarized light detection by a chiral organic semiconductor transistor. Nat Photonics 7:634–638
29. Miah MJ, Shahabuddin M, Karikomi M, Salim M, Nasuno E, Kato N, Iimura K (2016) Structural and photoelectrical characterization of thin films of a novel amphiphilic Oxa[9] helicene derivative. Bull Chem Soc Jpn 89:203–211
30. Miah MJ, Shahabuddin M, Kayes MN, Karikomi M, Nasuno E, Kato N, Iimura K (2016) Fabrication and characterization of molecular films of 11-Oxa[9]helicene and 9-Diethyleneglycoxy-11-oxa[9]helicene. Trans Mater Res Soc Jpn 41:151–154
31. Shane JY, Swager TM (1998) Porous shape persistent fluorescent polymer films: an approach to TNT sensory materials. J Am Chem Soc 120:5321–5322
32. Kim C, Marks TJ, Facchetti A, Schiavo M, Bossi A, Maiorana S, Licandro E, Todescato F, Toffanin S, Muccini M, Graiff C, Tiripicchio A (2009) Synthesis, characterization, and transistor response of Tetrathia-[7]-helicene precursors and derivatives. Org Electron 10:1511–1520
33. Sahasithiwat S, Mophuang T, Menbangpung L, Kamtonwong S, Sooksimuang T (2010) 3,12-Dimethoxy-7,8-dicyano-[5]helicene as a novel emissive material for organic light-emitting diode. Synth Met 160:1148–1152
34. Xu J, Nie G, Zhang S, Han X, Pu S, Shen L, Xiao O (2005) Electrosyntheses of Poly(2,3-benzofuran) films in boron trifluoride diethyl etherate containing poly(ethylene glycol). Eur Polym J 41:1654–1661
35. Rancic M, Trisovic N, Milcic M, Uscumlic G, Marinkovic A (2012) Substituent and solvent effects on intramolecular charge transfer of 5-arylidene-2, 4-thiazolidinediones. Spectrochim Acta Part A 86:500–507

Chapter 11
Stimuli-Responsive Polymer Micelles

Shin-ichi Yusa

Abstract Diblock copolymers (PNIPAM$_m$-PNVP$_n$) composed of poly(*N*-isopropyl acrylamide) (PNIPAM) and poly(*N*-vinyl-2-pyrrolidone) (PNVP) with well-defined block lengths were successfully prepared via organotellurium-mediated controlled radical polymerization (TERP). Thermo-responsive association properties in water of the diblock copolymers with different block lengths were characterized. All the diblock copolymers dissolve in water molecularly when the solution temperature is below an aggregation temperature (T_a) that is near a lower critical solution temperature (LCST) for the PNIPAM block. Comparing T_a between the two diblock copolymers of the same the PNIPAM block lengths (DP = 110), the diblock copolymer with a shorter PNVP block length tends to associate at a lower temperature. On the other hand, the association occurred at a higher temperature for the diblock copolymers with shorter PNIPAM block lengths. When the temperature was raised above T_a, PNIPAM$_m$-PNVP$_n$ formed aggregates due to hydrophobic interactions between the PNIPAM blocks. PNIPAM$_m$-PNVP$_n$ bounds to gold nanoparticles in water through coordination interaction of the PNVP block with Au. The polymer-coated gold nanoparticles indicated a temperature-dependent color change arising from a shift of the maximum wavelength of the plasmon band. PNIPAM$_m$-PNVP$_n$ was found to associate with fullerene (C_{60}) and thus C_{60} can be solubilized in water. The C_{60} complex formed a core–shell micelle-like aggregate comprising a C_{60}/PNVP hydrophobic core and a thermo-responsive PNIPAM shell. The hydrodynamic radius (R_h) of the C_{60} complex increased with increasing temperature, which was ascribed to the hydrophobic association between dehydrated PNIPAM shells above LCST. The generation of singlet oxygen by photosensitization by the C_{60} complex was confirmed from photo-oxidation of 9, 10-anthracenedipropionic acid. DNA was cleaved by visible light irradiation in the presence of the C_{60} complex. There may be a hope for a pharmaceutical application of the C_{60} complex to cancer photodynamic therapy (PDT).

S. Yusa (✉)
Department of Applied Chemistry, University of Hyogo,
2167 Shosha, 671-2280 Hyogo Himeji, Japan
e-mail: yusa@eng.u-hyogo.ac.jp

T. Kawai and M. Hashizume (eds.), *Stimuli-Responsive Interfaces*,
DOI 10.1007/978-981-10-2463-4_11

Keywords Organotellurium-mediated radical polymerization · Lower critical solution temperature · Gold nanoparticles · Photodynamic therapy · Fullerene

11.1 Introduction

Controlled radical polymerization of non-conjugated monomers, such as vinyl acetate, *N*-vinylformamide (NVF), and *N*-vinyl-2-pyrrolidone (NVP), is difficult because their propagating radicals apt to undergo various side reactions during the polymerization due to their high activity. It has been reported that organotellurium-mediated controlled radical polymerization (TERP) provides effective promoters for highly controlled radical polymerization for a wide range of monomers including non-conjugated ones [1–3]. Proposed mechanism for TERP is illustrated in Fig. 11.1. Condition for the polymerizations is the same as conventional radical polymerization except for the addition of an organotellurium compound. The mechanism includes a chain transfer between propagating radicals and chain transfer agents. There is a transition state for TERP [4, 5]. Side reactions of the propagating radicals are prevented, because the rate of the chain transfer is faster than that of the propagation. Hence, the polymerization proceeds in a living mechanism.

Poly(*N*-vinyl-2-pyrrolidone) (PNVP) has been widely used in personal-care, agricultural, and many other industrial products because of its low toxicity and desirable performance. PNVP is soluble in a large number of organic solvents, and it is also soluble in water. PNVP is a biocompatible material and thus used for various pharmaceutical applications [6]. PNVP shows a characteristic property of adhesion to various materials such as metals [7].

Water-soluble stimuli-responsive polymer micelles formed from amphiphilic diblock copolymers are actively studied as a nanometer-sized vehicle for drug delivery system (DDS) and other purposes [8]. These amphiphilic diblock copolymers composed of a permanently hydrophilic block and a stimuli-responsive (i.e., "smart") block that undergoes a change from water-soluble to a water-insoluble state by physical and chemical external stimuli such as temperature pH, and ionic strength. Thermo-responsive diblock copolymers that undergo

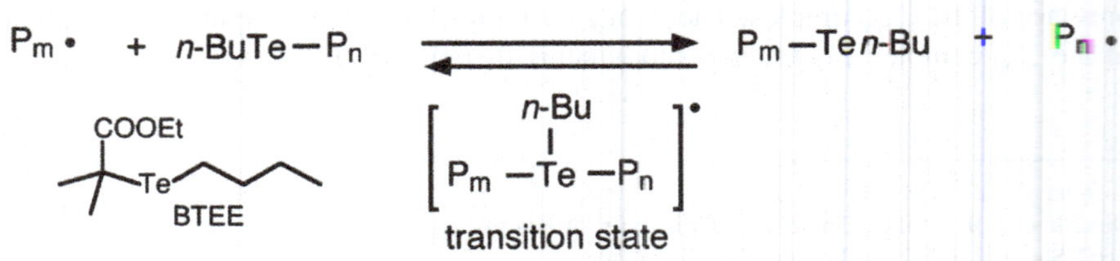

Fig. 11.1 Proposed mechanism for TERP and chemical structure of ethyl-2-methyl-2-butyltelanilpropionate (BTEE)

micelle formation and dissociation depending on the temperature are particularly attractive materials, because there is no need of any additive to the solution. Poly(*N*-isopropylacrylamide) (PNIPAM) shows a thermally reversible phase transition, i.e., a lower critical solution temperature (LCST), and it has been cited in the literature as a promising material for a number of biological applications including DDS [9]. There have been a number of reports dealing with PNIPAM containing thermo-responsive diblock copolymers [10]. Diblock copolymers (PNIPAM$_m$-PNVP$_n$) composed of thermo-responsive PNIPAM block and biocompatible PNVP block is a good candidate for a thermo-responsive drug delivery vehicle, because the diblock copolymer can form a micelle-like aggregate above LCST for the PNIPAM block with the biocompatible PNVP blocks forming an outer shell layer (Fig. 11.2).

PNVP is known to stabilize gold nanoparticles due to multiple coordination of the pendent amide residues with the surface of gold [11]. Gold nanoparticles may

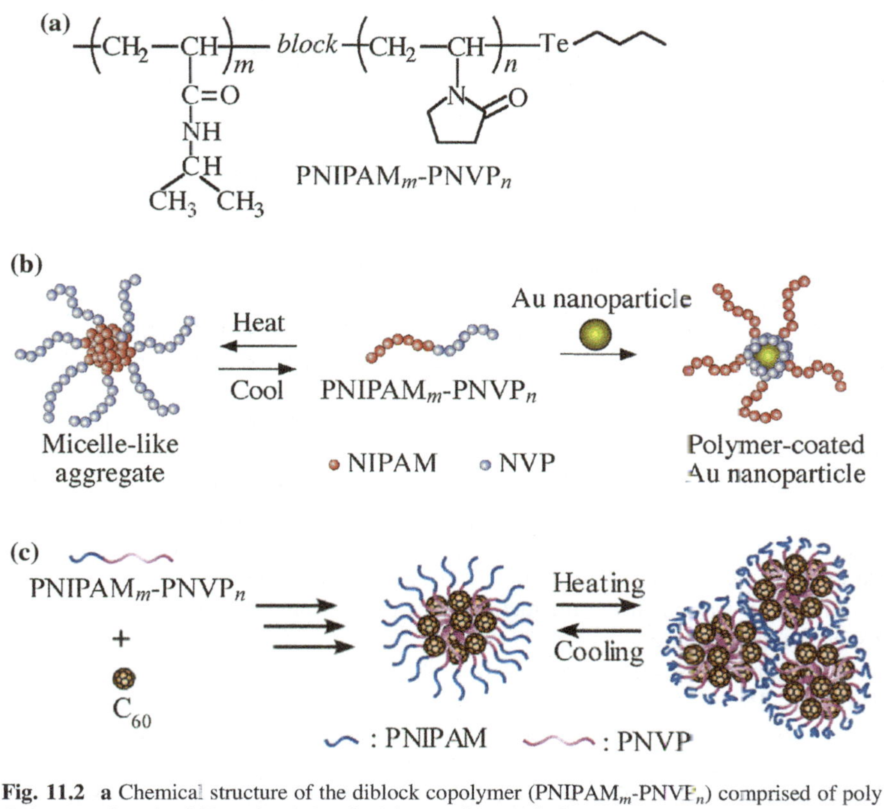

Fig. 11.2 **a** Chemical structure of the diblock copolymer (PNIPAM$_m$-PNVP$_n$) comprised of poly (*N*-isopropylacrylamide) and poly(*N*-vinyl-2-pyrrolidone). **b** Schematic illustration of the micellization of PNIPAM$_m$-PNVP$_n$ as a function of temperature and the formation of polymer-coated gold nanoparticle due to coordinated interaction between the PNVP block and gold nanoparticle. **c** Conceptual illustration of complexation of PNIPAM$_m$-PNVP$_n$ by association with C_{60} and thermo-responsive behavior of the C_{60} complex

be effectively coated with PNIPAM$_m$-PNVP$_n$ (Fig. 11.2b), and each of the polymer-coated gold nanoparticle may be separately dissolved in water when temperature is below LCST. However, at high temperatures above LCST, the gold nanoparticles may associate with each other due to the hydrophobic interactions between the dehydrated PNIPAM blocks. Visible absorption spectra of the surface plasmon band for gold nanoparticles are sensitive to their own size. Well-dispersed gold nanoparticles indicate an absorption band around 520 nm with a pink color due to the surface plasmon resonance. The association of the gold nanoparticles induces a color change of the solution from pink to bluish-purple, which may be applied to colloidal sensors [12].

The biological properties of fullerene (C_{60}) have attracted attention with regard to the development of novel pharmaceuticals [13]. C_{60} and its derivatives have been reported about their enzyme inhabitation [14], antiviral activity [15], DNA cleavage [16], radical quenching activity [17], and photodynamic therapy (PDT) [18]. PDT is a well-established anticancer therapy that kills cancer cells by photochemically generated cytotoxic reactive oxygen species using photosensitizer such as C_{60}, porphyrins, and phthalocyanine derivatives [19]. To protect harmful aftereffects i.e. PDT-induced photosensitivity disorder, the photosensitizer should selectively be accumulated in cancer cells. If C_{60} is utilized for a photosensitizer of PDT, C_{60} is to be solubilized in water and to be accumulated selectively in the target point. However, the biological application of C_{60} is limited due to its poor solubility in water [20]. To improve the solubility of C_{60} in water, several water-soluble derivatives have been synthesized [21]. Other solubilization methods for C_{60} have been explored using cyclodextrins [22], calixerenes [23], micelles [24], liposomes [25], and PNVP [26].

As C_{60} can be solubilized in water by association with PNVP, C_{60} may be bound to the PNVP block in PNIPAM$_m$-PNVP$_n$ in water at temperatures below LCST (Fig. 11.2c). In such a case, the PNVP block may become sufficiently hydrophobic for the block copolymer to form a micelle-like complex, an aggregate of the C_{60}-bound PNVP blocks forming a core and the hydrated PNIPAM blocks forming a shell. Thus, C_{60} can be solubilized in water by complexation with PNIPAM$_m$-PNVP$_n$. Upon heating to a temperature above LCST, the thermo-responsive outer shell made up of PNIPAM blocks collapse to form a hydrophobic layer surrounding the core. The C_{60} complex is expected to be applied to PDT, because C_{60} can be solubilized in water and the interactions between cell and the micelle can be controlled by temperature.

In the present work, we synthesized thermo-responsive diblock copolymers PNIPAM$_m$-PNVP$_n$ by TERP. The thermo-responsive properties were characterized. Furthermore, we found that gold nanoparticles were stabilized with PNIPAM$_m$-PNVP$_n$ and that the polymer-coated gold nanoparticles exhibited a color change due to their association and dissociation responding to a temperature change. The C_{60} complex was prepared, and its thermo-responsive behavior in water was studied. In addition, we describe some results of dye bleaching by singlet oxygen generated by photoirradiation of the C_{60} complex in water. The cleavage of DNA photosensitized by the C_{60} complex was also reported.

11.2 Synthesis of Thermo-Responsive Diblock Copolymers by TERP

First, we performed solution polymerization of NIPAM in *N, N*-dimethylformamide at 60 °C by TERP. At different time intervals, a small portion of the polymerization mixture was sampled out with a syringe and subjected to measurement of molecular weight by gel permeation chromatography (GPC) and conversion by ^{1}H NMR. Figure 11.3a shows time–conversion plots for the solution polymerization. After 2 h, a monomer conversion of ca. 100 % is reached. Figure 11.3b shows conversion–number-average molecular weight (M_n) and conversion–molecular weight distribution (M_w/M_n) plots. M_n for NIPAM by TERP increases linearly with increasing conversion. All values of M_n are close to the theoretical values predicted for a living mechanism. The values of M_w/M_n are reasonably small, suggesting that the polymerization of NIPAM by TERP progressed in accordance with a living mechanism.

Next, we applied the TERP technique to one-pot synthesis of diblock copolymers of PNIPAM and PNVP. NIPAM was polymerized for 3 h, followed by the addition of NVP to the polymerization mixture of NIPAM. Figure 11.4a shows the

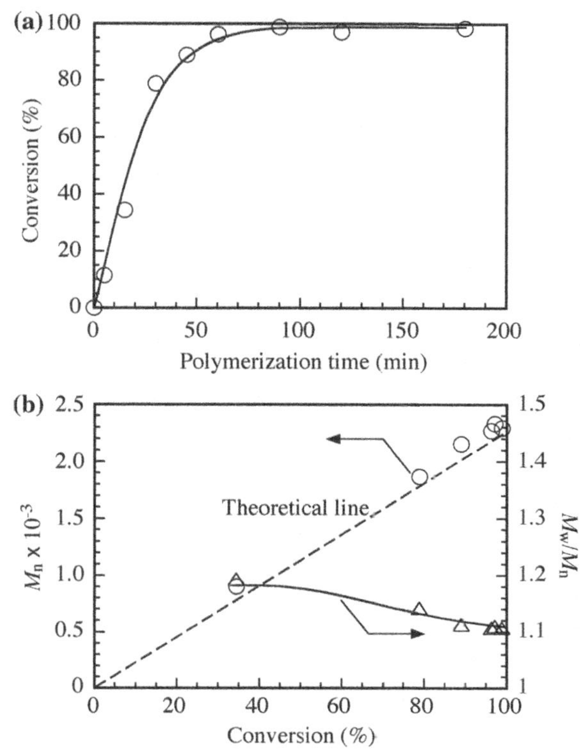

Fig. 11.3 **a** Time–conversion plots for the polymerization of NIPAM by TERP. **b** Dependence of M_n (*circle*) and M_w/M_n (*triangle*) on monomer conversion in the polymerization of NIPAM at 60 °C by TERP. The *broken line* represents the *theoretical line*

time–conversion profile for the block copolymerization of NVP by TERP in the presence of a PNIPAM macropromoter. After 3 h, the monomer conversion reached ca. 80 %. A time evolution of normalized GPC elution profiles observed during the block copolymerization of NVP in the presence of the PNIPAM macropromoter is shown in Fig. 11.4b. The molecular weight of the diblock copolymer increases progressively with polymerization time. These GPC elusion profiles are unimodal, suggesting no side reactions. Figure 11.4c shows M_n and M_w/M_n estimated from GPC measurement for the diblock copolymers of PNIPAM and PNVP as a function of the monomer conversion during the polymerization of NVP. M_n increases almost linearly with the conversion. All values of M_n are close to the theoretical values predicted for a living mechanism. The values of M_w/M_n are nearly constant at ca. 1.1 independent of the conversion. These observations indicate that the block copolymerization of NVP is well controlled. The number-average degrees of polymerization (DP) of the PNIPAM and PNVP blocks were estimated from GPC and ^{1}H NMR as listed in Table 11.1. Three samples with different block lengths are denoted as PNIPAM$_{110}$-PNVP$_{53}$, PNIPAM$_{110}$-PNVP$_{234}$, and PNIPAM$_{76}$-PNVP$_{219}$. The molecular weight distribution is narrow with M_w/M_n values ranging from 1.09 to 1.18 as estimated from GPC data.

11.3 Thermo-Responsive Behavior of the Diblock Copolymers

It is known that PNIPAM shows an LCST of ca. 32 °C, i.e., PNIPAM dissolves in water at room temperature, but it precipitates above LCST. It is expected that the diblock copolymer of PNIPAM and PNVP behaves as a thermo-responsive polymer. Both the PNIPAM and PNVP blocks are soluble in water at room temperature, whereas polymer micelles may be formed at temperatures above LCST for the PNIPAM block, the dehydrated PNIPAM blocks forming a core and hydrophilic PNVP blocks forming shells.

Thermo-responsive association behavior of AB diblock copolymers comprised of the hydrophilic block and PNIPAM block can be observed by ^{1}H NMR measurement [27]. Figure 11.5 shows the comparison of ^{1}H NMR spectra for PNIPAM$_{110}$-PNVP$_{234}$ in D_2O at 25 and 60 °C. The resonance peak at 1.5–2.6 ppm is attributed to the sum of the main chain protons and pendent methylene protons of PNVP. The broad resonance band centered at 3.3 ppm can be assigned as the pendent methylene protons in PNVP. The resonance bands at 1.2 and 3.9 ppm are attributed to the methyl and methine protons in the pendent isopropyl group in PNIPAM. The resonance peaks at 1.5–2.6 and 3.3 ppm due to PNVP at 60 °C are slightly larger than those at 25 °C, suggesting that the motions of the PNVP blocks increase with increasing temperature. On the other hand, the resonance peak intensity around 1.2 ppm, that is due to the NIPAM pendent methyl protons, at 60 °C

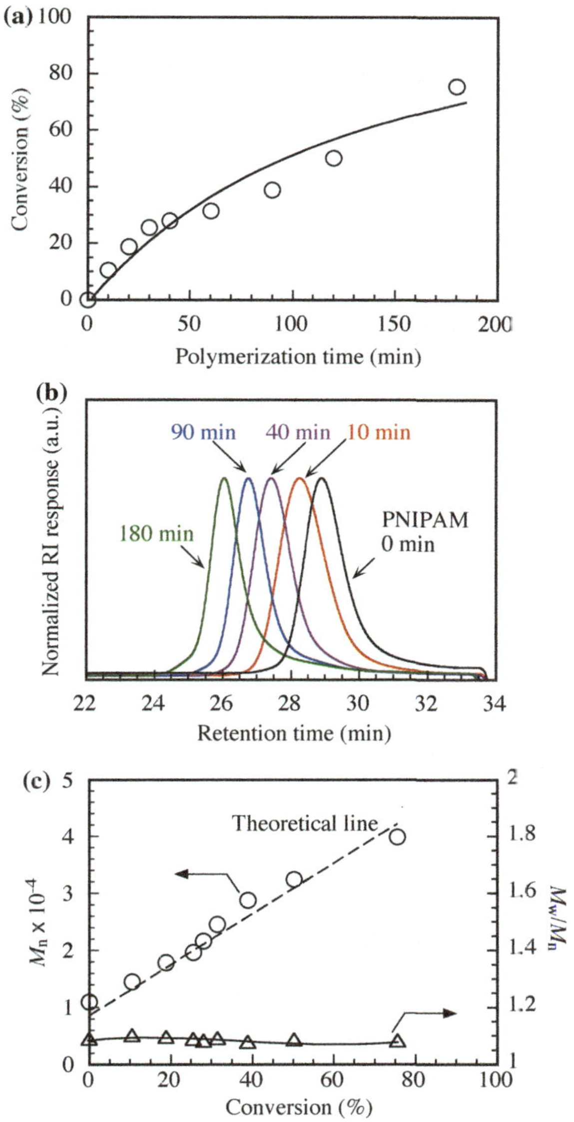

Fig. 11.4 **a** Time–conversion plots for block copolymerization of NVP in the presence of the PNIPAM macropromoter in DMF at 60 °C by TERP. **b** Normalized GPC elution profiles with time during the synthesis of PNIPAM$_m$-PNVP$_n$ diblock copolymer. **c** Dependence of M_n (*circle*) and M_w/M_n (*triangle*) on monomer conversion in the block copolymerization of NVP in the presence of the PNIPAM macropromoter at 60 °C by TERP. The *broken line* represents the *theoretical line*

Table 11.1 Compositions and molecular weights of the polymers

Sample code	DP PNIPAM[a]	DP PNVP[b]	M_n^b(NMR) × 10^{-4}	M_n^c(GPC) × 10^{-4}	M_w/M_n^c
$PNIPAM_{110}$	110		1.24	1.26	1.13
$PNIPAM_{76}$	76		0.86	1.09	1.09
$PNIPAM_{110}$-$PNVP_{53}$	110	53	1.83	1.70	1.09
$PNIPAM_{110}$-$PNVP_{234}$	110	234	3.85	2.63	1.11
$PNIPAM_{76}$-$PNVP_{219}$	76	219	3.29	3.67	1.15
$PNIPAM_{95}$-$PNVP_{240}$	95	240	3.74	4.55	1.18

[a]Estimated from the conversion determined by ^{1}H NMR and the initial ratio of NIPAM to BTEE
[b]Estimated by ^{1}H NMR for the purified diblock copolymers
[c]Estimated by GPC eluted with a DMF solution containing a 10-mM LiBr

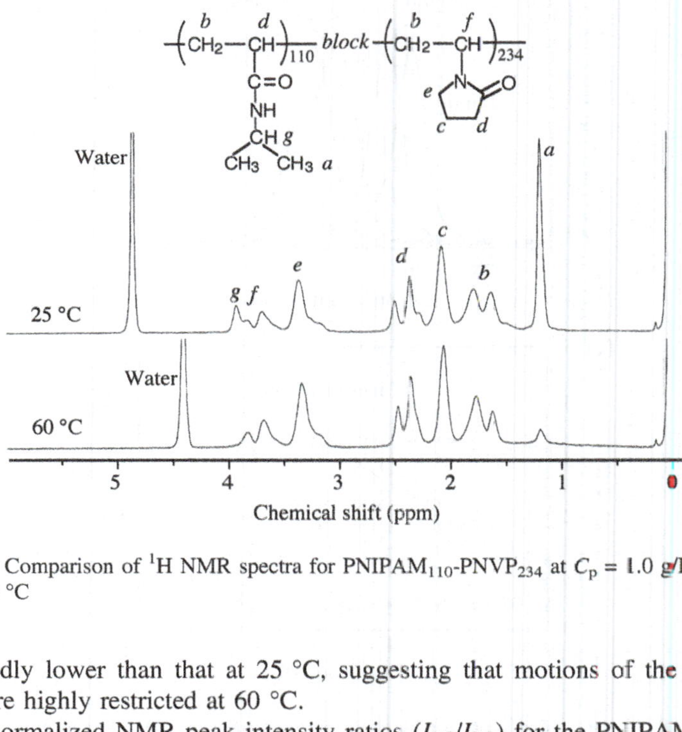

Fig. 11.5 Comparison of ^{1}H NMR spectra for $PNIPAM_{110}$-$PNVP_{234}$ at C_p = 1.0 g/L in D_2O at 25 and 60 °C

is markedly lower than that at 25 °C, suggesting that motions of the PNIPAM blocks are highly restricted at 60 °C.

The normalized NMR peak intensity ratios ($I_{1.2}/I_{3.3}$) for the PNIPAM pendent methyl proton at 1.2 ppm and the PNVP methylene protons at 3.3 ppm were measured at varying temperatures. In Fig. 11.6, the $I_{1.2}/I_{3.3}$ ratios for the diblock copolymers in D_2O are plotted as a function of temperature. The $I_{1.2}/I_{3.3}$ ratios are normalized with the $I_{1.2}/I_{3.3}$ value at 25 °C. The normalized $I_{1.2}/I_{3.3}$ ratio for

PNIPAM$_{110}$-PNVP$_{234}$ is practically constant at temperatures below 40 °C; however, the $I_{1.2}/I_{3.3}$ ratio starts to decrease with increasing temperature from 40 to 55 °C. The decrease in $I_{1.2}/I_{3.3}$ above 40 °C is due to a restricted motion of the PNIPAM blocks as a result of the formation of a hydrophobic microdomain. The temperature for the onset of a decrease in $I_{1.2}/I_{3.3}$ for the diblock copolymer with the shorter PNIPAM block length, PNIPAM$_{76}$-PNVP$_{219}$, is higher than those for the block copolymers with longer PNIPAM block lengths. This observation suggests that the micellization occurs at a higher temperature for the diblock copolymers with shorter PNIPAM block lengths. In the case of the same PNIPAM block lengths (DP = 110), the diblock copolymers with shorter PNVP block lengths tend to associate at lower temperatures.

The diblock copolymers in water undergo a transition from a molecularly dissolved unimer state at low temperatures to an association state above an aggregation temperature (T_a). As shown in Fig. 11.7, values of percent transmittance (%T) monitored at 600 nm for an aqueous solution of PNIPAM$_{110}$-PNVP$_{234}$ at polymer concentration (C_p) = 1.0 g/L are 100 % below 40 °C. The %T value decreases with increasing temperature, reaching a smaller value of 75 % above 45 °C. The turbid solution above T_a became clear again when the solution was cooled below T_a. The T_a values for the diblock copolymers were estimated from a break in the %T versus temperature plot. The T_a values for PNIPAM$_{110}$-PNVP$_{53}$, PNIPAM$_{110}$-PNVP$_{234}$, and PNIPAM$_{76}$-PNVP$_{219}$ are estimated to be 38.5, 40.6, and 44.0 °C, respectively (Table 11.2). Comparing the block copolymers of the same PNIPAM block lengths (i.e., PNIPAM$_{110}$), T_a is higher for longer PNVP block length. Comparing the diblock copolymers of similar PNVP block lengths (i.e., PNVP$_{234}$ and PNVP$_{219}$), T_a is higher for shorter PNIPAM block length. These observations are consistent with the aforementioned observations in ^{1}H NMR spectra. McCormick et al. [28] reported the thermo-responsive micellization of poly (*N*, *N*-dimethylacrylamide)-*block*-poly(*N*-isopropylacrylamide) (PDMA-PNIPAM)

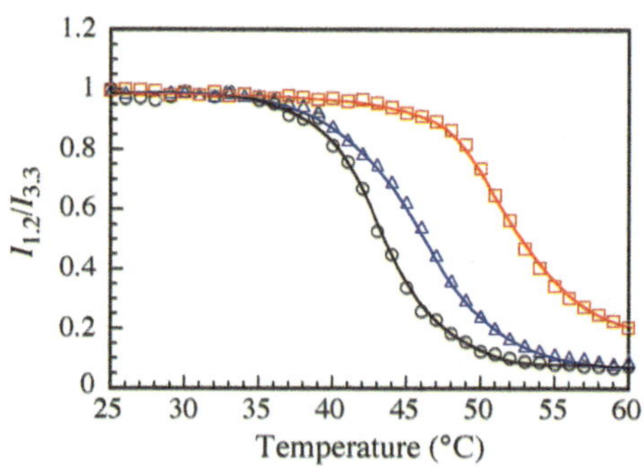

Fig. 11.6 Plots of normalized NMR signal intensity ratio of $I_{1.2}/I_{3.3}$ for PNIPAM$_{110}$-PNVP$_{53}$ (*circle, black*), PNIPAM$_{110}$-PNVP$_{234}$ (*triangle, blue*), and PNIPAM$_{76}$-PNVP$_{219}$ (*square, red*) at C_p = 1.0 g/L in D_2O as a function of temperature, where $I_{1.2}$ and $I_{3.3}$ are NMR signal intensities at 1.2 and 3.3 ppm, respectively

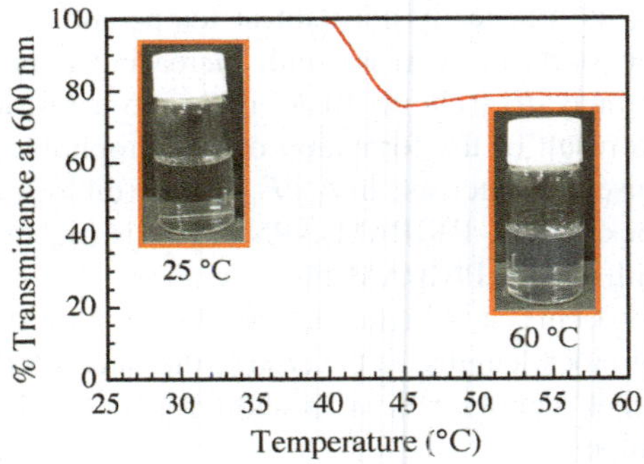

Fig. 11.7 Percent transmittance (%*T*) at 600 nm for an aqueous solution of PNIPAM$_{110}$-PNVP$_{234}$ at polymer concentration (C_p) = 1.0 g/L as a function of temperature. The temperature was changed with a heating rate of 0.5 °C/min. *Inserts* show digital photographs for aqueous solution of PNIPAM$_{110}$-PNVP$_{234}$ at 25 and 60 °C

Table 11.2 Dynamic light scattering (DLS) and static light scattering (SLS) data for the micelle-like aggregate formed from the diblock copolymers at 60 °C

Sample code	R_h^a (nm)	R_g^a (nm)	R_g/R_h	$M_w \times 10^{-7}$	N_{agg}^b	T_a^c (°C)
PNIPAM$_{110}$-PNVP$_{53}$	86.9	243	2.7	50.1	27,000	38.[illegible]
PNIPAM$_{110}$-PNVP$_{234}$	69.8	55.3	0.79	2.36	808	40.[illegible]
PNIPAM$_{76}$-PNVP$_{219}$	82.8	77.6	0.94	1.26	298	44.[illegible]

[a]Estimated by DLS and SLS in water at 60 °C, where C_p is fixed at 1.0 g/L
[b]Aggregation numbers at 60 °C estimated from M_w of the aggregates determined by SLS and M_w of the corresponding unimers determined by GPC
[c]Association temperature (T_a) in water determined by %*T* at 600 nm

with different block lengths in water. Our results are consistent with their findings that the association temperature for PDMA-PNIPAM increases with decreasing PNIPAM block length.

The thermo-responsive association of PNIPAM$_m$-PNVP$_n$ in water was confirmed by dynamic light scattering (DLS) measurement. The hydrodynamic radius (R_h) values for the diblock copolymers are plotted in Fig. 11.8a as a function of the temperature. Below 30 °C, the R_h values for the diblock copolymers are small on the order of ca. 10 nm, because the diblock copolymers dissolve molecularly in water. As the temperature is increased, the R_h value starts to increase around T_a, reaching a maximum value and then decreases with a further increase in temperature.

The temperature-dependent inter-molecular association of the diblock copolymers was evidenced by light scattering intensity data. In Fig. 11.8b, the scattering intensities for aqueous solutions of the diblock copolymers are plotted against temperature. As the temperature is increased, the scattering intensity starts to increase at a temperature fairly close to T_a estimated from %*T* (Table 11.2). The

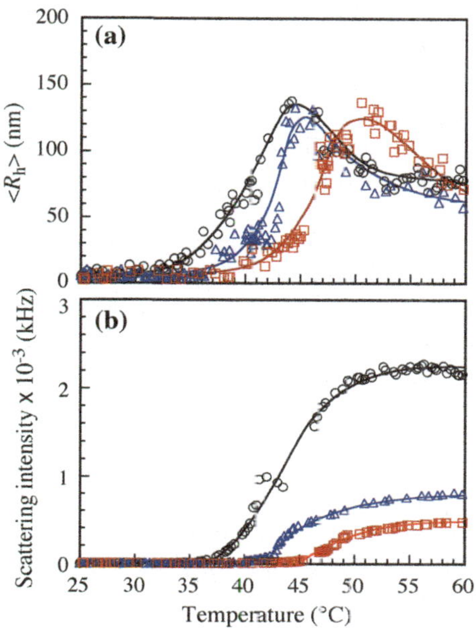

Fig. 11.8 **a** Intensity-average hydrodynamic radius (R_h) and **b** scattering intensity for $PNIPAM_{110}$-$PNVP_{53}$ (*circle, black*), $PNIPAM_{110}$-$PNVP_{234}$ (*triangle, blue*), and $PNIPAM_{76}$-$PNVP_{219}$ (*square, red*) at C_p = 1.0 g/L in water as a function of temperature

scattering intensities are nearly constant at high temperature, suggesting that the aggregation numbers (N_{agg}) are practically constant. N_{agg} is constant, while R_h decreases with increasing above a certain temperature. This observation suggests that the core of the polymer micelle becomes more compact with constant N_{agg} due to further dehydration of the PNIPAM blocks with increasing temperature. The temperature at which the light scattering intensity starts to increase cannot completely agree with the temperature at which the ^{1}H NMR peak intensity ratio starts to decrease (Fig. 11.6), because the two techniques measure a different matter, i.e., the light scattering and ^{1}H NMR techniques measure size and motional restriction, respectively. When the temperature was increased from 25 to 60 °C and subsequently decreased back to 25 °C, reversible changes in the R_h value and scattering intensity were observed.

It may be assumed that at a higher temperature the diblock copolymers form spherical core–shell micelles with an aggregate of collapsed PNIPAM blocks forming a micelle core and permanently water-soluble PNVP blocks forming a shell. The micelle size, molecular weight, and N_{agg} estimated from light scattering data at 60 °C are summarized in Table 11.2. The value of N_{agg} was calculated from the ratio of weight–average molecular weight (M_w) for the micelle at 60 °C by SLS and M_w for the unimer state estimated by GPC. The N_{agg} values for $PNIPAM_{110}$-$PNVP_{234}$ and $PNIPAM_{76}$-$PNVP_{219}$ were found to be 808 and 298, respectively; the micelle size of $PNIPAM_{110}$-$PNVP_{234}$ is smaller than that of $PNIPAM_{76}$-$PNVP_{219}$. This observation suggests that longer PNIPAM blocks tend to form a micelle core with a larger N_{agg}, because the longer PNIPAM block is more hydrophobic at temperatures above T_a. The N_{agg} value for $PNIPAM_{110}$-$PNVP_{53}$ was found to be 27,000, which is much larger than those for $PNIPAM_{110}$-$PNVP_{234}$ and $PNIPAM_{76}$-$PNVP_{219}$. This N_{agg} value for $PNIPAM_{110}$-$PNVP_{53}$ is obviously too large for a simple core–shell structure. Kim et al. [29] reported the formation of large aggregates of individual core–shell micelles formed from several kinds of amphiphilic diblock copolymers in water. Depending on the relative lengths of hydrophilic and hydrophobic blocks, a micelle core formed from hydrophobic blocks is not always completely protected from the exposure to the bulk water phase. Therefore, multicore structures may be formed by association of individual micelles, when the shell of the polymer micelle cannot completely prevent the hydrophobic interaction between the exposed hydrophobic cores of the micelles. Because the hydrophilic block length of $PNIPAM_{110}$-$PNVP_{53}$ is much shorter than those of the other two diblock copolymers, the individual simple core–shell micelles formed from $PNIPAM_{110}$-$PNVP_{53}$ may associate to form multicore inter-micellar aggregates.

A R_g/R_h ratio is an important parameter that depends on the shape and polydispersity of the object. The theoretical value of R_g/R_h for the homogeneous rigid sphere is 0.778, and it increases for a less dense structure and polydisperse solution [30]. The R_g/R_h ratios for $PNIPAM_{110}$-$PNVP_{234}$ and $PNIPAM_{76}$-$PNVP_{219}$ micelles are 0.79 and 0.94, respectively, suggesting that these diblock copolymers form spherical micelles at 60 °C. On the other hand, R_g/R_h for $PNIPAM_{110}$-$PNVP_{53}$ was found to be a much larger value of 2.7. Taken together with the N_{agg} data discussed above, it is suggested that $PNIPAM_{110}$-$PNVP_{53}$ forms elongated multicore aggregates formed by association of individual core–shell micelles.

11.4 Coating of Gold Nanoparticles with the Diblock Copolymer

It is known that PNVP can be attached to the surface of various metals. We looked into the capability of $PNIPAM_m$-$PNVP_n$ coating gold nanoparticles. We anticipate that, at room temperature, each of the polymer-coated gold nanoparticles is separately dissolved in water without aggregation. However, at temperatures above LCST for the PNIPAM blocks, the gold nanoparticles may associate with each other due to the hydrophobic interactions between the dehydrated PNIPAM blocks.

Visible absorption spectra of the surface plasmon band for gold nanoparticles are known to be sensitive to the size of the gold nanoparticles. When gold nanoparticles are close to each other, the plasmon band around 520 nm shifts toward longer wavelengths. Therefore, one can expect that the color of the polymer-coated gold nanoparticles is pink at room temperature, and the color changes to bluish-purple at a higher temperature. To prepare polymer-coated gold nanoparticles, we just mixed $PNIPAM_{110}$-$PNVP_{53}$ and gold nanoparticles in a 0.2 mM NaCl aqueous solution. Figure 11.9 shows the comparison of UV–visible absorption spectra for the $PNIPAM_{110}$-$PNVP_{53}$ coated gold nanoparticles at 25 and 50 °C. At 25 °C, the

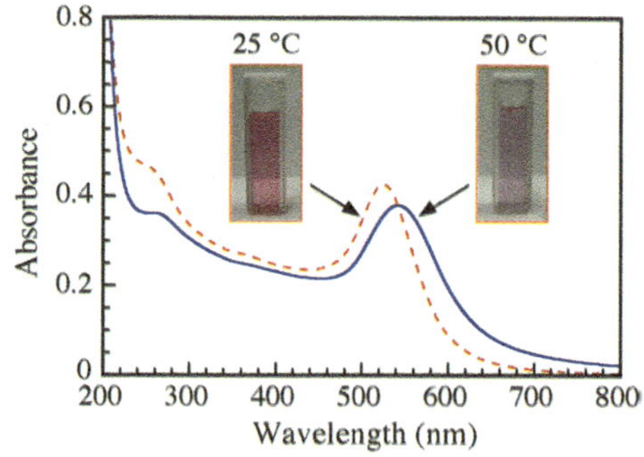

Fig. 11.9 UV–Vis absorption spectra for $PNIPAM_{110}$-$PNVP_{53}$ coated gold nanoparticles in 20 mM NaCl aqueous solutions at 25 (*broken line*) and 50 °C (*solid line*). *Inserts* show digital photographs of $PNIPAM_{110}$-$PNVP_{53}$ coated gold nanoparticles in 20 mM NaCl aqueous solutions at 25 and 50 °C

surface plasmon band for the polymer-coated gold nanoparticles can be observed at 525 nm; however, at 50 °C, the absorption maximum shifts to a longer wavelength of 542 nm. The color of the polymer-coated gold nanoparticles is pink at 25 °C. On the other hand, the pink color changes into bluish-purple at 50 °C. These observations indicate that heat-induced association of the $PNIPAM_{110}$-$PNVP_{53}$ coated gold nanoparticles can be controlled.

11.5 Complex Formation of the Thermo-responsive Diblock Copolymers with Fullerene

A typical procedure for the preparation of sample solution of the C_{60}/$PNIPAM_m$-$PNVP_n$ complex is as follows: Known amounts of $PNIPAM_m$-$PNVP_n$ and C_{60} powders were mixed in an agate mortar and then vigorously mulled for 30 min. Water was added to the mixture. Insoluble matters were removed by centrifugation. All sample solutions were filtered prior to measurements. The final C_p was 1.0 g/L. The C_{60} complex was prepared at C_p = 1.0 g/L, and the concentration of C_{60} was 0.06 g/L unless otherwise noted.

Figure 11.10a shows UV–Vis absorption spectra for the C_{60}/$PNIPAM_{95}$-$PNVP_{240}$ complex with varying amounts of added C_{60} in water at 25 °C. The spectra show absorption bands at 261 and 340 nm. The significant tailing of the absorption band up to around 700 nm was observed, which may be attributed to interactions between fullerenes [31]. It is known that the electron-deficient double bonds of C_{60} interact with the pendent amide groups in PNVP due to charge transfer interaction [32]. The charge transfer and hydrophobic interactions induce the C_{60} complex formation comprising a C_{60}/PNVP core. The aqueous solution of the C_{60} complex is brownish yellow in contrast to a violet solution of C_{60} in

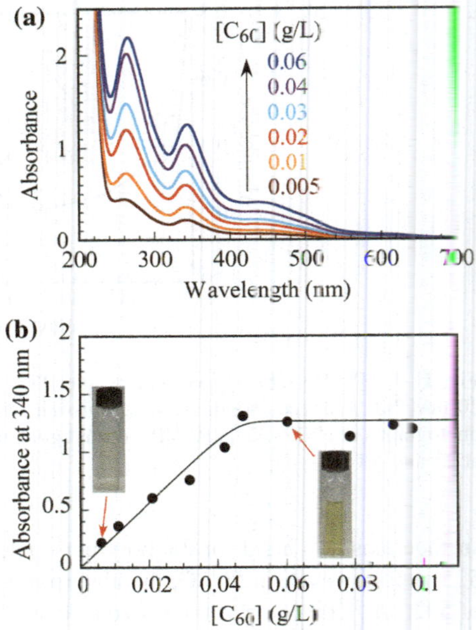

Fig. 11.10 **a** UV–Vis absorption spectra for an aqueous solution of C_{60}/PNIPAM$_{95}$-PNVP$_{240}$ complex with varying amounts of added C_{60} ([C_{60}]) at 25 °C. The concentration of the block copolymer (C_p) was fixed at 1.0 g/L. **b** Absorbance at 340 nm for the C_{60}/PNIPAM$_{95}$-PNVP$_{240}$ complex in water at 25 °C plotted as a function of [C_{60}]. *Inserts* show digital photographs for the aqueous solutions of [C_{60}] = 0.01 and 0.06 g/L

Table 11.3 Solubilized amounts of C_{60}, lower critical solution temperature (LCST), and hydrodynamic radius (R_h) for the C_{60} complex in water

Sample code	Solubilized C_{60}^{a} (g/L)	LCST[b] (°C)	R_h (nm) 25 °C	R_h (nm) 60 °C
C_{60}/PNIPAM$_{76}$-PNVP$_{219}$	0.022	47.6	62.8	80.8
C_{60}/PNIPAM$_{95}$-PNVP$_{240}$	0.020	41.9	78.5	108
C_{60}/PNIPAM$_{110}$-PNVP$_{149}$	0.026	40.7	67.9	94.7

[a]Estimated by UV–Vis absorption spectra at C_p = 1.0 g/L
[b]Estimated from temperature dependence of percent transmittance

n-hexane [33]. Figure 11.10b illustrates a solubility curve for C_{60} obtained based on the absorbance at 340 nm. The absorbance at 340 nm increases almost linearly with an increase in the amount of added C_{60} ([C_{60}]), reaching a plateau at [C_{60}] near 0.05 g/L. The maximum concentration of C_{60} in the presence of PNIPAM$_{95}$-PNVP$_{240}$ in water was calculated to be 0.020 g/L at C_p = 1.0 g/L using a molar absorption coefficient (ε) value of 42,700 L/(mol cm) at 332 nm reported for C_{60} solubilized by γ-CD in aqueous solution [34]. In the case of PNIPAM$_{75}$-PNVP$_{219}$ and PNIPAM$_{110}$-PNVP$_{149}$, the maximum concentrations of C_{60} in water were estimated to be 0.022 and 0.026 g/L, respectively, as summarized in Table 11.3. As shown in Table 11.3, the maximum concentration of C_{60} in water is almost constant around 0.02 g/L independent of the PNVP block length. It is known that the ε value of fullerenes is strongly dependent on the aggregation state [35]. Therefore, the maximum concentration of C_{60} in the text is an apparent value.

The diblock copolymers and the C_{60} complexes are soluble in water at room temperature, and the solutions exhibit LCST. Figure 11.11 shows a typical example of the temperature dependence of $\%T$ at 700 nm for aqueous solutions of $PNIPAM_{95}$-$PNVP_{240}$ and the C_{60}/$PNIPAM_{95}$-$PNVP_{240}$ complex. At 40 °C or lower, the $\%T$ values for the aqueous solutions of $PNIPAM_{95}$-$PNVP_{240}$ and the C_{60} complex are approximately 100 and 95 %, respectively. The $\%T$ value for $PNIPAM_{95}$-$PNVP_{240}$ decreases with increasing temperature, reaching a smaller value of ca. 92 % at a temperature near 44 °C. The LCST values for the C_{60} complex estimated from a break in the $\%T$ versus temperature plot are given in Table 11.3. The presence of LCST for the diblock copolymers without C_{60} indicates that the PNIPAM blocks are dehydrated and collapsed above LCST into a core–shell-type polymer micelle, the PNIPAM blocks forming a core. In the case of the C_{60} complex in water, an LCST was observed around 41.9 °C. LCST values for the C_{60} complexes were found to be slightly higher than those for the free diblock copolymers, and LCST decreases with increasing PNIPAM block length (Table 11.3). This observation suggests that LCST of the C_{60} complex may be controlled by the chain length of the PNIPAM block. The C_{60} complexes have higher LCST values comparing with the corresponding diblock copolymers.

In an attempt to obtain information as to how C_{60} is integrated into the diblock copolymer micelles, we measured DLS. Figure 11.12 shows the comparison of R_h distributions for free $PNIPAM_{95}$-$PNVP_{240}$ at 25 °C and the C_{60}/$PNIPAM_{95}$-$PNVP_{240}$ complex at 25 and 60 °C in water. All the distributions are unimodal. The R_h values for free $PNIPAM_{95}$-$PNVP_{240}$ and for its C_{60} complex at 25 °C were estimated to be 7.6 and 78.5 nm, respectively. The small R_h value of 7.6 nm corresponds to the size of the free diblock copolymer molecularly dissolved in water at 25 °C. In the case of the C_{60} complex at 25 °C, C_{60} is incorporated into a complex where the C_{60}-bound PNVP blocks form a core. It should be noted that the

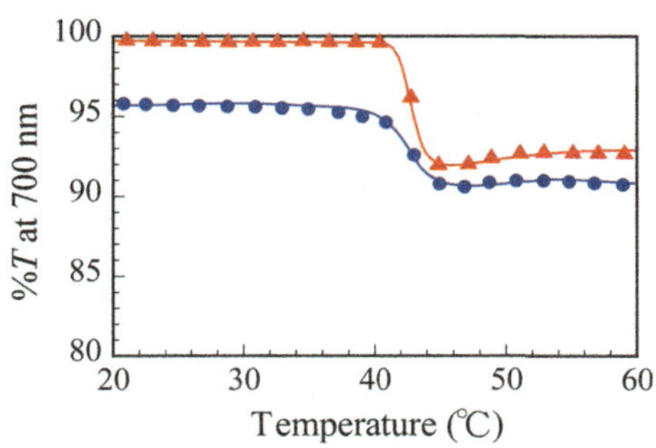

Fig. 11.11 Percent transmittance ($\%T$) at 700 nm as a function of temperature for an aqueous solution of free $PNIPAM_{95}$-$PNVP_{240}$ at C_p = 1.0 g/L (*dark triangle, red*) and the C_{60}/$PNIPAM_{95}$-$PNVP_{240}$ complex (*dark circle, blue*). The temperature was changed with a heating rate of 0.5 °C/min

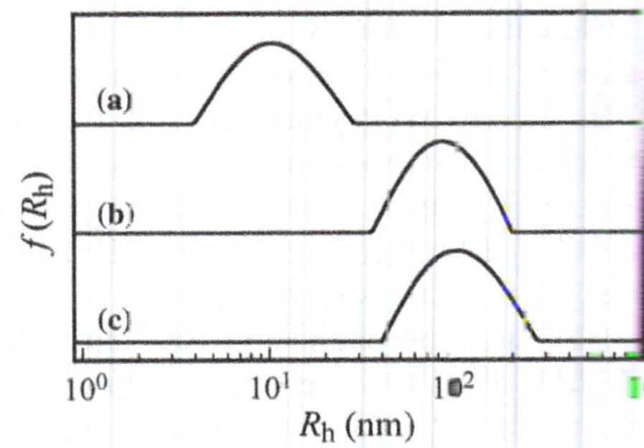

Fig. 11.12 Hydrodynamic radius (R_h) distributions for free PNIPAM$_{95}$-PNVP$_{240}$ at C_p = 1.0 g/L at 25 °C **a** and the C_{60}/PNIPAM$_{95}$-PNVP$_{240}$ complex in water at 25 **b** and 60 °C **c**

C_{60} complex gives a unimodal distribution at 25 °C with R_h value of 78.5 nm. However, when the concentration of added C_{60} is reduced by one-sixth (i.e., [C_{60}] = 0.01 g/L) while the polymer concentration keeping constant (i.e., C_p = 1.0 g/L), the distribution becomes bimodal where one component is due to free diblock copolymer with R_h of 6.7 nm and the other is due to the C_{60} complex with R_h of 67.8 nm. It can be concluded that under a condition where C_p = 1.0 g/L and [C_{60}] = 0.06 g/L, virtually all the diblock copolymer chains are used to form complexes with C_{60} and no free polymer remains. When the temperature was raised to 60 °C, R_h of the C_{60} complexes increased significantly. In Table 11.3 are compared R_h values at 25 and 60 °C for the C_{60} complexes with the diblock copolymers of different block lengths. For all the C_{60} complexes, the R_h values at 60 °C are larger than those at 25 °C. The increase in R_h at 60 °C can be attributed to the aggregation of the complexes caused by hydrophobic associations between the hydrophobic outer layers.

It is important to emphasize that the type of core–shell micelle-like complex formed from the C_{60}-bound block copolymers, where PNIPAM blocks remain as the shell component even after dehydrated above LCST, is completely different from core–shell micelles formed from the free block copolymers at temperatures above LCST. As discussed in our earlier paper [36], in micelles formed by the free block copolymers at temperatures above LCST, dehydrated PNIPAM blocks form a core and PNVP blocks, which remain to be hydrophilic independent of temperature, form a shell. In fact, such a micelle is formed by free PNIPAM$_{95}$-PNVP$_{240}$ at 60 °C whose R_h is 50.0 nm. There is no sign of the formation of such PNIPAM$_{95}$-PNVP$_{240}$ micelle when the C_{60}/PNIPAM$_{95}$-PNVP$_{240}$ complex was heated to 60 °C as far as we can see from DLS data. This indicates that C_{60} remains to be bound to the PNVP block even at this high temperature without undergoing thermal dissociation.

Figure 11.13a shows a plot of R_h for the C_{60}/PNIPAM$_{95}$-PNVP$_{240}$ complex in pure water as a function of temperature. Below LCST, the C_{60} complex exists in water with an R_h value on the order of 80 nm. As temperature is increased, R_h for the C_{60} complex starts to increase somewhat abruptly at a temperature near 40 °C reaching a maximum value of ca. 110 nm at ca. 45 °C. A similar profile is shown in Fig. 11.13b in which the scattering intensity for the C_{60}/PNIPAM$_{95}$-PNVP$_{240}$

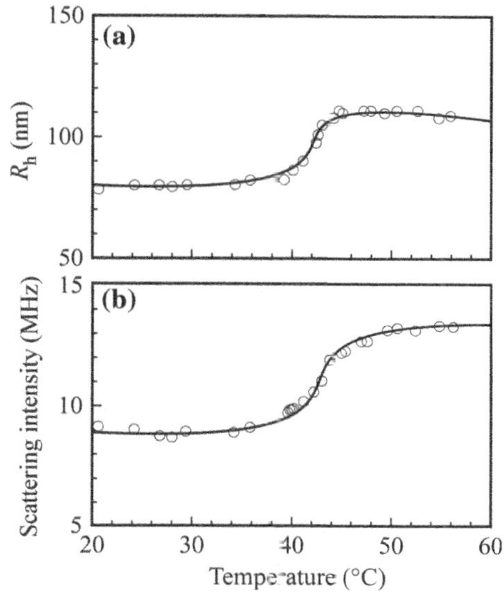

Fig. 11.13 **a** Plots of hydrodynamic radius (R_h) and **b** scattering intensity for the C_{60}/PNIPAM$_{95}$-PNVP$_{240}$ complex as a function of temperature in water

Table 11.4 Static light scattering (SLS) data for C_{60}/PNIPAM$_{95}$-PNVP$_{240}$ complex in water at 25 and 60 °C

Temperature (°C)	$M_w \times 10^{-7}$	R_g (nm)	$A_2 \times 10^6$ (mol mL/g^2)	N_{agg}^a	R_g/R_h
25	7.56	81.5	1.63	2000	1.04
60	10.7	75.8	1.09	2800	0.70

[a]Aggregation numbers (N_{agg}) for C_{60}/PNIPAM$_{95}$-PNVP$_{240}$ complex, defined as the total number of polymer chains forming one micelle

complex in pure water is plotted against temperature. The increase in R_h and the scattering intensity at a certain temperature is a clear manifestation that the C_{60} complexes form inter-complex aggregates at temperatures above LCST due to the dehydration of the PNIPAM block chains in the shell.

The SLS data for the C_{60} complex at 25 and 60 °C are summarized in Table 11.4. Values of M_w, R_g, and A_2 for the complex were estimated from SLS. A value of N_{agg} for the C_{60}/PNIPAM$_{95}$-PNVP$_{240}$ complex was calculated from apparent M_w for the complex estimated by SLS and M_w for a single polymer chain estimated from GPC to be 2000 at 25 °C (Table 11.4). Based on this N_{agg} value, taken together with the absorbance of a C_{60} complex solution, the number of C_{60} molecules incorporated in one C_{60}/PNIPAM$_{95}$-PNVP$_{240}$ complex at 25 °C was calculated to be ca. 2000. This value coincides with the N_{agg} value, implying that each PNVP block sequence associates with one C_{60} molecule in the C_{60}/PNIPAM$_{95}$-PNVP$_{240}$ complex. The association of the C_{60} complex at 60 °C is reflected in the larger values of M_w and N_{agg} for the C_{60} complex at 60 °C than those at 25 °C.

As shown in Table 11.4, R_g value at 60 °C is smaller than that at 25 °C, R_g = 81.5 nm at 25 °C decreasing to 75.8 nm at 60 °C. The R_g/R_h ratio for the C_{60} complex at 25 and 60 °C is 1.04 and 0.70, respectively (Table 11.4). This finding indicates that the PNIPAM shells collapse into a thin layer due to dehydration of the PNIPAM chain above LCST. At 60 °C, the micelle may be a homogeneous sphere, which is consistent with the angular dependence DLS data. The A_2 value for the C_{60} complex at 60 °C is smaller than that at 25°C, suggesting that the inter-complex association decreases the solubility in water at a higher temperature due to an increased dehydration of the PNIPAM chain.

11.6 Generation of Singlet Oxygen from the Fullerene Complex

Fullerenes are known to be photoexcited to the singlet state first and then efficiently converted to the triplet state that can convert molecular oxygen to singlet oxygen ($^1O_2$) with quantum yield of nearly unity [37]. The photo-oxidation of 9, 10-anthracenedipropionic acid (ADPA) to its endoperoxide derivative by singlet oxygen is usually used for the detection of $^1O_2$ generated by photosensitizers [38].

The ADPA bleaching experiment was performed to prove that the C_{60}/PNIPAM$_{95}$-PNVP$_{240}$ complex is photoactive and thus able to produce $^1O_2$ by light irradiation. The formation of the endoperoxide of ADPA by $^1O_2$ can be detected by loss of absorption of PDPA. Mixed solutions of the C_{60} complex and ADPA were irradiated at a selected wavelength that was selectively absorbed by C_{60}. In Fig. 11.14a are shown changes in the absorption spectra for ADPA in the presence of the C_{60} complex in D_2O by irradiation of visible light ($\geq$420 nm). The visible absorption spectra were corrected by subtracting the C_{60} absorption. The absorbance of ADPA progressively decreased with irradiation time. The reaction was followed at 400 nm by monitoring the decrease of the absorbance of ADPA in the presence of the C_{60} complex with the irradiation time. Figure 11.14b shows the progress of the photoreaction ($A(t)/A(0)$) of the ADPA with light irradiation time, where $A(t)$ and $A(0)$ are the absorbance at 400 nm at irradiation time t and 0 min, respectively. After 4 h of photosensitization, $A(t)/A(0)$ of the ADPA in the presence of the C_{60} complex in D_2O and in pure water decreased to 0.66 and 0.74, respectively. The most effective pathway for singlet oxygen to be deactivated to the ground state is by transferring its electronic energy to vibrational oscillators in the neighboring solvent molecules. It is known that the lifetime of $^1O_2$ in D_2O is longer than that in water [39]. On the other hand, $A(t)/A(0)$ of ADPA without the C_{60} complex in D_2O was constant during light irradiation, indicating that PDPA is stable toward irradiation at $\geq$420 nm. Therefore, it is concluded that $^1O_2$ can be generated by light irradiation of the C_{60} complex.

Nakamura et al. reported that DNA was cleaved by fullerene derivatives under photoirradiation [40]. The DNA cleavage is due to $^1O_2$ from fullerene generated by

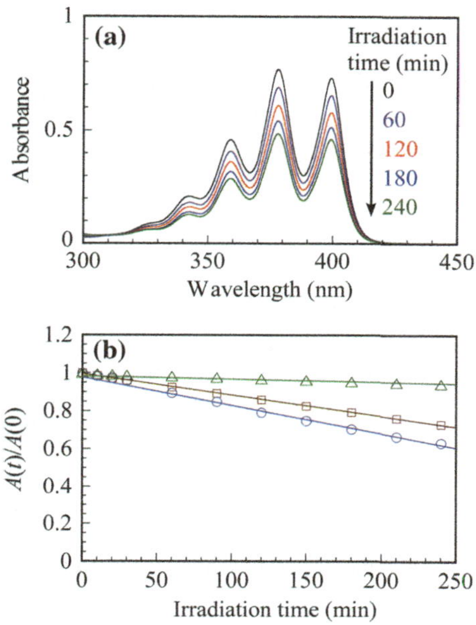

Fig. 11.14 (a) Change in UV–Vis absorption spectra of ADPA in the presence of the C_{60}/PNIPAM$_{95}$-PNVP$_{240}$ complex in D_2O caused by irradiation of visible light ($\geq$420 nm). The irradiation times are indicated in the figure. **b** Absorbance of ADPA at 400 nm as a function of irradiation time in the absence (*triangle, green*) and in the presence of the C_{60}/PNIPAM$_{95}$-PNVP$_{240}$ complex in D_2O (*circle, blue*). For the latter, data obtained in H_2O in place of D_2O are also plotted (*square, brown*)

the photoirradiation. We studied the interaction potential of the C_{60}/PNIPAM$_{95}$-PNVP$_{240}$ complex with biological molecules by photo-induced cleavage ability of the micelle toward pBR322 plasmid DNA. Supercoiled DNA (Form I) can be cleaved into nicked DNA (Form II) and linear DNA (Form III). We used plasmid DNA containing >70 % supercoiled DNA (Form I) for the photocleavage studies. The cleaving activity of the C_{60} complex on supercoiled DNA was examined under various conditions with visible light (>420 nm) irradiation. To evaluate the ability for the photo-induced DNA cleavage by the C_{60} complex, we monitored the formation of Form II and Form III by agarose gel electrophoresis (Fig. 11.15). The photocleavage experiments were performed in the presence of NADH, because the natural reducing agent, i.e., NADH is present in vivo. No effective DNA cleavage occurred in the absence of NADH [41]. Supercoiled DNA was completely cleaved into Form II and Form III in the presence of the C_{60} complex under visible light irradiation for 6 h. Under the same conditions as lane 1, except in the absence of C_{60} and the C_{60} complex, supercoiled DNA was scarcely converted to Form II under visible light irradiation for 6 h. Supercoiled DNA cleavage cannot be observed in the presence of the C_{60} complex incubated in the dark for 6 h, which is the same as the control experiment in the absence of the micelle incubated in the

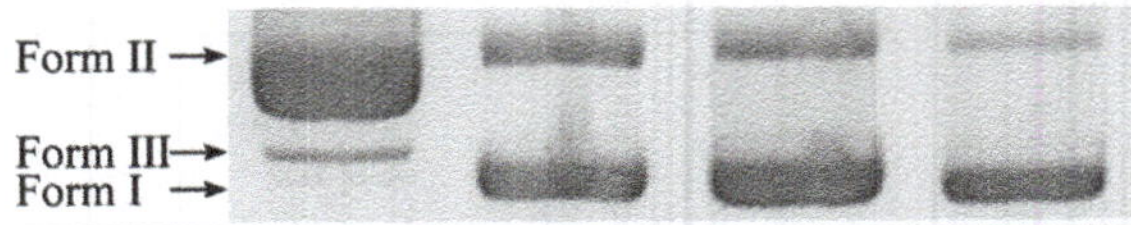

Lane	1	2	3	4
C_{60}	+	-	-	+
Polymer	+	+	-	+
Light	+	+	+	-

Fig. 11.15 Gel electrophoretic images of supercoiled plasmid pBR322 DNA nicking by the C_{60}/PNIPAM$_{95}$-PNVP$_{240}$ complex in the presence of NADH. The mixtures were analyzed on a 1.0 wt % agarose gel and photographed with ethidium bromide staining. *Lane 1* shows 0.9 g/L of the C_{60}/PNIPAM$_{95}$-PNVP$_{240}$ complex incubated under visible light irradiation for 6 h. *Lane 2* shows 0.9 g/L of PNIPAM$_{95}$-PNVP$_{240}$ incubated under visible light irradiation for 6 h. *Lane 3* shows no chemicals incubated under visible light irradiation for 6 h. *Lane 4* shows 0.9 g/L of the C_{60}/PNIPAM$_{95}$-PNVP$_{240}$ incubated in the *dark* for 6 h

dark. These observations indicate that the water-soluble C_{60}/PNIPAM$_{95}$-PNVP$_{240}$ complex has an ability to cleave DNA under visible light irradiation.

11.7 Conclusions

Thermo-responsive diblock copolymers with different block lengths, PNIPAM$_m$-PNVP$_n$, were prepared by one-pot synthesis using TERP. All the diblock copolymers dissolve in water molecularly below T_a; however, heat-induced micellization occurs above T_a due to the association of dehydrated PNIPAM blocks. T_a depends on the composition of the diblock copolymers. The diblock copolymer with longer PNVP and shorter PNIPAM blocks shows a higher T_a, while the diblock copolymer with shorter PNVP and longer PNIPAM blocks shows a lower T_a. Light scattering data suggested that PNIPAM$_{110}$-PNVP$_{234}$ and PNIPAM$_{76}$-PNVP$_{219}$ form spherical core–shell micelles with N_{agg} = 808 and 298, respectively, above T_a. By contrast, PNIPAM$_{110}$-PNVP$_{53}$ forms a much larger aggregate with N_{agg} = 27,000, which may be a multicore aggregate formed by the association of individual core–shell micelles.

Gold nanoparticles can be coated with PNIPAM$_{110}$-PNVP$_{53}$ because of coordinated interaction between the surface of gold nanoparticles and the PNVP blocks. The maximum wavelength of the plasmon band for the polymer-coated gold nanoparticles in a 20 mM NaCl aqueous solution at 25 °C shifted to a longer wavelength at 50 °C, suggesting that association behavior of the polymer-coated gold nanoparticles can be controlled by temperature.

The thermo-responsive diblock copolymers can solubilize C_{60} in water through binding to the PNVP block. Based on the characterization of the C_{60} complex, the formation of a micelle-like complex comprising a C_{60}/PNVP core and a thermo-responsive PNIPAM shell was proposed. The C_{60} complex with R_h = 60–80 nm below LCST (<40 °C) is expected to be accumulated to tumor by passive targeting EPR effect. Below LCST, the thermo-responsive shell of the C_{60} complex is hydrophilic; however, above LCST inter-complex association was observed due to the hydrophobic interaction of a dehydrated PNIPAM shell that collapsed in the outer layer of the micelle. It may be expected that the inter-cellular uptake of the C_{60} complex may be thermally controlled by changing temperature due to the hydrophilic/hydrophobic switching properties of the micelle. The C_{60} complex was found to generate singlet oxygen and cleave DNA by visible light irradiation. The C_{60} complex can be expected to be applied to PDT.

References

1. Yamago S, Ikeda K, Yoshida J (2002) J Am Chem Soc 124:2874–2875
2. Yamago S, Ray B, Iida K, Yoshida J, Tada T, Yoshizawa K, Kwak Y, Goto A, Fukuda T (2004) J Am Chem Soc 126:13908–13909
3. Yamago S (2006) J Polym Sci Part A Polym Chem 44:1–12
4. Yamago S, Miyazoe H, Goto R, Hashidume M, Sawazaki T, Yoshida J (2001) J Am Chem Soc 123:3697–3705
5. Kwak Y, Tezuka M, Goto A, Fukuda T, Yamago S (2007) Macromolecules 40:1881–1885
6. Haaf F, Sanner A, Straub F (1985) Polym J 17:143–152
7. Einaga H, Harada M (2005) Langmuir 21:2578–2584
8. McCormick CL, (ed) (2001) Stimuli-responsive water soluble and amphiphilic polymers; ACS Symp. Ser. 780. The American Chemical Society, Washington, DC
9. Kikuchi A, Okano T (2005) J Controlled Release 101:69–84
10. Artoçaréna M, Heise B, Ishaya S, Laschewsky A (2002) J Am Chem Soc 124:3787–3793
11. Tsunoyama H, Sakurai H, Negishi Y, Tsukuda T (2005) J Am Chem Soc 127:9374–9375
12. Ishii T, Otsuka H, Kataoka K, Nagasaki Y (2004) Langmuir 20:561–564
13. Jensen AW, Wilson SR, Schuster DI (1996) Bioorganic Med Chem 4:767–779
14. Nakamura E, Isobe H (2003) Acc Chem Res 36:807–815
15. Schinazi RF, Sijbesma R, Srdanov G, Hill CL, Wudl F (1993) Antimicrob Agents Chemother 37:1707–1710
16. Sera N, Tokiwa H, Miyata N (1996) Carcinogenesis 17:2163–2169
17. Chiang LY, Lu FJ, Lin JT (1995) Free radical scavenging activity of water-soluble fullerenols. Chem Commun 1283–1284
18. Tabata Y, Ikeda Y (1999) Pure Appl Chem 71:2047–2053
19. Ali H, van Lier JE (1999) Chem Rev 99:2379–2450
20. Ruoff RS, Tse DS, Malhotra R, Lorents DC (1993) J Phys Chem 97:3379–3383
21. Chiang LY, Wang LY, Swirczewski JW, Soled S, Cameron S (1994) J Org Chem 59: 3960–3968
22. Andersson T, Nilsson K, Sundahl M, Westman G, Wennerstrom O (1992) C 60 embedded in γ-cyclodextrin: a water-soluble fullerene. Chem Commun, 604–606
23. Atwood JL, Koutsantonis GA, Raston CL (1994) Nature 368:229–231
24. Hungerbuehler H, Guldi DM, Asmus KD (1993) J Am Chem Soc 115:3386–3387
25. Ikeda A, Sato T, Kitamura K, Nishiguchi K, Sasaki Y, Kikuchi J, Ogawa T, Yogo K, Takeya T (2005) Org Biomol Chem 3:2907–2909

26. Yamakoshi YN, Yagami T, Fukuhara K, Sueyoshi S, Miyata N (1994) Solubilization of fullerenes into water with polyvinylpyrrolidone applicable to biological tests. Chem Commun, 517–518
27. Yusa S, Shimada Y, Mitsukami Y, Yamamoto T, Morishima Y (2004) Macromolecules 37:7507–7513
28. Convertine AJ, Lokitz BS, Vasileva Y, Myrick LJ, Scales CW, Lowe AB, McCormick CL (2006) Macromolecules 39:1724–1730
29. Kim C, Lee SC, Shin JH, Yoon JS, Kwon IC, Jeong SY (2000) Macromolecules 33 7448–7452
30. Konishi T, Yoshizaki T, Yamakawa H (1991) Macromolecules 24:5614–5622
31. Pantos GD, Wietor JL, Sanders JKM (2007) Angew Chem Int Ed 46:2238–2240
32. Ungurenasu C, Airinei A (2000) J Med Chem 43:3186–3188
33. Hare JP, Kroto HW, Taylor R (1991) Chem Phys Lett 177:394–398
34. Ikeda A, Doi Y, Hashizume M, Kikuchi J, Konishi T (2007) J Am Chem Soc 129:4140–4141
35. Li C, Hu J, Yin J, Liu S (2009) Macromolecules 42:5007–5016
36. Yusa S, Yamago S, Sugahara M, Morikawa S, Yamamoto T, Morishima Y (2007) Macromolecules 40:5907–5915
37. Arbogast JW, Foote CS (1991) J Am Chem Soc 113:8886–8889
38. Diwu ZJ, Lown JW (1992) J Photochem Photobiol A Chem 64:273–287
39. Rodgers MAJ, Snowden PT (1982) J Am Chem Soc 104:5541–5543
40. Tokuyama H, Yamago S, Nakamura E, Shiraki T, Sugiura Y (1993) J Am Chem Soc 115:7918–7919
41. Yamakoshi Y, Sueyoshi S, Fukuhara K, Miyata N (1998) J Am Chem Soc 120:12363–12364

Chapter 12
Design of Biomimetic Interfaces at the Dendrimer Periphery and Their Applications

Chie Kojima

Abstract Dendrimers have well-defined and highly branched shapes, which are different from linear polymers. Dendrimers have inner space and a lot of functional groups at the periphery, which are useful for encapsulation and conjugation, respectively. In this chapter, artificial proteins such as collagen and elastin were designed by attaching the model peptides at the dendrimer periphery. The collagen-mimetic and the elastin-mimetic dendrimers exhibited thermosensitive properties similar to natural collagen and elastin. Collagen-mimetic dendrimers formed thermosensitive hydrogels, whose sensitivity was regulated by the sequences of the peptide. Due to the loading ability of dendrimers, collagen-mimetic and elastin-mimetic dendrimers worked as temperature-controllable carriers. And, dendritic peptidomimetics were prepared by conjugating amino acids to the terminal of dendron with fluorescent core. The clustering effects of the conjugated peptides and amino acids worked to induce some functions. Thus, dendrimers can work as a functional platform of bioactive molecules.

Keywords Dendrimer · Collagen · Elastin · Peptide · Thermoresponsive · Hydrogel · Drug delivery

12.1 Introduction

Proteins are supreme stimuli-responsive biomaterials. Some respond to temperature, pH, and specific ions and change the conformation. The functions such as the enzymatic activities and the protein binding properties are controlled by these stimuli. Consequently, proteins construct sophisticated systems in our body. Commercially available proteins are produced by extraction from animals or using recombinant technology. Mass production can be problematic because of concerns regarding

C. Kojima (✉)
Department of Applied Chemistry, Graduate School of Engineering,
Osaka Prefecture University, Osaka, Japan
e-mail: kojima@chem.osakafu-u.ac.jp

T. Kawai and M. Hashizume (eds.), *Stimuli-Responsive Interfaces*,
DOI 10.1007/978-981-10-2463-4_12

animal ethics and low productivity from recombinant technology. Additionally, deleterious pathogens can contaminate animal extracts. Thus, chemically synthesized proteins are more desirable for application as functional materials.

Because proteins are macromolecules in which various amino acids are linked via amide bonds, they are categorized as polymers in chemistry. However, their properties are different from those of synthetic polymers. Sequence and molecular weight are precisely controlled in proteins, but not in synthetic polymers. Thus, natural proteins cannot be chemically synthesized at the present. A short fragment of a protein is called a peptide. Peptides can be chemically synthesized by the solid-phase method. However, the functions of peptides are limited and different from those of whole proteins.

A dendrimer is a synthetic polymer with a highly branched structure. The name comes from the Greek word for tree and reflects their structure [1, 2]. Although synthetic polymers are synthesized by polymerization, dendrimers are synthesized by stepwise protocols. Thus, the sizes and molecular structures of dendrimers are controllable [2–7]. Moreover, dendrimers can contain interior space and multiple functional groups at the periphery. Thus, dendrimers can work as a scaffold of bioactive molecules such as peptides [2, 6–9]. Recently, we proposed that the conjugation of the model peptides to dendrimers is applicable to producing artificial proteins. We focused on collagen and elastin, which are components of the extracellular matrix, because they have repeat sequences and thermosensitive properties. In this chapter, the design and synthesis of collagen and elastin mimetics using dendrimers and model peptides, and their applications are shown. Additionally, dendritic peptidomimetics are also designed by conjugating amino acids to the dendrimer periphery.

12.2 Dendrimers

Dendrimers were first reported by Dr. Tomalia in 1985 [1]. Dendrimers are composed of a core, building units, and terminal groups (Fig. 12.1) [2–7]. Some dendrimers are synthesized by repeated reaction cycles starting from the core in a divergent method.

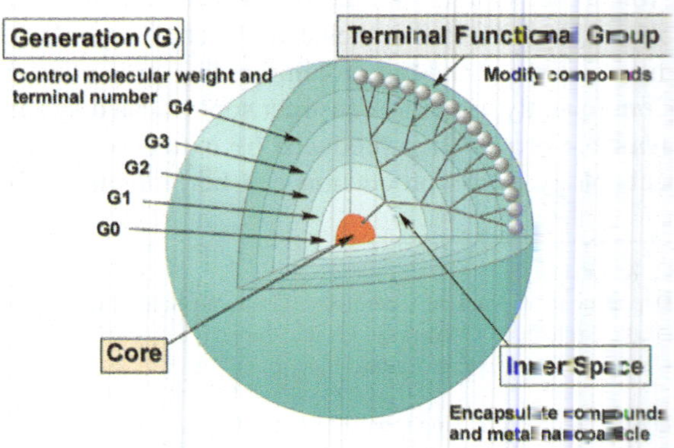

Fig. 12.1 Overview of the dendrimer structure

Table 12.1 Properties of ethylenediamine core PAMAM dendrimers of different generations

Generation	Terminal group	Molecular weight (Da)	Diameter (nm)
G1	8	1,430	2.2
G2	16	3,256	2.9
G3	32	6,909	3.6
G4	64	14,215	4.5
G5	128	28,826	5.4
G6	256	58,048	6.7
G7	512	116,493	8.1
G8	1024	233,383	9.7
G9	2048	467,162	11.4
G10	4096	934,720	13.5

A reaction cycle is referred to as a generation (G), which determines the molecular weight and the number of terminal groups. The most widely studied dendrimers are polyamidoamine (PAMAM) dendrimers, which are commercially available. The number of terminal groups, the molecular weights, and the sizes of PAMAM dendrimers with ethylenediamine cores are shown in Table 12.1. The number of terminal groups doubles with each generation, and the molecular weight increases stepwise. The diameters of the dendrimer from G1 to G10 range from 2.2 to 13.5 nm [2–4]. These sizes are similar to some proteins. For example, the sizes of insulin, cytochrome c, and hemoglobin are 3.0, 4.5, and 5.5 nm, respectively, and correlate with those of the G3, G4, and G5 PAMAM dendrimers [2–4]. Because dendrimers have well-controlled spherical structures, they are useful for designing protein mimetics. Dendrimers also possess characteristic cores and inner spaces; these are relatively isolated from their external environments, especially in large generation dendrimers [2, 6, 7, 10, 11]. Drug molecules and metal nanoparticles have been encapsulated in dendrimers, making them potential drug carriers, catalysts, and nanoreactors [2, 6–9, 11].

The terminal groups of dendrimers have been widely used for functionalization. Because it is easy to add reactive groups at the dendrimer termini, various compounds can be linked to the dendrimer periphery. In some cases, the modification is beyond the addition of compounds to the dendrimer. The modified moieties exhibit clustering effects. For example, the binding affinity of the sugar- and peptide-modified dendrimers is much larger than those of the intact molecules [2, 5, 12, 13]. Peptides linked to the surface of a dendrimer can also induce higher-order structures [12]. These clustering effects are unique to dendrimers and are useful for designing artificial proteins. Thus, collagen model peptides and elastin model peptides were conjugated to the termini of PAMAM dendrimers to produce collagen and elastin mimetics (Fig. 12.2). Amino acids were also conjugated to the dendrimer termini to produce peptidomimetics. The details are given below.

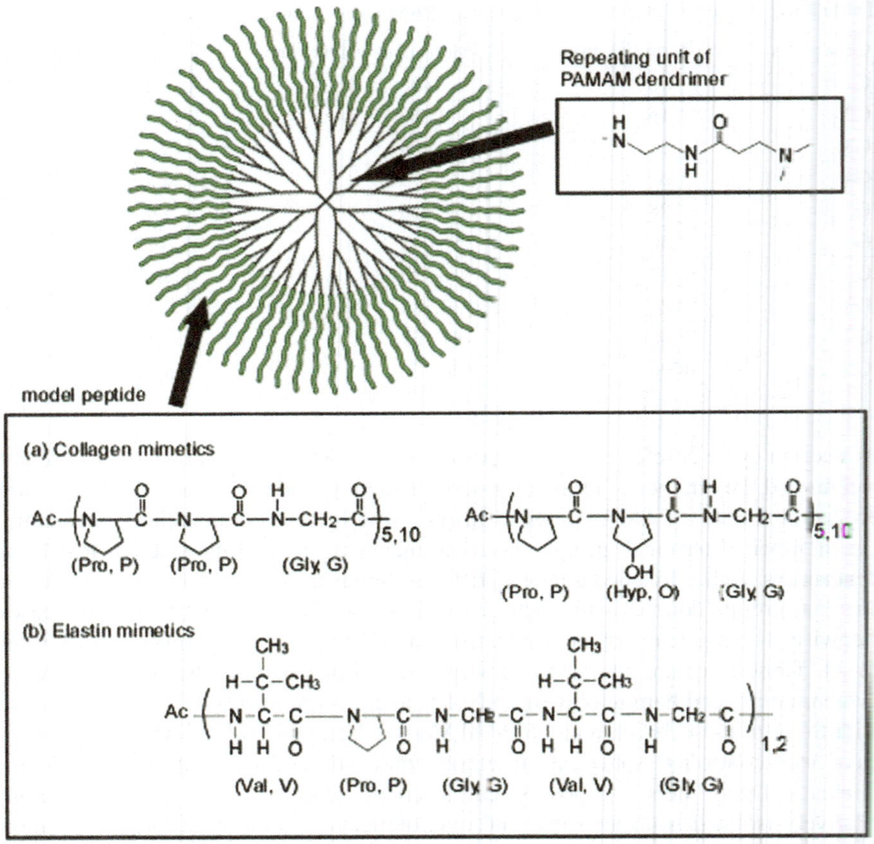

Fig. 12.2 Structure of collagen-mimetic dendrimers (a) and elastin-mimetic dendrimers (b)

12.3 Collagen Mimetics

Collagen is the most abundant protein in mammals. Collagen is composed of glycine-proline-(hydroxy) proline [Gly–Pro–Pro (Hyp)] repeats that form a triple helix in a temperature-dependent manner. The triple helical structure is formed at low temperature, but it dissociates above the melting point. Thermal denaturation of collagen at high temperature results in the formation of gelatin [14–15]. The melting point is largely affected by the amount of hydroxyproline (Hyp). The melting temperature increases with increasing Hyp. The composition and the thermal properties of collagen are different in each species. For example, porcine collagen has more Hyp and a higher melting temperature than that of fish collagen [14, 16, 17].

Both collagen and gelatin form hydrogels, which are useful for biomedical applications such as tissue engineering and drug delivery [15–17]. However, their gelation mechanisms are different [18]. In collagen gels, triple helixes self-assemble into fibrils and then fibers, leading to the formation of a gel network. Because the dehydration around triple helical structures influences self-association, collagen gels are formed upon heating. In gelatin gels, however, partially formed triple helixes play a role in cross-linking gelatin molecules. The triple helix forms efficiently at low temperature, so gelatin gels are formed by cooling. Thus, the temperature-dependent sol–gel transitions of collagen and gelatin are opposite. The biological properties and applications of collagen- and gelatin-type gels differ because of the different thermal properties. Collagen-type gels could be applied as injectable hydrogels that could be prepared at body temperature or by local heating systems. Gelatin-type gels are useful for temperature-inducible drug release.

At present, most collagen materials are extracted from animals such as pigs and fishes. Although porcine collagen has similar properties to that of human collagen, it is possibly contaminated by dangerous pathogens. Fish collagen is safer than porcine collagen, but its thermal properties, especially the melting temperature, are different from those of human collagen because of lower amount of Hyp. Hyp is produced by the post-modification of proline (Pro). Because controlling the post-modification is problematic, collagens are difficult to produce using recombinant technology. Collagen model peptides were previously synthesized, but it is difficult for short collagen peptides to form a stable triple helix. Recently, knotted collagen peptides have been shown to effectively form triple helixes and hydrogels as well as to self-assemble [16, 17]. However, these artificial collagen gels exhibited a gelatin-like thermal property; the hydrogel formed at low temperature and were dissolved at high temperature [19–21]. Our paper on a collagen model peptide-conjugated dendrimer was the first report on an artificial collagen-like gel that forms upon heating [22].

12.3.1 Synthesis and Characterization of Collagen-Mimetic Dendrimers

(Pro-Pro-Gly)$_{5,10}$ (PPG5 and PPG10) and (Pro-Hyp-Gly)$_{5,10}$ (POG5 and POG10) are commercially available. These peptides were acetylated and reacted with PAMAM dendrimers of different generations using condensation agents. PPG5 were reacted using HATU (1-[bis(dimethylamino)methylene]-1H-1,2,3-triazolo[4,5-b] pyridinium 3-oxid hexafluorophosphate) in organic solvent, and others were reacted using DMT-MM (4-(4,6-dimethoxy-1,3,5-triazin-2-yl)-4-methylmorpholinium chloride) in aqueous media. The synthesized compounds were identified by ^{1}H nuclear magnetic resonance (NMR), fluorescamine assay, and high-performance liquid chromatography (HPLC). In the ^{1}H NMR spectra, except for the spectrum of the POG10-conjugated dendrimer, signals derived from both the PAMAM dendrimer and collagen model peptides were observed. The number of bound peptide was calculated from the integral ratio of the peptide to the dendrimer and was similar to that of the terminal groups of the dendrimer. This result indicated that these peptides were attached to all the terminal amino groups of the dendrimer [23–25]. However, POG10 peptide signals were observed in the NMR spectrum of the POG10-conjugated dendrimer, but the PAMAM dendrimer signals were not. The HPLC chromatogram of the POG10-conjugated dendrimer showed no significant peak corresponding to the free peptide, indicating that no free peptides were contained in the dendrimer compound. These results suggest that the dendrimer was covered with POG10 peptides that diminished the dendrimer signals [22]. These collagen-mimetic dendrimers were also analyzed by a fluorescamine assay. This assay is used to determine primary amino groups and detects the unreacted amino groups of the dendrimers [27]. Less than 5 % of the terminal amino groups were detected in these collagen-mimetic dendrimers. Thus, the reactivity of the peptides was calculated as higher than 95 %. These results indicate that terminal groups of the dendrimer were almost fully modified with the collagen model peptides [22–26

Circular dichroism (CD) analyses are useful for estimating protein secondary structure. Collagen model peptides form a collagen-like triple helix, in which three left-handed polyproline II-like helixes intertwine to form a right-handed supercoiled helix [16, 17]. The triple helical structure shows a characteristic CD spectrum containing a positive Cotton effect around 225 nm and a negative Cotton effect around 200 nm [28]. CD spectra of collagen-mimetic dendrimers were measured at varying temperatures. All the samples showed characteristic collagen CD patterns, even though the molar ellipticity at 225 nm ($[\theta]_{225}$) varied [22–26]. The $[\theta]_{225}$ of dendrimers with longer POG10/PPG10 peptides was higher than those with the shorter POG5/PPG5 peptides. This result suggested that elongation of the peptide chain promoted the triple helix formation. The $[\theta]_{225}$ of the POG10-conjugated dendrimer was higher than the PPG10-conjugated dendrimer, because Hyp is indispensable for the enhancement of triple helicity [16, 17]. Therefore, both the peptide chain length and the peptide sequence in the collagen-mimic dendrimer are important for the triple helical structure to form efficiently. The ratio of positive and negative peaks is a standard indicator for evaluating the efficiency of the collagen-like triple helical structure [28]. The ratio of positive and negative peak of the POG10-conjugated dendrimer was estimated to be 0.12 and that of natural collagen one (porcine type IA collagen) was 0.11 under our conditions [22, 24]. This indicates that the POG10-conjugated dendrimer is comparable in this respect to natural collagen. ^{1}H NMR spectrometry and X-ray diffraction analyses were also performed to confirm the formation of the collagen-like triple helical structure [25].

The thermal stability of the collagen-like triple helix was also investigated by measuring $[\theta]_{225}$ during heating and subsequent cooling processes [22–26]. The $[\theta]_{225}$ of PPG5 and POG5 peptides and these short peptides-conjugated dendrimers showed a gradual decrease with increasing temperature, and complete recovery after cooling. However, the $[\theta]_{225}$ of PPG10 and POG10 peptides decreased sharply around 33 and 60 °C, respectively, and was partially recovered by cooling. The marked change in $[\theta]_{225}$ corresponds to the melting of the triple helical structure and indicates denaturation of these peptides. The melting of the triple helical structure in the PPG10- and POG10-conjugated dendrimers was observed around 45 and 75 °C, which was much higher than those of the peptides. These results suggested that the triple helical structures of the PPG10- and POG10-conjugated dendrimers were more thermally stable than those of the corresponding peptides because of the clustering effect. The high thermal stability of the triple helical structure in the POG10-conjugated dendrimer, compared with that in the PPG10-conjugated dendrimer, was affected by the Hyp content, because the hydroxyl groups formed hydrogen bonds to enhance the stability [16, 17].

12.3.2 Thermoresponsive Hydrogels

We prepared hydrogels using these collagen mimetics. Aqueous suspensions (15 wt %) of these collagen model peptides and the collagen-mimetic dendrimers were prepared and incubated at different temperatures. The solutions containing these peptides and PPG5- and POG5-conjugated dendrimers did not change into the hydrogel at any temperature. On the other hand, the PPG10- and POG10-conjugated dendrimers formed temperature-sensitive hydrogels (Fig. 12.3). Interestingly, the sensitivity was opposite [22, 25]. The PPG10-conjugated dendrimer formed the hydrogel at a lower temperature, and the hydrogel dissolved above 45 °C [25]. The manner of gelling was gelatin-like. The POG10-conjugated dendrimer suspension formed a hydrogel at 40 °C upon heating. Because this gel was formed by heating, the POG10-conjugated dendrimer formed a collagen-like hydrogel. Natural collagen gel extracted from mammals denatures to form a gelatin sol around 45 °C, at which the triple helical structure is disturbed. However, the transition from the gel state to the sol state was not observed up to 80 °C in the hydrogel made from the POG10-conjugated dendrimer [22]. This finding may be because the triple helical structure was closely packed at the dendrimer surface.

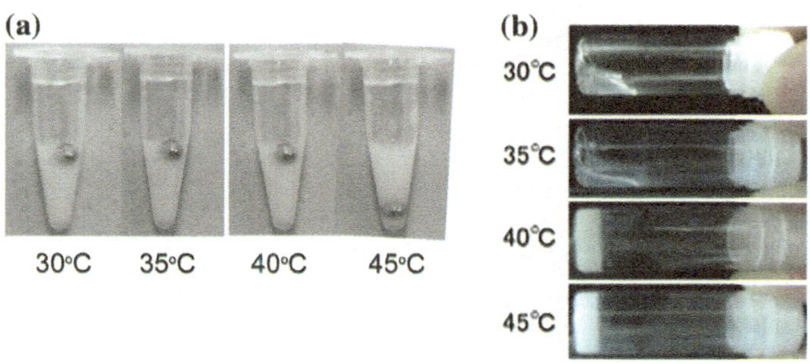

Fig. 12.3 Thermosensitive hydrogels made from PPG10-conjugated (**a**) and POG10-conjugated dendrimers (**b**)

Thus, the POG10-conjugated dendrimer showed thermal resistance superior to that of natural collagen.

12.3.3 Thermoresponsive Drug Delivery

Because of the drug encapsulation properties of dendrimers, rose bengal was chosen as a model drug, and functions of the collagen-mimetic dendrimers as a drug vehicle was evaluated. Interestingly, the release of rose bengal from the PPG5-conjugated dendrimer increased at higher temperatures. The formation of a collagen-like triple helix at lower temperature may improve the binding properties of rose bengal to the dendrimer because of the enhanced hydrophobic interaction and the suppressed permeability of rose bengal at the surface [23] (Fig. 12.4). Other collagen-mimetic dendrimers could also encapsulate rose bengal and release it in a temperature-dependent manner [26]. As shown in the previous section, the POG10- and the PPG10-conjugated dendrimers formed temperature-dependent hydrogels, which are also useful as smart drug containers [22, 25].

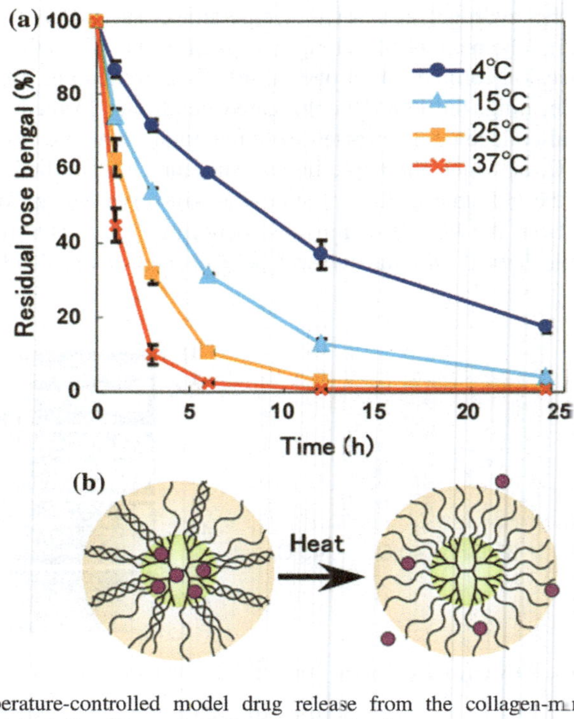

Fig. 12.4 Temperature-controlled model drug release from the collagen-mimetic dendrimer. **a** Experimental data. **b** Possible controlled release mechanism

12.4 Elastin Mimetics

Elastin is one of the main components of extracellular matrix and has been used as a temperature-sensitive biomaterial. Elastin is composed of repeat sequences such as valine-proline-glycine-valine-glycine (Val-Pro-Gly-Val-Gly, VPGVG) [29–31]. This peptide is called an elastin-like peptide and is used to prepare elastin-mimetic materials. Previous reports showed that elastin-like polypeptides induced temperature-dependent structure change from random coil to β-turn and/or β-spiral, leading to coacervation because of hydrophobic interactions [29–31]. Most elastin-mimetic polymers are produced using recombinant protein technology or by grafting elastin-like peptide at synthetic polymers [32, 33]. Elastin-like peptides have been conjugated to the PAMAM dendrimers to produce elastin mimetics. They exhibited temperature-dependent properties [34–36]. The temperature-sensitive properties and their applications are shown in this section.

12.4.1 Synthesis and Characterization

Boc (tert-butoxycarbonyl)-protected elastin-like peptide containing one and two repeats [VPGVG (ELP1) and $(VPGVG)_2$ (ELP2)] were synthesized, and reacted with PAMAM dendrimers of G3, G4 and G5 using HATU. The Boc group was deprotected, followed by acetylation. ^{1}H NMR spectra of elastin-mimetic dendrimers showed signals derived from the PAMAM dendrimer, the peptide unit, and the acetyl unit. The integral ratios indicated that all the termini of the dendrimers were modified with the peptides and the acetyl group. These elastin-mimetic dendrimers were characterized by CD measurements, solution transmittance assay, and differential scanning calorimetry (DSC) to determine the phase transition mechanism of the elastin-mimetic dendrimers and the influence of peptide lengths and dendrimer generations on the temperature dependency.

CD spectra of the elastin-mimetic dendrimers showed a negative Cotton effect that corresponds to a β-turn structure (218 nm) and random coil (196 nm), and the ratios were changed by heating [37]. The strong signal of the random coil structure was observed at low temperatures, and the signal of the β-turn structure increased at higher temperatures in all elastin mimetics. The dendrimer generation and the peptide length did not affect the negative Cotton effect at 218 nm, suggesting that the peptide lengths as well as the dendrimer generations did not influence the formation of the β-turn structure. Because these elastin-mimetic dendrimers were similar to elastin-like peptides, they can work as an artificial elastin-mimetic material with temperature-dependent conformation change [35, 36].

The solution containing elastin-mimetic dendrimers became turbid by heating, but the corresponding peptides did not. The change in the transmittance of the solution was monitored with increasing temperature. The drastic transmittance change at a specific temperature, called the phase transition temperature, was observed in all elastin-mimetic dendrimers. The average phase transition temperature of ELP1-G4, ELP2-G3, ELP2-G4, and ELP2-G5 were 48, 39, 38, and 30 °C, respectively. These results suggest that longer peptides and larger dendrimer generation resulted in the decreased phase transition temperature. Elastin mimetics can indicate a salt-dependent thermosensitive property [37, 38]. A linear relationship was observed in all elastin-mimetic dendrimers, consistent with previous observations in other elastin mimetics studies [35–38]. Because PAMAM dendrimers have many tertiary amines, it was expected that the elastin-mimetic dendrimers would show pH sensitivity. A higher phase transition temperature was observed at low pH in all elastin-mimetic dendrimers. When the inner tertiary amines were protonated, the phase transition temperature became higher because of the increased hydrophilicity [35, 36].

Temperature-dependent polymers such as poly(N-isopropylacrylamide) (poly (NIPAM)) exhibit an endothermal peak around the phase transition temperature at which the dehydration of bound water to the hydrophobic moiety of the polymer occurred [39]. Elastin mimetics showed an endothermal reaction corresponding to dehydration [40]. The thermal behavior of elastin-mimetic dendrimers was examined by DSC. ELP2-G3, ELP-G4, and ELP-G5 exhibited endothermal peaks at 38, 36, and 32 °C, respectively, which correspond to their phase transition temperatures [36]. This observation was consistent with those of previous elastin mimetics [40]. The area of the endothermal signal in DSC corresponds to ΔH. Normalized to dendrimer, larger endothermal peaks were observed in larger generation dendrimers. The larger generation dendrimer contains large amounts of peptide, which is a cause of a large endothermal signal. However, when normalized to peptide unit (VPGVG), the endothermal peaks of all ELP2-conjugated dendrimers look similar, suggesting that the peptide induced the similar dehydration. On the other hand, ELP1-G4 did not show any significant peaks under our conditions, suggesting that dehydration did not occur around the phase transition temperature [35]. Dendrimers conjugating NIPAM units at the termini did not show any significant endothermal signals [39]. Short peptide units at the surface of the dendrimer may provide only small amounts of bound water.

The possible mechanism of the temperature dependency of the elastin-mimetic dendrimers is shown in Fig. 12.5. The conformation of ELP at the termini of elastin-mimetic dendrimers changed from random coil to β-turn by heating (step 1); dehydration occurred (step 2), which led to aggregation (step 3); consequently, the turbidity rapidly increased around the phase transition temperature (step 4) [36]. Compared with ELP1-G4 and ELP2-G4, the effects of peptide length to the phase transition behavior are discussed. The conformation change was not significantly affected by the peptide length; however, drastic dehydration occurred in ELP2-G4, but not in ELP1-G4. Accordingly, aggregation was promoted and the phase transition temperature was significantly decreased when the peptide length increased. The balance of hydrophobicity is a key factor for temperature-dependent polymers [39]. The molecular weight of elastin mimetics affected phase transition behaviors [30, 31]. Thus, the phase transition temperature decreased when the hydrophobic ratio and the molecular weight were increased. ELP2-G3, ELP2-G4, and ELP2-G5

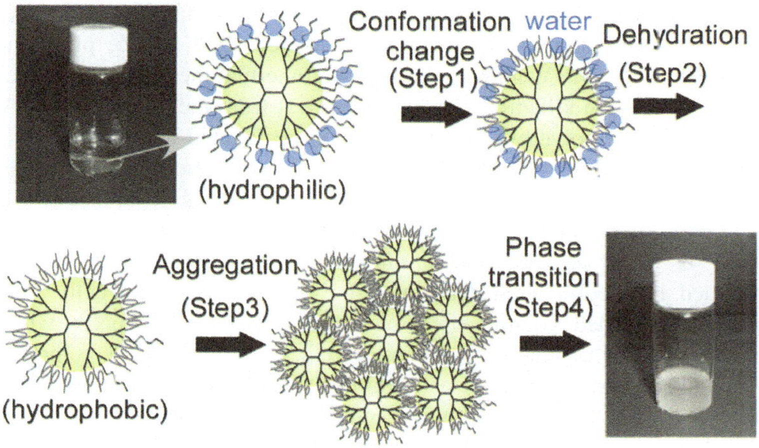

Fig. 12.5 Possible phase transition mechanism of the elastin-mimetic dendrimers. Reproduced from Ref. [36] by permission of John Wiley & Sons Ltd

were compared to understand the effects of dendrimer generation. Conformation change and dehydration were not affected by dendrimer generation. However, aggregation and phase transition temperature were affected by dendrimer generation. Even though the hydrophobicity balance was unchanged in the different-generation dendrimers, the molecular weights increased. The decreased phase transition temperature may be related to the increased molecular weight. In conclusion, both peptide length and dendrimer generation affected the thermosensitivity of the elastin-mimetic dendrimers. ELP2-G4 and ELP2-G3 exhibited thermosensitivity around body temperature, which is useful for biomedical applications [36].

12.4.2 Thermoresponsive Carrier of Bioactive Molecules

Thermosensitive polymers have been used for temperature-dependent drug delivery systems, because they can contribute to temperature-controlled drug release and cellular uptake [9, 41, 42]. The elastin-mimetic dendrimers are applicable as temperature-sensitive carriers of bioactive molecules. A model drug, rose bengal, was encapsulated in the elastin-mimetic dendrimers. The larger the dendrimer, the higher the encapsulation ability was. However, the thermosensitive properties did not affect the release profiles, which may be because rose bengal stably interacted with the core PAMAM dendrimer, and the peptide conformation change did not affect the complex stability [35].

Gold nanoparticles (AuNPs) have attracted attention because of their interesting shape- and size-dependent photosensitive properties, which are different from bulk gold and gold atoms [43]. Moreover, AuNPs are biocompatible and have photoinduced, heat-generating properties because of localized surface plasmon resonance. Therefore, AuNPs can be applied to photothermal therapy, in which cells at the affected tissues are killed by photoinduced heat from the administrated AuNPs [43–45]. Previously, polyethylene glycol (PEG)-conjugated dendrimers have been used as carriers for AuNPs. AuNP-loaded PEG-conjugated dendrimers exhibit photothermal properties and photoinduced cytotoxicity, suggesting that they can be applied to photothermal therapy [46, 47]. However, seeded growth of AuNPs in the dendrimer is necessary to obtain effective photocytotoxicity. Because the AuNP-loaded PEG-conjugated dendrimer did not associate with cells efficiently owing to the excluded volume effect of PEG chains, larger AuNPs with more efficient photothermogenic properties were required [47]. Instead of PEG-conjugated dendrimers, the elastin-mimetic dendrimers were used as a carrier of AuNP [48]. The AuNP-loaded elastin-mimetic dendrimers were prepared from mixtures of $HAuCl_4$ and the elastin-mimetic dendrimers at a mole ratio of 55/1; the subsequent reduction of the Au (III) ions was performed using $NaBH_4$. Localized surface plasmon resonance derived from the AuNP was observed after the AuNP loading. The average diameter of AuNP in ELP2-G4 was estimated to be 1.5 nm. The solution temperature of the AuNP-loaded elastin-mimetic dendrimer increased 2.7 °C, when green laser light (ca. 1 W, ϕ 3 mm) was irradiated for 5 min. The AuNP-loaded PEG-conjugated dendrimer exhibited a similar photothermogenic property, indicating that the elastin-mimetic dendrimers did not affect the property of loaded AuNP [48]. CD spectra of the AuNP-loaded elastin-mimetic dendrimers were similar to the elastin-mimetic dendrimers, indicating that the loaded AuNP did not interfere with the conformation change of the elastin-mimetic dendrimer. The hydrodynamic diameters of the AuNP-loaded elastin-mimetic dendrimers were measured at different temperatures by dynamic light scattering. The diameter of the AuNP-loaded elastin-mimetic dendrimer started to increase at 35 °C, reaching around 500 nm at temperatures of 45 °C or more. This increase in the diameter indicated that the AuNP-loaded elastin-mimetic dendrimer formed aggregates because of the phase transition process [48].

The photocytotoxicity of the AuNP-loaded elastin-mimetic dendrimer was examined using our single cell photoirradiation system. The viability of HeLa cells treated with AuNP-loaded elastin-mimetic dendrimer (300 μM Au) at 37 °C was determined to be 92 % when the cells were cultured in the absence of light. This result indicated that the AuNP-loaded elastin-mimetic dendrimer showed a high biocompatibility. HeLa cells were then incubated with AuNP-loaded elastin-mimetic dendrimer and PEG-conjugated dendrimer for 6 h at 37 °C or 25 °C, and photoirradiated for 1 min. A live/dead assay was then performed using calcein-AM (green) and PI (red) to stain the live and dead cells respectively (Fig. 12.6). In the phase contrast images, more black dots were observed for the cells treated with the AuNP-loaded-elastin-mimetic dendrimer at 37 °C than there were for the cells treated with the same material at 25 °C. Furthermore, fewer black dots were observed for the cells treated with the AuNP-loaded PEG-conjugated dendrimer than those treated with the AuNP-loaded elastin-mimetic dendrimer.

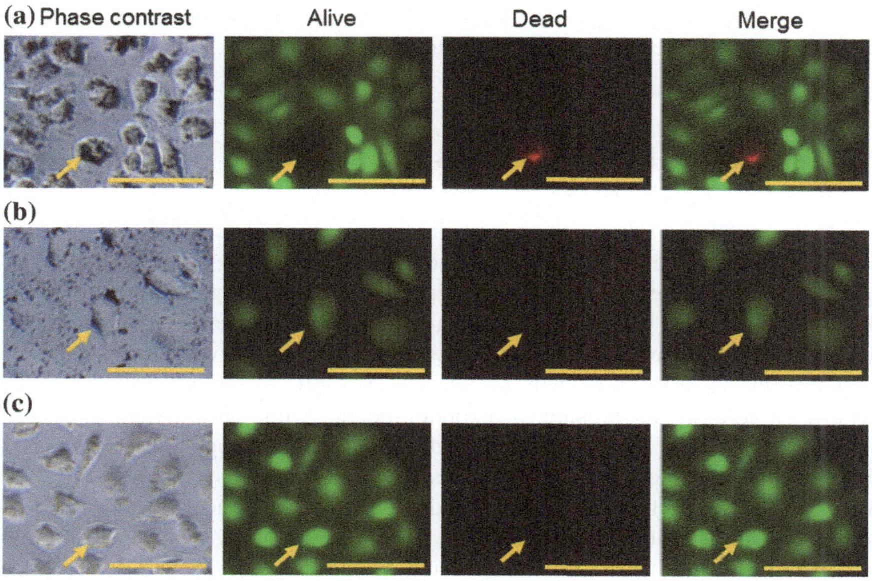

Fig. 12.6 Temperature-controlled photocytotoxicity using the AuNP-loaded elastin-mimetic dendrimers. **a** 37 °C. **b** 25°C. Photocytotoxicity of the AuNP-loaded PEG-conjugated dendrimer at 37 °C was also shown. *Arrows* indicate the photoirradiated cells. Reprinted from Ref. [48], Copyright 2015, with permission from Elsevier

This result suggested that the hydrophobic AuNP-loaded elastin-mimetic dendrimer associates more efficiently with the HeLa cells than the hydrophilic AuNP-loaded PEG-conjugated dendrimer and the AuNP-loaded elastin-mimetic dendrimer prior to the phase transition. Photoirradiated cells treated with the AuNP-loaded elastin-mimetic dendrimer at 37 °C became red, indicating cell death. However, those treated with the AuNP-loaded elastin-mimetic dendrimer at 25 °C and the AuNP-loaded PEG-conjugated dendrimer remained viable after the photoirradiation. These results indicate that photoirradiation with a laser for only 1 min can induce photocytotoxicity against a single target cell using the AuNP-loaded elastin-mimetic dendrimer above the phase transition temperature. Thus, the photocytotoxicity of the AuNP-loaded elastin-mimetic dendrimers can be controlled by temperature [48]. As described, the AuNP-loaded elastin-mimetic dendrimers worked as a dual stimuli- (photo- and temperature-) sensitive nanoparticle. The dual stimuli-sensitive dendrimer showed the cytotoxicity only when both light and heat were added. This indicated that the dual stimuli-sensitive dendrimer provided "AND" gate, which are useful for improving selectivity of the therapeutic effects. The AuNP-loaded dendrimer prepared in the current study responded to visible light, which cannot penetrate deeply into the tissue. In contrast, gold nanorods respond to near infrared light, which can penetrate more deeply into the tissues [49,

50]. Thus, the loading of gold nanorods onto elastin-mimetic dendrimers may be used to overcome the issue of light penetration.

12.5 Dendritic Peptidomimetics

As shown in the previous sections, artificial proteins were successfully prepared by conjugating the model peptides to the dendrimer periphery. Next, artificial peptides (peptidomimetics) were designed and prepared by conjugating amino acids to the dendrimer periphery [51]. First, alanine and phenylalanine were conjugated to the termini of a dendrimer at the different ratios. The bound ratio was almost the same as that in the in feed ratio. Then, proline (P) and arginine (R) were bound to the dendrimer termini to prepare a peptidomimetic with proline-rich sequence. Proline-rich peptides (PxxPxR) are bound to the Src Homology 3 (SH3) domain; this interaction is indispensable for the signal transduction system [52–54]. The dissociation constant (K_d) of the dendritic peptidomimetic with proline-rich sequence was estimated using isothermal titration calorimetry analysis. The binding experiment was also performed using a conventional proline-rich peptide (QVPSRPNRAP), which bound to the SH3 domain shown in the previous report [54]. K_d of the dendritic peptidomimetic was 4.72 mM, which was similar to that of the conventional proline-rich peptide (1.51 mM). This result suggests that the dendritic peptidomimetic is a potential peptidomimetic. However, further improvement to enhance the binding affinity is necessary.

Dendrimers are asymmetric: The core and the terminal groups are different. Using a part of a dendrimer, the dendron, different molecules can be bound to the core and the termini of the single dendritic molecule. Our dendritic peptidomimetic contained a fluorescent dye and amino acids at the core and the termini of the dendron, respectively (Fig. 12.7). Using fluorescent dendritic peptidomimetics, binding experiments were performed by the pull-down assay coupled with a fluorescent intensity measurement. This peptidomimetic can be used in combinatorial peptide libraries to determine protein interactions for the design of new drugs and to further develop biological physiology. Combinatorial peptide libraries designed using solid-phase peptide synthesis have traditionally been used to identify bioactive peptides [55, 56]. There are, however, two major difficulties associated with synthetic peptide libraries. First, short peptides exhibit low levels of affinity because of their overall structural flexibility [57, 58]. Clustering effects of the modified compounds at the dendrimer termini improved the binding affinity of dendritic peptidomimetics, as shown in Fig. 12.7 [2, 6, 12, 13]. Second, synthesizing diverse peptides is challenging [55, 56, 59]. In the dendritic peptidomimetic, a virtually one-step synthesis using the twenty natural amino acids (conjugation of amino acids to the dendrimer) can enable preparing a variety of peptidomimetics. For example, the reaction of the eight terminal groups from generation 3 (G3) of a dendron with twenty amino acids would yield a library of 8^{20} (2.56×10^{10}) peptides, which is much greater than that achievable with a conventional synthetic

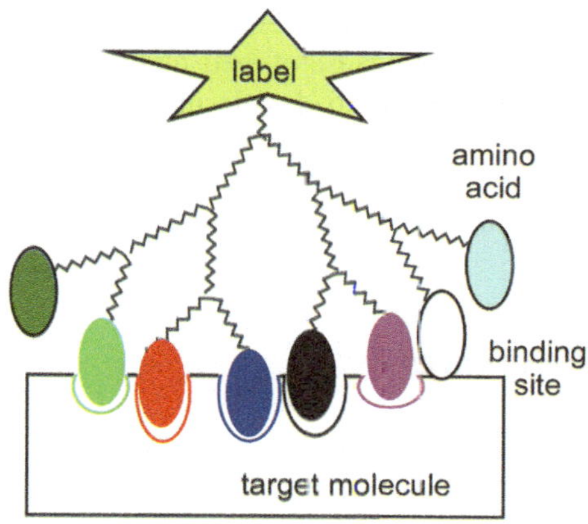

Fig. 12.7 Design of dendritic peptidomimetics. Reprinted from Ref. [51]

peptide library. Although it is difficult to exert control over the configuration of each amino acid in the compound, peptidomimetics of this type may be useful as a preliminary screening tool to determine which amino acids are important for the interactions.

12.6 Conclusions

Collagen mimetics and elastin mimetics are artificial proteins that were successfully prepared by conjugating the model peptides at dendrimer termini. Peptidomimetics were prepared by conjugating amino acids at dendrimer termini. The clustering effects of modified compounds are present at the dendrimer termini; these effects led to mimetic properties to corresponding natural proteins and peptides. The POG10-conjugated dendrimer is highly thermally resistant, which is superior to natural collagen. Thermal resistance is a useful property for easy handling. Moreover, the physicochemical properties were able to be regulated by peptide length and sequence as well as the dendrimer generation. Thus, suitable artificial proteins can be designed.

Research on dendrimer-based artificial proteins is now focused on their applications. The uniform structure of dendrimers contributes to their reproducibility. Because collagen and elastin are major components of extracellular matrices, these mimetics can be used as cell scaffolds. Collagen and elastin mimetics show temperature-sensitive properties. Thus, they can be used as smart biomaterials. Temperature-dependent delivery systems are a possible application. Because dendrimers are well-controlled nanosized particles, they can benefit from the enhanced permeability and retention (EPR) effect. The EPR effect is a phenomenon that macromolecules accumulate in tumor tissues, which is caused by leakage from angiogenic vessels and lack of lymphatic tissues [60]. Therefore, dendrimers have been used in cancer therapies for drug delivery and diagnosis. The large number of terminal groups and presence of inner space are useful in designing multifunctional dendrimers. Because of their multifunctionality, dendrimer-based artificial proteins and peptides can load molecules such as drugs, imaging probes, ligands, and other stimuli-responsive moieties in a single molecule. Thus, dendrimers are useful as a platform of multifunctional unimolecular micelles [61]. Dendrimers with biomimetic interfaces are widely applicable in biomedical applications.

References

1. Tomalia DA, Baker H, Dewald J, Hall M, Kallos G, Martin S, Roeck J, Ryder J, Smith P (1985) A new class of polymers: starburst-dendritic macromolecules. Polym J 17(1):117–132
2. Fréchet JMJ, Tomalia DA (2001) Dendrimers and other dendritic polymers. John Wiley & Sons, 688
3. Tomalia DA, Naylor AM, Goddard WA III (1990) Starbust dendrimers: molecular-level control of size, shape, surface chemistry, topology, and flexibility from atoms to macroscopic matter. Angew Chem Int Ed 29(2):138–175
4. Svenson S, Tomalia DA (2005) Dendrimers in biomedical applications–reflections on the field. Adv Drug Deliv Rev 57(15):2106–2129
5. Matthews OA, Shipway AN, Stoddart JF (1998) Dendrimers—branching out from curiosities into new technologies. Prog Polym Sci 23(1):1–56
6. Newkome GR, Moorefield CN, Vogtle F (2001) Dendrimers and dendrons: concepts, syntheses, applications. Wiley-VCH 623
7. Astruc D, Boisselier E, Ornelas C (2010) Dendrimers designed for functions: from physical, photophysical, and supramolecular properties to applications in sensing, catalysis, molecular electronics, photonics, and nanomedicine. Chem Rev 110(4):1857–1959
8. Lee CC, MacKay JA, Fréchet JMJ, Szoka FC (2005) Designing dendrimers for biological applications. Nat Biotechnol 23(12):1517–1526
9. Kojima C (2010) Design of stimuli-responsive dendrimers. Expert Opin Drug Deliv 7 (3):307–319
10. Nishiyama N, Morimoto Y, Jang WD, Kataoka K (2009) Design and development of dendrimer photosensitizer-incorporated polymeric micelles for enhanced photodynamic therapy. Adv Drug Deliv Rev 61(4):327–328
11. Crooks RM, Zhao M, Sun L, Chechik V, Yeung LK (2001) Dendrimer-encapsulated metal nanoparticles: synthesis, characterization, and applications to catalysis. Acc Chem Res 34 (3):181–190
12. Crespo L, Sanclimens G, Pons M, Giralt E, Royo M, Albericio F (2005) Peptide and amide bond-containing dendrimers. Chem Rev 105(5):1663–1681
13. Sadler K, Tam JP (2002) Peptide dendrimers: applications and synthesis. J Biotechnol 90(3–4):195–229
14. Gelse K, Poschl E, Aigner T (2003) Collagens—structure, function, and biosynthesis. Adv Drug Deliv Rev 55(12):1531–1546
15. Wallace DG, Rosenblatt J (2003) Collagen gel systems for sustained delivery and tissue engineering. Adv Drug Deliv Rev 55(12):1631–1649
16. Koide T (2005) Triple helical collagen-like peptides: engineering and applications in matrix biology. Connect Tissue Res 46(3):131–141

17. Fields GB (2010) Synthesis and biological applications of collagen-model triple-helical peptides. Org Biomol Chem 8(6):1237–1258
18. Djabourov M, Lechaire JP, Gaill F (1993) Structure and rheology of gelatin and collagen gels. Biorheology 30(3–4):191–205
19. Yamazaki CM, Asada S, Kitagawa K, Koide T (2008) Artificial collagen gels via self-assembly of de novo designed peptides. Biopolymers 90(6):816–823
20. Wang AY, Foss CA, Leong S, Mo X, Pomper MG, Yu SM (2008) Spatio-temporal modification of collagen scaffolds mediated by triple helical propensity. Biomacromolecules 9 (7):1755–1763
21. Werten MW, Teles H, Moers AP, Wolbert EJ, Sprakel J, Eggink G, de Wolf F (2009) Precision gels from collagen-inspired triblock copolymers. Biomacromolecules 10 (5):1106–1113
22. Kojima C, Suehiro T, Tada T, Sakamoto Y, Waku T, Tanaka N (2011) Preparation of heat-induced artificial collagen gels based on collagen-mimetic dendrimers. Soft Matter 7 (19):8991–8997
23. Kojima C, Tsumura S, Harada A, Kono K (2009) A collagen-mimic dendrimer capable of controlled release. J Am Chem Soc 131(17):6052–6053
24. Suehiro T, Kojima C, Tsumura S, Harada A, Kono K (2010) Higher order structure of short collagen model peptides attached to dendrimers and linear polymers. Biopolymers 93 (7):640–648
25. Suehiro T, Tada T, Waku T, Tanaka N, Hongo C, Yamamoto S, Nakahira A, Kojima C (2011) Temperature-dependent higher order structures of the $(Pro\text{-}Pro\text{-}Gly)_{10}$-modified dendrimer. Biopolymers 95(4):270–277
26. Kojima C, Suehiro T (2011) Improvement of temperature-responsive drug release from collagen-mimic dendrimers. Chem Lett 40(11):1249–1251
27. Yuba E, Kojima C, Harada A, Koiwai K, Kono K (2008) Gene delivery to dendritic cells mediated by complexes of lipoplexes and pH-sensitive fusogenic polymer-modified liposomes. J Control Rel 130(1):77–83
28. Khew ST, Yang QJ, Tong YW (2008) Enzymatically crosslinked collagen-mimetic dendrimers that promote integrin-targeted cell adhesion. Biomaterials 29(20):3034–3045
29. Tatham AS, Shewry PR (2000) Elastomeric proteins: biological roles, structures and mechanisms. Trends Biochem Sci 25(11):567–571
30. Urry DW (1993) Molecular machines: how motion and other functions of living organisms can result from reversible chemical changes. Angew Chem Int Ed Engl 32(6):819–841
31. Urry DW, Urry KD, Szaflarski W, Nowicki M (2010) Elastic-contractile model proteins: physical chemistry, protein function and drug design and delivery. Adv Drug Deliv Rev 62 (15):1404–1455
32. Ayres L, Vos MRJ, Adams PJHM, Shklyarevskiy IO, van Hest JCM (2003) Elastin-based side-chain polymers synthesized by ATRP. Macromolecules 36(16):5967–5973
33. McDaniel JR, Callahan DJ, Chilkoti A (2010) Drug delivery to solid tumors by elastin-like polypeptides. Adv Drug Deliv Rev 62(15):1456–1467
34. Koga T, Iimura M, Higashi N (2012) Novel peptide-shelled dendrimer with dramatically changeable thermo-responsive character. Macromol Biosci 12(8):1043–1047
35. Kojima C, Irie K (2013) Synthesis of temperature-dependent elastin-like peptide-modified dendrimer for drug delivery. Biopolym Peptide Sci 100(6):714–721
36. Kojima C, Irie K, Tada T, Tanaka N (2014) Temperature-sensitive elastin-mimetic dendrimers: effect of peptide length and dendrimer generation to temperature sensitivity. Biopolymers 101(6):603–612
37. Nuhn H, Klok HA (2008) Secondary structure formation and LCST behavior of short elastin-like peptides. Biomacromolecules 9(10):2755–2763
38. Reguera J, Urry DW, Parker TM, McPherson DT, Rodríguez-Cabello JC (2007) Effect of NaCl on the exothermic and endothermic components of the inverse temperature transition of a model elastin-like polymer. Biomacromolecules 8(2):354–358

39. Haba Y, Kojima C, Harada A, Kono K (2007) Comparison of thermosensitive properties between poly(amidoamine) dendrimers having peripheral N-isopropylamide groups and linear polymers with the same groups. Angew Chem Int Ed 46(1–2):234–237
40. Yamaoka T, Tamura T, Seto Y, Tada T, Kunugi S, Tirrell DA (2003) Mechanism for the phase transition of a genetically engineered elastin model peptide $(VPGIG)_{40}$ in aqueous solution. Biomacromolecules 4(6):1680–1685
41. Akimoto J, Nakayama M, Okano T (2014) Temperature-responsive polymeric micelles for optimizing drug targeting to solid tumors. J Control Rel 193:2–8
42. Kono K (2001) Thermosensitive polymer-modified liposomes. Adv Drug Deliv Rev 53 (3):307–319
43. Daniel MC, Astruc D (2004) Gold nanoparticles: assembly, supramolecular chemistry, quantum-size-related properties, and applications toward biology, catalysis, and nanotechnology. Chem Rev 104(1):293–346
44. Llevot A, Astruc D (2012) Applications of vectorized gold nanoparticles to the diagnosis and therapy of cancer. Chem Soc Rev 41(1):242–257
45. Govorov AO, Richardson HH (2007) Generating heat with metal nanoparticles. Nano today 2 (1):30–38
46. Haba Y, Kojima C, Harada A, Ura T, Horinaka H, Kono K (2007) Preparation of poly (ethylene glycol)-modified poly(amido amine) dendrimers encapsulating gold nanoparticles and their heat-generating ability. Langmuir 23(10):5243–5246
47. Umeda Y, Kojima C, Harada A, Horinaka H, Kono K (2010) PEG-attached PAMAM dendrimers encapsulating gold nanoparticles: growing gold nanoparticles in the dendrimers for improvement of their photothermal properties. Bioconjugate Chem 21(8):1559–1564
48. Fukushima D, UH Sk, Sakamoto Y, Nakase I, Kojima C (2015) Dual stimuli-sensitive dendrimers: photothermogenic gold nanoparticle-loaded thermo-responsive elastin-mimetic dendrimers. Colloid Surf B-Biointerfaces 132, 155–160
49. Hu M, Chen J, Li ZY, Au L, Hartland GV, Li X, Marquez M, Xia Y (2006) Gold nanostructures: engineering their plasmonic properties for biomedical applications. Chem Soc Rev 35(11):1084–1094
50. Chen H, Shao L, Li Q, Wang J (2013) Gold nanorods and their plasmonic properties. Chem Soc Rev 42(7):2679–2724
51. Kojima C, Fukada H, Inui T (2013) Synthesis and binding properties of peptidomimetics based on a dendritic polymer. Polymer J 45(3):339–345
52. Owen DJ, Wigge P, Vallis Y, Moore JDA, Evans PR, McMahon HT (1998) Crystal structure of the amphiphysin-2 SH3 domain and its role in the prevention of dynamin ring formation. EMBO J 17(18):5273–5285
53. Wigge P, McMahon HT (1998) The amphiphysin family of proteins and their role in endocytosis at the synapse. Trends Neurosci 21(8):339–344
54. Kojima C, Hashimoto A, Yabuta I, Hirose M, Hashimoto S, Kanaho Y, Sumimoto H, Ikegami T, Sabe H (2004) Regulation of bin1 SH3 domain binding by phosphoinositides. EMBO J 23(22):4413–4422
55. Falciani C, Lozzi L, Pini A, Bracci L (2005) Bioactive peptides from libraries. Chem Biol 12 (4):417–426
56. Gentilucci L, Tolomelli A, Squassabia F (2006) Peptides and peptidomimetics in medicine, surgery and biotechnology. Curr Med Chem 13(20):2449–2466
57. Olson GL, Bolin DR, Bonner MP, Bos M, Cook CM, Fry DC, Graves BJ, Hatada M, Hill DE, Kahn M, Madison VS, Rusiecki VK, Sarabu R, Sepinwall J, Vincent GP, Voss ME (1993) Concepts and progress in the development of peptide mimetics. J Med Chem 36 (21):3039–3049

58. Pastor MT, Mora P, Ferrer-Montiel A, Perez-Paya E (2004) Design of bioactive and structurally well-defined peptides from conformationally restricted libraries. Biopolymers 76 (4):357–365
59. Lam KC, Salmon SE, Hersh EM, Hruby VJ, Kazmierski WM, Knapp RJ (1991) A new type of synthetic peptide library for identifying ligand-binding activity. Nature 354(6348):82–84
60. Matsumura Y, Maeda H (1986) A new concept for macromolecular therapeutics in cancer chemotherapy: mechanism of tumoritropic accumulation of proteins and the antitumor agent smancs. Cancer Res 46(12):6387–6392
61. Majoros IJ, Thomas TP, Mehta CB, Baker JR Jr (2005) PAMAM dendrimer-based multifunctional engineered synthetic nanoscale bioconjugates for cancer therapy. J Med Chem 48(19):5892–5899

Chapter 13
Stimuli-Responsive Polymer Materials for Creation of Biointerfaces

Hidenori Otsuka and Daisuke Matsukuma

Abstract In this review, we introduce the biorecognition-driven stimuli-responsive surface and hydrogels. The first attention focuses on recent advances in the development of functionalizable antifouling coatings and their applications in label-free optical biosensors. Approaches to the development of antifouling coatings, ranging from self-assembled monolayers and PEG derivatives to low-fouling polymer brushes and polymerized gels, are reviewed. Preparation of antifouling coatings and the functionalization of antifouling coatings with bioreceptors are introduced, and the application example of biofunctional coating with fouling properties is discussed. Special attention is given to biofunctional coatings for label-free bioanalysis of blood plasma and serum for medical diagnostics. The following focus is fed light on the biorecognition-based stimuli-responsive hydrogels. We will discuss on peptides and proteins recognition system of stimuli-responsive hybrid hydrogel composed of synthetic polymers and biopolymers.

Keywords Biorecognition · Stimuli-responsive hydrogel · Peptide · Protein

13.1 Introduction

To sustain life and maintain biological function, nature requires selectively tailored molecular assemblies and interfaces that provide a specific chemical function and structure, and which change in their environment. Synthetic polymer systems (Fig. 13.1) including copolymers and polymer composites are often prepared as two-dimensional (2D) (films) and three-dimensional (3D) (particulates and their assemblies) stimuli-responsive structures for a broad range of applications, such as drug delivery, tissue engineering, medical diagnostics, and bioseparations [1]. This

H. Otsuka (✉) · D. Matsukuma
Department of Applied Chemistry, Faculty of Science Division I,
Tokyo University of Science, Tokyo, Japan
e-mail: h.otsuka@rs.kagu.tus.ac.jp

T. Kawai and M. Hashizume (eds.), *Stimuli-Responsive Interfaces*,
DOI 10.1007/978-981-10-2463-4_13

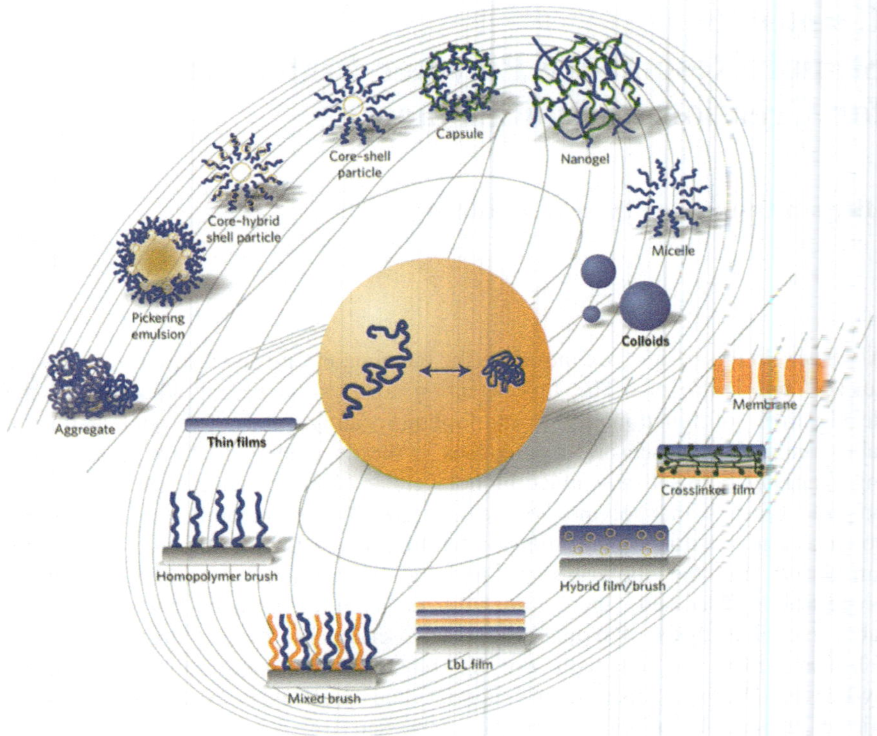

Fig. 13.1 'Galaxy' of nanostructured stimuli-responsive polymer materials. These materials rely on the phase behavior of macromolecule assemblies in thin films (polymer brushes, multilayered films made of different polymers, hybrid systems that combine polymers and particles, thin films of polymer networks, and membranes that are thin films with channels/pores), and nanoparticles (micelles, nanogels, capsules and vesicles, core–shell particles, hybrid particle-in-particle structures, and their assemblies in solutions and at interfaces in emulsions and foams) (data from Nat. Mater. (2010), 9, 101–113)

article focuses on stimuli-responsive macromolecular nanostructures that are capable of conformational and chemical changes on receiving an external signal. These changes are accompanied by variations in the physical properties of the polymer. The signal is derived from changes in the materials' environment, such as biological stimuli (e.g., pH, reduction–oxidation, enzymes, glucose) and externally applied triggers (e.g., temperature, light, solvent quality).

Furthermore, the integration of copolymer architectures (e.g., linear, branched, graft, block) and polymer composites into stimuli-responsive materials design enables exquisite control over the locations of responsive sites within self-assembled nanostructures. The combination of new synthesis techniques and well-defined copolymer self-assembly has facilitated substantial developments in stimuli-responsive materials in recent years. Developments in controlled polymerization techniques such as reversible addition-fragmentation chain-transfer (RAFT) polymerization, atom transfer radical polymerization (ATRP), nitroxide-mediated radical polymerization (NMP), and ring-opening metathesis polymerization (ROMP) allow for the synthesis of polymers with well-defined architectures, reproducible molecular weights, low polydispersities, and end-group fidelity. Many of these polymerizations proceed under mild conditions and are tolerant to functional groups, enabling the direct polymerization of responsive monomers or the incorporation of responsive end-groups through the use of functional chain-transfer agents (CTAs) or initiators. In this review, we discuss several methods that have been employed to synthesize self-assembling and stimuli-responsive copolymers and their surfaces for biomedical applications.

13.2 Low-Fouling Coatings for Label-Free Biosensing

Label-free optical affinity biosensors responding to biological recognition events enable direct real-time observation of interactions between target analytes and the counter receptors and rapid detection of chemical and biological species [1]. By combining the specific analyte targeting of carbohydrates with the multivalent structure and change of solution color as a consequence of plasmonic interactions with the aggregation of metal nanoparticles, glyconanoparticles with noble metals in the nanocore have been used extensively for the development of bioanalytical assays. This system is based on the assay of optical color change responding to specific biological recognition signal existing on the metal nanoparticles. These biosensors have become an important research tool in molecular biology and hold vast potential for bioanalytical applications in numerous fields, including medical diagnostics, environmental monitoring, food safety, and security.

In contrast with routinely used label-based bioanalytical methods, for example, enzyme-linked immunosorbent assays (ELISAs) or fluorescence biosensors, label-free biosensors are able to detect target analytes directly and therefore do not require the use of multistep assays or additional reagents (e.g., secondary antibodies specific to the analyte) [2–4]. However, because label-free optical affinity biosensors are based on the measurement of refractive index changes caused by the binding of analyte molecules to biorecognition elements immobilized on the surface of the sensor, their performance may be compromised by interfering effects producing refractive index changes unrelated to analyte binding. Potentially strong interferences can be caused by non-specific adsorption of molecules from a tested sample to the sensor surface; this effect is sometimes referred to as 'fouling.' These fouling effects may be particularly severe during the analysis of complex real-world samples, e.g., blood plasma or serum.

To suppress non-specific adsorption to the sensor surface, many research groups have pursued the development of coatings that enable the immobilization of biorecognition elements while also providing a low-fouling background [3, 5–11]. To further preserve the biological activity of the immobilized biorecognition elements, hydrophilic low-fouling surfaces based on poly(ethylene glycol) (PEG) and its derivatives have been a popular choice [12]. There have been several reports of the successful attachment of biorecognition elements to some of these coatings [8, 11]; however, a deterioration of the fouling properties of the coatings was observed after the attachment of biorecognition elements. Therefore, there is a continuing search for functionalizable coatings that can provide a high degree of resistance to non-specific adsorption from complex real-world samples [13, 14].

This review focuses on recent advances in the development of functionalizable antifouling coatings and their applications in label-free optical biosensors. It covers aspects related to coating materials and architectures, and methods used for the characterization of their fouling properties. Special attention is given to the application of label-free affinity biosensors with advanced functional coatings in medical diagnostics.

In any label-free optical affinity biosensors, the ability to avoid non-specific interactions between the surface and the protein-containing media is crucial in order to generate unbiased experimental outcomes. Advances in surface chemistry have made possible the synthesis of so-called non-fouling surfaces that significantly reduce or eliminate the non-specific adsorption of proteins and other biomolecules from biological fluids such as cell culture media. Several types of native molecules have the ability to reduce the adsorption of proteins at surfaces, e.g., carbohydrates such as agarose and mannitol as well as albumin [15, 16]. Due to their limited efficiency and stability, a number of synthetic materials have been developed [17]. The most widely used system is poly(ethylene glycol) or PEG with the monomeric repeating unit [-CH_2-CH_2-O-]- [also known as poly(ethylene oxide) or PEO]. The factors governing protein resistance of a PEG-graft copolymer were recently investigated in detail, with quantitative information provided on the interfacial architecture of PEG chains and their influence on protein resistance [18–20]. Several theories have been proposed to explain PEG's antifouling behavior, including its large excluded volume, osmotic repulsion, high molecular mobility, lack of protein binding sites, and high hydrophilicity. Protein-resistant PEGylated surfaces are often portrayed as near-liquid assemblies of highly mobile molecules and oligomeric segments that offer few binding sites to most proteins, as well as very short PEG–protein interaction times. Many different PEG surface-immobilization strategies have been successfully applied (Fig. 13.2). The simplest and least stable PEGylated surface is made by the adsorption of PEG homopolymer directly onto a substrate. Such studies have been performed on various substrate materials, including glass and polymer. In all cases, the amount of adsorbed protein is reduced compared with unmodified surfaces. Because of its amphoteric nature, PEG adsorbs only weakly onto most surfaces and alternate strategies incorporate endgroups or copolymer blocks, which preferentially adsorb onto hydrophobic surfaces. A good example is the family of triblock copolymers of PEO–PPO–PEO [21–24], known as Pluronics. The hydrophobic PPO block—formed by substituting a methyl group for one of PEO's hydrogens—adsorbs onto the surface, leaving the PEO segments to move freely in the aqueous solution (Fig. 13.2a). Surfaces coated with this copolymer show reduced adsorption of

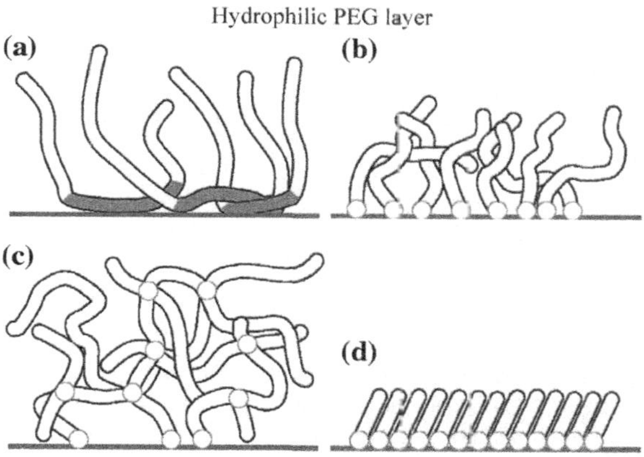

Fig. 13.2 Schematics of the different methods for attaching PEG molecules to a surface: **a** adsorbed copolymers, **b** grafted macromolecules, **c** crosslinked thin-film hydrogels, and **d** self-assembled monolayers of short oligomers or macromolecules

albumin, globulin, and fibrinogen, as well as reduced adhesion of platelets and macrophages. By changing the ratio of the PEO and PPO block lengths, the amphiphilic character of the block copolymer can be varied, which affects the hydrophobicity of the surfaces, as well as protein adsorption and cell adhesion [25, 26].

We have reported novel approaches for the construction of functionalized poly (ethylene glycol) (PEG) layer on surfaces using heterobifunctional PEG/polylactide (PLA) block copolymers [27, 28]. One of the objectives in the study is to investigate the effect of the variation in PEG chain length on surface properties. This includes assays on protein adsorption and cellular attachment to get a biochemical insight into the behavior of tethered PEG under biological conditions. For this purpose, numerous acetal-PEG/PLAs with different lengths of both PEG and PLA were synthesized. Molecular weights (MW) of PEG/PLA segments were abbreviated as follows: PEG/PLA(0.65/11.0, 1.8/7.0, 3.3/5.4, 5.0/4.6, 8.7/6.9) where the numbers in parenthesis denote the MW of the PEG segments and PLA segments in kg/mol, respectively. The PEG-brushed layer was constructed on the silanized glass surface by the spin coating of 4 % (w/v) solution of PLA solution/toluene, followed by the 2 % (w/v) solution of acetal-PEG/PLA/toluene. The wettability of the surface covered with PEG/PLA block copolymers was estimated both in air and in water by contact angle measurements (Table 13.1).

Protein adsorption on PEG/PLA surfaces was then estimated by using bovine serum albumin (BSA) as a model protein. On PLA surfaces, BSA adsorbed significantly, while on PEG-coated surfaces BSA adsorption is clearly decreased, mainly due to the steric stabilization by PEG chains. Moreover, minimum adsorption was obtained at a medium PEG chain length, i.e., PEG/PLA(3.3/5.4); we note that this surface revealed the maximum hysteresis in dynamic contact angle measurement. Protein adsorption may be related to the hysteresis observed in the dynamic contact angle, which is likely to depend on particular surface properties such as the density and mobility of tethered PEG chains on the surface.

Table 13.1 Molecular weights of PEG/PLA block copolymers and contact angle analysis

Sample[a]	PEG			PLA	Static Contact Angle[b]	
	Mn	Mw	Mw/Mn	Mn	Dry[c]	Wet[d]
1 PLA	–	–	–	20000	80.2	69.6
2 PEG/PLA (0.65/11.5)	650	720	1.10	11470	78.9	51.4
3 PEG/PLA (1.8/7.0)	1880	1930	1.03	7020	75.6	48.7
4 PEG/PLA (3.3/5.4)	3340	3470	1.04	5410	59.7	40.2
5 PEG/PLA (5.0/4.6)	5050	5210	1.03	4640	47.2	37.7
6 PEG/PLA (8.7/6.9)	8730	8810	1.01	6940	47.9	32.6

[a]Molecular weights (MW) of PEG/PLA segments were abbreviated as follows: PEG/PLA(1.8/7.0, 5.0/4.6, 8.7/6.9) where the numbers in parenthesis denote the MW of the PEG segments and PLA segments in kg/mol, respectively

[b]The values (degree) after hydration in PBS for 24 h

[c]The values in dry state were measured by a sessile droplet technique, where a water droplet was placed on the film surface

[d]The values in wet state were measured by a captive bubble technique, where a sample film was immersed in water and a small air bubble was placed on the film from the bottom using a curved needle

To enhance coating stability, many approaches have been studied to covalently bind PEG molecules to surfaces. These produce coatings that can vary from a monolayer to thin-film hydrogels (Fig. 13.2b, c). Two general approaches are used. One exploits the reactivity of the hydroxyl endgroup, which can be bound to surfaces such as activated silica. In order to increase the number of surface silanol groups, silica can be exposed to water plasmas. Silanols can then react with PEG hydroxyl groups to create an ester linkage to the activated surface. Surfaces modified in such a fashion are more hydrophilic, smoother, and demonstrate a reduced propensity for albumin adsorption. A second grafting approach replaces PEG's hydroxyl endgroup with more reactive groups. The type of replacement group is determined by the target surface or the chemical reaction used in the process. Gombotz et al., for example, used allylamine plasma glow discharge to introduce amine groups on the surface of poly(ethylene terephthalate), which were subsequently reacted with amine-terminated PEO using cyanuric chloride chemistry. A significant reduction in the adsorption of albumin and fibrinogen was achieved, despite an incomplete surface coverage. Similar results are found using silane chemistry to bind PEG molecules to the surface of Si and silica.

Another widely used PEG-chemistry-based approach relies on oligo-EG or PEG-modified alkanethiolate self-assembled monolayers (SAMs) [29–31] (Fig. 13.2d). The thiol group couples to Au and other transition metals, which can be deposited as ultrathin films on substrates. Although these PEG-modified SAMs show substantial reduction of protein adsorption, they still adsorb significant amounts of serum proteins. Moreover, they tend to oxidize under ambient conditions restricting their use to short-term cell culture studies [32]. Bearinger et al. [33] proposed an attractive alternative for modification of gold surfaces based on

PPS–PEG diblock or PEG–PPS–PEG triblock copolymers with poly(propylene sulfide) (PPS) as the central block (that binds to gold surfaces) and PEG-grafted chains. Other polymer architectures such as gels and polymeric SAMs have also been successfully utilized to link PEG chains to surfaces. For example, Healy and coworkers developed a gel-like interpenetrating polymeric network (IPN) of poly (acrylamide) and poly(ethylene glycol) [P(AAm-co-EG)] [34, 35], while Toner et al. used a poly(ethylene glycol) diacrylate (PEG-DA) hydrogel [36].

We have studied the surface properties of the PEG coating in detail to estimate protein adsorption on PEGylated surfaces. Three types of PEG immobilization [PEG5k(3), PEG5k(1)/2k(3), PEG5k(2)/2k(4)] were performed, as shown in SPR sensorgrams (Fig. 13.3). After first treatment with PEG5 k, the sensor surface was washed under running buffer to remove non-covalently adsorbed PEG. The sensor chip was then treated again with a solution of PEG5k. This cycle of adsorption/rinsing of PEG5k was repeated several times. Eventually, the total SPR angle shift was amplified by increasing the number of treatment cycles to three, indicating that repetitive treatment with PEG5k was effective in increasing the density of PEG [PEG5k(3)]. Notably, this trend became even more significant following additional treatment of the PEG5k surface with shorter PEG (PEG2k), as shown in Fig. 13.3b, c. We planned to increase the surface brush density by PEG2k, retaining the PEG5k brush surface character. Sensorgrams showed a number of interesting findings. First, immobilization of long-chain PEG [PEG5k (1)] increased markedly with changes in SPR angle (Fig. 13.3a–c). However, the extent of the shift decreased with the second injection of long-chain PEG [PEG5k (2)] (Fig. 13.3a, c), and little change was seen on the third injection of long-chain PEG (PEG5k(3); Fig. 13.3a). On the other hand, immobilization of short-chain PEG [PEG2k(1)] after long-chain PEG resulted again in marked changes (Fig. 13.3b, c). These results suggested that long-chain PEG5k can hardly penetrate into the preconstructed longer PEG-brushed layer due to its exclusion volume effect, while short-chain PEG2k appreciably filled the gap in the preconstructed longer PEG layer. It should be noted that SPR sensorgrams showed a steep increase curve in PEG2k(1), as shown in Fig. 13.3b, c, indicating the importance of a short underbrushed PEG layer in increasing the PEG chain density.

Non-specific protein adsorption from the culture medium for HUVEC was estimated on each PEG-coated surface to estimate the cytophobicity of PEGylated surfaces, because the adsorbed proteins are responsible for subsequent cell adhesion. On bare gold as a control, the SPR angle shift due to the non-specific adsorption of protein was 2927.4 RU, when serum-containing cell (HUVEC) culture medium (EBM-2) was passed over the surface. In contrast, PEG-coated surfaces clearly reduced protein adsorption (Fig. 13.4). Figure 13.4 also shows a comparison of protein adsorption on the three types of PEG surface. The PEG5k(2)/2k(4) surface showed greatest degree of inhibition of protein adsorption from the medium, suggesting that the inhibitory effect of non-specific protein adsorption was the highest for this surface among those studied. These results indicate that PEG surfaces with higher immobilized PEG chain density have greater ability to repel proteins. Based on these results, it was concluded that shorter PEG, *viz.* an

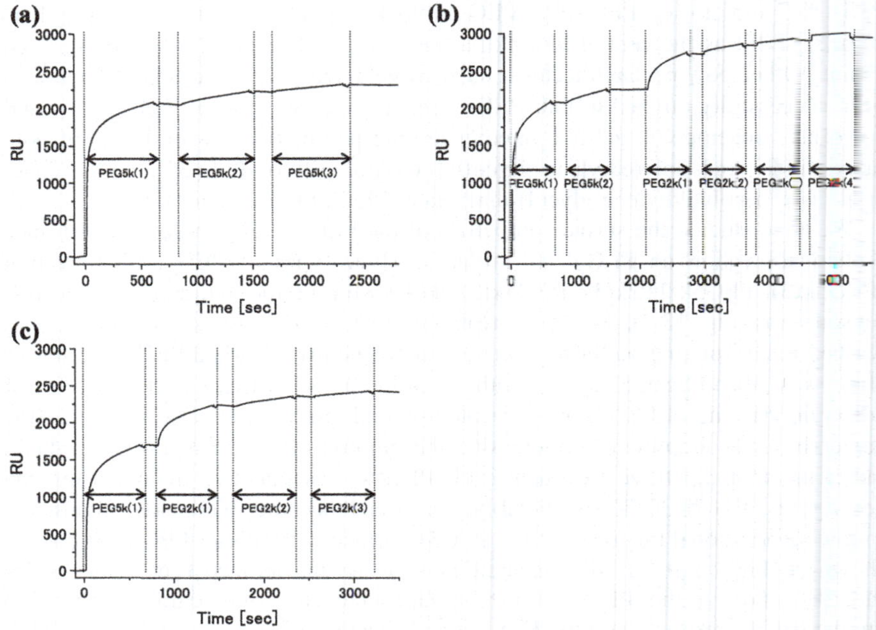

Fig. 13.3 Sensorgrams of PEG immobilization on gold surfaces. **a** PEG5k(3), **b** PEG5k(1)/2k(3), and **c** PEG5k(2)/2k(4). Flow rate, 10 mL/min; running buffer, PBS (0.15 M, pH 7.4, containing 1 M NaCl); sample, 0.01 mg/mL of PEG (M_w: 5k or 2k)/PBS (0.15 M, pH 7.4, containing 1 M NaCl) solution; sample injection, 100 mL for each time point (data from *Langmuir*, 2007, 23, 6609–6703)

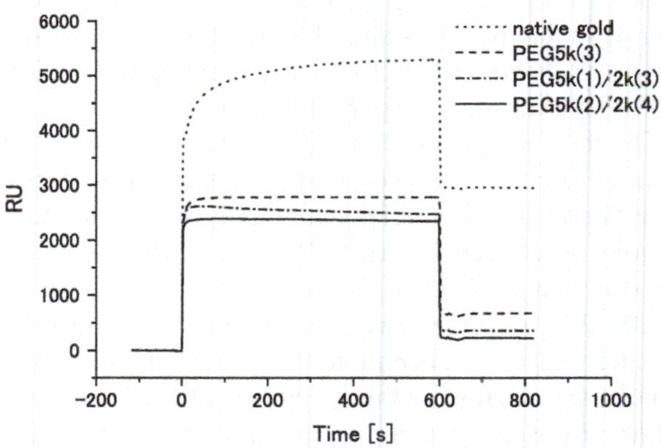

Fig. 13.4 Sensorgrams of injection of serum-containing cell culture medium (EBM-2 medium to culture HUVEC) on native gold and each PEGylated surfaces. Flow rate, 10 mL/min; running buffer, PBS (pH 7.4, 0.15 M); sample injection volume, 100 mL (data from *Langmuir*, 2007, 23, 6609–6703)

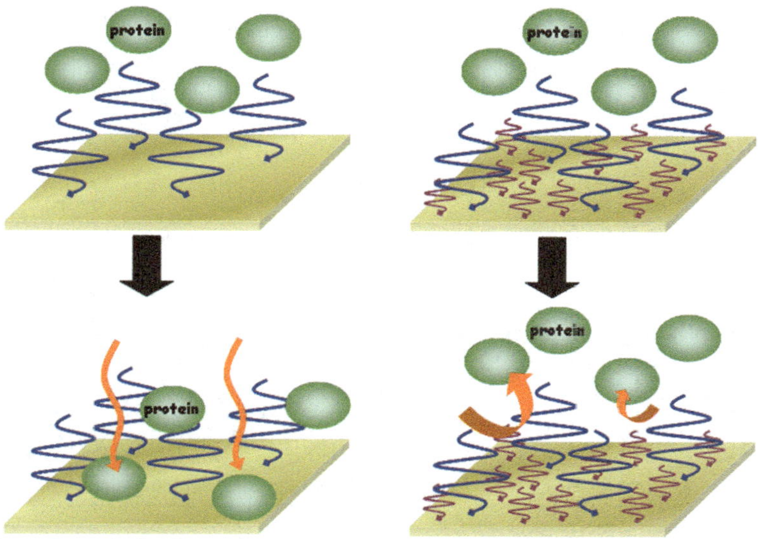

Fig. 13.5 Schematic representation to compare protein adsorption, depending on PEG density

underbrushed layer to increase the PEG surface density, played a substantial role in minimizing non-specific protein adsorption (Fig. 13.5). Other workers have also proposed that PEG mixtures which are polydisperse with respect to molecular weight are more efficacious than single molecular weights. Mixed PEGs were shown to have greatest efficacy in steric stabilization of colloidal particles and in protein repellency. The PEG5k(2)/2k(4) surface with the highest PEG chain density was expected to have the highest cytophobicity. In this way, a completely rejection of cultured cells with long-term viability was obtained. This clearly indicated the importance of a short underbrushed PEG layer in minimizing non-specific protein adsorption for long-term maintenance of the non-fouling surface.

13.3 Label-Free Biosensing by Glyconanoparticles: Colorimetric Bioassays

As shown in Fig. 13.6, we have previously reported a quantitative and reversible lectin-induced association of gold nanoparticles modified with α-Lactosyl-ω-mercapto-poly(ethylene glycol) [28]. Gold nanoparticles (1–10 nm size range) were prepared with an appreciably narrow size distribution by in situ reduction of

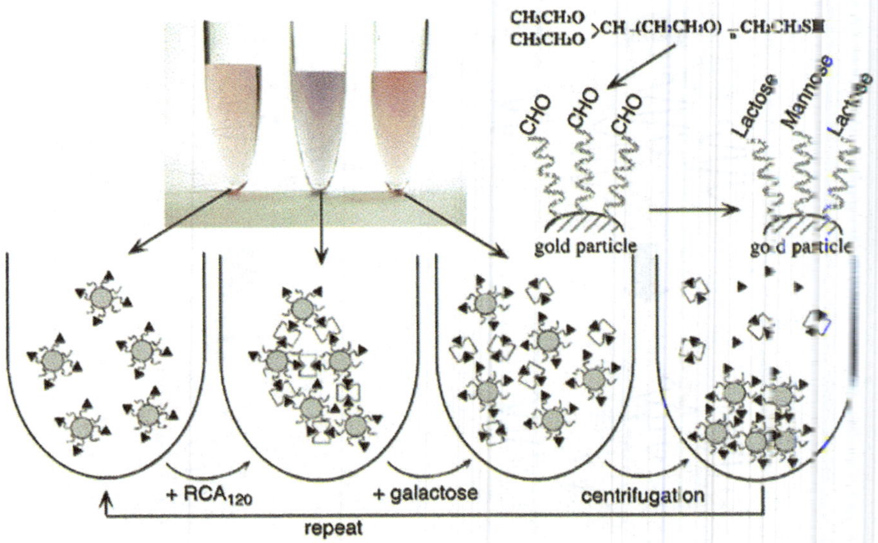

Fig. 13.6 Schematic representation of the reversible aggregation–dispersion behavior of lac-PEGylated gold nanoparticles by sequential addition of RCA120 lectin and galactose with actual concomitant change in color from *pinkish red* → *purple* → *pinkish red* (data from Otsuka et al. [28])

$HAuCl_4$ in the presence of heterobifunctional poly(ethylene glycol) (PEG) derivatives containing both mercapto and acetal groups (α-acetal-ω-mercapto-PEG). The α-acetal-PEG layers formed on gold nanoparticles impart appreciable stability to the nanoparticles in aqueous solutions with elevated ionic strength and also in serum-containing medium due to a non-fouling character. The PEG acetal terminal group was converted to aldehyde by gentle acid treatment, followed by the reaction with *p*-aminophenyl-*β*-D-lactopyranoside (Lac) in the presence of $(CH_3)_2NHBH_3$. Lac-conjugated gold nanoparticles exhibited selective aggregation when exposed to *Recinus communis* agglutinin (RCA_{120}) a bivalent lectin specifically recognizing the *β*-D-galactose residue, inducing significant changes in the absorption spectrum with concomitant visible color change from pinkish red to purple. Aggregation of the Lac-functionalized gold nanoparticles by the RCA_{120} lectin was reversible, recovering the original dispersed phase and color by addition of excess galactose. Further, the degree of aggregation was proportional to lectin concentration, allowing the system to be utilized to quantitate lectin concentration with nearly the same sensitivity as ELISA. This simple yet highly effective, derivatization of gold nanoparticles with heterobifunctional PEG provides a convenient method to construct various colloidal sensor systems currently applied in bioassays and biorecognition.

The new stable and glyco-derivatized gold nanorods (GNRs) were further prepared using developed amphiphilic block glycopolymers poly(lactose$_m$-b-pyridine$_n$

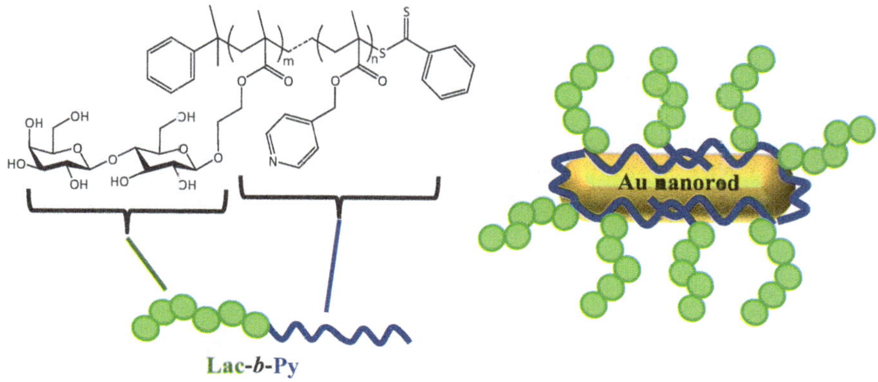

Fig. 13.7 Amphiphilic block glycopolymers poly(lactose$_m$-b-pyridine$_n$) to engineer self-organized, densely packed glycopolymeric recognition layers and generate multivalent binding cavity onto gold nanorod surface

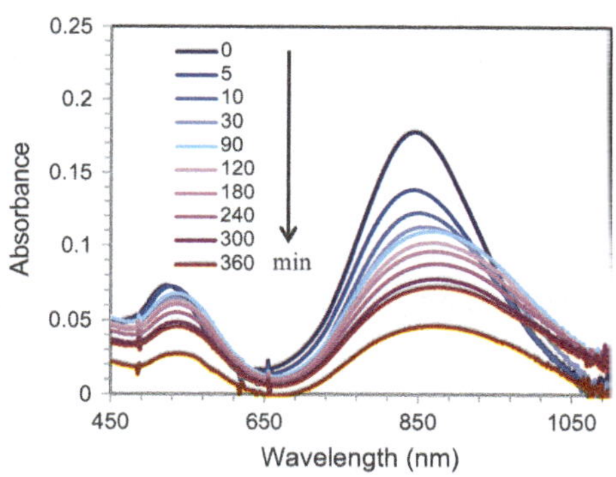

Fig. 13.8 UV–Vis absorption spectra of poly(Lac$_{23}$-b-Py$_{36}$)-coated GNRs with time after the addition of a very small concentration (1 ng/mL) of RCA$_{120}$ lection (data from Otsuka et al. [37])

to engineer self-organized, densely packed glycopolymeric recognition layers and generate multivalent binding cavity onto gold surface (Fig. 13.7). We reported a rapid and ultrasensitive biosensing for the weak affinity of protein–carbohydrate binding employing combination of longitudinal SPR of GNRs and cluster glycoside effect. Thus, a very small amount of lectin (100 pg/mL = 8.3×10^{-13} M) was detected by the aggregation of GNRs [37].

Reaction of these lactose-derivatized GNRs, originally functionalized with densely packed glycopolymeric recognition layers, with bivalent galactose-binding lectin (RCA$_{120}$) [38] was followed as a function of time through optical changes in the SPR band in the UV–Vis spectrum. Figure 13.8 shows representative UV–Vis spectra recorded with time after the addition of a very small concentration

(1 ng/mL) of RCA_{120} lectin in 0.15 M PBS to GNR colloidal suspension. The intensity at longitudinal SPR band observed at 840 nm was dramatically reduced accompanied with a peak redshift to much longer wavelength in a short time (ca. 10 min.). Broadening of the SPR peak was also confirmed. These are attributed to distance-dependent changes in the optical properties of three-dimensionally aggregated GNRs [39, 40] cross-linked by extremely small amount of added RCA_{120} lectin through cluster glycoside effect of lactose residues on the gold surface.

13.4 Biorecognition-Based Stimuli-Responsive Hydrogels

Self-assembly is present everywhere in nature. It is defined as the spontaneous association and/or organization of numerous individual units into rational and well-defined structures without external participation [41–44]. It is well known that most of biomacromolecules such as peptides and proteins are self-assemble into individual three-dimensional structure triggered by the molecular biorecognition. They serve as not only functional molecules in biological processes but also one of the building blocks for the body. Because the molecular biorecognition is the focal point of all biological events, it forms the basis for the design of precisely defined smart systems, including targeted therapeutics, bioimaging, stimuli-sensitive and self-assembled biomaterials, and biosensors. The biorecognition-based self-assembly system permits the design of materials programmed at nanometer to micrometer scales and is an efficient tool to fabricate the functional biomaterials [45].

Hydrogels are three-dimensional cross-linked water-soluble polymer networks and can swell but do not dissolve in water. They are commonly used in clinical practice and experimental medicine for biological applications because hydrogels can be formulated in a variety of physical forms, including slabs, microparticles, nanoparticles, coatings, and films. Their classification may be based on the source (natural and/or synthetic macromolecules), on the nature of cross-linking (physical or covalent), and on the nature of the network structure (homogeneous/heterogeneous network, interpenetrating network, and double network) [46–48]. With the development of synthetic procedure, hydrogels with regulated and/or complex network have emerged to improve physicochemical and biological functions.

Hydrogels are known to be sensitive to external stimuli such as temperature, pH, ion concentration, solvent composition, electric potential, magnetism, and light [49]. These parameters cause a change in phase, shape, mechanics, and permeability that subsequently can be reversed to return the material to its original state. Stimuli-sensitive hydrogels have been integrated with biomacromolecules such as peptides and proteins. These hydrogels show to swell/collapse in the presence of a target molecule. Particularly, hybrid hydrogels consist of self-assembled biomacromolecules have attracted extensive research interest [44, 50, 51]. Hybrid

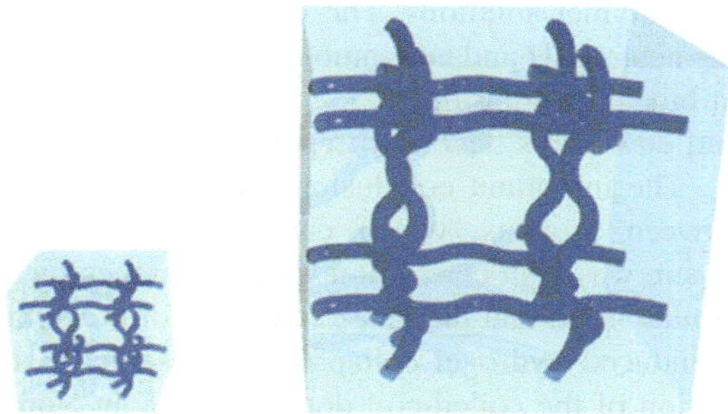

Fig. 13.9 Stimuli that can cause reversible swelling and collapse (shrinking) of hydrogels

hydrogels composed from at least two distinct classes of macromolecules, for example, synthetic and biological macromolecules interconnected either covalently or non-covalently. Tunable network structure of hydrogels in response to biological stimuli leads to dramatic changes in their permeability and mechanical properties (Fig. 13.9). They have been studied as innovative carriers for drug delivery and as scaffolds for tissue engineering [52, 53].

In this topic, we focus on stimuli-responsive hydrogels in response to molecular biorecognitions. Various biological stimuli that can change in physicochemical properties of hydrogel have been proposed. It is a good example as biorecognition-based hybrid hydrogel system formed through specific protein–protein, protein–peptide, and protein–polysaccharide interactions. We shall only discuss on peptides and proteins recognition system of stimuli-responsive hybrid hydrogel consists of biomacromolecules.

13.4.1 Hydrogel Formed by Self-assembly Biomacromolecules

Tirrell and coworkers have reported a reversible hydrogel based on self-assembling of artificial proteins [54]. This is the first example of a self-assembling hybrid hydrogel combined biorecognition of a homodimer-forming α-helical motif with synthetic polymer. They used the proteins consist of terminal leucine zipper domains flanking a central, flexible, water-soluble polyelectrolyte segment. Coiled-coil aggregates formation of the terminal domains in near-neutral aqueous solutions triggers formation of a three-dimensional polymer network with the polyelectrolyte segment retaining solvent and preventing precipitation of the chain. Dissociation of the coiled-coil aggregates through elevation of pH or temperature causes dissolution of the gel and a return to the viscous behavior that is characteristic of polymer solutions. The mild conditions for hydrogelation can be controlled (near-neutral pH and near-ambient temperature) and suggest the potential applications in bioengineering field.

Kopecek and coworkers have also reported studies on polymer–protein hybrid hydrogels [55]. The important conclusion was that the formation of a coiled-coil interaction between proteins, which can be imposed onto the hydrogelation (Fig. 13.10). There was high correlation with the temperatures between the melting point of the peptide motif and the phase transition of the hybrid hydrogel structure. A temperature-induced hydrogel collapse was observed that corresponded to the structural transition of the coiled-coil domains from an elongated helix to random coil. At that time, this is a new temperature-response system for stimuli-responsive hydrogels. That can be also tuned over a wide temperature range by assembling gels with coiled-coils that have different melting temperatures. Kopecek group use the genetically engineered protein motifs, which attached to the functional copolymer composed of the hydrophilic *N*-(2-hydroxypropyl)-methacrylamide (HPMA) and the metal complexation monomer *N*-(*N*,*N*-dicarboxymethylaminopropyl) methacrylamide (DAMA). The Ni^{2+} ion coordination-mediated grafting structure was formed between the pendant metal-chelating ligand and the His tag, and the coiled-coil interaction anchored on the side chain of the polymers. Hydrogels could be obtained through the aggregation of coiled-coil structure between grafted proteins. The temperature stability and the swelling property of resulting hydrogels were tunable by using different kinds of coiled-coil proteins. This kind of novel hydrogel system with an easy anchoring strategy, well-characterized synthetic polymer backbones, and tunable property of coiled-coil proteins could be used for incorporation and delivery of cellular and molecular species.

The interactions of proteins and glycosaminoglycans are critical matter for mediating the multiple responses of the extracellular matrix (ECM). Kiick and coworkers focused on the interaction between proteins and glycosaminoglycans and had reported the fabrication such a kind of hydrogel [56]. The non-covalently associated hydrogel networks were produced via the interaction of a low molecular weight heparin-terminated star-shaped poly(ethylene glycol) (PEG-LMWH) and a dimeric protein, heparin-binding vascular endothelial growth factor (VEGF). The erosion of the hydrogel and the release profile of VEGF could be controlled by the addition of VEGF receptor. The hydrogel remained intact in PBS solution without the VEGF receptor. However, its erosion could be visibly observed in the presence of a VEGF receptor. The release profile of encapsulated VEGF could also be enhanced in the presence of a VEGF receptor [57]. The bioactivity of released VEGF was assessed by a cell-proliferation assay. When porcine aortic endothelial cells (PAE DKR) were encapsulated in the hydrogel, the very low number of dead cells was confirmed by the confocal microscope observation. It is considered the lack of cytotoxicity of these materials. This concept can be also applied for generating non-covalent, responsive, and injectable hydrogels [58, 59]. This study by the Kiick group offered a novel mechanism for targeted delivery and erosion via the release of therapeutically protein cross-linkers in response to cell surface receptors.

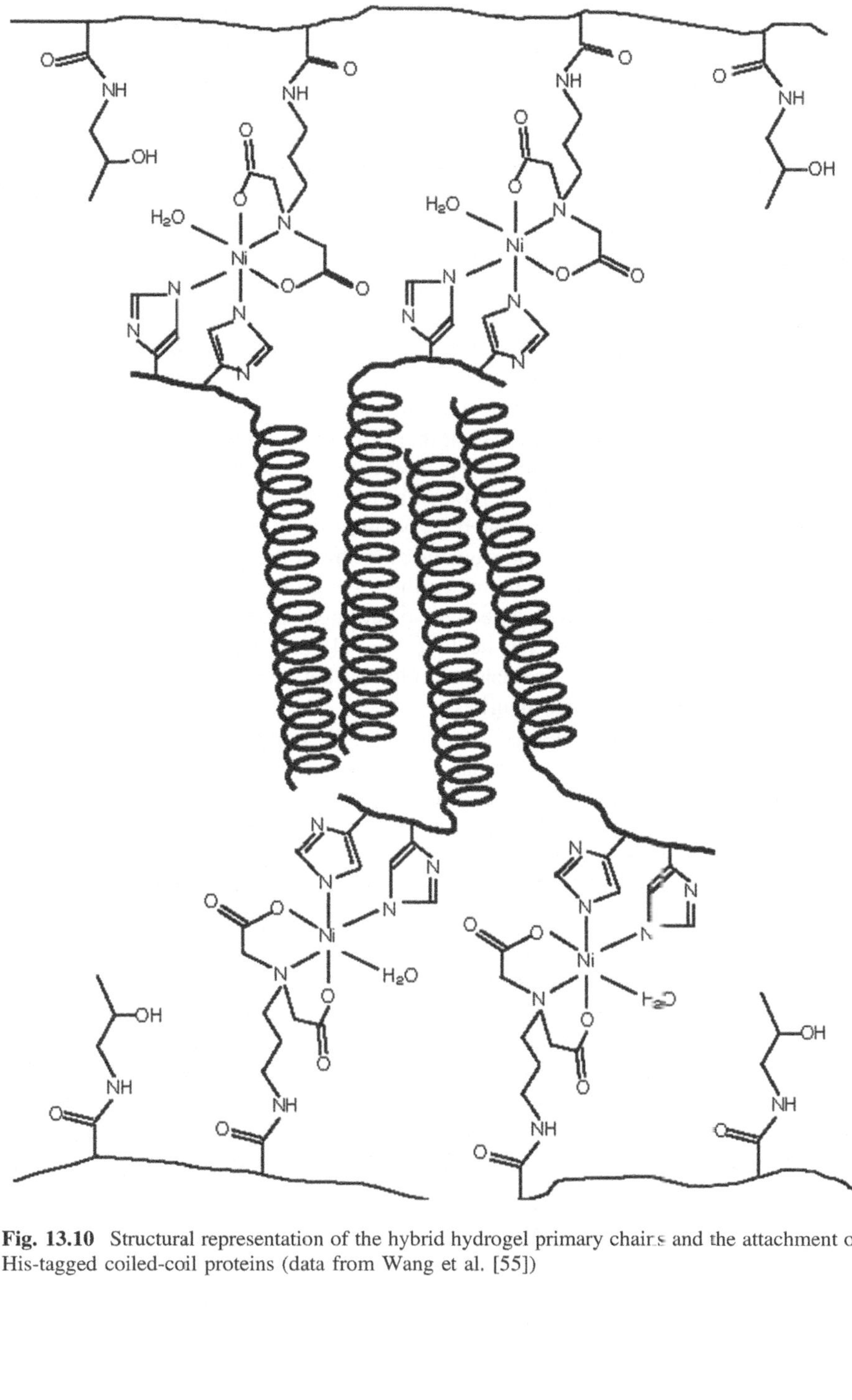

Fig. 13.10 Structural representation of the hybrid hydrogel primary chains and the attachment of His-tagged coiled-coil proteins (data from Wang et al. [55])

As mentioned above, Most of hydrogels are protein–polymer hybrid hydrogel systems. In contrast to these hydrogel systems, the fully genetically encoded and bacterial-produced protein-based hydrogel systems have been also investigated. Heilshorn and coworkers reported a simple hydrogel system based on protein–protein interactions with easily tunable their physical properties [60]. They reported the development of a two-component, molecular-recognition gelation strategy that enables cell encapsulation without environmental triggers. Instead, the two components, which contain multiple repeats of WW and proline-rich peptide domains, undergo a sol–gel phase transition upon simple mixing and heteroassembly of the peptide domains. These results demonstrate use of the WW and proline-rich domains in protein-engineered materials and expand the library of peptides successfully designed into engineered proteins. Because both of these association domains are normally found intracellularly, their molecular recognition is not disrupted by the presence of additional biomolecules in the extracellular milieu, thereby enabling reproducible encapsulation of multiple cell types, including PC-12 neuronal-like cells, human umbilical vein endothelial cells, and murine adult neural stem cells. Precise variations in the molecular-level design of the two components including (i) the frequency of repeated association domains per chain and (ii) the association energy between domains enable tailoring of the hydrogel viscoelasticity to achieve plateau shear moduli ranging from ~9 to 50 Pa. Because of the transient physical crosslinks that form between association domains, these hydrogels are shear-thinning, injectable, and self-healing. Neural stem cells encapsulated in the hydrogels form stable three-dimensional cultures that continue to self-renew, differentiate, and sprout extended neurites.

Specific protein–peptide interactions can also be applied for the formation of hydrogels. Due to the ease of synthesis of peptides, easy incorporation of peptides into different kinds of molecules such as polymers, and a large pool of specific protein–peptide interactions with different dissociation constants, the property of protein-based hydrogels based on specific protein–peptide interactions can be well manipulated and is highly tunable. Regan and coworkers reported a stimuli-responsive smart gel based on the specific protein–peptide interaction between a 34-amino acid tetratricopeptide repeat (TPR) and penta-peptide [61, 62]. The TPR protein arranges into a tandem repeating unit with a spacer. The penta-peptide was grafted to four-armed PEG providing multivalent interactions to the TPR protein. The simple mixing system led to the formation of hydrogels. Since the affinity of the protein–peptide interaction would decrease with increasing ionic strength, the gel would erode in high-ionic-strength buffer solutions. Therefore, the release profiles of encapsulated proteins or small molecules could be controlled by the ionic strength of the release buffer. The hydrogel system would be well-suited for cell culture and tissue engineering applications.

13.4.2 Hydrogel Response to Biomacromolecules

The described above works are focused on the formation of higher order architecture or specific interaction of biomacromolecules to produce hydrogels. Different strategy is the use of specific interaction between proteins as a responsive signal. Biomacromolecules fixed in hydrogel matrix can respond to active species and exhibit the change of physicochemical properties of hydrogel.

Miyata and coworkers have introduced a reversible binding between antigen and antibody into a semi-interpenetrating network (semi-IPN) hydrogel to produce a reversible antigen-responsive hydrogel (Fig. 13.11) [63]. They used rabbit immunoglobulin G (rabbit IgG) and goat antirabbit IgG (GAR IgG) as the antigen and antibody, respectively. Both the antigen and antibody were chemically

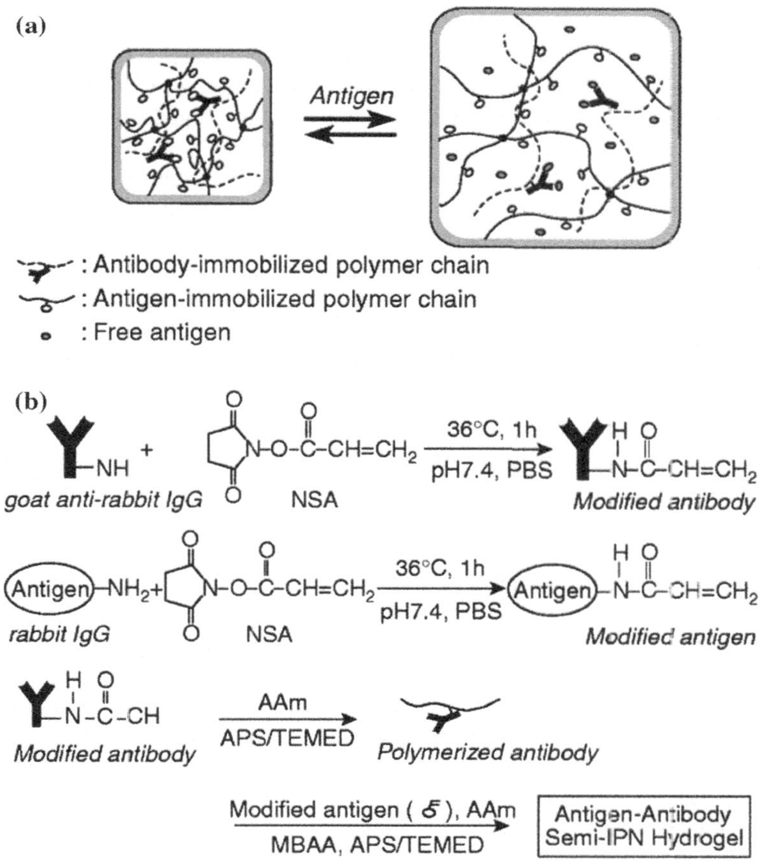

Fig. 13.11 Strategy for the preparation of an antigen-responsive hydrogel (data from Miyata et al. [63])

modified by *N*-succinimidylacrylate, respectively, and the copolymerization of the resultant vinyl (GAR IgG) with acrylamide (AAm) was performed. An antigen–antibody semi-IPN hydrogel could be obtained by the copolymerization of the vinyl (rabbit IgG), acrylamide, and *N,N*-methylenebisacrylamide as a cross-linker in the presence of the polymerized GAR IgG.

The antigen–antibody semi-IPN hydrogel could show a rapid swelling behavior following the addition of free rabbit IgG. It is because that the free rabbit IgG exhibit a stronger binding affinity to GAR IgG than polymerized rabbit IgG. Therefore, the addition of free native rabbit IgG can exchange to IgG grafted to the semi-IPN hydrogel, and induce the dissociation of the complex. The swelling of hydrogel was reversible, which meant that the hydrogel would shrink upon contact with buffer solutions without rabbit IgG. The reversible swelling change of the hydrogel could be applied to control the permeation of a protein drug through a membrane fabricated from the hydrogel. Alternately immersion of the hydrogel in phosphate buffer with and without rabbit IgG could control the penetration of protein drug through the hydrogel membrane. This hydrogel system might be applied for the drug delivery in response to specific antigens.

Daunert and coworkers have reported the hybrid hydrogel response to ligand–receptor interaction [64]. Reversible specific interactions between proteins and their ligands could also be used to produce responsive hydrogels. Calmodulin (CaM), a calcium-binding protein, has been selected as the biological recognition element in the stimuli-responsive hydrogel because it undergoes a large conformational change on binding calcium, certain peptides, and the phenothiazine class of drugs (antipsychotics). CaM undergoes two conformational changes, one in the presence of Ca^{2+} and the other in the presence of phenothiazines. Specifically, CaM goes from its 'dumbbell' conformation in the presence of Ca^{2+}, to a more constrictive conformation when bound to phenothiazine. When Ca^{2+} is removed from CaM, the protein changes from its bound conformation to its native conformation [65].

Daunert group has covalently incorporated CaM and a ligand of CaM (phenothiazine (PTZ)) into a hydrogel network (Fig. 13.12) [64]. The addition of Ca^{2+} ions induced a conformational change of CaM and initiated the binding between CaM and PTZ. This binding introduced non-covalent cross-linkings to the hydrogel, resulting in shrinkage of the hydrogel. Upon the addition of ethylene glycol-bis (β-aminoethyl ether)-*N,N,N′,N′*-tetraacetic acid (EGTA) to remove Ca^{2+}, PTZ would be released from the CaM binding site and the conformation of CaM would be changed, resulting in hydrogel swelling. The obtained hydrogel could be used to control the permeation of vitamin B12. Moreover, they demonstrated that their hydrogel could serve as a microactuator in microfluidics. Their hydrogel was integrated into a plunger microactuator configuration and also could act as a gate to control the flow rate. The hydrogel was able to prevent the release of blue dextran while in a swollen state, and then deswell in the presence of a stimulus in order to open the microactuator and allow the solution to be released.

Weber and coworkers proposed the stimuli-sensing hydrogel prepared from protein–protein interaction. They used the protein conjugated polymer that the genetically engineered bacterial gyrase subunit B (GyrB) with a hexahistidine tag

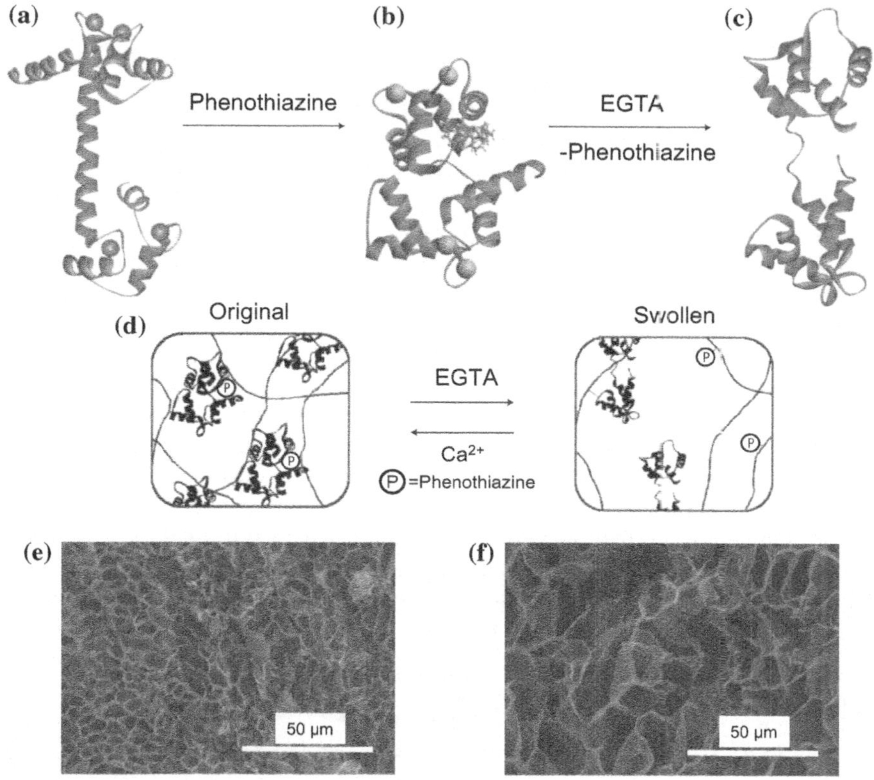

Fig. 13.12 Swelling of stimuli-responsive hydrogel (data from Ehrick et al. [64])

through the chelating of polyacrylamide functionalized with nitrilotriacetic acid-Ni^{2+} [66]. GyrB could be dimerized by the addition of the aminocoumarin antibiotic coumermycin, resulting in hydrogel formation. Addition of increasing concentrations of clinically validated novobiocin (Albamycin) dissociated the GyrB subunits, thereby resulting in dissociation of the hydrogel. The hydrogel was stable when incubated in PBS buffer solution for 24 days. However, it would dissociate into homogeneous solutions upon contact with PBS solution containing 1 mM of novobiocin within 2 days. The release profile of entrapped therapeutic agents such as the human vascular endothelial growth factor (hVEGF) could be controlled by the addition of different concentrations of novobiocin. These hydrogel systems developed by Daunert and Weber groups have a great potential for the spatiotemporally controlled delivery of therapeutic agents [64, 66].

Hybrid hydrogels formed by covalent strategies are mechanically stronger than those formed by non-covalent ones. However, physical protein-based hydrogels formed by non-covalent strategies have several advantages [67]. The simple mixing pathway for preparing physical hydrogels is biocompatible to cells and therapeutic agents. They are injectable due to the non-covalent interactions between components used for hydrogelations.

Otsuka and coworkers have reported a simple and one-pot fabrication of biocompatible hydrogels functioning as elastin-like polypeptide (ELP) [68, 69] matrix in terms of thermo-responsibility and structure-proved bioavailability, without complicated pretreatment and synthesis (Fig. 13.13) [70]. Elastin is an extracellular matrix protein, which is found predominantly in the connective tissue of artery, skin, lung, and ligament [71, 72]. The important role of elastin is the formulation of elastic tissues through self-assembly into fibrous form. However, the self-assembled elastin molecule is water insoluble, and it is difficult to use for creation of biofunctional materials.

Otsuka group developed the fabrication of novel biocompatible hydrogels with three-dimensionally arranged self-assembly structure of ELP. This concept is very simple without complicated pretreatment and synthesis, and obtained hydrogels behave as ELP-like hydrogels in terms of thermo-responsibility and structure-proved bioavailability. To achieve this purpose, a one-pot method was developed to prepare the composite hydrogels formed by untreated ELP (E) interpenetrating in alginate (A)-Ca^{2+} hydrogel as host three-dimensional network (denoted as AE gel). AE gels were prepared by spontaneous gelation of alginate-Ca^{2+} hydrogel in

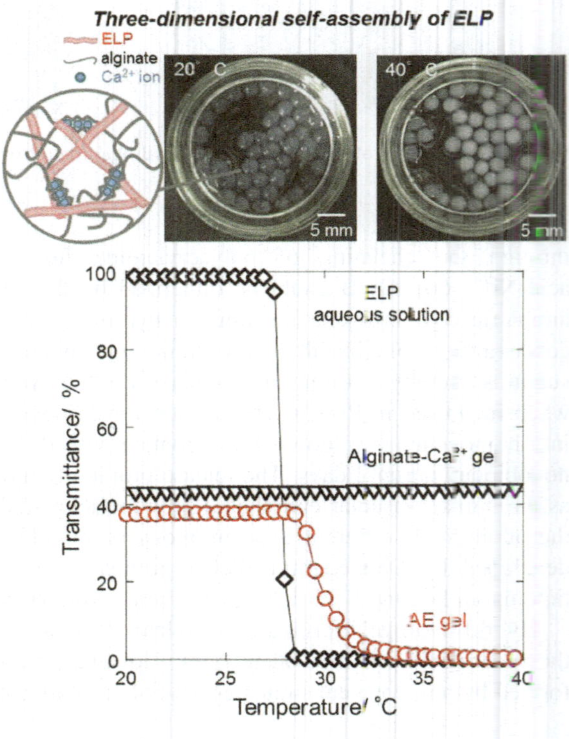

Fig. 13.13 *Upper* photograph of AE gel (alginate/ELP = 1.0/1.0 wt%) at 20 and 40 °C. *Lower* transmittance curves of ELP polymer, alginate-Ca^{2+} gel, and AE gel as a function of temperature (data from Matsukuma and Otsuka [70])

spherical bead shape coexisted with ELP polymer. Because they used the ELP without functional groups in primary structure and thus reaction with calcium ions, it was suggested that AE gels formed semi-IPN structure interpenetrated with ELP by physical entrapment in alginate-Ca^{2+} network.

Thermo-responsive phase-transition behaviors involving the formation of specific secondary structure of ELP were characterized as the fundamental investigation at molecular-scale dynamics of this novel smart hydrogels. It can be seen that homogeneous opaque AE gel beads at 20 °C (below LCST of ELP) turn into white turbidity at 40 °C (above LCST). This is due to the thermal phase transition of ELP, because the transmittance did not change for alginate-Ca^{2+} gel as a function of temperature. Furthermore, during the temperature sweep, the optical change of white turbidity uniformly spread in AE gel beads without the volume of spherical beads significantly changed. It is suggested that the thermal phase transition of ELP occurred three-dimensionally in alginate-Ca^{2+} network of AE gel. Phase-transition kinetics as a function of temperature was next varied with ELP aqueous solution in a similar manner, to demonstrate that AE gel formed semi-IPN structure interpenetrated with ELP in alginate network by physical entrapment. Actually, the thermo-responsive behavior of AE gel was observed repeatedly for at least 1 week, whereas during 1 month stocked in water, it has gradually diminished due to release of ELP from AE gel.

Temperature dependency of secondary structure of ELP was investigated by circular dichroism (CD) spectrum measurement. At low temperature, the spectrum gave a typical pattern of disordered random coil structure with negative cotton effect maximum at 196–198 nm. The CD value at 196 nm decreased with increasing temperature, while a positive peak in the region between 206 and 212 nm was observed, indicating the characteristic transition into a type II β-turn structure [73, 74]. Significantly, similar transition curves were observed between with and without alginate. From these results, it can be concluded that the thermo-responsive conformational character of ELP is not disturbed by alginate polymer in AE gel system. On the other hand, the thermal phase transition of AE gel was affected the alginate-Ca^{2+} network. The LCST of AE gels was shifted to lower temperature upon cooling process, and its negative hysteresis was noticeable when the ELP concentration and temperature scan rate increased. Considering there was no specific interaction between alginate and ELP from CD spectrum measurement, this temperature hysteresis of ELP was attributed to the presence of alginate-Ca^{2+} network. In other words, it was conceivable that thermo-reversible three-dimensional self-assembly of ELP was physically restricted by alginate-Ca^{2+} network, which should also help to disturb the aggregation of the ELP chains in AE gel. From these results, we can conclude that thermo-reversible self-assembled structure of ELP was arranged three- and molecule-dimensionally in alginate-Ca^{2+} hydrogel.

The AE gel was simply prepared by the addition of ELP polymer to sodium alginate followed by calcium cross-linking. The similar phase transitions of AE gel with ELP polymer were confirmed in both LCST and molecular conformation. The thermo-reversible self-assembly of ELP in three-dimension was formed in obtained composite AE hydrogel. This is the new strategy of the simple fabrication of ELP-like hydrogels without sacrificing the thermo-responsibility and biofunctionality of ELP. These smart hydrogels with dynamically convertible functional domain are useful for drug delivery and tissue-engineered scaffold in the field of biomedical application. The tissue differentiation and cellular matrix accumulation of mesenchymal stem cells (MSCs) will be investigated after cell encapsulation in an engineered AE hydrogel as an injectable matrix for delivery of cell-based therapeutics.

13.5 Summary and Outlook

In this review, we introduce the stimuli-responsive hybrid material system driven by molecular biorecognition. Today, there are many additional strategies for the creation of smart hybrid surface (2D) and hydrogels (3D) through biorecognition-based self-assembly system [52, 75]. The synergistic combination of two types of structures may produce new materials that possess unprecedented levels of structural organization and novel properties. The combination of synthetic polymers and peptide or protein segments into self-assembled hybrid materials provides an opportunity to combine and superimpose the properties of the component molecules. These materials are playing an increasingly important part in a diverse range of applications, such as drug delivery, diagnostics, tissue engineering, and 'smart' optical systems, as well as biosensors, microelectromechanical systems, coatings, and textiles. We review recent advances and challenges in the developments toward applications of stimuli-responsive polymeric materials that are self-assembled from nanostructured building blocks.

References

1. Sun Y-S (2013) Optical biosensors for label-free detection of biomolecular interactions. Instrum Sci Technol 42(2):109–127
2. Fan X, White IM, Shopova SI, Zhu H, Suter JD, Sun Y (2008) Sensitive optical biosensors for unlabeled targets: a review. Anal Chim Acta 620(1–2):8–26
3. Cretich M, Damin F, Pirri G, Chiari M (2006) Protein and peptide arrays: recent trends and new directions. Biomol Eng 23(2–3):77–88
4. Phelan ML, Nock S (2003) Generation of bioreagents for protein chips. Proteomics 3 (11):2123–2134
5. Banerjee I, Pangule RC, Kane RS (2011) Antifouling coatings: recent developments in the design of surfaces that prevent fouling by proteins, bacteria, and marine organisms. Adv Mater 23(6):690–718
6. Barbey R, Lavanant L, Paripovic D, Schuwer N, Sugnaux C, Tugulu S, Klok HA (2009) Polymer brushes via surface-initiated controlled radical polymerization: synthesis, characterization, properties, and applications. Chem Rev 109(11):5437–5527

7. Ligler FS, Taitt CAR (2002) Optical biosensors: present and future. Elsevier, Amsterdam, Oxford
8. Rodriguez-Emmenegger C, Brynda E, Riedel T, Houska M, Subr V, Alles AB, Hasan E, Gautrot JE, Huck WT (2011) Polymer brushes showing non-fouling in blood plasma challenge the currently accepted design of protein resistant surfaces. Macromol Rapid Commun 32(13):952–957
9. Rodriguez-Emmenegger C, Brynda E, Riedel T, Sedlakova Z, Houska M, Alles AB (2009) Interaction of blood plasma with antifouling surfaces. Langmuir 25(11):6328–6333
10. Rodriguez-Emmenegger C, Kylian O, Houska M, Brynda E, Artemenko A, Kousal J, Alles AB, Biederman H (2011) Substrate-independent approach for the generation of functional protein resistant surfaces. Biomacromolecules 12(4):1058–1066
11. Klok HA, Barbey R, Lavanant L, Paripovic D, Schuwer N, Sugnaux C, Tugulu S (2009) Polymer brushes via surface-initiated controlled radical polymerization: synthesis, characterization, properties, and applications. Chem Rev 109(11):5437–5527
12. Blattler TM, Pasche S, Textor M, Griesser HJ (2006) High salt stability and protein resistance of poly(L-lysine)-g-poly(ethylene glycol) copolymers covalently immobilized via aldehyde plasma polymer interlayers on inorganic and polymeric substrates. Langmuir 22(13):5760–5769
13. Calvo KR, Liotta LA, Petricoin EF (2005) Clinical proteomics: from biomarker discovery and cell signaling profiles to individualized personal therapy. Biosci Rep 25(1–2):107–125
14. Homola J (2008) Surface plasmon resonance sensors for detection of chemical and biological species. Chem Rev 108(2):462–493
15. Luk YY, Kato M, Mrksich M (2000) Self-assembled monolayers of alkanethiolates presenting mannitol groups are inert to protein adsorption and cell attachment. Langmuir 16:9604–9608
16. Nelson CM, Raghavan S, Tan JL, Chen CS (2003) Degradation of micropatterned surface by celldependent and independent processes. Langmuir 19:1493–1499
17. Ratner BD, Bryant SJ (2004) Biomaterials: Where we have been and where we are going. Ann Rev Biomed Eng 6:41–75
18. Pasche S, Vörös J, Griesser HJ, Spencer ND, Textor M (2005) Effects of Ionic strength and surface charge on protein adsorption at PEGylated surfaces. J Phys Chem B 109:17545–17552
19. Pasche SSM, Paul De, Vörös J, Spencer ND, Textor M (2003) Poly(L-lysine)-graft-poly (ethylene glycol) assembled monolayers on niobium oxide surfaces: a quantitative study of the influence of polymer interfacial architecture on resistance to protein adsorption by ToF-SIMS and in situ OWLS. Langmuir 19:9216–9225
20. Wagner MS, Pasche S, Castner DG, Textor M (2004) Characterization of poly(L-lysine)-graft-poly (ethylene glycol) assembled monolayers on niobium pentozide substrates using time-of-flight secondary ion mass spectrometry and multivariate analysis. Anal Chem 76:1483–1492
21. Booth C, Attwood D (2000) Effects of block architecture and composition on the association properties of poly(oxyalkylene) copolymers in aqueous solution. Macromol Rapid Commun 21:501–527
22. Lopes JR, Loh W (2000) Investigation of self-assembly and Micelle polarity for a wide range of ethylene oxide−propylene oxide−ethylene oxide block copolymers in water. Langmuir 14:750–756
23. Kozlov MY, Melik-Nubarov NS, Batrakova EV, Kabanov AV (2000) Relationship between pluronic block copolymer structure, critical micellization concentration and partitioning coefficients of low molecular mass solutions. Macromolecules 33:3305–3313
24. Thissen H, Hayes JP, Kingshott P, Johnson G, Harvey EC, Griesser HJ (2002) Nanometer thickness laser ablation for spatial control of cell attachment. Smart Mater Struct 11:792–799
25. Neff JA, Tresco PA, Caldwell KD (1999) Surface modification for controlled studies of cell-ligand interactions. Biomaterials 20:2377–2393
26. Liu VA, Jastromb WE, Bhatia SN (2002) Engineering protein and cell adhesivity using PEO terminated triblock polymers. J Biomed Mater Res 60:126–134

27. Otsuka H, Nagasaki Y, Kataoka K (2000) Surface characterization of functionalized polylactide through the coating with heterobifunctional poly(ethylene glycol)-polylactide block copolymer. Biomacromolecules 1:39–48
28. Otsuka H, Akiyama Y, Nagasaki Y, Kataoka K (2001) Quantitative and reversible lectin-induced association of gold nanoparticles modified with α-Lactosyl-ω-mercapto-(ethylene glycol). J Am Chem Soc 123:8226–8230
29. Ostuni E, Chapman RG, Liang MN, Meluleni G, Pier G, Ingber DE (2001) Self-assembled monolayers that resist the adsorption of proteins and the adhesion of bacterial and mammalian cells. Langmuir 17:6336–6343
30. Ostuni E, Chapman RG, Holmlin RE, Takayama S, Whitesides GM (2001) A survey of structure-property relationships of surfaces that resist the adsorption of protein. Langmuir 17:5605–5620
31. Schwendel D, Dahint R, Herrwerth S, Schloerholz M, Eck W, Grunze M (2001) Temperature dependence of the protein resistance of poly-and oligo(ethylene glycol)-terminated alkanethiolate monolayers. Langmuir 17:5717–5720
32. Schoenfisch MH, Pemberton JE (1998) Air stability of alkanethiol self-assembled monolayers on silver and gold surfaces. J Am Chem Soc 120:4502–4513
33. Bearinger JP, Terrettaz S, Michel R, Tirelli N, Vogel H, Textor M (2003) Chemisorbed poly (propylene sulphide)-based copolymers resist biomolecular interactions. Nat Mater 2:259–264
34. Barber TA, Golledge SL, Castner DG, Healy KE (2003) Peptide-modified p (AAm-co-EG/AAc) IPNs grafted to bulk titanium modulate osteoblast behavior in vitro. J Biomed Mater Res Part A 64A:38–47
35. Bearinger JP, Castner DG, Healy KE (1998) Biomolecular modification of p(AAm-co-EG/AA) IPNs supports osteoblast adhesion and phenotypic expression. J Biomater Sci Poly Ed 9:629–652
36. Revzin A, Tompkins RG, Toner M (2003) Surface engineering with poly(ethylene glycol) photolithography to create high-density cell arrays on glass. Langmuir 19:9855–9862
37. Otsuka H, Muramatsu Y, Matsukuma D (2015) Gold nanorods functionalized with self-assembled glycopolymers for ultrasensitive detection of proteins. Chem Lett 44:132–134
38. Sweeney EC, Tonevitsky AG, Temiakov DE, Agapov II, Saward S, Palmer RA (1997) Proteins 28:586
39. Huang H-Y, Chen W-F, Kuo P-L (2005) J Phys Chem B 109:24288
40. Gao B, Rozin MJ, Tao AR (2013) Nanoscale 5:5677
41. Kyle S, Aggeli A, Ingham E, McPherson MJ (2009) Production of self-assembling biomaterials for tissue engineering. Trends Biotechnol 27(7):423–433
42. Ryan DM, Nilsson BL (2012) Self-assembled amino acids and dipeptides as non-covalent hydrogels for tissue engineering. Polym Chem 3(1):18–33
43. Woolfson DN, Mahmoud ZN (2010) More than just bare scaffolds: towards multi-component and decorated fibrous biomaterials. Chem Soc Rev 39(9):3464–3479
44. Kopecek J, Yang J (2012) Smart self-assembled hybrid hydrogel biomaterials. Angew Chem Int Ed 51(30):7396–7417
45. Kyle S, Aggeli A, Ingham E, McPherson MJ (2010) Recombinant self-assembling peptides as biomaterials for tissue engineering. Biomaterials 31(36):9395–9405
46. Kopecek J, Yang J (2007) Hydrogels as smart biomaterials. Polym Int 56(9):1078–1098
47. Johnson JA, Turroa NJ, Kobersteinb JT, Mark JE (2010) Some hydrogels having novel molecular structures. Prog Polym Sci 35(3):332–337
48. Hoare TR, Kohane DS (2008) Hydrogels in drug delivery: progress and challenges. Polymer 49(8):1993–2007
49. Qui Y, Park K (2012) Environment-sensitive hydrogels for drug delivery. Adv Drug Deliver Rev 64:49–60
50. Wang H, Shi Y, Wang L, Yang Z (2013) Recombinant proteins as cross-linkers for hydrogelations. Chem Soc Rev 42(3):891–901
51. Liao X, Chen G, Jianga M (2013) Hydrogels locked by molecular recognition aiming at responsiveness and functionality. Polym Chem 4(6):1733–1745

52. Siegel RA (2014) Stimuli sensitive polymers and self regulated drug delivery systems: a very partial review. J Control Release 190:337–351
53. King WJ, Murphy WL (2011) Bioinspired conformational changes: an adaptable mechanism for bio-responsive protein delivery. Polym Chem 2(3):476–491
54. Petka WA, Harden JL, McGrath KP, Wirtz D, Tirrell DA (1998) Reversible hydrogels from self-assembling artificial proteins. Science 281:389–392
55. Wang C, Stewart RJ, Kopecek J (1999) Hybrid hydrogels assembled from synthetic polymers and coiled-coil protein domains. Nature 397:417–420
56. Yamaguchi N, Zhang L, Chae BS, Palla CS, Furst EM, Kiick KL (2007) Growth factor mediated assembly of cell receptor-responsive hydrogels. J Am Chem Soc 129(11):3040–3041
57. Neufeld G, Cohen T, Gengrinovitch S, Poltorak Z (1999) Vascular endothelial growth factor (VEGF) and its receptors. FASEB J 13(1):9–22
58. Nie T, Baldwin A, Yamaguchi N, Kiick KL (2007) Production of heparin-functionalized hydrogels for the development of responsive and controlled growth factor delivery systems. J. Control. Release 122:287–296
59. Kiick KL (2008) Peptide- and protein-mediated assembly of heparinized hydrogels. Soft Matter 4:29–37
60. Foo CTSWP, Lee JS, Mulyasasmita W, Parisi-Amon A, Heilshorn SC (2009) Two-component protein-engineered physical hydrogels for cell encapsulation. PNAS 106 (52):22067–22072
61. Grove TZ, Osuji CO, Forster JD, Dufresne ER, Regan L (2010) Stimuli-responsive smart gels realized via modular protein design. J Am Chem Soc 132(40):14024–14026
62. Grove TZ, Forster J, Pimienta G, Dufresne E, Regan L (2012) A modular approach to the design of protein-based smart gels. Biopolymers 97(7):508–517
63. Miyata T, Asami N, Uragami T (1999) A reversibly antigen-responsive hydrogel. Nature 399:766–769
64. Ehrick JD, Deo SK, Browning TW, Bachas LG, Madou MJ, Daunert S (2005) Genetically engineered protein in hydrogels tailors stimuli-responsive characteristics. Nature Mater 4 (4):298–302
65. Douglass PM, Salins LLE, Dikici E, Daunert S (2002) Class-selective drug detection: fluorescently-labeled calmodulin as the biorecognition element for phenothiazines and tricyclic antidepressants. Bioconjugate Chem 13(6):1186–1192
66. Ehrbar M, Schoenmakers R, Christen EH, Fussenegger M, Weber W (2008) Drug-sensing hydrogels for the inducible release of biopharmaceuticals. Nature Mater 7:800–804
67. DiMarco RL, Heilshorn SC (2012) Multifunctional materials through modular protein engineering. Adv Mater 24(29):3923–3940
68. Simnick AJ, Lim DW, Chow D, Chilkoti A (2007) Biomedical and biotechnological applications of elastin-like polypeptides. Polym Rev 47(1):121–154
69. MacEwan SR, Chilkoti A (2010) Elastin-like polypeptides: biomedical applications of tunable biopolymers. Pept Sci 94(1):60–77
70. Matsukuma D, Otsuka H (2015) Thermo-responsive reversible phase-transition of alginate based semi-IPN gel through self-assembly of interpenetrated elastin-like polypeptide. Chem Lett 44(1):26–28
71. Vrhocski B, Weiss AS (1998) Biochemistry of tropoelastin. Eur J Biochem 258:1–18
72. Bellingham CM, Lillie MA, Gosline JM, Wright GM, Starcher BC, Bailey AJ, Woodhouse KA, Keeley FW (2003) Recombinant human elastin polypeptides self-assemble into biomaterials with elastin-like properties. Biopolymers 70:445–455
73. Nuhn H, Klok HA (2008) Secondary structure formation and LCST behavior of short elastin-like peptides. Biomacrolecules 9:2755–2763
74. Koga T, Iimura M, Higashi N (2012) Novel peptide-shelled dendrimer with dramatically changeable thermo-responsive character. Macromol Biosci 12:1043–1047
75. Banta S, Wheeldon IR, Blenner M (2010) Protein engineering in the development of functional hydrogels. Annu Rev Biomed Eng 12:167–186

Chapter 14
Stimuli-Responsive Adhesion for 3D Fabrication of Hydrogels

Taka-Aki Asoh, Masatoshi Kato, Yasuyuki Tsuboi and Akihiko Kikuchi

Abstract Hydrogels have numerous applications in tissue engineering, drug delivery system, and as soft actuators, and then, adhesion of hydrogels has attracted attention because it is one of the simplest processes for building 3D constructs. For the fabrication of on-demand hydrogel materials, a stimulus-responsive adhesive method that can work well under physiological conditions is required. This chapter summarizes adhesion of hydrogels, particularly focusing on electrophoretic adhesion of hydrogels for fabrication of functional 3D materials. Oppositely charged and the same-charged hydrogels adhered to each other by DC and AC electrophoretic adhesion, respectively. Nonionic hydrogel also adhered by combination of AC electrophoretic adhesion and specific interactions. To apply this adhesive method to build a 3D material, the rapid fabrication of hydrogel constructs was achieved by the electrophoretic adhesion of microgel to each other. The electrophoretic adhesion of hydrogels will be useful as powerful method for the fabrication of novel 3D hydrogel constructs by bottom-up technology.

Keywords 3D fabrication · Hydrogel · Adhesion · Electrophoresis

T.-A. Asoh (✉)
Advanced Research Institute for Natural Science and Technology,
Osaka City University, 3-3-138 Sugimoto Sumiyoshi-ku, Osaka-shi 558-8585, Japan
e-mail: asoh@sci.osaka-cu.ac.jp

T.-A. Asoh · M. Kato · Y. Tsuboi
Department of Chemistry, Osaka City University, 3-3-138 Sugimoto Sumiyoshi-ku, Osaka-shi 558-8585, Japan

A. Kikuchi
Department of Materials Science and Technology, Tokyo University of Science, 6-3-1 Niijuku Katsushika-ku, Tokyo 125-8585, Japan

T. Kawai and M. Hashizume (eds.), *Stimuli-Responsive Interfaces*,
DOI 10.1007/978-981-10-2463-4_14

14.1 General

Hydrogels are similar to the organs in the living tissue, and therefore, three-dimensional (3D) hydrogels with chemical and/or microstructure regulation have numerous applications in tissue engineering [1], drug delivery system [2], and as soft actuators [3–5]. 3D printing technology of materials has used preparation of prototyping tool in research and manufacturing of next-generation specific devices. Key technology is the ability to precisely control structure and material properties in 3D and tailor-made. Additive manufacturing (AM) of polymeric materials is based on fused deposition modeling of thermoplastic particles or on photo-polymerization of photosensitive monomers and/or macromonomers [6]. On the other hand, hydrogels are prepared by chemical or physical cross-linking of water-soluble polymers or polymerization of monomer in the presence of cross-linkers. Precursors of hydrogels are often one-sort solution, and then, AM technology is limited in their ability to 3D print soft materials such as hydrogels, although soft tissue-like hydrogels were prepared by direct writing using hydrogel precursor ink into support bath such as thermoreversible sol–gel solution or within a hydrogel reservoir [7–9]. Therefore, novel method to create a 3D hydrogel material is required such as adhesion of building block. In this chapter, we provide an overview of research on stimuli-responsive adhesion of hydrogels for fabrication of 3D hydrogels.

14.2 Adhesion of Hydrogels

Adhesion of hydrogels has therefore attracted attention because it is one of the most important processes for building 3D constructs, and then, the adhesion of hydrogels is required to develop functional devices using biocompatible or stimuli-responsive hydrogels. However, solidified glues will limit the softness of adhesive coated surfaces and/or restrict mass transfer between adhered two hydrogels. Therefore, adhesion of gels by using interactions between two hydrogels, such as metal–ligand [10], electrostatic [11–14], host–guest [15], and organic–inorganic [16–19] interactions, has been reported (Fig. 14.1). In particular, utilizing electrostatic interactions is also effective to create a hydrogel in aqueous media at ambient temperature, and the aggregation of oppositely charged micro- or nanoparticles has been used to form scaffolds in cell cultures [20–22].

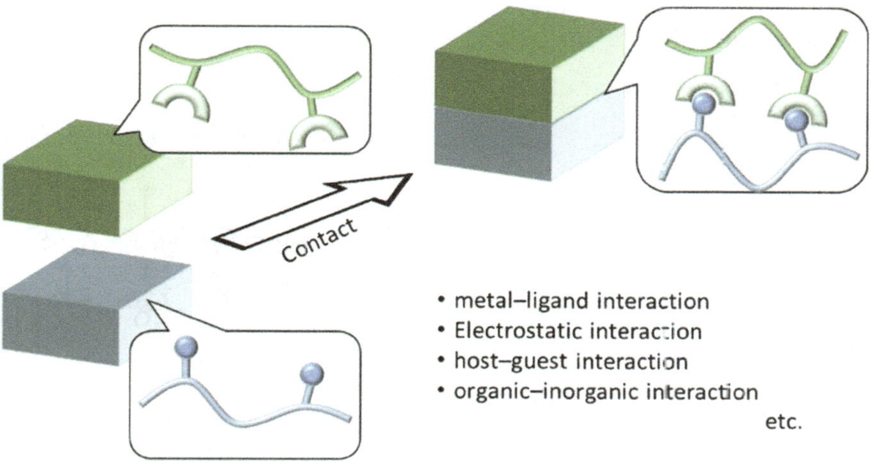

Fig. 14.1 Driving force of adhesion of hydrogels

14.3 Electrophoretic Adhesion (DC and AC Electrophoresis)

Although adhesion by interactions between two hydrogels has been reported, spontaneous adhesion may be disadvantageous for the preparation of hydrogels, because it is difficult to control the configuration, shape, elasticity, and handling. Moreover, hydrogel materials with a specific spatial configuration are difficult to prepare with the spontaneous adhesive method. For the fabrication of on-demand hydrogel materials, a stimulus-responsive adhesive method that can work well under physiological conditions is required. Therefore, a novel stimuli-responsive adhesion system is required for the rapid and concise fabrication of hydrogel constructs. We reported electrophoretic adhesion of hydrogels [23–28], which was a technique using electrophoretic manipulation of polyelectrolyte to form an adhesive layer at the interface of the two gels for adhesion (Fig. 14.2). Two types of gels, which consisting of a cationic or anionic polymers, showed unique adhesion to each other upon electric field impression. The two gels in contact adhered when a direct-current (DC) electric field was applied with a cationic gel on the anode and an anionic gel on the cathode [23, 24, 26] (Fig. 14.2a). Adhesion of the two gels was not observed by opposite electric field or with pressure bonding. It is well known that negative- and positive-charged polymers move to the anode and cathode during electrophoresis, respectively. Therefore, polyion complexes as adhesive layers were formed at the interface of the two gels by electrophoresis of polyelectrolyte, and then, hydrogel adheres each other. Detachment of the adhered hydrogels was also possible by simply applying the inverse voltage to deconstruct the polyion complex. The adhered gels were detached when an electric field was applied in the case of cationic gel on the cathode and anionic gel on the anode,

Fig. 14.2 **a** DC and **b** AC electrophoretic adhesion of hydrogels. The cationic and anionic polymers are shown in *orange* and *blue*, respectively. *Arrows* indicate movement of polyanion and polycation

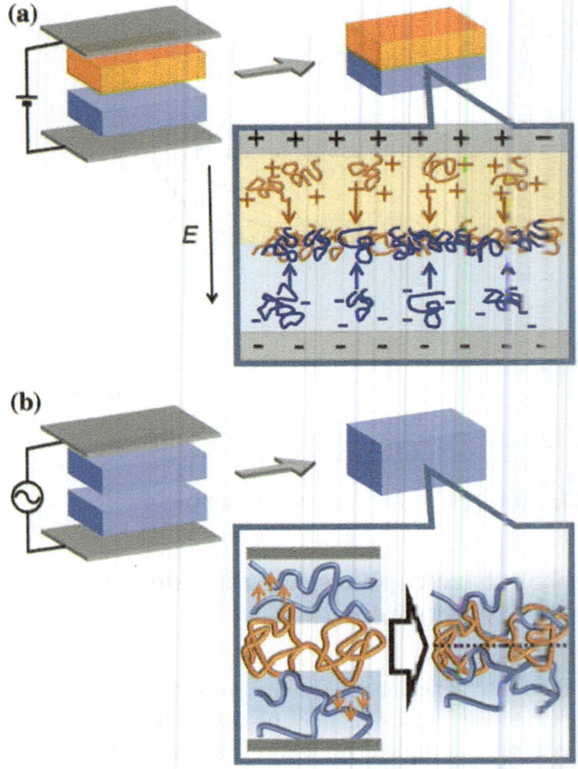

respectively. The same-charged hydrogels were adhered to each other by alternating-current (AC) electrophoretic adhesion [24, 27] (Fig. 14.2b). Cationic polymer solution was dropped onto one anionic gel surface, followed by placement of the other anionic gels. They adhered to each other when the AC electric field was applied with a square wave. Adhesion of the two gels utilizing cationic polymer as binders was not observed when the DC electric field was applied. The adhesive strength of gels decreased with increasing frequency and increased with increasing periodicity. These results indicate that the polyion complex as adhesive layers seemed to be formed between the hydrogel interfaces by oscillation of the cationic polymers in response to an AC electric field. Therefore, the stepwise formation of the polyion complex of anionic and cationic polymers by cationic polymer oscillation at the interface of the two anionic hydrogels was the driving force for the adhesion.

14.4 Adhesion of Stimuli-Responsive Hydrogels

Stimuli-responsive polymers, which change their physical and/or chemical properties in response to environmental signals, have gained much attention not only because of their unique properties, but also because of their potential for significant technological and biomedical applications. Thermoresponsive hydrogels such as poly(*N*-isopropylacrylamide) (PNIPAAm) hydrogels show a reversible volume phase transition in response to an temperature change and then can be used in soft actuators [5]. To develop functional devices using stimuli-responsive hydrogels such as soft actuators, the adhesion of PNIPAAm hydrogels is required. Therefore, adhesion of PNIPAAm hydrogels was investigated. Usually, solidified glues will limit the softness of adhesive coated surfaces and/or restrict mass transfer between adhered two hydrogels. However, when the cationic and anionic PNIPAAm gels were adhered to each other by electrophoresis, the adhered areas shrank and swelled with increasing and decreasing temperature, respectively, without detachment of the hydrogels, and the PNIPAAm/PNIPAAm bi-gels were quite stable in water during repetitive swelling and shrinking cycles [23, 26]. It was clear that the strongly jointed areas of the bi-gels showed flexibility without any inhibition of the hydrogel responsiveness, although the adhesive area of the gel surface changed over 18-fold [26]. As shown in Fig. 14.3, thermoresponsive bending gels [23] and uniaxial movement soft actuator [26] were simply prepared by adhesion of PNIPAAm gel and non-stimuli-responsive poly(acrylamide) (PAAm) gels. The bilayer gels of PNIPAAm/PAAm bent in an arc (Fig. 14.3a) and multilayer gels of PNIPAAm/PAAm showed uniaxial movement (Fig. 14.3b) in response to temperature without detachment, because PAAm does not show any temperature responsiveness, but PNIPAAm shrinks. Thus, the present adhesion method provides not only relative motion with the swelling/deswelling of the gels, but also the

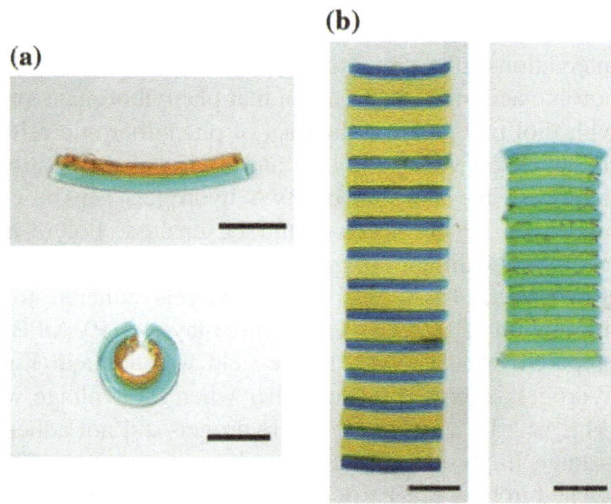

Fig. 14.3 Preparation of **a** PNIPAAm/poly (acrylamide) (PAAm) bilayerd hydrogels, **b** (PAAm/PNIPAAm) 13PAAm multilayer hydrogels at 20 and 40 °C. All scale bars are 5 mm. Reproduced from Asoh et al. [23, 26] from the Royal Society of Chemistry

ability to withstand bending stress without detachment. These adhered materials will be useful as smart soft actuators that can work in water.

14.5 Adhesion of Biodegradable Hydrogels

Biodegradable hydrogels have numerous applications in tissue engineering and drug delivery systems. The biodegradable hydrogels were also adhered to each other by electrophoretic adhesion [24, 27]. Chemical structure of polymers is shown in Fig. 14.4a. Gelatin hydrogels were prepared by chemically cross-linking gelatin type A. The cationic or anionic gels were prepared by addition of chitosan hydrochloride or sodium poly(γ-glutamate) (PGA) to the gelatin solution, respectively. The cured gelatin with chitosan or PGA by cooling was immersed into 1-ethyl-3-(3-dimethylaminopropyl)carbodiimide hydrochloride (EDC) solutions. The adhesion interface of the gels showed the ability to withstand the stress of a 180° twist without detachment (Fig. 14.4b), and the adhesion of the two gels was quite stable in water. Increasing the electrophoresis time resulted in stronger adhesion between the two gels, which clearly supports the hypothesis that increasing voltage increases gel interaction [24]. Two sodium dextran sulfate (DS) gels [27], which are prepared by chemical cross-linking of gelatin type A with DS, were adhered by AC electrophoretic adhesion with quaternized chitosan (TMC) binders (Fig. 14.4c).

14.6 Adhesion of Nonionic Hydrogels

Poly(vinyl alcohol) (PVA) is a well-known biocompatible and non-toxic material, often studied in the biomaterial field [29]. The adhesion of PVA hydrogels may be a convenient method for the construction of 3D scaffolds for cell culture. To study interactions of the hydroxy groups of PVA, we focused our attention on phenylboronic acid. It is well known that phenylboronate groups show reversible bonding with diol units around the pKa of phenylboronic acid, and phenylboronate is used for glucose sensing [30–33] and hydrogel preparation [34] in the field of biomaterials. If the adhesion of PVA hydrogels can be controlled, biocompatible and non-toxic 3D constructs could be created. Poly(3-methacrylamidophenylboronic acid-co-*N*,*N*-dimethylacrylamide) (PBA) was prepared by radical copolymerization (Fig. 14.5a). PVA gels and PBA gels adhered to each other by DC voltage impression (Fig. 14.5b), and three-layered PVA/PBA/PVA hydrogels were also prepared when an AC electric field was applied (Fig. 14.5c) [28]. The two PVA hydrogels adhered to each other when AC voltage was applied with square wave (Fig. 14.5d), while two PVA hydrogels did not adhere by pressure bonding or DC voltage impression. When a voltage was impressed to a PBA copolymer sandwiched in between PVA hydrogels, the negatively charged PBA copolymer moved

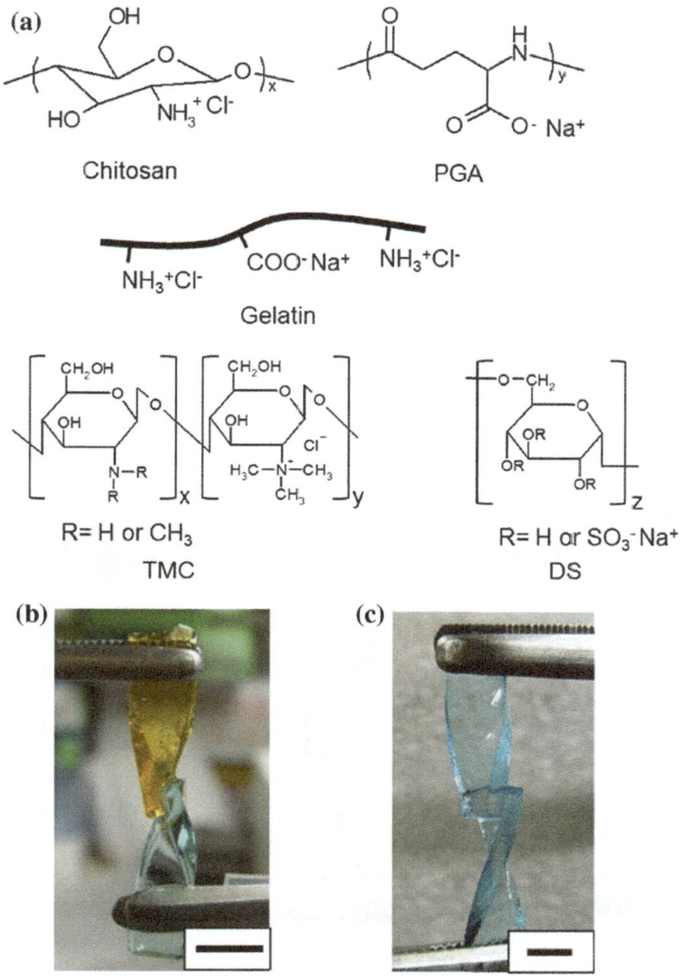

Fig. 14.4 **a** Chemical structure of PGA, Chitosan, gelatin, TMC, and DS. Images of the adhered **b** Chitosan/PGA and **c** DS gels, twisted 180°. The PGA and DS chitosan gels were stained by the cationic dye methylene blue, and the chitosan gels were stained by the anionic dye methyl orange. All scale bars are 5 mm. Reproduced from Asoh et al. [24, 27] from the Royal Society of Chemistry and the Elsevier

to the anode by electrophoresis, and the PBA units reacted with diols of the PVA hydrogel interface. Continuously, partially bonded PBA copolymers on the PVA surface move to the opposite side of PVA hydrogels by inverse voltage impression, and then, two PVA hydrogels adhered to each other utilizing PBA copolymers. These results are in good agreement with a study using electrostatic interaction and indicate that both the mobility of the PBA copolymers and the interaction of PBA

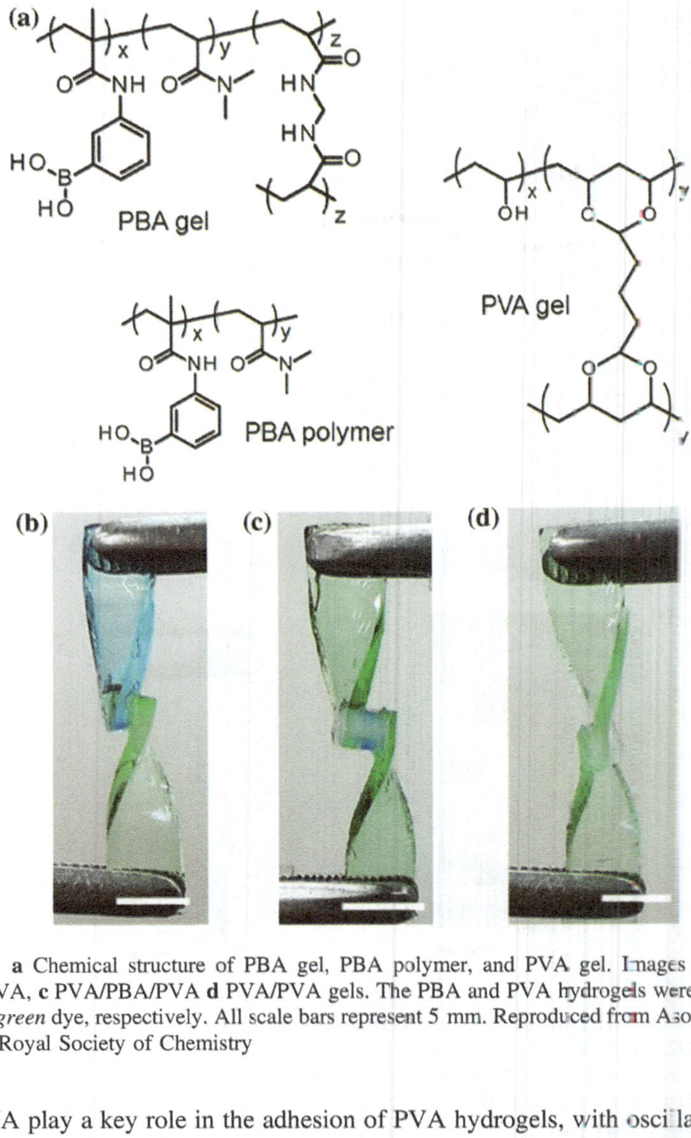

Fig. 14.5 **a** Chemical structure of PBA gel, PBA polymer, and PVA gel. Images of adhered **b** PBA/PVA, **c** PVA/PBA/PVA **d** PVA/PVA gels. The PBA and PVA hydrogels were stained by *blue* and *green* dye, respectively. All scale bars represent 5 mm. Reproduced from Asoh et al. [28] from the Royal Society of Chemistry

with PVA play a key role in the adhesion of PVA hydrogels, with oscillating PBA copolymers stepwise interacting with diols at the interface of the PVA hydrogels during AC electrophoresis. The adhered PVA interface was stable under physiological conditions, but detachment was observed when sugar and acid were added in excess. Detached gels re-adhered under the same conditions, indicating that the adhesion of the hydrogels exhibits repeatability characteristics.

14.7 Free-Form Fabrication of Hydrogel Adhesion

To apply this electrophoretic adhesion of hydrogels to build a material, the rapid fabrication of hydrogel constructs will be achieved by electric field impression to a precursor [25, 27]. Poly(vinyl alcohol) (PVA) microgels with poly(diallyldimethylammonium chloride) (PDDA) or sodium poly(styrene sulfonate) (PSS) were used as the building blocks of the hydrogel constructs. A pasty product was formed when the same amounts of cationic and anionic microgel powders were swelled in water, but dispersed in an excess amount of water. This result is expedient for configuring and molding the shape of the gels. The microgel paste was gelled by electric field impression, and obtained construct was stable in water after the electric field removal and was elastic and durable enough to handle with forceps (Fig. 14.6a) [25]. As shown in Fig. 14.6b, the cationic and anionic microgel contacted each other three-dimensionally. Hydrogel constructs which are built by the adhesion of microgel as the minimum constituent can form another network. The reconstruction of hydrogel can be achieved by the re-adhesion of microbeads. At first, the microgel paste molded into a strip shape, and electrophoresis was carried out to prepare a hydrogel construct. Next, the strip-shaped constructs turned back into a pasty material by grinding to obliterate the adhesion interface of the microgels (Fig. 14.6c[i]). After molding the pasty microgels into a triangle shape, the construct was re-gelled by electrophoresis (Fig. 14.6c[ii]). Thereafter, a square-shaped hydrogel was prepared by electrophoresis after grinding the triangle-shaped hydrogel (Fig. 14.6c[iii] and [iv]). In all steps, the pasty microgels could be molded into any shape, but the hydrogel constructs were handled by forceps and were stable in water without leakage of the microgels.

We also prepared hydrogel constructs by the adhesion of biodegradable microgels (Fig. 14.7a) [27]. Anionic microgel powders (DS microgels) were swelled in water, and then, cationic TMC droplets were added into the swelled microgels. After molding a square shape, AC voltage was applied to a paste composed of DS microgels and TMC. The microgels adhered to each other and formed an elastic gel-like material by AC electric field application on the x-, y-, and z-axes. The constructs were also durable enough to be handled with forceps (Fig. 14.7b). Through the microscopic observations of the hydrogel constructs, we confirmed that the microgels had adhered to each other (Fig. 14.7c). Microgels formed 3D networks through the adhesion of the microgels by AC electrophoresis of TMC at the microgel interfaces.

3D fabrication based on stimuli-responsive adhesion control, especially electrophoretic adhesion of hydrogels, will be useful as powerful method for the fabrication of novel 3D hydrogel constructs with precisely controlled structure and material properties.

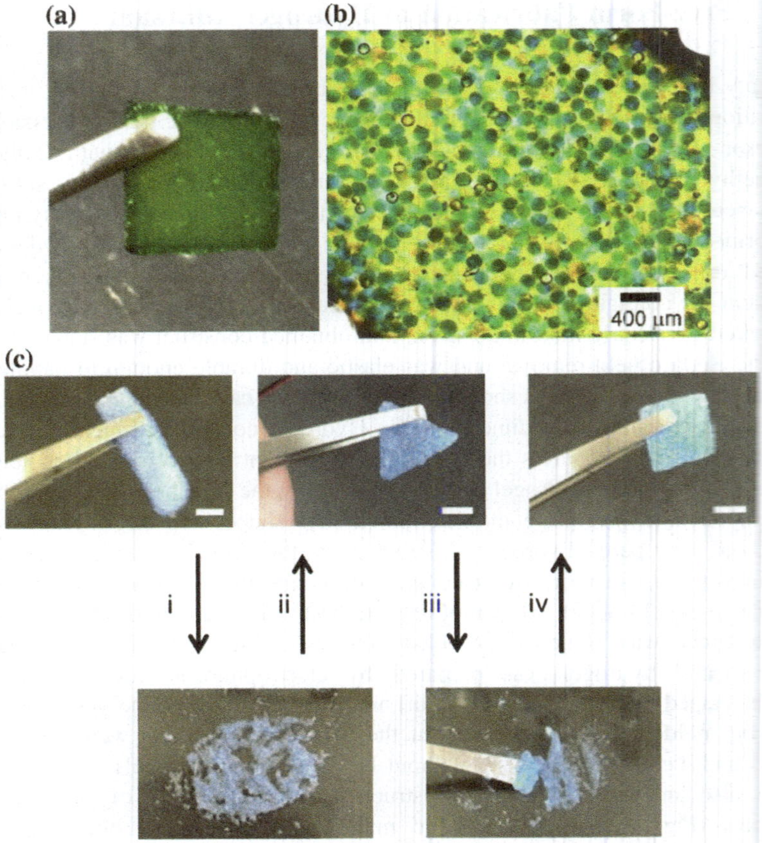

Fig. 14.6 **a** Macro- and **b** microscopic observations of the hydrogel constructs. The anionic and cationic microbeads were stained by the cationic dye methylene blue and the anionic dye methyl orange, respectively. The *square-shaped* hydrogel constructs were 10 mm × 10 mm, and the thickness was 1 mm. **c** Macroscopic images of the reconstruction of hydrogels. The anionic microbeads were stained by methylene blue. All scale bars are 5 mm. Reproduced from Asoh et al. [25] from the Royal Society of Chemistry

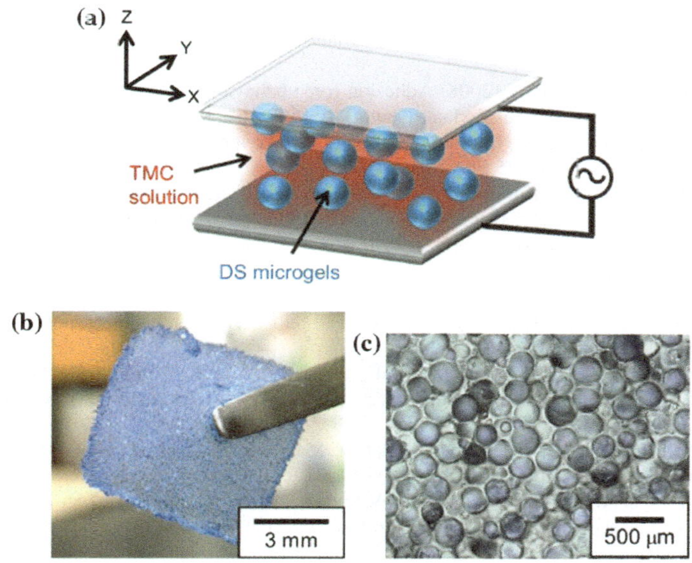

Fig. 14.7 **a** Preparation of hydrogel constructs composed of DS microgels with TMC by AC voltage impression. **b** Macro- and **c** microscopic images of the hydrogel constructs prepared by DS30 microgel adhesion. The DS microgels were stained by methylene blue. Reproduced from Asoh et al. [27] from the Elsevier

References

1. Lee KY, Mooney DJ (2001) Hydrogels for tissue engineering. Chem Rev 101:1869–1880
2. Vermonden T, Censi R, Hennink WE (2012) Hydrogels for protein delivery. Chem Rev 112:2853–2888
3. Osada Y, Okuzaki H, Hori H (1992) A polymer gel with electrically driven motility. Nature 355:242–244
4. Hu Z, Zhang X, Li Y (1995) Synthesis and application of modulated polymer gels. Science 269:525–527
5. Asoh T, Matsusaki M, Kaneko T, Akashi M (2008) Fabrication of temperature-responsive bending hydrogels with a nanostructured gradient. Adv Mater 20:2080–2083
6. Derby B (2012) Printing and prototyping of tissues and scaffolds. Science 338:921–926
7. Hinton TJ, Jallerat Q, Palchesko RN, Park JH, Grodzicki MS, Sue HJ, Ramadan MH, Hudson AR, Feinberg AW (2015) Sci Adv 1:e1500758
8. Wu W, DeConinck A, Lewis JA (2011) Omnidirectional printing of 3d microvascular networks. Adv Mater 23:H178–H183
9. Kolesky D, Truby RL, Gladman AS, Busbee TA, Homan KA, Lewis JA (2014) Adv Mater 26:3124–3130
10. Holten-Andersen N, Harrington MJ, Birkedal H, Lee BP, Messersmith PB, Lee KYC, Waite JH (2011) pH-induced metal-ligand cross-links inspired by mussel yield self-healing polymer networks with near-covalent elastic moduli. Proc Natl Acad Sci USA 108:2651–2655

11. Tamagawa H, Nagato F, Umemoto S, Okui N, Popovic S, Taya M (2002) The influence of salt-linkages on the adhesion and hardness variance of hydrogels. Bull Chem Soc Jpn 75:383–388
12. Tamagawa H, Takahashi Y (2008) Adhesion force behavior between two gels attached with an electrolytic polymer liquid. Mater Chem Phys 107:164–170
13. Abe H, Hara Y, Maeda S, Hashimoto S (2014) Surface modification method for adhesion of gels. Chem Lett 43:243–245
14. Sun TL, Kurokawa T, Kuroda S, Ihsan AB, Akasaki T, Sato K, Haque MA, Nakajima T, Gong JP (2013) Physical hydrogels composed of polyampholytes demonstrate high toughness and viscoelasticity. Nat Mater 12:932–937
15. Harada A, Kobayashi R, Takashima Y, Hashidzume A, Yamaguchi H (2011) Macroscopic self-assembly through molecular recognition. Nat Chem 3:34–37
16. Haraguchi K, Uyama K, Tanimoto H (2011) Self-healing in nanocomposite hydrogels. Macromol Rapid Commun 32:1253–1258
17. Wang Q, Mynar JL, Yoshida M, Lee E, Lee M, Okuro K, Kinbara K, Aida T (2010) High-water-content mouldable hydrogels by mixing clay and a dendritic molecular binder. Nature 463:339–343
18. Rosel S, Prevoteau A, Elzière P, Hourdet D, Marcellan A, Leibler L (2014) Nanoparticle solutions as adhesives for gels and biological tissues. Nature 505:382–385
19. Meddahi-Pellé A, Legrand A, Marcellan A, Louedec L, Letourneur D, Leibler L (2014) Organ repair, hemostasis, and in vivo bonding of medical devices by aqueous solutions of nanoparticles. Angew Chem Int Ed 53:6369–6373
20. Van Tomme SR, van Steenbergen MJ, De Smedt SC, van Nostrum CF, Hennink WE (2005) Self-gelling hydrogels based on oppositely charged dextran microspheres. Biomaterials 26:2129–2135
21. Wang Q, Wang L, Detamore MS, Berkland C (2008) Biodegradable colloidal gels as moldable tissue engineering scaffolds. Adv Mater 20:236–239
22. Wang H, Hansen, MB, M. Löwik DWP, van Hest JCM, Li Y, Jansen JA Leeuwenburgh SC (2011) Oppositely charged gelatin nanospheres as building blocks for injectable and biodegradable gels. Adv Mater 23:H119–H124
23. Asoh T (2010) Kikuchi a electrophoretic adhesion of stimuli-responsive hydrogels. Chem Commun 46:7793–7795
24. Asoh T, Kawai W, Kikuchi A (2012) Electrophoretic adhesion of biodegradable hydrogels through the intermediary of oppositely charged polyelectrolytes. Soft Matter 8:1923–1927
25. Asoh T, Kikuchi A (2012) Rapid fabrication of reconstructible hydrogels by electrophoretic microbead adhesion. Chem Commun 48:10019–10021
26. Asoh T, Kawamura E, Kikuchi A (2013) Stabilization of electrophoretically adhered gel-interfaces to construct multi-layered hydrogels. RSC Adv 3:7947–7952
27. Asoh T, Kawai W, Kikuchi A (2014) Alternating-current electrophoretic adhesion of biodegradable hydrogel utilizing intermediate polymers. Colloids Surf B 123:742–746
28. Asoh T, Takaishi K, Kikuchi A (2015) Adhesion of Poly(vinyl alcohol) hydrogels by electrophoretic manipulation of phenylboronic acid copolymer. J Mater Chem B 3:6740–6745
29. Peppas NA, Hoffman AS (2013) Chapter I.2.5 Hydrogels. In: Ratner BD, Hoffman AS, Schoen FJ, Lemons JE (ed) Biomaterials science, 3rd edn. Elsevier, Academic Press. pp 166–179
30. Kikuchi A, Suzuki K, Okabayashi O, Hoshino H, Kataoka K, Sakurai Y, Okano T (1996) Glucose-sensing electrode coated with polymer complex gel containing phenylboronic acid. Anal Chem 68:823–828
31. Kataoka K, Miyazki H, Bunya M, Okano T, Sakurai Y (1998) Totally synthetic polymer gels responding to external glucose concentration: their preparation and application to on–off regulation of insulin release. J Am Chem Soc 120:12694–12695
32. Alexeev VL, Sharma AC, Goponenko AV, Das S, Lednev IK, Wilcox CS, Finegold DN, Asher SA (2003) High ionic strength glucose-sensing photonic crystal. Anal Chem 75:2316–2323

33. Matsumoto A, Ishii T, Nishida J, Matsumoto H, Kataoka K, Miyahara Y (2012) A synthetic approach toward a self-regulated insulin delivery system. Angew Chem Int Ed 51:2124–2128
34. Konno T, Ishihara K (2007) Temporal and spatially controllable cell encapsulation using a water-soluble phospholipid polymer with phenylboronic acid moiety. Biomaterials 28:1770–1777

Chapter 15
Media-Responsive Swelling and Material Release Properties of Polysaccharide Composite Films

Mineo Hashizume and Kazutoshi Iijima

Abstract Free-standing, water-insoluble thin films made of polyion complexes of natural polysaccharides, for example, chondroitin sulfate and chitosan, are successfully obtained by utilizing hot press techniques. These films exhibit sufficient mechanical properties to be used as structural materials in the dry state. On the other hand, the films swell to some extent and soften in water, but are not dissolved out, although they are made of water-soluble polysaccharides. The degree of swelling is controllable depending on the species and pH of the media, in addition to the polysaccharide species. When small dye molecules are used as model drugs, dye-loaded films are obtained using both pre- and post-loading methods. These dye-loaded films release the dye molecules when immersed in aqueous media. The kinetics of the dye release is affected by the species and pH of the solutions, which indicate that it is correlated with the swelling behavior of the films. These results support that the free-standing polysaccharide composite films have promise to be used as biocompatible film materials such as plasters or wound dressings having media-responsive drug release abilities.

Keywords Thin film · Polysaccharide · Polyion complex · Controlled release · pH responsive

15.1 Introduction

Polymer materials and their composites are one of the promising candidates to be used as stimuli-responsive materials [1–6]. For example, poly(*N*-isopropylacrylamide) and its derivatives exhibit volume phase transition behaviors based on their temperature-dependent change of hydration states. Photo-responsive polymers usually bare photo-responsive units to induce conformational change of the

---

M. Hashizume (✉) · K. Iijima
Department of Industrial Chemistry, Faculty of Engineering,
Tokyo University of Science, Tokyo, Japan
e-mail: mhashizu@ci.kagu.tus.ac.jp

T. Kawai and M. Hashizume (eds.), *Stimuli-Responsive Interfaces*,
DOI 10.1007/978-981-10-2463-4_15

polymers by photoirradiation. In some cases, photoirradiation induces chemical reactions that produce chemical species, which brings specific functions to the polymer materials. Media pH is another stimulus to induce conformational and/or functional changes of the polymers, where the pH changes result in the change of local and net charges of the polymers. Recently, such stimuli-responsive polymers are also called as "smart polymers". Some of these systems are described in other chapters of this book. Certain kinds of stimuli-responsive polymers can be utilized for the application of drug delivery system (DDS) because the required stimuli can be generated naturally or artificially in vivo [7, 8]. Along this line, various kinds of synthetic and semi-synthetic polymers have been designed and used as various states such as monomers, micelles, colloidal particles, films, coating layers, and bulk solids.

On the other hand, we have been investigating utilization of biomass or commonly used polymers for preparation of structural materials to be used as biomaterials. We especially have been focused on utilization of natural polysaccharides because of their potential use in biomedical applications [9–12] and their relatively well-established industrial productions. As a result of the investigation, we have demonstrated that free-standing films can be obtained by hot pressing of polyion complexes (PICs) of natural polysaccharides (Fig. 15.1) [13, 14]. Most of the previous studies for the fabrication of structural materials made of water-soluble polysaccharides have been conducted accompanying with cross-linking [15–18] or chemical modification [19, 20] of the polysaccharides to gain water insolubility of the resulting materials, whereas our free-standing films are made of water-soluble polysaccharides without using such treatments. The PICs of natural polysaccharides have been studied by several groups [21–25]: however, most of them have handled the PICs as bulk solids or cast films and no film fabrication process analogous to

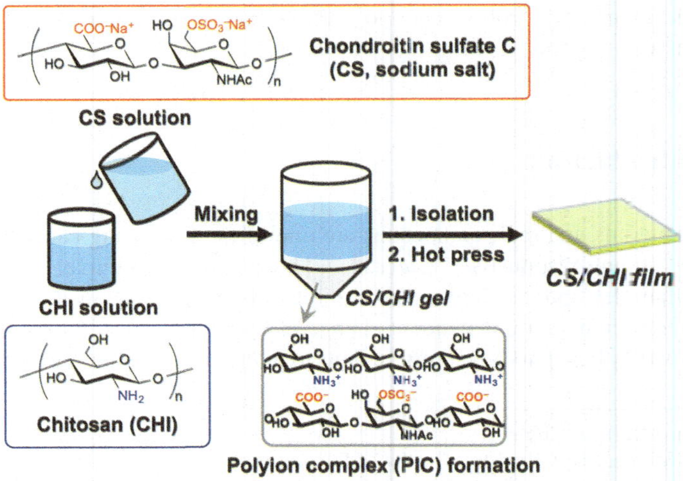

Fig. 15.1 Chemical structures of polysaccharides and schematic illustration of preparation of free-standing polysaccharide composite films

ours has not been employed. The layer-by-layer (LbL) assembly [26–28] may be regarded as a technique for preparation of a special type of PICs. Using this technique, nanometer-thickness of free-standing films made of CHI and alginate have been prepared and their potential use as a nanoplaster has been demonstrated [29]. On the other hand, our free-standing films have thicknesses of several tenths of micrometers, which results in the fact that the films are robust to be treated easily by hand.

During the physical characterization of the polysaccharide composite films, we found that these films exhibited some kinds of media-responsive properties. In this chapter, we describe the media-responsive swelling and material release properties of the films, in addition to the basics of the preparation and other physical characterization of the films.

15.2 Preparation of Free-Standing Polysaccharide Composite Films

The standard procedure for the preparation of our free-standing polysaccharide composite films is schematically shown in Fig. 15.1. Briefly, the aqueous solutions of glycosaminoglycans (GAGs) such as chondroitin sulfate C (CS), heparin (HEP), and hyaluronic acid (HYA) are added dropwise to the chitosan (CHI) solutions (generally in dilute acetic acid) to form PICs of the polysaccharides. An example of the sample for the case of the PIC of CS and CHI is shown in Fig. 15.2a (hereafter PICs made of CS, HEP, HYA and CHI are denoted as CS/CHI, HEP/CHI, and HYA/CHI, respectively). The volumes of the GAG solutions needed to complete the PIC formation with the given amounts of CHI solutions are not identical to the theoretical values as estimated from the number of acidic functional groups of GAG molecules in the solution. This is understandable when considering the fact that the reactions (PIC formation) occur between macromolecules: that is, some of ionic functional groups of the polysaccharides might be sterically hindered to interact with oppositely charged functional groups of the counterpart polysaccharides. The hydrated CS/CHI PICs in the solutions are centrifuged, washed with distilled water, and then isolated as dried PICs using lyophilization (Fig. 15.2b). Usually, the isolation yields of the PICs are *ca.* 80–85 % as calculated from the volume of the solutions used for the PIC formation. In some cases, longer times of centrifugation are used instead of lyophilization to isolate PICs, where the resulting PICs are not completely dehydrated. The dried PICs are firstly pressed using a commercial hot press apparatus (typically 120 °C, 20 MPa) (Fig. 15.2c and then small portions of water are added to make the PICs swollen (Fig. 15.2d). When water-containing PICs obtained using longer times of centrifugations, these steps can be omitted. Incorporation of water in the PICs is necessary to increase their moldability. The swollen PICs are hot pressed at a lower pressure (nearly 0 MPa) to be molded and to evaporate the water simultaneously (Fig. 15.2e) and then folded into small pieces

Fig. 15.2 Morphologies of the CS/CHI PICs at each preparation step. **a** Formed in the mixed solution, **b** isolated by lyophilization, **c** after the first hot pressing, **d** after swollen by a small portion of water, **e** after the second pressing, **f** after folded and placed in the hole of the PET spacer, and **g** final morphology of the film after the third pressing

and placed in the rectangular holes provided by the PET spacers (50 or 100 μm thickness) that control film thickness (Fig. 15.2f). The folded PICs are again hot pressed (20 MPa) and finally free-standing CS/CHI films are obtained (Fig. 15.2g). Using the conditions almost the same as to that of CS/CHI films, free-standing HEP/CHI and HYA/CHI films are also obtained [13, 14]. The sizes of the films are limited by the size of the hot press apparatus. When using an apparatus having a 15 cm × 20 cm size of hot press stage, the films of 5 cm × 5 cm sizes are reproducibly obtained.

Very recently, we have found out that roll press techniques are also available to obtain free-standing polysaccharide composite films. The detail is currently under investigation in our group. We have also demonstrated that the polysaccharide composite fibers can be obtained by spinning of the PICs formed at the interfaces of the CS and the CHI solutions [30]. This approach (PIC formation at the solution interface) can be combined with microfluidic technologies: that is, the PIC formation at the interface of co-axial flows supplied by microfluidic devices can be utilized to continuous formation of the PIC fibers or tubes. We experimentally confirm that such an approach can be realized.

The obtained polysaccharide composite films (CS/CHI films) are transparent (Fig. 15.3a), which support that no domain structures that scatter visible lights form inside the films. The origin of pale yellowish color of the films probably comes from decomposition of impurities by heat treatments. The films are robust, elastic and bendable without cracking (Fig. 15.3b). Microscopic observations of the films using scanning electron microscopy (SEM) support smooth surfaces and dense inner structures of the films (Fig. 15.3c). Both the HEP/CHI films and the HYA/CHI films also have basically similar macroscopic and microscopic

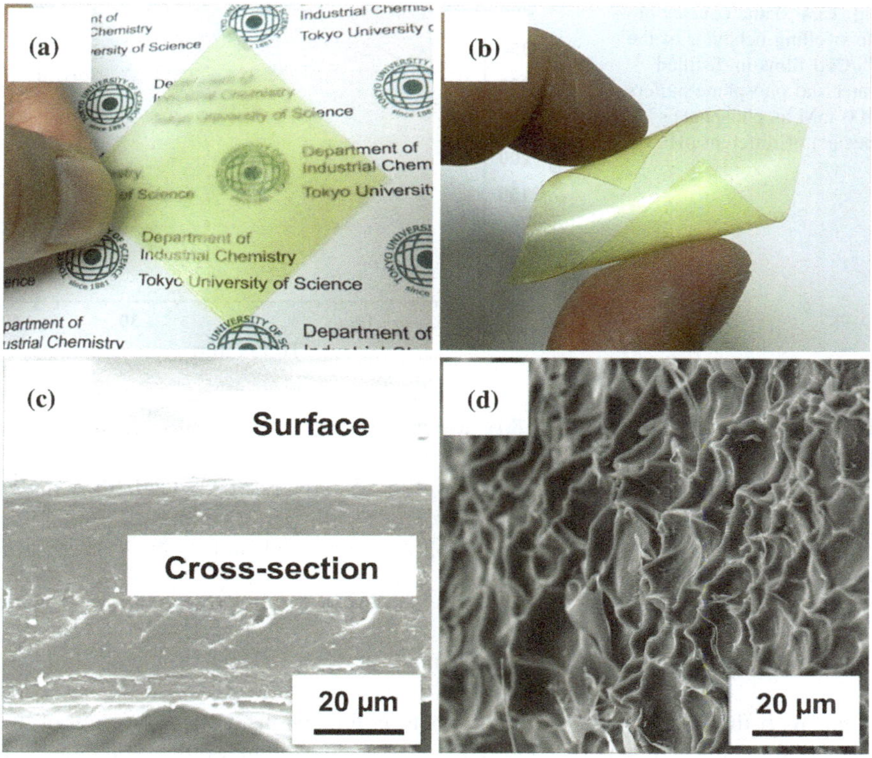

Fig. 15.3 **a**, **b** Macroscopic views of the CS/CHI films. **c**, **d** SEM images of the CS/CHI film (**c**) and the film after incubation in cell culture medium (**d**)

morphologies to that of CS/CHI films. Tensile strengths of these polysaccharide composite films in the dry state are *ca.* 50–100 MPa depending on the film components, for example, the species and molecular weight of the polysaccharides, and the preparative conditions [14]. The results show that these films possess sufficient mechanical strength to be used as structural materials.

15.3 Swelling Behaviors of the Films

The polysaccharide composite films have sufficient mechanical strength in the dry state, whereas they become soft and much stretchable when subjected to high humidity. The films swell in water and buffer solutions to some degree, but not dissolved out. The SEM images of the CS/CHI films lyophilized after immersion in buffers, for example, a cell culture medium (D-MEM, serum free), show that the

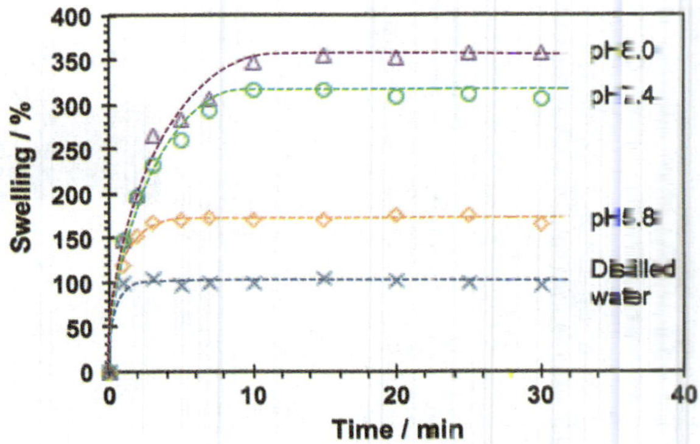

Fig. 15.4 Time courses of the swelling behavior of the CS/CHI films in distilled water and phosphate buffers (100 mM as phosphate species) of different pH

film surfaces are porous (Fig. 15.3d), which supports that the films are swollen in these solutions.

Figure 15.4 is the summary of the swelling behaviors of CS/CHI films under various media conditions. Here, the degree of swelling ("Swelling" in Fig. 15.4) and the weight loss (%) are defined as follows (Eqs. 15.1 and 15.2):

$$\text{Degree of swelling}\,(\%) = (W_S - W_I)/W_I \times 100 \tag{15.1}$$

$$\text{Weight loss}\,(\%) = (W_I - W_F)/W_I \times 100 \tag{15.2}$$

where W_I is the initial weight of the film before immersion, W_S is the weight of the swelled film after immersion, and W_F is the final weight of the film after the immersion experiments followed by drying in air, respectively. According to Eq. 15.1, 100 % of swelling corresponds to the situation where the film incorporates the same amount of water as to its dried weight. The results in Fig. 15.4 show that the swellings are moderately fast and reach equilibrium within 20 min for all cases. They also show that the degree of swelling is affected by the media species. The films in phosphate buffers swell much more than that in distilled water that showed approximately 100 % of swelling. The difference supposedly comes from the difference in ionic concentrations. Among the buffers, the swelling degrees of the films increase with the increase of the pH of the buffers. The CS/CHI films swell *ca.* 350 % at pH 8.0. These pH-dependent swelling properties are presumably due to the nature of ionic functional groups of the polysaccharides. The degrees of the dissociation of the ester sulfate groups (pK_a = 2.6 [31] or smaller) and the carboxylic acid groups of CS (pK_a = 4.6 [31] or 3.6 in the solution of 0.1 M of ion concentration [32]) are not significantly affected by the pH change from 5.8 to 8.0. On the other hand, the pK_a of the ammonium groups of CHI is 6.5 [33]. Therefore, the increase of the swelling degree by changing the media pH from 5.8 to 8.0 should be brought by the contribution of CHI because the amount of ammonium groups of CHI is decreased by this pH change. Dissociation of protons of the ammonium groups of CHI can decrease intermolecular interactions between CS and

CHI molecules, and between CS or CHI molecule themselves in the films, which must result in the change, probably increase, of the swelling. Proton dissociation can also affect the polymer conformations in the hydration state, which can be another factor to affect the swelling degree. Along the examination of the swelling behavior of the polysaccharide composite films, we also have demonstrated that the swelling degree is different depending on the polysaccharide species [14].

The increase of the swelling degree of the films might result in the increase of weight loss of the film, as defined by Eq. 15.2, after the swelling experiments. In the case of distilled water, the weight losses are up to *ca.* 10 %, whereas they become larger in the case of phosphate buffers of higher pH. The preliminary examinations indicate that the increase of the solution temperature also tends to increase the weight loss of the films. The weight loss of the film corresponds to the degradation of the film. Therefore, it is an important factor to design of the films as the carrier for sustained drug release.

These evaluations shown in this section demonstrate that the swelling degree and the weight loss of the polysaccharide composite films in aqueous media are controllable by the selection of several parameters such as the species and molecular weight of the polysaccharides and the species of the media.

15.4 Material Loading and Release Properties of the Films

We have recently found that the model drug-loaded polysaccharide composite films can be prepared and the loaded molecules can be released from the film [34]. For example, methylene blue (MB), a cationic dye molecule as a model drug, is successfully loaded in CS/CHI films in a concentration-dependent manner (Fig. 15.5a). Here, the MB solution is firstly mixed with the CHI (same charge) solution, and the resulting the MB-CHI mixed solution is then mixed with the CS (opposite charge) solution to form CS/CHI PICs incorporating MB molecules. The PICs are then formed into free-standing films using hot pressing. For the case of anionic model drugs, their solutions should be premixed with CS solutions before PIC formation to prevent formation of ion complexes between the model drug and CHI. In addition to the pre-loading method as described above, we have also confirmed that the MB-loaded CS/CHI films can also be obtained by immersion of the unmodified films into the MB solutions (the post-loading method): that is, the film swells and incorporate MB molecules. The model drug loading does not significantly affect macroscopic (except for the color of the film) and microscopic morphologies and physical properties of the films such as mechanical strength. These results are interesting because incorporation of charged molecules inside the films might interfere or assist PIC formation.

When these model drug-loaded films are immersed in aqueous solutions, they release the model drugs to the solutions. Figure 15.5 shows the example of the model drug release experiments for the case of the MB-loaded films. After the release experiments by film immersion, the solutions turn colored by MB, whereas

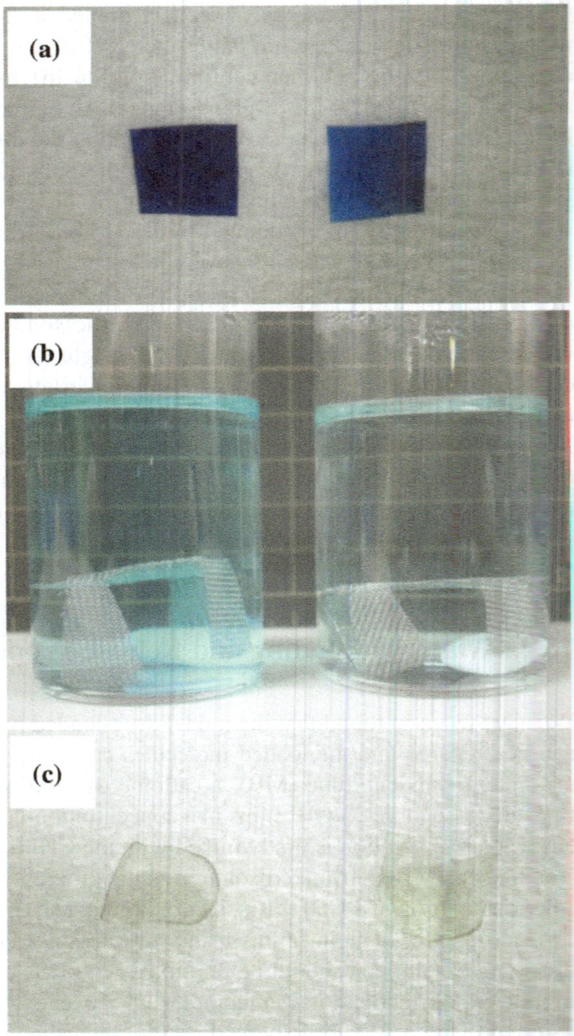

Fig. 15.5 Macroscopic views of the model drug (MB)-loaded CS/CHI films before (**a**) and after (**c**) the release experiments and that of the solutions after the release experiments (**b**). The *left film* in the images was prepared from the PIC that obtained using the MB-CHI premixed solution (10 mL of 1×10^{-3} M MB solutions + 10 mL of CHI solutions (1 wt% in dilute acetic acid)) and the required volume (approximately 7 mL) of the CS solutions (2 wt% in water). The *right film* was obtained using the same procedure except for that the concentration of the MB solution was 5×10^{-4} M

the films become decolorized that corresponds to MB release (Fig. 15.5b, c). Figure 15.5c also shows that simple (careless) drying of the films after immersion in solutions under the ambient conditions results in partial shrinking of the films.

The use of dye molecules such as MB makes it possible to monitor the model drug release profiles of the films spectroscopically. Such evaluations demonstrate that the release profiles of MB from the CS/CHI films depend on the media species and pH (Fig. 15.6). The MB release in distilled water is slow and lasts for over hours, whereas that in phosphate buffers reaches equilibrium in approximately 30 min (Fig. 15.6, inset). In addition, the media pH affects initial MB release rates. The estimated initial release rates obtained from the release profiles are 0.08, 3.8,

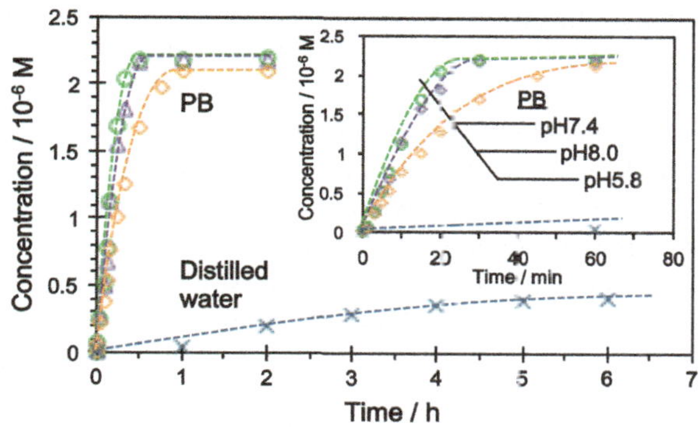

Fig. 15.6 MB release profiles of the MB-loaded CS/CHI films in distilled water and in phosphate buffers (pH 5.8, 7.4, and 8.0) at room temperature (approximately 25 °C). The inset shows the initial stage of the profiles. The films were prepared using the same procedure as to that for the film shown in the left image of Fig. 15.5

5.5, 5.3 nmol min^{-1} for the MB release in water and phosphate buffers of pH 5.8, 7.4, 8.0, respectively. Under the same ion concentrations, the increase of media pH results in the increase of initial release rates but such increases are saturated at higher pH. It is supposed that such differences mainly come from the difference in the swelling degrees of these media as described in the previous section. The pH-dependent swelling and corresponding drug release were also observed for the case of PIC nanoparticles consisted of HYA and polyethyleneimine [35]. However, other factors also need to be considered to explain the detail of the MB release properties from the films. For example, the swelling of the films in distilled water reaches equilibrium in approximately 30 min, whereas the time of MB releases necessary to reach equilibrium is much longer. Electrostatic interactions, both attractive and repulsive interactions, between MB and polysaccharides are an important considerable factor because they are affected by ionic concentrations and pH of the solutions. Interestingly, the releases of anionic model drug molecules from the films also show similar media and pH dependencies even though they possess opposite charges to MB [34]. Further evaluations are currently being conducted to prove the detailed release mechanism. Similar to the swelling degree, the (model) drug release behaviors of the films should be controllable by polysaccharide species and molecular weight, and media conditions.

15.5 Conclusion

The series of evaluations confirm that free-standing composite films prepared from PICs of oppositely charged natural polysaccharides possess media-responsive swelling and material release properties. In addition to small model drugs, very recently we have found out that some kinds of proteins can also be loaded in the films, which support the availability of the films as the carriers for protein drugs. Because the polysaccharides used in this study are intrinsically biocompatible, these drug-loaded polysaccharide composite films have potential application as plasters and implantable film materials having drug release ability such as wound dressings and the films used for cancer therapies. These films can also be used as cell culture scaffolds that continuously supply growth factors. Degradation of the films under the conditions of higher ion concentrations and/or higher temperatures, and biodegradation of the polysaccharide molecules under physiological conditions can be utilized as the other stimuli for the controlled release of the drugs from the films. We believe that the prove of the detailed release mechanism for the films will guide the tailor-made design of the films, for example, the choice of suitable species and molecular weight of the polysaccharides, to realize programmed release of the target drugs under the situation where the films used.

Acknowledgments This work was partly supported by JSPS KAKENHI Grant Numbers 22550155, 25410178, and a grant from The Association for the Progress of New Chemistry (ASPRONC), Japan.

References

1. Stuart MAC, Huck WTS, Genzer J, Muller M, Ober C, Stamm M, Sukhorukov GB, Szleifer I, Tsukruk VV, Urban M, Winnik F, Zauscher S, Luzinov I, Minko S (2010) Emerging applications of stimuli-responsive polymer materials. Nat Mater 9:101–113
2. Theato P, Sumerlin BS, O'Reilly RK, Epps III TH (eds) (2013) Stimuli responsive themed issue. Chem Soc Rev 42(17):7045–7486
3. Ganesh VA, Baji A, Ramakrishna S (2014) Smart functional polymers—a new route towards creating a sustainable environment. RSC Adv 4:53352–53364
4. Guragain S, Bastakoti BP, Malgras V, Nakashima K, Yamauchi Y (2015) Multi-stimuli-responsive polymeric materials. Chem Eur J 21:13164–13174
5. Chen JK, Chang CJ (2014) Fabrications and applications of stimulus-responsive polymer films and patterns on surfaces: a review. Materials 7:805–875
6. Meng H, Li G (2013) A review of stimuli-responsive shape memory polymer composites. Polymer 54:2199–2221
7. Ganta S, Devalapally H, Shahiwala A, Amiji M (2008) A review of stimuli-responsive nanocarriers for drug and gene delivery. J Controlled Release 126:187–204
8. Molina M, Asadian-Birjand M, Balach J, Bergueiro J, Miceli E, Calderón M (2015) Stimuli-responsive nanogel composites and their application in nanomedicine. Chem Soc Rev 44:6161–6186
9. Rinaudo M (2008) Main properties and current applications of some polysaccharides as biomaterials. Polym Int 57:397–430
10. Iijima K, Hashizume M (2015) Application of polysaccharides as structural materials. Trends Glycosci Glycotechnol 27:67–79
11. Czaja WK, Young DJ, Kawecki M, Brown RM Jr (2007) The future prospects of microbial cellulose in biomedical applications. Biomacromolecules 8:1–12
12. Ravi Kumer MNV, Muzzarelli RAA, Muzzarelli C, Sashiwa H, Domb AJ (2004) Chitosan chemistry and pharmaceutical perspectives. Chem Rev 104:6017–6084
13. Hashizume M, Kobayashi H, Ohashi M (2011) Preparation of free-standing films of natural polysaccharides using hot press technique and their surface functionalization with biomimetic apatite. Colloids Surf B 88:534–538
14. Hashizume M, Ohashi M, Kobayashi H, Tsuji Y, Iijima K (2015) Preparation of free-standing polysaccharide composite films: improved preparation and physical properties. Colloids Surf A 483:18–24

15. Sintov A, Di-Capua N, Rubinstein A (1995) Cross-linked chondroitin sulphate: characterization for drug delivery purposes. Biomaterials 16:473–478
16. Kubo K, Kuroyanagi Y (2004) Development of cultured dermal substitute composed of a spongy matrix of hyaluronic acid and atelo-collagen combined with fibroblasts: cryopreservation. Artif Organs 28:182–188
17. Vercruysse KP, Marecak DM, Marecek JF, Prestwich GD (1997) Synthesis and in vitro degradation of new polyvalent hydrazide cross-linked hydrogels of hyaluronic acid. Bioconjugate Chem 8:686–694
18. Luo Y, Kirker KR, Prestwich GD (2000) Cross-linked hyaluronic acid hydrogel films: new biomaterials for drug delivery. J Controlled Release 69:169–184
19. Hunt JA, Joshi HN, Stella VJ, Topp EM (1990) Diffusion and drug release in polymer films prepared from ester derivatives of hyaluronic acid. J Controlled Release 12:159–169
20. Campoccia D, Doherty P, Radice M, Brun P, Abatangelo G, Williams DF (1998) Semisynthetic resorbable materials from hyaluronan esterification. Biomaterials 19:2101–2127
21. Chen WB, Wang LF, Chen JS, Fan SY (2005) Characterization of polyelectrolyte complexes between chondroitin sulfate and chitosan in the solid state. J Biomed Mater Res 75A:128–137
22. Kikuchi Y, Noda A (1976) Polyelectrolyte complexes of heparin and chitosan. J Appl Polym Sci 20:2561–2563
23. Peniche C, Fernández M, Rodríguez G, Parra J, Jimenez J, López Bravo A, Gómez D, San Román J (2007) Cell supports of chitosan/hyaluronic acid and chondroitin sulphate system. Morphology and biological behaviour. J Mater Sci Mater Med 18:1719–1726
24. Meng X, Tian F, Yang J, He CN, Xing N, Li F (2010) Chitosan and alginate polyelectrolyte complex membranes and their properties for wound dressing application. J Mater Sci Mater Med 21:1751–1759
25. Fajardo AR, Lopes LC, Valente AJM, Rubira AF, Muniz EC (2011) Effect of stoichiometry and pH on the structure and properties of chitosan/chondroitin sulfate complexes. Colloid Polym Sci 289:1739–1748
26. Decher G (1997) Fuzzy nanoassemblies: toward layered polymeric multicomponents. Science 277:1232–1237
27. Decher G, Schlenoff JB (eds) (2012) Multilayer thin films, 2nd edn. Wiley-VCH, Weinheim
28. Ariga K, Yamauchi Y, Rydzek G, Ji Q, Yonamine Y, Wu KCW, Hill JP (2013) Layer-by-layer nanoarchitectonics: invention, innovation, and evolution. Chem Lett 43:36–68
29. Fujie T, Okamura Y, Takeoka S (2007) Ubiquitous transference of a free-standing polysaccharide nanosheet with the development of a nano-adhesive plaster. Adv Mater 19:3549–3553
30. Iijima K, Yuyama K, Asaine K, Irie K, Hashizume M (2014) Preparation of chondroitin sulfate/chitosan composite fibers by spinning from aqueous solution interfaces. Kobunshi Ronbunshu 71:11–16
31. Larsson B, Nilson M, Tjälve H (1981) The binding of inorganic and organic cations and H^+ to cartilage in vitro. Biochem Pharmacol 30:2963–2970
32. Scordilis-Kelley C, Osteryoung JG (1996) Voltammetric studies of counterion transport in solutions of chondroitin sulfate. J Phys Chem 100:797–804
33. Domard A (1987) pH and c.d. measurements on a fully deacetylated chitosan: application to Cu^{II}–polymer interactions. Int J Biol Macromol 9:98–104
34. Hashizume M, Murata Y, Iijima K, Shibata T (2016) Drug loading and release behaviors of freestanding polysaccharide composite films. Polymer J 48:545–550
35. Chen JX, Wang M, Tian HH, Chen JH (2015) Hyaluronic acid and polyethyleneimine self-assembled polyion complexes as pH-sensitive drug carrier for cancer therapy. Colloids Surf B 134:81–87

Chapter 16
Stimuli-Responsive Thin Films Composed of Photochromic Compounds to Construct Surface Relief

Takashi Ubukata

Abstract Photochromic compounds are very interesting and potentially useful for optical data storage, optical switching, optical filter, and so on; because their chemical structures are reversibly changed by light stimuli. In recent years, unique photoinduced surface relief (SR) formation caused by lateral material transfer has been reported and has attracted great attention. Patterned light irradiation construct a large SR structure on a thin film containing azobenzene that is one of the representative photochromic compounds. This chapter focuses on new classes of SR formation using photochromic compounds other than the azobenzene derivatives are demonstrated.

Keywords Thin film · Surface relief · Photochromic compounds · Azobenzene · Spiropyran · Spirooxazine · Hexaarylbiimidazole · Diarylethene

16.1 Introduction

Surface fine processing is one of the important techniques in various technology fields. Current lithography-based fine processing techniques enable the application of elaborate processing through a multistep process triggered by local photoreactions. In contrast, in 1995, two research groups independently reported that patterned relief structures are formed on the surface of an azobenzene-functionalized polymer thin film in the single step of the process of patterned light irradiation [1, 2]. These features can be reversibly erased and reformed because they are based on the transfer of the polymer. This phenomenon has drawn intense attention because of both fundamental scientific interest and the possible application of this effect as a new fine

T. Ubukata (✉)
Department of Advanced Materials Chemistry,
Yokohama National University, Yokohama, Japan
e-mail: ubukata@ynu.ac.jp

T. Kawai and M. Hashizume (eds.), *Stimuli-Responsive Interfaces*,
DOI 10.1007/978-981-10-2463-4_16

processing technique that differs from lithography methods in its operational principles and characteristics. In this chapter, after outlining this phenomenon, recent advances in materials are introduced, focusing on photochromic compounds

16.2 Surface Relief Formed on Azobenzene Containing Materials

The formation of patterned relief structures is found in polymer thin films that are modified by the azobenzene group. The photoisomerization [3] of azobenzene between the rod-like *trans*-form and the bent *cis*-form is considered to be important to form patterned relief structures (Fig. 16.1). Both of the papers that first described relief formation phenomenon [1, 2] in 1995 reported that the interference exposure to an amorphous thin film of polymer with azobenzene side groups leads to the formation of patterned relief structures that match with the period of the interference of laser. This structure is called "Surface Relief Grating" (SRG) and is still widely used. In addition to interference exposure [4], condensing laser [5], exposure through a photomask [6, 7], and near field light [8–10] have been shown to form various patterned relief structures that are collectively referred to as the surface relief (SR). Henceforth, in this chapter, these structures will be referred to as SR. The shape of the SR structure was stable under the temperature below the glass transition temperature, T_g of the materials. In contrast, the system returns to the original flat thin film surface upon heating to the temperature higher than the T_g [1, 2]. Furthermore, it was found that under the appropriate uniform light irradiation conditions, the SR structure reverts to the original flat thin film surface [11] (Fig. 16.2).

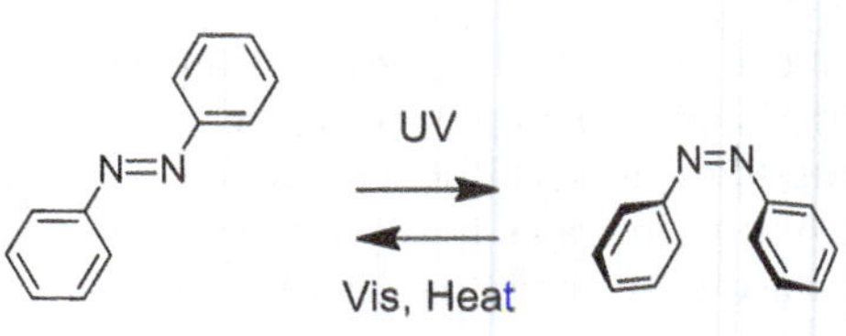

Fig. 16.1 Interconversion of azobenzene

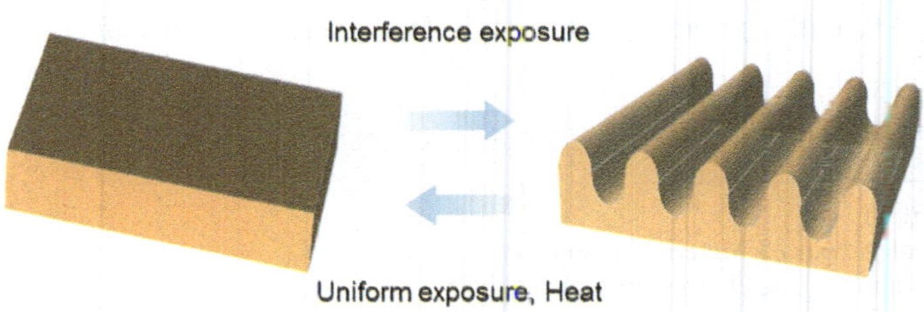

Fig. 16.2 Schematic illustration of reversible photoformation of surface relief

For SR formation by light, a continuous wave argon ion laser with a wavelength of 488 or 514 mm is normally used. The grating period is adjustable according to the interference conditions of the laser and periods of several hundred nm to several μm have been usually selected. Because of the fact that high light intensity is not required for SR formation, an intensity of several mW cm^{-2} to several hundred mW cm^{-2} is sufficient for the formation of the SR with a height difference of up to approximately 2 μm [12]. In most cases, the regions of the surface exposed to strong light have indented shapes and the movement of the substances occurs from the exposed region to the unexposed region. Because the SR-forming capacity greatly depends on the polarization state of the laser light [5], it is considered that SR formation occurs via an optical process as opposed to the thermal process.

Since the first reports of photogeneration of SRs on the films of amorphous polymers containing azobenzene [13], various material systems have been developed such as liquid crystalline polymers [6, 14–22], sol–gel materials [9, 23, 24], supramolecules [12, 25–27], dendritic materials [28, 29], amorphous low molecular weight materials [30–36], and single crystal surface [37]. On the other hand, these findings might have been difficult to guide the consensus for photogeneration mechanism in SR. Firstly, classifying each system then elucidating the predominant interactions may be necessary to understand the phenomena. Since the details of SR including the proposed formation mechanism have been summarized in several reviews [38–41] and publication [42], the reader can be referred to these references. In the next section, photogeneration of SR using materials other than azobenzene that have scarcely been discussed in the aforementioned reviews and publication will be introduced.

16.3 Surface Relief Formed on Films without Azobenzene

The SR formation discussed thus far is a phenomenon that has been implicitly understood to refer to the thin films that contain azobenzene. However, recently, it has been found that a surface shape similar to that of SR can be formed on the thin films that do not contain azobenzene by patterned light irradiation. In these cases, photopolymerization and photo-cross-linking reaction were found to play a key role. Including these photopolymerization and photo-cross-linking induced SR formation; hereinafter, the designation of SR shall be used for these materials as well.

SR with a period of 400 μm and a height difference of 1.5 μm was formed by the ultraviolet (UV) light irradiation to a spin-coating film of UV-curing type optical adhesive agent through a photomask followed by irradiation with the same amount of UV light all over the surface [43]. Furthermore, SR formation with a period of 667 nm and a height difference of 120 nm was observed by performing the laser interference exposure with a wavelength of 257 mm for the thin film of dipeptide having two thymine groups that are known to be photodimerizable [44]. This SR was stable even when heated at a temperature of 383 K for 16 h.

Kawatsuki, Ono, and others found that SR was obtained on the liquid crystalline polymer thin film that has a crosslinkable cinnamoyl side chain by heat treatment at a temperature of 423 K after interference exposure of a He–Cd laser or UV irradiation through a photomask [45–47]. Because the crosslinking proceeds in the exposed regions, and polymer diffusion from unexposed to exposed regions is induced by the subsequent heating, the SR structure where the exposed regions become the protrusions is formed. The same phenomenon has also been observed in the liquid crystalline polymer that has a crosslinkable styrylpyridine side chain [48]. A method for forming an SR structure through heating after such patterned exposure has been drawing attention recently and is known as the photoembossing method [49, 50]. Using the photoembossing method, SR structure is constructed on a photopolymer blend comprising a binder polymer, a multifunctional acrylate, and small percent of photoinitiator. These shape changes are thought to occur via the polymerization reaction of monomers in the exposure regions and diffusion of the monomers in the unexposed regions by the heating.

On the other hand, it has also been reported that on the thin films of polyphenylene vinylene [51] and polybutadiene [52], the polymer forms an SR with a height difference of several tens of nm by the interference exposure of a laser beam with wavelengths that can exhibit light absorption (405 and 244 nm, respectively). More recently, SR structures are found on the general purpose polystyrene films which are irradiation with UV light through a photomask under the control of the temperature of the film [53, 54]. Figure 16.3 shows the SR structure formed on the polystyrene film observed by atomic force microscopy (AFM). Sensitivity to form the SR structure can be improved by the addition of a benzophenone derivative to the polystyrene film [55].

The described SR formation by the materials that do not contain azobenzene is anticipated to diminish the availability of erasing/reformation, which is one of the main advantages of the SR formation techniques found in systems where azobenzene is used. Researchers have been attempting to obtain the photoformation of SR by using compounds other than azobenzene that exhibit photochromism for the purpose of reversible relief formation. In the next section, the strategies involving photoformation of the SR using photochromic compounds other than azobenzene shall be discussed.

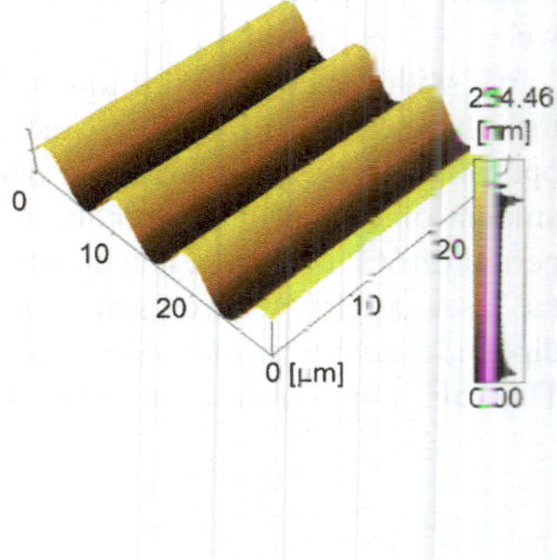

Fig. 16.3 AFM image of polystyrene ($M_w = 3.3 \times 10^3$) film after exposed to high-pressure mercury lamp (>290 nm) with 0.175 W cm^{-2} for 1.5 h through a photomask at 379 K

16.4 Surface Relief Formed on Polymer Thin Films Doped with Photochromic Molecules

Photochromism is referred to as "a reversible transformation of a single chemical species being induced in one or both directions by electromagnetic radiation between two states having different distinguishable absorption spectra" [56]. Upon photoirradiation, photochromic compounds change their structures accompanying with their colour change. Therefore, their physical and chemical properties, as well as absorption spectra, may be changed by photochromic transformation [57, 58]. Azobenzene, which have ability to form SR, is one of the representative photochromic compounds. In this sections, it is shown that SR structures formed on polymer films containing photochromic compounds that are not an azobenzene-related compound.

16.4.1 *Spiropyran*

In the beginning, spiropyrans (SPs) are employed because they are representative thermally reversible photochromic compounds such as azobenzenes. SPs show a large structural change upon photoirradiation; an approximately 90° rotation of one half of the molecule during the ring opening and closing reactions between the spiro form (SP form) and the merocyanine form (MC form) (Fig. 16.4). In this section, the formation of the SR structure using SP is presented [59].

SP shown in (Fig. 16.4) was synthesized by the condensation of a nitrosalicylaldehyde with an *N*-alkylindolenium bromide [60]. The spin-coated films were prepared from the chloroform solutions of each SPs and poly(methyl methacrylate) (PMMA) on a cleaned glass substrate. The thickness of films was 0.3 μm, when the film was composed of 60 % of SP and 40 % of PMMA in weight.

The wavelength of the light for irradiation longer than 350 nm was taken out from a high-pressure mercury lamp by a long pass filter and water filter. The light was collimated to obtain a homogeneous light intensity (0.1 W cm^{-2}) over the entire area to be exposed. Patterned light irradiation to form SR structures was performed through a photomask attached to the film. The photomask was composed of stripes with the same widths of the transparent and opaque lines. After the film

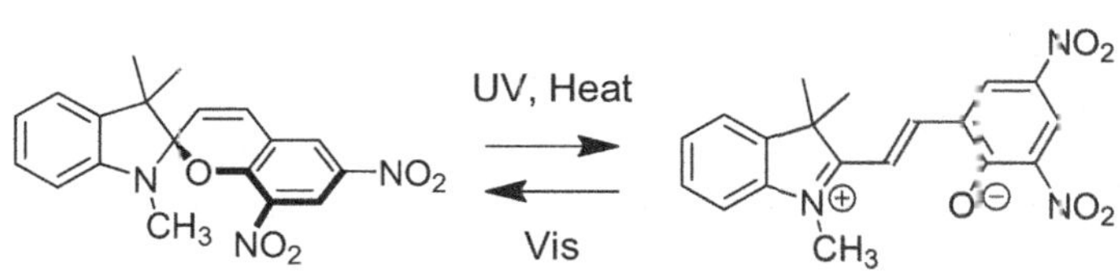

Fig. 16.4 Interconversion of dinitrospiropyran

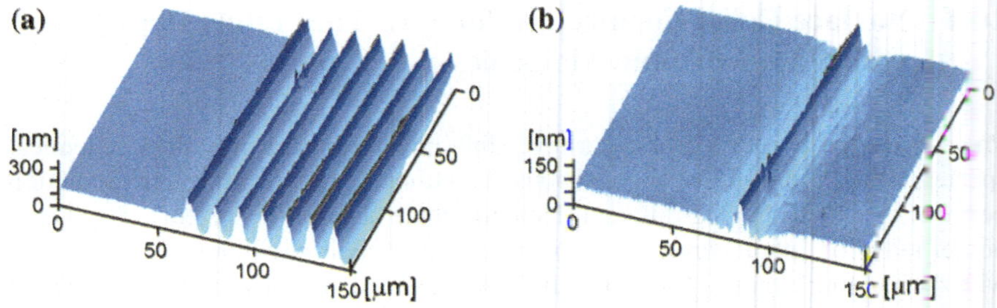

Fig. 16.5 AFM images of SP-doped PMMA films (60 wt% of SP) after exposed to high-pressure mercury lamp (>350 nm) with 0.1 W cm^{-2} for 2 h through a photomask at room temperature. The photomask was composed of stripes with the same widths (6.25 μm) of the *transparent* and *opaque lines* **a** and one *slit line* with 6.25-μm width **b**

was exposed to the irradiation light for 2 h, the surface of the film was observed through a differential interference microscopy and AFM.

Figure 16.5 shows the SR structure formed on the SP-doped PMMA film. When the light was shaded completely, only a flat surface was observed as shown in the left-hand side of Fig. 16.5a. On the other hand, when the film was irradiated with patterned light, regular surface modulation was produced, and its spatial period was coincident with that of the photomask as shown in the right-hand side of Fig. 16.5a. When the cross-sectional topography was examined, it was found that the top of the convex is higher than the initial surface level and base of depression is lower than the initial level. This result clearly shows that the lateral material transfer occurred at the film surface. In order to investigate the direction of material transfer, an irradiation experiment through a photomask possessing one slit line with 6.25-μm width was performed (Fig. 16.5b). In the case of the SP-doped PMMA film, irradiation produced a significant topographic maximum, indicating that polymer transfer occurred from the shaded areas to the exposed areas. The direction of mass transfer was opposite to that observed for almost azobenzene containing film.

16.4.2 Diarylethene

Azobenzene and spiropyran can be classified as the thermally reversible organic photochromic compounds that exhibit structural change thermally as well as photochemically. However, there have been no reports on SR formation with thermally irreversible photochromic compounds, classified into the other counterpart. In this section, it is shown that SR formation on polymer films containing diarylethene (DAE) known as one of the representatives of the thermally irreversible photochromic compounds (Fig. 16.6) [61].

Syntheses of DAE was carried out with standard procedures reported [62]. Optically transparent films, about 0.4 μm thick, were prepared by spin-coating

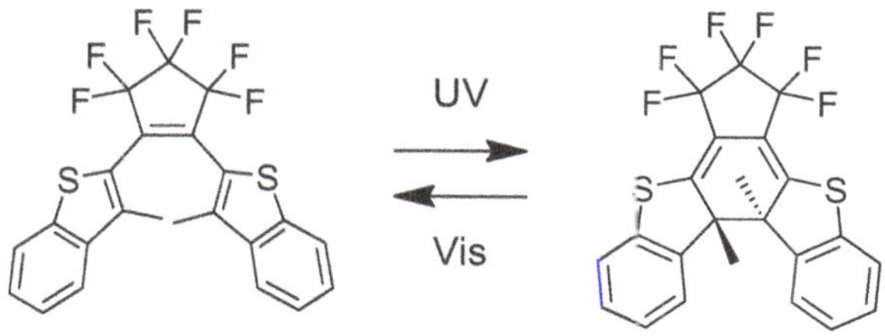

Fig. 16.6 Interconversion of diarylethene

method onto a cleaned glass plate, from the chloroform solution of each DAE and PMMA, 60 and 40 wt%, respectively.

Figure 16.7a shows the SR structure produced by the patterned light irradiation (>350 nm) to the PMMA film doped with DAE through a photomask. The regular surface modulation was produced, and its spatial period was coincident with that of the photomask. In order to investigate the direction of material transfer, an irradiation experiment through a photomask possessing one slit with 4-μm width was performed. In the case of the film doped with DAE, irradiation produced a significant topographic maximum, indicating that material transfer occurred from the shaded areas to the exposed areas (Fig. 16.7b).

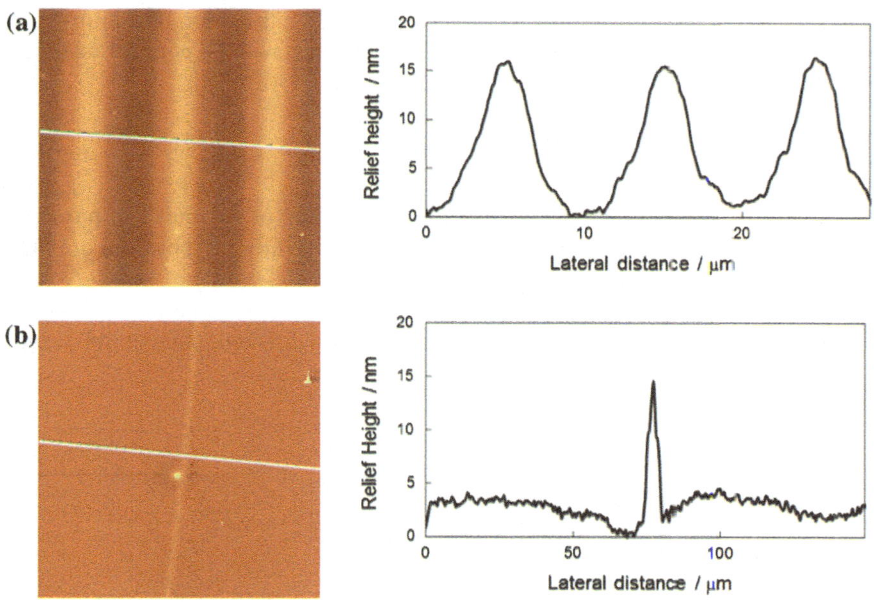

Fig. 16.7 AFM images of DAE-doped PMMA films (60 wt% of DAE) after exposed to high-pressure mercury lamp (>350 nm) with 0.1 W cm^{-2} for 2 h through a photomask at room temperature. The photomask was composed of stripes with the same widths (4 μm) of the *transparent* and *opaque lines* **a** and one *slit line* with 4-μm width **b**. The figures on the *right* side indicate the cross-sectional height profiles for the *solid lines* shown in the AFM images

The same phenomenon has also been observed in the PMMA film doped with the overcrowded alkene that shows specific rotational motion to the double bond by light and heat [63]. Unfortunately, reversibility was not recognized in the SRs formed in the films described above. It may be ascribed to the decomposition of these photochromic compounds in the PMMA by the UV light irradiation. Reversible SRG formation is anticipated using the other photochromic compounds with high durability against UV light.

16.5 Surface Relief Formed on Amorphous Thin Films Composed of Low Molecular Weight Photochromic Compounds

As shown in the former section, SR structures could be constructed on the mixed films composed of PMMA and photochromic compounds other than the azobenzenes using patterned UV light. However, reversible SRG formation has not been attained in these films due to low durability of the photochromic compounds against the UV light. In addition, entanglement of polymer chain in matrix should suppress efficient transfer of photochromic compounds in these systems. In order to avoid such difficulties occurring in polymer systems, low molecular weight photochromic compounds have been used. In this section, it is described that SR structure formed on the amorphous films composed of low molecular weight photochromic compounds without polymer matrix.

16.5.1 Hexaarylbiimidazole

SR structure formed on amorphous thin films consisting of derivative of hexaarylbiimidazoles (HABIs) has been reported [64]. The photochromic behaviour of HABI can be attributed to photoinduced homolytic reversible cleavage of the C–N bond between the imidazole rings to form a pair of 2,4,5-triphenylimidazolyl radicals (TPIRs). Here, 2,2′-di(*ortho*-chlorophenyl)-4,4′,5,5′-tetraphenylbiimidazole, *o*-Cl-HABI is used (Fig. 16.8).

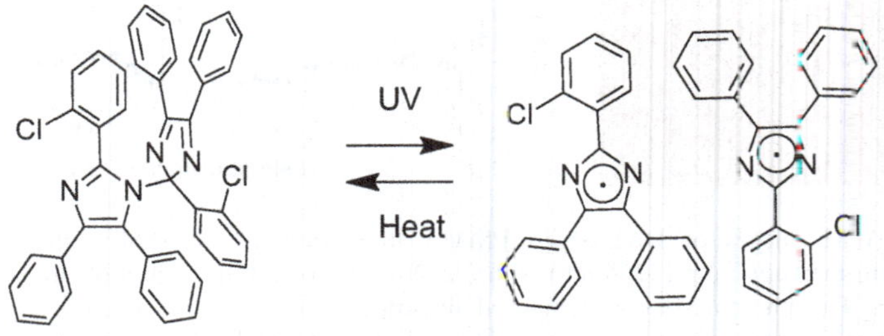

Fig. 16.8 Interconversion of *o*-Cl-HABI

Transparent amorphous thin films (ca. 120 nm thickness) were prepared by spin-coating on a clean glass substrate from *o*-Cl-HABI benzene solution. The first differential scanning calorimetry (DSC) scan of *o*-Cl-HABI microcrystals displayed the melting point (462 K), whereas no peak due to amorphism was observed in cooling process. Therefore, the second DSC heating scan displayed no peak for the melting point of *o*-Cl-HABI microcrystals. The T_g of *o*-Cl-HABI films was detected to be around 366 K by DSC. The photochromic behaviour of the *o*-Cl-HABI amorphous thin film is confirmed by absorption spectra before and after UV light irradiation at 365 nm. Before irradiation, the main absorption band is at 270 nm. After irradiation, new bands appear at 350 and 550 nm, ascribed to *o*-Cl-TPIR, produced by photoinduced homolytic dissociation. Then, under dark conditions, *o*-Cl-TPIRs thermally recombine to reproduce the dimer form, *o*-Cl-HABI within in ca. 6–10 s in an amorphous film at 383 K. This photochromic behaviour is in good agreement with that observed in *o*-Cl-HABI acetonitrile solution. Patterned light irradiation (365 nm, 22 mW cm^{-2}) was performed through a photomask composed of stripes with the 2 μm widths of transparent and opaque lines at the controlled temperature (383 K) under argon atmosphere.

Figure 16.9 shows SR structures formed by patterned UV light irradiation for 20 min on the film through the photomask. Regular surface modulation was achieved with a spatial period of 2 μm, in accordance with the photomask pattern, and observed peak-to-trough modulation depth of ca. 30 nm (Fig. 16.9). To investigate the direction of mass transfer, the film was irradiated through a photomask with just one slit line. The cross-sectional topography shows that the convex is higher and the concave is lower than the initial surface level, which clearly demonstrates that lateral mass transfer occurred on the film surface. Based on these results, driving force of the material transfer is presented as follows. On the surface of the irradiated areas, *o*-Cl-HABIs are readily cleaved photochemically into TPIRs; in contrast, on the surface of the shaded areas, almost all molecules are present in their dimer form, *o*-Cl-HABI. Mass transfer induced as a consequence of the radical dissociative photochromic reaction on the film surface can be explained as follows.

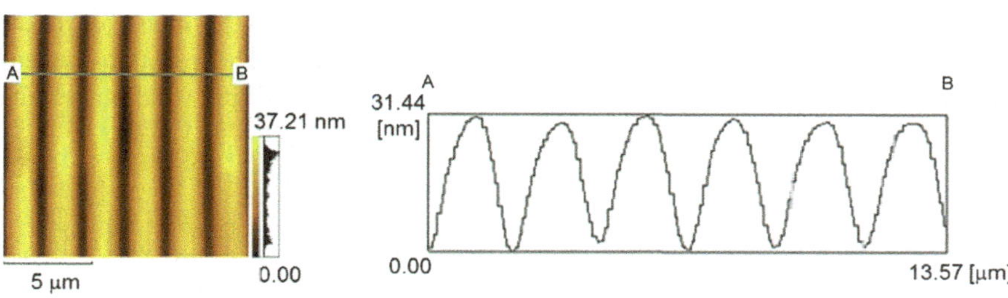

Fig. 16.9 AFM images of an amorphous thin film of *o*-Cl-HABI after exposed to UV (366 nm) light irradiation with 22 mW cm^{-2} for 20 min through a photomask at 383 K under argon atmosphere. The photomask was composed of stripes with the same widths (1 μm) of the *transparent* and *opaque line*. The figure on the *right* side indicate the cross-sectional height profile for the *solid lines* shown in the AFM image

Photochromic reaction of *o*-Cl-HABI increases the amount of *o*-Cl-TPIRs in the irradiated area, which causes the chemical potential of the irradiated area to differ from that of the shaded area and in turn drives the mass transfer. Molecular diffusion can be induced by chemical potential gradients. That is, the material moved from the irradiated areas to the shaded areas.

16.5.2 Spirooxazine

Reversible SR structures are reported on amorphous films composed only of highly durable spirooxazine (SO) molecules by spatially patterned UV light exposure due to the mass transfer [65]. It is well known that SO derivatives are one of the most widely studied classes of representative thermally reversible photochromic compounds, which exhibit extraordinarily high resistance to photodegradation [66]. Because the electrons localize in each ring, the stable spiro (SP) form absorbs light whose wavelength is shorter than the visible region. Upon UV light irradiation, the SP form undergoes bond cleavage between the spiro carbon atom and the oxygen atom to produce the planar merocyanine (MC) form (Fig. 16.10). Because of the delocalization of electrons in the planar MC form, the MC form can absorb the light in the visible regions and becomes coloured. When UV irradiation is discontinued and the coloured MC form is kept in the dark, it rapidly reverts to the initial SP form. Such back reaction is accelerated by visible light irradiation. Thus, SO molecules undergo thermally reversible photochromism between the colourless closed SP form and coloured open MC form.

The amorphous thin film of SO was prepared by spin-coating from the chloroform solution on a cleaned glass or a quartz substrate. The film was transparent and isotropic as revealed by the polarized optical microscopy and UV-visible absorption spectroscopy for several hours. The thickness of the film was ca. 120 nm. Before SR experiments, photochromic property of the film was investigated. The SO film exhibited photochromism between colourless SP form and coloured MC form. Upon irradiation with UV (366 nm) light, the colourless film changed to blue due to the generation of the coloured form, and a new absorption band in the visible region (λ_{max} = 612 nm) was detected. The reaction system reached the photostationary state within 70 s. When UV light irradiation was

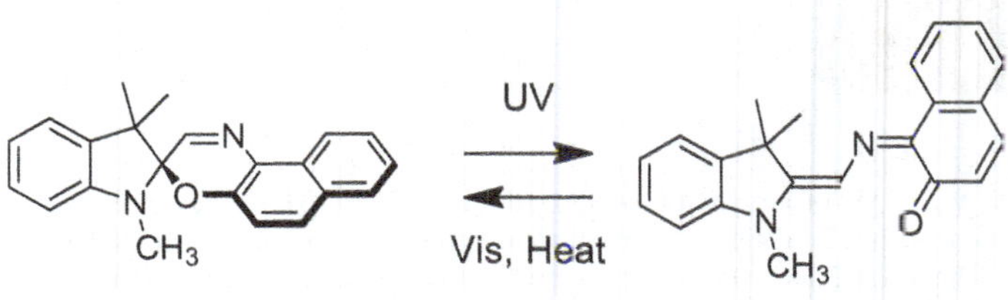

Fig. 16.10 Interconversion of spirooxazine

ceased, the original absorption spectrum was recovered completely, indicating that the backward ring-closing reaction took place thermally with half-life period of ca. 40 s. In order to check the durability of the SO film against the UV light, the visible absorption band ascribed to the open form was monitored during UV light irradiation (12.5 mW cm^{-2}). Although the MC form disappeared by the UV irradiation for 1 h under the aerial condition, it was held almost completely under the nitrogen atmosphere. Thus, the following experiment using UV light was performed under the nitrogen atmosphere. It should be noted that the stability of amorphous nature of the SO film was dramatically improved by the preceding UV light irradiation to the film.

Figure 16.11 shows the SR structures of SO films produced by the patterned UV light irradiation through a grating photomask with periods of 5 μm. When the film was irradiated with 10 mW cm^{-2} for 20 min, regular surface modulation was produced, and its spatial period was coincident with that of the photomask. In order to investigate the formation behaviour of SRG, an irradiation experiment through a photomask possessing one slit line with 2.5-μm width was performed. When the cross-sectional topography was examined, it is found that the top of the convex irradiated is higher and bottom of depression is lower than the initial surface level. This result clearly shows that the lateral material transfer occurred from the shaded areas to the irradiated areas. The phenomena of such material transfer are also observed on films of other SO derivatives [67]. By heating the film at 328 K for 20 min, the produced SR structure disappeared completely. Such reversible SR formation on alternate irradiation with patterned UV light and heating could be repeated at least 10 cycles, though the SR formation efficiency decreased gradually.

The reversible process was also confirmed by naked eyes (Fig. 16.12). When the SO film was irradiated with uniform UV light to improve the amorphous property of the thin film. A blue film was obtained because of the generation of the MC form (Fig. 16.12a), and the colour of the film changed from blue to colourless immediately by thermal back isomerization (Fig. 16.12b). In this step, no morphological change was observed, and only flat surface was obtained. The SO film was photoinscribed with spatially patterned UV light using the photomask. Irradiated part of

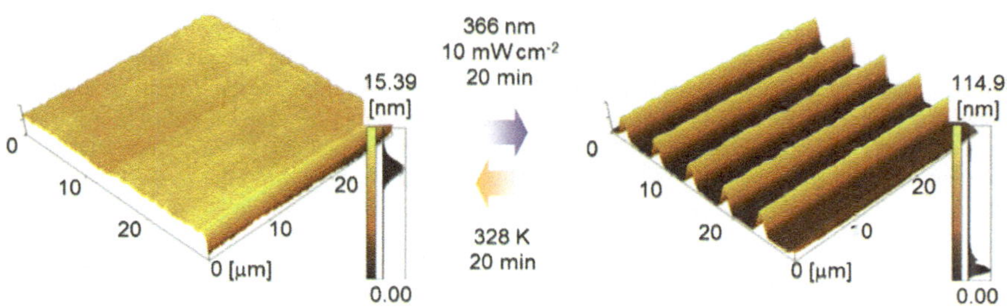

Fig. 16.11 Reversible construction of SR on an amorphous thin film of spirooxazine by patterned UV (366 nm) light irradiation with 10 mW cm^{-2} for 20 min through a photomask at room temperature and heating at 328 K for 20 min under nitrogen atmosphere

the film was coloured (Fig. 16.12c), and clear iridescent colour due to the interference emerged at the part irradiated with periodic pattern after quick thermal back isomerization (Fig. 16.12d). By heating the film at 328 K for 20 min, the produced SR structure disappeared completely (Fig. 16.12e). The film was again irradiated with UV light through the photomask placed perpendicularly to the first patterned light setup, aiming to prepare different SR structure from the first SR. The SR structure with the corresponding period was reformed on the SO film in the similar manner to the first photoinscription (Fig. 16.12f), and clear iridescent colour was appeared after quick thermal back isomerization (Fig. 16.12g).

As far as we know, this is the first reversible SR formation using a photochromic compound other than the azobenzene-related compound. Moreover, since the colourless SP form of SO is stable in the film, the SO film can be an excellent candidate for the photonic devices. In addition, the SP form of SO in the SR can be changed freely to the coloured MC form by irradiating with UV light without the destruction of SRG structure. These properties are useful when this material is used as optical devices.

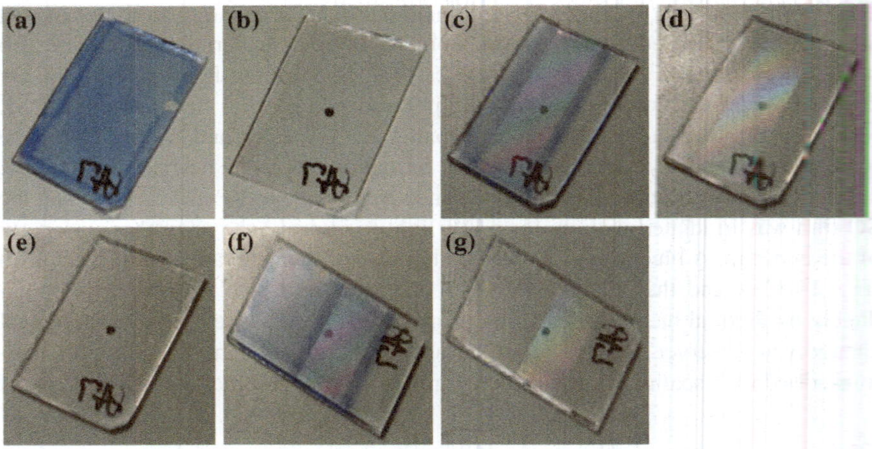

Fig. 16.12 Photographic images on an amorphous thin film of spirooxazine in the process of reversible construction of SR. Immediately after **a** and 10 min after **b,** the treatment of uniform irradiation with UV (366 nm) light (13.8 mW cm^{-2}) for 30 min under nitrogen atmosphere. Immediately after **c** and 10 min after **d,** the treatment of patterned irradiation with UV (366 nm) light (13.8 mW cm^{-2}) for 18 min at 299 K under nitrogen atmosphere. Immediately after **e,** the treatment of heating at 328 K for 20 min. Immediately after **f** and 10 min after **g,** the treatment of patterned irradiation with UV (366 nm) light (13.8 mW cm^{-2}) for 18 min at 299 K under nitrogen atmosphere

16.6 Conclusions

Two decades have passed since first discovery of photoinduced SR formation due to unusual material transfer on azobenzene-functionalized films. Intensive studies as to photoinduced SR have been developed in view of both fundamental science and practical applications. We are engaged in the construction of novel SR formation system that does not have azobenzene moiety in thin films. SR structures are formed on polymer thin films doped with photochromic molecules such as spiropyran and diarylethene upon irradiation with spatially patterned UV light. Low molecular weight photochromic compounds are used as SR-forming materials instead of the polymer films doped with photochromic compounds. SR structure formed on the spirooxazine amorphous thin film could be erased by heating and be reformed by a spatially patterned UV light, and this process could be repeated. Promotions of these research areas are anticipated to provide wide and new opportunities for material processing and creation of new class of photonic devices.

References

1. Rochon P, Batalla E, Natansohn A (1995) Optically induced surface gratings on azoaromatic polymer films. Appl Phys Lett 66(2):136–138
2. Kim DY, Tripathy SK, Li L, Kumar J (1995) Laser-induced holographic surface relief gratings on nonlinear optical polymer films. Appl Phys Lett 66(10):1166–1168
3. Rau H (2003) Azo compounds. In: Dürr H, Bouas-Laurent H (eds) Photochromism; Molecules and Systems, Rev edn. Elsevier, Amsterdam, pp 165–192
4. Viswanathan NK, Balasubramanian S, Li L, Tripathy SK, Kumar J (1999) A detailed investigation of the polarization-dependent surface-relief-grating formation process on azo polymer films. Jpn J Appl Phys 38(10):5928–5937
5. Bian S, Williams JM, Kim DY, Li L, Balasubramanian S, Kumar J, Tripathy S (1999) Photoinduced surface deformations on azobenzene polymer films. J Appl Phys 86(8):4498–4508
6. Holme NCR, Nikolova L, Hvilsted S, Rasmussen PH, Berg RH, Ramanujam PS (1999) Optically induced surface relief phenomena in azobenzene polymers. Appl Phys Lett 74 (4):519–521
7. Fiorini C, Prudhomme N, de Veyrac G, Maurin I, Raimond P, Nunzi JM (2000) Molecular migration mechanism for laser induced surface relief grating formation. Synth Met 11(1–3):121–125
8. Ikawa T, Mitsuoka T, Hasegawa M, Tsuchimori M, Watanabe O, Kawata Y, Egami C, Sugihara O, Okamoto N (2000) Optical near field induced change in viscoelasticity on an azobenzene-containing polymer surface. J Phys Chem B 104(39):9055–9058
9. Landraud N, Peretti J, Chaput F, Lampel G, Boilot JP, Lahlil K, Safarov VI (2001) Near-field optical patterning on azo-hybrid sol–gel films. Appl Phys Lett 79(27):4562–4564
10. Fukuda T, Sumaru K, Kimura T, Matsuda H, Narita Y, Inoue T, Sato F (2001) Observation of optical near-field as photo-induced surface relief formation. Jpn J Appl Phys 40(8B):L900–L902
11. Jiang XL, Li L, Kumar J, Kim DY, Tripathy SK (1998) Unusual polarization dependent optical erasure of surface relief gratings on azobenzene polymer films. Appl Phys Lett 72 (20):2502–2504

12. Kulikovska O, Goldenberg LM, Stumpe J (2007) Supramolecular azobenzene-based materials for optical generation of microstructures. Chem Mater 19(13):3343–3521
13. Natansohn A, Rochon P (1999) Photoinduced motions in azobenzene-based amorphous polymers: possible photonic devices. Adv Mater 11(16):1387–1391
14. Ubukata T, Seki T, Ichimura K (2000) Surface relief gratings in host-guest supramolecular materials. Adv Mater 12(22):1675–1678
15. Ubukata T, Seki T, Ichimura K (2002) Surface relief grating in hybrid films composed of azobenzene polymer and liquid crystal molecule. Colloids Surf A 198–200:113–117
16. Ubukata T, Hara M, Ichimura K, Seki T (2004) Phototactic transport motions of polymer film for micropatterning and alignment of functional materials. Adv Mater 16(3):220–223
17. Zettsu N, Ubukata T, Seki T, Ichimura K (2001) Soft crosslinkable azo polymer for rapid surface relief formation and persistent fixation. Adv Mater 13(22):1693–1697
18. Zettsu N, Fukuda T, Matsuda H, Seki T (2003) Unconventional polarization characteristic of rapid photoinduced material motion in liquid crystalline azobenzene polymer films. Appl Phys Lett 83(24):4960–4962
19. Zettsu N, Seki T (2004) Highly efficient photogeneration of surface relief structure and its immobilization in cross-linkable liquid crystalline azobenzene polymers. Macromolecules 37 (23):8692–8698
20. Ubukata T, Higuchi T, Zettsu N, Seki T, Hara M (2005) Spontaneous motion observed in highly sensitive surface relief formation. Colloids Surf A 257–258:123–126
21. Zettsu N, Ogasawara T, Arakawa R, Nagano S, Ubukata T, Seki T (2007) Highly photosensitive surface relief gratings formation in a liquid crystalline azobenzene polymer: new implications for the migration process. Macromolecules 40(13):4607–4613
22. Isayama J, Nagano S, Seki T (2010) Phototriggered mass migrating motions in liquid crystalline azobenzene polymer films with systematically varied thermal properties. Macromolecules 43(9):4105–4112
23. Darracq B, Chaput F, Lahlil K, Roosen G, Lévy Y, Boilot JP (1998) Photoinscription of surface relief gratings on azo-hybrid gels. Adv Mater 10(14):1133–1136
24. Nishizawa K, Nagano S, Seki T (2009) Novel liquid crystalline organic-inorganic hybrid for highly sensitive photoinscriptions. Chem Mater 21(13):2624–2631
25. Gao J, He Y, Xu H, Song B, Wang Z, Wang X (2007) Azobenzene-containing supramolecular polymer films for laser-induced surface relief gratings. Chem Mater 19 (1):14–17
26. Zettsu N, Ogasawara T, Mizoshita N, Nagano S, Seki T (2008) Photo-triggered surface relief grating formation in supramolecular liquid cystalline polymer systems with detachable azobenzene units. Adv Mater 20(3):516–521
27. Priimagi A, Cavallo G, Forni A, Gorynsztejn-Leben M, Kaivola M, Metrangolo P, Milani R, Shishido A, Pilati T, Resnati G, Terraneo G (2012) Halogen bonding versus hydrogen bonding in driving self-assembly and performance of light-responsive supramolecular polymers. Adv Funct Mater 22(12):2572–2579
28. Li W, Dohi T, Hara M, Nagano S, Haba O, Yonetake K, Seki T (2012) Phototriggered mass migration consorted with surface dewetting in thin films of a liquid crystalline azobenzene-containing dendrimer. Macromolecules 45(16):6618–6627
29. Koskela JE, Liljeström V, Lim J, Simanek EE, Ras RHA, Priimagi A, Kostiainen MA (2012) Light-fuelled transport of large dendrimers and proteins. J Am Chem Soc 136(19):6850–6853
30. Fuhrmann T, Tsutsui T (1999) Synthesis and properties of a hole-conducting, photopatternable molecular glass. Chem Mater 11(8):2226–2232
31. Nakano H, Takahashi T, Kadota T, Shirota Y (2002) Formation of a surface relief grating using a novel azobenzene-based photochromic amorphous molecular material. Adv Mater 14 (16):1157–1160
32. Ando H, Takahashi T, Nakano H, Shirota Y (2003) Comparative studies of the formation of surface relief grating. Amorphous molecular material vs. vinyl polymer. Chem Lett 32 (8):710–711

33. Ueda H, Tanino T, Ando H, Nakano H, Shirota Y (2004) Significant effect of molecular structure on surface relief grating formation for novel azobenzene-based photochromic amorphous molecular materials. Chem Lett 33(9):1152–1153
34. Nakano H, Tanino T, Takahashi T, Ando H, Shirota Y (2008) Relationship between molecular structure and photoinduced surface relief grating formation using azobenzene-based photochromic amorphous molecular materials. J Mater Chem 18(2):242–246
35. Kim MJ, Seo EM, Vak D, Kim DY (2003) Photodynamic properties of azobenzene molecular films with triphenylamines. Chem Mater 15(21):4021–4027
36. Ishow E, Lebon B, He Y, Wang X, Bouteiller L, Galmiche L, Nakatani K (2006) Structural and photoisomerization cross studies of polar photochromic monomeric glasses forming surface relief gratings. Chem Mater 18(5):1261–1267
37. Nakano H, Tanino T, Shirota Y (2005) Surface relief grating formation on a single crystal of 4-(dimethylamino)azobenzene. Appl Phys Lett 87(6):061910
38. Viswanathan NK, Kim DY, Bian S, Williams J, Liu W, Li L, Samuelson L, Kumar J, Tripathy SK (1999) Surface relief structures on azo polymer films. J Mater Chem 9(9):1941–1955
39. Yager KG, Barrett CJ (2001) All-optical patterning of azo polymer films. Curr Opin Solid State Mat Sci 5(6):487–494
40. Natansohn A, Rochon P (2002) Photoinduced motions in azo-containing polymers. Chem Rev 102(11):4139–4175
41. Seki T (2014) Meso- and microscopic motions in photoresponsive liquid crystalline polymer films. Macromol Rapid Commun 35(3):271–290
42. Sekkat Z, Knoll W (eds) (2002) Photoreactive organic thin films. Academic Press, California
43. Park JH, Yoon TY, Lee WJ, Lee SD (2002) Multi-domain liquid crystal display with self-aligned 4-domains on surface relief gratings of photopolymer. Mol Cryst Liq Cryst 375 (1):433–440
44. Ramanujam PS, Berg RH (2004) Photodimerization in dipeptides for high capacity optical digital storage. Appl Phys Lett 85(10):1665–1667
45. Kawatsuki N, Hasegawa T, Ono H, Tamoto T (2003) Formation of polarization gratings and surface relief gratings in photocrosslinkable polymer liquid crystals by polarization holography. Adv Mater 15(12):991–994
46. Ono H, Emoto A, Kawatsuki N, Hasegawa T (2003) Self-organized phase gratings in photoreactive polymer liquid crystals. Appl Phys Lett 82(9):1359–1361
47. Kawatsuki N, Fujio K, Hasegawa T, Emoto A, Ono H (2006) Surface relief formation with molecular orientation in photoreactive liquid crystalline polymer film. J Photopolym Sci Technol 19(2):151–156
48. Yamaki S, Nakagawa M, Morino S, Ichimura K (2000) Surface relief gratings generated by a photocrosslinkable polymer with styrylpyridine side chains. Appl Phys Lett 76(18):2520–2522
49. Sánchez C, de Gans BJ, Kozodaev D, Alexeev A, Escuti MJ, van Heesch C, Bel T, Schubert US, Bastiaansen CWM, Broer DJ (2005) Photoembossing of periodic relief structures using polymerization-induced diffusion: a combinatorial study. Adv Mater 17 (21):2567–2571
50. Hermans K, Wolf FK, Perelaer J, Janssen RAJ, Schubert US, Bastiaansen CWM, Broer DJ (2007) High aspect ratio surface relief structures by photoembossing. Appl Phys Lett 91 (17):174103
51. Krebs FC, Ramanujam PS (2006) Holographic recording in a series of conjugated polymers. Opt Mater 28(4):350–354
52. Taguenang JM, Kassu A, Govindarajalu G, Dokhanian M, Sharma A, Ruffin PB, Brantley C (2006) Fabrication and characterization of a diffraction-grating transducer in thin polybutadiene rubber film for sensing dynamical strain. Appl Opt 45(27):6903–6909
53. Ubukata T, Moriya Y, Yokoyama Y (2012) Facile one-step photopatterning of polystyrene films. Polym J 44(9):966–972

54. Katzenstein JM, Janes DW, Cushen JD, Hira NB, McGuffin DL, Prisco NA, Ellison CJ (2012) Patterning by photochemically directing the Marangoni effect. ACS Macro Lett 1 (10):1150–1154
55. Ubukata T, Yamamoto S, Moriya Y, Fujii S, Yokoyama Y (2012) Photo-triggered surface relief of polystyrene films—highly photo-sensitive formation by the addition of a benzophenone derivative. J Photopolym Sci Tech 25(5):675–678
56. Bouas-Laurent H, Dürr H (2001) Organic photochromism. Pure Appl Chem 73(4):639–665
57. Crano JC, Guglielmetti RJ (1999) Organic photochromic and thermochromic compounds. Plenum Press, New York
58. Feringa BL, Browne R (eds) (2011) Molecular switches, 2nd edn. Wiley, Weinheim
59. Ubukata T, Takahashi K, Yokoyama Y (2007) Photoinduced surface relief structures formed on polymer films doped with photochromic spiropyrans. J Phys Org Chem 20(11):981–984
60. Yokoyama Y, Hara W, Inoue T, Ubukata T, Sakomura M, Tukada H (2005) Negative photochromism of a spiropyran in a Langmuir-Blodgett film. Chem Lett 34(12):1622–1623
61. Ubukata T, Yamaguchi S, Yokoyama Y (2007) Photoinduced surface relief structures formed on polymer films mixed with diarylethenes. Chem Lett 36(10):1224–1225
62. Irie M (2000) Diarylethenes for memories and switches. Chem Rev 100(5):1685–1716
63. Okano K, Ogino S, Kawamoto M, Yamashita T (2011) Mass migration on a polymer surface caused by photoinduced molecular rotation. Chem Commun 47(43):11891–11893
64. Kikuchi A, Harada Y, Yagi M, Ubukata T, Yokoyama Y, Abe J (2010) Photoinduced diffusive mass transfer in *o*-Cl-HABI amorphous thin films. Chem Commun 46(13):2262–2264
65. Ubukata T, Fujii S, Yokoyama Y (2009) Reversible phototriggered micromanufacturing using amorphous photoresponsive spirooxazine film. J Mater Chem 19(21):3373–3377
66. Chu NYC (2003) 4n + 2 system: spirooxazines. In: Dürr H, Bouas-Laurent H (eds) Photochromism; molecules and systems, Rev edn. Elsevier, Amsterdam, pp 493–509
67. Ubukata T, Fujii S, Arimatsu K, Yokoyama Y (2012) Phototriggered micromanufacturing using photoresponsive amorphous spirooxazine films. J Mater Chem 22(29):14410–14417

Chapter 17
Electric Field-Induced Arrangement of Colloidal Materials in Microfluidic Devices

Masahiro Motosuke

Abstract In this chapter, colloidal materials arrangement by microfluidic approach utilizing an electric field-induced interfacial flow is presented. A microfluidic platform is suitable for precise control of behavior of colloidal particles. Electrokinetics in the platform has been growing owing to its versatility and simplicity in fabrication, implementation and handling. Alternating-current electroosmosis (ACEO), which is nonlinear electrokinetic phenomena resulting in advent motion of fluid generated by an interaction between AC electric field and an induced diffuse charge layer on the electrode surface, has a potential for a particle control in a microfluidic channel. The arrangement of the colloidal particles is achieved by the ACEO flow in a microfluidic device with a parallel double-gap electrode. The experimental results show that concentrated particles exist on the electrode surface within 5 μm, and that three-dimensional concentration of particle with the concentration factor as large as 700 is attained. As a result, total concentration efficiency was 98.4 % for all the particles in the fluidic device. Also, this approach can be used for nanoparticle concentration. Examples of gold nanoparticles collection from 20 to 100 nm are presented here. The ACEO-based arrangement technique provides us to ensure the novel functionality for colloidal materials.

Keywords Microfluidics · Electrokinetics · Alternating-current electroosmosis (ACEO) · Flow-based particle handling

17.1 Introduction

Controlled arrangement of colloidal materials has been tremendously important for assembly of functional materials, enhancement or flexible control of surface reaction, improved detection of target materials and so on. Recently, microfluidics is an emerging field attracting considerable attention and broadening the application in

M. Motosuke (✉)
Department of Mechanical Engineering, Tokyo University of Science, Tokyo, Japan
e-mail: mot@rs.tus.ac.jp

T. Kawai and M. Hashizume (eds.), *Stimuli-Responsive Interfaces*,
DOI 10.1007/978-981-10-2463-4_17

various fields such as biology, chemistry, life science and engineering [1–3]. The biological research and clinical diagnostics in microfluidic system have been progressed with the aid of the particle handling, sorting, separation and alignment techniques. Various methods for the microfluidic particle control relying on a wide variety of principles have been studied and reported [4, 5]. Electrophoresis [6], dielectrophoresis [7, 8], magnetophoresis [9], acoustic pressure [10], hydrodynamic force [11–13], optical force [14, 15] and surface affinity [16, 17] are utilized. The particle control technique is quite important in plenty of microfluidic applications. For example, focusing particles into a tight stream in the bulk flow in a microchannel is an essential step for counting, detecting and sorting them resulting in signal with high signal-to-noise ratio such as microfluidic cytometer [18–21]. A preconcentration greatly enhances the detection performance of rare samples in the reagent [22, 23]. Also, particle alignment in microfluidic system can be useful for high functionality [24, 25].

Microfluidic particle handling techniques can be classified into two categories based on their driving force to control particles: particle force and fluidic force. The former uses force acting on the particle to manipulate particles in the fluid, and the latter relies on advective transport. The particle force acting on a spherical particle F_p is basically proportional to the volume of the particle and is described as follows:

$$F_\mathrm{p} \propto d^3 \tag{17.1}$$

where d is diameter of the particle. On the other hand, the fluidic force F_f is proportional to the projected area of the particle and then is presented as,

$$F_\mathrm{f} \propto d^2 \tag{17.2}$$

The resistance of the particle R moving the velocity v in the fluid with the viscosity η should be,

$$R = 3\pi\eta d \tag{17.3}$$

in the Stokes regime. Overall mobility of the particle, M_p or M_f, that corresponds to the manipulation velocity, can be evaluated by the value where the force is divided by the resistance, namely F_p/R or F_f/R, described as,

$$M_\mathrm{p} \propto \frac{F_\mathrm{p}}{R} = \frac{d^2}{3\pi\eta} \propto d^2 \tag{17.4}$$

$$M_\mathrm{f} \propto \frac{F_\mathrm{f}}{R} = \frac{d}{3\pi\eta} \propto d \tag{17.5}$$

Therefore, M_p is proportional to the square of the diameter, while M_f is proportional to the diameter. Figure 17.1 depicts the particle size dependence on M_p and M_f. Here, the mobility in the figure was normalized by the value for 1-μm

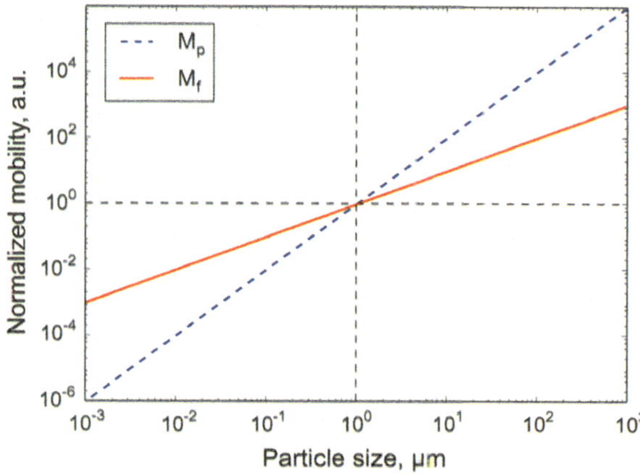

Fig. 17.1 Particle size dependence of mobility based on the particle force (M_p) and fluidic force (M_f) which is normalized by the value for 1-μm particle

particle. It is clear that the particle force is more sensitive in the target size rather the fluidic one. Thus, the particle force can be used as a separation technique based on the size-dependent mobility difference [7, 26, 27]. On the other hand, the fluidic force shows less sensitivity of the target size. This insensitivity can be important for controlling a particle group with size variation or nanoscopic materials. Furthermore, the fluid-force-based approach would provide particle manipulation which does not depend on surface charge or physicochemical properties of the particles because of the use of the fluid flow to control them.

In our study, an alternating-current electroosmosis (ACEO), one of the AC electrokinetic fluid flows, was exploited for the control of colloidal materials ranging its size from 10 μm to 10 nm. Electrokinetics is currently the mechanism of choice for generation of fluid motion and particle manipulation at microscale and nanoscale dimensions. AC electrokinetic approach introduces advantages over DC-based one like electrophoresis, such as low temperature rise due to Joule heating, less electrolysis of solution and less damage on target particles.

17.2 AC Electric Field-Induced Liquid Flow

17.2.1 Principle of ACEO

As mentioned above, the AC electric field application offers many advantages over DC approach. Among several AC electrokinetic phenomena, AC electroosmosis (ACEO), which is a field-induced fluid flow induced by an interaction of the electric double layer and the tangential component of the electric field [28], is one of the promising techniques of particle manipulation which does not depend on electrical properties of the particle because the transportation of particle by ACEO utilizes the fluidic force. ACEO was firstly reported the fluidic motion under AC electric field imposed on co-planar electrode in a channel with an electrolyte [29]. Then, a theory was proposed based on linear coupling of induced charged layer above the electrode and ions in liquid [30]. The driving force of ACEO flow is the ion movement on an electrode subjected to AC electric field. Instantaneous surface charge induced by the electric field application interacts with the small horizontal component of the electric field to generate a fluid motion across the surface of electrode. A schematic model of ACEO around a co-planar electrode pair with a gap is shown in Fig. 17.2.

The application of an AC voltage to a pair of co-planar electrodes generates a steady (nonzero time-averaged) fluid flow with a velocity that depends both on the applied potential and on frequency. A simple circuit model based on an array of resistors and capacitors gives a reasonable correlation with experimentally observed values of velocity in a limited condition but provides us qualitative description of the flow field of ACEO. The time-averaged expression for the slip velocity is given by

$$u_{\mathrm{slip}} \propto -\frac{\varepsilon}{4\eta}\Lambda\frac{\partial|\Delta\phi_{\mathrm{DL}}|^2}{\partial x} \tag{17.6}$$

where $\Delta\phi_{\mathrm{DL}}$ represents the voltage drop across the double layer and Λ is an empirical parameter that accounts for the ratio of the voltage drop across the diffuse layer to the total voltage drop across the double layer, and is given by

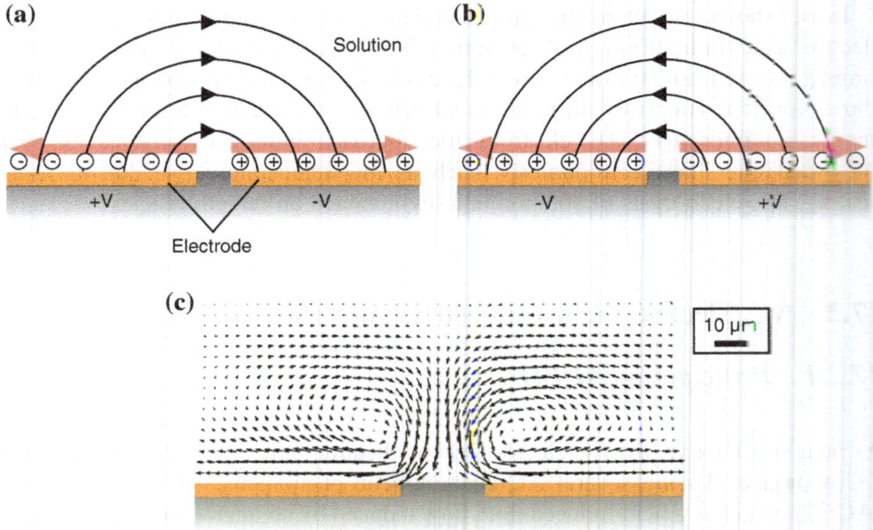

Fig. 17.2 Schematic of alternating-current electroosmosis (ACEO) on symmetric co-planar electrodes due to the capacitive charging mechanism under AC electric field. As the field sign switches, **a** ↔ **b**, the sign of induced charge layers also switches, resulting in uniform flow direction **c**

$$\Lambda = \frac{C_S}{C_S + C_D} \tag{17.7}$$

where C_S and C_D are the capacitances per unit of area of the Stern and diffuse layer, respectively. And $\Delta\phi_{DL}$ can be simply expressed based on simple circuit model as

$$\phi_{DL} = \frac{V}{2 + i\pi\Lambda C_D \omega x/\sigma} \tag{17.8}$$

Therefore, the slip velocity induced by ACEO is

$$u_{ACEO} = \Lambda \frac{\varepsilon V^2}{8\eta x} \frac{\Omega^2}{(1 + \Omega^2)} \tag{17.9}$$

with a nondimensional frequency Ω given by

$$\Omega = \frac{\omega \Lambda C_D \pi x}{2\sigma} \tag{17.10}$$

This simple description of ACEO phenomenon is valid only at low voltage condition, i.e., $V \ll 25$ mV in a typical condition [31] that is much smaller than practical range of several V, and the theory which can quantitatively describe the ACEO phenomenon has not been established despite tremendous efforts to describe nonlinear charge dynamics [32–34]. Therefore, the experimental investigation of ACEO is still strongly needed to obtain its performance in application. The unique flow structure induced by ACEO enables particles to concentrate into a certain location in a microchannel with electrode arrays.

17.2.2 Induced ACEO Velocity with Its Parameter Dependence

In this study, the induced ACEO velocity above the electrode array was measured with its parameter dependence.

A microfluidic device used in this study has a polydimethylsiloxane (PDMS; Sylgard 184, Dow Corning Toray, Japan) microchannel fabricated by the soft lithography [35, 36]. The channel scale is 500 μm wide and 50 μm high in typical conditions. The PDMS channel was bonded to a glass substrate with indium tin oxide (ITO) electrode (Geomatec, Japan) using oxygen plasma (YHS-360, SAKIGAKE-Semiconductor, Japan). ITO has an optical transparency in visible wavelength so that an optical access through the substrate becomes possible. The electrode was sputtered on the glass substrate, and the patterning was performed by conventional wet etching process. The thickness of ITO film was approximately 90 nm. The electrode has a spacing of 25 μm. As a source of AC electric field, a function generator (AFG3022B, Tektronix, USA) was connected to the electrodes and it applied AC electric field into the device with the voltage of 0.4–4.0 V_{pp} and at the frequency of 0.1–100 kHz.

The working fluid was KCl aqueous solution (Wako Pure Chemical Industries, Japan). Electrical conductivity of the solution was set at 4.4 mS/m. Tracer particle used here is fluorescent polystyrene particles with a diameter of 1.0 μm (FluoSpheres carboxylate-modified microspheres, Invitrogen, USA). Volume fraction of particles in the working fluid was approximately 0.01 % to avoid particle–particle interaction.

As the measurement method of ACEO velocity, the micron-resolution particle image velocimetry (PIV) was employed in this study [37, 38]. This technique is based on the volume illumination and detection of the movement of fluorescent particles in the focal plane of an objective lens. A continuous mercury lamp (Intensilight, Nikon, Japan) is used as an illumination source. The filter cube consisting of an exciter filter, an emitter filter and a dichroic mirror is optimized to perform effective excitation and collection of the fluorescent light for particles. The excitation and emission wavelengths of the fluorescent particles are 505 and 515 nm, respectively. The time interval between two images in a pair recorded by a sCMOS (scientific complementary metal–oxide–semiconductor) camera (ORCA-Flash2.8, Hamamatsu Photonics, Japan) was typically 30 ms. The time-average processing between 200 image pairs was performed to reduce the scattering of the measured data due to the random Brownian motion of particles. Since the nonuniformity of particle density in the image used for PIV could affect the correlation-based analytical procedure, the velocity measurements were taken only during the initial stage of concentration, typically less than 10 s after applying voltage, to prevent the measurement error owing to the nonuniformity.

As a preliminary measurement of the ACEO velocity, evaluation of effective measurement depth through the ITO electrode was performed. The focal plane of the objective lens (40×, NA 0.60) was set at the surface of the electrode, and then, the time-dependent particle displacement parallel to the wall in quiescent fluid condition was measured using a tracking procedure based on the pattern matching procedure using Gaussian distribution. The time revolution of the mean square displacement of 1-μm particles is shown in Fig. 17.3. The similar measurement results obtained at the focal place far from the substrate (25 μm from the bottom surface) are also shown as comparison. The stochastic fluctuation of the particle in bulk condition (focal position z = 25 μm) indicates good agreement with the following Stokes–Einstein equation in infinite fluid within 1.6 %

$$\langle \Delta x^2 \rangle = 2D_{\infty} t = \frac{k_B T}{6\pi r \eta} t \tag{17.11}$$

where D_{∞} is diffusion coefficient in bulk liquid, t is time, k_B is Boltzmann's constant, T is absolute temperature, and r is particle radius. The near-wall particle

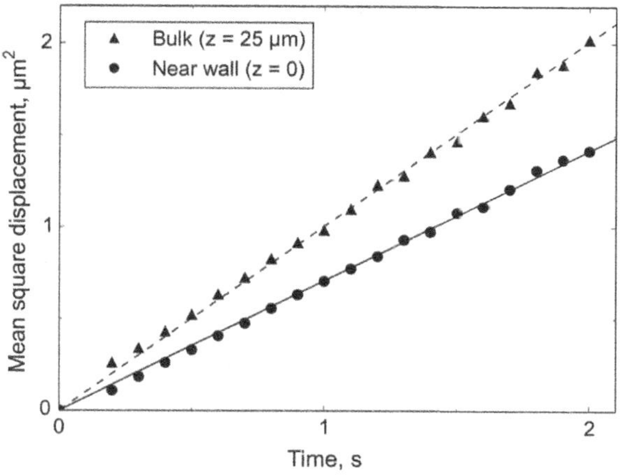

Fig. 17.3 Mean square displacement of 1-μm particle due to Brownian motion. In bulk condition (z = 25 μm), measured diffusion coefficient matches Stokes–Einstein relationship. Hindered diffusion occurs in near-wall case (z = 0 μm) indicating that the actual measurement height in this experiment is 1.1 μm from the bottom substrate

fluctuation shows hindered motion due to the existence of the boundary wall. The correction factor of diffusion coefficient parallel to the wall at the distance of particle center from the wall h is presented as follows [39].

$$D^* = \frac{D_{\parallel}}{D_{\infty}} = 1 - \frac{9}{16}\left(\frac{r}{h}\right) + \frac{1}{8}\left(\frac{r}{h}\right)^3 - \frac{45}{256}\left(\frac{r}{h}\right)^3 - \frac{1}{16}\left(\frac{r}{h}\right)^5 \tag{17.12}$$

where D^* and $D_{\parallel}$ are corrected diffusion coefficient and diffusion coefficient parallel to the wall, respectively. From the fitted experimental data by the above equation, the effective measurement height was estimated to be 1.1 μm from the bottom surface. This value is reasonable considering the effective height of 1.6 μm based on the measurement depth in the micro-PIV setup in the present study [40, 41].

As mentioned in the previous section, ACEO has strong parameter dependence on the induced velocity. Here, ACEO velocity on the ITO electrode surface was measured with the focal plane at the same height as the above diffusion measurement. Figure 17.4a depicts the induced velocity distribution on one side of co-planar electrode pair at the applied voltages from 0.4 to 2.0 V_{pp} under the frequency of 1 kHz. ACEO velocity shows its peak around 12.5 μm from the gap center. This position corresponds to the edge of electrode (hald of the gap width of 25 μm). The voltage dependence at several locations is shown in Fig. 17.4b. The velocity at each distance increases with square power of the applied voltage (2.16, 1.93 and 1.96 for 12.5, 22.5 and 32.5 μm from the gap center, respectively). This trend agrees well with the ACEO theory as in Eq. (17.9). Frequency dependence on ACEO velocity is presented in Fig. 17.5. Here, the applied voltage was fixed at 2.0 V_{pp}. Under low frequency condition, the tangential fluid flow remains farther from the edge instead of showing some decay in the peak velocity. Through the velocity measurements, ACEO behavior with strong dependence of the electric field condition was confirmed.

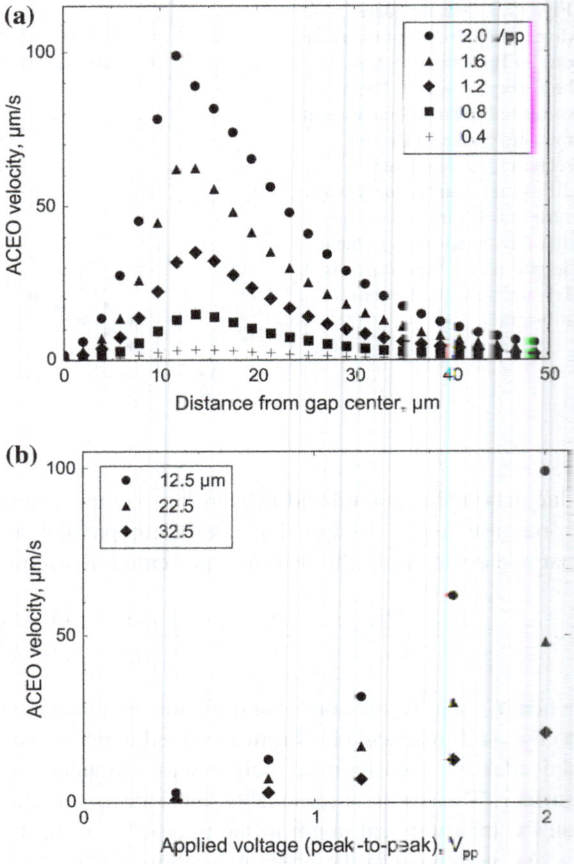

Fig. 17.4 ACEO velocity on ITO electrode surface at an applied frequency of 1 kHz. **a** Velocity distribution at different voltages and **b** voltage dependence at several distances from the gap center

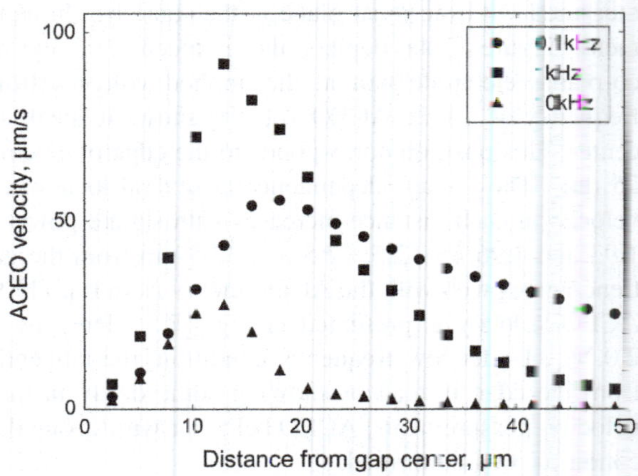

Fig. 17.5 ACEO velocity on ITO electrode surface at an applied frequency of 1 kHz. **a** Velocity distribution at different voltages and **b** voltage dependence at several distances from the gap center

17.3 Particle Arrangement by ACEO

17.3.1 Collection Mechanism of Particles

The tangential ACEO flow transports suspended particles along the electrode surface, and then, the particles are accumulated at a certain location on the electrode. A pair of accumulated lines of particles is observed along the stream in the microchannel. The particle accumulation process at the entrance (upstream region) of the electrode is shown in Fig. 17.6. Here, a half of the field is presented in this figure due to the symmetric behavior. In typical conditions, the accumulated location converges within 1 mm from the electrode entrance. The converged accumulated location of the particle is sensitive to the input parameters of AC electric field. Figure 17.7 shows accumulated particles (1-μm fluorescent polystyrene beads) at 60 s after applying AC electric field with frequencies from 0.2 to 5.0 kHz at the voltage of 4.0 V_{pp}. Low applied frequency provides further concentrated location for the particles. Figure 17.7b is the fluorescent intensity distribution at the half area from the channel center. It is found that integrals of intensities over the distance direction are almost the same values in all the cases. Therefore, it can be stated that ACEO concentrates particles from inner area toward a specific location depending on the applied frequencies. The insight about this accumulation phenomenon can be useful for optimized design of ACEO-based particle concentrator.

Because the direction of ACEO flow along the electrode surface is tangential, transported particles experience no forces normal to the surface except for their spontaneous diffusion. Under an assumption that the transportation of colloidal particles occurs as far as the convective motion by ACEO overcomes the inherent diffusive motion of the particles, particles are concentrated at the balanced position of the advective force and diffusive one. Figure 17.8 shows the comparison between experimental results and estimated concentration location for 1-μm and 500-nm particles under different frequencies of the applied voltage. In this

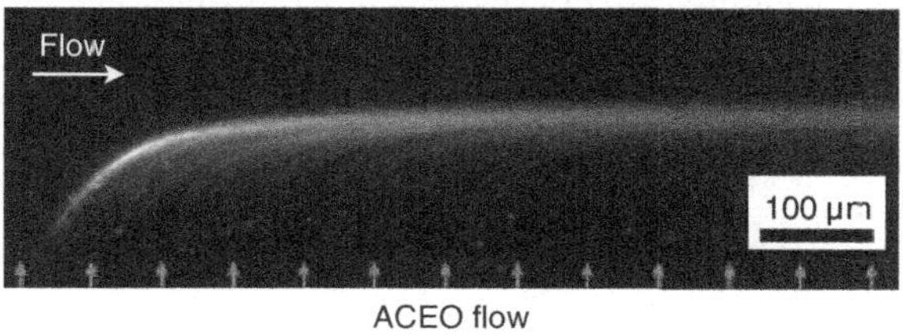

Fig. 17.6 Particle accumulation process at the entrance of the electrode. Particles are collected at a certain location from the electrode edge. In this figure, one-half of the electrode pair is shown because of its geometrical symmetry. The accumulation location of the particles converges within 1 mm in the standard experimental condition

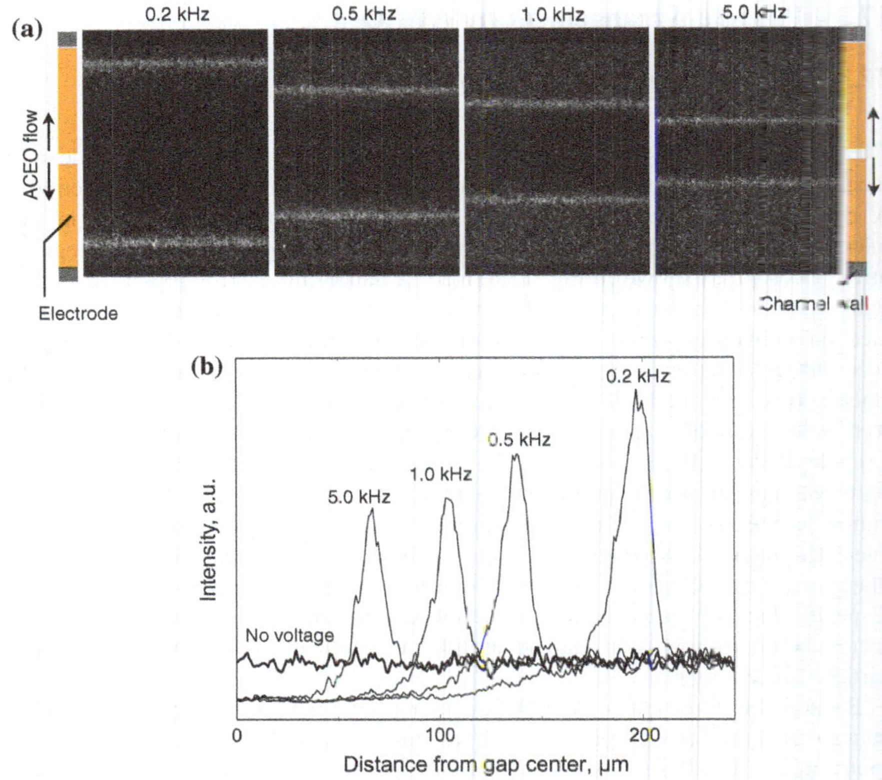

Fig. 17.7 Behaviors of particle concentration under different frequencies with a voltage of [illegible] V_{pp}. **a** Particle image of fluorescent beads and **b** fluorescent intensity profile at half of the channel

estimation, the accumulation locations were calculated from a balance of diffusive velocity of the particle using the hindered diffusion coefficient parallel to the wall in Eq. (17.12) and the measured ACEO velocity. The diffusive velocity V_D is given by the following equation.

$$V_D = \frac{\sqrt{\langle \Delta x^2 \rangle |_{t=t_u}}}{t_u} = \sqrt{\frac{2D_{\|}}{t_u}} \tag{17.13}$$

where t_u is a unit of time. $D_{\|}$ was determined by particle tracking without the applied voltage. The validity of this theoretical model can be confirmed by a good agreement between the calculated and experimental collection location. Therefore, particle concentration in unlimited boundary by ACEO can be explained by this balance model.

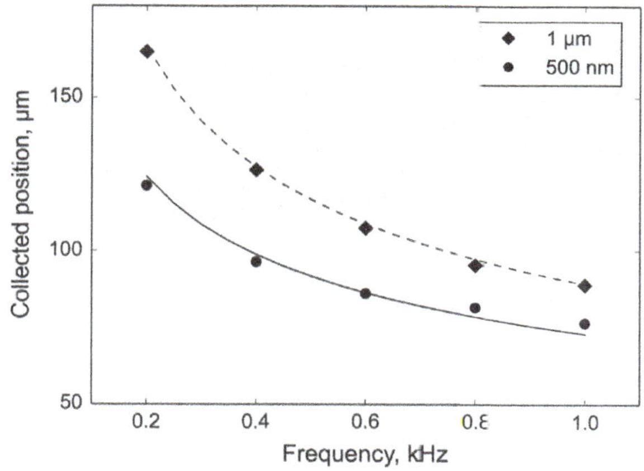

Fig. 17.8 Experimental results and theoretical estimation of collected position under different frequencies and particle sizes of 1 μm and 500 nm in diameter. *Lines* shown in the figure correspond calculated positions

17.3.2 Particle Collection in Interfered ACEO Flow

When the electrode has multiple patterns of electrode array, we need to consider an interaction of ACEO flow and resultant particle collection. As in Fig. 17.9, the particle collection site depends on the electrode geometry. In this study, a double-gap electrode geometry is used as the simplest case to investigate the effect of interfered ACEO flow on the particle collection process.

In the double-gap electrode device, ACEO flows from each gap face each other. Therefore, particles are transported and accumulated from each gap toward the center electrode. The accumulation line of particles at the entrance region in different frequencies of AC electric field is shown in Fig. 17.10. There are separated and merged conditions of two symmetric accumulation lines of particles, one is separated condition (17.2 KHz) and the other is merged one (0.5 kHz). In separated condition (Fig. 17.10a), particles migrate resulting in two separated lines even in the downstream region. Meanwhile, the two accumulation lines merge at the center

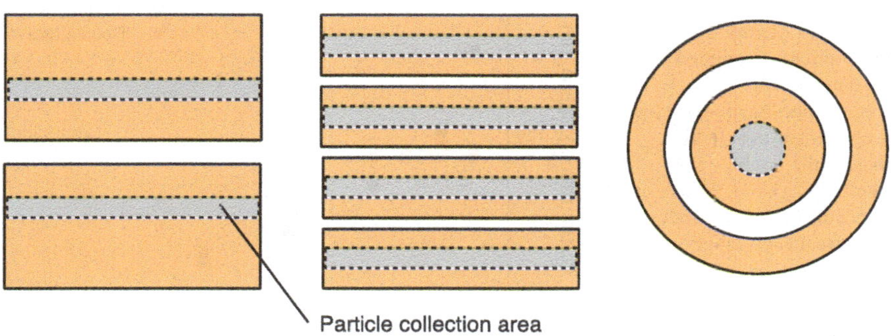

Fig. 17.9 Particle collection by ACEO with different electrode geometries. Interaction of ACEO flow plays a key role in the particle accumulation behavior

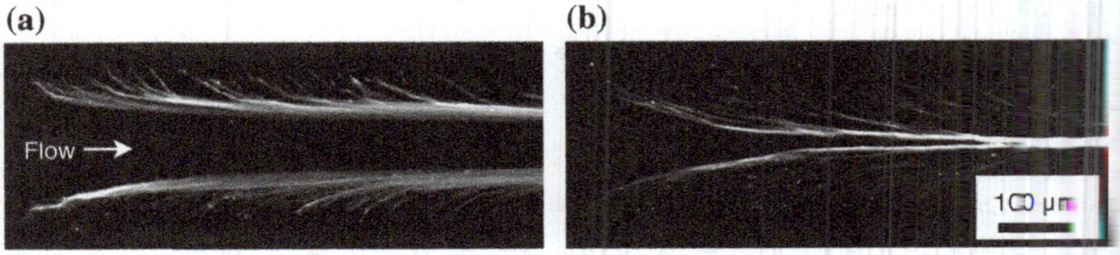

Fig. 17.10 Symmetric accumulation lines of particles at the entrance of double-gap electrode pattern under the applied voltage of 6 V_{pp} and bulk flow velocity of 0.2 mm/s. **a** Separated condition (17.2 kHz) and **b** merged condition (0.5 kHz)

of the center electrode under a certain condition in which particles are highly concentrated into a single zone as in Fig. 17.10b. The accumulation line in the separated condition is quite similar to that in the single-gap case without any interaction of ACEO flow. A comparison of the accumulation process in the single-gap and double-gap electrode system under different applied frequencies is depicted in Fig. 17.11. The accumulation trend of the double-gap system in this case changes at the frequency of 1 kHz. In low frequency region of less than 1 kHz, the accumulation location agrees with that of the single-gap system. On the other hand, in higher frequency region, particles are accumulated at the center of the electrode without frequency dependence. This would be attributed to the interference of two facing ACEO flow creating a stagnation area around the electrode center. It is worth noting that we can estimate the particle collection process of the double-gap system, and even multiple-gap one like comb-type electrode or more complicated, based on the collection performance of the single-gap system. When the half width of the center electrode in the double-gap system is wider than the convergence distance of the particle accumulation in the single-gap one, particles are collected into separated lines. And when the half width of the center electrode of the double-gap system is less than the single-gap accumulation distance, particles are collected into a single line at the electrode center. This estimation is useful to

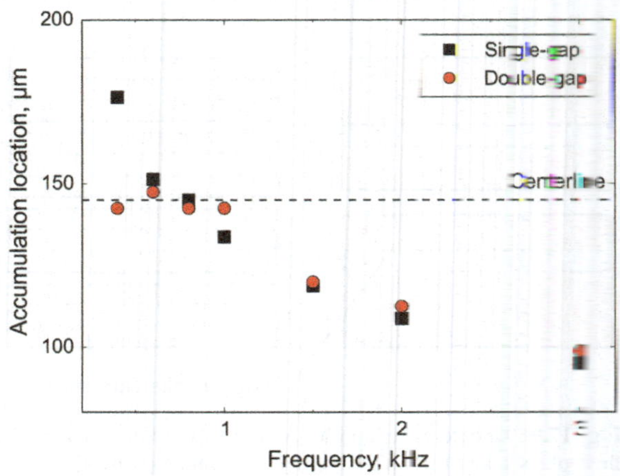

Fig. 17.11 Accumulation locations in single-gap and double-gap electrode systems under difference applied frequencies. The accumulation locations show an agreement over 1 kHz, whereas the double-gap system collect the particle at the center of the electrode in lower frequency range

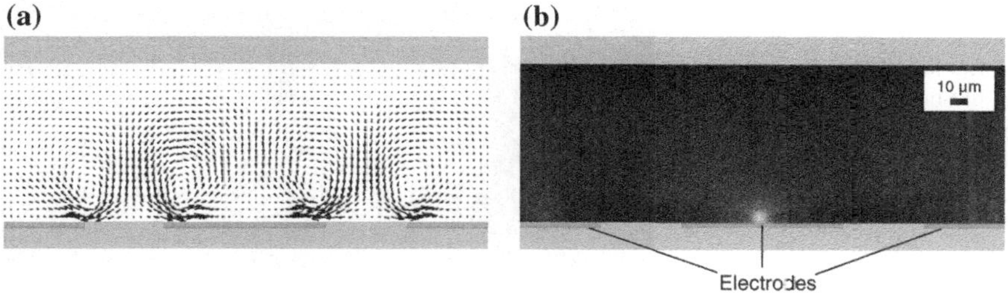

Fig. 17.12 Cross-sectional view of particle concentration by ACEO of double-gap electrode system in merged condition. **a** Velocity field measured by lateral micro-PIV. A stagnation area is observed above the center of electrode and **b** fluorescent particle distribution observed by the lateral view setup that is same as that in the velocity measurement. Particles are concentrated in the vicinity of the electrode surface within a height less than 5 μm

design the electrode geometry for a focusing or arrangement microfluidic system for colloidal materials because the concentrating performance can be predicted a priori by data obtained from the single-gap device.

The relation between ACEO velocity field and particle collection in the double-gap system is indicated in Fig. 17.12. In Fig. 17.12a, velocity field in a plane perpendicular to the substrate is depicted. These velocity fields were measured by lateral micro-PIV using a small mirror prism that changed the optical path of a microscope from vertical to lateral. A stagnation area is observed above the center of electrode. Figure 17.12b shows a cross-sectional image of the fluorescent intensity profile observed by the same lateral microscope setup as the velocity measurement. The results clearly indicate that most of particles suspended in the solution can be concentrated three-dimensionally on the surface area, less than 5 μm from the electrode surface, where the area corresponds to the stagnation area of the interfering ACEO flow. Note that the concentration factor of the 1-μm particles achieves 700. The highly efficient particle concentration would contribute to arrange colloidal materials and to enhance the surface reaction.

This velocity result also implies that particles are transported from bulk area above the both gaps to the surface area. Other experimental results show that the maximum collection rate of the particle in the double-gap system is equivalent to the coverage rate of the center electrode in the channel. For example, a double-gap system with a 300-μm center electrode in 500-μm-wide microchannel can collect particles as much as 60 %. Experimental result showing the collection rate of 56.4 % supports this simple explanation. In order to overcome this limitation, we need to adjust the initial concentration distribution of particles by using more sophisticated way, e.g., a utilization of sheath flow to prevent the particle flowing outer area of the center electrode or a cascade electrode system with the same purpose as the sheath flow approach. Then, collection rated of 98.4 % is confirmed [42]. This insight can be applied in a microfluidic particle concentrator or filtration as in Fig. 17.13. ACEO would benefit to extract particles in microchannel to a branch without clogging of filter in the system.

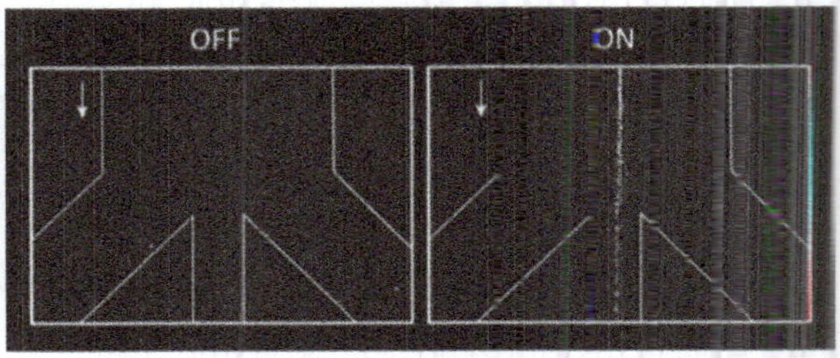

Fig. 17.13 An example of particle extraction using ACEO flow. This ACEO-based microfluidic system works as concentrator of filtration without clogging

17.3.3 Nanoparticle Arrangement

The electrokinetic flow can also be useful for the arrangement of nanoparticles [43]. Gold nanoparticles with diameters of 20, 50 and 100 nm (EMGC20, 50 and 100, BBI Solutions, USA) were used in this study. In contrast to fluorescent particle in the previous section, gold nanoparticle is invisible by a normal microscope. Therefore, the surface plasmon resonance induces strong absorption of light in a specific visible spectrum was used for visualization and quantification of particle collection [44, 45]. Prior to microfluidic experiments, calibration curve of gold nanoparticle aqueous solution for each diameter was obtained using a spectrometer. Then, the concentration distribution of gold nanoparticle can be visualized by an illumination of light in absorption band. A filtered LED was used in the present experiment. Figure 17.14 depicts the concentration of 100-nm gold nanoparticles with 8 V_{pp} application. A collected zone of nanoparticles can be visualized as a dark area in the figure, only when the electric field is applied ('ON' in Fig. 17.14). The size-dependent concentration factor calculated based on the calibration curves are presented in Fig. 17.15. The concentration factor shows positive dependence both in applied voltage and in particle size. For 100-nm particles with 10 V_{pp}, the

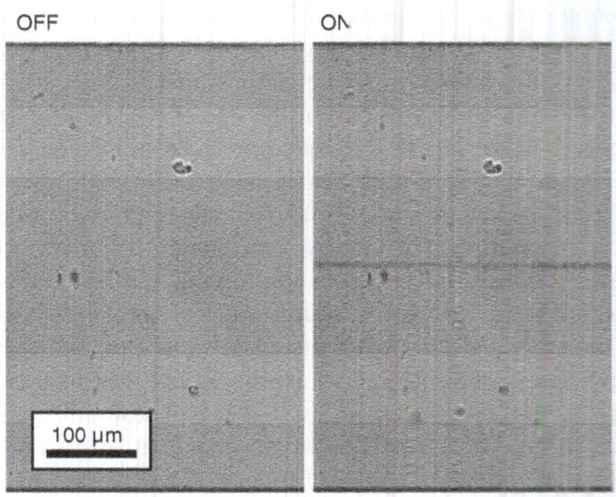

Fig. 17.14 Concentration of gold nanoparticles (100 nm). Nanoparticles are visualized by light absorption due to the surface plasmon resonance

Fig. 17.15 Concentration factors of gold nanoparticles with different sizes from 20 to 100 nm

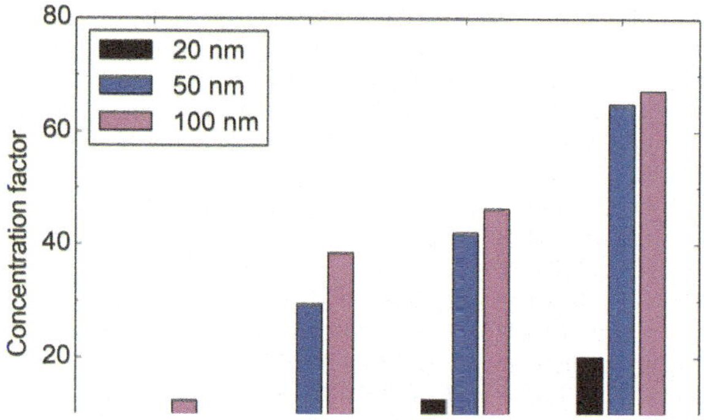

concentration factor approaches almost 70. Taking into account that there is also three-dimensional collection effect as in Fig. 17.12, the local maximum factor can be more than this value. In overall condition, 20-nm particles indicate relatively low factor. This is due to stronger diffusion and higher traceability of the particles to the advective motion of fluid. The factor for smaller particles can be improved by optimization of electrode array geometry and operating condition considering the balance of advection and diffusion.

References

1. Whitesides GM (2006) The origins and the future of microfluidics. Nature 442:368–373
2. Neethirajan S, Kobayashi I, Nakajima M, Wu D, Nandagopaldand S, Lin F (2011) Microfluidics for food, agriculture and biosystems industries. Lab Chip 11:1574–1586
3. Yeo LY, Chang HC, Chan PPY, Friend JR (2011) Microfluidic devices for bioapplications. Small 7:12–48
4. Pamme N (2007) Continuous flow separations in microfluidic devices. Lab Chip 7:1644–1659
5. Sajeesh P, Sen AK (2014) Particle separation and sorting in microfluidic device: a review. Microfluid Nanofluid 17:1–52
6. Yalcin SE, Sharma Qian AS, Joo SW, Baysal O (2010) Manipulating particles in microfluidics by floating electrodes. Electrophoresis 31:3711–3718
7. Gascoyne PRC, Vykoukal J (2002) Particle separation by dielectrophoresis. Electrophoresis 23:1973–1983
8. Dong L, Huang JP, Yu KW (2004) Theory of dielectrophoresis in colloidal suspensions. J Appl Phys 95:8321–8326
9. Liu C, Lagae L, Borghs G (2007) Manipulation of magnetic particles on chip by magnetophoretic actuation and dielectrophoretic levitation. Appl Phys Lett 90:184109
10. Lairell T, Petersson F, Nilsson A (2007) Chip integrated strategies for acoustic separation and manipulation of cells and particles. Chem Soc Rev 36:492–506
11. Yamada M, Nakashima M, Seki M (2004) Pinched flow fractionation: continuous size separation of particle utilizing a laminar flow profile in a pinched microchannel. Anal Chem 76:5465–5471
12. Kuntaegowdanahalli SS, Bhagat AS, Kumar G, Papautsky I (2009) Inertial microfluidics for continuous particle separation in spiral microchannels. Lab Chip 9:2973–2980

13. Martel JM, Toner M (2013) Particle focusing in curved microfluidic channels. Sci Rep 3:3340
14. Hart SJ, Terray AV (2003) Refractive-index-driven separation of colloidal polymer particles using optical chromatography. Appl Phys Lett 83:5316–5318
15. Kotari H, Motosuke M (2015) Simple applications of microparticle transportation by tender optical scattering force. Microfluid Nanofluid 18:549–558
16. Nagrath S, Sequist LV, Maheswaran S, Bell DW, Irimia D, Ulkus L, Smith MR, Kwak EL, Digumarthy S, Muzikansky A, Ryan P, Balis UJ, Tompkins RG, Haber DA, Toner M (2007) Isolation of rare circulating tumour cells in cancer patients by microchip technology. Nature 450:1235–1239
17. Bose S, Singh R, Hanewich-Hollatz M, Chong Shen, Lee CH, Dorfman DM, Karp JM, Karnik R (2013) Affinity flow fractionation of cells via transient interactions with asymmetric molecular patterns. Sci Rep 3:2329
18. Chung TD, Kim HC (2007) Recent advances in miniaturized microfluidic flow cytometry for clinical use. Electrophoresis 28:4511–4520
19. Ateya DA, Erickson JS, Howell PB Jr, Hilliard LR, Golden JP, Ligler FS (2008) The good, the bad, and the tiny: a review of microflow cytometry. Anal Bioanal Chem 391:1485–1498
20. Godin J, Chen C, Cho SH, Qiao W, Tsai F, Lo Y (2008) Microfluidics and photonics for bio-system-on-a-chip: a review of advancements in technology towards a microfluidic flow cytometry chip. J Biophoton 1:355–376
21. Sun T, Morgan H (2010) Single-cell microfluidic impedance cytometry: a review. Microfluid Nanofluid 8:423–443
22. Lin CC, Hsu JL, Lee GB (2011) Sample preconcentration in microfluidic devices. Microfluid Nanofluid 10:481–511
23. Giordano BC, Burgi DS, Hart SJ, Terray A (2012) On-line sample pre-concentration in microfluidic devices: a review. Anal Chim Acta 718:11–24
24. Wood CD, Evans SD, Cunningham JE, O'Rorke R, Wälti C, Davies AG (2008) Alignment of particles in microfluidic systems using standing surface acoustic waves. Appl Phys Lett 92:044104
25. Chen HH, Sun B, Tran KK, Shen H, Gao D (2009) A microfluidic manipulator for enrichment and alignment of moving cells and particles. J Biomech Eng 131:074505
26. Petersson F, Åberg L, Swärd-Nilsson AM, Laurell T (2007) Free flow acoustophoresis: microfluidic-based mode of particle and cell separation. Anal Chem 79:5117–5123
27. Lenshofa A, Laurell T (2010) Continuous separation of cells and particles in microfluidic systems. Chem Soc Rev 39:1203–1217
28. Castellanos A, Ramos A, González A, Green NG, Morgan H (2003) Electrohydrodynamics and dielectrophoresis in microsystem: scaling laws. J Phys D 36:2584–2597
29. Ramos A, Morgan H, Green NG, Castellanos A (1998) AC electrokinetics: a review of forces in microelectrode structures. J. Phys. D 31:2338–2353
30. Ramos A, Morgan H, Green NG, Castellanos A (1999) AC electric-field-induced fluid flow in microelectrodes. J Colloid Interface Sci 217:420–422
31. Bazant MZ, Thornton K, Ajdari A (2004) Diffuse-charge dynamics in electrochemical systems. Phys Rev E 70:021506
32. Olesen LH, Bruus H, Ajdari A (2006) AC electrokinetic micropumps: the effect of geometrical confinement, Faradaic current injection, and nonlinear surface capacitance. Phys Rev E 73:056313
33. Gregersen MM, Olesen LH, Brask A, Hansen MF, Bruus H (2007) Flow reversal at low voltage and low frequency in a microfabricated AC electrokinetic pump. Phys Rev E 76:056305
34. Olesen LH, Bazant MZ, Bruus H (2010) A Strongly nonlinear dynamics of electrolytes in large ac voltages. Phys Rev E 82:011501
35. Xia Y, Whitesides G (1998) Soft lithography. Annu Rev Mater Sci 28:153–184
36. Sia S, Whitesides G (2003) Microfluidic devices fabricated in poly(dimethylsiloxane) for biological studies. 24:3563–3576

37. Santiago JG, Wereley ST, Meinhart CD, Beebe DJ, Adrian RJ (1998) A particle image velocimetry system for microfluidics 25:316–319
38. Wereley ST, Meinhart CD (2010) Recent advances in micro-particle image velocimetry. Ann Rev Fluid Mech 42:557–576
39. Goldman AJ, Cox RG, Brenner H (1967) Slow viscous motion of a sphere parallel to a plane wall—I motion through a quiescent fluid. Chem Eng Sci 22:637–651
40. Meinhart CD, Wereley ST, Gray MHB (2000) Volume illumination for two-dimensional particle image velocimetry. Meas Sci Technol 11:809–814
41. Olsen MG, Adrian RJ (2000) Out-of-focus effects on particle image visibility and correlation in microscopic particle image velocimetry. Exp Fluids 29:S166–S174
42. Motosuke M, Yamasaki K, Ishida A, Toki H, Honami S (2013) Improved particle concentration by cascade AC electroosmotic flow. Microfluid Nanofluid 14:1021–1030
43. Bhatt KH, Grego S, Velev OD (2005) An AC electrokinetic technique for collection and concentration of particles and cells on patterned electrodes. Langmuir 21:6603–6612
44. He YQ, Liu SP, Kong L, Liu ZF (2005) A study on the sizes and concentrations of gold nanoparticles by spectra of absorption, resonance Rayleigh scattering and resonance non-linear scattering. Spectrochim Acta A 61:2861–2866
45. Jain PK, Lee KS, El-Sayed IH, El-Sayed MA (2006) Calculated absorption and scattering properties of gold nanoparticles of different size, shape, and composition: applications in biological imaging and biomedicine. J Phys Chem B 110:7238–7248

The manufacturer's authorised representative in the EU is Springer Nature Customer Service Centre GmbH, Europaplatz 3, 69115 Heidelberg, Germany. If you have any concerns regarding our products, please contact ProductSafety@springernature.com

Printed and bound by CPI Group (UK) Ltd, Croydon, CR0 4YY

23/07/2026

02175027-0004

Mathematics for Industry

Volume 22

Aims & Scope

The meaning of "Mathematics for Industry" (sometimes abbreviated as MI or MfI) is different from that of "Mathematics in Industry" (or of "Industrial Mathematics"). The latter is restrictive: it tends to be identified with the actual mathematics that specifically arises in the daily management and operation of manufacturing. The former, however, denotes a new research field in mathematics that may serve as a foundation for creating future technologies. This concept was born from the integration and reorganization of pure and applied mathematics in the present day into a fluid and versatile form capable of stimulating awareness of the importance of mathematics in industry, as well as responding to the needs of industrial technologies. The history of this integration and reorganization indicates that this basic idea will someday find increasing utility. Mathematics can be a key technology in modern society.

The series aims to promote this trend by (1) providing comprehensive content on applications of mathematics, especially to industry technologies via various types of scientific research, (2) introducing basic, useful, necessary and crucial knowledge for several applications through concrete subjects, and (3) introducing new research results and developments for applications of mathematics in the real world. These points may provide the basis for opening a new mathematics oriented technological world and even new research fields of mathematics.

More information about this series at http://www.springer.com/series/13254

Kazuhiko Ozeki

Theory of Affine Projection Algorithms for Adaptive Filtering

Springer

Kazuhiko Ozeki
The University of Electro-Communications
Tokyo
Japan

ISSN 2198-350X ISSN 2198-3518 (electronic)
Mathematics for Industry
ISBN 978-4-431-56310-5 ISBN 978-4-431-55738-8 (eBook)
DOI 10.1007/978-4-431-55738-8

Springer Tokyo Heidelberg New York Dordrecht London

Softcover reprint of the hardcover 1st edition 2016

Printed on acid-free paper

Springer Japan KK is part of Springer Science+Business Media
(www.springer.com)

Preface

This book is concerned with theoretical aspects of the affine projection algorithm (APA). The APA is a natural generalization of the classical, normalized least-mean-squares (NLMS) algorithm. It exploits multiple regressors for updating the coefficients of an adaptive filter, while the NLMS algorithm uses only the current, single regressor. The APA improves the convergence rate of an adaptive filter over the NLMS algorithm, especially for correlated input signals. By adjusting the number of regressors used, one can make a trade-off between the convergence rate and the computational complexity.

After the birth of the APA in the middle of the 1980s, several adaptation algorithms that also exploit multiple regressors were put forward independently. Most of them are now recognized as variants of the APA, constituting a family of APAs. Subsequently, many efforts were made to investigate the convergence behavior of the APA. We have also seen developments of techniques to reduce the computational complexity of the APA. Furthermore, we now have the kernel APA (KAPA), which extends the APA by the kernel trick so that it is applicable to identification of nonlinear systems. Recently, variable parameter APAs of various kinds are being studied. In this way, a tiny seed planted around 30 years ago has now grown, cherished by many researchers, into a big tree with lots of branches. The publication of this book was motivated by such a situation.

The book traces the above-mentioned history of development of the APA. In introducing the APA, the idea of affine projection, from which the name of the algorithm comes, plays an essential role. Actually, the APA emerged from the affine projection interpretation of the NLMS algorithm. Such a geometrical view enables us to see many of the important properties of the APA. Thus, to understand the principle of the APA, Chap. 3 is most fundamental.

I have tried to make the book self-contained and the mathematical proofs easily traceable by the reader as much as possible. The prerequisite is an elementary knowledge of calculus, linear algebra, probability, and statistics. Mathematical facts that can be described independently of the main body of the book are mostly put into appendices for readability. Since the theory of the APA relies on diverse mathematical materials, this made the appendices, especially in Chap. 3, unusually voluminous.

The APA is still developing. Computational problems will be overcome with the advancement of algorithms and digital technology. I hope this book will give a perspective of the APA world, and provide a basis for its further development.

In writing this book, I benefited from numerous papers that contributed to the development of the APA. Many parts of the book are reviews of those contributions. I wish to thank the authors of those papers. My thanks also go to the staff of Springer Japan, for their unfailing courtesy, encouragement, and help.

Yokohama
May 2015

Kazuhiko Ozeki

Contents

Symbols

$x \stackrel{\Delta}{=} y$	x is defined as y
$\{x;\, C(x)\}$	The set of x satisfying the condition $C(x)$
$x \in S$	x is an element of the set S
$X \subseteq Y$ or $Y \supseteq X$	X is a subset of Y
$S \cup T$	Union of the sets S and T
$S \cap T$	Intersection of the sets S and T
S^c	Complement of the set S
$\emptyset$	Empty set
$\mathbb{R}$	Real number system
$\mathbb{C}$	Complex number system
$\overline{x}$	Complex conjugate of the complex number x
$\mathbb{R}^n$	Cartesian product of n $\mathbb{R}$'s
$\mathbb{C}^n$	Cartesian product of n $\mathbb{C}$'s
$\{x(k)\}$	Signal that takes the value $x(k)$ at time k
$f : S \to T$	f is a mapping from the set S to the set T
$\max_{x \in S} f(x)$	Maximum of the function $f(x)$ on the set S
$\mathrm{argmax}_{x \in S} f(x)$	Maximum point of the function $f(x)$ on the set S
$\min_{x \in S} f(x)$	Minimum of the function $f(x)$ on the set S
$\mathrm{argmin}_{x \in S} f(x)$	Minimum point of the function $f(x)$ on the set S
fg	Composite mapping of f and g
sgn	Signum function
$\nabla_x f(x)$	Gradient of the function $f(x)$, x being a vector
$\mathrm{diag}\,[a_1, \ldots, a_n]$	Diagonal matrix with diagonal components $a_1, \ldots, a_n$
I_n	$n \times n$ identity matrix. Subscript n may be omitted
$\mathrm{span}\,(S)$	Vector space spanned by the set S
$\mathrm{cl}\,(S)$	Closure of the set S
$\mathcal{R}(A)$	Range space of the matrix A
$\mathcal{N}(A)$	Null space of the matrix A
x^t	Transpose of the vector x
A^t	Transpose of the matrix A

$\operatorname{rank} A$	Rank of the matrix A
$\dim \mathbb{V}$	Dimension of the vector space $\mathbb{V}$
A^+	Moore-Penrose pseudoinverse of the matrix A
$\langle x, y \rangle$	Inner product of the vectors x and y
$\|\|x\|\|$	Euclidean norm of the vector x
$\|\|x\|\|_\Sigma$	Norm of the vector x weighted with the positive definite matrix Σ
$x \perp y$	The vectors x and y are orthogonal
$x \perp \mathbb{H}$	The vector x is orthogonal to the vector space $\mathbb{H}$
$\mathbb{G} \perp \mathbb{H}$	The vector spaces $\mathbb{G}$ and $\mathbb{H}$ are orthogonal
$\mathbb{G} \oplus \mathbb{H}$	Direct sum of the vector spaces $\mathbb{G}$ and $\mathbb{H}$
$\mathbb{V}^\perp$	Orthogonal complement of the vector space $\mathbb{V}$
$P_\mathbb{V}$	Orthogonal projection onto the linear subspace $\mathbb{V}$
$\det A$	Determinant of the matrix A
$\operatorname{tr}[A]$	Trace of the matrix A
$\operatorname{vec}\{A\}$	Vector formed by concatenating the columns of the matrix A
vec^{-1}	Inverse operation of vec
Π	Affine subspace
P_Π	Affine projection onto the affine subspace Π
$A \otimes B$:	Tensor product, or Kronecker product of the matrices A and B
$A \boxtimes B$	Schur product, or Hadamard product of the matrices A and B
$\mathrm{E}\{\cdot\}$	Expectation of a random variable, a random vector, or a random matrix
$\mathrm{V}(\cdot)$	Variance of a random variable
$\mathrm{Cov}(x)$	Covariance matrix of the random vector x
$O(f(n))$	Big-O of $f(n)$
P iff Q	P if and only if Q

Chapter 1
Introduction

1.1 Adaptive Filter and Its Application Areas

An *adaptive filter* is a kind of learning machine that automatically changes its transfer function [1, 2]. As illustrated in Fig. 1.1, an adaptive filter has two key components: a *transversal filter*, or *finite impulse response (FIR) filter* [3, p. 155] to process the input signal, and a mechanism to control the coefficients of the transversal filter. It updates the coefficient vector $w_k = (w_{k,0}, w_{k,1}, \ldots, w_{k,n-1})^T$ each time the new signal $x(k)$ arrives so that the magnitude of the error $e(k)$ between the output signal $z(k)$ from the filter and the desired signal $d(k)$ decreases. Thus, the coefficient vector w_k is time varying, k being the time-index. The integer n is the filter length.

There is a long history of research on adaptive filters, and various applications have been reported. In order to see how adaptive filters are applied to real-world problems, let us take up some examples before proceeding to the subject of this book, the adaptation algorithm.

The first example is *system identification* [4]. In Fig. 1.2, we want to make a replica of the unknown system using an adaptive filter. In this figure, the desired signal $d(k)$ is the output of the unknown system, the observation noise being neglected for simplicity. If adaptation is successful to make the magnitude of error $e(k)$ sufficiently small, we can identify the unknown system with the transversal filter as far as the input–output relation is concerned. This scheme covers a wide variety of problems including *echo cancellation* [1, p. 327], [2, p. 323], and *active noise control* [5]. Acoustic echo cancellation using an adaptive filter will be discussed in the last chapter.

The second example is *channel equalization* [2, p. 171]. As shown in Fig. 1.3, the input to the adaptive filter is the output $y(k)$ of the unknown channel in this case. The symbol Δ denotes the time required for the test signal $x(k)$ to go through the channel and the adaptive filter. The desired signal is the delayed version $x(k - \Delta)$ of the test signal $x(k)$. If adaptation is successful (and if the test signal $x(k)$ is appropriately chosen), the adaptive filter acquires the inverse characteristic of the channel, equalizing the channel.

K. Ozeki, *Theory of Affine Projection Algorithms for Adaptive Filtering*,
Mathematics for Industry 22, DOI 10.1007/978-4-431-55738-8_1

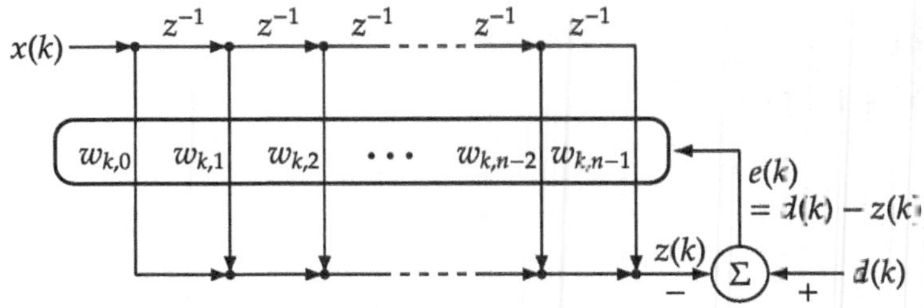

Fig. 1.1 Basic configuration of adaptive filter

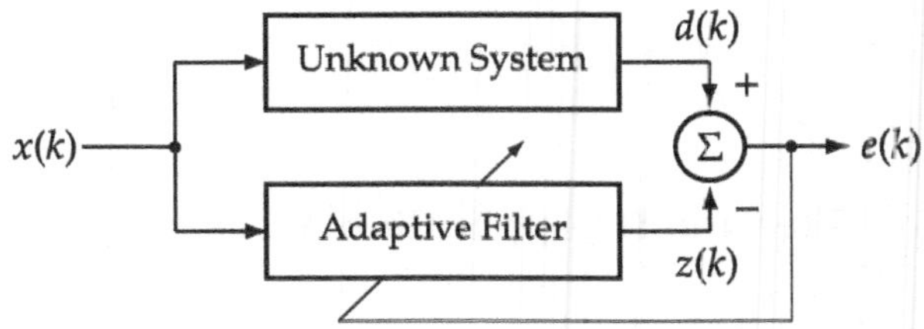

Fig. 1.2 System identification using adaptive filter

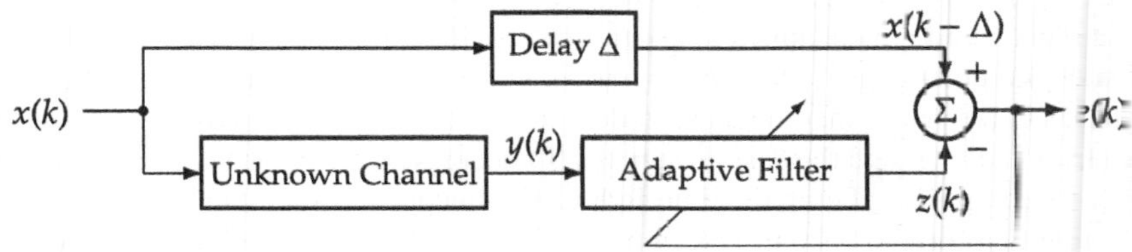

Fig. 1.3 Channel equalization using adaptive filter

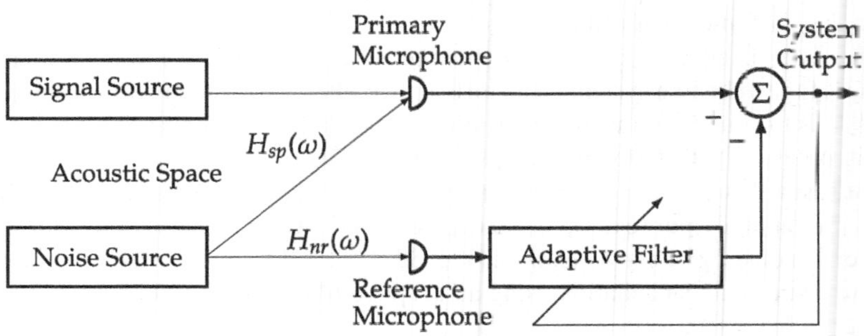

Fig. 1.4 Acoustic noise cancellation using adaptive filter

Figure 1.4 shows yet another example, *acoustic noise cancellation* [5]. For simplicity, we assume that the reference microphone picks up only the sound from the noise source, and that adaptation is done so that the adaptive filter approximates the transfer function $H_{sp}(\omega)H_{nr}(\omega)^{-1}$, where $H_{sp}(\omega)$ and $H_{nr}(\omega)$ are the transfer functions from the noise source to the primary microphone and from the noise source to the reference microphone, respectively. Then, the signal from the noise source is cancelled, and only the signal from the signal source comes out from the system output.

Besides the examples above, applications to such problems as *antenna beamforming* [2, p. 101], *howling suppression* [7], and feedback cancellation in *hearing aids* [8, 9] have also been reported.

1.2 Basic Scheme for Adaptation

In an adaptive filter, the coefficient vector is updated as

$$w_{k+1} = w_k + \mu \Delta w_k, \tag{1.1}$$

where Δw_k is a vector to give the direction and the magnitude of correction added to the current coefficient vector w_k to obtain the new coefficient vector w_{k+1} as shown in Fig. 1.5. The parameter μ is called the *step-size*. The vector Δw_k is a function of the current coefficient vector w_k, the current and the past input signal $x(k), x(k-1), \ldots$ and the current and the past desired signal $d(k), d(k-1), \ldots$:

$$\Delta w_k = f(w_k; x(k), x(k-1), \ldots; d(k), d(k-1), \ldots). \tag{1.2}$$

The choice of the function f determines the performance and the computational complexity of an adaptation algorithm. In classical adaptation algorithms such as the least-mean-squares (LMS) algorithm and the normalized least-mean-squares (NLMS) algorithm, f is a simple function of $w_k, x(k), x(k-1), \ldots, x(k-(n-1))$, and $d(k)$, where n is the length of the transversal filter. These algorithms have low complexity and are easy to implement. However, the speed of convergence is limited especially for correlated input signals. By using a more elaborate function, it is possible to improve the convergence rate. In fact, the affine projection algorithm (APA) [10], the main subject of this book, is such an algorithm. The APA is a natural generalization of the NLMS algorithm, exploiting $x(k), x(k-1), \ldots, x(k-(n-1)-(p-1))$ and $d(k), d(k-1), \ldots, d(k-(p-1))$ for updating the coefficient

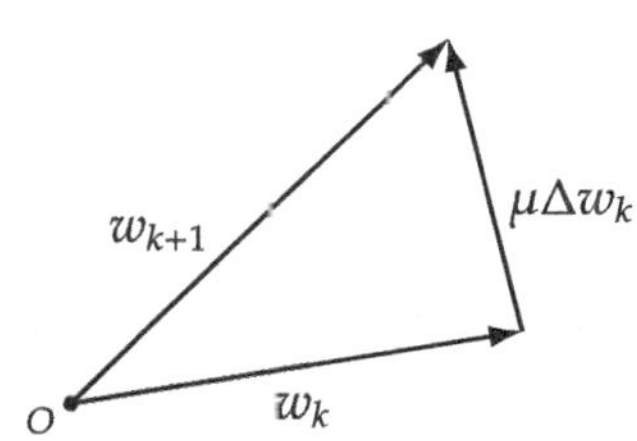

Fig. 1.5 Update of the coefficient vector w_k. The vector Δw_k is a function of w_k, the past input signal, and the past desired signal. The parameter μ is the step-size

vector, where the positive integer p is a parameter called the projection order. When $p = 1$, the APA reduces to the NLMS algorithm. As p increases, the convergence rate improves in exchange for increased computational complexity. One can adjust the value of p to make a trade-off between the convergence rate and the computational complexity.

After the original APA appeared, many variants were put forward independently. They now form a family of affine projection algorithms. In the following chapters, we will see how the original APA emerged as a natural generalization of the NLMS algorithm, and how it evolved to one of the standard algorithms in the field of adaptive filtering.

1.3 Organization of the Book

The book is organized as follows in the subsequent chapters.

Chapter 2 reviews two classical adaptation algorithms: the least-mean-squares (LMS) algorithm and its normalized version, the NLMS algorithm. The progress from the LMS algorithm to the NLMS algorithm is an important step toward the APA, the main subject of this book.

Chapter 3 gives the formulation of the APA. Starting with the initial motivation, it develops the basic APA (B-APA) and the regularized APA (R-APA). It also covers the decorrelating property of the APA, together with the formal similarities between the recursive least-squares (RLS) algorithm and the APA. In introducing the APA, the affine projection structure of the algorithm is emphasized. In fact, it was an affine projection interpretation of the NLMS algorithm that inspired the APA. Such a geometrical view gives a deep insight into the properties of the APA.

Chapter 4 surveys adaptation algorithms that were put forward after the appearance of the APA, and are now recognized as variants of the APA. It gives proofs that these algorithms are essentially equivalent to the APA, together with their unified formulation as a family of APAs.

Chapter 5 focuses on the convergence behavior of the APA. After showing the basic convergence behavior of the APA, it reviews two analysis examples based on simplifying assumptions on regressors: the discrete orientation assumption and the autoregressive regressor assumption. It also reviews a general treatment for the analysis of the R-APA based on the energy conservation relation and the weighted energy conservation relation.

Chapter 6 describes two methods of reducing the computational complexity of the APA: the fast affine projection (FAP) algorithm and the block exact fast affine projection (BEFAP) algorithm. It is shown that these algorithms significantly reduce the amount of computation necessary to implement the APA.

Chapter 7 is for the kernel APA (KAPA). It first explains how the kernel trick is used for extending a linear model to a nonlinear model by taking the kernel perceptron as an example. Then, it describes the way the APA is extended to the KAPA by the kernel trick so that it is applicable to identification of nonlinear systems.

Chapter 8, the last chapter of this book, gives an overview of recent topics on variable parameter APAs such as the variable step-size APA (VSS-APA), the proportionate APA (PAPA), the evolving order APA (E-APA), and the variable regularized APA (VR-APA).

1.4 Remarks on Notation

Throughout the book the following notation is employed unless otherwise noted. A vector space is denoted by a capital blackboard font, such as $\mathbb{G}, \mathbb{H}, \mathbb{V}$. The real number system is denoted by $\mathbb{R}$. The complex number system is denoted by $\mathbb{C}$. The Cartesian product $\mathbb{R}^n$ is considered a vector space over $\mathbb{R}$. The Cartesian product $\mathbb{C}^n$ is considered a vector space over $\mathbb{C}$. Scalars and vectors are denoted by normal font letters, such as $a, b, c, x, y, z, \alpha, \beta, \gamma$. Boldface letters are reserved for random variables and random vectors as stated below.

Signals that appear in this book take real values. The symbol $\{x(k)\}$ denotes a signal that takes the value $x(k)$ at time k.

A vector is normally a column vector. The transpose of a vector x is denoted by x^t. A vector $x \in \mathbb{R}^n$ may be expressed as $x = (x_0, x_1, \ldots, x_{n-1})^t$, where x_i is the ith component of x. An indexed vector x_k is expressed as $x_k = (x_{k,0}, x_{k,1}, \ldots, x_{k,n-1})^t$. The inner product of two vectors x, y is denoted by $\langle x, y\rangle$. For $x, y \in \mathbb{R}^n$, $\langle x, y\rangle \triangleq x^t y = \sum_{i=0}^{n-1} x_i y_i$, and for $x, y \in \mathbb{C}^n$, $\langle x, y\rangle \triangleq x^t \overline{y} = \sum_{i=0}^{n-1} x_i \overline{y}_i$. The Euclidean norm $\|x\|$ of a vector x is defined by $\|x\| \triangleq \sqrt{\langle x, x\rangle}$.

Matrices are denoted by normal capital letters, such as A, B, C, X, Y, Z. An $m \times n$ matrix whose (i, j)th component equals a_{ij} may simply be represented as $A = [a_{ij}]$. The transpose of a matrix A is denoted by A^t just as in the case of a vector. A column vector $x = (x_0, x_1, \ldots, x_{n-1})^t$ may also be denoted, by using matrix notation, as $x = [x_0\ x_1\ \cdots\ x_{n-1}]^t$. If $x_0, x_1, \ldots, x_{n-1}$ are m-dimensional vectors, then $[x_0\ x_1\ \cdots\ x_{n-1}]$ is an $m \times n$ matrix.

Random variables and random vectors are denoted by boldface letters, such as $\boldsymbol{x}, \boldsymbol{y}, \boldsymbol{z}$. Random matrices are denoted by boldface capital letters, such as $\boldsymbol{X}, \boldsymbol{Y}, \boldsymbol{Z}$.

The integers n and p are reserved for the filter length and the projection order, respectively.

References

1. Haykin, S.: Adaptive Filter Theory. Prentice-Hall, Upper Saddle River (2002)
2. Sayed, A.H.: Adaptive Filters. Wiley, Hoboken (2008)
3. Oppenheim, A.V., Schafer, R.W.: Digital Signal Processing. Prentice-Hall, Englewood Cliffs (1975)
4. Nagumo, J., Noda, A.: A learning method for system identification. IEEE Trans. Autom. Control AC **12**(3), 282–287 (1967)

5. Kuo, S.M., Morgan, D.R.: Active noise control: a tutorial review. Proc. IEEE **87**(6), 943–973 (1999)
6. Widrow, B., Glover Jr, J.R., McCool, J.M., Kaunitz, J., Williams, C.S., Hearn, R.H., Zeidler, J.R., Dong Jr, E., Goodlin, R.C.: Adaptive noise cancelling: principles and applications. Proc. IEEE **63**(12), 1692–1716 (1975)
7. Umeda, T., Ozeki, K.: Suppression of howling between microphones and monitoring loudspeakers—an application of the adaptive filter. NHK Laboratories Note. 321 (1985)
8. Lee, S., Kim, I., Park, Y.: Approximated affine projection algorithm for feedback cancellation in hearing aids. Comput. Meth. Programs Biomed. **87**(3), 254–261 (2007)
9. Rotaru, M., Albu, F., Coanda, H.: A variable step size modified decorrelated NLMS algorithm for adaptive feedback cancellation in hearing aids. In: Proceedings of ISETC 2012, pp. 263–266. Timisoara, 15–16 Nov (2012)
10. Ozeki, K., Umeda, T.: An adaptive filtering algorithm using an orthogonal projection to an affine subspace and its properties. IEICE Trans. **J67-A**(2), 126–132 (1984) (Also in Electron. Commun. Jpn. **67-A**(5), 19–27 (1984))

Chapter 2
Classical Adaptation Algorithms

2.1 Least Mean-Squares (LMS) Algorithm

We review two classical adaptation algorithms in this chapter. The first one is the *least-mean-squares (LMS) algorithm*. In this algorithm, the coefficient vector $w_k \in \mathbb{R}^n$ is updated as

$$\begin{aligned} w_{k+1} &= w_k + \mu \Delta w_k, \\ \Delta w_k &= e(k) x_k, \end{aligned} \tag{2.1}$$

where

$$x_k \stackrel{\triangle}{=} (x(k), x(k-1), \ldots, x(k-(n-1)))^t$$

is the signal vector,

$$e(k) \stackrel{\triangle}{=} d(k) - z(k) = d(k) - \langle x_k,\ w_k \rangle$$

is the error signal, and μ is a parameter called the *step-size*. The integer n is the filter length throughout the book. The step-size can be time varying as $\mu(k)$. We will treat such a *variable step-size* case in the last chapter. The expression $\langle x_k,\ w_k \rangle \stackrel{\triangle}{=} x_k^t w_k$ means the inner product of x_k and w_k. The signal vector is also called a *regressor*.

How to time-index the coefficient vector is a matter of convention. In the literature, we sometimes encounter a formulation like

$$\begin{aligned} w_k &= w_{k-1} + \mu \Delta w_{k-1}, \\ \Delta w_{k-1} &= e(k) x_k, \\ e(k) &= d(k) - \langle x_k,\ w_{k-1} \rangle \end{aligned} \tag{2.2}$$

K. Ozeki, *Theory of Affine Projection Algorithms for Adaptive Filtering*,
Mathematics for Industry 22, DOI 10.1007/978-4-431-55738-8_2

instead of (2.1). However, (2.1) and (2.2) are essentially the same. The important point is that the time-index of the coefficient vector is increased by 1 after it is updated, showing that the update is performed sample-by-sample.

The motivation of the LMS algorithm can be explained by an argument based on the steepest-descent method (Appendix 1: "Steepest-Descent Method"). Let us assume that $x(k)$ and $d(k)$ are realizations of random variables $\boldsymbol{x}(k)$ and $\boldsymbol{d}(k)$, respectively. Then, $z(k)$ and $e(k)$ are also realizations of random variables defined by

$$\boldsymbol{z}(k) \triangleq \langle \boldsymbol{x}_k,\ w_k \rangle = \boldsymbol{x}_k^t w_k,$$

$$\boldsymbol{e}(k) \triangleq \boldsymbol{d}(k) - \boldsymbol{z}(k),$$

where $\boldsymbol{x}_k = (\boldsymbol{x}(k), \boldsymbol{x}(k-1), \dots, \boldsymbol{x}(k-(n-1)))^t$. In this stochastic framework, we can consider the mean-square error between $\boldsymbol{d}(k)$ and $\boldsymbol{z}(k)$:

$$\mathrm{E}\left\{\boldsymbol{e}^2(k)\right\} = \mathrm{E}\left\{(\boldsymbol{d}(k) - \boldsymbol{z}(k))^2\right\} = \mathrm{E}\left\{(\boldsymbol{d}(k) - \boldsymbol{x}_k^t w_k)^2\right\}. \tag{2.3}$$

We differentiate (2.3) with respect to w_k to obtain the gradient of the mean-square error at w_k:

$$\begin{aligned}
\nabla_{w_k}\mathrm{E}\left\{\boldsymbol{e}^2(k)\right\} &= \nabla_{w_k}\mathrm{E}\left\{\boldsymbol{d}^2(k) - 2\boldsymbol{d}(k)\boldsymbol{x}_k^t w_k + (\boldsymbol{x}_k^t w_k)^2\right\} \\
&= \nabla_{w_k}\left(\mathrm{E}\left\{\boldsymbol{d}^2(k)\right\} - 2\mathrm{E}\{\boldsymbol{d}(k)\boldsymbol{x}_k^t\}w_k + w_k^t\mathrm{E}\{\boldsymbol{x}_k\boldsymbol{x}_k^t\}w_k\right) \\
&= -2\mathrm{E}\{\boldsymbol{d}(k)\boldsymbol{x}_k\} + 2\mathrm{E}\{\boldsymbol{x}_k\boldsymbol{x}_k^t\}w_k \\
&= -2(\mathrm{E}\{\boldsymbol{d}(k)\boldsymbol{x}_k\} - \mathrm{E}\{\boldsymbol{x}_k\boldsymbol{x}_k^t\}w_k).
\end{aligned}$$

However, the expectations $\mathrm{E}\{\boldsymbol{d}(k)\boldsymbol{x}_k\}$ and $\mathrm{E}\{\boldsymbol{x}_k\boldsymbol{x}_k^t\}$ are not known. Therefore, we employ *instantaneous approximation*: we replace $\mathrm{E}\{\boldsymbol{d}(k)\boldsymbol{x}_k\}$ with the realization $d(k)x_k$, and $\mathrm{E}\{\boldsymbol{x}_k\boldsymbol{x}_k^t\}$ with $x_k x_k^t$. Then, the gradient is approximated by

$$\begin{aligned}
\nabla_{w_k}\mathrm{E}\left\{\boldsymbol{e}^2(k)\right\} &\approx -2(d(k)x_k - x_k x_k^t w_k) \\
&= -2x_k(d(k) - x_k^t w_k) \\
&= -2x_k(d(k) - z(k)) \\
&= -2e(k)x_k.
\end{aligned}$$

Since a small change of w_k along the direction $-\nabla_{w_k}\mathrm{E}\left\{\boldsymbol{e}^2(k)\right\}$ decreases $\mathrm{E}\left\{\boldsymbol{e}^2(k)\right\}$, if we replace w_k with $w_{k+1} = w_k + \mu e(k)x_k$ for a small positive μ, then $\mathrm{E}\left\{\boldsymbol{e}^2(k+1)\right\}$ will be smaller than $\mathrm{E}\left\{\boldsymbol{e}^2(k)\right\}$. This is just the LMS algorithm (2.1). The origin of the name "least-mean-squares (LMS)" algorithm will be apparent. The LMS algorithm is also called the *Widrow-Hoff algorithm* after its originators [1]. The geometrical meaning of the LMS algorithm will become clearer in the next section.

2.2 Normalized LMS (NLMS) Algorithm

Let us consider the system identification problem in Fig. 2.1. We assume that the unknown system is represented by a transversal filter with the coefficient vector $w^o = (w_0^o, w_1^o, \ldots, w_{n-1}^o)^t$. We also assume $\{v(k)\} = 0$[1] throughout this chapter. This is a more specific version of Fig. 1.2. In the LMS algorithm, the effective magnitude of the step-size μ depends on the volume of the input signal as explained below. If the input signal $\{x(k)\}$ is multiplied by a scalar a, then other signals are also multiplied by a. Let the new signals be denoted by

$$\begin{aligned}
\tilde{x}(k) &\stackrel{\triangle}{=} ax(k), \\
\tilde{x}_k &\stackrel{\triangle}{=} (\tilde{x}(k), \tilde{x}(k-1), \ldots, \tilde{x}(k-(n-1)))^t \\
&= a(x(k), x(k-1), \ldots, x(k-(n-1)))^t \\
&= ax_k, \\
\tilde{y}(k) &\stackrel{\triangle}{=} \langle \tilde{x}_k, w^o \rangle = a\langle x_k, w^o \rangle = ay(k), \\
\tilde{d}(k) &\stackrel{\triangle}{=} \tilde{y}(k) = ay(k) = ad(k), \\
\tilde{z}(k) &\stackrel{\triangle}{=} \langle \tilde{x}_k, w_k \rangle = \langle ax_k, w_k \rangle = a\langle x_k, w_k \rangle = az(k), \\
\tilde{e}(k) &\stackrel{\triangle}{=} \tilde{d}(k) - \tilde{z}(k) = ad(k) - az(k) = ae(k).
\end{aligned}$$

For the new signals, $\mu \Delta w_k$ can be written as

$$\begin{aligned}
\mu \Delta w_k &= \mu \tilde{e}(k) \tilde{x}_k \\
&= (a^2 \mu) e(k) x_k. \qquad (2.4)
\end{aligned}$$

Thus, multiplying a scalar a to $\{x(k)\}$ has the same effect as multiplying a^2 to the step-size μ. This phenomenon makes it difficult to set an optimal value for μ: just

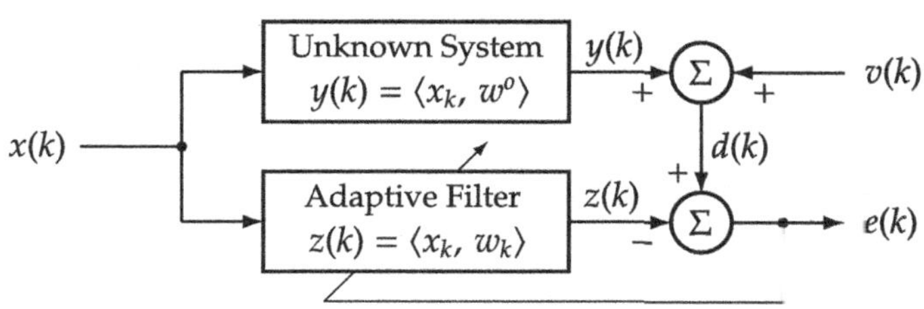

Fig. 2.1 Identification of unknown linear system

[1] This means $v(k) = 0$ for $k \geq 0$.

changing the gain of the amplifier for the input signal affects the effective value of μ.

In the *normalized LMS (NLMS) algorithm*, the coefficient vector is updated as

$$w_{k+1} = w_k + \mu \Delta w_k,$$
$$\Delta w_k = e(k)\frac{x_k}{\|x_k\|^2}. \tag{2.5}$$

Note that Δw_k is normalized by $\|x_k\|^2$. It is obvious from the above arguments that Δw_k is invariant under multiplication of a scalar to the input signal $\{x(k)\}$. Thus, in the NLMS algorithm, μ has a definite meaning that is independent of the volume of the signal $\{x(k)\}$.

The NLMS algorithm can be interpreted from a geometrical point of view. If $\mu = 1$, then as shown in Fig. 2.2, w_{k+1} is the affine projection (Chap. 3, Appendix 1: "Affine Projection") of w_k onto Π_k: $w_{k+1} = P_{\Pi_k} w_k$, where Π_k is the hyperplane defined by

$$\Pi_k \stackrel{\triangle}{=} \{w;\ d(k) - \langle x_k,\ w\rangle = 0\},$$

and P_{Π_k} denotes the affine projection onto Π_k. In fact,

$$\begin{aligned}\langle x_k,\ w_{k+1}\rangle &= x_k^t\left(w_k + e(k)\frac{x_k}{\|x_k\|^2}\right)\\ &= x_k^t w_k + e(k)\\ &= d(k).\end{aligned}$$

This shows $w_{k+1} \in \Pi_k$. Also, since x_k is orthogonal to Π_k by Theorem 3.12,

$$w_{k+1} - w_k = \frac{e(k)}{\|x_k\|^2}x_k$$

is orthogonal to Π_k.

If $\mu \neq 1$, w_{k+1} is a point somewhere on the straight line connecting w_k and $P_{\Pi_k} w_k$, the location of which depends on μ as illustrated in Fig. 2.2. In fact,

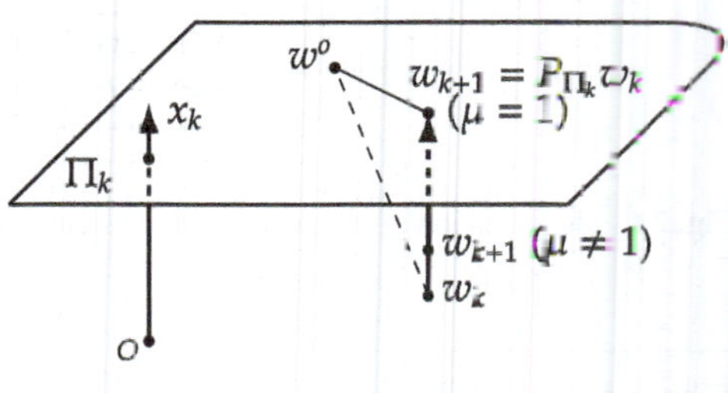

Fig. 2.2 Geometrical interpretation of the NLMS algorithm. If $\{v(k)\} = 0$, $w^o \in \Pi_k$

$$\begin{aligned} w_{k+1} - w_k &= \mu \Delta w_k \\ &= \mu(w_k + \Delta w_k - w_k) \\ &= \mu(P_{\Pi_k} w_k - w_k). \end{aligned}$$

Thus, the step-size μ determines the ratio

$$\frac{\|w_{k+1} - w_k\|}{\|P_{\Pi_k} w_k - w_k\|} = |\mu|. \tag{2.6}$$

Since the hyperplane Π_k is invariant under the change of the volume of the input signal $\{x(k)\}$, the vector w_{k+1} is also invariant. In the LMS algorithm, this is not true. Although w_{k+1} is also a point on the straight line connecting w_k and $P_{\Pi_k} w_k$, the location depends not only on μ but also on the volume of the signal $\{x(k)\}$ even if its waveform is kept unchanged.

Under the present assumption that $\{v(k)\} = 0$, the coefficient vector w^o of the unknown system satisfies $d(k) - \langle x_k, w^o \rangle = 0$. Thus, w^o is an element of Π_k as shown in Fig. 2.2.

Theorem 2.1 *If* $0 < \mu < 2$, *then* $\|w_{k+1} - w^o\| \leq \|w_k - w^o\|$. *If* $\mu \leq 0$ *or* $\mu \geq 2$, *then* $\|w_{k+1} - w^o\| \geq \|w_k - w^o\|$.

Proof In the triangle $\{w^o, w_k, P_{\Pi_k} w_k\}$, the vectors $(w^o - P_{\Pi_k} w_k)$ and $(P_{\Pi_k} w_k - w_k)$ are orthogonal as shown in Fig. 2.2. Therefore, by the Pythagorean theorem,

$$\begin{aligned} \|w^o - w_k\|^2 &= \|w^o - P_{\Pi_k} w_k\|^2 + \|P_{\Pi_k} w_k - w_k\|^2 \\ &= \|w^o - P_{\Pi_k} w_k\|^2 + \|\Delta w_k\|^2. \end{aligned} \tag{2.7}$$

Also, in the triangle $\{w^o, w_{k+1}, P_{\Pi_k} w_k\}$, the vectors $(w^o - P_{\Pi_k} w_k)$ and $(P_{\Pi_k} w_k - w_{k+1})$ are orthogonal. Therefore,

$$\begin{aligned} \|w^o - w_{k+1}\|^2 &= \|w^o - P_{\Pi_k} w_k\|^2 + \|P_{\Pi_k} w_k - w_{k+1}\|^2 \\ &= \|w^o - P_{\Pi_k} w_k\|^2 + \|(1 - \mu)\Delta w_k\|^2 \\ &= \|w^o - P_{\Pi_k} w_k\|^2 + (1 - \mu)^2 \|\Delta w_k\|^2. \end{aligned} \tag{2.8}$$

From (2.7) and (2.8), we have

$$\|w^o - w_k\|^2 - \|w^o - w_{k+1}\|^2 = \mu(2 - \mu)\|\Delta w_k\|^2.$$

Since

$$\mu(2 - \mu) \begin{cases} > 0, & \text{if } 0 < \mu < 2; \\ \leq 0, & \text{otherwise}, \end{cases}$$

we have proved the theorem. □

This theorem guarantees that if $0 < \mu < 2$, then $\|w^o - w_k\|^2$ decreases monotonously. Since a bounded, monotone sequence always converges, $\lim_{k\to\infty} \|w^o - w_k\|^2$ exists. However, this does not mean $\lim_{k\to\infty} w_k = w^o$.

For arbitrary μ, we have from (2.8)

$$\|w^o - P_{\Pi_k} w_k\| \leq \|w^o - w_{k+1}\|,$$

with equality only if $\mu = 1$ unless $\Delta w_k = 0$. Therefore, $\mu = 1$ is the best choice for fast convergence. However, in the presence of the observation noise $\{v(k)\}$, we have to consider a trade-off between the convergence rate and the adaptation error. In such a case, $\mu = 1$ may not be the best choice.

In the above argument, we see that the meaning of the step-size μ in the NLMS algorithm is quite different from that in the LMS algorithm. In the LMS algorithm, μ is a small positive number for approximating differential with finite difference. In the NLMS algorithm on the other hand, μ is a parameter to prevent over-fitting to noisy data. In this sense, it may be more appropriate to call it the *relaxation factor* [2].

The progress from the LMS algorithm to the NLMS algorithm, also known as the *learning method* [3], is an important step toward the affine projection algorithm (APA) [2], the main subject of the subsequent chapters.

Appendix 1: Steepest-Descent Method

Let f be a real differentiable function defined on $\mathbb{R}^n$. The *steepest-descent method* searches for $\text{argmin}_x f(x)$ by iterating the following computation starting from an initial point x_0:

$$x_{k+1} = x_k - \mu \nabla_x f(x_k), \tag{2.9}$$

where $\mu > 0$ is the step-size. The step-size may depend on the iteration index k as $\mu(k)$.

This algorithm is motivated by the following fact. Let $x(t) \in \mathbb{R}^n$ be a function of t, and consider a differential equation

$$\frac{\mathrm{d}x(t)}{\mathrm{d}t} = -\nabla_x f(x(t)). \tag{2.10}$$

Using the chain rule for differentiation and (2.10), we have

$$\begin{aligned}\frac{\mathrm{d}f(x(t))}{\mathrm{d}t} &= (\nabla_x f(x(t)))^t \frac{\mathrm{d}x(t)}{\mathrm{d}t} \\ &= -(\nabla_x f(x(t)))^t \nabla_x f(x(t)) \\ &= -\|\nabla_x f(x(t))\|^2 \\ &\leq 0.\end{aligned}$$

with equality only if $\nabla_x f(x(t)) = 0$. This shows that if the curve $x(t)$ is a solution of (2.10), $f(x(t))$ is a decreasing function of t.

The differential equation (2.10) can be approximated by a difference equation as

$$\frac{x(t + \Delta t) - x(t)}{\Delta t} \approx -\nabla_x f(x(t)).$$

If we let $x_k \triangleq x(t)$, $x_{k+1} \triangleq x(t + \Delta t)$, and $\mu \triangleq \Delta t$, we obtain (2.9). Under a certain condition, the vector sequence $x_0, x_1, x_2, \ldots$ converges to a local minimum point. If the initial point x_0 and the step-size μ are appropriately chosen, the sequence converges to $\mathrm{argmin}_x f(x)$.

References

1. Widrow, B., Hoff, M.E.Jr.: Adaptive switching circuits. IRE WESCON Conv. Rec. Pt.**4**, 96–104 (1960)
2. Ozeki, K., Umeda, T.: An adaptive filtering algorithm using an orthogonal projection to an affine subspace and its properties. IEICE Trans. **J67-A**(2), 126–132 (1984) (Also in Electron. Commun. Jpn. 67-A(5), 19–27 (1984))
3. Nagumo, J., Noda, A.: A learning method for system identification. IEEE Trans. Autom. Control **AC-12**(3), 282–287 (1967)

Chapter 3
Affine Projection Algorithm

3.1 Motivation of Affine Projection Algorithm

Let us consider the system identification problem in Fig. 2.1. We assume $\{v(k)\} = 0$ for a while. If we run the NLMS algorithm with $\mu = 1$ starting from $k = 0$, then, regardless of the initial coefficient vector w_0, we have $w_k \in \Pi_{k-1}$ for $k \geq 1$. This situation is illustrated in Fig. 3.1. In the figure, we see that as the angle between Π_k and Π_{k-1} approaches 0 or π, Δw_k tends to 0, which will result in slow convergence. The angle between Π_k and Π_{k-1} equals, by definition, the angle between x_k and x_{k-1}. Let the angle be denoted by θ. Then,

$$\cos\theta = \frac{\langle x_k,\ x_{k-1}\rangle}{\|x_k\| \cdot \|x_{k-1}\|},$$

where $\langle x_k,\ x_{k-1}\rangle = x_k^t x_{k-1}$ is the inner product of x_k and x_{k-1}. The quantity $r \stackrel{\triangle}{=} \langle x_k,\ x_{k-1}\rangle/(\|x_k\| \cdot \|x_{k-1}\|)$ is the first-order autocorrelation coefficient of the input signal $\{x(k)\}$ calculated for $x_k = (x(k), x(k-1), \ldots, x(k-(n-1)))^t$ and $x_{k-1} = (x(k-1), x(k-2), \ldots, x(k-1-(n-1)))^t$. If the signal $\{x(k)\}$ is a realization of a white noise, r is expected to take a very small value for sufficiently large n, and θ will be near $\pm\pi/2$. However, if $\{x(k)\}$ is a correlated signal, r will go away from 0 toward ± 1, and θ may approach 0 or π, resulting in slow convergence. The same problem arises in the case of $\mu \neq 1$, and in the LMS algorithm as well because the direction of Δw_k is the same in the LMS algorithm and in the NLMS algorithm.

The problem above will be avoided by projecting w_k, not onto Π_k, but onto $\Pi_k \cap \Pi_{k-1}$ as in Fig. 3.2: $w_{k+1} = P_{\Pi_k \cap \Pi_{k-1}} w_k$ [1]. Since $\Pi_k \cap \Pi_{k-1}$ is independent of the angle θ, the convergence rate will not be affected even if r approaches ± 1.

The hyperplanes Π_k and Π_{k-1} are $(n-1)$-dimensional affine subspaces (Appendix 1: "Affine Projection"). Therefore, by Theorem 3.10, $\Pi_k \cap \Pi_{k-1}$ is also an affine subspace, and the mapping $P_{\Pi_k \cap \Pi_{k-1}}$ is the affine projection onto $\Pi_k \cap \Pi_{k-1}$.

K. Ozeki, *Theory of Affine Projection Algorithms for Adaptive Filtering*, Mathematics for Industry 22, DOI 10.1007/978-4-431-55738-8_3

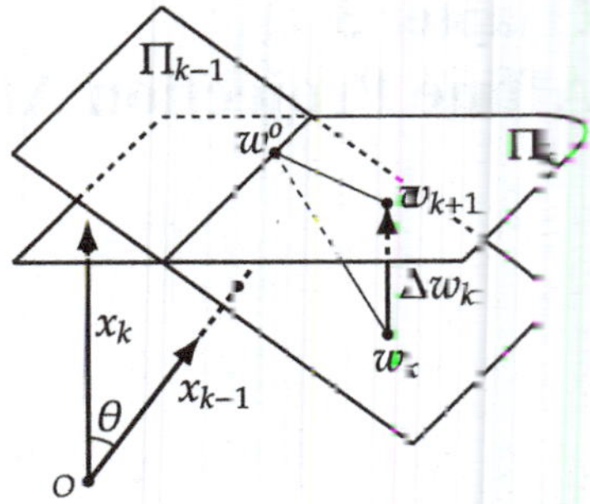

Fig. 3.1 Update of coefficient vector in the NLMS algorithm

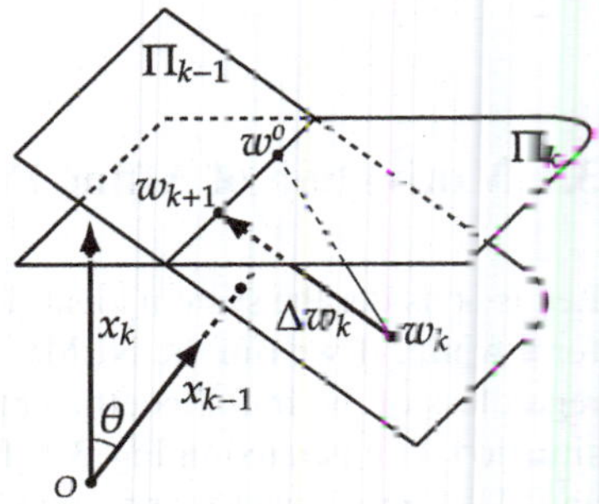

Fig. 3.2 Affine projection of w_k onto $\Pi_k \cap \Pi_{k-1}$

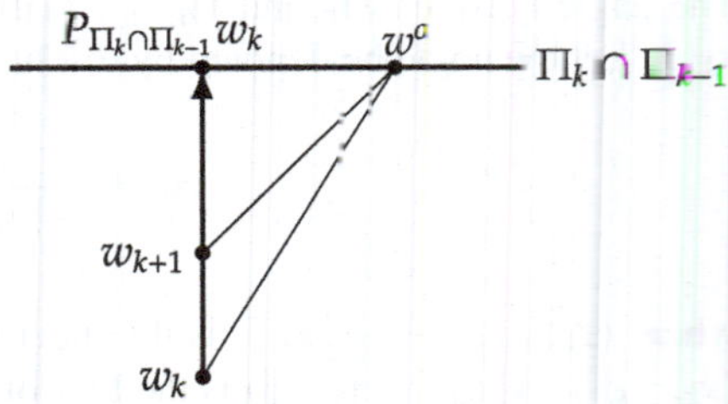

Fig. 3.3 Right-angled triangles $\{w^o, w_k, P_{\Pi_k \cap \Pi_{k-1}} w_k\}$ and $\{w^o, w_{k+1}, P_{\Pi_k \cap \Pi_{k-1}} w_k\}$

Introducing the step-size μ, we can write this new update equation as

$$\begin{aligned} w_{k+1} &= w_k + \mu \Delta w_k, \\ \Delta w_k &= P_{\Pi_k \cap \Pi_{k-1}} w_k - w_k. \end{aligned} \tag{3.1}$$

As shown in Fig. 3.3, the triangle $\{w^o, w_k, P_{\Pi_k \cap \Pi_{k-1}} w_k\}$ is right-angled. The triangle $\{w^o, w_{k+1}, P_{\Pi_k \cap \Pi_{k-1}} w_k\}$ is right-angled as well. Therefore, Theorem 2.1 also holds for this new algorithm.

When $\mu = 1$, the coefficient vector w_k is on the hyperplane Π_{k-1} for $k \geq 1$ regardless of the initial coefficient vector w_0. In this case, the update of the coefficient vector is performed by the following series equations:

$$\begin{aligned} u_k &= \langle x_k, x_{k-1}\rangle / \|x_{k-1}\|^2 x_{k-1}, \\ x_k' &= x_k - u_k, \\ e(k) &= d(k) - \langle w_k, x_k\rangle, \\ \Delta w_k &= e(k) x_k' / \|x_k'\|^2, \\ w_{k+1} &= w_k + \Delta w_k. \end{aligned} \tag{3.2}$$

Note that the vector u_k is the orthogonal projection of x_k onto the subspace spanned by the single vector x_{k-1}. The vector x_k', which is orthogonal to x_{k-1}, determines the direction of Δw_k.

Theorem 3.1 *Let w_{k+1} in the NLMS algorithm (2.5) be denoted by $w_{k+1}^{(1)}$, and w_{k+1} in the new algorithm (3.1) by $w_{k+1}^{(2)}$. If $0 \le \mu \le 2$, then $\|w^o - w_{k+1}^{(1)}\| \ge \|w^o - w_{k+1}^{(2)}\|$.*

Proof Because the vector $P_{\Pi_k} w_k - w_k$ and the affine subspace Π_k are orthogonal, $P_{\Pi_k} w_k - w_k$ and $\Pi_k \cap \Pi_{k-1}$ are orthogonal. Also, $P_{\Pi_k \cap \Pi_{k-1}} w_k - w_k$ and $\Pi_k \cap \Pi_{k-1}$ are orthogonal. Therefore, the vector

$$P_{\Pi_k \cap \Pi_{k-1}} w_k - P_{\Pi_k} w_k = (P_{\Pi_k \cap \Pi_{k-1}} w_k - w_k) - (P_{\Pi_k} w_k - w_k)$$

and the affine subspace $\Pi_k \cap \Pi_{k-1}$ are orthogonal. Hence, the vector $P_{\Pi_k \cap \Pi_{k-1}} w_k - P_{\Pi_k} w_k$ and the vector $P_{\Pi_k \cap \Pi_{k-1}} w_k - w^o$ are orthogonal. From this fact, we have

$$\|P_{\Pi_k} w_k - w^o\| \ge \|P_{\Pi_k \cap \Pi_{k-1}} w_k - w^o\| \tag{3.3}$$

as can be seen in Fig. 3.4. Now, let

$$\begin{aligned} &a \triangleq \|w^o - w_k\|, \quad b \triangleq \|w^o - P_{\Pi_k} w_k\|, \quad c \triangleq \|w^o - P_{\Pi_k \cap \Pi_{k-1}} w_k\|, \\ &d \triangleq \|w_k - P_{\Pi_k} w_k\|, \quad e \triangleq \|w_k - P_{\Pi_k \cap \Pi_{k-1}} w_k\|. \end{aligned}$$

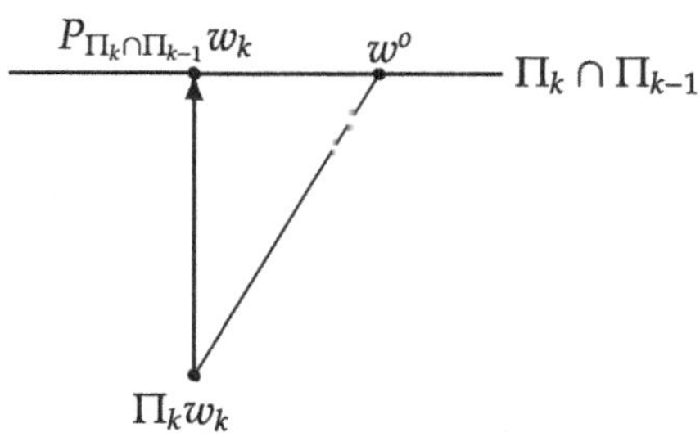

Fig. 3.4 Right-angled triangle $\{w^o, P_{\Pi_k \cap \Pi_{k-1}} w_k, P_{\Pi_k} w_k\}$

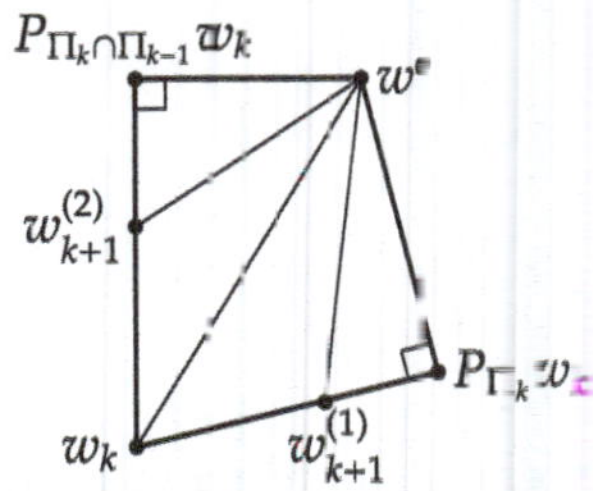

Fig. 3.5 Comparison of $\|w^o - w_{k+1}^{(1)}\|$ and $\|w^o - w_{k+1}^{(2)}\|$

In Fig. 3.5, we see that

$$\begin{aligned} \|w^o - w_{k+1}^{(1)}\|^2 &= b^2 + \|w_{k+1}^{(1)} - P_{\Pi_k} w_k\|^2 \\ &= b^2 + (1-\mu)^2 d^2 \\ &= b^2 + (1-\mu)^2 (a^2 - b^2) \\ &= \mu(2-\mu) b^2 + (1-\mu)^2 a^2. \end{aligned} \tag{3.4}$$

Similarly,

$$\begin{aligned} \|w^o - w_{k+1}^{(2)}\|^2 &= c^2 + \|w_{k+1}^{(2)} - P_{\Pi_k \cap \Pi_{k-1}} w_k\|^2 \\ &= c^2 + (1-\mu)^2 e^2 \\ &= c^2 + (1-\mu)^2 (a^2 - c^2) \\ &= \mu(2-\mu) c^2 + (1-\mu)^2 a^2. \end{aligned} \tag{3.5}$$

Noting $b \geq c$ by (3.3), and $\mu(2-\mu) \geq 0$ by the given condition, we have from (3.4) and (3.5)

$$\begin{aligned} \|w^o - w_{k+1}^{(1)}\| - \|w^o - w_{k+1}^{(2)}\| &= \mu(2-\mu)(b^2 - c^2) \\ &\geq 0. \end{aligned}$$

□

3.2 Basic Affine Projection Algorithm (B-APA)

Theorem 3.1 suggests that if we use more regressors $x_k, x_{k-1}, \ldots, x_{k-(p-1)}$, and make the affine projection of w_k onto $\Pi_k \cap \Pi_{k-1} \cap \cdots \cap \Pi_{k-(p-1)}$ to generalize the update equation (3.1) as

$$\begin{aligned} w_{k+1} &= w_k + \mu \Delta w_k, \\ \Delta w_k &= P_{\Pi_k \cap \Pi_{k-1} \cap \cdots \cap \Pi_{k-(p-1)}} w_k - w_k, \end{aligned} \tag{3.6}$$

then the convergence rate of the adaptive filter will be further improved. This is the original form of the *affine projection algorithm (APA)* [1]. The positive integer p is the *projection order* of the algorithm. Throughout the book, we assume $p < n$. The NLMS algorithm and the algorithm (3.1) are special cases of the APA with $p = 1$ and $p = 2$, respectively.

Since $(P_{\Pi_k \cap \Pi_{k-1} \cap \cdots \cap \Pi_{k-(p-1)}} w_k - w_k) \perp (P_{\Pi_k \cap \Pi_{k-1} \cap \cdots \cap \Pi_{k-(p-1)}} w_k - w^o)$, Theorem 2.1 also holds for the APA.

Theorem 3.1 is generalized for the APA as follows:

Theorem 3.2 *Let the coefficient vector w_{k+1} in the update equation* (3.6) *for the APA with order p be denoted by $w_{k+1}^{(p)}$. If $0 \le \mu \le 2$, and $q < p$, then $\|w^o - w_{k+1}^{(q)}\| \ge \|w^o - w_{k+1}^{(p)}\|$.*

Proof It suffices to show $\|w^o - w_{k+1}^{(p-1)}\| \ge \|w^o - w_{k+1}^{(p)}\|$. In a similar way to prove (3.3), we can easily show that

$$\|P_{\Pi_k \cap \cdots \cap \Pi_{k-(p-2)}} w_k - w^o\| \ge \|P_{\Pi_k \cap \cdots \cap \Pi_{k-(p-1)}} w_k - w^o\|.$$

Also, we see that

$$(P_{\Pi_k \cap \cdots \cap \Pi_{k-(p-1)}} w_k - w_k) \perp (P_{\Pi_k \cap \cdots \cap \Pi_{k-(p-1)}} w_k - w^o),$$

and

$$(P_{\Pi_k \cap \cdots \cap \Pi_{k-(p-2)}} w_k - w_k) \perp (P_{\Pi_k \cap \cdots \cap \Pi_{k-(p-2)}} w_k - w^o).$$

By using these facts, the proof goes exactly in parallel with that of Theorem 3.1. □

The affine subspace $\Pi_k \cap \cdots \cap \Pi_{k-(p-1)}$ is the set of solutions of a system of linear equations

$$\langle x_{k-j},\ w\rangle = d(k-j) \quad (j = 0, 1, \ldots, p-1), \tag{3.7}$$

w being the unknown. By using the matrix notation

$$X_k \triangleq \left[x_k\ x_{k-1}\ \cdots\ x_{k-(p-1)}\right]^t,$$

and

$$d_k \triangleq (d(k), d(k-1), \ldots, d(k-(p-1)))^t,$$

Eq. (3.7) is rewritten as

$$X_k w = d_k. \tag{3.8}$$

The $p \times n$ matrix X_k is called the *regressor block* in this book.

Now, let $\Pi^{(k)} \triangleq X_k^+ d_k + \mathcal{N}(X_k)$, where X_k^+ is the Moore–Penrose pseudoinverse of X_k (Appendix 1: "Affine Projection"). Under the present assumption that

$\{v(k)\} = 0$, (3.8) has a solution w^o. Therefore, by Corollary 3.1 (Appendix 1: "Affine Projection"), $\Pi^{(k)} = \Pi_k \cap \cdots \cap \Pi_{k-(p-1)}$. As a matter of fact, $\Pi^{(k)}$ has a more general meaning. There are various cases where (3.8) has no solution. For example, the true order of the unknown transversal filter may not be equal to n. Also, the observation noise $\{v(k)\}$ may not be negligible. Even in such cases, (3.8) has a least-squares solution, and by Theorem 3.8, the affine subspace $\Pi^{(k)}$ gives the set of least-squares solutions of (3.8).

By Theorem 3.13 (Appendix 1: "Affine Projection"), the affine projection of w_k onto $\Pi^{(k)}$ is given by

$$P_{\Pi^{(k)}} w_k = w_k + X_k^+ (d_k - X_k w_k),$$

which is the solution of (3.8) that is closest to w_k when (3.8) has at least one solution, or the least-squares solution of (3.8) that is closest to w_k when (3.8) has no solution. By using this formula, the algorithm (3.6) is modified as

$$\begin{aligned} w_{k+1} &= w_k + \mu \Delta w_k, \\ \Delta w_k &= P_{\Pi^{(k)}} w_k - w_k = X_k^+ (d_k - X_k w_k). \end{aligned} \tag{3.9}$$

This algorithm is meaningful, unlike the algorithm (3.6), even when Eq. (3.8) has no solution.

In practice, computing the pseudoinverse in a general case is not simple. Therefore, we make here a moderate assumption that the regressors $x_k, x_{k-1}, \ldots, x_{k-(p-1)}$ are linearly independent. Then, rank $X_k = p$, and so by Theorem 3.7 (Appendix 1: "Affine Projection"), $X_k^+ = X_k^t (X_k X_k^t)^{-1}$. Note that the Gramian matrix $X_k X_k^t$ is nonsingular by Theorem A.8 (Appendix A). Substituting this equation into (3.9), we have

$$w_{k+1} = w_k + \mu X_k^t (X_k X_k^t)^{-1} e_k, \tag{3.10}$$

where e_k is the error vector defined by

$$\begin{aligned} e_k &\triangleq d_k - X_k w_k \\ &= \begin{bmatrix} d(k) - \langle x_k, w_k \rangle \\ d(k-1) - \langle x_{k-1}, w_k \rangle \\ \vdots \\ d(k-(p-1)) - \langle x_{k-(p-1)}, w_k \rangle \end{bmatrix}. \end{aligned} \tag{3.11}$$

Note that $e_k \neq (e(k), e(k-1), \ldots, e(k-(p-1))^t$. The adaptation algorithm using (3.10) is referred to as the *basic APA (B-APA)* henceforth.

Under the assumption that regressors $x_k, x_{k-1}, \ldots, x_{k-(p-1)}$ are linearly independent, Eq. (3.8) has a solution even in the presence of the observation noise. Therefore, $\Pi^{(k)} = \Pi_k \cap \cdots \cap \Pi_{k-(p-1)}$.

3.3 Regularized Affine Projection Algorithm (R-APA)

Although the Gramian matrix $X_k X_k^t$ is invertible under the present assumption that the regressors $x_k, x_{k-1}, \ldots, x_{k-(p-1)}$ are linearly independent, numerical inversion of $X_k X_k^t$ becomes difficult if the condition number of $X_k X_k^t$, cond $(X_k X_k^t)$, is large (Appendix 2: " Condition Number").[1] One way of avoiding this problem is to add a diagonal matrix δI to $X_k X_k^t$, where δ is a small positive number and I is the $p \times p$ identity matrix.

The matrix $X_k X_k^t$ is obviously symmetric. Since $X_k X_k^t$ is nonsingular, the eigenvalues $\lambda_1, \lambda_2, \ldots, \lambda_p$ are all positive by Theorem A.6 and Corollary A.2 (Appendix A). Thus, by (3.68),

$$\text{cond}\,(X_k X_k^t) = \frac{\max_k \lambda_k}{\min_k \lambda_k}.$$

On the other hand, if v_k is the eigenvector corresponding to λ_k,

$$(X_k X_k^t + \delta I)v_k = (\lambda_k + \delta)v_k,$$

which shows that the eigenvalues of $X_k X_k^t + \delta I$ are given by $\lambda_k + \delta$ $(k = 1, 2, \ldots, p)$. Therefore,

$$\text{cond}\,(X_k X_k^t + \delta I) = \frac{\max_k(\lambda_k + \delta)}{\min_k(\lambda_k + \delta)} = \frac{\max_k \lambda_k + \delta}{\min_k \lambda_k + \delta}.$$

Thus, by adding δI to $X_k X_k^t$, the distribution of the relative magnitude of eigenvalues is flattened, which leads to

$$\text{cond}\,(X_k X_k^t + \delta I) < \text{cond}\,(X_k X_k^t).$$

This inequality shows that inverting $X_k X_k^t + \delta I$ is easier than inverting $X_k X_k^t$ numerically. Even if the regressors $x_k, x_{k-1}, \ldots, x_{k-(p-1)}$ are linearly dependent, $X_k X_k^t + \delta I$ is still invertible.

This technique, adding δI to $X_k X_k^t$ to stabilize the numerical inversion process, is referred to as *regularization*, and δ is called the *regularization factor*. The coefficient update equation with regularization is stated as

$$w_{k+1} = w_k + \mu X_k^t (X_k X_k^t + \delta I)^{-1} e_k, \qquad (3.12)$$

where e_k is the error vector defined by (3.11).

The adaptation algorithm using (3.12) is called the *regularized APA (R-APA)*. This is a variant of the APA often described as the standard APA in textbooks [2–4]. Although regularization makes w_{k+1} for $\mu = 1$ only approximately equal to the affine projection of w_k onto $\Pi^{(k)}$, it is a practical way of avoiding numerical difficulty.

---

[1]Numerical inversion of a matrix A is based on solving a linear equation of the type $Ax = b$. If cond (A) is large, a small error in A or b can result in a large error in the solution.

3.4 Alternative Approaches to Derivation of Update Equations

3.4.1 Update Equation as Solution of an Optimization Problem

In the foregoing arguments, it is seen that the affine projection of w_k onto $\Pi^{(k)}$ is the solution of the minimization problem

$$\min_{w} \|w - w_k\|^2$$

subject to the constraints

$$\begin{aligned} d(k) - x_k^t w &= 0, \\ d(k-1) - x_{k-1}^t w &= 0, \\ &\vdots \\ d(k-(p-1)) - x_{k-(p-1)}^t w &= 0. \end{aligned} \tag{3.13}$$

We can use the method of Lagrange multipliers (Appendix 3: "The Method of Lagrange Multipliers") to solve this problem. The Lagrange function for this problem is given by

$$L(w, \lambda_0, \lambda_1, \ldots, \lambda_{p-1}) = \|w - w_k\|^2 + \sum_{i=0}^{p-1} \lambda_i (d(k-i) - x_{k-i}^t w).$$

We differentiate L with respect to w and set it equal to 0, obtaining

$$\begin{aligned} w &= w_k + \sum_{i=0}^{p-1} (\lambda_i/2) x_{k-i} \\ &= w_k + X_k^t \lambda, \end{aligned} \tag{3.14}$$

where

$$\lambda \stackrel{\triangle}{=} (\lambda_0/2, \lambda_1/2, \ldots, \lambda_{p-1}/2)^t.$$

Multiplying X_k to both sides of (3.14) from the left, and using the constraints (3.13), we have

$$X_k X_k^t \lambda = e_k.$$

If we assume that the regressors $x_k, x_{k-1}, \ldots, x_{k-(p-1)}$ are linearly independent, then the Gramian matrix $X_k X_k^t$ is nonsingular by Theorem A.8 (Appendix A). Thus, we have

$$\lambda = (X_k X_k^t)^{-1} e_k.$$

Substituting this equation into (3.14), we obtain the solution as

$$w = w_k + X_k^t (X_k X_k^t)^{-1} e_k.$$

Using this solution for w_{k+1}, and introducing the step-size μ, we finally have an update equation

$$w_{k+1} = w_k + \mu X_k^t (X_k X_k^t)^{-1} e_k,$$

which is just the update equation (3.10) for the B-APA. The method of Lagrange multipliers is also used for deriving the *set-membership affine projection algorithm (SM-APA)* [5].

The update equation can also be derived from other optimization problems. For example, let us look for an n-dimensional vector Δw that minimizes a function

$$L(\Delta w) \stackrel{\triangle}{=} \mu \|e_k - X_k \Delta w\|^2 + (1-\mu)\|X_k \Delta w\|^2 + \delta\|\Delta w\|^2,$$

where μ and δ are real parameters [6]. The first two terms can be looked upon as linear interpolation of two quadratic forms. The last term is for regularization. By straightforward calculation, $L(\Delta w)$ can be rewritten as

$$L(\Delta w) = \mu \left(\|e_k\|^2 - 2\Delta w^t X_k^t e_k\right) + \Delta w^t X_k^t X_k \Delta w + \delta\|\Delta w\|^2.$$

Differentiating $L(\Delta w)$ with respect to Δw by the use of (A.17) (Appendix A), and setting it equal to 0, we have an equation

$$\begin{aligned} \frac{\partial L}{\partial \Delta w}(\Delta w) &= 2\left(-\mu X_k^t e_k + (X_k^t X_k + \delta I_n)\Delta w\right) \\ &= 0. \end{aligned}$$

Since the matrix $X_k^t X_k$ is nonnegative definite, $X_k^t X_k + \delta I_n$ is nonsingular for $\delta > 0$. Hence,

$$\Delta w = \mu (X_k^t X_k + \delta I_n)^{-1} X_k^t e_k. \tag{3.15}$$

In the matrix inversion lemma (A.5) (Appendix A), replace $A = \delta I_n$, $B = -X_k^t$, $C = X_k$, $D = I_p$. Then, we have

$$(X_k^t X_k + \delta I_n)^{-1} X_k^t = X_k^t (X_k X_k^t + \delta I_p)^{-1}. \tag{3.16}$$

Note that $X_k X_k^t + \delta I_p$ is also nonsingular. From (3.15) and (3.16),

$$\Delta w = \mu X_k^t (X_k X_k^t + \delta I_p)^{-1} e_k.$$

Using this vector as an increment from w_k, we have an update equation

$$w_{k+1} = w_k + \mu X_k^t (X_k X_k^t + \delta I_p)^{-1} e_k,$$

which is exactly the update equation (3.12) for the R-APA.

3.4.2 Derivation of Update Equation by Instantaneous Approximation

Just as the LMS algorithm is derived by the steepest descent method (Chap. 2, Appendix 1: "Steepest Descent Method") for solving a minimization problem with the use of instantaneous approximation of expectations, the R-APA is derived by Newton's method (Appendix 4: "Newton's Method") with the use of instantaneous approximation of expectations. To see this, let us assume that X_k and d_k are realizations of a random matrix $\boldsymbol{X}_k$ and a random vector $\boldsymbol{d}_k$, respectively. In this stochastic framework, let us consider a function of the n-dimensional vector w

$$\begin{aligned} f(w) &\triangleq \mathrm{E}\left\{\|\boldsymbol{d}_k - \boldsymbol{X}_k w\|^2\right\} \\ &= \mathrm{E}\left\{\boldsymbol{d}_k^t \boldsymbol{d}_k\right\} - 2\mathrm{E}\{\boldsymbol{d}_k \boldsymbol{X}_k\} w + w^t \mathrm{E}\left\{\boldsymbol{X}_k^t \boldsymbol{X}_k\right\} w, \end{aligned}$$

and apply Newton's method to minimize it. For that purpose, we calculate the gradient $\nabla_w f(w_k)$ and the Hessian matrix $\nabla_w^2 f(w_k)$ to obtain

$$\begin{aligned} \nabla_w f(w_k) &= -2\mathrm{E}\left\{\boldsymbol{X}_k^t \boldsymbol{d}_k\right\} + 2\mathrm{E}\left\{\boldsymbol{X}_k^t \boldsymbol{X}_k\right\} w_k, \\ \nabla_w^2 f(w_k) &= 2\mathrm{E}\left\{\boldsymbol{X}_k^t \boldsymbol{X}_k\right\}. \end{aligned}$$

Since we know neither $\mathrm{E}\left\{\boldsymbol{X}_k^t \boldsymbol{d}_k\right\}$ nor $\mathrm{E}\left\{\boldsymbol{X}_k^t \boldsymbol{X}_k\right\}$, we resort to instantaneous approximation

$$\begin{aligned} \nabla_w f(w_k) &\approx -2X_k^t d_k + 2X_k^t X_k w_k \\ &= -2X_k^t e_k, \\ \nabla_w^2 f(w_k) &\approx 2X_k^t X_k, \end{aligned}$$

where expectations are replaced with realizations. Applying (3.71) in Newton's method, we obtain

$$w_{k+1} = w_k + \mu (X_k^t X_k + \delta I)^{-1} X_k^t e_k. \tag{3.17}$$

By using (3.16), this equation is rewritten as

$$w_{k+1} = w_k + \mu X_k^t (X_k X_k^t + \delta I)^{-1} e_k,$$

which is the same as the update equation (3.12) for the R-APA.

3.5 Decorrelating Property of B-APA

The B-APA has a decorrelating property in the following sense [7]. Let us introduce linear predictor coefficients $\alpha_1, \alpha_2, \ldots, \alpha_{p-1}$ to predict $x(k-j)$ by a linear combination

$$\sum_{i=1}^{p-1} \alpha_i x(k-j-i)$$

for $j = 0, 1, \ldots, n-1$ (Appendix 5: "Linear Prediction"). Then, we can make n equations

$$\begin{aligned} r(k) &= x(k) - \sum_{i=1}^{p-1} \alpha_i x(k-i) \\ r(k-1) &= x(k-1) - \sum_{i=1}^{p-1} \alpha_i x(k-1-i) \\ &\vdots \\ r(k-(n-1)) &= x(k-(n-1)) - \sum_{i=1}^{p-1} \alpha_i x(k-(n-1)-i), \end{aligned} \tag{3.18}$$

where $r(k), r(k-1), \ldots, r(k-(n-1))$ are prediction errors. By using the vector and matrix notation, (3.18) can be rewritten in a compact form as

$$r_k = x_k - \tilde{X}_k^t \alpha,$$

where

$$\begin{aligned} r_k &\overset{\triangle}{=} (r(k), r(k-1), \ldots, r(k-(n-1)))^t, \\ \alpha &\overset{\triangle}{=} (\alpha_1, \alpha_2, \ldots, \alpha_{p-1})^t, \\ \tilde{X}_k &\overset{\triangle}{=} \left[x_{k-1}\ x_{k-2} \cdots x_{k-(p-1)}\right]^t. \end{aligned}$$

The least-squares estimate of α is obtained by solving the minimization problem

$$\underset{\alpha}{\operatorname{argmin}} \|r_k\|^2 = \underset{\alpha}{\operatorname{argmin}} \left\| x_k - \tilde{X}_k^t \alpha \right\|^2, \tag{3.19}$$

which is equivalent to solving the normal equation

$$\tilde{X}_k \tilde{X}_k^t \alpha = \tilde{X}_k x_k. \tag{3.20}$$

Let us assume that the regressors $x_{k-1}, x_{k-2}, \ldots, x_{k-(p-1)}$ are linearly independent. Then, by Theorem A.8 (Appendix A), the Gramian matrix $\tilde{X}_k \tilde{X}_k^t$ is nonsingular, so that the solution of (3.20) is expressed as

$$\hat{\alpha} = (\tilde{X}_k \tilde{X}_k^t)^{-1} \tilde{X}_k x_k. \tag{3.21}$$

Now let us define the error vector

$$\phi_k \stackrel{\triangle}{=} x_k - \tilde{X}_k^t \hat{\alpha},$$

and consider a coefficient update equation

$$w_{k+1} = w_k + \frac{\phi_k}{\|\phi_k\|^2} e(k), \tag{3.22}$$

where, as usual,

$$e(k) = d(k) - x_k^t w_k. \tag{3.23}$$

Equation (3.22) has a similar form as the update equation for the NLMS algorithm, x_k being replaced with the error vector ϕ_k. However, we have to note that x_k in (3.23) is not replaced with ϕ_k.

Because $\tilde{X}_k^t \hat{\alpha}$ is the orthogonal projection of x_k onto the linear subspace spanned by $x_{k-1}, x_{k-2}, \ldots, x_{k-(p-1)}$ (Appendix 1: "Affine Projection"), we have

$$\phi_k^t x_i = 0 \quad (i = k-1, k-2, \ldots, k-(p-1)). \tag{3.24}$$

Also,

$$\begin{aligned} \phi_k^t \phi_k &= \phi_k^t \left(x_k - \tilde{X}_k^t \hat{\alpha} \right) \\ &= \phi_k^t x_k - \phi_k^t \tilde{X}_k^t \hat{\alpha} \\ &= \phi_k^t x_k. \end{aligned} \tag{3.25}$$

Using (3.24) and (3.25), we can show that w_{k+1} is the affine projection of w_k onto $\Pi_k \cap \Pi_{k-1} \cap \cdots \cap \Pi_{k-(p-1)}$ provided $w_k \in \Pi_{k-1} \cap \Pi_{k-2} \cap \cdots \cap \Pi_{k-(p-1)}$. In fact, since

$$w_{k+1} - w_k = \frac{\phi_k}{\|\phi_k\|^2} e(k) \tag{3.26}$$

is a linear combination of $x_k, x_{k-1}, \ldots, x_{k-(p-1)}$, we have

$$(w_{k+1} - w_k) \perp \Pi_k \cap \Pi_{k-1} \cap \cdots \cap \Pi_{k-(p-1)}$$

by Theorem 3.12 (Appendix 1: "Affine Projection"). Furthermore,

$$
\begin{aligned}
x_k^t w_{k+1} &= x_k^t w_k + \frac{x_k^t \phi_k}{\|\phi_k\|^2} e(k) \\
&= x_k^t w_k + \frac{\phi_k^t \phi_k}{\|\phi_k\|^2} e(k) \\
&= x_k^t w_k + e(k) \\
&= d(k),
\end{aligned}
\tag{3.27}
$$

and for $i = k-1, k-2, \ldots, k-(p-1)$,

$$
\begin{aligned}
x_i^t w_{k+1} &= x_i^t w_k + \frac{x_i^t \phi_k}{\|\phi_k\|^2} e(k) \\
&= x_i^t w_k \\
&= d(i).
\end{aligned}
\tag{3.28}
$$

Equations (3.27) and (3.28) show

$$
w_{k+1} \in \Pi_k \cap \Pi_{k-1} \cap \cdots \cap \Pi_{k-(p-1)}.
$$

Thus, we see that (3.22) is equivalent to the update equation for the B-APA with $\mu = 1$.

The error vector ϕ_k is nearly orthogonal to other error vectors. This is most evident when the input signal $\{x(k)\}$ is a realization of an *autoregressive (AR) process*. An autoregressive process is a sequence of random variables $\{\boldsymbol{x}(k)\}$ defined by a stochastic equation

$$
\boldsymbol{x}(k) = \sum_{i=1}^{p-1} \alpha_i \boldsymbol{x}(k-i) + \boldsymbol{\xi}(k),
\tag{3.29}
$$

where $\{\boldsymbol{\xi}(k)\}$ is a white noise: for $-\infty < i, j < \infty$,

$$
\begin{aligned}
\mathrm{E}\{\boldsymbol{\xi}(i)\} &= 0, \\
\mathrm{E}\{\boldsymbol{\xi}(i)\boldsymbol{\xi}(j)\} &= \begin{cases} \sigma_\xi^2, & \text{if } i = j; \\ 0, & \text{otherwise.} \end{cases}
\end{aligned}
$$

If the input signal $\{x(k)\}$ is a realization of the AR process $\{\boldsymbol{x}(k)\}$, then the optimum linear predictor coefficients $\hat{\alpha}(1), \hat{\alpha}(2), \ldots, \hat{\alpha}(p-1)$ are estimates of the AR coefficients in (3.29), and the error signal $\{\phi(k)\}$ defined by

$$
\phi(k) \stackrel{\triangle}{=} x(k) - \sum_{i=1}^{p-1} \hat{\alpha}_i x(k-i)
$$

is the realization $\{\tilde{\xi}(k)\}$ of the estimate of the white noise $\{\xi(k)\}$. Therefore, we can expect that the error vector

$$\begin{aligned}\phi_k &= (\phi(k), \phi(k-1), \ldots, \phi(k-(n-1)))^t \\ &= (\tilde{\xi}(k), \tilde{\xi}(k-1), \ldots, \tilde{\xi}(k-(n-1)))^t\end{aligned}$$

has the property that for $i \neq j$,

$$\phi_i^t \phi_j = \sum_{l=0}^{n-1} \tilde{\xi}(i-l)\tilde{\xi}(j-l) \approx 0.$$

Even when the input signal is not a realization of an AR process, the error signal has a flatter spectrum and becomes closer to the white noise than the original input signal. Thus, we can view the B-APA ($\mu = 1$) as something close to the NLMS algorithm with the regressors replaced with modified regressors that are decorrelated and still have the same information to update the coefficient vector as the original regressors. The adaptation algorithm using (3.22) is referred to as the *decorrelation APA (D-APA)*.

An early paper [8] reported on an update equation that is essentially the same as (3.22). The aim of the paper is to extend the NLMS algorithm using a single vector that represents the information contained in multiple regressors $x_k, x_{k-1}, \ldots, x_{k-(p-1)}$. In the method, the regressors $x_k, x_{k-1}, \ldots, x_{k-(p-1)}$ are first orthogonalized, starting with $x_{k-(p-1)}$, by the Gram–Schmidt orthogonalization procedure (Appendix 1: "Affine Projection"):

$$\begin{aligned}
x'_{k-(p-1)} &\triangleq x_{k-(p-1)}, \\
z_{k-(p-1)} &\triangleq x'_{k-(p-1)}/\|x'_{k-(p-1)}\|, \\
x'_{k-(p-1)+1} &\triangleq x_{k-(p-1)+1} - \langle x_{k-(p-1)+1}, z_{k-(p-1)}\rangle z_{k-(p-1)}, \\
z_{k-(p-1)+1} &\triangleq x'_{k-(p-1)+1}/\|x'_{k-(p-1)+1}\|, \\
&\vdots \\
x'_k &\triangleq x_k - \sum_{j=k-1}^{k-(p-1)} \langle x_k, z_j\rangle z_j.
\end{aligned}$$

Then, the coefficient vector w_k is updated as

$$w_{k+1} = w_k + \frac{x'_k}{\|x'_k\|^2} e(k), \tag{3.30}$$

where $e(k) = d(k) - x_k^t w_k$ as usual. Because $\sum_{j=k-1}^{k-(p-1)} \langle x_k, z_j \rangle z_j$ is the orthogonal projection of x_k onto the subspace spanned by the regressors $x_{k-1}, x_{k-2}, \ldots, x_{k-(p-1)}$, Eq. (3.30) gives the same update equation as (3.22). The update equation (3.2) is a special case of (3.30) for $p = 2$.

From a present-day point of view, we can say that the paper presented a method of computing w_{k+1} in (3.10), which is only applicable to the case of $\mu = 1$. Although the paper does not mention the affine projection, it is a remarkable work that tried to generalize the NLMS algorithm nearly 10 years ahead of the appearance of the APA. This algorithm is referred to as the *Hinamoto-Maekawa algorithm (HMA)* in this book.

3.6 Formal Similarities Between APA and RLS

Let us assume that all the past regressors $x_k, x_{k-1}, \ldots, x_0$ and the desired signal values $d(k), d(k-1), \ldots, d(0)$ are available, and try to estimate the coefficient vector w by the least-squares method using these data. We first define an objective function $\mathcal{E}(w)$ by

$$\mathcal{E}(w) \triangleq \sum_{i=0}^{k} (x_i^t w - d(i))^2 + \delta \|w\|^2,$$

where the second term on the right-hand side is for regularization. In order to minimize $\mathcal{E}(w)$, we differentiate it with respect to w:

$$\frac{\partial \mathcal{E}(w)}{\partial w} = 2\left\{ \left(\sum_{i=0}^{k} x_i x_i^t \right) w - \sum_{i=0}^{k} x_i d(i) + \delta w \right\}.$$

Setting this expression equal to zero, we obtain the normal equation

$$\left(\sum_{i=0}^{k} x_i x_i^t + \delta I_n \right) w = \sum_{i=0}^{k} x_i d(i). \tag{3.31}$$

We employ the solution of this equation as the coefficient vector at $k+1$:

$$w_{k+1} \triangleq \left(\sum_{i=0}^{k} x_i x_i^t + \delta I_n \right)^{-1} \left(\sum_{i=0}^{k} x_i d(i) \right).$$

Now, let

$$P_k \triangleq \left(\sum_{i=0}^{k} x_i x_i^t + \delta I_n \right)^{-1}.$$

Then,

$$P_k^{-1} = \sum_{i=0}^{k-1} x_i x_i^t + \delta I_n + x_k x_k^t$$
$$= P_{k-1}^{-1} + x_k x_k^t.$$

Substitution of $A = P_{k-1}^{-1}$, $B = -x_k$, $C = x_k^t$, $D = 1$ into the matrix inversion lemma (A.4) (Appendix A) immediately leads to

$$P_k = (P_{k-1}^{-1} + x_k x_k^t)^{-1}$$
$$= P_{k-1} - \frac{P_{k-1} x_k x_k^t P_{k-1}}{1 + x_k^t P_{k-1} x_k}. \tag{3.32}$$

Note that $(1 + x_k^t P_{k-1} x_k)$ is a scalar. Thus, we obtain

$$w_{k+1} = \left(P_{k-1} - \frac{P_{k-1} x_k x_k^t P_{k-1}}{1 + x_k^t P_{k-1} x_k}\right)\left(\sum_{i=0}^{k-1} x_i d(i) + x_k d(k)\right)$$
$$= P_{k-1} \sum_{i=0}^{k-1} x_i d(i) - \frac{P_{k-1} x_k x_k^t P_{k-1}}{1 + x_k^t P_{k-1} x_k} \sum_{i=0}^{k-1} x_i d(i)$$
$$+ P_{k-1} x_k d(k) - \frac{P_{k-1} x_k x_k^t}{1 + x_k^t P_{k-1} x_k} P_{k-1} x_k d(k)$$
$$= w_k - \frac{P_{k-1}}{1 + x_k^t P_{k-1} x_k} x_k x_k^t w_k + P_{k-1} x_k \left(1 - \frac{x_k^t P_{k-1} x_k}{1 + x_k^t P_{k-1} x_k}\right) d(k)$$
$$= w_k + \frac{P_{k-1}}{1 + x_k^t P_{k-1} x_k} x_k (d(k) - x_k^t w_k). \tag{3.33}$$

If we compute P_k directly by a standard method like the Cholesky decomposition, then the complexity is $O(n^3)$ per sample. However, if we use the recursion (3.32), then it is reduced to $O(n^2)$. Equation (3.33), together with (3.32), is referred to as the *recursive least-squares (RLS) algorithm*.[2]

It is noted that if we let

$$\tilde{X}_k \triangleq \left[x_k \; x_{k-1} \; \cdots \; x_0\right]^t,$$

then

$$P_{k-1} = (\tilde{X}_{k-1}^t \tilde{X}_{k-1} + \delta I_n)^{-1}.$$

[2]There are some variants of RLS. The present type is called the "prewindowed RLS" algorithm [9].

Therefore, (3.33) is rewritten as

$$w_{k+1} = w_k + \tilde{\mu}(k)(\tilde{X}_{k-1}^t \tilde{X}_{k-1} + \delta I_n)^{-1} x_k e(k), \tag{3.34}$$

where $\tilde{\mu}(k) \triangleq (1 + x_k^t P_{k-1} x_k)^{-1}$, and $e(k) \triangleq d(k) - x_k^t w_k$.

Now, we turn to the update equation (3.17) for the R-APA. Let us assume $\mu \approx 1$ and $\delta \approx 0$. Then, the R-APA becomes close to the B-APA, so that

$$e_k = \begin{bmatrix} d(k) - x_k^t w_k \\ d(k-1) - x_{k-1}^t w_k \\ \vdots \\ d(k-(p-1)) - x_{k-(p-1)}^t w_k \end{bmatrix} \approx \begin{bmatrix} e(k) \\ 0 \\ \vdots \\ 0 \end{bmatrix}. \tag{3.35}$$

Therefore, $X_k^t e_k \approx x_k e(k)$. By using this approximate equation and (3.16), the update equation (3.17) for the R-APA is rewritten as

$$w_{k+1} = w_k + \mu(X_k^t X_k + \delta I_n)^{-1} x_k e(k). \tag{3.36}$$

Thus, we see formal similarities between the RLS algorithm (3.34) and the R-APA (3.36) [4, p. 255].

Appendix 1: Affine Projection

Orthogonal Projection

The *inner product* $\langle x, y \rangle$ of vectors $x, y \in \mathbb{R}^n$ is defined by $\langle x, y \rangle \triangleq x^t y$. The Euclidean *norm* $\|x\|$ of $x \in \mathbb{R}^n$ is defined by $\|x\| \triangleq \sqrt{\langle x, x \rangle}$. Vectors x and y are said to be *orthogonal* and denoted by $x \perp y$ iff[3] $\langle x, y \rangle = 0$. A vector x and a linear subspace $\mathbb{V}$ of $\mathbb{R}^n$ are said to be orthogonal and denoted by $x \perp \mathbb{V}$ iff $x \perp y$ for every element y in $\mathbb{V}$. Linear subspaces $\mathbb{V}$ and $\mathbb{W}$ are said to be orthogonal and denoted by $\mathbb{V} \perp \mathbb{W}$ iff any element of $\mathbb{V}$ and any element of $\mathbb{W}$ are orthogonal.

For a linear subspace $\mathbb{V}$ of $\mathbb{R}^n$, $\mathbb{V}^\perp \triangleq \{x \in \mathbb{R}^n \,;\, x \perp \mathbb{V}\}$ is called the *orthogonal complement* of $\mathbb{V}$. The set $\mathbb{V}^\perp$ is a linear subspace of $\mathbb{R}^n$. It is easy to verify that $V \perp V^\perp$ and that $V \cap V^\perp = \{0\}$.

Let $\mathbb{V}$ be a linear subspace of $\mathbb{R}^n$ and x an element of $\mathbb{R}^n$. An element $x_0 \in \mathbb{V}$ is called the *orthogonal projection* of x onto $\mathbb{V}$ iff $(x - x_0) \perp \mathbb{V}$. Such x_0 is uniquely determined for x. In fact, suppose that there are two orthogonal projections x_0 and x_0' of x. Then, for any $y \in \mathbb{V}$, $\langle x - x_0, y \rangle = 0$ and $\langle x - x_0', y \rangle = 0$. From these equations,

[3]This is the abbreviation of the phrase *if and only if*, which is very often used in mathematics.

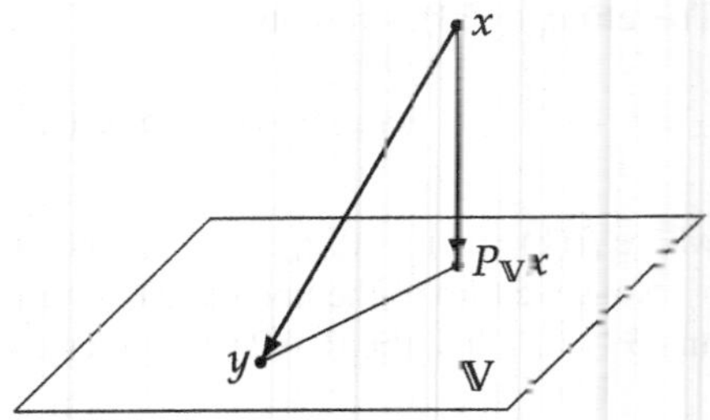

Fig. 3.6 Minimum distance property of orthogonal projection

we have $\langle x_0 - x_0', y\rangle = 0$. Let $y \triangleq x_0 - x_0' \in \mathbb{V}$. Then $\langle x_0 - x_0', x_0 - x_0'\rangle = 0$, which leads to $x_0 = x_0'$. The mapping that maps x to x_0 is also called the orthogonal projection, and denoted by $P_{\mathbb{V}}$. This is a linear mapping from $\mathbb{R}^n$ onto $\mathbb{V}$.

As illustrated in Fig. 3.6, the orthogonal projection has the minimum distance property. That is, if y is an element of $\mathbb{V}$, then $\|P_{\mathbb{V}}x - x\| \le \|y - x\|$ with equality iff $y = P_{\mathbb{V}}(x)$. This is a direct consequence of the Pythagorean theorem:

$$\begin{aligned}\|y - x\|^2 &= \|y - P_{\mathbb{V}}x\|^2 + \|P_{\mathbb{V}}x - x\|^2 \\ &\ge \|P_{\mathbb{V}}x - x\|^2.\end{aligned}$$

Any element $x \in \mathbb{R}^n$ is decomposed as

$$x = P_{\mathbb{V}}x + P_{\mathbb{V}^\perp}x. \tag{3.37}$$

In fact, let $y \triangleq x - P_{\mathbb{V}}x$. Then y is an element of $\mathbb{V}^\perp$. Since $x - y = P_{\mathbb{V}}x \in V$, $(x - y) \perp V^\perp$. Hence, $y = P_{\mathbb{V}^\perp}x$.

The decomposition (3.37) is unique in the sense that if $x = y_1 + y_2$, $y_1 \in \mathbb{V}$, $y_2 \in \mathbb{V}^\perp$, then $y_1 = P_{\mathbb{V}}x$ and $y_2 = P_{\mathbb{V}^\perp}x$. Therefore, as in Fig. 3.7, $\mathbb{R}^n$ is represented as the (orthogonal) *direct sum* of $\mathbb{V}$ and $\mathbb{V}^\perp$:

$$\mathbb{R}^n = \mathbb{V} \oplus \mathbb{V}^\perp.$$

Since $P_{\mathbb{V}}$ is a linear mapping from $\mathbb{R}^n$ to $\mathbb{R}^n$, it can be represented by a real matrix. The following theorem is well known [10, p. 149].

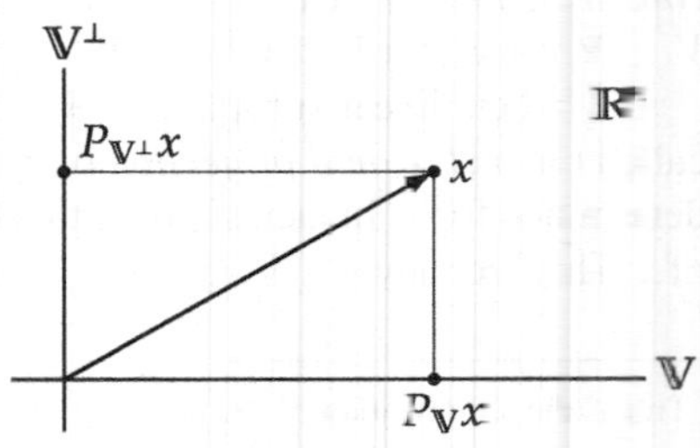

Fig. 3.7 Orthogonal decomposition of $\mathbb{R}^n$

Theorem 3.3 *A matrix A is an orthogonal projection iff $A = A^t$ (symmetric) and $A^2 = A$ (idempotent). When these conditions are met, $A = P_{\mathcal{R}(A)}$ and $I - A = P_{\mathcal{R}(A)^\perp}$, where $\mathcal{R}(A)$ is the range space of A, i.e., the linear subspace spanned by the columns of A.*

Theorem 3.4 *If A is an orthogonal projection, then its eigenvalues are 0 or 1.*

Proof If λ is an eigenvalue of A, and x the corresponding eigenvector (Appendix A), then $Ax = \lambda x$. The eigenvector x can be uniquely decomposed as $x = x_1 + x_2,\ x_1 \in \mathcal{R}(A),\ x_2 \in \mathcal{R}(A)^\perp$. Therefore, $Ax = x_1$ and $Ax = \lambda x_1 + \lambda x_2$, from which we have $(\lambda - 1)x_1 + \lambda x_2 = 0$. By the uniqueness of orthogonal decomposition, $(\lambda - 1)x_1 = 0$ and $\lambda x_2 = 0$. Thus, if $x_1 \neq 0$, then $\lambda = 1$ (and $x_2 = 0$). If $x_1 = 0$ then $x_2 \neq 0$ since $x \neq 0$, which implies $\lambda = 0$. □

Gram–Schmidt Orthogonalization Procedure

Theorem 3.5 *Let $\{x_1, x_2, \ldots, x_m\}$ $(m \leq n)$ be a set of linearly independent vectors in $\mathbb{R}^n$. From this set, construct a new set of vectors $\{x'_1, x'_2, \ldots, x'_m\}$ as follows: Define x'_1 by*

$$x'_1 \triangleq x_1/\|x_1\|, \tag{3.38}$$

and define x'_l for $l = 2, 3, \ldots, m$ recursively as

$$y_l \triangleq x_l - \sum_{k=1}^{l-1} \langle x_l,\ x'_k \rangle x'_k,$$

$$x'_l \triangleq y_l/\|y_l\|. \tag{3.39}$$

Then,

$$\langle x'_j,\ x'_k \rangle = 0 \quad (1 \leq j < k \leq m),$$

and the sets of vectors $\{x_1, x_2, \ldots, x_l\}$ and $\{x'_1, x'_2, \ldots, x'_l\}$ span the same linear subspace for any l $(1 \leq l \leq m)$.

Proof First note that each x'_k is normalized so that $\|x'_k\| = 1$. Let us show, by mathematical induction on l, that for $l = 2, 3, \ldots, m$, the vector x'_l and each of $x'_1, x'_2, \ldots, x'_{l-1}$ are orthogonal. For $l = 2$, this is true. In fact,

$$\begin{aligned}
\langle x'_2,\ x'_1 \rangle &= \langle (x_2 - \langle x_2,\ x'_1 \rangle x'_1)/\|y_2\|,\ x'_1 \rangle \\
&= (\langle x_2,\ x'_1 \rangle - \langle x_2, x'_1 \rangle \langle x'_1, x'_1 \rangle)/\|y_2\| \\
&= (\langle x_2,\ x'_1 \rangle - \langle x_2, x'_1 \rangle)/\|y_2\| \\
&= 0.
\end{aligned}$$

Next, assume that for $l \geq 3$,

x_2' and x_1' are orthogonal,
x_3' and each element of $\{x_1', x_2'\}$ are orthogonal,

$\vdots$

x_{l-1}' and each element of $\{x_1', x_2', \ldots, x_{l-2}'\}$ are orthogonal.

Then, for $1 \leq j \leq l-1$,

$$\begin{aligned}
\langle x_l', x_j' \rangle &= \left\langle \frac{1}{\|y_l\|}\left(x_l - \sum_{k=1}^{l-1} \langle x_l, x_k' \rangle x_k' \right), x_j' \right\rangle \\
&= \frac{1}{\|y_l\|}\left(\langle x_l, x_j' \rangle - \sum_{k=1}^{l-1} \langle x_l, x_k' \rangle \langle x_k', x_j' \rangle \right) \\
&= \frac{1}{\|y_l\|}(\langle x_l, x_j' \rangle - \langle x_l, x_j' \rangle) \\
&= 0.
\end{aligned}$$

This shows that x_l' and each element of $\{x_1', x_2', \ldots, x_{l-1}'\}$ are orthogonal.

From the recursion (3.38) and (3.39), it is obvious that each x_l' is a linear combination of $x_1, x_2, \ldots, x_l$. Also from the recursion,

$$\begin{aligned}
x_1 &= \|x_1\| x_1', \\
x_l &= \sum_{k=1}^{l-1} \langle x_l, x_k' \rangle x_k' + \|y_l\| x_l',
\end{aligned}$$

which shows that each x_l is a linear combination of $x_1', x_2', \ldots, x_l'$. Therefore, $\{x_1, x_2, \ldots, x_l\}$ and $\{x_1', x_2', \ldots, x_l'\}$ span the same linear subspace. □

The recursion in (3.38) and (3.39) is referred to as the *Gram–Schmidt orthogonalization procedure*. Note that $\sum_{k=1}^{l-1} \langle x_l, x_k' \rangle x_k'$ in (3.39) is the orthogonal projection of x_l onto the linear subspace spanned by $\{x_1', x_2', \ldots, x_{l-1}'\}$.

If normalization is not performed, the Gram–Schmidt orthogonalization procedure is simply written as

$$\begin{aligned}
x_1' &\triangleq x_1, \\
x_l' &\triangleq x_l - \sum_{k=1}^{l-1} (\langle x_l, x_k' \rangle / \|x_k'\|^2) x_k' \quad (l = 2, 3, \ldots, m).
\end{aligned} \tag{3.40}$$

Moore–Penrose Pseudoinverse

An $m \times n$ real matrix X can be considered as a mapping from $\mathbb{R}^n$ to $\mathbb{R}^m$. The *range space* of X, denoted by $\mathcal{R}(X)$, is defined by

$$\mathcal{R}(X) \stackrel{\triangle}{=} \{Xv;\ v \in \mathbb{R}^n\}, \tag{3.41}$$

which is a linear subspace of $\mathbb{R}^m$ spanned by the columns of X. The *null space* of X, denoted by $\mathcal{N}(X)$, is defined by

$$\mathcal{N}(X) \stackrel{\triangle}{=} \{v \in \mathbb{R}^n;\ Xv = 0\}, \tag{3.42}$$

which is also a linear subspace of $\mathbb{R}^n$.

Let $\mathcal{N}(X)^\perp$ be the orthogonal complement of $\mathcal{N}(X)$ in $\mathbb{R}^n$:

$$\mathcal{N}(X)^\perp \stackrel{\triangle}{=} \{v \in \mathbb{R}^n;\ \langle v,\ w\rangle = 0\ \ (w \in \mathcal{N}(X))\}.$$

Likewise, let $\mathcal{R}(X)^\perp$ be the orthogonal complement of $\mathcal{R}(X)$ in $\mathbb{R}^m$:

$$\mathcal{R}(X)^\perp \stackrel{\triangle}{=} \{v \in \mathbb{R}^m;\ \langle v,\ w\rangle = 0\ \ (w \in \mathcal{R}(X))\}.$$

Then, $\mathbb{R}^n$ and $\mathbb{R}^m$ are represented respectively as the direct sum

$$\begin{aligned}\mathbb{R}^n &= \mathcal{N}(X)^\perp \oplus \mathcal{N}(X),\\ \mathbb{R}^m &= \mathcal{R}(X) \oplus \mathcal{R}(X)^\perp.\end{aligned}$$

Denote by $X|_{\mathcal{N}(X)^\perp}$ the restriction of X, as a mapping, on $\mathcal{N}(X)^\perp$. $X|_{\mathcal{N}(X)^\perp}$ is a one-to-one mapping from $\mathcal{N}(X)^\perp$ onto $\mathcal{R}(X)$. In fact, if $X|_{\mathcal{N}(X)^\perp}v = X|_{\mathcal{N}(X)^\perp}w$ for $v, w \in \mathcal{N}(X)^\perp$, then $X|_{\mathcal{N}(X)^\perp}(v - w) = X(v - w) = 0$. Therefore, $v - w \in \mathcal{N}(X)^\perp \cap \mathcal{N}(X) = \{0\}$. Hence $v = w$, i.e., $X|_{\mathcal{N}(X)^\perp}$ is a one-to-one mapping. Furthermore, for $w \in \mathcal{R}(X)$, there exists $v \in \mathbb{R}^n$ such that $Xv = w$. The element v can be decomposed as $v = v_1 + v_2$, $v_1 \in \mathcal{N}(X)^\perp$, $v_2 \in \mathcal{N}(X)$. Since $Xv_2 = 0$,

$$X|_{\mathcal{N}(X)^\perp}v_1 = Xv_1 = Xv_1 + Xv_2 = X(v_1 + v_2) = Xv = w.$$

This shows that $X|_{\mathcal{N}(X)^\perp}$ is a mapping from $\mathcal{N}(X)^\perp$ onto $\mathcal{R}(X)$. Thus, we see that the mapping $X|_{\mathcal{N}(X)^\perp}$ has the inverse

$$(X|_{\mathcal{N}(X)^\perp})^{-1} :\ \mathcal{R}(X) \longrightarrow\ \mathcal{N}(X)^\perp.$$

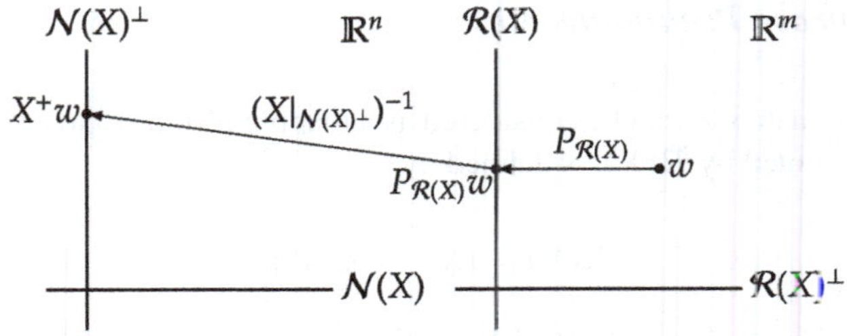

Fig. 3.8 Formation of the Moore–Penrose pseudoinverse X^+

Now, $w \in \mathbb{R}^m$ can be uniquely decomposed as $w = w_1 + w_2, \quad w_1 \in \mathcal{R}(X), \quad w_2 \in \mathcal{R}(X)^\perp$. Let $P_{\mathcal{R}(X)}$ be the orthogonal projection from $\mathbb{R}^m$ onto $\mathcal{R}(X)$: $P_{\mathcal{R}(X)} w \overset{\triangle}{=} w_1$. The composite mapping $X^+ \overset{\triangle}{=} (X|_{\mathcal{N}(X)^\perp})^{-1} P_{\mathcal{R}(X)}$ is called the *Moore–Penrose pseudoinverse* of X. This is a linear mapping from $\mathbb{R}^m$ to $\mathbb{R}^n$. The formation of X^+ is illustrated in Fig. 3.8.

Lemma 3.1 *(1)* $X^+X = P_{\mathcal{N}(X)^\perp}$.
(2) $XX^+ = P_{\mathcal{R}(X)}$.
(3) If X is nonsingular, then $X^+ = X^{-1}$.

Proof (1) Let us decompose $v \in \mathbb{R}^n$ as $v = v_1 + v_2$, $v_1 \in \mathcal{N}(X)^\perp$, $v_2 \in \mathcal{N}(X)$. Then,

$$\begin{aligned} X^+Xv &= X^+X(v_1 + v_2) \\ &= X^+(Xv_1) \\ &= (X|_{\mathcal{N}(X)^\perp})^{-1} P_{\mathcal{R}(X)}(Xv_1) \\ &= (X|_{\mathcal{N}(X)^\perp})^{-1} P_{\mathcal{R}(X)}(X|_{\mathcal{N}(X)^\perp} v_1) \\ &= (X|_{\mathcal{N}(X)^\perp})^{-1}(X|_{\mathcal{N}(X)^\perp} v_1) \\ &= v_1 \\ &= P_{\mathcal{N}(X)^\perp} v, \end{aligned}$$

which shows $X^+X = P_{\mathcal{N}(X)^\perp}$.
(2) Let us decompose $w \in \mathbb{R}^m$ as $w = w_1 + w_2$, $w_1 \in \mathcal{R}(X)$, $w_2 \in \mathcal{R}(X)^\perp$. Then,

$$\begin{aligned} XX^+w &= X(X|_{\mathcal{N}(X)^\perp})^{-1} P_{\mathcal{R}(X)}(w_1 + w_2) \\ &= X(X|_{\mathcal{N}(X)^\perp})^{-1} w_1 \\ &= X|_{\mathcal{N}(X)^\perp}(X|_{\mathcal{N}(X)^\perp})^{-1} w_1 \\ &= w_1 \\ &= P_{\mathcal{R}(X)} w, \end{aligned}$$

which shows $XX^+ = P_{\mathcal{R}(X)}$.

(3) If X is an $n \times n$ nonsingular matrix, then $\mathcal{N}(X)^{\perp} = \mathbb{R}^n$, and $\mathcal{R}(X) = \mathbb{R}^n$. Therefore, $P_{\mathcal{R}(X)}$ equals the identity mapping I_n on $\mathbb{R}^n$. Hence,

$$X^{+} = (X|_{\mathcal{N}(X)^{\perp}})^{-1} P_{\mathcal{R}(X)} = X^{-1} I_n = X^{-1}.$$

□

Lemma 3.2 *If* rank $X = m$, *then* $XX^{+} = I_m$.

Proof Because $\dim \mathcal{R}(X) = \text{rank } X = m$, we have $\mathcal{R}(X) = \mathbb{R}^m$. Thus, in view of Lemma 3.1(2),

$$XX^{+} = P_{\mathcal{R}(X)} = P_{\mathbb{R}^m} = I_m.$$

□

Lemma 3.3 $\mathcal{R}(X^t) = \mathcal{N}(X)^{\perp}$.

Proof Note that X^t is a mapping from $\mathbb{R}^m$ to $\mathbb{R}^n$. Because there exists a one-to-one linear mapping $X|_{\mathcal{N}(X)^{\perp}}$ from $\mathcal{N}(X)^{\perp}$ onto $\mathcal{R}(X)$,

$$\dim \mathcal{N}(X)^{\perp} = \dim \mathcal{R}(X) = \text{rank } X = \text{rank } X^t. \tag{3.43}$$

For arbitrary $w \in \mathbb{R}^m$ and $v \in \mathcal{N}(X)$,

$$\langle X^t w, v\rangle = \langle w, Xv\rangle = 0.$$

Therefore, $X^t w \in \mathcal{N}(X)^{\perp}$, which shows that

$$\mathcal{R}(X^t) \subseteq \mathcal{N}(X)^{\perp}. \tag{3.44}$$

On the other hand, by (3.43),

$$\dim \mathcal{R}(X^t) = \text{rank } X^t = \dim \mathcal{N}(X)^{\perp}. \tag{3.45}$$

From (3.44) and (3.45), we conclude that $\mathcal{R}(X^t) = \mathcal{N}(X)^{\perp}$. □

Theorem 3.6 *(1)* $(X^{+})^{+} = X$.
(2) $(X^t)^{+} = (X^{+})^t$.
(3) $X(X^t X)^{+} X^t X = X$.
(4) $X(X^t X)^{+} X^t = P_{\mathcal{R}(X)}$.

Proof (1) By definition of the Moore–Penrose pseudoinverse,

$$(X^{+})^{+} = (X^{+}|_{\mathcal{N}(X^{+})^{\perp}})^{-1} P_{\mathcal{R}(X^{+})}.$$

Since $\mathcal{R}(X^+) = \mathcal{N}(X)^\perp$ and $\mathcal{N}(X^+)^\perp = (\mathcal{R}(X)^\perp)^\perp = \mathcal{R}(X)$, we have

$$\begin{aligned}(X^+)^+ &= (X^+|_{\mathcal{R}(X)})^{-1} P_{\mathcal{N}(X)^\perp} \\ &= ((X|_{\mathcal{N}(X)^\perp})^{-1})^{-1} P_{\mathcal{N}(X)^\perp} \\ &= X|_{\mathcal{N}(X)^\perp} P_{\mathcal{N}(X)^\perp}.\end{aligned}$$

Let v be an arbitrary element in $\mathbb{R}^n$, and decompose it as $v = v_1 + v_2$, where $v_1 \in \mathcal{N}(X)^\perp$ and $v_2 \in \mathcal{N}(X)$. Then,

$$\begin{aligned}(X^+)^+ v &= X|_{\mathcal{N}(X)^\perp} P_{\mathcal{N}(X)^\perp} v \\ &= X|_{\mathcal{N}(X)^\perp} v_1 \\ &= X v_1.\end{aligned}$$

On the other hand,

$$Xv = X(v_1 + v_2) = X v_1.$$

Therefore, $(X^+)^+ v = Xv$ for any $v \in \mathbb{R}^n$, i.e., $(X^+)^+ = X$.

(2) The matrix $X|_{\mathcal{N}(X)^\perp}$ is a one-to-one mapping from $\mathcal{N}(X)^\perp$ onto $\mathcal{R}(X)$, and $X^t|_{\mathcal{N}(X^t)^\perp}$ is a one-to-one mapping from $\mathcal{N}(X^t)^\perp$ onto $\mathcal{R}(X^t)$. In view of Lemma 3.3, $\mathcal{N}(X)^\perp = \mathcal{R}(X^t)$ and $\mathcal{N}(X^t)^\perp = \mathcal{R}(X)$. That is, the domain of $X|_{\mathcal{N}(X)^\perp}$ coincides with the range of $X^t|_{\mathcal{N}(X^t)^\perp}$, and the domain of $X^t|_{\mathcal{N}(X^t)^\perp}$ coincides with the range of $X|_{\mathcal{N}(X)^\perp}$ as shown in the following diagram:

$$\begin{array}{ccc} \mathcal{N}(X)^\perp & \xrightarrow{X|_{\mathcal{N}(X)^\perp}} & \mathcal{R}(X) \\ \| & & \| \\ \mathcal{R}(X^t) & \xleftarrow[X^t|_{\mathcal{N}(X^t)^\perp}]{} & \mathcal{N}(X^t)^\perp \end{array}$$

Let us define

$$\mathcal{S} \triangleq \mathcal{N}(X)^\perp = \mathcal{R}(X^t),$$

and

$$\mathcal{T} \triangleq \mathcal{R}(X) = \mathcal{N}(X^t)^\perp.$$

Then, we have

$$(X|_{\mathcal{S}})^t = X^t|_{\mathcal{T}}. \tag{3.45}$$

In fact, let v and w be arbitrary elements in $\mathcal{S}$ and in $\mathcal{T}$, respectively. Then,

$$\begin{aligned}\langle (X|_{\mathcal{S}})^t w, v\rangle &= \langle w, X|_{\mathcal{S}} v\rangle \\ &= \langle w, Xv\rangle,\end{aligned}$$

and

$$\begin{aligned}\langle X^t|_{\mathcal{T}}w, v\rangle &= \langle X^t w, v\rangle \\ &= \langle w, Xv\rangle.\end{aligned}$$

Therefore, for arbitrary $v \in \mathcal{S}$ and $w \in \mathcal{T}$,

$$\langle (X|_{\mathcal{S}})^t w, v\rangle = \langle X^t|_{\mathcal{T}}w, v\rangle,$$

which is equivalent to (3.46).

We also have

$$((X|_{\mathcal{S}})^t)^{-1} = ((X|_{\mathcal{S}})^{-1})^t. \tag{3.47}$$

In fact, let v and w be arbitrary elements in $\mathcal{S}$ and $\mathcal{T}$, respectively, and let $u \triangleq ((X|_{\mathcal{S}})^t)^{-1}v$. Then,

$$\langle ((X|_{\mathcal{S}})^t)^{-1}v, w\rangle = \langle u, w\rangle, \tag{3.48}$$

and

$$\begin{aligned}\langle ((X|_{\mathcal{S}})^{-1})^t v, w\rangle &= \langle v, (X|_{\mathcal{S}})^{-1}w\rangle \\ &= \langle (X|_{\mathcal{S}})^t u, (X|_{\mathcal{S}})^{-1}w\rangle \\ &= \langle u, X|_{\mathcal{S}}(X|_{\mathcal{S}})^{-1}w\rangle \\ &= \langle u, w\rangle.\end{aligned} \tag{3.49}$$

Combining (3.48) and (3.49),

$$\langle ((X|_{\mathcal{S}})^t)^{-1}v, w\rangle = \langle ((X|_{\mathcal{S}})^{-1})^t v, w\rangle,$$

which is equivalent to (3.47).

Now, to prove $(X^t)^+ = (X^+)^t$, it suffices to show that for arbitrary $v \in \mathbb{R}^n$ and $w \in \mathbb{R}^m$,

$$\langle (X^t)^+ v, w\rangle = \langle (X^+)^t v, w\rangle.$$

This equation is equivalent to

$$\langle (X^t)^+ v, w\rangle = \langle v, X^+ w\rangle. \tag{3.50}$$

Let v be decomposed as $v = v_1 + v_2$, where $v_1 \in \mathcal{S}$ and $v_2 \in \mathcal{S}^\perp$. Then, using (3.46) and (3.47), we have

$$\begin{aligned}(X^t)^+ v &= (X^t|_{\mathcal{T}})^{-1}P_{\mathcal{S}}v \\ &= (X^t|_{\mathcal{T}})^{-1}v_1 \\ &= ((X|_{\mathcal{S}})^t)^{-1}v_1 \\ &= ((X|_{\mathcal{S}})^{-1})^t v_1.\end{aligned}$$

Therefore, if we decompose w as $w = w_1 + w_2$, where $w_1 \in \mathcal{T}$ and $w_2 \in \mathcal{T}^\perp$, we obtain (3.50) in the following way:

$$\begin{aligned}\langle (X^t)^+ v, w\rangle &= \langle ((X|_{\mathcal{S}})^{-1})^t v_1, w\rangle \\ &= \langle ((X|_{\mathcal{S}})^{-1})^t v_1, w_1\rangle \\ &= \langle v_1, (X|_{\mathcal{S}})^{-1} w_1\rangle \\ &= \langle v_1, X^+ w\rangle \\ &= \langle v, X^+ w\rangle.\end{aligned}$$

(3) Let $Y \triangleq X(X^tX)^+X^tX - X$. Since $((X^tX)^+)^t = (X^tX)^-$ by (2) above, we have

$$\begin{aligned}Y^tY &= (X(X^tX)^+X^tX - X)^t(X(X^tX)^+X^tX - X) \\ &= (X^tX(X^tX)^+X^t - X^t)(X(X^tX)^+X^tX - X) \\ &= X^tX(X^tX)^+X^tX(X^tX)^+X^tX \\ &\quad - X^tX(X^tX)^+X^tX - X^tX(X^tX)^+X^tX + X^tX.\end{aligned}$$

By Lemma 3.1(2), $X^tX(X^tX)^+ = P_{\mathcal{R}(X^tX)}$. Therefore,

$$\begin{aligned}X^tX(X^tX)^+X^tX &= P_{\mathcal{R}(X^tX)}X^tX \\ &= X^tX,\end{aligned}$$

and

$$\begin{aligned}X^tX(X^tX)^+X^tX(X^tX)^+X^tX &= P_{\mathcal{R}(X^tX)}P_{\mathcal{R}(X^tX)}X^tX \\ &= X^tX.\end{aligned}$$

Hence, $Y^tY = 0$, from which $Y = 0$ is concluded.
(4) By (3) above, $X(X^tX)^+X^tu = u$ for any $u \in \mathcal{R}(X)$. Moreover, $X(X^tX)^+X^t v = 0$ for any $v \in \mathcal{R}(X)^\perp$, since $X^tv = 0$. Therefore, $X(X^tX)^+X^t = P_{\mathcal{R}(X)}$. □

Theorem 3.7 *If* rank $X = m$, *then* $X^+ = X^t(XX^t)^{-1}$.

Proof Let x_k be the kth column of X^t. Then,

$$XX^t = \begin{bmatrix} \langle x_1, x_1\rangle & \langle x_1, x_2\rangle & \cdots & \langle x_1, x_m\rangle \\ \langle x_2, x_1\rangle & \langle x_2, x_2\rangle & \cdots & \langle x_2, x_m\rangle \\ \vdots & \vdots & \ddots & \vdots \\ \langle x_m, x_1\rangle & \langle x_m, x_2\rangle & \cdots & \langle x_m, x_m\rangle \end{bmatrix}.$$

If rank X = rank $X^t = m$, the vectors $x_1, x_2, \ldots, x_m$ are linearly independent. Therefore, by Theorem A.8 (Appendix A), $\det(XX^t) \neq 0$, so that the Gramian matrix XX^t has the inverse $(XX^t)^{-1}$. For arbitrary $w \in \mathbb{R}^m$, let

$$v \triangleq X^+ w. \tag{3.51}$$

Because $v \in \mathcal{N}(X)^\perp$, Lemma 3.3 guarantees the existence of $z \in \mathbb{R}^m$ such that

$$v = X^t z. \tag{3.52}$$

Combining (3.51) and (3.52), and using Lemma 3.2, we have

$$XX^t z = Xv = XX^+ w = w.$$

Since the Gramian matrix XX^t has the inverse $(XX^t)^{-1}$,

$$z = (XX^t)^{-1} w. \tag{3.53}$$

Substitution of (3.53) into (3.52) yields

$$v = X^t (XX^t)^{-1} w. \tag{3.54}$$

Comparison of (3.51) and (3.54) leads to

$$X^+ = X^t (XX^t)^{-1}.$$

□

Theorem 3.8 *Let X be an $m \times n$ matrix, and y an element of $\mathbb{R}^m$. The Moore–Penrose pseudoinverse gives a least-squares solution of the linear equation*

$$Xv = y. \tag{3.55}$$

That is, $v \triangleq X^+ y$ minimizes $\|Xv - y\|^2$. Moreover, $X^+ y - \mathcal{N}(X) \triangleq \{X^+ y + w\,;\ w \in \mathcal{N}(X)\}$ is the set of least-squares solutions of (3.55). *Therefore, $X^+ y$ gives the minimum norm least-squares solution of* (3.55).

Proof Because $\{Xv\,;\ v \in \mathbb{R}^n\} = \mathcal{R}(X)$, the quantity $\|Xv - y\|^2$ is minimized iff $Xv = P_{\mathcal{R}(X)} y$. This is attained for $v = X^+ y$, since $XX^+ = P_{\mathcal{R}(X)}$. If $v = X^+ y + w$, $w \in \mathcal{N}(X)$, then

$$\begin{aligned} Xv &= X(X^+ y + w) \\ &= XX^+ y + Xw \\ &= XX^+ y \\ &= P_{\mathcal{R}(X)} y. \end{aligned}$$

Therefore, any element in $X^+y + \mathcal{N}(X)$ is a least-squares solution of (3.55). Conversely, suppose $v = X^+y + w$, $w \in \mathbb{R}^n$, is a least-squares solution of (3.55). Decomposition of w as $w = w_1 + w_2$, $w_1 \in \mathcal{N}(X)^\perp$, $w_2 \in \mathcal{N}(X)$ leads to

$$\begin{aligned} Xv &= XX^+y + Xw_1 + Xw_2 \\ &= XX^+y + Xw_1. \end{aligned}$$

Because $Xv = P_{\mathcal{R}(X)}y = XX^+y$, $Xw_1 = 0$. Therefore, $w_1 \in \mathcal{N}(X)^\perp \cap \mathcal{N}(X) = \{0\}$. This shows that $v = X^+y + w_2 \in X^+y + \mathcal{N}(X)$.

Since $X^+y \perp \mathcal{N}(X)$, X^+y is the minimum norm element in $X^+y + \mathcal{N}(X)$. □

Suppose a linear equation has at least one solution. Then, a least-squares solution is a solution, and vice versa. Thus, we have the following corollary:

Corollary 3.1 *If* (3.55) *has a solution,* $X^+y + \mathcal{N}(X)$ *gives the set of solutions, and* X^+y *gives the minimum norm solution.*

Corollary 3.2 *For any* $m \times n$ *matrix* X,

$$X^+ = (X^tX)^+X^t \tag{3.56}$$

$$= X^t(XX^t)^+. \tag{3.57}$$

Proof Let us prove (3.56) first. By Theorem 3.8, $v = X^+y$ gives the minimum norm least-squares solution of (3.55). On the other hand, v minimized $\|Xv - y\|^2$ iff $(Xv - y) \perp X$, that is,

$$X^t(Xv - y) = X^tXv - X^ty = 0.$$

The minimum norm solution v' of this equation is given by $v' = (X^tX)^+X^ty$. Because the minimum norm least-squares solution is unique, $v = v'$. Thus, $X^+y = (X^tX)^+X^ty$ for any $y \in \mathbb{R}^m$, which shows (3.56).

Replace X with X^t in (3.56):

$$(X^t)^+ = (XX^t)^+X. \tag{3.58}$$

By Theorem 3.6(2), $((X^t)^+)^t = X^+$, and $((XX^t)^+)^t = (XX^t)^+$. Therefore, taking the transpose of both sides of (3.58), we immediately obtain (3.57). □

Affine Projection

A subset Π of $\mathbb{R}^n$ is called an *affine subspace* of $\mathbb{R}^n$ iff there exists an element $a \in \Pi$ such that

$$\Pi - a \triangleq \{x - a\,;\, x \in \Pi\}$$

is a linear subspace of $\mathbb{R}^n$. The element a is called the *origin* of Π, and $\Pi - a$ the linear subspace *associated* with Π. The dimension of Π is defined as $\dim \Pi \triangleq \dim(\Pi - a)$.

Theorem 3.9 *Let Π be an affine subspace of $\mathbb{R}^n$, and a its origin. Then, for any $b \in \Pi$,*

$$\Pi - a = \Pi - b.$$

That is, any element of Π can be chosen as its origin, and the linear subspace associated with Π is independent of the choice of the origin.

Proof If we denote $\mathbb{V} \triangleq \Pi - a$, Π is represented as $\Pi = \mathbb{V} + a$. Then, noting $b - a \in \mathbb{V}$, we have $\Pi = \mathbb{V} - (b - a) + b = \mathbb{V} + b$. This shows $\Pi - a = \mathbb{V} = \Pi - b$. □

Theorem 3.10 *If Π_1 and Π_2 are affine subspaces of $\mathbb{R}^n$ satisfying $\Pi_1 \cap \Pi_2 \neq \emptyset$, then $\Pi_1 \cap \Pi_2$ is an affine subspace of $\mathbb{R}^n$.*

Proof For arbitrarily chosen $a \in \Pi_1 \cap \Pi_2$, let

$$\mathbb{V}_1 \triangleq \Pi_1 - a, \quad \mathbb{V}_2 \triangleq \Pi_2 - a.$$

Then, it is shown that

$$\Pi_1 \cap \Pi_2 - a = \mathbb{V}_1 \cap \mathbb{V}_2. \tag{3.59}$$

In fact, if $v \in \Pi_1 \cap \Pi_2 - a$, there exists $w \in \Pi_1 \cap \Pi_2$ such that $v = w - a$. Because $w \in \Pi_1$, we have $v \in \mathbb{V}_1$. Also, because $w \in \Pi_2$, we have $v \in \mathbb{V}_2$. Therefore, $v \in \mathbb{V}_1 \cap \mathbb{V}_2$. Conversely, if $v \in \mathbb{V}_1 \cap \mathbb{V}_2$, then $v \in \mathbb{V}_1$. Therefore, there exists $w_1 \in \Pi_1$ such that

$$v = w_1 - a. \tag{3.60}$$

Also, because $v \in \mathbb{V}_2$, there exists $w_2 \in \Pi_2$ such that

$$v = w_2 - a. \tag{3.61}$$

From (3.60) and (3.61), $w_1 = w_2$. Let $w \triangleq w_1 = w_2$. Then, $w \in \Pi_1 \cap \Pi_2$ and $v = w - a$. Therefore, $v \in \Pi_1 \cap \Pi_2 - a$.

Because $\mathbb{V}_1 \cap \mathbb{V}_2$ is a linear subspace of $\mathbb{R}^n$, (3.59) shows that $\Pi_1 \cap \Pi_2$ is an affine subspace. □

We can immediately generalize the above theorem.

Corollary 3.3 *If $\Pi_1, \Pi_2, \ldots, \Pi_p$ are affine subspaces of $\mathbb{R}^n$ satisfying $\Pi_1 \cap \Pi_2 \cap \cdots \cap \Pi_p \neq \emptyset$, then $\Pi_1 \cap \Pi_2 \cap \cdots \cap \Pi_p$ is an affine subspace of $\mathbb{R}^n$.*

Let Π be an affine subspace of $\mathbb{R}^n$, and $\mathbb{V}$ the associated linear subspace. An element $v \in \mathbb{R}^n$ and Π are said to be *orthogonal* and denoted by $v \perp \Pi$ iff $v \perp \mathbb{V}$.

An element $v' \in \Pi$ is called the *affine projection* of $v \in \mathbb{R}^n$ onto Π iff $(v - v') \perp \Pi$. The affine projection $v' \in \Pi$ is uniquely determined for v. In fact, suppose v' and v'' are both affine projections of v. Then, by definition, $v' - v \in \mathbb{V}^\perp$. Also, $v'' - v \in \mathbb{V}^\perp$. Since $\mathbb{V}^\perp$ is a linear subspace of $\mathbb{R}^n$, $v' - v'' \in \mathbb{V}^\perp$. On the other hand, there exists $w' \in \mathbb{V}$ such that $v' = w' + a$, where a is the origin of $\mathbb{V}$. In the same way, there exists $w'' \in \mathbb{V}$ such that $v'' = w'' + a$. Hence, $v' - v'' = w' - w'' \in \mathbb{V}$. Thus, we have $v' - v'' \in \mathbb{V} \cap \mathbb{V}^\perp = \{0\}$, from which $v' = v''$ is concluded.

The mapping that maps $v \in \mathbb{R}^n$ to its affine projection $v' \in \Pi$ is also called the affine projection onto Π, and denoted by P_Π. Just as in the case of orthogonal projection onto a linear subspace, $P_\Pi v$ is characterized as the unique element $v' \in \Pi$ that minimizes $\|v' - v\|$.

Now, given $x_1, x_2, \ldots, x_m \in \mathbb{R}^n$ and $y_1, y_2, \ldots, y_m \in \mathbb{R}$ $(m \leq n)$, let us consider a system of linear equations

$$\langle x_k, v \rangle = y_k \quad (k = 1, 2, \ldots, m), \tag{3.62}$$

where $v \in \mathbb{R}^n$ is the unknown vector. Using the matrix and vector notation

$$X \triangleq [x_1\ x_2\ \cdots\ x_m]^t,$$
$$y \triangleq (y_1, y_2, \ldots, y_m)^t,$$

we can rewrite (3.62) as

$$Xv = y. \tag{3.63}$$

Note that X is an $m \times n$ matrix, and y an m-dimensional vector.

By Theorem 3.8, $\Pi \triangleq X^+ y + \mathcal{N}(X)$ gives the set of least-squares solutions of (3.63). If (3.63) has at least one solution, then, by Corollary 3.1, Π gives the set of solutions. Note that Π is an affine subspace of $\mathbb{R}^n$, with the origin $X^+ y$, and the associated linear subspace $\mathcal{N}(X)$.

Theorem 3.11 rank X + dim $\Pi = n$.

Proof As stated in "Moore-Penrose Pseudoinverse" above, there is a one-to-one linear mapping from $\mathcal{N}(X)^\perp$ onto $\mathcal{R}(X)$. Therefore, $\dim \mathcal{R}(X) = \dim \mathcal{N}(X)^\perp$, so that

$$\begin{aligned} \operatorname{rank} X + \dim \Pi &= \dim \mathcal{R}(X) + \dim \mathcal{N}(X) \\ &= \dim \mathcal{N}(X)^\perp + \dim \mathcal{N}(X) \\ &= \dim \mathbb{R}^n \\ &= n. \end{aligned}$$

□

If X is composed of a single, nonzero row, then rank $X = 1$. Therefore, by Theorem 3.11, dim $\Pi = n - 1$. Such an affine subspace is called a *hyperplane*.

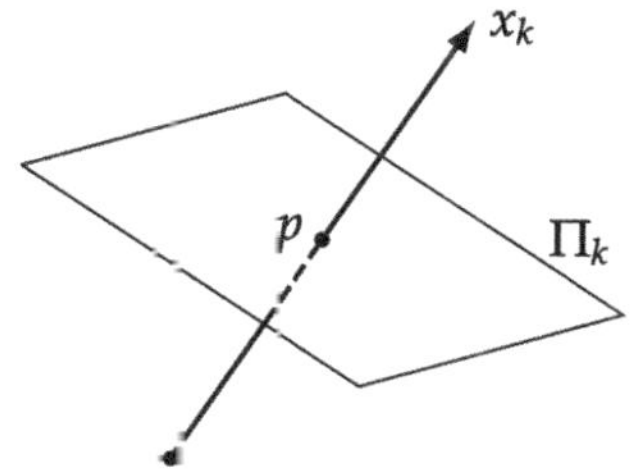

Fig. 3.9 Hyperplane determined by the equation $\langle x_k, v\rangle = y_k$. $p = y_k x_k/\|x_k\|^2$

If $x_k \neq 0$, each equation $\langle x_k, v\rangle = y_k$ in (3.62) determines a hyperplane. Let it be denoted by $\Pi_k = \{v \in \mathbb{R}^n ; \langle x_k, v\rangle = y_k\}$. As illustrated in Fig. 3.9, Π_k is a hyperplane that is orthogonal to x_k and passes through the point $p = y_k x_k/\|x_k\|^2$. If (3.63) has a solution, the set of solutions Π is the intersection of such hyperplanes: $\Pi = \Pi_1 \cap \Pi_2 \cap \cdots \cap \Pi_m$.

The angle between two hyperplanes Π_i and Π_j is defined to be the angle between x_i and x_j.

Theorem 3.12 *A vector $v \in \mathbb{R}^n$ and $\Pi = X^+y + \mathcal{N}(X)$ are orthogonal iff v is a linear combination of $x_1, x_2, \ldots, x_m$.*

Proof Let $\mathbb{V}$ be the linear subspace spanned by $\{x_1, x_2, \ldots, x_m\}$. Then, by Lemma 3.3, $\mathbb{V} = \mathcal{R}(X^t) = \mathcal{N}(X)^\perp$, from which the theorem is obvious. □

Theorem 3.13 *For $\Pi = X^+y + \mathcal{N}(X)$, the affine projection P_Π is given by*

$$P_\Pi v = v + X^+(y - Xv) \quad (v \in \mathbb{R}^n). \tag{3.64}$$

Proof Let $w \triangleq v + X^+(y - Xv)$. We first show that w is an element of Π. Note that w can be rewritten as $w = X^+y + (v - X^+Xv)$. By Lemma 3.1, $XX^+ = P_{\mathcal{R}(X)}$. Since $Xv \in \mathcal{R}(X)$,

$$\begin{aligned} X(v - X^+Xv) &= Xv - XX^+Xv \\ &= Xv - P_{\mathcal{R}(X)}Xv \\ &= Xv - Xv \\ &= 0. \end{aligned}$$

Therefore, we have $v - X^+Xv \in \mathcal{N}(X)$. Hence, $w = X^+y + (v - X^+Xv) \in X^+y + \mathcal{N}(X) = \Pi$. Furthermore, because $(w - v) = X^+(y - Xv) \in \mathcal{R}(X^+) = \mathcal{N}(X)^\perp$, we have $(w - v) \perp \mathcal{N}(X)$. Hence, by definition, $(w - v) \perp \Pi$. Thus we have shown $w = P_\Pi v$. □

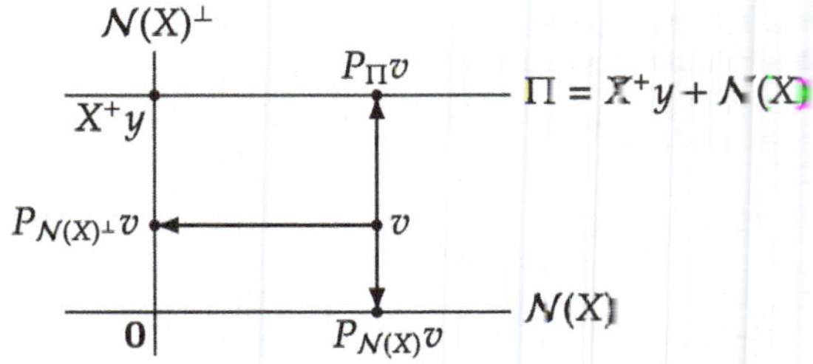

Fig. 3.10 Geometrical meaning of Theorem 3.13

Figure 3.10 illustrates the geometrical meaning of Theorem 3.13. In the figure, we see two interpretations of $P_{\Pi}v$:

$$P_{\Pi}v = v + (X^{+}y - P_{\mathcal{N}(X)^{\perp}}v) \quad \text{and} \quad P_{\Pi}v = X^{+}y + P_{\mathcal{N}(X)}v.$$

Appendix 2: Condition Number

Natural Norm of a Matrix

The *natural norm* $\|A\|$ of an $n \times n$ matrix A is defined by

$$\|A\| \triangleq \sup_{x \neq 0} \frac{\|Ax\|}{\|x\|}.$$

The natural norm is also called the induced norm, induced from the vector norm used in the definition.

Lemma 3.4 *The natural norm has the following properties:*

(1) $\|A\| \geq 0$, *and* $\|A\| = 0$ *implies* $A = 0$.
(2) For a scalar c, $\|cA\| = |c|\,\|A\|$.
(3) $\|A + B\| \leq \|A\| + \|B\|$.
(4) $\|AB\| \leq \|A\| \cdot \|B\|$.
(5) $\|I\| = 1$.

Proof (1) Since $\|Ax\|/\|x\| \geq 0$, $\|A\| \geq 0$ is obvious. If $A \neq 0$, there exists $\tilde{x} \neq 0$ such that $A\tilde{x} \neq 0$. Since $\|A\tilde{x}\|/\|\tilde{x}\| > 0$, $\|A\| > 0$.
(2) Since $\|cAx\| = |c|\,\|Ax\|$,

$$\begin{aligned}\|cA\| &= \sup_{x\neq 0} \|cAx\|/\|x\| \\ &= |c| \sup_{x\neq 0} \|Ax\|/\|x\| \\ &= |c|\,\|A\|.\end{aligned}$$

(3) By the triangle inequality for the vector norm, we have

$$\begin{aligned}\|A+B\| &= \sup_{x\neq 0} \frac{\|(A+B)x\|}{\|x\|} \\ &\leq \sup_{x\neq 0} \frac{\|Ax\|+\|Bx\|}{\|x\|} \\ &\leq \sup_{x\neq 0} \frac{\|Ax\|}{\|x\|} + \sup_{x\neq 0} \frac{\|Bx\|}{\|x\|} \\ &= \|A\| + \|B\|.\end{aligned}$$

(4) First note that for any vector y,

$$\|Ay\| \leq \|A\| \cdot \|y\|. \tag{3.65}$$

In fact, if $y = 0$, this is obvious. If $y \neq 0$, then

$$\|Ay\|/\|y\| \leq \sup_{x\neq 0} \|Ax\|/\|x\| = \|A\|,$$

which leads to (3.65). Using this inequality, we have

$$\begin{aligned}\|AB\| &= \sup_{x\neq 0} \frac{\|ABx\|}{\|x\|} \\ &\leq \sup_{x\neq 0} \frac{\|A\| \cdot \|Bx\|}{\|x\|} \\ &= \|A\| \sup_{x\neq 0} \frac{\|Bx\|}{\|x\|} \\ &= \|A\| \cdot \|B\|.\end{aligned}$$

(5) is obvious from the definition of the natural norm. □

Lemma 3.5 *If A is an $n \times n$ Hermitian matrix with eigenvalues $\lambda_1, \lambda_2, \ldots, \lambda_n$ (Appendix A), then*

$$\|A\| = \max_k |\lambda_k|.$$

Proof Let $\{x_1, x_2, \ldots, x_n\}$ be the orthonormal basis of $\mathbb{C}^n$ comprised of eigenvectors corresponding to the eigenvalues $\lambda_1, \lambda_2, \ldots, \lambda_n$. Then, any vector $x \in \mathbb{C}^n$ is represented as $x = \sum_{k=1}^{n} c_k x_k$ with some complex coefficients c_k's. Using this representation, we have

$$\begin{aligned} \|Ax\|^2 &= \left\| \sum_{k=1}^{n} c_k \lambda_k x_k \right\|^2 \\ &= \sum_{k=1}^{n} |\lambda_k|^2 |c_k|^2 \\ &\le \max_k |\lambda_k|^2 \sum_{i=1}^{n} |c_i|^2. \end{aligned}$$

We also have

$$\begin{aligned} \|x\|^2 &= \left\| \sum_{k=1}^{n} c_k x_k \right\|^2 \\ &= \sum_{i=1}^{n} |c_i|^2. \end{aligned}$$

Thus, if $x \neq 0$,

$$\frac{\|Ax\|}{\|x\|} \le \max_k |\lambda_k|. \tag{3.66}$$

On the other hand, let $m \triangleq \text{argmax}_k |\lambda_k|$. Then,

$$\frac{\|Ax_m\|}{\|x_m\|} = \frac{|\lambda_m| \cdot \|x_m\|}{\|x_m\|} = |\lambda_m| = \max_k |\lambda_k|. \tag{3.67}$$

From (3.66) and (3.67),

$$\|A\| = \sup_{x \neq 0} \frac{\|Ax\|}{\|x\|} = \max_k |\lambda_k|.$$

□

Condition Number

The *condition number* $\kappa(A)$ of a nonsingular matrix A is defined by

$$\kappa(A) \triangleq \|A\| \cdot \|A^{-1}\|.$$

By (4) and (5) of Lemma 3.4, it is easily verified that $\kappa(A) \geq 1$.
If A is a Hermitian matrix with eigenvalues $\lambda_1, \lambda_2, \ldots, \lambda_n$, then

$$\kappa(A) = \frac{\max_k |\lambda_k|}{\min_k |\lambda_k|}. \tag{3.68}$$

This is immediately proved by Lemma 3.5 if we note that the eigenvalues of A^{-1} are $1/\lambda_1, 1/\lambda_2, \ldots, 1/\lambda_n$.

Theorem 3.14 *Let us consider a linear equation*

$$Ax = b, \tag{3.69}$$

where A is an $n \times n$ nonsingular matrix, and $b \in \mathbb{C}^n$.

(1) If the solution x changes to $x + \Delta x$ when b changes to $b + \Delta b$, the following inequality holds:

$$\frac{\|\Delta x\|}{\|x\|} \leq \kappa(A) \frac{\|\Delta b\|}{\|b\|}.$$

(2) Let ΔA be an $n \times n$ matrix satisfying $\|A^{-1}\Delta A\| < 1$. If the solution x changes to $x + \Delta x$ when A changes to $A + \Delta A$, the following inequality holds:

$$\frac{\|\Delta x\|}{\|x\|} \leq \frac{\kappa(A)}{1 - \|A^{-1}\Delta A\|} \frac{\|\Delta A\|}{\|A\|}.$$

Proof (1) From (3.69) and $A(x + \Delta x) = b + \Delta b$, we have $A\Delta x = \Delta b$. Therefore,

$$\|\Delta x\| = \|A^{-1}\Delta b\| \leq \|A^{-1}\| \cdot \|\Delta b\|.$$

On the other hand, $\|b\| = \|Ax\| \leq \|A\| \cdot \|x\|$. Thus,

$$\begin{aligned} \frac{\|\Delta x\|}{\|x\|} &\leq \|A^{-1}\| \frac{\|\Delta b\|}{\|x\|} \\ &\leq \|A^{-1}\| \cdot \|A\| \frac{\|\Delta b\|}{\|b\|} \\ &= \kappa(A) \frac{\|\Delta b\|}{\|b\|}. \end{aligned}$$

(2) Since $(A + \Delta A)(x + \Delta x) = b = Ax$, we have

$$(A + \Delta A)\Delta x = -(\Delta A)x.$$

From the assumption that $\|A^{-1}\Delta A\| < 1$, we have for an arbitrary nonzero vector y,

$$\|A^{-1}(\Delta A)y\| < \|y\|.$$

Therefore,

$$\begin{aligned}\|(I + A^{-1}\Delta A)y\| &\geq \|Iy\| - \|A^{-1}(\Delta A)y\| \\ &> \|y\| - \|y\| \\ &= 0.\end{aligned}$$

This shows $(I + A^{-1}\Delta A)y \neq 0$, so that $I + A^{-1}\Delta A$ is nonsingular. Hence, $A - \Delta A = A(I + A^{-1}\Delta A)$ is also nonsingular. Now, let $B \triangleq A^{-1}\Delta A$ and $C \triangleq (I - B)^{-1}$. Then, since $I = C + BC$,

$$\begin{aligned}\|C\|(1 - \|B\|) &= \|C\| - \|C\| \cdot \|B\| \\ &\leq \|C\| - \|BC\| \\ &\leq \|C + BC\| \\ &= \|I\| \\ &= 1.\end{aligned}$$

From these facts,

$$\begin{aligned}\frac{\|\Delta x\|}{\|x\|} &= \frac{\| - (A + \Delta A)^{-1}(\Delta A)x\|}{\|x\|} \\ &= \frac{\| - CA^{-1}(\Delta A)x\|}{\|x\|} \\ &\leq \|C\| \cdot \|A^{-1}\| \cdot \|\Delta A\| \\ &= \|A\| \cdot \|A^{-1}\| \cdot \|C\| \frac{\|\Delta A\|}{\|A\|} \\ &\leq \frac{\kappa(A)}{1 - \|B\|} \frac{\|\Delta A\|}{\|A\|} \\ &= \frac{\kappa(A)}{1 - \|A^{-1}\Delta A\|} \frac{\|\Delta A\|}{\|A\|}.\end{aligned}$$

□

Theorem 3.14 shows that if $\kappa(A)$ is close to 1, then the relative change in the solution x is comparable to the relative change in A and b. However, if $\kappa(A)$ is large, this is not guaranteed. In fact, if $\kappa(A)$ is large, there are cases where a small relative change in A or b causes quite a large relative change in the solution x. Because of this, if the condition number $\kappa(A)$ is large, the linear equation (3.69) is said to be *ill-conditioned*.

Appendix 3: The Method of Lagrange Multipliers

The *method of Lagrange multipliers* is a method of solving constrained optimization problems. Let $f(x), g_0(x), g_1(x), \ldots, g_{m-1}(x)$ be real differentiable functions defined on a domain D in $\mathbb{R}^n$, and let us consider the problem of minimizing $f(x)$ subject to constraints $g_k(x) = 0 \ \ (k = 0, 1, \ldots, m-1)$. From these functions we construct the *Lagrangian function* as

$$L(x, \lambda_0, \lambda_1, \ldots, \lambda_{m-1}) \triangleq f(x) + \sum_{k=0}^{m-1} \lambda_k g_k(x),$$

where $\lambda_k \in \mathbb{R} \ \ (k = 0, 1, \ldots, m-1)$ are variables called the *Lagrange multipliers*. We denote the *Jacobian matrix* of the constraining functions at x by $J(x)$:

$$J(x) \triangleq \left[\frac{\partial g_i}{\partial x_j}(x) \right].$$

A point $x^* \in D$ is called a *regular point* iff it satisfies the constraints and $\operatorname{rank} J(x^*) = m$. The latter condition is equivalent to the statement that the vectors $\nabla_x g_k(x^*) \ (k = 0, 1, \ldots, m-1)$ are linearly independent.

The following theorem is well known in the field of nonlinear programming [11, p. 300].

Theorem 3.15 *Let x^* be a regular point. If x^* minimizes $f(x)$, then there exist $\lambda_k^* \ (k = 0, 1, \ldots, m-1)$ such that*

$$\frac{\partial L}{\partial x}(x^*, \lambda_0^*, \lambda_1^*, \ldots, \lambda_{m-1}^*) = 0. \tag{3.70}$$

Solving (3.70) with the use of the constraining equations, we can find x^* that is a candidate for $\operatorname{argmin}_x f(x)$ under the constraints.

Appendix 4: Newton's Method

A twice continuously differentiable function f defined on $\mathbb{R}^n$ can be approximated by a quadratic function on a neighborhood of each point $x^o \in \mathbb{R}^n$ as

$$f(x) \approx f(x^o) + (\nabla_x f(x^o))^t (x - x^o) + \frac{1}{2}(x - x^o)^t \nabla_x^2 f(x^o)(x - x^o).$$

If the Hessian matrix $\nabla_x^2 f(x^o)$ is positive definite, the right-hand side of this equation is minimized by solving the equation

$$\nabla_x \left\{ f(x^o) + (\nabla_x f(x^o))^t (x - x^o) + \frac{1}{2}(x - x^o)^t \nabla_x^2 f(x^o)(x - x^o) \right\} = 0.$$

By using the formula (A.17) (Appendix A), this equation can be rewritten as

$$\nabla_x f(x^o) + \nabla_x^2 f(x^o)(x - x^o) = 0,$$

from which the minimum point x^m is obtained as

$$x^m = x^o - (\nabla_x^2 f(x^o))^{-1} \nabla_x f(x^o).$$

Motivated by this fact, *Newton's method* searches for $\text{argmin}_x f(x)$ by iterating the following computation starting from an initial point x_0:

$$x_{k+1} = x_k - \mu(\delta I + \nabla_x^2 f(x_k))^{-1} \nabla_x f(x_k), \tag{3.71}$$

where $\delta > 0$, and I is the $n \times n$ identity matrix. The *regularization* term δI is added so that the algorithm does not fail even if $\nabla_x^2 f(x_k)$ is not invertible. Just as in the case of the steepest descent algorithm, an index dependent step-size $\mu(k)$ may be used. Under a certain condition, the vector sequence $x_0, x_1, x_2, \ldots$ converges to a local minimum point. If the initial point x_0 and the step-size μ are appropriately chosen, the sequence converges to $\text{argmin}_x f(x)$.

Appendix 5: Linear Prediction

Let $\{x(k)\}$ be a real signal. We make a linear combination $\hat{x}(j) \triangleq \sum_{i=1}^m f_i x(j-i)$ of $x(j-1), x(j-2), \ldots, x(j-m)$, where $f_1, f_2, \ldots, f_m$ are real coefficients. Let us determine the coefficients so that the function

$$L_f(f_1, f_2, \ldots, f_m) \triangleq \sum_{j=k}^{k-(n-1)} (x(j) + \hat{x}(j))^2$$

is minimized. That is, we want to determine the coefficients so that $-\hat{x}(j)$ becomes the least-mean-squares estimate of $x(j)$ on the interval $j = k, k-1, \ldots, k-(n-1)$ [12]. In order to minimize L_f, we differentiate it with respect to each f_l, and set it equal to zero. Because

$$\frac{\partial L_f}{\partial f_l}(f_1, f_2, \ldots, f_m) = 2 \left(\sum_{j=k}^{k-(n-1)} x(j)x(j-l) + \sum_{i=1}^{m} f_i \sum_{j=k}^{k-(n-1)} x(j-i)x(j-l) \right),$$

we obtain a linear equation

$$\begin{bmatrix} r_{1,1} & r_{1,2} & \cdots & r_{1,m} \\ r_{2,1} & r_{2,2} & \cdots & r_{2,m} \\ \vdots & \vdots & \ddots & \vdots \\ r_{m,1} & r_{m,2} & \cdots & r_{m,m} \end{bmatrix} \begin{bmatrix} f_1 \\ f_2 \\ \vdots \\ f_m \end{bmatrix} + \begin{bmatrix} r_{0,1} \\ r_{0,2} \\ \vdots \\ r_{0,m} \end{bmatrix} = 0, \tag{3.72}$$

where

$$r_{s,t} \triangleq \sum_{j=k}^{k-(n-1)} x(j-s)x(j-t).$$

Note that $r_{s,t}$ is symmetric: $r_{s,t} = r_{t,s}$. Equation (3.72) is called the *normal equation*. We assume that the matrix

$$R_l \triangleq \begin{bmatrix} r_{1,1} & r_{1,2} & \cdots & r_{1,m} \\ r_{2,1} & r_{2,2} & \cdots & r_{2,m} \\ \vdots & \vdots & \ddots & \vdots \\ r_{m,1} & r_{m,2} & \cdots & r_{m,m} \end{bmatrix}$$

is nonsingular. The solution of (3.72) is called the *forward linear predictor*. We denote it by the same symbol as the unknown variable:

$$(f_1, f_2, \ldots, f_m)^t = -R_l^{-1}(r_{0,1}, r_{0,2}, \ldots, r_{0,m})^t. \tag{3.73}$$

Each f_i is referred to as a *forward linear predictor coefficient* or simply *linear predictor coefficient*.[4]

Now let us calculate the forward *residual power* E_f defined by

$$E_f \triangleq \min L_f(f_1, f_2, \ldots, f_m).$$

The function $L_f(f_1, f_2, \ldots, f_m)$ is expressed as

$$\begin{aligned} L_f(f_1, f_2, \ldots, f_m) &= \sum_{j=k}^{k-(n-1)} \left(x(j) + \sum_{i=1}^{m} f_i x(j-i) \right)^2 \\ &= r_{0,0} + 2(r_{0,1}, r_{0,2}, \ldots, r_{0,m})(f_1, f_2, \ldots, f_m)^t \\ &\quad + (f_1, f_2, \ldots, f_m) R_{1m} (f_1, f_2, \ldots, f_m)^t. \end{aligned} \tag{3.74}$$

[4]We sometimes use a formulation that the sum of $(x(j) - \hat{x}(j))^2$ is to be minimized. In this formulation the sign of f_i is reversed.

$$
\begin{array}{ccccc}
f_m & f_{m-1} & \cdots & f_1 & \\
\times & \times & \cdots & \times & \\
x(j-m) & x(j-m+1) & \cdots & x(j-1) & x(j) \\
 & \times & \cdots & \times & \times \\
 & b_1 & \cdots & b_{m-1} & b_m
\end{array}
$$

Fig. 3.11 Forward prediction and backward prediction. In the forward prediction, $x(j)$ is predicted by using the past data. In the backward prediction on the other hand, $x(j-m)$ is predicted, or estimated, by using the future data

Substitution of (3.73) into (3.74) yields

$$E_f = r_{0,0} + (r_{0,1}, r_{0,2}, \ldots, r_{0,m})(f_1, f_2, \ldots, f_m)^t.$$

Thus, we see that the *forward residual filter*

$$f \triangleq (f_0, f_1, f_2, \ldots, f_m)^t \quad (f_0 = 1)$$

satisfies the *augmented normal equation*

$$Rf = (E_f, 0, \ldots, 0)^t, \tag{3.75}$$

where

$$R \triangleq \begin{bmatrix} r_{0,0} & r_{0,1} & \cdots & r_{0,m} \\ r_{1,0} & r_{1,1} & \cdots & r_{1,m} \\ \vdots & \vdots & \ddots & \vdots \\ r_{m,0} & r_{m,1} & \cdots & r_{m,m} \end{bmatrix}. \tag{3.76}$$

We can also consider backward prediction as in Fig. 3.11. For a linear combination $\tilde{x}(j-m) \triangleq \sum_{i=1}^{m} b_i x(j-m+i)$ of $x(j-m+1), x(j-m+2), \ldots, x(j)$, we determine the coefficients $b_1, b_2, \ldots, b_m$ so that the function

$$L_b(b_1, b_2, \ldots, b_m) \triangleq \sum_{j=k}^{k-(n-1)} (x(j-m) + \tilde{x}(j-m))^2$$

is minimized. Differentiating L_b with respect to each b_l, and setting it equal to zero, we obtain the normal equation

$$\begin{bmatrix} r_{0,0} & r_{0,1} & \cdots & r_{0,m-1} \\ r_{1,0} & r_{1,1} & \cdots & r_{1,m-1} \\ \vdots & \vdots & \ddots & \vdots \\ r_{m-1,0} & r_{m-1,1} & \cdots & r_{m-1,m-1} \end{bmatrix} \begin{bmatrix} b_m \\ b_{m-1} \\ \vdots \\ b_1 \end{bmatrix} + \begin{bmatrix} r_{0,m} \\ r_{1,m} \\ \vdots \\ r_{m-1,m} \end{bmatrix} = 0. \tag{3.77}$$

The solution of (3.77), denoted by the same symbol $(b_m, b_{m-1}, \ldots, b_1)^t$ as the unknown variable, is called the *backward linear predictor*. Just as in the case of the forward residual power, the backward residual power

$$E_b \triangleq \min L_b(b_1, b_2, \ldots, b_m)$$

is calculated as

$$E_b = r_{m,m} + (r_{m,0}, r_{m,1}, \ldots, r_{m,m-1})(b_m, b_{m-1}, \ldots, b_1)^t.$$

Therefore, the *backward residual filter*

$$b \triangleq (b_m, b_{m-1}, \ldots, b_0)^t \quad (b_0 = 1)$$

satisfies the augmented normal equation in the backward case

$$Rb = (0, \ldots, 0, E_b)^t. \tag{3.78}$$

Theorem 3.16 *Assume that the matrix R in* (3.76) *is nonsingular. Then,*

$$R^{-1} = \begin{bmatrix} 0 & \mathbf{0}^t \\ \mathbf{0} & R_l^{-1} \end{bmatrix} + \frac{ff^t}{E_f}, \tag{3.79}$$

where **0** *is the m-dimensional zero vector.*

Similarly,

$$R^{-1} = \begin{bmatrix} R_u^{-1} & \mathbf{0} \\ \mathbf{0}^t & 0 \end{bmatrix} + \frac{bb^t}{E_b}, \tag{3.80}$$

where R_u is the $m \times m$ sub-matrix of R consisting of the upper left corner of R:

$$R_u \triangleq \begin{bmatrix} r_{0,0} & r_{0,1} & \cdots & r_{0,m-1} \\ r_{1,0} & r_{1,1} & \cdots & r_{1,m-1} \\ \vdots & \vdots & \ddots & \vdots \\ r_{m-1,0} & r_{m-1,1} & \cdots & r_{m-1,m-1} \end{bmatrix}.$$

Proof First note R_l and R_u are nonsingular by the assumption that R is nonsingular. Let us show (3.79). Defining $r \triangleq (r_{1,0}, r_{2,0}, \ldots, r_{m,0})^t$, we can express R as

$$R = \begin{bmatrix} r_{0,0} & r^t \\ r & R_l \end{bmatrix}.$$

By the formula (A.1) (Appendix A) and (3.75),

$$\begin{aligned}
R\left(\begin{bmatrix}0 & \mathbf{0}^t\\ \mathbf{0} & R_l^{-1}\end{bmatrix}+\frac{ff^t}{E_f}\right) &= \begin{bmatrix}r_{0,0} & r^t\\ r & R_l\end{bmatrix}\begin{bmatrix}0 & \mathbf{0}^t\\ \mathbf{0} & R_l^{-1}\end{bmatrix}+R\frac{ff^t}{E_f}\\
&= \begin{bmatrix}r_{0,0}\cdot 0+r^t\cdot\mathbf{0} & r_{0,0}\cdot\mathbf{0}^t+r^tR_l^{-1}\\ r\cdot 0+R_l\cdot\mathbf{0} & r\cdot\mathbf{0}^t+R_lR_l^{-1}\end{bmatrix}+\frac{1}{E_f}Rff^t\\
&= \begin{bmatrix}0 & r^tR_l^{-1}\\ \mathbf{0} & I\end{bmatrix}+\frac{1}{E_f}(E_f,0,\ldots,0)^t f^t\\
&= \begin{bmatrix}0 & r^tR_l^{-1}\\ \mathbf{0} & I\end{bmatrix}+(1,0,\ldots,0)^t f^t\\
&= \left[\begin{array}{c|cccc}0 & -f_1 & -f_2 & \cdots & -f_m\\ \hline 0 & & & & \\ \vdots & & & I & \\ 0 & & & & \end{array}\right]+\begin{bmatrix}1 & f_1 & f_2 & \cdots & f_m\\ 0 & 0 & 0 & \cdots & 0\\ \vdots & \vdots & \vdots & \ddots & \vdots\\ 0 & 0 & 0 & \cdots & 0\end{bmatrix}\\
&= \begin{bmatrix}1 & \mathbf{0}^t\\ \mathbf{0} & I\end{bmatrix}\\
&= I.
\end{aligned}$$

This shows that

$$\begin{bmatrix}0 & \mathbf{0}^t\\ \mathbf{0} & R_l^{-1}\end{bmatrix}+\frac{ff^t}{E_f}=R^{-1}.$$

Equation (3.80) can be shown in a similar way. □

References

1. Ozeki, K., Umeda, T.: An adaptive filtering algorithm using an orthogonal projection to an affine subspace and its properties. IEICE Trans. **J67-A**(2), 126–132 (1984) (Also in Electron. Commun. Jpn. **67-A**(5), 19–27 (1984))
2. Haykin, S.: Adaptive Filter Theory. Prentice-Hall, Upper Saddle River (2002)
3. Sayed, A.H.: Adaptive Filters. Wiley, Hoboken (2008)
4. Haykin, S., Widrow, B. (eds.): Least-Mean-Square Adaptive Filters. Wiley, Hoboken (2003)
5. Werner, S., Diniz, P.S.R.: Set-membership affine projection algorithm. IEEE Signal Process. Lett. **8**(8), 231–235 (2001)
6. Morgan, D.R., Kratzer, S.G.: On a class of computationally efficient, rapidly converging, generalized NLMS algorithms. IEEE Signal Process. Lett. **3**(8), 245–247 (1996)
7. Rupp, M.: A family of adaptive filter algorithms with decorrelating properties. IEEE Trans. Signal Process. **46**(3), 771–775 (1998)
8. Hinamoto, T., Maekawa, S.: Extended theory of learning identification. J. IEEJ-C **95**(10), 227–234 (1975)

9. Cioffi, J.M., Kailath, T.: Windowed fast transversal filters adaptive algorithms with normalization. IEEE Trans. Acoust. Speech Signal Process. **ASSP–33**(3), 607–625 (1985)
10. Satake, I.: Linear Algebra. Marcel Dekker, New York (1975)
11. Luenberger, D.G.: Linear and Nonlinear Programming. Addison-Wesley, Menlo Park (1989)
12. Markel, J.D., Gray, A.H.: Linear Prediction of Speech. Springer, Berlin (1976)

Chapter 4
Family of Affine Projection Algorithms

4.1 Variants of Basic Affine Projection Algorithm

After the publication of the paper on the basic affine projection algorithm (B-APA) [1], several adaptive filtering algorithms that also exploit multiple regressors were put forward. Most of them can be considered variants of B-APA, though they were developed independently from different perspectives [2]. We have already discussed the R-APA and the D-APA. Let us give a brief review of other variant algorithms, which now constitute a family of affine projection algorithms.

4.2 Partial-Rank Algorithm (PRA)

While the coefficient vector is updated sample-by-sample in the B-APA as well as in the R-APA, it is possible to modify the algorithm so that the coefficient vector is updated only once every r samples. By denoting the coefficient vector by $\tilde{w}_k$ in this scheme, the update equation is expressed as

$$\tilde{w}_{k+1} = \tilde{w}_{k+1-r} + \mu X_k^t (X_k X_k^t + \delta I)^{-1} \tilde{e}_k, \tag{4.1}$$

where

$$\begin{aligned} X_k &= \left[x_k \; x_{k-1} \; \cdots \; x_{k-(p-1)}\right]^t, \\ x_k &= (x(k), x(k-1), \ldots, x(k-(n-1)))^t, \\ \tilde{e}_k &= d_k - X_k \tilde{w}_{k+1-r}, \\ d_k &= (d(k), d(k-1), \ldots, d(k-(p-1)))^t. \end{aligned}$$

It is assumed that the coefficient vector is kept constant for a period of r samples: $\tilde{w}_{k+1-r} = \tilde{w}_{k+1-r+1} = \cdots = \tilde{w}_k$. For this reason, we also have $\tilde{e}_k = d_k - X_k \tilde{w}_k$.

K. Ozeki, *Theory of Affine Projection Algorithms for Adaptive Filtering*, Mathematics for Industry 22, DOI 10.1007/978-4-431-55738-8_4

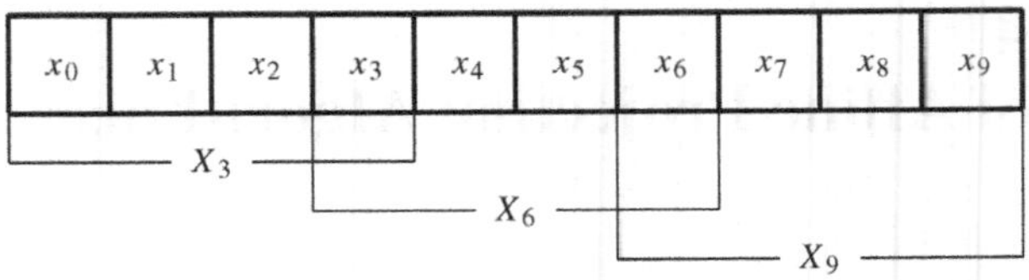

Fig. 4.1 Example of the use of regressors for $p = 4$ and $r = 3$

The new coefficient vector $\tilde{w}_{k+1}$ is then updated by using the regressor block X_{k+r} to obtain $\tilde{w}_{k+1+r}$. When $r = 1$, the update equation (4.1) coincides with (3.12) for the R-APA. Reducing the frequency of updating coefficient vectors in this way does not affect the basic properties of the APA, because $\tilde{w}_{k+1}$ for $\mu = 1$ is still the (approximate) affine projection of $\tilde{w}_{k+1-r}$ onto $\Pi_k \cap \Pi_{k-1} \cap \cdots \cap \Pi_{k-(p-1)}$. It is obvious that the computational complexity for updating the coefficient vector is reduced by a factor of $1/r$ per sample compared with that of the R-APA.

Figure 4.1 exemplifies the use of regressors for $p = 4$ and $r = 3$. If $r < p$, there is an overlap of regressors in adjacent regressor blocks X_k and X_{k+r}. If $r > p$, some regressors are skipped and not used. The algorithm with $r = p$ is called the *partial-rank algorithm (PRA)*. The PRA was originally developed within the context of spatial processing for adaptive beamforming [2, 3]. This algorithm was rediscovered a couple of times. The *block orthogonal projection algorithm (BOPA)* [4] and the *general optimum block adaptive (GOBA) algorithm* [5] are such algorithms. The same algorithm was also presented in [6].[1]

4.3 NLMS Algorithm with Orthogonal Correction Factors (NLMS-OCF)

In the B-APA, we can take "sparse" regressors $x_k, x_{k-D}, x_{k-2D}, \ldots, x_{k-(p-1)D}$ for an integer $D \geq 1$ instead of $x_k, x_{k-1}, x_{k-2}, \ldots, x_{k-(p-1)}$. Let us assume, for simplicity, that the new regressors are linearly independent. Applying the Gram–Schmidt orthogonalization procedure without normalization (3.40) (Chap. 3, Appendix 1: "Affine Projection") starting from x_k, we obtain a sequence of orthogonal vectors

$$\tilde{x}_k, \tilde{x}_{k-D}, \tilde{x}_{k-2D}, \ldots, \tilde{x}_{k-(p-1)D},$$

[1] In this paper, a sample-by-sample algorithm was also developed, which is the same as the APA.

which satisfies

(1) $\tilde{x}_{k-iD}$ is a linear combination of $x_k, x_{k-D}, \ldots, x_{k-iD}$ $(0 \le i \le p-1)$, (4.2)

(2) $x_{k-jD} \perp \tilde{x}_{k-iD}$ $(0 \le j < i)$ for $1 \le i \le p-1$, (4.3)

(3) $x^t_{k-iD}\tilde{x}_{k-iD} = \|\tilde{x}_{k-iD}\|^2$ $(0 \le i \le p-1)$. (4.4)

Using this vector sequence, let us define a vector

$$\tilde{\Delta} w_k \stackrel{\triangle}{=} \sum_{i=0}^{p-1} \tilde{\mu}(k-iD)\tilde{x}_{k-iD}, \tag{4.5}$$

where

$$\tilde{\mu}(k-iD) \stackrel{\triangle}{=} \frac{\tilde{e}(k-iD)}{\|\tilde{x}_{k-iD}\|^2},$$
$$\tilde{e}(k-iD) \stackrel{\triangle}{=} d(k-iD) - x^t_{k-iD}v^i_{k+1},$$
$$v^i_{k+1} \stackrel{\triangle}{=} w_k + \sum_{j=0}^{i-1} \tilde{\mu}(k-jD)\tilde{x}_{k-jD}.$$

Then, $w_k + \tilde{\Delta} w_k$ is the affine projection of w_k onto $\Pi_k \cap \Pi_{k-D} \cap \Pi_{k-2D} \cap \cdots \cap \Pi_{k-(p-1)D}$. In fact, since $\tilde{\Delta} w_k$ is a linear combination of x_k, x_{k-D}, x_{k-2D}, $\ldots$, $x_{k-(p-1)D}$ due to (4.2), we have $\tilde{\Delta} w_k \perp \Pi_k \cap \Pi_{k-D} \cap \Pi_{k-2D} \cap \cdots \cap \Pi_{k-(p-1)D}$ by Theorem 3.12. Furthermore, by (4.3) and (4.4),

$$\begin{aligned}
x^t_{k-jD}(w_k + \tilde{\Delta} w_k) &= x^t_{k-jD}w_k + \sum_{i=0}^{p-1} \tilde{\mu}(k-iD)x^t_{k-jD}\tilde{x}_{k-iD} \\
&= x^t_{k-jD}\left(w_k + \sum_{i=0}^{j-1} \tilde{\mu}(k-iD)\tilde{x}_{k-iD}\right) + \tilde{\mu}(k-jD)x^t_{k-jD}\tilde{x}_{k-jD} \\
&= x^t_{k-jD}v^j_{k+1} + \tilde{\mu}(k-jD)\|\tilde{x}_{k-jD}\|^2 \\
&= d(k-jD) - \tilde{e}(k-jD) + \tilde{e}(k-jD) \\
&= d(k-jD) \quad (0 \le j \le p-1),
\end{aligned}$$

which shows $w_k + \tilde{\Delta} w_k \in \Pi_{k-jD}$ for $0 \le j \le p-1$, that is, $w_k + \tilde{\Delta} w_k \in \Pi_k \cap \Pi_{k-D} \cap \Pi_{k-2D} \cap \cdots \cap \Pi_{k-(p-1)D}$. Thus, introducing the step-size parameter μ, we have a coefficient update equation

$$w_{k+1} = w_k + \mu\tilde{\Delta} w_k \tag{4.6}$$

that is equivalent to the B-APA with a modified regressor block

$$\tilde{X}_k \triangleq \left[x_k \; x_{k-D} \; x_{k-2D} \; \cdots \; x_{k-(p-1)D}\right]^t .$$

When $D = 1$, (4.6) coincides with (3.10). The adaptation algorithm using (4.6) is referred to as the *NLMS algorithm with orthogonal correction factors (NLMS-OCF)* [7].

4.4 Binormalized Data-Reusing LMS Algorithm (BNDR-LMS)

In a conference paper [8], an algorithm named the *binormalized data-reusing LMS algorithm (BNDR-LMS)* was presented. In this algorithm, the coefficient vector is updated as

$$w_{k+1} = w_k + \mu(Ax_k + Bx_{k-1}), \tag{4.7}$$

where μ is the step-size parameter, and

$$A = \frac{(d(k) - x_k^t w_k)\|x_{k-1}\|^2 - (d(k-1) - x_{k-1}^t w_k)x_k^t x_{k-1}}{\|x_k\|^2\|x_{k-1}\|^2 - (x_k^t x_{k-1})^2},$$

$$B = \frac{(d(k-1) - x_{k-1}^t w_k)\|x_k\|^2 - (d(k) - x_k^t w_k)x_k^t x_{k-1}}{\|x_k\|^2\|x_{k-1}\|^2 - (x_k^t x_{k-1})^2}.$$

Equation (4.7) is a special case of (3.10) for the B-APA [9]. In fact, when $p = 2$, the update equation (3.10) is written as

$$w_{k+1} = w_k + \mu \Delta w_k, \tag{4.8}$$

where

$$\begin{aligned} \Delta w_k &= X_k^t (X_k X_k^t)^{-1} e_k, \\ X_k &= \left[x_k \; x_{k-1}\right]^t, \\ e_k &= (d(k) - x_k^t w_k, d(k-1) - x_{k-1}^t w_k)^t. \end{aligned}$$

Since

$$X_k X_k^t = \begin{bmatrix} \|x_k\|^2 & x_k^t x_{k-1} \\ x_{k-1}^t x_k & \|x_{k-1}\|^2 \end{bmatrix} = \begin{bmatrix} \|x_k\|^2 & x_k^t x_{k-1} \\ x_k^t x_{k-1} & \|x_{k-1}\|^2 \end{bmatrix}$$

we have

$$(X_k X_k^t)^{-1} = \frac{1}{\det(X_k X_k^t)} \begin{bmatrix} \|x_{k-1}\|^2 & -x_k^t x_{k-1} \\ -x_k^t x_{k-1} & \|x_k\|^2 \end{bmatrix},$$
$$\det(X_k X_k^t) = \|x_k\|^2 \|x_{k-1}\|^2 - (x_k^t x_{k-1})^2.$$

Therefore,

$$\begin{aligned} \Delta w_k &= \frac{[x_k \ x_{k-1}]}{\det(X_k X_k^t)} \begin{bmatrix} \|x_{k-1}\|^2 & -x_k^t x_{k-1} \\ -x_k^t x_{k-1} & \|x_k\|^2 \end{bmatrix} \begin{bmatrix} d(k) - x_k^t w_k \\ d(k-1) - x_{k-1}^t w_k \end{bmatrix} \\ &= \frac{[x_k \ x_{k-1}]}{\det(X_k X_k^t)} \begin{bmatrix} (d(k) - x_k^t w_k)\|x_{k-1}\|^2 - (d(k-1) - x_{k-1}^t w_k) x_k^t x_{k-1} \\ -(d(k) - x_k^t w_k) x_k^t x_{k-1} + (d(k-1) - x_{k-1}^t w_k)\|x_k\|^2 \end{bmatrix} \\ &= A x_k + B x_{k-1}, \end{aligned}$$

which shows the equivalence of (4.7) and (4.8).

4.5 Unified Formulation

In order to describe the variants of the B-APA in a unified manner, let us consider the following model: First we take up a sequence of sparse regressors $x_k, x_{k-D}, x_{k-2D}, \ldots, x_{k-(p-1)D}$, and make a regressor block as

$$\tilde{X}_k \triangleq [x_k \ x_{k-D} \ x_{k-2D} \ \ldots \ x_{k-(p-1)D}]^t .$$

Based on this regressor block, we define a coefficient update equation as

$$\tilde{w}_{k+1} = \tilde{w}_{k+1-r} + \mu \tilde{X}_k^t (\tilde{X}_k \tilde{X}_k^t + \delta I)^{-1} \tilde{e}_k, \tag{4.9}$$

where

$$\tilde{e}_k \triangleq \begin{bmatrix} d(k) - x_k^t \tilde{w}_{k+1-r} \\ d(k-D) - x_{k-D}^t \tilde{w}_{k+1-r} \\ d(k-2D) - x_{k-2D}^t \tilde{w}_{k+1-r} \\ \vdots \\ d(k-(p-1)D) - x_{k-(p-1)D}^t \tilde{w}_{k+1-r} \end{bmatrix}$$

The coefficient vector $\tilde{w}_k$ is updated once every r samples, and kept constant for a period of r samples: $w_{k+1-r} = w_{k+1-r+1} = \cdots = w_k$.

Table 4.1 Family of affine projection algorithms

Algorithm	p	μ	δ	r	D
NLMS	$p = 1$	$\mu > 0$	$\delta = 0$	$r = 1$	$D = 1$
BNDR-LMS	$p = 2$	$\mu > 0$	$\delta = 0$	$r = 1$	$D = 1$
HMA	$n > p \geq 1$	$\mu = 1$	$\delta = 0$	$r = 1$	$D = 1$
D-APA	$n > p \geq 1$	$\mu = 1$	$\delta = 0$	$r = 1$	$D = 1$
B-APA	$n > p \geq 1$	$\mu > 0$	$\delta = 0$	$r = 1$	$D =$ [illegible]
R-APA	$n > p \geq 1$	$\mu > 0$	$\delta > 0$	$r = 1$	$D =$ [illegible]
PRA	$n > p \geq 1$	$\mu > 0$	$\delta > 0$	$r = p$	$D =$ [illegible]
NLMS-OCF	$n > p \geq 1$	$\mu > 0$	$\delta = 0$	$r = 1$	$D \geq$ [illegible]

NLMS The normalized LMS algorithm, *BNDR-LMS* The binormalized data-reusing LMS algorithm, *HMA* The Hinamoto-Maekawa algorithm, *D-APA* The decorrelation APA, *B-APA* The basic APA, *R-APA* The regularized APA, *PRA* The partial-rank algorithm, *NLMS-OCF* The normalized LMS algorithm with orthogonal correction factors

Equation (4.9) encompasses a wide range of update equations in the APA family. By setting the values of the parameters appropriately, various algorithms are obtained including the ones shown in Table 4.1 [10], which are all considered to be variants of the B-APA.

References

1. Ozeki, K., Umeda, T.: An adaptive filtering algorithm using an orthogonal projection to an affine subspace and its properties. IEICE Trans. **J67-A**(2), 126–132 (1984) (Also in Electron. Commun. Jpn. **67-A**(5), 19–27 (1984))
2. Morgan, D.R., Kratzer, S.G.: On a class of computationally efficient, rapidly converging generalized NLMS algorithms. IEEE Signal Process. Lett. **3**(8), 245–247 (1996)
3. Kratzer, S.G., Morgan, D.R.: The partial-rank algorithm for adaptive beamforming. In: Proceedings of SPIE 0564, Real-Time Signal Processing VIII, vol. 564, pp. 9–14. San Diego, CA 20 Aug 1985
4. Furukawa, T., Kubota, H., Tsujii, S.: Orthogonal projection algorithm for block adaptive signal processing and its some properties. Trans. IEICE **J71-A**(12), 2138–2146 (1988)
5. Wang, T., Wang, C.-L.: On the optimum design of the block adaptive FIR digital filter. IEEE Trans. Signal Process. **41**(6), 2131–2140 (1993)
6. Montazeri, M., Duhamel, P.: A set of algorithms linking NLMS and block RLS algorithms. IEEE Trans. Signal Process. **43**(2), 444–453 (1995)
7. Sankaran, S.G., Beex, A.A.: Normalized LMS algorithm with orthogonal correction factors. In: Proceedins of Thirty-First Asilomar Conference Signals, System, Computer, pp. 1670–1673. Pacific Grove, CA, 2–5 Nov 1997
8. Apolinário, J.A.Jr., de Campos, M.L.R., Diniz, P.S.R.: The binormalized data-reusing LMS algorithm. In: Procedings of XV Simpósio Brasileiro de Telecomunicações, pp. 77–80. Recife, 8–11 Sept 1997
9. de Campos, M.L.R., Diniz, P.S.R., Apolinário Jr, J.A..: On normalized data-reusing and affine-projections algorithms. In: Proceedings of ICECS'99, pp. 843–846. Pafos, 5–8 Sep 1999
10. Shin, H.-C., Sayed, A.H.: Mean-square performance of a family of affine projection algorithms. IEEE Trans. Signal Process. **52**(1), 90–102 (2004)

Chapter 5
Convergence Behavior of APA

5.1 Measures of Convergence Behavior

Needless to say, the behavior of an adaptive filter depends on the input signal $\{x(k)\}$. However, since there are infinitely many variations of signals, it is difficult to draw general, useful conclusions on the behavior of an adaptive filter taking the waveform of each signal into consideration, especially when the observation noise is not negligible. Therefore, we resort to a statistical approach, where the input signal $\{x(k)\}$ is regarded as a realization of a stochastic process $\{\boldsymbol{x}(k)\}$. Accordingly, a regressor block X_k is considered as a realization of a stochastic regressor block $\boldsymbol{X}_k$. Other signals and the coefficient vector sequence $\{w_k\}$ are also treated as realizations of corresponding stochastic processes as shown in Fig. 5.1, which is a stochastic version of Fig. 2.1. In the analysis, we are concerned with the behavior of the error signal $\boldsymbol{e}(k)$ and the coefficient vector $\boldsymbol{w}_k$. Even in this stochastic framework, rigorous analysis in a general case is intractable because of complicated correlations among variables. Therefore, we make the following assumptions throughout this chapter:

(A-1) The input signal $\{\boldsymbol{x}(k)\}$ is stationary up to necessary order of moments with $\mathrm{E}\{\boldsymbol{x}(k)\} = 0$ and $\mathrm{E}\left\{\boldsymbol{x}_k \boldsymbol{x}_k^t\right\} = R_x$.
(A-2) The observation noise $\{\boldsymbol{v}(k)\}$ is independent and identically distributed (i.i.d.) with $\mathrm{E}\{\boldsymbol{v}(k)\} = 0$ and $\mathrm{E}\left\{\boldsymbol{v}^2(k)\right\} = \sigma_v^2$.
(A-3) The input signal $\{\boldsymbol{x}(k)\}$ and the observation noise $\{\boldsymbol{v}(k)\}$ are statistically independent.
(A-4) Statistical dependence of $\boldsymbol{w}_k$ on $\{\boldsymbol{v}(k), \boldsymbol{v}(k-1), \boldsymbol{v}(k-2), \ldots\}$ is negligible.
(A-5) Statistical dependence of $\boldsymbol{w}_k$ on $\boldsymbol{X}_k$ is negligible.

The assumptions (A-1)–(A-3) are moderate ones, which may actually occur in the real world, whereas (A-4) and (A-5) are strong assumptions.

We start with defining measures of convergence. The first measure we take up here is the *mean-square error (MSE)* defined by

$$\mathrm{MSE} \triangleq \lim_{k\to\infty} \mathrm{E}\left\{\boldsymbol{e}^2(k)\right\}.$$

K. Ozeki, *Theory of Affine Projection Algorithms for Adaptive Filtering*,
Mathematics for Industry 22, DOI 10.1007/978-4-431-55738-8_5

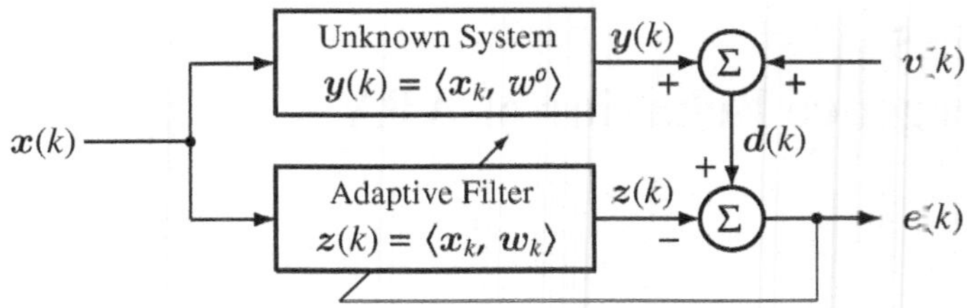

Fig. 5.1 Stochastic version of Fig. 2.1. The signals and the noise, as well as the coefficient vector sequence of the adaptive filter, are stochastic processes

We note

$$\begin{aligned} e(k) &= d(k) - x_k^t w_k \\ &= x_k^t (w^o - w_k) + v(k) \\ &= x_k^t \tilde{w}_k + v(k), \end{aligned}$$

where $\tilde{w}_k = w^o - w_k$ is the coefficient error vector. Using the assumptions (A-2)–(A-4), we have

$$\begin{aligned} \mathrm{E}\left\{e^2(k)\right\} &= \mathrm{E}\left\{(x_k^t \tilde{w}_k)^2\right\} + \mathrm{E}\left\{v^2(k)\right\} + 2\mathrm{E}\left\{x_k^t \tilde{w}_k\right\} \cdot \mathrm{E}\left\{v(k)\right\} \\ &= \mathrm{E}\left\{(x_k^t \tilde{w}_k)^2\right\} + \sigma_v^2. \end{aligned} \tag{5.1}$$

We define the *excess mean-square error (EMSE)* by

$$\mathrm{EMSE} \triangleq \lim_{k \to \infty} \mathrm{E}\left\{(x_k^t \tilde{w}_k)^2\right\}.$$

From (5.1), we have

$$\mathrm{MSE} = \mathrm{EMSE} + \sigma_v^2.$$

The *misadjustment* $\mathcal{M}$ of an adaptive filter is defined by

$$\mathcal{M} \triangleq \frac{\mathrm{EMSE}}{\sigma_v^2}.$$

The plot of $\mathrm{E}\left\{e^2(k)\right\}$ versus k is referred to as the *learning curve*. If we define the *a priori error*

$$e_a(k) \triangleq x_k^t \tilde{w}_k,$$

then the EMSE is rewritten as

$$\mathrm{EMSE} = \lim_{k \to \infty} \mathrm{E}\left\{e_a^2(k)\right\}.$$

In parallel with the a priori error $\boldsymbol{e}_a(k)$, the *a posteriori error* is defined by

$$\boldsymbol{e}_p(k) \stackrel{\triangle}{=} \boldsymbol{x}_k^t \tilde{\boldsymbol{w}}_{k+1}.$$

In addition to the EMSE, we are also interested in the behavior of the coefficient vector $\boldsymbol{w}_k$ or the coefficient error vector $\tilde{\boldsymbol{w}}_k$.

The adaptive filter is said to converge in the mean iff

$$\lim_{k\to\infty} \mathrm{E}\{\boldsymbol{w}_k\} = w^o,$$

or equivalently,

$$\lim_{k\to\infty} \mathrm{E}\{\tilde{\boldsymbol{w}}_k\} = 0,$$

for any initial coefficient vector $\boldsymbol{w}_0$.

If $\tilde{\boldsymbol{w}}_k$ converges in the mean-square, the quantity defined by

$$\mathrm{MSD} \stackrel{\triangle}{=} \lim_{k\to\infty} \mathrm{E}\left\{\|\tilde{\boldsymbol{w}}_k\|^2\right\}$$

is referred to as the *mean-square deviation (MSD)*.

An adaptive algorithm is said to be *stable* if $\tilde{\boldsymbol{w}}_k$ converges in the mean, and $\mathrm{E}\left\{\|\tilde{\boldsymbol{w}}_k\|^2\right\}$ remains bounded as $k \to \infty$ for any initial coefficient vector $\boldsymbol{w}_0$.

Besides the stochastic convergence measures above, a nonstochastic measure of performance called the *normalized misalignment*, expressed in dB as

$$20 \log_{10} \frac{\|\tilde{w}_k\|}{\|w^o\|},$$

is often used in simulation experiments.

5.2 Fundamental Behavior of B-APA

5.2.1 Shuffling Property

In terms of the coefficient error vector $\tilde{w}_k$, the update equation (3.10) for the B-APA is rewritten as

$$\tilde{w}_{k+1} = \left(I - \mu X_k^t (X_k X_k^t)^{-1} X_k\right) \tilde{w}_k - \mu X_k^t (X_k X_k^t)^{-1} v_k, \tag{5.2}$$

where

$$v_k \stackrel{\triangle}{=} (v(k), v(k-1), \ldots, v(k-(p-1)))^t.$$

Obviously, the matrix $X_k^t(X_k X_k^t)^{-1} X_k$ plays the key role as to the convergence of the coefficient error vector.

By Theorem 3.6 (Chap. 3, Appendix 1: "Affine Projection"),

$$X_k^t(X_k X_k^t)^{-1} X_k = X_k^t(X_k X_k^t)^{+} X_k = P_{\mathcal{R}(X_k^t)}. \tag{5.3}$$

That is, the matrix $A_k \overset{\triangle}{=} X_k^t(X_k X_k^t)^{-1} X_k$ is the orthogonal projection onto the subspace spanned by the columns of X_k^t. Therefore, if the space $\mathcal{R}(X_k^t)$ is kept unchanged as k increases, $\tilde{w}_k$ will not behave favorably even when $\{v(k)\} = 0$ In fact, in such a case, once $\tilde{w}_k$ falls in $\mathcal{R}(X_k^t)^{\perp}$, then from (5.2),

$$\tilde{w}_k = \tilde{w}_{k+1} = \tilde{w}_{k+2} = \cdots$$

eternally. This shows that in order for $\tilde{w}_k$ to have a favorable convergence property, some kind of "shuffling" is necessary for the distribution of X_k.

Lemma 5.1 *The eigenvalues of A_k are* 1 *or* 0. *The multiplicity of* 1 *is p, and that of* 0 *is $(n - p)$.*

Proof Because A_k is an orthogonal projection, the first part of the lemma comes from Theorem 3.4 (Chap. 3, Appendix 1: "Affine Projection"). As to the second part, we have by Formula A.7 (Appendix A),

$$\begin{aligned} \operatorname{tr}[A_k] &= \operatorname{tr}\left[X_k^t(X_k X_k^t)^{-1} X_k\right] \\ &= \operatorname{tr}\left[X_k X_k^t(X_k X_k^t)^{-1}\right] \\ &= \operatorname{tr}\left[I_p\right] \\ &= p. \end{aligned}$$

Thus, in view of (A.23) (Appendix A), the multiplicity of 1 is p, and that of 0 is $(n - p)$. □

The eigenspace of A_k corresponding to the eigenvalue 0 is denoted by S_0, and that corresponding to the eigenvalue 1 by S_1. Any vector $\phi \in \mathbb{R}^n$ is uniquely decomposed as $\phi = \phi_0 + \phi_1,\ \phi_0 \in S_0,\ \phi_1 \in S_1$.

In the following arguments, the regressor block X_k is treated as a random matrix $\boldsymbol{X}_k$. In accordance with this, A_k is also treated as a random matrix $\boldsymbol{A}_k = \boldsymbol{X}_k^t(\boldsymbol{X}_k \boldsymbol{X}_k^t)^{-1} \boldsymbol{X}_k$. Together with $\boldsymbol{A}_k$, ϕ_0 and ϕ_1 become random vectors denoted by $\boldsymbol{\phi}_0$ and $\boldsymbol{\phi}_1$, respectively. The matrix $\boldsymbol{A}_k$ is said to have the *shuffling property* iff for any $\phi \in \mathbb{R}^n$,

$$P(\boldsymbol{\phi}_0 \neq 0 \wedge \boldsymbol{\phi}_1 \neq 0) > 0. \tag{5.4}$$

Lemma 5.2 *If the matrix $\boldsymbol{A}_k$ has the shuffling property, then* $0 < \lambda < 1$ *for any eigenvalue λ of* $\mathrm{E}\{\boldsymbol{A}_k\}$.

Proof Let $\phi \in \mathbb{R}^n$ be a vector satisfying $\|\phi\| = 1$, and let

$$\Omega_0 \triangleq \{\boldsymbol{\phi}_1 = 0\}, \quad \Omega_1 \triangleq \{\boldsymbol{\phi}_0 = 0\}, \quad \Omega_2 \triangleq \{\boldsymbol{\phi}_0 \neq 0 \wedge \boldsymbol{\phi}_1 \neq 0\}.$$

The stochastic events $\Omega_0, \Omega_1, \Omega_2$ constitute a partition of the underlying probability space. Since $\boldsymbol{A}_k\phi = 0 \cdot \phi$ on Ω_0, and $\boldsymbol{A}_k\phi = 1 \cdot \phi$ on Ω_1, we have

$$\phi^t \boldsymbol{A}_k \phi = \begin{cases} \phi^t(0 \cdot \phi) = 0 & \text{on } \Omega_0; \\ \phi^t(1 \cdot \phi) = 1 & \text{on } \Omega_1; \\ \|\boldsymbol{\phi}_1\|^2 & \text{on } \Omega_2. \end{cases}$$

Thus,

$$\begin{aligned} \phi^t \mathrm{E}\{\boldsymbol{A}_k\}\phi &= \mathrm{E}\left\{\phi^t \boldsymbol{A}_k \phi\right\} \\ &= \int_{\Omega_0} \phi^t \boldsymbol{A}_k \phi \mathrm{d}P + \int_{\Omega_1} \phi^t \boldsymbol{A}_k \phi \mathrm{d}P + \int_{\Omega_2} \phi^t \boldsymbol{A}_k \phi \mathrm{d}P \\ &= P(\Omega_1) + \int_{\Omega_2} \|\boldsymbol{\phi}_1\|^2 \mathrm{d}P, \end{aligned} \tag{5.5}$$

where P is the probability measure on the underlying probability space. Since $\|\boldsymbol{\phi}_1\|^2 > 0$ on Ω_2, and $P(\Omega_2) > 0$, $\int_{\Omega_2} \|\boldsymbol{\phi}_1\|^2 \mathrm{d}P > 0$. From (5.5), we see

$$\phi^t \mathrm{E}\{\boldsymbol{A}_k\}\phi > 0. \tag{5.6}$$

Also, since $\|\boldsymbol{\phi}_1\|^2 < 1$ on Ω_2,

$$\int_{\Omega_2} \|\boldsymbol{\phi}_1\|^2 \mathrm{d}P < P(\Omega_2).$$

Therefore,

$$\begin{aligned} \phi^t \mathrm{E}\{\boldsymbol{A}_k\}\phi &= P(\Omega_1) + \int_{\Omega_2} \|\boldsymbol{\phi}_1\|^2 \mathrm{d}P \\ &< P(\Omega_1) + P(\Omega_2) \\ &\leq 1. \end{aligned} \tag{5.7}$$

Now, let $\lambda_{\min}$ and $\lambda_{\max}$ be the smallest and largest eigenvalues of $\mathrm{E}\{\boldsymbol{A}_k\}$, respectively. Then, it is easily verified that

$$\begin{aligned} \lambda_{\min} &= \min_{\|\phi\|=1} \phi^t \mathrm{E}\{\boldsymbol{A}_k\}\phi, \\ \lambda_{\max} &= \max_{\|\phi\|=1} \phi^t \mathrm{E}\{\boldsymbol{A}_k\}\phi. \end{aligned} \tag{5.8}$$

From Eqs. (5.6), (5.7) and (5.8), we finally have

$$0 < \lambda_{\min} \le \lambda_{\max} < 1.$$

□

Since $\mathrm{tr}\,[A_k] = p$, $\mathrm{tr}\,[\mathrm{E}\,\{A_k\}] = \mathrm{E}\,\{\mathrm{tr}\,[A_k]\} = p$. Because $\mathrm{tr}\,[A_k]$ is equal to the sum of its eigenvalues, we expect that each eigenvalue of $\mathrm{E}\,\{A_k\}$ is equal to p/n if some kind of uniformity is assumed on the distribution of A_k. We say that A_k has the *complete shuffling property* iff $\mathrm{E}\,\{A_k\} = (p/n)I$.

Example 5.1 Let $n = 3$, and $p = 1$ or $p = 2$. If the orientation of $\mathcal{R}(A_k)$ is uniformly distributed, then A_k has the complete shuffling property.

Proof First, let $p = 1$. In this case, $\mathcal{R}(A_k)$ is a one-dimensional subspace spanned by a unit vector u, which is parametrized by

$$u = (\sin\theta\cos\varphi, \sin\theta\sin\varphi, \cos\theta)^t$$

as illustrated in Fig. 5.2. Since the orthogonal projection of a vector a onto the subspace spanned by a unit vector u is given by uu^ta, the matrix $A_k = P_{\mathcal{R}(A_k)}$ is given by

$$\begin{aligned} A_k &= \begin{bmatrix} \sin\theta\cos\varphi \\ \sin\theta\sin\varphi \\ \cos\theta \end{bmatrix} (\sin\theta\cos\varphi, \sin\theta\sin\varphi, \cos\theta) \\ &= \begin{bmatrix} \sin^2\theta\cos^2\varphi & \sin^2\theta\sin\varphi\cos\varphi & \sin\theta\cos\theta\cos\varphi \\ \sin^2\theta\sin\varphi\cos\varphi & \sin^2\theta\sin^2\varphi & \sin\theta\cos\theta\sin\varphi \\ \sin\theta\cos\theta\cos\varphi & \sin\theta\cos\theta\sin\varphi & \cos^2\theta \end{bmatrix}. \end{aligned}$$

Integrating this expression with a probability density function $p(\theta, \varphi) = \sin\theta/(4\pi)$ $(0 \le \theta \le \pi,\ 0 \le \varphi \le 2\pi)$, which makes the distribution of the orientation uniform, we have

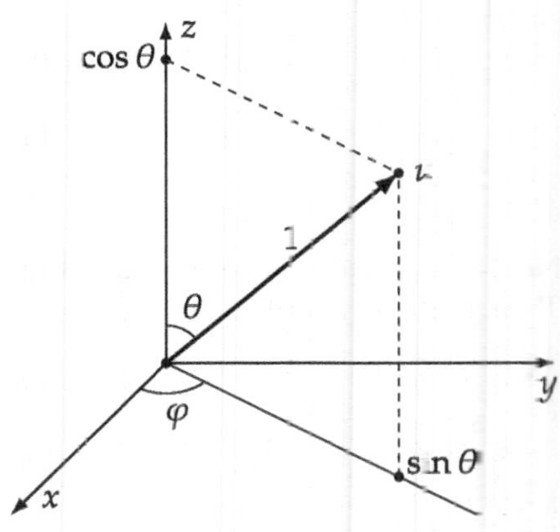

Fig. 5.2 Representation of a unit vector u in $\mathbb{R}^3$ by the polar coordinates. $u = (\sin\theta\cos\varphi, \sin\theta\sin\varphi, \cos\theta)^t$

$$\mathrm{E}\{\boldsymbol{A}_k\} = \begin{bmatrix} a_{11} & a_{12} & a_{13} \\ a_{21} & a_{22} & a_{23} \\ a_{31} & a_{32} & a_{33} \end{bmatrix},$$

where

$$\begin{aligned}
a_{11} &= \frac{1}{4\pi}\int_0^{\pi}\int_0^{2\pi} \sin^3\theta\cos^2\varphi\,\mathrm{d}\theta\mathrm{d}\varphi,\\
a_{12} &= \frac{1}{4\pi}\int_0^{\pi}\int_0^{2\pi} \sin^3\theta\sin\varphi\cos\varphi\,\mathrm{d}\theta\mathrm{d}\varphi,\\
a_{13} &= \frac{1}{4\pi}\int_0^{\pi}\int_0^{2\pi} \sin^2\theta\cos\theta\cos\varphi\,\mathrm{d}\theta\mathrm{d}\varphi,\\
a_{21} &= \frac{1}{4\pi}\int_0^{\pi}\int_0^{2\pi} \sin^3\theta\sin\varphi\cos\varphi\,\mathrm{d}\theta\mathrm{d}\varphi.\\
a_{22} &= \frac{1}{4\pi}\int_0^{\pi}\int_0^{2\pi} \sin^3\theta\sin^2\varphi\,\mathrm{d}\theta\mathrm{d}\varphi,\\
a_{23} &= \frac{1}{4\pi}\int_0^{\pi}\int_0^{2\pi} \sin^2\theta\cos\theta\sin\varphi\,\mathrm{d}\theta\mathrm{d}\varphi\\
a_{31} &= \frac{1}{4\pi}\int_0^{\pi}\int_0^{2\pi} \sin^2\theta\cos\theta\cos\varphi\,\mathrm{d}\theta\mathrm{d}\varphi,\\
a_{32} &= \frac{1}{4\pi}\int_0^{\pi}\int_0^{2\pi} \sin^2\theta\cos\theta\sin\varphi\,\mathrm{d}\theta\mathrm{d}\varphi,\\
a_{33} &= \frac{1}{4\pi}\int_0^{\pi}\int_0^{2\pi} \sin\theta\cos^2\theta\,\mathrm{d}\theta\mathrm{d}\varphi.
\end{aligned}$$

It is immediately seen that all the off-diagonal components are equal to 0. Using the formulas

$$\begin{aligned}
\int_0^{\pi} \sin^3\theta\,\mathrm{d}\theta &= \frac{4}{3},\\
\int_0^{\pi} \sin\theta\cos^2\theta\,\mathrm{d}\theta &= \frac{2}{3},\\
\int_0^{2\pi} \sin^2\varphi\,\mathrm{d}\varphi &= \int_0^{2\pi} \cos^2\varphi\,\mathrm{d}\varphi = \pi,
\end{aligned}$$

we finally have

$$\mathrm{E}\{\boldsymbol{A}_k\} = \begin{bmatrix} \frac{1}{3} & 0 & 0 \\ 0 & \frac{1}{3} & 0 \\ 0 & 0 & \frac{1}{3} \end{bmatrix} = (p/n)\boldsymbol{I}.$$

Therefore, $\boldsymbol{A}_k$ has the complete shuffling property.

Next, let $p = 2$. In this case, $\mathcal{R}(A_k)$ is a 2-dimensional subspace whose orientation is determined by its normal, which coincides with $\mathcal{R}(A_k)^\perp$. By Theorem 3.3 (Chap 3, Appendix 1: "Affine Projection"),

$$A_k = P_{\mathcal{R}(A_k)} = I - P_{\mathcal{R}(A_k)^\perp}.$$

Since $\mathcal{R}(A_k)^\perp$ is a one-dimensional subspace, the above argument for $p = 1$ applies. Hence,

$$\mathrm{E}\left\{P_{\mathcal{R}(A_k)^\perp}\right\} = (1/3)I,$$

which leads to

$$\begin{aligned}\mathrm{E}\{A_k\} &= \mathrm{E}\left\{I - P_{\mathcal{R}(A_k)^\perp}\right\} \\ &= I - \mathrm{E}\left\{P_{\mathcal{R}(A_k)^\perp}\right\} \\ &= (2/3)I \\ &= (p/n)I.\end{aligned}$$

This shows that the matrix A_k has the complete shuffling property for $p = 2$ as well. □

5.2.2 Stability and Mean-Square Deviation

The stochastic version of (5.2) is written as

$$\tilde{w}_{k+1} = \left(I - \mu X_k^t(X_k X_k^t)^{-1} X_k\right)\tilde{w}_k - \mu X_k^t(X_k X_k^t)^{-1} v_k. \qquad (5.9)$$

Based on this equation, we prove the following theorems:

Theorem 5.1 *Suppose that the matrix* $A_k \triangleq X_k^t(X_k X_k^t)^{-1}X_k$ *has the shuffling property. Then,* $\tilde{w}_k$ *converges to* 0 *in the mean if* $0 < \mu < 2$. *It does not if* $\mu \leq 2$.

Proof Taking expectations of both sides of (5.9), and using the assumptions (A-2), (A-3), and (A-5), we obtain

$$\mathrm{E}\{\tilde{w}_{k+1}\} = (I - \mu\mathrm{E}\{A_k\})\mathrm{E}\{\tilde{w}_k\}. \qquad (5.10)$$

Let $\lambda_1, \lambda_2, \ldots, \lambda_n$ be the eigenvalues of $\mathrm{E}\{A_k\}$. Then, the eigenvalues $\rho_1, \rho_2, \ldots, \rho_n$ of $(I - \mu\mathrm{E}\{A_k\})$ are given by

$$\rho_i = 1 - \mu\lambda_i \quad (1 \leq i \leq n).$$

Because we assume that the matrix A_k has the shuffling property, $0 < \lambda_i < 1$ $(1 \le i \le n)$ by Lemma 5.2. Therefore, if $0 < \mu < 2$, then $|\rho_i| < 1$ $(1 \le i \le n)$. Also, if $\mu \le 0$, then $\rho_i \ge 1$ $(1 \le i \le n)$. Using these facts, we immediately draw the conclusion by Theorem 5.9 (Appendix 1: "Stability of Linear Systems"). □

Theorem 5.2 *Suppose the matrix $A_k = X_k^t(X_kX_k^t)^{-1}X_k$ has the shuffling property. Then,* $\mathrm{E}\left\{\|\tilde{w}_k\|^2\right\}$ *remains bounded as $k \to \infty$ if $0 < \mu < 2$. It does not if $\mu < 0$ or $2 \le \mu$.*

Proof From (5.9), we obtain

$$\begin{aligned}
\|\tilde{w}_{k+1}\|^2 &= \left((I-\mu A_k)\tilde{w}_k - \mu X_k^t(X_kX_k^t)^{-1}v_k\right)^t \left((I-\mu A_k)\tilde{w}_k - \mu X_k^t(X_kX_k^t)^{-1}v_k\right)\\
&= \tilde{w}_k^t(I-\mu A_k)^2\tilde{w}_k - \mu\tilde{w}_k^t(I-\mu A_k)X_k^t(X_kX_k^t)^{-1}v_k\\
&\quad - \mu v_k^t(X_kX_k^t)^{-1}X_k(I-\mu A_k)\tilde{w}_k + \mu^2 v_k^t(X_kX_k^t)^{-1}X_kX_k^t(X_kX_k^t)^{-1}v_k\\
&= \tilde{w}_k^t\{I-\mu(2-\mu)A_k\}\tilde{w}_k - \mu\tilde{w}_k^t(I-\mu A_k)X_k^t(X_kX_k^t)^{-1}v_k\\
&\quad - \mu v_k^t(X_kX_k^t)^{-1}X_k(I-\mu A_k)\tilde{w}_k + \mu^2 v_k^t(X_kX_k^t)^{-1}v_k.
\end{aligned}$$

Taking expectation of both sides of the above equation, and using the assumptions (A-2), (A-3), and (A-5), we have

$$\begin{aligned}
\mathrm{E}\left\{\|\tilde{w}_{k+1}\|^2\right\} &= \mathrm{E}\left\{\tilde{w}_k^t\left[I-\mu(2-\mu)A_k\right]\tilde{w}_k\right\} + \mu^2\mathrm{E}\left\{v_k^t(X_kX_k^t)^{-1}v_k\right\}\\
&= \mathrm{E}\left\{\tilde{w}_k^t\left[I-\mu(2-\mu)\mathrm{E}\{A_k\}\right]\tilde{w}_k\right\} + \mu^2\mathrm{E}\left\{\mathrm{tr}\left[v_kv_k^t(X_kX_k^t)^{-1}\right]\right\}\\
&= \mathrm{E}\left\{\tilde{w}_k^t\left[I-\mu(2-\mu)\mathrm{E}\{A_k\}\right]\tilde{w}_k\right\} + \mu^2\mathrm{tr}\left[\mathrm{E}\left\{v_kv_k^t\right\}\mathrm{E}\left\{(X_kX_k^t)^{-1}\right\}\right]\\
&= \mathrm{E}\left\{\tilde{w}_k^t\left[I-\mu(2-\mu)\mathrm{E}\{A_k\}\right]\tilde{w}_k\right\} + \mu^2\sigma_v^2\mathrm{tr}\left[\mathrm{E}\left\{(X_kX_k^t)^{-1}\right\}\right]. \quad (5.11)
\end{aligned}$$

Let $\lambda_{\min}$ be the smallest eigenvalue of $\mathrm{E}\{A_k\}$. Since we assume A_k has the shuffling property, $0 < \lambda_{\min} < 1$ by Lemma 5.2. If $0 < \mu < 2$, then $\mu(2-\mu) > 0$, so that

$$\mathrm{E}\left\{\tilde{w}_k^t\left[I-\mu(2-\mu)\mathrm{E}\{A_k\}\right]\tilde{w}_k\right\} \le (1-\mu(2-\mu)\lambda_{\min})\mathrm{E}\left\{\|\tilde{w}_k\|^2\right\}.$$

Therefore, we have from (5.11)

$$\mathrm{E}\left\{\|\tilde{w}_{k+1}\|^2\right\} \le (1-\mu(2-\mu)\lambda_{\min})\mathrm{E}\left\{\|\tilde{w}_k\|^2\right\} + \mu^2\sigma_v^2\mathrm{tr}\left[\mathrm{E}\left\{(X_kX_k^t)^{-1}\right\}\right].$$

Let $\rho \triangleq 1-\mu(2-\mu)\lambda_{\min}$, and $c \triangleq \sigma_v^2\mathrm{tr}\left[\mathrm{E}\left\{(X_kX_k^t)^{-1}\right\}\right]$. Then,

$$\mathrm{E}\left\{\|\tilde{w}_k\|^2\right\} \le \rho^k\mathrm{E}\left\{\|\tilde{w}_0\|^2\right\} + \mu^2 c\sum_{i=0}^{k-1}\rho^i.$$

Since $0 < \rho < 1$, the right-hand side converges to $\mu^2 c/(1-\rho)$ as $k \to \infty$. This shows that $\mathrm{E}\left\{\|\tilde{\boldsymbol{w}}_k\|^2\right\}$ remains bounded.

If $\mu < 0$ or $2 \le \mu$, then $\mu(2-\mu) \le 0$, and so

$$\mathrm{E}\left\{\tilde{\boldsymbol{w}}_k^t \left[I - \mu(2-\mu)\mathrm{E}\left\{\boldsymbol{A}_k\right\}\right]\tilde{\boldsymbol{w}}_k\right\} \ge (1-\mu(2-\mu)\lambda_{\min})\mathrm{E}\left\{\|\tilde{\boldsymbol{w}}_k\|^2\right\}.$$

Thus,

$$\mathrm{E}\left\{\|\tilde{\boldsymbol{w}}_{k+1}\|^2\right\} \ge \rho \mathrm{E}\left\{\|\tilde{\boldsymbol{w}}_k\|^2\right\} + \mu^2 c.$$

Since $\rho \ge 1$, and $\mu^2 c > 0$, this inequality shows that $\mathrm{E}\left\{\|\tilde{\boldsymbol{w}}_k\|^2\right\}$ diverges as $k \to \infty$. □

From Theorems 5.1 and 5.2, we have the following result:

Corollary 5.1 *If the matrix $\boldsymbol{A}_k$ has the shuffling property, the B-APA is stable iff $0 < \mu < 2$.*

The stability condition $0 < \mu < 2$ is one of the facts we could anticipate from the geometrical investigation of the APA in Chap. 3.

Theorem 5.3 *Suppose that the matrix $\boldsymbol{A}_k$ has the complete shuffling property. If $0 < \mu < 2$, then $\mathrm{E}\left\{\|\tilde{\boldsymbol{w}}_k\|^2\right\}$ converges as $k \to \infty$, and MSD is given by*

$$\mathrm{MSD} = \frac{n}{p}\frac{\mu}{2-\mu}\sigma_v^2 tr\left[\mathrm{E}\left\{(\boldsymbol{X}_k\boldsymbol{X}_k^t)^{-1}\right\}\right].$$

Proof From the assumption that $\boldsymbol{A}_k$ has the complete shuffling property,

$$\mathrm{E}\left\{\boldsymbol{A}_k\right\} = (p/n)I.$$

Therefore, from (5.11),

$$\mathrm{E}\left\{\|\tilde{\boldsymbol{w}}_{k+1}\|^2\right\} = \rho \mathrm{E}\left\{\|\tilde{\boldsymbol{w}}_k\|^2\right\} + \mu^2 c,$$

where

$$\rho \triangleq 1 - \mu(2-\mu)(p/n),$$

and

$$c \triangleq \sigma_v^2 \mathrm{tr}\left[\mathrm{E}\left\{(\boldsymbol{X}_k\boldsymbol{X}_k^t)^{-1}\right\}\right].$$

Thus,

$$\mathrm{E}\left\{\|\tilde{\boldsymbol{w}}_k\|^2\right\} = \rho^k \mathrm{E}\left\{\|\tilde{\boldsymbol{w}}_0\|^2\right\} + \mu^2 c \sum_{i=0}^{k-1}\rho^i. \tag{5.12}$$

Since $0 < \rho < 1$, this equation leads to

$$\begin{aligned}\text{MSD} &= \lim_{k\to\infty} \text{E}\left\{\|\tilde{\boldsymbol{w}}_k\|^2\right\} \\ &= \frac{\mu^2 c}{1-\rho} \\ &= \frac{n}{p}\frac{\mu}{2-\mu}\sigma_v^2 \text{tr}\left[\text{E}\left\{(\boldsymbol{X}_k\boldsymbol{X}_k^t)^{-1}\right\}\right]. \end{aligned} \tag{5.13}$$

□

If an approximation

$$\text{tr}\left[\text{E}\left\{(\boldsymbol{X}_k\boldsymbol{X}_k^t)^{-1}\right\}\right] \approx \text{E}\left\{\frac{p}{\|\boldsymbol{x}_k\|^2}\right\}$$

is used, (5.13) is rewritten as

$$\text{MSD} = \frac{\mu\sigma_v^2 n}{2-\mu}\text{E}\left\{\frac{1}{\|\boldsymbol{x}_k\|^2}\right\}. \tag{5.14}$$

From (5.12) and (5.14), we make the following observations:

(1) Because $\mu = 1$ maximizes $\mu(2-\mu)$, it minimizes ρ. Therefore, $\mu = 1$ maximizes the convergence rate for $\text{E}\left\{\|\tilde{\boldsymbol{w}}_k\|^2\right\}$. This has already been expected in the geometrical investigation of the APA in Chap. 3.
(2) For $0 < \mu < 2$, ρ is a decreasing function of p/n. Therefore, increasing p/n makes the convergence rate higher. We have already seen in Chap. 3 that increasing the projection order p can improve the convergence rate.
(3) The squared norm $\|\boldsymbol{x}_k\|^2$ tends to be proportional to n. Therefore, very roughly speaking, MSD is independent of n and p.
(4) MSD is approximately proportional to $\sigma_v^2/\text{E}\left\{x^2(k)\right\}$, the inverse of the S/N ratio.
(5) Within the range $0 < \mu < 2$, decreasing μ makes MSD smaller. Therefore, although $\mu = 1 - a$ and $\mu = 1 + a$ for $0 \le a < 1$ give the same convergence rate, it is wise to choose the value of μ within the range $0 < \mu \le 1$.

5.3 Discrete Orientation Assumption for Regressors

In this subsection, we review the paper [1] with slight modifications. In the work, which is considered to be an extension of [2] to the case of the APA, the convergence behavior of the B-APA is analyzed based on simplifying assumptions that

(B-1) the regressors $\{\boldsymbol{x}_k\}$ are independent and identically distributed (i.i.d.),

and

(B-2) the regressor $\boldsymbol{x}_k$ is represented by the product of three quantities:

$$\boldsymbol{x}_k = \boldsymbol{s}(k)\boldsymbol{r}(k)\boldsymbol{u}_k, \tag{5.15}$$

where $\boldsymbol{s}(k)$ and $\boldsymbol{r}(k)$ are random variables, and $\boldsymbol{u}_k$ is a random vector having the same dimension as $\boldsymbol{x}_k$. It is assumed that $\boldsymbol{s}(k)$, $\boldsymbol{r}(k)$, $\boldsymbol{u}_k$ are independent, and each of $\{\boldsymbol{s}(k)\}$, $\{\boldsymbol{r}(k)\}$, $\{\boldsymbol{u}_k\}$ is i.i.d. with distributions

$$P\{\boldsymbol{s}(k) = \pm 1\} = \frac{1}{2},$$

$\boldsymbol{r}(k)$ and $\|\boldsymbol{x}_k\|$ have the same probability distribution,

$$P\{\boldsymbol{u}_k = v_i\} = p_i \triangleq \frac{\lambda_i}{\sum_{j=1}^{n} \lambda_j} = \frac{\lambda_i}{\operatorname{tr}[R_x]},$$

where λ_i's are eigenvalues of the covariance matrix $R_x = \mathrm{E}\left\{\boldsymbol{x}_k\boldsymbol{x}_k^t\right\}$, and v_i is the normalized eigenvector corresponding to λ_i.

Note that $(p_1, p_2, \ldots, p_n)$ defines a probability distribution. By Theorem A.5 (Appendix A), the matrix R_x can be diagonalized as

$$R_x = V\Lambda V^t,$$

where

$$V = \begin{bmatrix} v_1 \ v_2 \ldots v_n \end{bmatrix},$$
$$\Lambda = \operatorname{diag}[\lambda_1, \lambda_2, \ldots, \lambda_n].$$

The representation (5.15) is consistent with the assumption (A-1). In fact,

$$\begin{aligned}
\mathrm{E}\{\boldsymbol{s}(k)\boldsymbol{r}(k)\boldsymbol{u}_k\} &= \mathrm{E}\{\boldsymbol{s}(k)\}\mathrm{E}\{\boldsymbol{r}(k)\}\mathrm{E}\{\boldsymbol{u}_k\} \\
&= 0 \cdot \mathrm{E}\{\boldsymbol{r}(k)\}\mathrm{E}\{\boldsymbol{u}_k\} \\
&= 0,
\end{aligned}$$

and

$$\begin{aligned}
\mathrm{E}\left\{(\boldsymbol{s}(k)\boldsymbol{r}(k)\boldsymbol{u}_k)(\boldsymbol{s}(k)\boldsymbol{r}(k)\boldsymbol{u}_k)^t\right\} &= \mathrm{E}\left\{\boldsymbol{s}^2(k)\right\}\mathrm{E}\left\{\boldsymbol{r}^2(k)\right\}\mathrm{E}\left\{\boldsymbol{u}_k\boldsymbol{u}_k^t\right\} \\
&= \mathrm{E}\left\{\|\boldsymbol{x}_k\|^2\right\}\sum_{i=1}^{n} v_i v_i^t \frac{\lambda_i}{\operatorname{tr}[R_x]}
\end{aligned}$$

$$
\begin{aligned}
&= \sum_{i=1}^{n} \lambda_i v_i v_i^t \\
&= V \Lambda V^t \\
&= R_x .
\end{aligned}
$$

The assumption (B-2), which says that $\boldsymbol{x}_k$ can take only one of finite number of orientations, was first introduced in [2]. The assumptions (B-1) and (B-2), though unrealistic in ordinary situations, make the analysis tractable, and yield results that help us to see fundamental properties of the B-APA.

Let us consider the stochastic version of the update equation (3.10) for the B-APA:

$$\boldsymbol{w}_{k+1} = \boldsymbol{w}_k + \mu \boldsymbol{X}_k^t (\boldsymbol{X}_k \boldsymbol{X}_k^t)^{-1} \boldsymbol{e}_k. \tag{5.16}$$

Note that the coefficient vector $\boldsymbol{w}_k$ and the error vector $\boldsymbol{e}_k$ are also random vectors. If the regressors $\boldsymbol{x}_k, \boldsymbol{x}_{k-1}, \ldots, \boldsymbol{x}_{k-(p-1)}$ are orthogonal to each other, then

$$\boldsymbol{X}_k \boldsymbol{X}_k^t = \mathrm{diag}\,[\boldsymbol{x}_k^t \boldsymbol{x}_k, \boldsymbol{x}_{k-1}^t \boldsymbol{x}_{k-1}, \ldots, \boldsymbol{x}_{k-(p-1)}^t \boldsymbol{x}_{k-(p-1)}].$$

Therefore, (5.16) is rewritten as

$$\boldsymbol{w}_{k+1} = \boldsymbol{w}_k + \mu \sum_{i=0}^{p-1} \boldsymbol{\mu}^{(i)} \boldsymbol{x}_{k-i}, \tag{5.17}$$

where

$$\boldsymbol{\mu}^{(i)} \triangleq \frac{\boldsymbol{e}_{k,i}}{\boldsymbol{x}_{k-i}^t \boldsymbol{x}_{k-i}} = \frac{\boldsymbol{d}(k-i) - \boldsymbol{x}_{k-i}^t \boldsymbol{w}_k}{\boldsymbol{x}_{k-i}^t \boldsymbol{x}_{k-i}}.$$

According to the assumption (B-2), we expect that many of the regressors are orthogonal to each other. However, there are chances for some of them to become parallel. To avoid this, we select an orthogonal set of regressors

$$\{\boldsymbol{x}_{k-j_1}, \boldsymbol{x}_{k-j_2}, \ldots, \boldsymbol{x}_{k-j_J}\} \subseteq \{\boldsymbol{x}_k, \boldsymbol{x}_{k-1}, \ldots, \boldsymbol{x}_{k-(p-1)}\},$$

by removing $\boldsymbol{x}_{k-i}$ if it is parallel to some of $\boldsymbol{x}_k, \boldsymbol{x}_{k-1}, \ldots, \boldsymbol{x}_{k-(i-1)}$, and revise (5.17) as

$$\boldsymbol{w}_{k+1} = \boldsymbol{w}_k + \mu \sum_{j \in J_k} \boldsymbol{\mu}^{(j)} \boldsymbol{x}_{k-j}, \tag{5.18}$$

where

$$J_k \triangleq \{j_1, j_2, \ldots, j_J\} \subseteq \{0, 1, \ldots, p-1\}.$$

For example, let $n = 16$ and $p = 8$. If $\boldsymbol{u}_k, \boldsymbol{u}_{k-1}, \ldots, \boldsymbol{u}_{k-7}$ are realized as

$$\begin{aligned}&\boldsymbol{u}_k = \boldsymbol{\nu}_3,\ \boldsymbol{u}_{k-1} = \boldsymbol{\nu}_1,\ \boldsymbol{u}_{k-2} = \boldsymbol{\nu}_{15},\ \boldsymbol{u}_{k-3} = \boldsymbol{\nu}_3,\\&\boldsymbol{u}_{k-4} = \boldsymbol{\nu}_5,\ \boldsymbol{u}_{k-5} = \boldsymbol{\nu}_{10},\ \boldsymbol{u}_{k-6} = \boldsymbol{\nu}_5,\ \boldsymbol{u}_{k-7} = \boldsymbol{\nu}_8.\end{aligned} \tag{5.19}$$

then $J_k = \{0, 1, 2, 4, 5, 7\}$. Since which ν_i is realized is a matter of probability, J_k is a random variable that takes a value in the set of all the subsets of $\{0, 1, \ldots, p-1\}$. The probability that the realization (5.19) occurs is given by $p_1 p_3^2 p_5^2 p_8 p_{10} p_{15}$. However, it is possible that other realizations also produce the same J_k. Therefore, the probability of the event $J_k = \{0, 1, 2, 4, 5, 7\}$ is given by the sum of the probabilities of all such realizations that produce the event.

Equation (5.18) is rewritten in terms of the coefficient error vector $\tilde{\boldsymbol{w}}_k$ as

$$\begin{aligned}\tilde{\boldsymbol{w}}_{k+1} &= w^o - \boldsymbol{w}_{k+1}\\&= w^o - \boldsymbol{w}_k - \mu \sum_{j\in J_k} \mu^{(j)} \boldsymbol{x}_{k-j}\\&= \tilde{\boldsymbol{w}}_k - \mu \sum_{j\in J_k} \frac{e_{k,j}}{\boldsymbol{x}_{k-j}^t \boldsymbol{x}_{k-j}} \boldsymbol{x}_{k-j}\\&= \tilde{\boldsymbol{w}}_k - \mu \sum_{j\in J_k} \frac{\boldsymbol{x}_{k-j}^t \tilde{\boldsymbol{w}}_k + v(k-j)}{\boldsymbol{x}_{k-j}^t \boldsymbol{x}_{k-j}} \boldsymbol{x}_{k-j}\\&= \left[I - \mu \sum_{j\in J_k} \frac{\boldsymbol{x}_{k-j}\boldsymbol{x}_{k-j}^t}{\boldsymbol{x}_{k-j}^t \boldsymbol{x}_{k-j}} \right] \tilde{\boldsymbol{w}}_k - \mu \sum_{j\in J_k} \frac{v(k-j)\boldsymbol{x}_{k-j}}{\boldsymbol{x}_{k-j}^t \boldsymbol{x}_{k-j}},\end{aligned} \tag{5.20}$$

where we used the fact that

$$\begin{aligned}e_{k,j} &= d(k-j) - \boldsymbol{x}_{k-j}^t \boldsymbol{w}_k\\&= \boldsymbol{x}_{k-j}^t w^o + v(k-j) - \boldsymbol{x}_{k-j}^t \boldsymbol{w}_k\\&= \boldsymbol{x}_{k-j}^t \tilde{\boldsymbol{w}}_k + v(k-j).\end{aligned}$$

From (5.20), we can derive a recurrence equation for the covariance matrix of $\tilde{\boldsymbol{w}}_k$, $\mathrm{Cov}(\tilde{\boldsymbol{w}}_k)$, in the following way. The covariance matrix of $\tilde{\boldsymbol{w}}_{k+1}$ is expanded into four terms as

$$\begin{aligned}\mathrm{Cov}(\tilde{\boldsymbol{w}}_{k+1}) = \mathrm{E}&\left\{\left[I - \mu \sum_{j\in J_k} \frac{\boldsymbol{x}_{k-j}\boldsymbol{x}_{k-j}^t}{\boldsymbol{x}_{k-j}^t \boldsymbol{x}_{k-j}} \right] \tilde{\boldsymbol{w}}_k \tilde{\boldsymbol{w}}_k^t \left[I - \mu \sum_{l\in J_k} \frac{\boldsymbol{x}_{k-l}\boldsymbol{x}_{k-l}^t}{\boldsymbol{x}_{k-l}^t \boldsymbol{x}_{k-l}} \right]\right\}\\+\mathrm{E}&\left\{\left[\mu \sum_{j\in J_k} \frac{v(k-j)\boldsymbol{x}_{k-j}}{\boldsymbol{x}_{k-j}^t \boldsymbol{x}_{k-j}} \right] \left[\mu \sum_{l\in J_k} \frac{v(k-l)\boldsymbol{x}_{k-l}^t}{\boldsymbol{x}_{k-l}^t \boldsymbol{x}_{k-l}} \right]\right\}\end{aligned}$$

$$
-\mathrm{E}\left\{\left[I-\mu\sum_{j\in J_k}\frac{\boldsymbol{x}_{k-j}\boldsymbol{x}_{k-j}^t}{\boldsymbol{x}_{k-j}^t\boldsymbol{x}_{k-j}}\right]\tilde{\boldsymbol{w}}_k\left[\mu\sum_{l\in J_k}\frac{\boldsymbol{v}(k-l)\boldsymbol{x}_{k-l}^t}{\boldsymbol{x}_{k-l}^t\boldsymbol{x}_{k-l}^t}\right]\right\}
$$

$$
-\mathrm{E}\left\{\left[\mu\sum_{j\in J_k}\frac{\boldsymbol{v}(k-j)\boldsymbol{x}_{k-j}}{\boldsymbol{x}_{k-j}^t\boldsymbol{x}_{k-j}}\right]\tilde{\boldsymbol{w}}_k^t\left[I-\mu\sum_{l\in J_k}\frac{\boldsymbol{x}_{k-l}\boldsymbol{x}_{k-l}^t}{\boldsymbol{x}_{k-l}^t\boldsymbol{x}_{k-l}}\right]\right\}.
$$

By the assumption (A-5), the first term is rewritten as

$$
\begin{aligned}
&\mathrm{E}\left\{\left[I-\mu\sum_{j\in J_k}\frac{\boldsymbol{x}_{k-j}\boldsymbol{x}_{k-j}^t}{\boldsymbol{x}_{k-j}^t\boldsymbol{x}_{k-j}}\right]\tilde{\boldsymbol{w}}_k\tilde{\boldsymbol{w}}_k^t\left[I-\mu\sum_{l\in J_k}\frac{\boldsymbol{x}_{k-l}\boldsymbol{x}_{k-l}^t}{\boldsymbol{x}_{k-l}^t\boldsymbol{x}_{k-l}}\right]\right\}\\
&=\mathrm{E}\left\{\left[I-\mu\sum_{j\in J_k}\frac{\boldsymbol{x}_{k-j}\boldsymbol{x}_{k-j}^t}{\boldsymbol{x}_{k-j}^t\boldsymbol{x}_{k-j}}\right]\mathrm{Cov}(\tilde{\boldsymbol{w}}_k)\left[I-\mu\sum_{l\in J_k}\frac{\boldsymbol{x}_{k-l}\boldsymbol{x}_{k-l}^t}{\boldsymbol{x}_{k-l}^t\boldsymbol{x}_{k-l}}\right]\right\}.
\end{aligned}
$$

Also, using the assumptions (A-2) and (A-3), we have for $j\neq l$

$$
\begin{aligned}
&\mathrm{E}\left\{\left[\frac{\boldsymbol{v}(k-j)\boldsymbol{x}_{k-j}}{\boldsymbol{x}_{k-j}^t\boldsymbol{x}_{k-j}}\right]\left[\frac{\boldsymbol{v}(k-l)\boldsymbol{x}_{k-l}^t}{\boldsymbol{x}_{k-l}^t\boldsymbol{x}_{k-l}}\right]\right\}\\
&=\mathrm{E}\left\{\boldsymbol{v}(k-j)\boldsymbol{v}(k-l)\left[\frac{\boldsymbol{x}_{k-j}}{\boldsymbol{x}_{k-j}^t\boldsymbol{x}_{k-j}}\right]\left[\frac{\boldsymbol{x}_{k-l}^t}{\boldsymbol{x}_{k-l}^t\boldsymbol{x}_{k-l}}\right]\right\}\\
&=\mathrm{E}\{\boldsymbol{v}(k-j)\}\mathrm{E}\{\boldsymbol{v}(k-l)\}\mathrm{E}\left\{\left[\frac{\boldsymbol{x}_{k-j}}{\boldsymbol{x}_{k-j}^t\boldsymbol{x}_{k-j}}\right]\left[\frac{\boldsymbol{x}_{k-l}^t}{\boldsymbol{x}_{k-l}^t\boldsymbol{x}_{k-l}}\right]\right\}\\
&=0.
\end{aligned}
$$

Therefore, the second term is rewritten as

$$
\begin{aligned}
&\mathrm{E}\left\{\left[\mu\sum_{j\in J_k}\frac{\boldsymbol{v}(k-j)\boldsymbol{x}_{k-j}}{\boldsymbol{x}_{k-j}^t\boldsymbol{x}_{k-j}}\right]\left[\mu\sum_{l\in J_k}\frac{\boldsymbol{v}(k-l)\boldsymbol{x}_{k-l}^t}{\boldsymbol{x}_{k-l}^t\boldsymbol{x}_{k-l}}\right]\right\}\\
&=\mathrm{E}\left\{\mu^2\sum_{j\in J_k}\boldsymbol{v}^2(k-j)\frac{\boldsymbol{x}_{k-j}\boldsymbol{x}_{k-j}^t}{\|\boldsymbol{x}_{k-j}\|^2\boldsymbol{x}_{k-j}^t\boldsymbol{x}_{k-j}}\right\}.
\end{aligned}
$$

It is easy to see that the third and the fourth terms vanish by the assumptions (A-2)–(A-4).

Using the assumption (B-2), we have

$$\begin{aligned}\frac{\boldsymbol{x}_{k-j}\boldsymbol{x}_{k-j}^t}{\boldsymbol{x}_{k-j}^t\boldsymbol{x}_{k-j}} &= \frac{(\boldsymbol{s}(k-j)\boldsymbol{r}(k-j)\boldsymbol{u}_{k-j})(\boldsymbol{s}(k-j)\boldsymbol{r}(k-j)\boldsymbol{u}_{k-j})^t}{(\boldsymbol{s}(k-j)\boldsymbol{r}(k-j)\boldsymbol{u}_{k-j})^t(\boldsymbol{s}(k-j)\boldsymbol{r}(k-j)\boldsymbol{u}_{k-j})} \\ &= \frac{\boldsymbol{s}^2(k-j)\boldsymbol{r}^2(k-j)\boldsymbol{u}_{k-j}\boldsymbol{u}_{k-j}^t}{\boldsymbol{s}^2(k-j)\boldsymbol{r}^2(k-j)\|\boldsymbol{u}_{k-j}\|^2} \\ &= \boldsymbol{u}_{k-j}\boldsymbol{u}_{k-j}^t. \end{aligned} \tag{5.21}$$

Thus, we can express $\text{Cov}(\tilde{\boldsymbol{w}}_{k+1})$ as

$$\begin{aligned}\text{Cov}(\tilde{\boldsymbol{w}}_{k+1}) &= \text{E}\left\{\left[I-\mu\sum_{j\in J_k}\boldsymbol{u}_{k-j}\boldsymbol{u}_{k-j}^t\right]\text{Cov}(\tilde{\boldsymbol{w}}_k)\left[I-\mu\sum_{l\in J_k}\boldsymbol{u}_{k-l}\boldsymbol{u}_{k-l}^t\right]\right\} \\ &\quad + \text{E}\left\{\mu^2\sum_{j\in J_k}\boldsymbol{v}^2(k-j)\frac{1}{\boldsymbol{r}^2(k-j)}\boldsymbol{u}_{k-j}\boldsymbol{u}_{k-j}^t\right\} \\ &= \text{E}\left\{\left[I-\mu\sum_{j\in K_k}\nu_j\nu_j^t\right]\text{Cov}(\tilde{\boldsymbol{w}}_k)\left[I-\mu\sum_{l\in K_k}\nu_l\nu_l^t\right]\right\} \\ &\quad + \mu^2\sigma_v^2\text{E}\left\{1/r^2\right\}\text{E}\left\{\sum_{j\in K_k}\nu_j\nu_j^t\right\}, \end{aligned} \tag{5.22}$$

where $\text{E}\left\{1/r^2\right\} \triangleq \text{E}\left\{1/\boldsymbol{r}^2(k)\right\}$, for all k, and $K_k \subseteq \{1, 2, \ldots, n\}$ is the set of indices of ν_i's that appear as realizations of $\{\boldsymbol{u}_{k-j};\ j \in J_k\}$. In other words, K_k is the set of indices of ν_i's that appear at least once as realizations of $\boldsymbol{u}_k, \boldsymbol{u}_{k-1}, \ldots, \boldsymbol{u}_{k-(p-1)}$. In the example (5.19), $K_k = \{1, 3, 5, 8, 10, 15\}$. This is a random variable that takes a value in the set of all the subsets of $\{1, 2, \ldots, n\}$. For $i \in \{1, 2, \ldots, n\}$, the set $\{i \in K_k\}$ is the event that ν_i appears at least once as realizations of $\boldsymbol{u}_k, \boldsymbol{u}_{k-1}, \ldots, \boldsymbol{u}_{k-(p-1)}$. Since it is the complement of the event that ν_i does not appear at all as realizations of $\boldsymbol{u}_k, \boldsymbol{u}_{k-1}, \ldots, \boldsymbol{u}_{k-(p-1)}$,

$$P(i \in K_k) = 1 - (1 - p_i)^p. \tag{5.23}$$

Now we define

$$\gamma_{k,i} \triangleq \nu_i^t\text{Cov}(\tilde{\boldsymbol{w}}_k)\nu_i,$$

which is the ith diagonal component of $V^t\text{Cov}(\tilde{\boldsymbol{w}}_k)V$ for $i = 1, 2, \ldots, n$. Using (5.22), and noting

$$\nu_i^t\sum_{j\in K_k}\nu_j\nu_j^t = \begin{cases}\nu_i^t, & \text{if } i \in K_k; \\ 0, & \text{otherwise,}\end{cases}$$

we can calculate $\gamma_{k+1,i}$ as

$$\begin{aligned}
\gamma_{k+1,i} &= \nu_i^t \mathrm{Cov}(\tilde{\boldsymbol{w}}_{k+1})\nu_i \\
&= \mathrm{E}\left\{\nu_i^t \left[I - \mu \sum_{j\in K_k} \nu_j \nu_j^t\right] \mathrm{Cov}(\tilde{\boldsymbol{w}}_k) \left[I - \mu \sum_{l\in K_k} \nu_l \nu_l^t\right] \nu_i\right\} \\
&\quad + \mu^2 \sigma_v^2 \mathrm{E}\left\{\frac{1}{r^2}\right\} \mathrm{E}\left\{\nu_i^t \left[\sum_{j\in K_k} \nu_j \nu_j^t\right] \nu_i\right\} \\
&= \nu_i^t \mathrm{Cov}(\tilde{\boldsymbol{w}}_k)\nu_i + \mathrm{E}\left\{\nu_i^t \left[\mu \sum_{j\in K_k} \nu_j \nu_j^t\right] \mathrm{Cov}(\tilde{\boldsymbol{w}}_k) \left[\mu \sum_{l\in K_k} \nu_l \nu_l^t\right] \nu_i\right\} \\
&\quad - \mathrm{E}\left\{\nu_i^t \mathrm{Cov}(\tilde{\boldsymbol{w}}_k) \left[\mu \sum_{l\in K_k} \nu_l \nu_l^t\right] \nu_i\right\} - \mathrm{E}\left\{\nu_i^t \left[\mu \sum_{j\in K_k} \nu_j \nu_j^t\right] \mathrm{Cov}(\tilde{\boldsymbol{w}}_k)\nu_i\right\} \\
&\quad + \mu^2 \sigma_v^2 \mathrm{E}\left\{\frac{1}{r^2}\right\} \mathrm{E}\left\{\nu_i^t \left[\sum_{j\in K_k} \nu_j \nu_j^t\right] \nu_i\right\} \\
&= \nu_i^t \mathrm{Cov}(\tilde{\boldsymbol{w}}_k)\nu_i + \mu^2 \nu_i^t \mathrm{Cov}(\tilde{\boldsymbol{w}}_k)\nu_i \mathrm{E}\left\{\Xi_{i\in K_k}\right\} - 2\mu \nu_i^t \mathrm{Cov}(\tilde{\boldsymbol{w}}_k)\nu_i \mathrm{E}\left\{\Xi_{i\in K_k}\right\} \\
&\quad + \mu^2 \sigma_v^2 \mathrm{E}\left\{1/r^2\right\} \mathrm{E}\left\{\Xi_{i\in K_k}\right\},
\end{aligned}$$

where $\Xi_{i\in K_k}$ is the characteristic function of the event $\{i \in K_k\}$:

$$\Xi_{i\in K_k} = \begin{cases} 1, & \text{if } i \in K_k; \\ 0, & \text{otherwise.} \end{cases}$$

Since

$$\mathrm{E}\left\{\Xi_{i\in K_k}\right\} = P(i \in K_k),$$

we finally have

$$\begin{aligned}
\gamma_{k+1,i} &= \nu_i^t \mathrm{Cov}(\tilde{\boldsymbol{w}}_k)\nu_i + \mu^2 \nu_i^t \mathrm{Cov}(\tilde{\boldsymbol{w}}_k)\nu_i P(i \in K_k) \\
&\quad - 2\mu \nu_i^t \mathrm{Cov}(\tilde{\boldsymbol{w}}_k)\nu_i P(i \in K_k) + \mu^2 \sigma_v^2 \mathrm{E}\left\{1/r^2\right\} P(i \in K_k) \\
&= \gamma_{k,i}\{1 - \mu(2-\mu)P(i \in K_k)\} + \mu^2 \sigma_v^2 \mathrm{E}\left\{1/r^2\right\} P(i \in K_k) \\
&= (1 - \alpha\beta_i)\gamma_{k,i} + \mu^2 \sigma_v^2 \mathrm{E}\left\{1/r^2\right\} \beta_i,
\end{aligned} \tag{5.24}$$

where $\alpha = \mu(2-\mu)$, and by (5.23), $\beta_i = P(i \in K_k) = 1 - (1-p_i)^P$.

Lemma 5.3 *For each i, $\gamma_{k,i}$ converges if $0 < \mu < 2$, and when it converges,*

$$\lim_{k\to\infty} \gamma_{k,i} = \frac{\mu}{2-\mu} \sigma_v^2 \mathrm{E}\left\{1/r^2\right\}. \tag{5.25}$$

If $\mu < 0$ *or* $2 \le \mu$, $\gamma_{k,i}$ *does not remain bounded.*

Proof Since $\alpha \le 1$ and $0 < \beta_i \le 1$, we always have $1 - \alpha\beta_i \ge 0$. Moreover $1 - \alpha\beta_i < 1$ if $0 < \mu < 2$. Therefore, from (5.24), $\gamma_{k,i}$ converges if $0 < \mu < 2$. It is easy to see that (5.24) is rewritten as

$$\gamma_{k+1,i} - \frac{\mu}{2-\mu}\sigma_v^2 \mathrm{E}\left\{1/r^2\right\} = (1 - \alpha\beta_i)\left(\gamma_{k,i} - \frac{\mu}{2-\mu}\sigma_v^2 \mathrm{E}\left\{1/r^2\right\}\right),$$

from which (5.25) is obvious.

If $\mu < 0$ or $2 \le \mu$, then $(1 - \alpha\beta_i) \ge 1$ and $\mu^2\sigma_v^2\mathrm{E}\left\{1/r^2\right\}\beta_i > 0$. Therefore, $\gamma_{k,i}$ does not remain bounded. □

Theorem 5.4 *(1) The a priori error* $\boldsymbol{e}_a(k)$ *converges in the mean-square if* $0 < \mu < 2$.

(2) The coefficient error vector $\tilde{\boldsymbol{w}}_k$ *converges in the mean-square if* $0 < \mu < 2$. *If* $\mu < 0$ *or* $2 \le \mu$, $\mathrm{E}\left\{\|\tilde{\boldsymbol{w}}_k\|^2\right\}$ *does not remain bounded.*

(3) The coefficient error vector $\tilde{\boldsymbol{w}}_k$ *converges to* 0 *in the mean if* $0 < \mu < 2$. *It does not if* $\mu \le 0$.

Proof (1) Using the assumption (A-5), we obtain

$$\begin{aligned}
\mathrm{E}\left\{\boldsymbol{e}_a^2(k)\right\} &= \mathrm{E}\left\{(\tilde{\boldsymbol{w}}_k^t \boldsymbol{x}_k)^2\right\} \\
&= \mathrm{tr}\left[R_x \mathrm{Cov}(\tilde{\boldsymbol{w}}_k)\right] \\
&= \mathrm{tr}\left[V \Lambda V^t \mathrm{Cov}(\tilde{\boldsymbol{w}}_k)\right] \\
&= \mathrm{tr}\left[\Lambda V^t \mathrm{Cov}(\tilde{\boldsymbol{w}}_k) V\right] \\
&= \sum_{i=1}^n \lambda_i \gamma_{k,i}.
\end{aligned}$$

This shows $\mathrm{E}\left\{\boldsymbol{e}_a^2(k)\right\}$ converges if $\gamma_{k,i}$ converges for each i. Therefore, in view of Lemma 5.3, $\mathrm{E}\left\{\boldsymbol{e}_a^2(k)\right\}$ converges if $0 < \mu < 2$.

(2) Since V is an orthogonal matrix, $V^t = V^{-1}$. Therefore, by Formula A.8 (Appendix A),

$$\begin{aligned}
\mathrm{E}\left\{\|\tilde{\boldsymbol{w}}_k\|^2\right\} &= \mathrm{tr}\left[\mathrm{Cov}(\tilde{\boldsymbol{w}}_k)\right] \\
&= \mathrm{tr}\left[V^t \mathrm{Cov}(\tilde{\boldsymbol{w}}_k) V\right] \\
&= \sum_{i=1}^n \gamma_{k,i}.
\end{aligned}$$

This equation, together with Lemma 5.3, shows that $\mathrm{E}\left\{\|\tilde{\boldsymbol{w}}_k\|^2\right\}$ converges if $0 < \mu < 2$. If $\mu < 0$ or $2 \leq \mu$, again by Lemma 5.3, $\mathrm{E}\left\{\|\tilde{\boldsymbol{w}}_k\|^2\right\}$ does not remain bounded.
(3) From (5.20) and (5.21),

$$\begin{aligned}\mathrm{E}\left\{\tilde{\boldsymbol{w}}_{k+1}\right\} &= \mathrm{E}\left\{\left[I - \mu \sum_{j\in J_k} \boldsymbol{u}_{k-j}\boldsymbol{u}_{k-j}^t\right]\tilde{\boldsymbol{w}}_k\right\} \\ &= \mathrm{E}\left\{\tilde{\boldsymbol{w}}_k\right\} - \mathrm{E}\left\{\mu \sum_{j\in K_k} \nu_j \nu_j^t \tilde{\boldsymbol{w}}_k\right\}. \end{aligned} \tag{5.26}$$

Let us define a vector ρ_k by

$$\rho_k \triangleq V^t \mathrm{E}\left\{\tilde{\boldsymbol{w}}_k\right\}.$$

The ith component of ρ_k is expressed as

$$\rho_{k,i} = \nu_i^t \mathrm{E}\left\{\tilde{\boldsymbol{w}}_k\right\} = \mathrm{E}\left\{\nu_i^t \tilde{\boldsymbol{w}}_k\right\}.$$

Multiplying ν_i^t to both sides of (5.26) from the left, we have

$$\begin{aligned}\rho_{k+1,i} &= \rho_{k,i} - \mathrm{E}\left(\mu\nu_i^t \sum_{j\in K_k} \nu_j\nu_j^t \tilde{\boldsymbol{w}}_k\right) \\ &= \rho_{k,i} - \mathrm{E}\left\{\mu\nu_i^t \tilde{\boldsymbol{w}}_k \Xi_{i\in K_k}\right\} \\ &= \rho_{k,i} - \mu\mathrm{E}\left\{\nu_i^t \tilde{\boldsymbol{w}}_k\right\}\mathrm{E}\left\{\Xi_{i\in K_k}\right\} \\ &= \rho_{k,i} - \mu\mathrm{E}\left\{\nu_i^t \tilde{\boldsymbol{w}}_k\right\} P(i\in K_k) \\ &= (1-\mu\beta_i)\rho_{k,i}. \end{aligned} \tag{5.27}$$

Suppose $0 < \mu < 2$. Then, since $0 < \beta_i \leq 1$, we have $0 < \mu\beta_i < 2$. This leads to $|1-\mu\beta_i| < 1$, which implies $\lim_{k\to\infty}\rho_{k,i} = 0$. Therefore,

$$\begin{aligned}\lim_{k\to\infty}\mathrm{E}\left\{\tilde{\boldsymbol{w}}_k\right\} &= \lim_{k\to\infty}(V^t)^{-1}\rho_k \\ &= V\lim_{k\to\infty}\rho_k \\ &= 0. \end{aligned}$$

If $\mu \leq 0$, then $1-\mu\beta_i \geq 1$. Therefore, $\rho_{k,i}$ does not converge to 0. □

From (2) and (3) of the above theorem, we have the following corollaries:

Corollary 5.2 *The B-APA is stable iff* $0 < \mu < 2$.

This result is compatible with Corollary 5.1.

Corollary 5.3 *If* $0 < \mu < 2$, *the* EMSE *and the misadjustment* $\mathcal{M}$ *are given, respectively, by*

$$\mathrm{EMSE} = \frac{\mu}{2-\mu}\sigma_v^2 \mathrm{E}\left\{1/r^2\right\} tr\left[R_x\right], \tag{5.23}$$

and

$$\mathcal{M} = \frac{\mu}{2-\mu}\mathrm{E}\left\{1/r^2\right\} tr\left[R_x\right].$$

Proof By Lemma 5.3 and the proof of Theorem 5.4(1),

$$\begin{aligned}
\mathrm{EMSE} &= \lim_{k\to\infty} \mathrm{E}\left\{\boldsymbol{e}_a^2(k)\right\} \\
&= \sum_{i=1}^{n} \lambda_i \lim_{k\to\infty} \gamma_{k,i} \\
&= \sum_{i=1}^{n} \lambda_i \frac{\mu}{2-\mu}\sigma_v^2 \mathrm{E}\left\{1/r^2\right\} \\
&= \frac{\mu}{2-\mu}\sigma_v^2 \mathrm{E}\left\{1/r^2\right\} \mathrm{tr}\left[R_x\right].
\end{aligned}$$

Therefore,

$$\begin{aligned}
\mathcal{M} &= \frac{\mathrm{EMSE}}{\sigma_v^2} \\
&= \frac{\mu}{2-\mu}\mathrm{E}\left\{1/r^2\right\} \mathrm{tr}\left[R_x\right].
\end{aligned}$$

□

This corollary shows that the EMSE is independent of the projection order p. However, it should be noted that this conclusion is based on strong assumptions. In fact, a simulation result in [1] shows a slight dependence of the EMSE on p.

From Lemma 5.3 and the proof of (1) in Theorem 5.4, it is seen that $\mathrm{E}\left\{\boldsymbol{e}_a^2(k)\right\}$ is a weighted sum of exponential functions, where the convergence rate of each function is determined by a factor $1 - \alpha\beta_i$.

Since $\mu = 1$ gives the maximum α, it is the best choice as far as the convergence rate is concerned. However, choosing a small μ has a merit of obtaining small EMSE as seen in Lemma 5.3. Because $\beta_i = 1 - (1 - p_i)^p$ and $p_i > 0$, β_i increases as p increases, explaining the faster convergence of the APA over the NLMS algorithm. These results are compatible with the results in the preceding section.

Let us consider a situation where we can neglect the observation noise: $\sigma_v^2 = 0$, and $\gamma_{0,i} = c^2/\mathrm{tr}\,[R_x] > 0 \ (i = 1, 2, \ldots, n)$. Then, from (5.24), we have

$$\gamma_{k,i} = \frac{c^2}{\mathrm{tr}[R_x]}(1 - \alpha\beta_i)^k.$$

From this fact and the proof of Theorem 5.4(1),

$$\begin{aligned} \mathrm{E}\left\{e^2(k)\right\} &= \mathrm{E}\left\{e_a^2(k)\right\} \\ &= \sum_{i=1}^{n} \lambda_i \gamma_{k,i} \\ &= c^2 \sum_{i=1}^{n} p_i (1 - \alpha\beta_i)^k. \end{aligned}$$

If ν_i's are uniformly distributed, that is, $p_i = 1/n \ (i = 1, 2, \ldots, n)$, and $\mu = 1$, then

$$1 - \alpha\beta_i = (1 - 1/n)^p.$$

Therefore, we have

$$\mathrm{E}\left\{e^2(k)\right\} = c^2(1 - 1/n)^{kp},$$

which leads to

$$\begin{aligned} 10\log_{10} \mathrm{E}\left\{e^2(k)\right\} &= 10\log_{10} c^2 + 10\log_{10}(1 - 1/n)^{kp} \\ &= 10\log_{10} c^2 + (10/\log_e 10)kp\log_e(1 - 1/n) \\ &\approx 10\log_{10} c^2 - 4.343\frac{kp}{n}. \end{aligned}$$

Thus, the learning curve is linear in the logarithmic scale, and goes down by roughly 20 dB in every $5n/p$ samples. This result is compatible with the observation in [3] for the case of the NLMS algorithm. It is also noted that the convergence slows down when the filter length n is large.

The above analysis also applies, with necessary changes, to the case of sparse regressors $\boldsymbol{x}_k, \boldsymbol{x}_{k-D}, \boldsymbol{x}_{k-2D}, \ldots, \boldsymbol{x}_{k-(p-1)D}$ with a delay of D samples between regressors as employed in the NLMS-OCF.

The theory described in this section is based on strong assumptions, e.g., the statistical dependence of the coefficient vector on the past measurement noise is negligible. A recent paper [4] investigated the effect of the correlation between the coefficient vector and the past measurement noise on the MSE.

5.4 Autoregressive Regressor Assumption

5.4.1 Formulation

In the papers [5, 6], the convergence behavior of the B-APA is investigated when the input signal is an autoregressive (AR) process, and $\mu = 1$. This work uses the representation of the B-APA that was first presented in [3] to show the decorrelating property of the APA as discussed in Sect. 3.5. Let us start with a brief review of that section with a modification that all the signals are treated as random processes. Throughout this section, we assume $n \gg p$.

The input signal $\{x(k)\}$ is assumed to be a stationary random process determined by a difference equation

$$x(k) = \sum_{i=1}^{p-1} \alpha_i x(k-i) + \xi(k),$$

where $\{\xi(k)\}$ is a white noise with $\mathrm{E}\{\xi(k)\} = 0$, $\mathrm{E}\{\xi^2(k)\} = \sigma_\xi^2$, and scalars $\alpha_1, \alpha_2, \ldots, \alpha_{p-1}$ are linear predictor coefficients (Chap. 3, Appendix 5 "Linear Prediction"). From this equation, we have a matrix representation

$$\boldsymbol{x}_k = \tilde{\boldsymbol{X}}_k^t \alpha + \boldsymbol{\xi}_k, \tag{5.29}$$

where

$$\begin{aligned}
\boldsymbol{x}_k &\stackrel{\Delta}{=} (x(k), x(k-1), \ldots, x(k-(n-1)))^t, \\
\tilde{\boldsymbol{X}}_k &\stackrel{\Delta}{=} \left[\boldsymbol{x}_{k-1}\ \boldsymbol{x}_{k-2}\ \ldots\ \boldsymbol{x}_{k-(p-1)}\right]^t, \\
\alpha &\stackrel{\Delta}{=} (\alpha_1, \alpha_2, \ldots, \alpha_{p-1})^t, \\
\boldsymbol{\xi}_k &\stackrel{\Delta}{=} (\xi(k), \xi(k-1), \ldots, \xi(k-(n-1)))^t.
\end{aligned}$$

The stochastic process $\{x(k)\}$ is called an *autoregressive (AR) process*. It is the output of an *infinite impulse response (IIR) filter* [7, p. 18] for the input signal $\{\xi(k)\}$. Since we assume that $\{\xi(k)\}$ is of zero mean, $\{x(k)\}$ is also of zero mean.

In order to make calculation tractable, we assume the following:

(C-1) The statistical dependence between $\tilde{\boldsymbol{X}}_k$ and $\boldsymbol{\xi}_k$ is negligible.

If the matrix $\tilde{\boldsymbol{X}}_k \tilde{\boldsymbol{X}}_k^t$ is nonsingular almost everywhere,[1] the least-squares estimate of the linear predictor α is given by

$$\hat{\alpha}_k = (\tilde{\boldsymbol{X}}_k \tilde{\boldsymbol{X}}_k^t)^{-1} \tilde{\boldsymbol{X}}_k \boldsymbol{x}_k \tag{5.30}$$

---

[1] This means $P(\det(\tilde{\boldsymbol{X}}_k \tilde{\boldsymbol{X}}_k^t) \neq 0) = 1$.

as in (3.21). Now let us define the error vector by

$$\boldsymbol{\phi}_k \triangleq \boldsymbol{x}_k - \tilde{\boldsymbol{X}}_k^t \hat{\boldsymbol{\alpha}}_k. \tag{5.31}$$

The vector $\tilde{\boldsymbol{X}}_k^t \hat{\boldsymbol{\alpha}}_k$ is the orthogonal projection of $\boldsymbol{x}_k$ onto the subspace spanned by $\boldsymbol{x}_{k-1}, \boldsymbol{x}_{k-2}, \ldots, \boldsymbol{x}_{k-(p-1)}$, which are the column vectors of $\tilde{\boldsymbol{X}}_k^t$. Therefore, $\boldsymbol{\phi}_k$ is orthogonal to every column vector of $\tilde{\boldsymbol{X}}_k^t$. This holds for every realization of the input signal. Using $\boldsymbol{\phi}_k$, we have a stochastic version of (3.22)

$$\boldsymbol{w}_{k+1} = \boldsymbol{w}_k + \frac{\boldsymbol{\phi}_k}{\|\boldsymbol{\phi}_k\|^2} \boldsymbol{e}(k), \tag{5.32}$$

where

$$\begin{aligned} \boldsymbol{e}(k) &= \boldsymbol{d}(k) - \boldsymbol{x}_k^t \boldsymbol{w}_k \\ &= \boldsymbol{x}_k^t w^o + \boldsymbol{v}(k) - \boldsymbol{x}_k^t \boldsymbol{w}_k \\ &= \boldsymbol{x}_k^t \tilde{\boldsymbol{w}}_k + \boldsymbol{v}(k), \end{aligned}$$

and $\tilde{\boldsymbol{w}}_k \triangleq w^o - \boldsymbol{w}_k$ is the coefficient error vector. Equation (5.32) can be rewritten in terms of $\tilde{\boldsymbol{w}}_k$ as

$$\begin{aligned} \tilde{\boldsymbol{w}}_{k+1} &= w^o - \boldsymbol{w}_{k+1} \\ &= w^o - \boldsymbol{w}_k - \frac{\boldsymbol{\phi}_k}{\|\boldsymbol{\phi}_k\|^2} \boldsymbol{e}(k) \\ &= \tilde{\boldsymbol{w}}_k - \frac{\boldsymbol{\phi}_k}{\|\boldsymbol{\phi}_k\|^2} (\boldsymbol{x}_k^t \tilde{\boldsymbol{w}}_k + \boldsymbol{v}(k)) \\ &= \tilde{\boldsymbol{w}}_k - \frac{\boldsymbol{\phi}_k \boldsymbol{x}_k^t}{\|\boldsymbol{\phi}_k\|^2} \tilde{\boldsymbol{w}}_k - \frac{\boldsymbol{\phi}_k}{\|\boldsymbol{\phi}_k\|^2} \boldsymbol{v}(k). \end{aligned} \tag{5.33}$$

Let us prepare a couple of lemmas here for later use.

Lemma 5.4 *(1)* $\boldsymbol{\phi}_k^t \tilde{\boldsymbol{w}}_{k+1} = \boldsymbol{\phi}_k^t \tilde{\boldsymbol{w}}_k - \boldsymbol{x}_k^t \tilde{\boldsymbol{w}}_k - \boldsymbol{v}(k)$.
(2) $\boldsymbol{x}_k^t \tilde{\boldsymbol{w}}_{k+1} = -\boldsymbol{v}(k)$.
(3) $\tilde{\boldsymbol{X}}_k \tilde{\boldsymbol{w}}_{k+1} = \tilde{\boldsymbol{X}}_k \tilde{\boldsymbol{w}}_k = -\tilde{\boldsymbol{v}}_{k-1}$, *where*

$$\tilde{\boldsymbol{v}}_{k-1} \triangleq (\boldsymbol{v}(k-1), \boldsymbol{v}(k-2), \ldots, \boldsymbol{v}(k-(p-1)))^t.$$

(4) Equation (5.33) *is rewritten as*

$$\tilde{\boldsymbol{w}}_{k+1} = \tilde{\boldsymbol{w}}_k - \frac{\boldsymbol{\phi}_k \boldsymbol{\phi}_k^t}{\|\boldsymbol{\phi}_k\|^2} \tilde{\boldsymbol{w}}_k - \frac{\boldsymbol{\phi}_k}{\|\boldsymbol{\phi}_k\|^2} \boldsymbol{v}_f(k).$$

where $\{v_f(k)\}$ *is the filtered noise defined by*

$$v_f(k) \triangleq v(k) - \hat{\alpha}_k^t \tilde{v}_{k-1}.$$

Proof (1) This equation is directly obtained by multiplying ϕ_k^t to both sides of (5.33) from the left.
(2) Since $x_k^t = \phi_k^t + \hat{\alpha}_k^t \tilde{X}_k$ from (5.31), and $\tilde{X}_k \phi_k = 0$, we have from (5.33)

$$\begin{aligned}
x_k^t \tilde{w}_{k+1} &= (\phi_k^t + \hat{\alpha}_k^t \tilde{X}_k)\tilde{w}_{k+1} \\
&= (\phi_k^t + \hat{\alpha}_k^t \tilde{X}_k)\tilde{w}_k \\
&\quad - (\phi_k^t + \hat{\alpha}_k^t \tilde{X}_k)\frac{\phi_k x_k^t}{\|\phi_k\|^2}\tilde{w}_k \\
&\quad - (\phi_k^t + \hat{\alpha}_k^t \tilde{X}_k)\frac{\phi_k}{\|\phi_k\|^2} v(k) \\
&= \phi_k^t \tilde{w}_k + \hat{\alpha}_k^t \tilde{X}_k \tilde{w}_k - x_k^t \tilde{w}_k - v(k) \\
&= (\phi_k^t + \hat{\alpha}_k^t \tilde{X}_k - x_k^t)\tilde{w}_k - v(k) \\
&= -v(k).
\end{aligned}$$

(3) Since $\tilde{X}_k \phi_k = 0$, the first equality is obvious from (5.33). Because

$$\begin{aligned}
x_{k-1}^t w_k &= d(k-1) = x_{k-1}^t w^o + v(k-1), \\
x_{k-2}^t w_k &= d(k-2) = x_{k-2}^t w^o + v(k-2), \\
&\vdots \\
x_{k-(p-1)}^t w_k &= d(k-(p-1)) = x_{k-(p-1)}^t w^o + v(k-(p-1)),
\end{aligned}$$

we have

$$\begin{aligned}
x_{k-1}^t \tilde{w}_k &= -v(k-1), \\
x_{k-2}^t \tilde{w}_k &= -v(k-2), \\
&\vdots \\
x_{k-(p-1)}^t \tilde{w}_k &= -v(k-(p-1)),
\end{aligned}$$

that is,

$$\tilde{X}_k \tilde{w}_k = -(v(k-1), v(k-2), \ldots, v(k-(p-1)))^t = -\tilde{v}_{k-1}.$$

(4) Substitution of $\boldsymbol{x}_k^t = \boldsymbol{\phi}_k^t + \hat{\boldsymbol{\alpha}}_k^t \tilde{\boldsymbol{X}}_k$ into (5.33) with the use of (2) and (3) above yields

$$\begin{aligned}
\tilde{\boldsymbol{w}}_{k+1} &= \tilde{\boldsymbol{w}}_k - \frac{\boldsymbol{\phi}_k(\boldsymbol{\phi}_k^t + \hat{\boldsymbol{\alpha}}_k^t \tilde{\boldsymbol{X}}_k)}{\|\boldsymbol{\phi}_k\|^2}\tilde{\boldsymbol{w}}_k - \frac{\boldsymbol{\phi}_k}{\|\boldsymbol{\phi}_k\|^2}\boldsymbol{v}(k) \\
&= \tilde{\boldsymbol{w}}_k - \frac{\boldsymbol{\phi}_k\boldsymbol{\phi}_k^t}{\|\boldsymbol{\phi}_k\|^2}\tilde{\boldsymbol{w}}_k - \frac{\boldsymbol{\phi}_k}{\|\boldsymbol{\phi}_k\|^2}(\boldsymbol{v}(k) + \hat{\boldsymbol{\alpha}}_k^t \tilde{\boldsymbol{X}}_k \tilde{\boldsymbol{w}}_k) \\
&= \tilde{\boldsymbol{w}}_k - \frac{\boldsymbol{\phi}_k\boldsymbol{\phi}_k^t}{\|\boldsymbol{\phi}_k\|^2}\tilde{\boldsymbol{w}}_k - \frac{\boldsymbol{\phi}_k}{\|\boldsymbol{\phi}_k\|^2}(\boldsymbol{v}(k) - \hat{\boldsymbol{\alpha}}_k^t \tilde{\boldsymbol{v}}_{k-1}) \\
&= \tilde{\boldsymbol{w}}_k - \frac{\boldsymbol{\phi}_k\boldsymbol{\phi}_k^t}{\|\boldsymbol{\phi}_k\|^2}\tilde{\boldsymbol{w}}_k - \frac{\boldsymbol{\phi}_k}{\|\boldsymbol{\phi}_k\|^2}\boldsymbol{v}_f(k).
\end{aligned}$$

□

Lemma 5.5 *If the matrix* $\boldsymbol{A}_k \triangleq \tilde{\boldsymbol{X}}_k^t(\tilde{\boldsymbol{X}}_k\tilde{\boldsymbol{X}}_k^t)^{-1}\tilde{\boldsymbol{X}}_k$ *has the complete shuffling property, then* $R_\phi \triangleq \mathrm{E}\left\{\boldsymbol{\phi}_k\boldsymbol{\phi}_k^t\right\}$ *is given by*

$$R_\phi = \left(\frac{n-(p-1)}{n}\right)\sigma_\xi^2 I.$$

Proof Substitution of (5.29) into (5.30) yields

$$\begin{aligned}
\hat{\boldsymbol{\alpha}}_k &= (\tilde{\boldsymbol{X}}_k\tilde{\boldsymbol{X}}_k^t)^{-1}\tilde{\boldsymbol{X}}_k(\tilde{\boldsymbol{X}}_k^t\alpha + \boldsymbol{\xi}_k) \\
&= \alpha + (\tilde{\boldsymbol{X}}_k\tilde{\boldsymbol{X}}_k^t)^{-1}\tilde{\boldsymbol{X}}_k\boldsymbol{\xi}_k.
\end{aligned} \tag{5.34}$$

Therefore, from (5.31), we get

$$\begin{aligned}
\boldsymbol{\phi}_k &= (\tilde{\boldsymbol{X}}_k^t\alpha + \boldsymbol{\xi}_k) - \tilde{\boldsymbol{X}}_k^t(\alpha + (\tilde{\boldsymbol{X}}_k\tilde{\boldsymbol{X}}_k^t)^{-1}\tilde{\boldsymbol{X}}_k\boldsymbol{\xi}_k) \\
&= \boldsymbol{\xi}_k - \tilde{\boldsymbol{X}}_k^t(\tilde{\boldsymbol{X}}_k\tilde{\boldsymbol{X}}_k^t)^{-1}\tilde{\boldsymbol{X}}_k\boldsymbol{\xi}_k \\
&= (I - \boldsymbol{A}_k)\boldsymbol{\xi}_k.
\end{aligned} \tag{5.35}$$

Thus, by using the assumption (C-1),

$$\begin{aligned}
R_\phi &= \mathrm{E}\left\{\boldsymbol{\phi}_k\boldsymbol{\phi}_k^t\right\} \\
&= \mathrm{E}\left\{(I - \boldsymbol{A}_k)\boldsymbol{\xi}_k\boldsymbol{\xi}_k^t(I - \boldsymbol{A}_k)^t\right\} \\
&= \mathrm{E}\left\{(I - \boldsymbol{A}_k)\mathrm{E}\left\{\boldsymbol{\xi}_k\boldsymbol{\xi}_k^t\right\}(I - \boldsymbol{A}_k)^t\right\} \\
&= \sigma_\xi^2\mathrm{E}\left\{(I - \boldsymbol{A}_k)(I - \boldsymbol{A}_k)^t\right\} \\
&= \sigma_\xi^2\mathrm{E}\left\{I - \boldsymbol{A}_k\right\} \\
&= \left(\frac{n-(p-1)}{n}\right)\sigma_\xi^2 I.
\end{aligned}$$

□

We assume that the matrix $\boldsymbol{A}_k$ has the complete shuffling property hereafter in this section. Furthermore, we add here one more assumption:

(C-2) The signal $\{\boldsymbol{\xi}(k)\}$ is Gaussian.

Lemma 5.6 *The expectations of $\|\boldsymbol{\phi}_k\|^2$, $\|\boldsymbol{\phi}_k\|^{-2}$, $\|\boldsymbol{\phi}_k\|^4$, $\|\boldsymbol{\phi}_k\|^{-4}$ are given, respectively, by*

(1) $\mathrm{E}\left\{\|\boldsymbol{\phi}_k\|^2\right\} = \sigma_\xi^2 g$,

(2) $\mathrm{E}\left\{\|\boldsymbol{\phi}_k\|^{-2}\right\} = \left(\sigma_\xi^2(g-2)\right)^{-1}$,

(3) $\mathrm{E}\left\{\|\boldsymbol{\phi}_k\|^4\right\} = \sigma_\xi^4(g+2)g$,

(4) $\mathrm{E}\left\{\|\boldsymbol{\phi}_k\|^{-4}\right\} = \left(\sigma_\xi^4(g-2)(g-4)\right)^{-1}$,

where $g \triangleq n-(p-1)$.

Proof From (5.35), we have

$$\begin{aligned}
\|\boldsymbol{\phi}_k\|^2 &= \boldsymbol{\phi}_k^t\boldsymbol{\phi}_k \\
&= \boldsymbol{\xi}_k^t(I-\boldsymbol{A}_k)^t(I-\boldsymbol{A}_k)\boldsymbol{\xi}_k \\
&= \boldsymbol{\xi}_k^t(I-\boldsymbol{A}_k)\boldsymbol{\xi}_k \\
&= \boldsymbol{\xi}_k^t(T^t\Lambda T)\boldsymbol{\xi}_k,
\end{aligned}$$

where T is an orthogonal matrix, and $\Lambda = \mathrm{diag}\,[\lambda_1, \lambda_2, \ldots, \lambda_n]$, λ_i's being the eigenvalues of $I-\boldsymbol{A}_k$. By Lemma 5.1, $(n-(p-1))$ eigenvalues out of $\lambda_1, \lambda_2, \ldots, \lambda_n$ take the value 1, and the remaining $(p-1)$ ones take the value 0. Let $\tilde{\boldsymbol{\xi}}_k \triangleq T\boldsymbol{\xi}_k$. Then,

$$\begin{aligned}
y &\triangleq \|\boldsymbol{\phi}_k\|^2 \\
&= \tilde{\boldsymbol{\xi}}_k^t\mathrm{diag}\,[\lambda_1, \lambda_2, \ldots, \lambda_n]\tilde{\boldsymbol{\xi}}_k \\
&= \sum_{i=1}^{n}\lambda_i\tilde{\xi}^2(k-(i-1)).
\end{aligned}$$

The vector $\tilde{\boldsymbol{\xi}}_k$ is zero-mean Gaussian since it is a linear transformation of a zero-mean Gaussian vector $\boldsymbol{\xi}_k$. Moreover, by the assumption (C-1),

$$\begin{aligned}
\mathrm{E}\left\{\tilde{\boldsymbol{\xi}}_k\tilde{\boldsymbol{\xi}}_k^t\right\} &= \mathrm{E}\left\{T\boldsymbol{\xi}_k\boldsymbol{\xi}_k^tT^t\right\} \\
&= \sigma_\xi^2\mathrm{E}\left\{TT^t\right\} \\
&= \sigma_\xi^2 I.
\end{aligned}$$

Therefore, $\tilde{\xi}(k), \tilde{\xi}(k-1), \ldots, \tilde{\xi}(k-(n-1))$ are mutually statistically independent, and $\mathrm{E}\left\{\tilde{\xi}^2(i)\right\} = \sigma_\xi^2$. Thus, $\tilde{y} \triangleq y/\sigma_\xi^2$ has the chi-square distribution with $g \triangleq$

$n - (p - 1)$ degrees of freedom since it is the sum of squares of $(n - (p - 1))$ independent random variables each of which has the Gaussian distribution with zero mean and unit variance [8, p. 183]. From this fact, the probability density function of y is given by

$$f(y) = \frac{1}{2^{g/2}\sigma_\xi^g \Gamma(g/2)} y^{(g/2-1)} e^{-y/(2\sigma_\xi^2)} \quad (y \geq 0), \tag{5.36}$$

where Γ denotes the gamma function.[2] Using this probability density function, and noting $\Gamma(z+1) = z\Gamma(z)$, we can evaluate $\mathrm{E}\left\{\|\boldsymbol{\phi}_k\|^2\right\}$ as follows:

$$\begin{aligned}
\mathrm{E}\left\{\|\boldsymbol{\phi}_k\|^2\right\} &= \int_0^\infty y f(y)\,\mathrm{d}y \\
&= \frac{1}{2^{g/2}\sigma_\xi^g \Gamma(g/2)} \int_0^\infty y^{(g/2)} e^{-y/(2\sigma_\xi^2)}\,\mathrm{d}y \\
&= \frac{1}{2^{g/2}\sigma_\xi^g \Gamma(g/2)} \int_0^\infty y^{((g+2)/2-1)} e^{-y/(2\sigma_\xi^2)}\,\mathrm{d}y \\
&= \frac{1}{2^{g/2}\sigma_\xi^g \Gamma(g/2)} 2^{(g+2)/2}\sigma_\xi^{(g+2)} \Gamma((g+2)/2) \\
&= 2\sigma_\xi^2 \frac{\Gamma(g/2+1)}{\Gamma(g/2)} \\
&= 2\sigma_\xi^2 \frac{(g/2)\Gamma(g/2)}{\Gamma(g/2)} \\
&= \sigma_\xi^2 g.
\end{aligned}$$

Other expectations are evaluated in a similar way. □

5.4.2 Convergence of Mean Coefficient Vector

Theorem 5.5 *The mean coefficient vector* $\mathrm{E}\{\boldsymbol{w}_k\}$ *converges to* 0 *for any initial value* $\mathrm{E}\{\tilde{\boldsymbol{w}}_0\}$ *iff* $p \leq n-2$.

Proof By Lemma 5.4(4), together with the assumptions (A-2) and (A-3), we have

$$\mathrm{E}\{\tilde{\boldsymbol{w}}_{k+1}\} = \mathrm{E}\{\tilde{\boldsymbol{w}}_k\} - \mathrm{E}\left\{\frac{\boldsymbol{\phi}_k \boldsymbol{\phi}_k^t}{\|\boldsymbol{\phi}_k\|^2}\right\} \mathrm{E}\{\tilde{\boldsymbol{w}}_k\}. \tag{5.37}$$

---

[2]In the original paper [6], $\sigma_\phi = \sqrt{g/n}\sigma_\xi$ is used instead of σ_ξ in (5.36) However, under the assumption $n \gg p$, these two quantities coincide.

Each component of $\boldsymbol{\phi}_k\boldsymbol{\phi}_k^t$ involves only two components of $\boldsymbol{\phi}_k$, whereas $\|\boldsymbol{\phi}_k\|^2$ involves all the components. Therefore, $\boldsymbol{\phi}_k\boldsymbol{\phi}_k^t$ and $\|\boldsymbol{\phi}_k\|^2$ are only weakly dependent when n is large. Thus, we can make an approximation

$$\begin{aligned} \mathrm{E}\left\{\frac{\boldsymbol{\phi}_k\boldsymbol{\phi}_k^t}{\|\boldsymbol{\phi}_k\|^2}\right\} &\approx \mathrm{E}\left\{\boldsymbol{\phi}_k\boldsymbol{\phi}_k^t\right\}\mathrm{E}\left\{\|\boldsymbol{\phi}_k\|^{-2}\right\} \\ &= R_\phi \mathrm{E}\left\{\|\boldsymbol{\phi}_k\|^{-2}\right\}. \end{aligned} \tag{5.38}$$

Substitution of (5.38) into (5.37) with the use of Lemmas 5.5 and 5.6 leads to

$$\mathrm{E}\left\{\tilde{\boldsymbol{w}}_{k+1}\right\} = \left(1 - \frac{g}{n(g-2)}\right)\mathrm{E}\left\{\tilde{\boldsymbol{w}}_k\right\}.$$

Since we assume $n \gg p$, this is equivalent to

$$\mathrm{E}\left\{\tilde{\boldsymbol{w}}_{k+1}\right\} = \left(1 - \frac{1}{g-2}\right)\mathrm{E}\left\{\tilde{\boldsymbol{w}}_k\right\}.$$

This shows that $\mathrm{E}\left\{\tilde{\boldsymbol{w}}_k\right\}$ converges to 0 for any initial value $\mathrm{E}\left\{\tilde{\boldsymbol{w}}_0\right\}$ iff[3]

$$\left|1 - \frac{1}{g-2}\right| < 1,$$

which is equivalent to $g \geq 3$. This inequality is satisfied because we assume $n \gg p$. □

5.4.3 Behavior of Mean-Square Error

Using the formulas in Lemma 5.4, we can calculate $\boldsymbol{e}(k)$ as follows:

$$\begin{aligned} \boldsymbol{e}(k) &= \boldsymbol{x}_k^t\tilde{\boldsymbol{w}}_k + \boldsymbol{v}(k) \\ &= \left(\boldsymbol{\phi}_k^t\tilde{\boldsymbol{w}}_k - \boldsymbol{\phi}_k^t\tilde{\boldsymbol{w}}_{k+1} - \boldsymbol{v}(k)\right) + \boldsymbol{v}(k) \\ &= \left(\boldsymbol{x}_k^t - \hat{\boldsymbol{\alpha}}_k^t\tilde{\boldsymbol{X}}_k\right)\tilde{\boldsymbol{w}}_k - \left(\boldsymbol{x}_k^t - \hat{\boldsymbol{\alpha}}_k^t\tilde{\boldsymbol{X}}_k\right)\tilde{\boldsymbol{w}}_{k+1} \\ &= \boldsymbol{x}_k^t\tilde{\boldsymbol{w}}_k - \boldsymbol{x}_k^t\tilde{\boldsymbol{w}}_{k+1} \\ &= \left(\boldsymbol{\phi}_k^t + \hat{\boldsymbol{\alpha}}_k^t\tilde{\boldsymbol{X}}_k\right)\tilde{\boldsymbol{w}}_k + \boldsymbol{v}(k) \\ &= \boldsymbol{v}(k) + \hat{\boldsymbol{\alpha}}_k^t\tilde{\boldsymbol{X}}_k\tilde{\boldsymbol{w}}_k + \boldsymbol{\phi}_k^t\tilde{\boldsymbol{w}}_k \\ &= \boldsymbol{v}(k) + \hat{\boldsymbol{\alpha}}_k^t(-\tilde{\boldsymbol{v}}_{k-1}) + \boldsymbol{\phi}_k^t\tilde{\boldsymbol{w}}_k \\ &= \boldsymbol{v}_f(k) + \boldsymbol{\phi}_k^t\tilde{\boldsymbol{w}}_k. \end{aligned}$$

[3] A more accurate equation, taking the initialization error into consideration, is presented in [9].

Hence,

$$\begin{aligned}\mathrm{E}\left\{e^2(k)\right\} &= \mathrm{E}\left\{(v_f(k)+\boldsymbol{\phi}_k^t\tilde{\boldsymbol{w}}_k)^2\right\}\\ &= \mathrm{E}\left\{v_f^2(k)\right\} + 2\mathrm{E}\left\{\boldsymbol{\phi}_k^t\tilde{\boldsymbol{w}}_k v_f(k)\right\} + \mathrm{E}\left\{\tilde{\boldsymbol{w}}_k^t\boldsymbol{\phi}_k\boldsymbol{\phi}_k^t\tilde{\boldsymbol{w}}_k\right\}.\end{aligned}$$

We first focus on the second term. By (5.30) and (5.29),

$$\begin{aligned}\hat{\boldsymbol{\alpha}}_k &= (\tilde{\boldsymbol{X}}_k\tilde{\boldsymbol{X}}_k^t)^{-1}\tilde{\boldsymbol{X}}_k(\tilde{\boldsymbol{X}}_k^t\alpha + \boldsymbol{\xi}_k)\\ &= \boldsymbol{\alpha} + (\tilde{\boldsymbol{X}}_k\tilde{\boldsymbol{X}}_k^t)^{-1}\tilde{\boldsymbol{X}}_k\boldsymbol{\xi}_k.\end{aligned} \tag{5.39}$$

If we neglect the off-diagonal components of $\tilde{\boldsymbol{X}}_k\tilde{\boldsymbol{X}}_k^t$,

$$(\tilde{\boldsymbol{X}}_k\tilde{\boldsymbol{X}}_k^t)^{-1} \approx \mathrm{diag}\left[\frac{1}{\boldsymbol{x}_{k-1}^t\boldsymbol{x}_{k-1}}\ \frac{1}{\boldsymbol{x}_{k-2}^t\boldsymbol{x}_{k-2}}\ \cdots\ \frac{1}{\boldsymbol{x}_{k-(p-1)}^t\boldsymbol{x}_{k-(p-1)}}\right].$$

Roughly speaking, $\boldsymbol{x}_{k-i}^t\boldsymbol{x}_{k-i} = \sum_{j=0}^{n-1}x^2(k-i-j)$ is proportional to n. Therefore, when n is sufficiently large, each component of $(\tilde{\boldsymbol{X}}_k\tilde{\boldsymbol{X}}_k^t)^{-1}\tilde{\boldsymbol{X}}$ takes a small value. This means that $\hat{\boldsymbol{\alpha}}_k$ is only weakly dependent on the regressors. This in turn means that $v_f(k)$ is only weakly dependent on the regressors. Also, since the dependence of $\boldsymbol{\phi}_k$ on $\tilde{\boldsymbol{w}}_k$ is negligible by the assumption (A-5), it is only weakly dependent on $\tilde{\boldsymbol{w}}_k v_f(k)$. Because $\{x(k)\}$ is zero mean,

$$\begin{aligned}\mathrm{E}\{\boldsymbol{\phi}_k\} &= \mathrm{E}\{\boldsymbol{x}_k\} - \mathrm{E}\left\{\tilde{\boldsymbol{X}}_k^t\hat{\boldsymbol{\alpha}}_k\right\}\\ &= \mathrm{E}\{\boldsymbol{x}_k\} - \mathrm{E}\left\{\tilde{\boldsymbol{X}}_k^t\right\}\mathrm{E}\{\hat{\boldsymbol{\alpha}}_k\}\\ &= 0.\end{aligned}$$

Hence,

$$\mathrm{E}\left\{\boldsymbol{\phi}_k^t\tilde{\boldsymbol{w}}_k v_f(k)\right\} \approx \mathrm{E}\left\{\boldsymbol{\phi}_k^t\right\}\mathrm{E}\left\{\tilde{\boldsymbol{w}}_k v_f(k)\right\} = \boldsymbol{0},$$

which leads to

$$\begin{aligned}\mathrm{E}\left\{e^2(k)\right\} &= \mathrm{E}\left\{v_f^2(k)\right\} + \mathrm{E}\left\{\tilde{\boldsymbol{w}}_k^t\boldsymbol{\phi}_k\boldsymbol{\phi}_k^t\tilde{\boldsymbol{w}}_k\right\}\\ &= \mathrm{E}\left\{v_f^2(k)\right\} + \mathrm{E}\left\{\mathrm{tr}[\tilde{\boldsymbol{w}}_k^t\boldsymbol{\phi}_k\boldsymbol{\phi}_k^t\tilde{\boldsymbol{w}}_k]\right\}\\ &= \mathrm{E}\left\{v_f^2(k)\right\} + \mathrm{E}\left\{\mathrm{tr}\left[\boldsymbol{\phi}_k\boldsymbol{\phi}_k^t\tilde{\boldsymbol{w}}_k\tilde{\boldsymbol{w}}_k^t\right]\right\}\\ &= \mathrm{E}\left\{v_f^2(k)\right\} + \mathrm{tr}\left[\mathrm{E}\left\{\boldsymbol{\phi}_k\boldsymbol{\phi}_k^t\tilde{\boldsymbol{w}}_k\tilde{\boldsymbol{w}}_k^t\right\}\right]\\ &= \mathrm{E}\left\{v_f^2(k)\right\} + \mathrm{tr}\left[\mathrm{E}\left\{\boldsymbol{\phi}_k\boldsymbol{\phi}_k^t\right\}\mathrm{E}\left\{\tilde{\boldsymbol{w}}_k\tilde{\boldsymbol{w}}_k^t\right\}\right]\\ &= \mathrm{E}\left\{v_f^2(k)\right\} + \mathrm{tr}[R_\phi Q_k],\end{aligned} \tag{5.40}$$

where $Q_k \triangleq \mathrm{E}\left\{\tilde{\boldsymbol{w}}_k \tilde{\boldsymbol{w}}_k^t\right\}$. Noting that $\{\boldsymbol{v}(k)\}$ is a white noise, we can calculate the first term as follows:

$$\begin{aligned}
\mathrm{E}\left\{\boldsymbol{v}_f^2(k)\right\} &= \mathrm{E}\left\{\boldsymbol{v}^2(k)\right\} - 2\mathrm{E}\left\{\sum_{i=1}^{p-1} \hat{\boldsymbol{\alpha}}_{k,i}\boldsymbol{v}(k)\boldsymbol{v}(k-i)\right\} \\
&\quad + \mathrm{E}\left\{\sum_{i=1}^{p-1}\sum_{j=1}^{p-1} \hat{\boldsymbol{\alpha}}_{k,i}\boldsymbol{v}(k-i)\hat{\boldsymbol{\alpha}}_{k,j}\boldsymbol{v}(k-j)\right\} \\
&= \mathrm{E}\left\{\boldsymbol{v}^2(k)\right\} + \mathrm{E}\left\{\sum_{i=1}^{p-1} \hat{\boldsymbol{\alpha}}_{k,i}^2\right\}\mathrm{E}\left\{\boldsymbol{v}^2(k)\right\} \\
&= \left(1 + \mathrm{E}\left\{\hat{\boldsymbol{\alpha}}_k^t \hat{\boldsymbol{\alpha}}_k\right\}\right)\sigma_v^2,
\end{aligned}$$

where $\hat{\boldsymbol{\alpha}}_{k,i}$ is the ith component of $\hat{\boldsymbol{\alpha}}_k$. From (5.39), we have

$$\begin{aligned}
\hat{\boldsymbol{\alpha}}_k^t \hat{\boldsymbol{\alpha}}_k &= \alpha^t\alpha + \alpha^t(\tilde{\boldsymbol{X}}_k\tilde{\boldsymbol{X}}_k^t)^{-1}\tilde{\boldsymbol{X}}_k\boldsymbol{\xi}_k + \boldsymbol{\xi}_k^t\tilde{\boldsymbol{X}}_k^t(\tilde{\boldsymbol{X}}_k\tilde{\boldsymbol{X}}_k^t)^{-1}\alpha \\
&\quad + \boldsymbol{\xi}_k^t\tilde{\boldsymbol{X}}_k^t(\tilde{\boldsymbol{X}}_k\tilde{\boldsymbol{X}}_k^t)^{-1}(\tilde{\boldsymbol{X}}_k\tilde{\boldsymbol{X}}_k^t)^{-1}\tilde{\boldsymbol{X}}_k\boldsymbol{\xi}_k.
\end{aligned}$$

Using the assumption (C-1), we can calculate the expectation of $\hat{\boldsymbol{\alpha}}_k^t \hat{\boldsymbol{\alpha}}_k$ in the following way:

$$\begin{aligned}
\mathrm{E}\left\{\hat{\boldsymbol{\alpha}}_k^t \hat{\boldsymbol{\alpha}}_k\right\} &= \alpha^t\alpha + \mathrm{E}\left\{\boldsymbol{\xi}_k^t\tilde{\boldsymbol{X}}_k^t\left(\tilde{\boldsymbol{X}}_k\tilde{\boldsymbol{X}}_k^t\right)^{-1}\left(\tilde{\boldsymbol{X}}_k\tilde{\boldsymbol{X}}_k^t\right)^{-1}\tilde{\boldsymbol{X}}_k\boldsymbol{\xi}_k\right\} \\
&= \alpha^t\alpha + \mathrm{E}\left\{\mathrm{tr}\left[\boldsymbol{\xi}_k^t\tilde{\boldsymbol{X}}_k^t\left(\tilde{\boldsymbol{X}}_k\tilde{\boldsymbol{X}}_k^t\right)^{-1}\left(\tilde{\boldsymbol{X}}_k\tilde{\boldsymbol{X}}_k^t\right)^{-1}\tilde{\boldsymbol{X}}_k\boldsymbol{\xi}_k\right]\right\} \\
&= \alpha^t\alpha + \mathrm{E}\left\{\mathrm{tr}\left[\boldsymbol{\xi}_k\boldsymbol{\xi}_k^t\tilde{\boldsymbol{X}}_k^t\left(\tilde{\boldsymbol{X}}_k\tilde{\boldsymbol{X}}_k^t\right)^{-1}\left(\tilde{\boldsymbol{X}}_k\tilde{\boldsymbol{X}}_k^t\right)^{-1}\tilde{\boldsymbol{X}}_k\right]\right\} \\
&= \alpha^t\alpha + \mathrm{tr}\left[\mathrm{E}\left\{\boldsymbol{\xi}_k\boldsymbol{\xi}_k^t\tilde{\boldsymbol{X}}_k^t\left(\tilde{\boldsymbol{X}}_k\tilde{\boldsymbol{X}}_k^t\right)^{-1}\left(\tilde{\boldsymbol{X}}_k\tilde{\boldsymbol{X}}_k^t\right)^{-1}\tilde{\boldsymbol{X}}_k\right\}\right] \\
&= \alpha^t\alpha + \mathrm{tr}\left[\mathrm{E}\left\{\boldsymbol{\xi}_k\boldsymbol{\xi}_k^t\right\}\mathrm{E}\left\{\tilde{\boldsymbol{X}}_k^t\left(\tilde{\boldsymbol{X}}_k\tilde{\boldsymbol{X}}_k^t\right)^{-1}\left(\tilde{\boldsymbol{X}}_k\tilde{\boldsymbol{X}}_k^t\right)^{-1}\tilde{\boldsymbol{X}}_k\right\}\right] \\
&= \alpha^t\alpha + \sigma_\xi^2\mathrm{tr}\left[\mathrm{E}\left\{\tilde{\boldsymbol{X}}_k^t\left(\tilde{\boldsymbol{X}}_k\tilde{\boldsymbol{X}}_k^t\right)^{-1}\left(\tilde{\boldsymbol{X}}_k\tilde{\boldsymbol{X}}_k^t\right)^{-1}\tilde{\boldsymbol{X}}_k\right\}\right] \\
&= \alpha^t\alpha + \sigma_\xi^2\mathrm{E}\left\{\mathrm{tr}\left[\tilde{\boldsymbol{X}}_k^t\left(\tilde{\boldsymbol{X}}_k\tilde{\boldsymbol{X}}_k^t\right)^{-1}\left(\tilde{\boldsymbol{X}}_k\tilde{\boldsymbol{X}}_k^t\right)^{-1}\tilde{\boldsymbol{X}}_k\right]\right\} \\
&= \alpha^t\alpha + \sigma_\xi^2\mathrm{E}\left\{\mathrm{tr}\left[\left(\tilde{\boldsymbol{X}}_k\tilde{\boldsymbol{X}}_k^t\right)\left(\tilde{\boldsymbol{X}}_k\tilde{\boldsymbol{X}}_k^t\right)^{-1}\left(\tilde{\boldsymbol{X}}_k\tilde{\boldsymbol{X}}_k^t\right)^{-1}\right]\right\} \\
&= \alpha^t\alpha + \sigma_\xi^2\mathrm{E}\left\{\mathrm{tr}\left[\left(\tilde{\boldsymbol{X}}_k\tilde{\boldsymbol{X}}_k^t\right)^{-1}\right]\right\} \\
&= \alpha^t\alpha + \sigma_\xi^2\mathrm{tr}\left[\mathrm{E}\left\{(\tilde{\boldsymbol{X}}_k\tilde{\boldsymbol{X}}_k^t)^{-1}\right\}\right].
\end{aligned}$$

Thus, we get

$$\mathrm{E}\left\{v_f^2(k)\right\} = \left(1 + \alpha^t\alpha + \sigma_\xi^2 \mathrm{tr}\left[\mathrm{E}\left\{\left(\tilde{\boldsymbol{X}}_k\tilde{\boldsymbol{X}}_k^t\right)^{-1}\right\}\right]\right)\sigma_v^2. \tag{5.41}$$

Substitution of (5.41) into (5.40) leads to

$$\mathrm{E}\left\{e(k)^2\right\} = \left(1 + \alpha^t\alpha + \sigma_\xi^2 \mathrm{tr}\left[\mathrm{E}\left\{(\tilde{\boldsymbol{X}}_k\tilde{\boldsymbol{X}}_k^t)^{-1}\right\}\right]\right)\sigma_v^2 + \mathrm{tr}\left[R_\phi Q_k\right]. \tag{5.42}$$

Expansion of $\mathrm{E}\left\{\tilde{\boldsymbol{w}}_{k+1}\tilde{\boldsymbol{w}}_{k+1}^t\right\}$ using Lemma 5.4(4) yields nine terms:

$$\begin{aligned} Q_{k+1} &= \mathrm{E}\left\{\tilde{\boldsymbol{w}}_{k+1}\tilde{\boldsymbol{w}}_{k+1}^t\right\} \\ &= t_1 + t_2 + t_3 + t_4 + t_5 + t_6 + t_7 + t_8 + t_9, \end{aligned}$$

where

$$\begin{aligned}
t_1 &= \mathrm{E}\left\{\tilde{\boldsymbol{w}}_k\tilde{\boldsymbol{w}}_k^t\right\} = Q_k, \\
t_2 &= -\mathrm{E}\left\{\tilde{\boldsymbol{w}}_k\tilde{\boldsymbol{w}}_k^t\frac{\boldsymbol{\phi}_k\boldsymbol{\phi}_k^t}{\|\boldsymbol{\phi}_k\|^2}\right\}, \\
t_3 &= -\mathrm{E}\left\{\tilde{\boldsymbol{w}}_k\frac{\boldsymbol{\phi}_k^t}{\|\boldsymbol{\phi}_k\|^2}\boldsymbol{v}_f(k)\right\}, \\
t_4 &= -\mathrm{E}\left\{\frac{\boldsymbol{\phi}_k\boldsymbol{\phi}_k^t}{\|\boldsymbol{\phi}_k\|^2}\tilde{\boldsymbol{w}}_k\tilde{\boldsymbol{w}}_k^t\right\}, \\
t_5 &= \mathrm{E}\left\{\frac{\boldsymbol{\phi}_k\boldsymbol{\phi}_k^t}{\|\boldsymbol{\phi}_k\|^2}\tilde{\boldsymbol{w}}_k\tilde{\boldsymbol{w}}_k^t\frac{\boldsymbol{\phi}_k\boldsymbol{\phi}_k^t}{\|\boldsymbol{\phi}_k\|^2}\right\}, \\
t_6 &= \mathrm{E}\left\{\frac{\boldsymbol{\phi}_k\boldsymbol{\phi}_k^t}{\|\boldsymbol{\phi}_k\|^2}\tilde{\boldsymbol{w}}_k\frac{\boldsymbol{\phi}_k^t}{\|\boldsymbol{\phi}_k\|^2}\boldsymbol{v}_f(k)\right\}, \\
t_7 &= -\mathrm{E}\left\{\frac{\boldsymbol{\phi}_k}{\|\boldsymbol{\phi}_k\|^2}\boldsymbol{v}_f(k)\tilde{\boldsymbol{w}}_k^t\right\}, \\
t_8 &= \mathrm{E}\left\{\frac{\boldsymbol{\phi}_k}{\|\boldsymbol{\phi}_k\|^2}\boldsymbol{v}_f(k)\tilde{\boldsymbol{w}}_k^t\frac{\boldsymbol{\phi}_k\boldsymbol{\phi}_k^t}{\|\boldsymbol{\phi}_k\|^2}\right\}, \\
t_9 &= \mathrm{E}\left\{\frac{\boldsymbol{\phi}_k}{\|\boldsymbol{\phi}_k\|^2}\boldsymbol{v}_f^2(k)\frac{\boldsymbol{\phi}_k^t}{\|\boldsymbol{\phi}_k\|^2}\right\}.
\end{aligned}$$

As stated previously, $\hat{\boldsymbol{\alpha}}$ is represented as the sum of α and a fluctuation term that is only weakly dependent on the regressors $\boldsymbol{x}_k, \boldsymbol{x}_{k-1}, \ldots, \boldsymbol{x}_{k-(p-1)}$. Therefore, $\boldsymbol{\phi}_k = \boldsymbol{x}_k - \tilde{\boldsymbol{X}}_k\hat{\boldsymbol{\alpha}}$ is approximated as a linear function of the regressors, which implies that $\boldsymbol{\phi}_k$ is Gaussian. Since $\boldsymbol{\phi}_k$ is a zero-mean vector, the third moments of its components are equal to zero [10, p. 39]. In addition, for the same reason as we derived (5.38), we have an approximation

$$\mathrm{E}\left\{\frac{\boldsymbol{\phi}_k^t}{\|\boldsymbol{\phi}_k\|^2}\right\} \approx \mathrm{E}\left\{\boldsymbol{\phi}_k^t\right\}\mathrm{E}\left\{\frac{1}{\|\boldsymbol{\phi}_k\|^2}\right\}$$
$$= 0.$$

Also, since each component of $\mathrm{E}\left\{\boldsymbol{\phi}_k\boldsymbol{\phi}_k^t\mathrm{E}\left\{\tilde{\boldsymbol{w}}_k\boldsymbol{v}_f(k)\right\}\boldsymbol{\phi}_k^t\right\}$ is a weighted sum of the third moments of the components of $\boldsymbol{\phi}_k$, we have

$$\mathrm{E}\left\{\boldsymbol{\phi}_k\boldsymbol{\phi}_k^t\tilde{\boldsymbol{w}}_k\boldsymbol{v}_f(k)\boldsymbol{\phi}_k^t\right\} \approx \mathrm{E}\left\{\boldsymbol{\phi}_k\boldsymbol{\phi}_k^t\mathrm{E}\left\{\tilde{\boldsymbol{w}}_k\boldsymbol{v}_f(k)\right\}\boldsymbol{\phi}_k^t\right\}$$
$$= 0.$$

Using these facts, we have

$$t_3 = -\mathrm{E}\left\{\tilde{\boldsymbol{w}}_k\boldsymbol{v}_f(k)\right\}\mathrm{E}\left\{\frac{\boldsymbol{\phi}_k^t}{\|\boldsymbol{\phi}_k\|^2}\right\} = 0,$$

and

$$t_6 = \mathrm{E}\left\{\boldsymbol{\phi}_k\boldsymbol{\phi}_k^t\tilde{\boldsymbol{w}}_k\boldsymbol{v}_f(k)\boldsymbol{\phi}_k^t\right\}\mathrm{E}\left\{\frac{1}{\|\boldsymbol{\phi}_k^t\|^4}\right\} = 0.$$

In a similar way, we can show $t_7 = 0$ and $t_8 = 0$. The second term is calculated as

$$\begin{aligned} t_2 &= -\mathrm{E}\left\{\tilde{\boldsymbol{w}}_k\tilde{\boldsymbol{w}}_k^t\right\}\mathrm{E}\left\{\frac{\boldsymbol{\phi}_k\boldsymbol{\phi}_k^t}{\|\boldsymbol{\phi}_k\|^2}\right\} \\ &= -\mathrm{E}\left\{\tilde{\boldsymbol{w}}_k\tilde{\boldsymbol{w}}_k^t\right\}\mathrm{E}\left\{\boldsymbol{\phi}_k\boldsymbol{\phi}_k^t\right\}\mathrm{E}\left\{\frac{1}{\|\boldsymbol{\phi}_k\|^2}\right\} \\ &= -Q_kR_\phi/\left(\sigma_\xi^2(g-2)\right). \end{aligned}$$

Similarly, we have

$$t_4 = -R_\phi Q_k/(\sigma_\xi^2(g-2)).$$

Evaluation of t_5 and t_9 needs a little bit more calculation.

Lemma 5.7 *The following equality holds*[4]*:*

$$\mathrm{E}\left\{\boldsymbol{\phi}_k\boldsymbol{\phi}_k^t\tilde{\boldsymbol{w}}_k\tilde{\boldsymbol{w}}_k^t\boldsymbol{\phi}_k\boldsymbol{\phi}_k^t\right\} = \sigma_\xi^2(g/n)tr[Q_k]R_\phi.$$

Proof Since $\boldsymbol{\phi}_k^t\tilde{\boldsymbol{w}}_k\tilde{\boldsymbol{w}}_k^t\boldsymbol{\phi}_k$ is a scalar,

$$\begin{aligned} \mathrm{E}\left\{\boldsymbol{\phi}_k\boldsymbol{\phi}_k^t\tilde{\boldsymbol{w}}_k\tilde{\boldsymbol{w}}_k^t\boldsymbol{\phi}_k\boldsymbol{\phi}_k^t\right\} &= \mathrm{E}\left\{\boldsymbol{\phi}_k(\boldsymbol{\phi}_k^t\tilde{\boldsymbol{w}}_k\tilde{\boldsymbol{w}}_k^t\boldsymbol{\phi}_k)\boldsymbol{\phi}_k^t\right\} \\ &= \mathrm{E}\left\{(\boldsymbol{\phi}_k^t\tilde{\boldsymbol{w}}_k\tilde{\boldsymbol{w}}_k^t\boldsymbol{\phi}_k)\boldsymbol{\phi}_k\boldsymbol{\phi}_k^t\right\} \\ &= \mathrm{E}\left\{\mathrm{tr}\left[\boldsymbol{\phi}_k\boldsymbol{\phi}_k^t\tilde{\boldsymbol{w}}_k\tilde{\boldsymbol{w}}_k^t\right]\boldsymbol{\phi}_k\boldsymbol{\phi}_k^t\right\}. \end{aligned} \quad (5.43)$$

[4]This is different from the corresponding result in the original paper [6]. However, under the present assumption that $n \gg p$, the two results are approximately the same.

Denoting the ith component of $\boldsymbol{\phi}_k$ and $\tilde{\boldsymbol{w}}_k$ by $\boldsymbol{\phi}_{k,i}$ and $\tilde{\boldsymbol{w}}_{k,i}$, respectively, we can write the (i, j)th component of (5.43) as

$$\begin{aligned}
&\mathrm{E}\left\{\mathrm{tr}\left[\boldsymbol{\phi}_k\boldsymbol{\phi}_k^t\tilde{\boldsymbol{w}}_k\tilde{\boldsymbol{w}}_k^t\right]\boldsymbol{\phi}_{k,i}\boldsymbol{\phi}_{k,j}\right\} \\
&\quad = \mathrm{E}\left\{\left[\sum_{l=1}^{n}\sum_{m=1}^{n}\boldsymbol{\phi}_{k,l}\boldsymbol{\phi}_{k,m}\tilde{\boldsymbol{w}}_{k,m}\tilde{\boldsymbol{w}}_{k,l}\right]\boldsymbol{\phi}_{k,i}\boldsymbol{\phi}_{k,j}\right\}. \qquad (5.44)
\end{aligned}$$

By Lemma 5.5, $\mathrm{E}\left\{\boldsymbol{\phi}_{k,i}\boldsymbol{\phi}_{k,j}\right\} = 0 \quad (i \neq j)$. Also, $\boldsymbol{\phi}_k$ and $\tilde{\boldsymbol{w}}_k$ are only weakly dependent as stated previously. Therefore, among n^2 terms in the square parentheses in (5.44), at most $4n$ terms are statistically dependent on $\boldsymbol{\phi}_{k,i}\boldsymbol{\phi}_{k,j}$. This enables us to assume that the quantity in the square parentheses and $\boldsymbol{\phi}_{k,i}\boldsymbol{\phi}_{k,j}$ are statistically independent for large n. Thus,

$$\begin{aligned}
&\mathrm{E}\left\{\left[\sum_{l=1}^{n}\sum_{m=1}^{n}\boldsymbol{\phi}_{k,l}\boldsymbol{\phi}_{k,m}\tilde{\boldsymbol{w}}_{k,m}\tilde{\boldsymbol{w}}_{k,l}\right]\boldsymbol{\phi}_{k,i}\boldsymbol{\phi}_{k,j}\right\} \\
&= \mathrm{E}\left\{\sum_{l=1}^{n}\sum_{m=1}^{n}\boldsymbol{\phi}_{k,l}\boldsymbol{\phi}_{k,m}\tilde{\boldsymbol{w}}_{k,m}\tilde{\boldsymbol{w}}_{k,l}\right\}\mathrm{E}\left\{\boldsymbol{\phi}_{k,i}\boldsymbol{\phi}_{k,j}\right\} \\
&= \mathrm{E}\left\{\sum_{l=1}^{n}\boldsymbol{\phi}_{k,l}^2\mathrm{E}\left\{\tilde{\boldsymbol{w}}_{k,l}^2\right\}\right\}\mathrm{E}\left\{\boldsymbol{\phi}_{k,i}\boldsymbol{\phi}_{k,j}\right\} \\
&\quad + \left\{\sum_{l=1}^{n}\sum_{m=1,m\neq l}^{n}\mathrm{E}\left\{\boldsymbol{\phi}_{k,l}\boldsymbol{\phi}_{k,m}\right\}\mathrm{E}\left\{\tilde{\boldsymbol{w}}_{k,m}\tilde{\boldsymbol{w}}_{k,l}\right\}\right\}\mathrm{E}\left\{\boldsymbol{\phi}_{k,i}\boldsymbol{\phi}_{k,j}\right\} \\
&= \mathrm{E}\left\{\sum_{l=1}^{n}\boldsymbol{\phi}_{k,l}^2\mathrm{E}\{\tilde{\boldsymbol{w}}_{k,l}^2\}\right\}\mathrm{E}\left\{\boldsymbol{\phi}_{k,i}\boldsymbol{\phi}_{k,j}\right\}. \qquad (5.45)
\end{aligned}$$

Assuming

$$\mathrm{V}(\tilde{\boldsymbol{w}}_{k,1}) = \mathrm{V}(\tilde{\boldsymbol{w}}_{k,2}) = \cdots = \mathrm{V}(\tilde{\boldsymbol{w}}_{k,n}),$$

where

$$\mathrm{V}(\tilde{\boldsymbol{w}}_{k,l}) \stackrel{\triangle}{=} \mathrm{E}\left\{(\tilde{\boldsymbol{w}}_{k,l} - \mathrm{E}\left\{\tilde{\boldsymbol{w}}_{k,l}\right\})^2\right\}$$

is the variance of $\tilde{\boldsymbol{w}}_{k,l}$, we have

$$\begin{aligned}
\mathrm{E}\left\{\tilde{\boldsymbol{w}}_{k,l}^2\right\} &= \mathrm{E}\left\{\tilde{\boldsymbol{w}}_{k,l}\right\}^2 + \mathrm{V}(\tilde{\boldsymbol{w}}_{k,l}) \\
&= \mathrm{E}\left\{\tilde{\boldsymbol{w}}_{k,l}\right\}^2 + \frac{1}{n}\mathrm{tr}\left[\mathrm{E}\left\{(\tilde{\boldsymbol{w}}_k - \mathrm{E}\left\{\tilde{\boldsymbol{w}}_k\right\})(\tilde{\boldsymbol{w}}_k^t - \mathrm{E}\{\tilde{\boldsymbol{w}}_k^t\})\right\}\right] \\
&= \frac{1}{n}\left(\mathrm{tr}[Q_k] - \mathrm{E}\left\{\tilde{\boldsymbol{w}}_k^t\right\}\mathrm{E}\left\{\tilde{\boldsymbol{w}}_k\right\}\right) + \mathrm{E}\left\{\tilde{\boldsymbol{w}}_{k,l}\right\}^2.
\end{aligned}$$

Substitution of this expression into (5.45) with the use of Lemma 5.6(1) yields

$$
\begin{aligned}
&\mathrm{E}\left\{\sum_{l=1}^{n}\boldsymbol{\phi}_{k,l}^2\mathrm{E}\left\{\tilde{\boldsymbol{w}}_{k,l}^2\right\}\right\}\mathrm{E}\left\{\boldsymbol{\phi}_{k,i}\boldsymbol{\phi}_{k,j}\right\}\\
&=\mathrm{E}\left\{\sum_{l=1}^{n}\boldsymbol{\phi}_{k,l}^2\left[\frac{1}{n}\left(\mathrm{tr}\left[Q_k\right]-\mathrm{E}\left\{\tilde{\boldsymbol{w}}_k^t\right\}\mathrm{E}\left\{\tilde{\boldsymbol{w}}_k\right\}\right)+\mathrm{E}\left\{\tilde{\boldsymbol{w}}_{k,l}\right\}^2\right]\right\}\mathrm{E}\left\{\boldsymbol{\phi}_{k,i}\boldsymbol{\phi}_{k,j}\right\}\\
&=\left\{\frac{1}{n}\left(\mathrm{tr}\left[Q_k\right]-\mathrm{E}\left\{\tilde{\boldsymbol{w}}_k^t\right\}\mathrm{E}\left\{\tilde{\boldsymbol{w}}_k\right\}\right)\mathrm{E}\left\{\boldsymbol{\phi}_k^t\boldsymbol{\phi}_k\right\}+\sum_{l=1}^{n}\mathrm{E}\left\{\phi_{k,l}^2\right\}\mathrm{E}\left\{\tilde{\boldsymbol{w}}_{k,l}\right\}^2\right\}\\
&\quad\times\mathrm{E}\left\{\boldsymbol{\phi}_{k,i}\boldsymbol{\phi}_{k,j}\right\}\\
&=\left\{\frac{1}{n}\left(\mathrm{tr}\left[Q_k\right]-\mathrm{E}\left\{\tilde{\boldsymbol{w}}_k^t\right\}\mathrm{E}\left\{\tilde{\boldsymbol{w}}_k\right\}\right)\left(n\sigma_\phi^2\right)+\sigma_\phi^2\sum_{l=1}^{n}\mathrm{E}\{\tilde{\boldsymbol{w}}_{k,l}\}^2\right\}\mathrm{E}\left\{\boldsymbol{\phi}_{k,i}\boldsymbol{\phi}_{k,j}\right\}\\
&=\left(\frac{1}{n}\left(\mathrm{tr}\left[Q_k\right]-\mathrm{E}\left\{\tilde{\boldsymbol{w}}_k^t\right\}\mathrm{E}\left\{\tilde{\boldsymbol{w}}_k\right\}\right)\left(n\sigma_\phi^2\right)+\sigma_\phi^2\mathrm{E}\left\{\tilde{\boldsymbol{w}}_k^t\right\}\mathrm{E}\left\{\tilde{\boldsymbol{w}}_k\right\}\right)\mathrm{E}\left\{\boldsymbol{\phi}_{k,i}\boldsymbol{\phi}_{k,j}\right\}\\
&=\sigma_\xi^2(g/n)\mathrm{tr}\left[Q_k\right]\mathrm{E}\left\{\boldsymbol{\phi}_{k,i}\boldsymbol{\phi}_{k,j}\right\}.
\end{aligned}
\tag{5.46}
$$

Thus, we have shown that the (i, j)th component of $\mathrm{E}\left\{\boldsymbol{\phi}_k\boldsymbol{\phi}_k^t\tilde{\boldsymbol{w}}_k\tilde{\boldsymbol{w}}_k^t\boldsymbol{\phi}_k\boldsymbol{\phi}_k^t\right\}$ is given by (5.46), which is the (i, j)th component of $\sigma_\xi^2(g/n)\mathrm{tr}[Q_k]R_\phi$. □

Let us employ an approximation

$$
\begin{aligned}
t_5&=\mathrm{E}\left\{\frac{\boldsymbol{\phi}_k\boldsymbol{\phi}_k^t}{\|\boldsymbol{\phi}_k\|^2}\tilde{\boldsymbol{w}}_k\tilde{\boldsymbol{w}}_k^t\frac{\boldsymbol{\phi}_k\boldsymbol{\phi}_k^t}{\|\boldsymbol{\phi}_k\|^2}\right\}\\
&\approx\frac{1}{\mathrm{E}\left\{\|\boldsymbol{\phi}_k\|^4\right\}}\mathrm{E}\left\{\boldsymbol{\phi}_k\boldsymbol{\phi}_k^t\tilde{\boldsymbol{w}}_k\tilde{\boldsymbol{w}}_k^t\boldsymbol{\phi}_k\boldsymbol{\phi}_k^t\right\}.
\end{aligned}
$$

Then, by Lemmas 5.6(3) and 5.7, t_5 is evaluated as

$$
t_5=\frac{1}{n\sigma_\xi^2(g+2)}\mathrm{tr}[Q_k]R_\phi.
$$

In order to evaluate the last term t_9 we again use the approximations that $v_f(k)$ is independent of $\boldsymbol{\phi}_k$ and that $\|\boldsymbol{\phi}_k\|^2$ is independent of $\boldsymbol{\phi}_k\boldsymbol{\phi}_k^t$. Then,

$$
\begin{aligned}
t_9&=\mathrm{E}\left\{\frac{\boldsymbol{\phi}_k}{\|\boldsymbol{\phi}_k\|^2}\boldsymbol{v}_f^2(k)\frac{\boldsymbol{\phi}_k^t}{\|\boldsymbol{\phi}_k\|^2}\right\}\\
&\approx\mathrm{E}\left\{\boldsymbol{v}_f^2(k)\right\}\mathrm{E}\left\{\frac{1}{\|\boldsymbol{\phi}_k\|^4}\right\}\mathrm{E}\left\{\boldsymbol{\phi}_k\boldsymbol{\phi}_k^t\right\}.
\end{aligned}
$$

The first factor on the right-hand side is given by (5.41). The second factor is given by Lemma 5.6(4). Thus, we have

$$t_9 = \left(1 + \alpha^t \alpha + \sigma_\xi^2 \text{tr}\left[\text{E}\left\{(\tilde{\boldsymbol{X}}_k \tilde{\boldsymbol{X}}_k^t)^{-1}\right\}\right]\right) \frac{\sigma_v^2}{\sigma_\xi^4 (g-2)(g-4)} R_\phi.$$

Adding all of t_1 through t_9, and using Lemma 5.5, we obtain the following recursion for Q_k:

$$Q_{k+1} = c_1 Q_k + c_2 \text{tr}[Q_k] I + c_3 I, \tag{5.47}$$

where

$$\begin{aligned}
c_1 &= 1 - \frac{2g}{n(g-2)}, \\
c_2 &= \frac{g}{n^2(g+2)}, \\
c_3 &= \left(1 - \alpha^t \alpha + \sigma_\xi^2 \text{tr}\left[\text{E}\left\{(\tilde{\boldsymbol{X}}_k \tilde{\boldsymbol{X}}_k^t)^{-1}\right\}\right]\right) \frac{\sigma_v^2 g}{n\sigma_\xi^2 (g-2)(g-4)}.
\end{aligned}$$

Taking the trace of both sides of (5.47) yields the following recursion for $\text{tr}\,[Q_k]$:

$$\text{tr}[Q_{k+1}] = (c_1 + nc_2)\text{tr}[Q_k] + nc_3.$$

The closed form solution of this equation is given by

$$\text{tr}[Q_k] = (c_1 + nc_2)^k \text{tr}[Q_0] + nc_3 \sum_{i=0}^{k-1} (c_1 + nc_2)^i. \tag{5.48}$$

Substituting (5.48) into (5.42) with the use of $\text{tr}[R_\phi Q_k] = \sigma_\xi^2 (g/n) \text{tr}[Q_k]$, we can calculate the time evolution of $\text{E}\left\{e^2(k)\right\}$, and are able to draw the learning curve.

Now we focus on the steady-state performance. Since

$$c_1 + nc_2 = 1 - \frac{g(g+6)}{n(g+2)(g-2)},$$

we have $0 < c_1 + nc_2 < 1$ if

$$0 < \frac{g(g+6)}{n(g+2)(g-2)} < 1.$$

Under this condition, $\text{tr}[Q_k]$ converges as $k \to \infty$, giving

$$\begin{aligned}
\text{tr}[Q_\infty] &\triangleq \lim_{k\to\infty} \text{tr}[Q_k] \\
&= nc_3 \frac{1}{1-(c_1+nc_2)} \\
&= \left(1 + \alpha^t \alpha + \sigma_\xi^2 \text{tr}\left[\text{E}\left\{(\tilde{\boldsymbol{X}}_k \tilde{\boldsymbol{X}}_k^t)^{-1}\right\}\right]\right) \frac{n\sigma_v^2 (g-2)}{\sigma_\xi^2 (g-4)(g+6)}.
\end{aligned} \tag{5.49}$$

In practical situations,

$$0 < \frac{2g}{n(g-2)} < 1.$$

Hence,

$$0 < 1 - \frac{2g}{n(g-2)} < 1.$$

Equation (5.47) shows that Q_k converges under this condition, and that

$$Q_\infty = c_1 Q_\infty + c_2 \mathrm{tr}[Q_\infty] I + c_3 I,$$

where $Q_\infty \triangleq \lim_{k\to\infty} Q_k$. Solving this equation for Q_∞ with the use of (5.49), we get

$$Q_\infty = \left(1 + \alpha^t\alpha + \sigma_\xi^2 \mathrm{tr}\left[\mathrm{E}\left\{(\tilde{\boldsymbol{X}}_k \tilde{\boldsymbol{X}}_k^t)^{-1}\right\}\right]\right) \frac{\sigma_v^2 (g+2)}{\sigma_\xi^2 (g-4)(g+6)}.$$

Furthermore, substitution of (5.49) into (5.42) yields

$$\begin{aligned}\mathrm{MSE} &= \lim_{k\to\infty} \mathrm{E}\left\{\boldsymbol{e}(k)^2\right\} \\ &= \left(1 + \alpha^t\alpha + \sigma_\xi^2 \mathrm{tr}\left[\mathrm{E}\left\{(\tilde{\boldsymbol{X}}_k \tilde{\boldsymbol{X}}_k^t)^{-1}\right\}\right]\right)\left(1 + \frac{g(g+2)}{(g-4)(g+6)}\right)\sigma_v^2.\end{aligned}$$

5.5 Convergence Behavior of R-APA

In this section, we make a review of a work on convergence behavior of algorithms in the APA family (Chap. 4) [11]. We restrict ourselves to the R-APA case because the results are readily extended to other algorithms in the family.

5.5.1 Steady-State Performance

If $\mathrm{E}\left\{\|\tilde{\boldsymbol{w}}_k\|^2\right\}$ and $\mathrm{E}\left\{\boldsymbol{e}_a^2(k)\right\}$ converge, we can expect that the equations

$$\mathrm{E}\left\{\|\tilde{\boldsymbol{w}}_k\|^2\right\} = \mathrm{E}\left\{\|\tilde{\boldsymbol{w}}_{k+1}\|^2\right\} = \mathrm{E}\left\{\|\tilde{\boldsymbol{w}}_{k+2}\|^2\right\} = \cdots,$$

and

$$\mathrm{E}\left\{\boldsymbol{e}_a^2(k)\right\} = \mathrm{E}\left\{\boldsymbol{e}_a^2(k+1)\right\} = \mathrm{E}\left\{\boldsymbol{e}_a^2(k+2)\right\} = \cdots$$

hold approximately for large k. When an adaptive filter reaches equilibrium in this sense, it is said to be in the *steady state*. In this subsection, we present approximate expressions for the EMSE and the MSE at steady state.

5.5.1.1 Energy Conservation Relation

Let us define the a priori error vector $\boldsymbol{e}_{a,k}$ and the a posteriori error vector $\boldsymbol{e}_{p,k}$, respectively, by

$$\boldsymbol{e}_{a,k} \stackrel{\triangle}{=} \left(\boldsymbol{x}_k^t \tilde{\boldsymbol{w}}_k, \boldsymbol{x}_{k-1}^t \tilde{\boldsymbol{w}}_k, \ldots, \boldsymbol{x}_{k-(p-1)}^t \tilde{\boldsymbol{w}}_k\right)^t = \boldsymbol{X}_k \tilde{\boldsymbol{w}}_k,$$

and

$$\boldsymbol{e}_{p,k} \stackrel{\triangle}{=} \left(\boldsymbol{x}_k^t \tilde{\boldsymbol{w}}_{k+1}, \boldsymbol{x}_{k-1}^t \tilde{\boldsymbol{w}}_{k+1}, \ldots, \boldsymbol{x}_{k-(p-1)}^t \tilde{\boldsymbol{w}}_{k+1}\right)^t = \boldsymbol{X}_k \tilde{\boldsymbol{w}}_{k+1}.$$

Note that we have

$$\boldsymbol{e}_k = \boldsymbol{e}_{a,k} + \boldsymbol{v}_k. \tag{5.50}$$

There is a relation among $\tilde{\boldsymbol{w}}_{k+1}$, $\tilde{\boldsymbol{w}}_k$, $\boldsymbol{e}_{a,k}$, and $\boldsymbol{e}_{p,k}$ that is useful for performance analysis of the APA. In terms of $\tilde{\boldsymbol{w}}_k$, the stochastic version of the update equation (3.12) for the R-APA is written as

$$\tilde{\boldsymbol{w}}_{k+1} = \tilde{\boldsymbol{w}}_k - \mu \boldsymbol{X}_k^t (\boldsymbol{X}_k \boldsymbol{X}_k^t + \delta \boldsymbol{I})^{-1} \boldsymbol{e}_k. \tag{5.51}$$

Multiplying $\boldsymbol{X}_k$ to both sides of (5.51) from the left, we obtain

$$\boldsymbol{e}_{p,k} = \boldsymbol{e}_{a,k} - \mu \boldsymbol{X}_k \boldsymbol{X}_k^t (\boldsymbol{X}_k \boldsymbol{X}_k^t + \delta \boldsymbol{I})^{-1} \boldsymbol{e}_k. \tag{5.52}$$

Let us assume that $\boldsymbol{X}_k \boldsymbol{X}_k^t$ is nonsingular almost everywhere. We solve (5.52) for $(\boldsymbol{X}_k \boldsymbol{X}_k^t + \delta \boldsymbol{I})^{-1} \boldsymbol{e}_k$, and substitute it into (5.51), finding

$$\tilde{\boldsymbol{w}}_{k+1} + \boldsymbol{X}_k^t (\boldsymbol{X}_k \boldsymbol{X}_k^t)^{-1} \boldsymbol{e}_{a,k} = \tilde{\boldsymbol{w}}_k + \boldsymbol{X}_k^t (\boldsymbol{X}_k \boldsymbol{X}_k^t)^{-1} \boldsymbol{e}_{p,k}.$$

Taking squared norms of both sides of this equation, we have an equation

$$\|\tilde{\boldsymbol{w}}_{k+1}\|^2 + \boldsymbol{e}_{a,k}^t (\boldsymbol{X}_k \boldsymbol{X}_k^t)^{-1} \boldsymbol{e}_{a,k} = \|\tilde{\boldsymbol{w}}_k\|^2 + \boldsymbol{e}_{p,k}^t (\boldsymbol{X}_k \boldsymbol{X}_k^t)^{-1} \boldsymbol{e}_{p,k}, \tag{5.53}$$

which is referred to as the *energy conservation relation*. Note that neither the step-size μ nor the regularization parameter δ appears in (5.53). From this equation, we immediately obtain

$$\mathrm{E}\left\{\|\tilde{\boldsymbol{w}}_{k+1}\|^2\right\} + \mathrm{E}\left\{\boldsymbol{e}_{a,k}^t (\boldsymbol{X}_k \boldsymbol{X}_k^t)^{-1} \boldsymbol{e}_{a,k}\right\} = \mathrm{E}\left\{\|\tilde{\boldsymbol{w}}_k\|^2\right\} + \mathrm{E}\left\{\boldsymbol{e}_{p,k}^t (\boldsymbol{X}_k \boldsymbol{X}_k^t)^{-1} \boldsymbol{e}_{p,k}\right\}. \tag{5.54}$$

Now let us assume that the adaptive filter is in the steady state. Then, for large k we have from (5.54)

$$\mathrm{E}\left\{e^t_{a,k}(X_kX^t_k)^{-1}e_{a,k}\right\} = \mathrm{E}\left\{e^t_{p,k}(X_kX^t_k)^{-1}e_{p,k}\right\}. \tag{5.55}$$

Substitution of (5.52) into the right-hand side of (5.55) yields

$$\begin{aligned}\mathrm{E}\left\{e^t_{a,k}(X_kX^t_k)^{-1}e_{a,k}\right\} = \mathrm{E}\left\{e^t_{a,k}(X_kX^t_k)^{-1}e_{a,k}\right\} - \mu\mathrm{E}\left\{e^t_{a,k}B_ke_k\right\}\\ - \mu\mathrm{E}\left\{e^t_kB_ke_{a,k}\right\} + \mu^2\mathrm{E}\left\{e^t_kA_ke_k\right\},\end{aligned} \tag{5.56}$$

where

$$A_k \triangleq (X_kX^t_k + \delta I)^{-1}X_kX^t_k(X_kX^t_k + \delta I)^{-1},$$

and

$$B_k \triangleq (X_kX^t_k + \delta I)^{-1}.$$

Thus, at steady state, we have

$$\mu\mathrm{E}\left\{e^t_kA_ke_k\right\} = \mathrm{E}\left\{e^t_{a,k}B_ke_k\right\} + \mathrm{E}\left\{e^t_kB_ke_{a,k}\right\}. \tag{5.57}$$

5.5.1.2 Evaluation of EMSE

Using (5.50) together with the assumptions (A-3) and (A-4), we get from (5.57)

$$\mu\mathrm{E}\left\{e^t_{a,k}A_ke_{a,k}\right\} + \mu\mathrm{E}\left\{v^t_kA_kv_k\right\} = 2\mathrm{E}\left\{e^t_{a,k}B_ke_{a,k}\right\}. \tag{5.58}$$

In order to derive an expression for $\mathrm{E}\left\{e^2_a(k)\right\}$, we first note

$$\begin{aligned}\mathrm{E}\left\{v^t_kA_kv_k\right\} &= \mathrm{E}\left\{\mathrm{tr}\left[v^t_kA_kv_k\right]\right\}\\ &= \mathrm{E}\left\{\mathrm{tr}\left[v_kv^t_kA_k\right]\right\}\\ &= \mathrm{tr}\left[\mathrm{E}\left\{v_kv^t_kA_k\right\}\right]\\ &= \sigma^2_v\mathrm{tr}\left[\mathrm{E}\left\{A_k\right\}\right].\end{aligned} \tag{5.59}$$

In the same way, we have

$$\begin{aligned}\mathrm{E}\left\{e^t_{a,k}A_ke_{a,k}\right\} &= \mathrm{E}\left\{\mathrm{tr}\left[e^t_{a,k}A_ke_{a,k}\right]\right\}\\ &= \mathrm{E}\left\{\mathrm{tr}\left[e_{a,k}e^t_{a,k}A_k\right]\right\}\\ &= \mathrm{tr}\left[\mathrm{E}\left\{e_{a,k}e^t_{a,k}A_k\right\}\right].\end{aligned} \tag{5.60}$$

and

$$\begin{aligned}\mathrm{E}\left\{e_{a,k}^t B_k e_{a,k}\right\} &= \mathrm{E}\left\{\mathrm{tr}\left[e_{a,k}^t B_k e_{a,k}\right]\right\} \\ &= \mathrm{E}\left\{\mathrm{tr}\left[e_{a,k} e_{a,k}^t B_k\right]\right\} \\ &= \mathrm{tr}\left[\mathrm{E}\left\{e_{a,k} e_{a,k}^t B_k\right\}\right]. \end{aligned} \tag{5.61}$$

To simplify (5.60) and (5.61), we assume the following:

(D-1) The random vector $e_{a,k}$ is statistically independent of X_k at steady state.

This assumption is called the *separation principle* [12, p. 307]. By using this assumption, (5.60) and (5.61) are further rewritten, respectively, as

$$\begin{aligned}\mathrm{E}\left\{e_{a,k}^t A_k e_{a,k}\right\} &= \mathrm{tr}\left[\mathrm{E}\left\{e_{a,k} e_{a,k}^t A_k\right\}\right] \\ &= \mathrm{tr}\left[\mathrm{E}\left\{e_{a,k} e_{a,k}^t\right\}\mathrm{E}\left\{A_k\right\}\right], \end{aligned} \tag{5.62}$$

and

$$\begin{aligned}\mathrm{E}\left\{e_{a,k}^t B_k e_{a,k}\right\} &= \mathrm{tr}\left[\mathrm{E}\left\{e_{a,k} e_{a,k}^t B_k\right\}\right] \\ &= \mathrm{tr}\left[\mathrm{E}\left\{e_{a,k} e_{a,k}^t\right\}\mathrm{E}\left\{B_k\right\}\right]. \end{aligned} \tag{5.63}$$

To get an approximate expression for $\mathrm{E}\left\{e_a^2(k)\right\}$ from $\mathrm{E}\left\{e_{a,k} e_{a,k}^t\right\}$, the following lemma is useful.

Lemma 5.8 *Suppose* $\delta \approx 0$. *If off-diagonal components are neglected,* $\mathrm{E}\left\{e_{a,k} e_{a,k}^t\right\}$ *is given by*

$$\mathrm{E}\left\{e_{a,k} e_{a,k}^t\right\} = \mathrm{E}\left\{e_a^2(k)\right\} S + \mu^2 \sigma_v^2 T, \tag{5.64}$$

where S *and* T *are diagonal matrices given, respectively, by*

$$S \triangleq \mathit{diag}\left[1, (1-\mu)^2, \ldots, (1-\mu)^{2(p-1)}\right],$$

and

$$T \triangleq \mathit{diag}\left[0, 1, 1+(1-\mu)^2, \ldots, 1+(1-\mu)^2+\cdots+(1-\mu)^{2(p-2)}\right].$$

Proof Since we assume $\delta \approx 0$, we have from (5.51)

$$X_k \tilde{w}_{k+1} = X_k \tilde{w}_k - \mu e_k.$$

Substitution of (5.50) into this equation yields

$$X_k \tilde{w}_{k+1} = (1-\mu) X_k \tilde{w}_k - \mu v_k.$$

Replacing k with $k-1$, we have a series of equations

$$\begin{aligned} \boldsymbol{x}_{k-1}^t \tilde{\boldsymbol{w}}_k &= (1-\mu)\boldsymbol{x}_{k-1}^t \tilde{\boldsymbol{w}}_{k-1} - \mu \boldsymbol{v}(k-1), \\ \boldsymbol{x}_{k-2}^t \tilde{\boldsymbol{w}}_k &= (1-\mu)\boldsymbol{x}_{k-2}^t \tilde{\boldsymbol{w}}_{k-1} - \mu \boldsymbol{v}(k-2), \\ \boldsymbol{x}_{k-3}^t \tilde{\boldsymbol{w}}_k &= (1-\mu)\boldsymbol{x}_{k-3}^t \tilde{\boldsymbol{w}}_{k-1} - \mu \boldsymbol{v}(k-3), \\ &\vdots \\ \boldsymbol{x}_{k-p}^t \tilde{\boldsymbol{w}}_k &= (1-\mu)\boldsymbol{x}_{k-p}^t \tilde{\boldsymbol{w}}_{k-1} - \mu \boldsymbol{v}(k-p). \end{aligned}$$

Squaring both sides of these equations, taking expectations, and using the assumptions (A-3) and (A-4), we obtain

$$\begin{aligned} \mathrm{E}\left\{(\boldsymbol{x}_{k-1}^t \tilde{\boldsymbol{w}}_k)^2\right\} &= (1-\mu)^2 \mathrm{E}\left\{(\boldsymbol{x}_{k-1}^t \tilde{\boldsymbol{w}}_{k-1})^2\right\} + \mu^2\sigma_v^2, && (5.65) \\ \mathrm{E}\left\{(\boldsymbol{x}_{k-2}^t \tilde{\boldsymbol{w}}_k)^2\right\} &= (1-\mu)^2 \mathrm{E}\left\{(\boldsymbol{x}_{k-2}^t \tilde{\boldsymbol{w}}_{k-1})^2\right\} + \mu^2\sigma_v^2, && (5.66) \\ \mathrm{E}\left\{(\boldsymbol{x}_{k-3}^t \tilde{\boldsymbol{w}}_k)^2\right\} &= (1-\mu)^2 \mathrm{E}\left\{(\boldsymbol{x}_{k-3}^t \tilde{\boldsymbol{w}}_{k-1})^2\right\} + \mu^2\sigma_v^2, && (5.67) \\ &\vdots \\ \mathrm{E}\left\{(\boldsymbol{x}_{k-p}^t \tilde{\boldsymbol{w}}_k)^2\right\} &= (1-\mu)^2 \mathrm{E}\left\{(\boldsymbol{x}_{k-p}^t \tilde{\boldsymbol{w}}_{k-1})^2\right\} + \mu^2\sigma_v^2. \end{aligned}$$

In the same way,

$$\begin{aligned} \mathrm{E}\left\{(\boldsymbol{x}_{k-2}^t \tilde{\boldsymbol{w}}_{k-1})^2\right\} &= (1-\mu)^2 \mathrm{E}\left\{(\boldsymbol{x}_{k-2}^t \tilde{\boldsymbol{w}}_{k-2})^2\right\} + \mu^2\sigma_v^2, && (5.68) \\ \mathrm{E}\left\{(\boldsymbol{x}_{k-3}^t \tilde{\boldsymbol{w}}_{k-1})^2\right\} &= (1-\mu)^2 \mathrm{E}\left\{(\boldsymbol{x}_{k-3}^t \tilde{\boldsymbol{w}}_{k-2})^2\right\} + \mu^2\sigma_v^2, && (5.69) \\ \mathrm{E}\left\{(\boldsymbol{x}_{k-4}^t \tilde{\boldsymbol{w}}_{k-1})^2\right\} &= (1-\mu)^2 \mathrm{E}\left\{(\boldsymbol{x}_{k-4}^t \tilde{\boldsymbol{w}}_{k-2})^2\right\} + \mu^2\sigma_v^2, \\ &\vdots \\ \mathrm{E}\left\{(\boldsymbol{x}_{k-(p+1)}^t \tilde{\boldsymbol{w}}_{k-1})^2\right\} &= (1-\mu)^2 \mathrm{E}\left\{(\boldsymbol{x}_{k-(p+1)}^t \tilde{\boldsymbol{w}}_{k-2})^2\right\} + \mu^2\sigma_v^2. \end{aligned}$$

Also,

$$\begin{aligned} \mathrm{E}\left\{(\boldsymbol{x}_{k-3}^t \tilde{\boldsymbol{w}}_{k-2})^2\right\} &= (1-\mu)^2 \mathrm{E}\left\{(\boldsymbol{x}_{k-3}^t \tilde{\boldsymbol{w}}_{k-3})^2\right\} + \mu^2\sigma_v^2, && (5.70) \\ \mathrm{E}\left\{(\boldsymbol{x}_{k-4}^t \tilde{\boldsymbol{w}}_{k-2})^2\right\} &= (1-\mu)^2 \mathrm{E}\left\{(\boldsymbol{x}_{k-4}^t \tilde{\boldsymbol{w}}_{k-3})^2\right\} + \mu^2\sigma_v^2, \\ \mathrm{E}\left\{(\boldsymbol{x}_{k-5}^t \tilde{\boldsymbol{w}}_{k-2})^2\right\} &= (1-\mu)^2 \mathrm{E}\left\{(\boldsymbol{x}_{k-5}^t \tilde{\boldsymbol{w}}_{k-3})^2\right\} + \mu^2\sigma_v^2, \\ &\vdots \\ \mathrm{E}\left\{(\boldsymbol{x}_{k-(p+2)}^t \tilde{\boldsymbol{w}}_{k-2})^2\right\} &= (1-\mu)^2 \mathrm{E}\left\{(\boldsymbol{x}_{k-(p+2)}^t \tilde{\boldsymbol{w}}_{k-3})^2\right\} + \mu^2\sigma_v^2, \end{aligned}$$

and so on.

The first diagonal component of $\mathrm{E}\left\{\boldsymbol{e}_{a,k}\boldsymbol{e}_{a,k}^t\right\}$ is simply given by

$$\mathrm{E}\left\{(\boldsymbol{x}_k^t\tilde{\boldsymbol{w}}_k)^2\right\} = \mathrm{E}\left\{\boldsymbol{e}_a^2(k)\right\}.$$

From (5.65), the second diagonal component is given by

$$\mathrm{E}\left\{(\boldsymbol{x}_{k-1}^t\tilde{\boldsymbol{w}}_k)^2\right\} = (1-\mu)^2\mathrm{E}\left\{\boldsymbol{e}_a^2(k-1)\right\} + \mu^2\sigma_v^2.$$

From (5.66) and (5.68), the third component is calculated as

$$\begin{aligned}
\mathrm{E}\left\{(\boldsymbol{x}_{k-2}^t\tilde{\boldsymbol{w}}_k)^2\right\} &= (1-\mu)^2\mathrm{E}\left\{(\boldsymbol{x}_{k-2}^t\tilde{\boldsymbol{w}}_{k-1})^2\right\} + \mu^2\sigma_v^2, \\
&= (1-\mu)^2[(1-\mu)^2\mathrm{E}\left\{(\boldsymbol{x}_{k-2}^t\tilde{\boldsymbol{w}}_{k-2})^2\right\} + \mu^2\sigma_v^2] + \mu^2\sigma_v^2, \\
&= (1-\mu)^4\mathrm{E}\left\{\boldsymbol{e}_a^2(k-2)\right\} + \left(1+(1-\mu)^2\right)\mu^2\sigma_v^2.
\end{aligned}$$

In the same way, the fourth component is obtained from (5.67), (5.69), and (5.70) as

$$\mathrm{E}\left\{(\boldsymbol{x}_{k-3}^t\tilde{\boldsymbol{w}}_k)^2\right\} = (1-\mu)^6\mathrm{E}\left\{\boldsymbol{e}_a^2(k-3)\right\} + \left(1+(1-\mu)^2+(1-\mu)^4\right)\mu^2\sigma_v^2.$$

Continuing this process, and noting that equations

$$\mathrm{E}\left\{\boldsymbol{e}_a^2(k)\right\} = \mathrm{E}\left\{\boldsymbol{e}_a^2(k-1)\right\} = \mathrm{E}\left\{\boldsymbol{e}_a^2(k-2)\right\} = \cdots$$

hold at steady state, we finally obtain Eq. (5.64). □

In the above lemma, we see that if δ and σ_v^2 are small enough,

$$\mathrm{E}\left\{\boldsymbol{e}_{a,k}\boldsymbol{e}_{a,k}^t\right\} \approx \mathrm{E}\left\{\boldsymbol{e}_a^2(k)\right\} S.$$

By using this approximation, (5.62) is rewritten as

$$\mathrm{E}\left\{\boldsymbol{e}_{a,k}^t\boldsymbol{A}_k\boldsymbol{e}_{a,k}\right\} = \mathrm{E}\left\{\boldsymbol{e}_a^2(k)\right\}\mathrm{tr}\left[S\cdot\mathrm{E}\left\{\boldsymbol{A}_k\right\}\right]. \tag{5.71}$$

In the same way, (5.63) is rewritten as

$$\mathrm{E}\left\{\boldsymbol{e}_{a,k}^t\boldsymbol{B}_k\boldsymbol{e}_{a,k}\right\} = \mathrm{E}\left\{\boldsymbol{e}_a^2(k)\right\}\mathrm{tr}\left[S\cdot\mathrm{E}\left\{\boldsymbol{B}_k\right\}\right]. \tag{5.72}$$

If we define

$$\alpha \triangleq \mathrm{tr}\left[S\cdot\mathrm{E}\left\{A_k\right\}\right],$$

and

$$\beta \triangleq \operatorname{tr}\left[S \cdot \mathrm{E}\left\{B_k\right\}\right],$$

then (5.58) is expressed as

$$(2\beta - \mu\alpha)\mathrm{E}\left\{e_a^2(k)\right\} = \mu\sigma_v^2 \operatorname{tr}\left[\mathrm{E}\left\{\boldsymbol{A}_k\right\}\right].$$

Thus, we have at steady state

$$\mathrm{EMSE} = \frac{\mu\sigma_v^2 \operatorname{tr}\left[\mathrm{E}\left\{\boldsymbol{A}_k\right\}\right]}{(2\beta - \mu\alpha)}. \tag{5.73}$$

In the case of the B-APA, the regularization parameter δ is equal to 0. This implies $\boldsymbol{A}_k = \boldsymbol{B}_k$, so that $\alpha = \beta$. Thus, in this special case, we have from (5.73)

$$\mathrm{EMSE} = \frac{\mu\sigma_v^2}{(2-\mu)} \frac{\operatorname{tr}\left[\mathrm{E}\left\{\boldsymbol{A}_k\right\}\right]}{\operatorname{tr}\left[S \cdot \mathrm{E}\left\{\boldsymbol{A}_k\right\}\right]}. \tag{5.74}$$

We treat here the following two cases:

(I) The step-size μ is close to 0: In this case, $S \approx I$ by Lemma 5.8. Thus,

$$\mathrm{EMSE} = \frac{\mu\sigma_v^2}{2-\mu}. \tag{5.75}$$

(II) The step-size μ is close to 1: In this case, $S \approx \operatorname{diag}[1, 0, \ldots, 0]$ by Lemma 5.8. Therefore,

$$\mathrm{EMSE} = \frac{\mu\sigma_v^2}{(2-\mu)} \frac{\operatorname{tr}\left[\mathrm{E}\left\{\boldsymbol{A}_k\right\}\right]}{\mathrm{E}\left\{\boldsymbol{A}_k(1,1)\right\}},$$

where $\boldsymbol{A}_k(1,1)$ is the first diagonal component of $\boldsymbol{A}_k$. If we use approximations

$$\operatorname{tr}\left[\mathrm{E}\left\{\boldsymbol{A}_k\right\}\right] \approx \mathrm{E}\left\{\frac{p}{\|\boldsymbol{x}_k\|^2}\right\},$$

and

$$\frac{1}{\mathrm{E}\left\{\boldsymbol{A}_k(1,1)\right\}} \approx \operatorname{tr}\left[R_x\right],$$

we get another expression

$$\mathrm{EMSE} = \frac{\mu\sigma_v^2}{(2-\mu)} \mathrm{E}\left\{\frac{p}{\|\boldsymbol{x}_k\|^2}\right\} \operatorname{tr}\left[R_x\right]. \tag{5.76}$$

Once the EMSE is obtained, the MSE and the misadjustment $\mathcal{M}$ are given, respectively, by

$$\mathrm{MSE} = \mathrm{EMSE} + \sigma_v^2, \tag{5.77}$$

and

$$\mathcal{M} = \frac{\mathrm{EMSE}}{\sigma_v^2} = \frac{\mu}{(2-\mu)} \mathrm{E}\left\{ \frac{p}{\|\boldsymbol{x}_k\|^2} \right\} \mathrm{tr}\,[R_x] .$$

The factor $\mu\sigma_v^2/(2-\mu)$ seems to be essential for the steady-state performance of the APA. It appears in all of (5.13), (5.28) in Corollary 5.3, (5.75), and (5.76).

Equation (5.76) is similar to the EMSE given by (5.28). However, Eq. (5.28) lacks the projection order p.

5.5.2 *Transient Analysis*

5.5.2.1 Weighted Energy Conservation Relation

Let Σ be an $n \times n$ symmetric, positive-definite matrix. Using Σ as a weighting matrix, we can derive a weighted version of the energy conservation relation. We first define the *weighted a priori error vector* and the *weighted a posteriori error vector*, respectively, by

$$\boldsymbol{e}_{a,k}^{\Sigma} \triangleq \boldsymbol{X}_k \Sigma \tilde{\boldsymbol{w}}_k,$$

and

$$\boldsymbol{e}_{p,k}^{\Sigma} \triangleq \boldsymbol{X}_k \Sigma \tilde{\boldsymbol{w}}_{k+1}.$$

We also define the *weighted norm* $\|a\|_\Sigma$ of a vector a by

$$\|a\|_\Sigma^2 \triangleq a^t \Sigma a.$$

Multiplying $\boldsymbol{X}_k \Sigma$ to both sides of (5.51) from the left, we obtain

$$\boldsymbol{e}_{p,k}^{\Sigma} = \boldsymbol{e}_{a,k}^{\Sigma} - \mu \boldsymbol{X}_k \Sigma \boldsymbol{X}_k^t (\boldsymbol{X}_k \boldsymbol{X}_k^t + \delta I)^{-1} \boldsymbol{e}_k. \tag{5.78}$$

We assume that $\boldsymbol{X}_k \Sigma \boldsymbol{X}_k^t$ is nonsingular. Then, we can solve (5.78) for $(\boldsymbol{X}_k \boldsymbol{X}_k^t + \delta I)^{-1} \boldsymbol{e}_k$. Substitution of the result into (5.51) yields

$$\tilde{\boldsymbol{w}}_{k+1} + \boldsymbol{X}_k^t (\boldsymbol{X}_k \Sigma \boldsymbol{X}_k^t)^{-1} \boldsymbol{e}_{a,k}^{\Sigma} = \tilde{\boldsymbol{w}}_k + \boldsymbol{X}_k^t (\boldsymbol{X}_k \Sigma \boldsymbol{X}_k^t)^{-1} \boldsymbol{e}_{p,k}^{\Sigma}. \tag{5.79}$$

Taking the squared weighted norm of both sides of (5.79), we have

$$\|\tilde{\boldsymbol{w}}_{k+1}\|_\Sigma^2+\left(\boldsymbol{e}_{a,k}^\Sigma\right)^t(\boldsymbol{X}_k\Sigma\boldsymbol{X}_k^t)^{-1}\boldsymbol{e}_{a,k}^\Sigma=\|\tilde{\boldsymbol{w}}_k\|_\Sigma^2+\left(\boldsymbol{e}_{p,k}^\Sigma\right)^t(\boldsymbol{X}_k\Sigma\boldsymbol{X}_k^t)^{-1}\boldsymbol{e}_{p,k}^\Sigma, \tag{5.80}$$

which is the *weighted energy conservation relation*. When $\Sigma = I$, this equation reduces to (5.53) as a special case.

5.5.2.2 Recursion for Weighted Energy Conservation Relation

From (5.78), we can derive

$$\begin{aligned}\left(\boldsymbol{e}_{p,k}^\Sigma\right)^t(\boldsymbol{X}_k\Sigma\boldsymbol{X}_k^t)^{-1}\boldsymbol{e}_{p,k}^\Sigma &= \left(\boldsymbol{e}_{a,k}^\Sigma-\mu\boldsymbol{X}_k\Sigma\boldsymbol{X}_k^t(\boldsymbol{X}_k\boldsymbol{X}_k^t+\delta I)^{-1}\boldsymbol{e}_k\right)^t\\ &\quad\times(\boldsymbol{X}_k\Sigma\boldsymbol{X}_k^t)^{-1}\left(\boldsymbol{e}_{a,k}^\Sigma-\mu\boldsymbol{X}_k\Sigma\boldsymbol{X}_k^t(\boldsymbol{X}_k\boldsymbol{X}_k^t+\delta I)^{-1}\boldsymbol{e}_k\right)\\ &=\left(\boldsymbol{e}_{a,k}^\Sigma\right)^t(\boldsymbol{X}_k\Sigma\boldsymbol{X}_k^t)^{-1}\boldsymbol{e}_{a,k}^\Sigma\\ &\quad-\mu\left(\boldsymbol{e}_{a,k}^\Sigma\right)^t(\boldsymbol{X}_k\boldsymbol{X}_k^t+\delta I)^{-1}\boldsymbol{e}_k\\ &\quad-\mu\boldsymbol{e}_k^t(\boldsymbol{X}_k\boldsymbol{X}_k^t+\delta I)^{-1}\boldsymbol{e}_{a,k}^\Sigma\\ &\quad+\mu^2\boldsymbol{e}_k^t(\boldsymbol{X}_k\boldsymbol{X}_k^t+\delta I)^{-1}(\boldsymbol{X}_k\Sigma\boldsymbol{X}_k^t)(\boldsymbol{X}_k\boldsymbol{X}_k^t+\delta I)^{-1}\boldsymbol{e}_k.\end{aligned} \tag{5.81}$$

Equations (5.80) and (5.81) yield

$$\begin{aligned}\|\tilde{\boldsymbol{w}}_{k+1}\|_\Sigma^2 &= \|\tilde{\boldsymbol{w}}_k\|_\Sigma^2-\mu\left(\boldsymbol{e}_{a,k}^\Sigma\right)^t(\boldsymbol{X}_k\boldsymbol{X}_k^t+\delta I)^{-1}\boldsymbol{e}_k-\mu\boldsymbol{e}_k^t(\boldsymbol{X}_k\boldsymbol{X}_k^t+\delta I)^{-1}\boldsymbol{e}_{a,k}^\Sigma\\ &\quad+\mu^2\boldsymbol{e}_k^t(\boldsymbol{X}_k\boldsymbol{X}_k^t+\delta I)^{-1}(\boldsymbol{X}_k\Sigma\boldsymbol{X}_k^t)(\boldsymbol{X}_k\boldsymbol{X}_k^t+\delta I)^{-1}\boldsymbol{e}_k.\end{aligned} \tag{5.82}$$

Substituting $\boldsymbol{e}_k=\boldsymbol{e}_{a,k}+\boldsymbol{v}_k$ into (5.82), taking expectations of both sides, and using the assumptions (A-3) and (A-4), we obtain

$$\begin{aligned}\mathrm{E}\left\{\|\tilde{\boldsymbol{w}}_{k+1}\|_\Sigma^2\right\} &= \mathrm{E}\left\{\|\tilde{\boldsymbol{w}}_k\|_\Sigma^2\right\}-\mu\mathrm{E}\left\{\left(\boldsymbol{e}_{a,k}^\Sigma\right)^t(\boldsymbol{X}_k\boldsymbol{X}_k^t+\delta I)^{-1}\boldsymbol{e}_{a,k}\right\}\\ &\quad-\mu\mathrm{E}\left\{\boldsymbol{e}_{a,k}^t(\boldsymbol{X}_k\boldsymbol{X}_k^t+\delta I)^{-1}\boldsymbol{e}_{a,k}^\Sigma\right\}\\ &\quad+\mu^2\mathrm{E}\left\{\boldsymbol{e}_{a,k}^t\boldsymbol{A}_k^\Sigma\boldsymbol{e}_{a,k}\right\}+\mu^2\mathrm{E}\left\{\boldsymbol{v}_k^t\boldsymbol{A}_k^\Sigma\boldsymbol{v}_k\right\},\end{aligned} \tag{5.83}$$

where

$$\boldsymbol{A}_k^\Sigma \triangleq (\boldsymbol{X}_k\boldsymbol{X}_k^t+\delta I)^{-1}(\boldsymbol{X}_k\Sigma\boldsymbol{X}_k^t)(\boldsymbol{X}_k\boldsymbol{X}_k^t+\delta I)^{-1}.$$

Equation (5.83) is rewritten, by substituting $\boldsymbol{e}_{a,k}=\boldsymbol{X}_k\tilde{\boldsymbol{w}}_k$, and $\boldsymbol{e}_{a,k}^\Sigma=\boldsymbol{X}_k\Sigma\tilde{\boldsymbol{w}}_k$, as

$$\mathrm{E}\left\{\|\tilde{\boldsymbol{w}}_{k+1}\|_\Sigma^2\right\}=\mathrm{E}\left\{\|\tilde{\boldsymbol{w}}_k\|_{\Sigma'}^2\right\}+\mu^2\mathrm{E}\left\{\boldsymbol{v}_k^t\boldsymbol{A}_k^\Sigma\boldsymbol{v}_k\right\}, \tag{5.84}$$

where

$$\Sigma' = \Sigma - \mu \Sigma X_k^t (X_k X_k^t + \delta I)^{-1} X_k - \mu X_k^t (X_k X_k^t + \delta I)^{-1} X_k \Sigma \\ + \mu^2 X_k^t A_k^{\Sigma} X_k.$$

Under the assumption (A-5), $\tilde{w}_k$ and Σ' are statistically independent. Therefore, we have

$$\mathrm{E}\left\{\|\tilde{w}_k\|_{\Sigma'}^2\right\} = \mathrm{E}\left\{\|\tilde{w}_k\|_{\mathrm{E}\{\Sigma'\}}^2\right\}.$$

Thus, (5.84) is rewritten as

$$\mathrm{E}\left\{\|\tilde{w}_{k+1}\|_{\Sigma}^2\right\} = \mathrm{E}\left\{\|\tilde{w}_k\|_{\mathrm{E}\{\Sigma'\}}^2\right\} + \mu^2 \mathrm{E}\left\{v_k^t A_k^{\Sigma} v_k\right\}. \tag{5.85}$$

Let us define a matrix Φ by

$$\Phi \triangleq \mathrm{E}\{\Sigma'\} \\ = \Sigma - \mu \Sigma \mathrm{E}\left\{X_k^t (X_k X_k^t + \delta I)^{-1} X_k\right\} - \mu \mathrm{E}\left\{X_k^t (X_k X_k^t + \delta I)^{-1} X_k\right\} \Sigma \\ + \mu^2 \mathrm{E}\left\{X_k^t A_k^{\Sigma} X_k\right\}.$$

Then, we have

$$\mathrm{E}\left\{\|\tilde{w}_{k+1}\|_{\Sigma}^2\right\} = \mathrm{E}\left\{\|\tilde{w}_k\|_{\Phi}^2\right\} + \mu^2 \mathrm{E}\left\{v_k^t A_k^{\Sigma} v_k\right\}. \tag{5.86}$$

If we define

$$P_k \triangleq X_k^t (X_k X_k^t + \delta I)^{-1} X_k,$$

then

$$X_k^t A_k^{\Sigma} X_k = X_k^t (X_k X_k^t + \delta I)^{-1} (X_k \Sigma X_k^t)(X_k X_k^t - \delta I)^{-1} X_k \\ = P_k \Sigma P_k.$$

Thus, Φ is rewritten as

$$\Phi = \Sigma - \mu I \cdot \Sigma \cdot \mathrm{E}\{P_k\} - \mu \mathrm{E}\{P_k\} \cdot \Sigma \cdot I + \mu^2 \mathrm{E}\{P_k \Sigma P_k\}. \tag{5.87}$$

We use the "vec" notation here. By Theorem A.10 (Appendix A), Eq. (5.87) leads to

$$\mathrm{vec}\{\Phi\} = \mathrm{vec}\{\Sigma\} - \mu(\mathrm{E}\left\{P_k^t\right\} \otimes I)\mathrm{vec}\{\Sigma\} - \mu(I \otimes \mathrm{E}\{P_k\})\mathrm{vec}\{\Sigma\} \\ + \mu^2 \mathrm{E}\left\{P_k^t \otimes P_k\right\} \mathrm{vec}\{\Sigma\}. \tag{5.88}$$

By denoting

$$F \stackrel{\triangle}{=} I - \mu\left(\mathrm{E}\left\{\boldsymbol{P}_k^t\right\} \otimes I + I \otimes \mathrm{E}\{\boldsymbol{P}_k\}\right) + \mu^2 \mathrm{E}\left\{\boldsymbol{P}_k^t \otimes \boldsymbol{P}_k\right\},$$

and

$$\sigma \stackrel{\triangle}{=} \mathrm{vec}\{\Sigma\}, \quad \phi \stackrel{\triangle}{=} \mathrm{vec}\{\Phi\},$$

Eq. (5.88) is expressed as

$$\phi = F\sigma. \tag{5.89}$$

Note that F is an $n^2 \times n^2$ matrix.

The last term of (5.86) is calculated as

$$\begin{aligned}
\mathrm{E}\left\{\boldsymbol{v}_k^t \boldsymbol{A}_k^\Sigma \boldsymbol{v}_k\right\} &= \mathrm{E}\left\{\mathrm{tr}\left[\boldsymbol{v}_k \boldsymbol{v}_k^t \boldsymbol{A}_k^\Sigma\right]\right\} \\
&= \mathrm{tr}\left[\mathrm{E}\left\{\boldsymbol{v}_k \boldsymbol{v}_k^t \boldsymbol{A}_k^\Sigma\right\}\right] \\
&= \mathrm{tr}\left[\mathrm{E}\left\{\boldsymbol{v}_k \boldsymbol{v}_k^t\right\} \mathrm{E}\left\{\boldsymbol{A}_k^\Sigma\right\}\right] \\
&= \sigma_v^2 \mathrm{tr}\left[\mathrm{E}\left\{\boldsymbol{A}_k^\Sigma\right\}\right] \\
&= \sigma_v^2 \mathrm{tr}\left[\mathrm{E}\left\{(\boldsymbol{X}_k \boldsymbol{X}_k^t + \delta I)^{-1} (\boldsymbol{X}_k \Sigma \boldsymbol{X}_k^t)(\boldsymbol{X}_k \boldsymbol{X}_k^t + \delta I)^{-1}\right\}\right] \\
&= \sigma_v^2 \mathrm{E}\left\{\mathrm{tr}\left[(\boldsymbol{X}_k \boldsymbol{X}_k^t + \delta I)^{-1} (\boldsymbol{X}_k \Sigma \boldsymbol{X}_k^t)(\boldsymbol{X}_k \boldsymbol{X}_k^t + \delta I)^{-1}\right]\right\} \\
&= \sigma_v^2 \mathrm{E}\left\{\mathrm{tr}\left[\boldsymbol{X}_k^t (\boldsymbol{X}_k \boldsymbol{X}_k^t + \delta I)^{-2} \boldsymbol{X}_k \Sigma\right]\right\} \\
&= \sigma_v^2 \mathrm{tr}\left[\mathrm{E}\left\{\boldsymbol{X}_k^t (\boldsymbol{X}_k \boldsymbol{X}_k^t + \delta I)^{-2} \boldsymbol{X}_k\right\} \Sigma\right] \\
&= \sigma_v^2 \gamma^t \sigma,
\end{aligned} \tag{5.90}$$

where

$$\gamma \stackrel{\triangle}{=} \mathrm{vec}\left\{\mathrm{E}\left\{\boldsymbol{X}_k^t (\boldsymbol{X}_k \boldsymbol{X}_k^t + \delta I)^{-2} \boldsymbol{X}_k\right\}\right\},$$

and the relation

$$\mathrm{tr}[AB] = \mathrm{vec}\{A^t\}^t \mathrm{vec}\{B\}$$

is used. From (5.86) and (5.90), we obtain

$$\mathrm{E}\left\{\|\tilde{\boldsymbol{w}}_{k+1}\|^2_{\mathrm{vec}^{-1}\{\sigma\}}\right\} = \mathrm{E}\left\{\|\tilde{\boldsymbol{w}}_k\|^2_{\mathrm{vec}^{-1}\{F\sigma\}}\right\} + \mu^2 \sigma_v^2 \gamma^t \sigma. \tag{5.91}$$

Replacing σ with $F\sigma$ in (5.91), we have

$$\mathrm{E}\left\{\|\tilde{\boldsymbol{w}}_{k+1}\|^2_{\mathrm{vec}^{-1}\{F\sigma\}}\right\} = \mathrm{E}\left\{\|\tilde{\boldsymbol{w}}_k\|^2_{\mathrm{vec}^{-1}\{F^2\sigma\}}\right\} + \mu^2 \sigma_v^2 (\gamma^t F\sigma).$$

Repeating this operation, we get the following series of equations:

$$
\begin{aligned}
\mathrm{E}\left\{\|\tilde{\boldsymbol{w}}_{k+1}\|^2_{\mathrm{vec}^{-1}\{\sigma\}}\right\} &= \mathrm{E}\left\{\|\tilde{\boldsymbol{w}}_{k}\|^2_{\mathrm{vec}^{-1}\{F\sigma\}}\right\} + \mu^2\sigma_v^2(\gamma^t\sigma) \\
\mathrm{E}\left\{\|\tilde{\boldsymbol{w}}_{k+1}\|^2_{\mathrm{vec}^{-1}\{F\sigma\}}\right\} &= \mathrm{E}\left\{\|\tilde{\boldsymbol{w}}_{k}\|^2_{\mathrm{vec}^{-1}\{F^2\sigma\}}\right\} + \mu^2\sigma_v^2(\gamma^t F\sigma), \\
&\vdots \\
\mathrm{E}\left\{\|\tilde{\boldsymbol{w}}_{k+1}\|^2_{\mathrm{vec}^{-1}\{F^{(n^2-1)}\sigma\}}\right\} &= \mathrm{E}\left\{\|\tilde{\boldsymbol{w}}_{k}\|^2_{\mathrm{vec}^{-1}\{F^{n^2}\sigma\}}\right\} + \mu^2\sigma_v^2(\gamma^t F^{(n^2-1)}\sigma).
\end{aligned}
\tag{5.92}
$$

Now, let

$$
f_F(x) = x^{n^2} + c_{(n^2-1)}x^{(n^2-1)} + \cdots + c_1 x + c_0 \tag{5.93}
$$

be the characteristic polynomial of the matrix F:

$$
f_F(x) \triangleq \det(xI - F).
$$

Theorem A.1 (Appendix A) asserts $f_F(F) = 0$, that is

$$
F^{n^2} = -\sum_{i=0}^{n^2-1} c_i F^i, \tag{5.94}
$$

which leads to

$$
\mathrm{E}\left\{\|\tilde{\boldsymbol{w}}_{k}\|^2_{\mathrm{vec}^{-1}\{F^{n^2}\sigma\}}\right\} = -\sum_{i=0}^{n^2-1} c_i \mathrm{E}\left\{\|\tilde{\boldsymbol{w}}_{k}\|^2_{\mathrm{vec}^{-1}\{F^i\sigma\}}\right\}. \tag{5.95}
$$

Equations (5.92) and (5.95) are expressed in a compact form as in the next theorem.

Theorem 5.6 *For a matrix $\mathcal{F}$, and vectors $\mathcal{W}_k$, $\mathcal{Y}$ defined by*

$$
\mathcal{F} \triangleq \begin{bmatrix} 0 & 1 & \cdots & \cdots & 0 \\ 0 & 0 & 1 & \cdots & 0 \\ \vdots & \vdots & \vdots & \ddots & \vdots \\ 0 & 0 & 0 & \cdots & 1 \\ -c_0 & -c_1 & -c_2 & \cdots & -c_{n^2-1} \end{bmatrix},
$$

$$
\mathcal{W}_k \triangleq \left(\mathrm{E}\left\{\|\tilde{\boldsymbol{w}}_{k}\|^2_{vec^{-1}\{\sigma\}}\right\}, \mathrm{E}\left\{\|\tilde{\boldsymbol{w}}_{k}\|^2_{vec^{-1}\{F\sigma\}}\right\}, \ldots, \mathrm{E}\left\{\|\tilde{\boldsymbol{w}}_{k}\|^2_{vec^{-1}\{F^{(n^2-1)}\sigma\}}\right\}\right)^t,
$$

$$
\mathcal{Y} \triangleq \left(\gamma^t\sigma, \gamma^t F\sigma, \ldots, \gamma^t F^{(n^2-1)}\sigma\right)^t,
$$

a recurrence equation

$$\mathcal{W}_{k+1} = \mathcal{F}\mathcal{W}_k + \mu^2\sigma_v^2\mathcal{Y} \tag{5.96}$$

holds, where

$$\begin{aligned}
\gamma &= vec\left\{\mathrm{E}\left\{\boldsymbol{X}_k^t\left(\boldsymbol{X}_k\boldsymbol{X}_k^t + \delta I\right)^{-2}\boldsymbol{X}_k\right\}\right\},\\
\sigma &= vec\left\{\Sigma\right\},\\
F &= I - \mu\left(\mathrm{E}\left\{\boldsymbol{P}_k^t\right\}\otimes I + I \otimes \mathrm{E}\left\{\boldsymbol{P}_k\right\}\right) + \mu^2\mathrm{E}\left\{\boldsymbol{P}_k^t \otimes \boldsymbol{P}_k\right\},
\end{aligned}$$

and $c_0, c_1, \ldots, c_{n^2-1}$ *are the coefficients of the characteristic polynomial of* F *as in* (5.93).

5.5.2.3 Mean-Square Stability

In order to discuss the stability of the linear system (5.96), we prepare a lemma.

Lemma 5.9 *The characteristic polynomial of* $\mathcal{F}$ *coincides with that of* F.

Proof By expanding $\det(xI - \mathcal{F})$ with respect to the last row, we have

$$\begin{aligned}
f_{\mathcal{F}}(x) &= \det(xI - \mathcal{F})\\
&= \det\begin{bmatrix}
x & -1 & \cdots & \cdots & \cdots & 0\\
0 & x & -1 & \cdots & \cdots & 0\\
0 & 0 & x & -1 & \cdots & 0\\
\vdots & \vdots & \ddots & \ddots & \vdots & \vdots\\
0 & 0 & \cdots & \cdots & x & -1\\
c_0 & c_1 & c_2 & \cdots & \cdots & x + c_{(n^2-1)}
\end{bmatrix}\\
&= c_0(-1)^{(n^2+1)}(-1)^{(n^2-1)} + c_1(-1)^{(n^2+2)}x(-1)^{(n^2-2)}\\
&\quad + c_2(-1)^{(n^2+3)}x^2(-1)^{(n^2-3)} + \cdots + (x + c_{(n^2-1)})(-1)^{(n^2+n^2)}x^{(n^2-1)}\\
&= c_0 + c_1x + \cdots + c_{(n^2-1)}x^{(n^2-1)} + x^{n^2}\\
&= f_F(x).
\end{aligned}$$

□

According to Theorem 5.9 and Corollary 5.4 (Appendix 1: "Stability of Linear Systems"), the linear system (5.96) is asymptotically stable, and $\mathcal{W}_k$ converges, if

$$|\rho_i| < 1 \ \ (i = 1, 2, \ldots, n^2), \tag{5.97}$$

where ρ_i's are eigenvalues of $\mathcal{F}$. Convergence of $\mathcal{W}_k$ implies convergence of its first component $\mathrm{E}\left\{\|\tilde{\boldsymbol{w}}_k\|^2_{\mathrm{vec}^{-1}\{\sigma\}}\right\}$.

Theorem 5.7 *Under the assumption that* $\mathrm{E}\{\boldsymbol{P}_k\}$ *is positive definite, the linear system* (5.96) *is asymptotically stable, so that* $\mathrm{E}\left\{\|\tilde{\boldsymbol{w}}_k\|^2_{vec^{-1}\{\sigma\}}\right\}$ *converges, if*

$$0 < \mu < \min\left\{\frac{1}{\lambda_{\max}(C^{-1}D)}, \frac{1}{\lambda^+_{\max}(H)}\right\}, \tag{5.98}$$

where

$$\begin{aligned}
C &\triangleq \mathrm{E}\left\{\boldsymbol{P}^t_k\right\} \otimes I + I \otimes \mathrm{E}\{\boldsymbol{P}_k\}, \\
D &\triangleq \mathrm{E}\left\{\boldsymbol{P}^t_k \otimes \boldsymbol{P}_k\right\}, \\
H &\triangleq \begin{bmatrix} C/2 & -D/2 \\ I & 0 \end{bmatrix},
\end{aligned}$$

$\lambda_{\max}(C^{-1}D)$ *is the largest eigenvalue of the matrix* $C^{-1}D$, *and* $\lambda^+_{\max}(H)$ *is the largest positive eigenvalue of* H. *If there is no positive eigenvalue of* H, *the condition* (5.98) *becomes*

$$0 < \mu < \frac{1}{\lambda_{\max}(C^{-1}D)}.$$

Proof Let ρ_i's be the eigenvalues of F. Since F is symmetric, its eigenvalues are real. Therefore, in view of Lemma 5.9, it suffices to show that

$$-1 < \rho_i < 1 \ (i = 1, 2, \ldots, n^2).$$

Since we assume that $\mathrm{E}\{\boldsymbol{P}_k\} = \mathrm{E}\left\{\boldsymbol{P}^t_k\right\}$ is positive definite, $\mathrm{E}\left\{\boldsymbol{P}^t_k\right\} \otimes I$ and $I \otimes \mathrm{E}\{\boldsymbol{P}_k\}$ are positive definite by Corollary A.4 (Appendix A). Therefore, C is positive definite. Also, by Theorem A.9 (Appendix A) and by definition of the inner product in the tensor space, we have for any vector $r \otimes s$ of compatible size,

$$\begin{aligned}
(r \otimes s)^t \mathrm{E}\left\{\boldsymbol{P}^t_k \otimes \boldsymbol{P}_k\right\}(r \otimes s) &= \mathrm{E}\left\{(r \otimes s)^t(\boldsymbol{P}^t_k \otimes \boldsymbol{P}_k)(r \otimes s)\right\} \\
&= \mathrm{E}\left\{(r \otimes s)^t\left(\left(\boldsymbol{P}^t_k r\right) \otimes (\boldsymbol{P}_k s)\right)\right\} \\
&= \mathrm{E}\left\{\left(r^t \boldsymbol{P}^t_k r\right) \cdot \left(s^t \boldsymbol{P}_k s\right)\right\}.
\end{aligned}$$

As is easily verified, $\boldsymbol{P}_k = \boldsymbol{P}^t_k$ is nonnegative definite. Therefore, $r^t \boldsymbol{P}^t_k r \geq 0$ and $s^t \boldsymbol{P}_k s \geq 0$, which shows that $D = \mathrm{E}\{\boldsymbol{P}^t_k \otimes \boldsymbol{P}_k\}$ is nonnegative definite.

The condition $\rho_i < 1$ $(i = 1, 2, \ldots, n^2)$ is equivalent to the condition $F < I$, that is

$$C - \mu D > 0. \tag{5.99}$$

By Theorem A.7 (Appendix A), there exists a nonsingular matrix T such that

$$\begin{aligned}
T^t C T &= I, \\
T^t D T &= \Lambda,
\end{aligned}$$

where $\Lambda \triangleq \text{diag}\,[\lambda_1, \lambda_2, \ldots, \lambda_{n^2}]$, and λ_i's are the roots of $\det(\lambda C - D) = 0$. Since

$$I - \mu\Lambda = T^t(C - \mu D)T,$$

the matrix $(C - \mu D)$ is positive definite iff $I - \mu\Lambda$ is positive definite. Hence $(C - \mu D)$ is positive definite iff

$$\mu < \frac{1}{\max\{\lambda_1, \lambda_2, \ldots, \lambda_n\}}. \tag{5.100}$$

On the other hand, we have

$$\begin{aligned}\det(\lambda C - D) &= \det(C(\lambda I - C^{-1}D)) \\ &= \det C \cdot \det(\lambda I - C^{-1}D),\end{aligned}$$

which shows that λ_i's are the eigenvalues of $C^{-1}D$. Therefore, the inequality (5.100) is equivalent to

$$\mu < \frac{1}{\lambda_{\max}(C^{-1}D)}.$$

The condition $-1 < \rho_i$ $(i = 1, 2, \ldots, n^2)$ is equivalent to the condition $-I < F$. The following proof is due to [13]. If we define

$$G \triangleq 2I - \mu C + \mu^2 D,$$

then the condition $-I < F$ is equivalent to $G > 0$. The eigenvalues of G are equal to 2 for $\mu = 0$, and vary continuously as μ increases. Therefore, if G is not positive definite for $\mu = a > 0$, there must be a value of μ for which $\det G = 0$. Let μ_0 be the smallest value of $\mu > 0$ for which $\det G = 0$. Then, the condition $G > 0$ is satisfied for $\mu < \mu_0$. Now let us define a block matrix

$$K \triangleq \begin{bmatrix} 2I - \mu C & \mu D \\ -\mu I & I \end{bmatrix}.$$

By Formula A.5 (Appendix A), we see $\det G = \det K$. Since K is rewritten as

$$K \triangleq \begin{bmatrix} 2I & 0 \\ 0 & I \end{bmatrix} \left(\begin{bmatrix} I & 0 \\ 0 & I \end{bmatrix} - \mu \begin{bmatrix} C/2 & -D/2 \\ I & 0 \end{bmatrix} \right),$$

the condition $\det G = 0$ is equivalent to the condition $\det(I - \mu L) = 0$, where

$$L \triangleq \begin{bmatrix} C/2 & -D/2 \\ I & 0 \end{bmatrix}.$$

Thus, we see

$$\mu_0 = \frac{1}{\lambda_{\max}(L)^+}.$$

If there is no real, positive eigenvalue of L, then it never happens that $\det(I - \mu L) = 0$ for positive μ. Thus, the condition $-1 < \rho_i$ $(i = 1, 2, \ldots, n^2)$ does not impose any restriction on the range of μ. □

By setting $\Sigma = I$, the above theorem gives a sufficient condition for convergence of $\tilde{\boldsymbol{w}}_k$ in the mean-square as a special case.

5.5.2.4 Convergence in the Mean

Substituting $\boldsymbol{e}_k = \boldsymbol{X}_k \tilde{\boldsymbol{w}}_k + \boldsymbol{v}_k$ into (5.51), we obtain

$$\begin{aligned}
\tilde{\boldsymbol{w}}_{k+1} &= (I - \mu \boldsymbol{X}_k^t (\boldsymbol{X}_k \boldsymbol{X}_k^t + \delta I)^{-1} \boldsymbol{X}_k) \tilde{\boldsymbol{w}}_k - \mu \boldsymbol{X}_k^t (\boldsymbol{X}_k^t (\boldsymbol{X}_k \boldsymbol{X}_k^t + \delta I)^{-1} \boldsymbol{v}_k \\
&= (I - \mu \boldsymbol{P}_k) \tilde{\boldsymbol{w}}_k - \mu \boldsymbol{X}_k^t (\boldsymbol{X}_k^t (\boldsymbol{X}_k \boldsymbol{X}_k^t + \delta I)^{-1} \boldsymbol{v}_k.
\end{aligned}$$

By using the assumptions (A-2) and (A-3), this equation leads to

$$\mathrm{E}\{\tilde{\boldsymbol{w}}_{k+1}\} = (I - \mu \mathrm{E}\{\boldsymbol{P}_k\}) \mathrm{E}\{\tilde{\boldsymbol{w}}_k\}.$$

Hence, by Theorem 5.9 and Corollary 5.4 (Appendix 1: "Stability of Linear Systems"), the next theorem is obvious.

Theorem 5.8 *If*

$$0 < \mu < \frac{2}{\lambda_{\max}(\mathrm{E}\{\boldsymbol{P}_k\})},$$

then

$$\lim_{k \to \infty} \mathrm{E}\{\tilde{\boldsymbol{w}}_k\} = 0,$$

where $\lambda_{\max}(\mathrm{E}\{\boldsymbol{P}_k\})$ *is the largest eigenvalue of* $\mathrm{E}\{\boldsymbol{P}_k\}$.

5.5.2.5 Learning Curves

To draw the learning curve, we need to evaluate, for each k,

$$\begin{aligned}
\mathrm{E}\left\{e_a^2(k)\right\} &= \mathrm{E}\left\{(\boldsymbol{x}_k^t \tilde{\boldsymbol{w}}_k)^2\right\} \\
&= \mathrm{E}\left\{\|\tilde{\boldsymbol{w}}_k\|_{R_x}^2\right\} \\
&= \mathrm{E}\left\{\|\tilde{\boldsymbol{w}}_k\|_{\mathrm{vec}^{-1}\{r\}}^2\right\},
\end{aligned} \quad (5.101)$$

where $r \triangleq \text{vec}\{R_x\}$. From (5.91), we obtain a series of equations

$$\begin{aligned}
\mathrm{E}\left\{\|\tilde{\boldsymbol{w}}_k\|^2_{\text{vec}^{-1}\{r\}}\right\} &= \mathrm{E}\left\{\|\tilde{\boldsymbol{w}}_{k-1}\|^2_{\text{vec}^{-1}\{Fr\}}\right\} + \mu^2\sigma_v^2(\gamma^t r),\\
\mathrm{E}\left\{\|\tilde{\boldsymbol{w}}_{k-1}\|^2_{\text{vec}^{-1}\{Fr\}}\right\} &= \mathrm{E}\left\{\|\tilde{\boldsymbol{w}}_{k-2}\|^2_{\text{vec}^{-1}\{F^2r\}}\right\} + \mu^2\sigma_v^2(\gamma^t Fr).\\
&\vdots\\
\mathrm{E}\left\{\|\tilde{\boldsymbol{w}}_1\|^2_{\text{vec}^{-1}\{F^{(k-1)}r\}}\right\} &= \mathrm{E}\left\{\|\tilde{\boldsymbol{w}}_0\|^2_{\text{vec}^{-1}\{F^k r\}}\right\} + \mu^2\sigma_v^2(\gamma^t F^{(k-1)}r).
\end{aligned}$$

Hence,

$$\mathrm{E}\left\{\|\tilde{\boldsymbol{w}}_k\|^2_{\text{vec}^{-1}\{r\}}\right\} = \mathrm{E}\left\{\|\tilde{\boldsymbol{w}}_0\|^2_{\text{vec}^{-1}\{F^k r\}}\right\} + \mu^2\sigma_v^2\gamma^t(I + F + \cdots + F^{(k-1)})r. \tag{5.102}$$

By putting $f_k \triangleq F^k r$ and $g_{k-1} \triangleq \gamma^t((I+F+\cdots+F^{(k-1)})r$, Eq. (5.102) is rewritten as

$$\mathrm{E}\left\{\|\tilde{\boldsymbol{w}}_k\|^2_{\text{vec}^{-1}\{r\}}\right\} = \mathrm{E}\left\{\|\tilde{\boldsymbol{w}}_0\|^2_{\text{vec}^{-1}\{f_k\}}\right\} + \mu^2\sigma_v^2 g_{k-1}. \tag{5.103}$$

This equation, together with recurrence equations

$$f_0 = r, \quad f_k = F f_{k-1},$$

and

$$g_0 = \gamma^t r, \quad g_{k-1} = g_{k-2} + \gamma^t f_{k-1},$$

can be used to evaluate (5.101).

5.5.2.6 Steady-State Performance

We have already examined the steady-state performance of the R-APA in Sect. 5.5.1. We can also examine it using the recurrence equation (5.91). We assume that the values of μ is chosen so that the linear system (5.96) is asymptotically stable. Let us denote

$$\mathrm{E}\left\{\|\tilde{\boldsymbol{w}}_\infty\|^2_{\text{vec}^{-1}\{\sigma\}}\right\} \triangleq \lim_{k\to\infty} \mathrm{E}\left\{\|\tilde{\boldsymbol{w}}_k\|^2_{\text{vec}^{-1}\{\sigma\}}\right\}.$$

Then, taking the limit of both sides of (5.91), we have

$$\mathrm{E}\left\{\|\tilde{\boldsymbol{w}}_\infty\|^2_{\text{vec}^{-1}\{\sigma\}}\right\} = \mathrm{E}\left\{\|\tilde{\boldsymbol{w}}_\infty\|^2_{\text{vec}^{-1}\{F\sigma\}}\right\} + \mu^2\sigma_v^2(\gamma^t\sigma).$$

which is rewritten as

$$\mathrm{E}\left\{\|\tilde{\boldsymbol{w}}_\infty\|^2_{\mathrm{vec}^{-1}\{(I-F)\sigma\}}\right\} = \mu^2\sigma_v^2(\gamma^t\sigma).$$

If we choose $\sigma = (I-F)^{-1}\mathrm{vec}\,\{I\}$, then $(I-F)\sigma = \mathrm{vec}\,\{I\}$. Thus, we obtain

$$\begin{aligned}\mathrm{MSD} &= \mathrm{E}\left\{\|\tilde{\boldsymbol{w}}_\infty\|^2\right\}\\ &= \mu^2\sigma_v^2\gamma^t(I-F)^{-1}\mathrm{vec}\,\{I\}.\end{aligned}$$

Similarly, if we choose $\sigma = (I-F)^{-1}\mathrm{vec}\,\{R_x\}$, then $(I-F)\sigma = \mathrm{vec}\,\{R_x\}$. Therefore, we get

$$\begin{aligned}\mathrm{EMSE} &= \mathrm{E}\left\{\|\tilde{\boldsymbol{w}}_\infty\|^2_{R_x}\right\}\\ &= \mu^2\sigma_v^2\gamma^t(I-F)^{-1}\mathrm{vec}\,\{R_x\}.\end{aligned}$$

Appendix 1: Stability of Linear Systems

Let us consider a linear system

$$x_{k+1} = Ax_k + y_k, \tag{5.104}$$

where $\{x_k\}$ and $\{y_k\}$ are sequences of n-dimensional vectors, and A is an $n \times n$ matrix. The linear system (5.104) is said to be *asymptotically stable* iff $\lim_{k\to\infty} A^k = 0$.

Theorem 5.9 *Let $\lambda_1, \lambda_2, \ldots, \lambda_n$ be the eigenvalues of A. The linear system* (5.104) *is asymptotically stable iff*

$$|\lambda_i| < 1 \quad (i = 1, 2, \ldots, n). \tag{5.105}$$

Proof Let us assume for simplicity that eigenvalues are different from each other. Then, there exists a nonsingular matrix T such that

$$\begin{aligned}A &= T^{-1}\Lambda T,\\ \Lambda &= \mathrm{diag}\,[\lambda_1, \lambda_2, \ldots, \lambda_n]\end{aligned} \tag{5.106}$$

From this fact, we have

$$\begin{aligned}A^k &= T^{-1}\Lambda^k T,\\ \Lambda^k &= \mathrm{diag}\,[\lambda_1^k, \lambda_2^k, \ldots, \lambda_n^k].\end{aligned}$$

Thus, we have

$$\lim_{k\to\infty} A^k = T^{-1}(\lim_{k\to\infty} \Lambda^k)T,$$

from which the conclusion is obvious.

When the eigenvalues have multiplicity, we use the Jordan normal form [11, p. 174] instead of the diagonal form (5.106), and reach the same conclusion. □

Corollary 5.4 *Let y_k be a constant vector: $y_k = y$. If the eigenvalues $\lambda_1, \lambda_2, \ldots, \lambda_n$ are different from each other, and the condition* (5.105) *is met, then*

$$\lim_{k\to\infty} x_k = T^{-1} diag\left[\frac{1}{1-\lambda_1}, \frac{1}{1-\lambda_2}, \ldots, \frac{1}{1-\lambda_n}\right] Ty.$$

Proof It is easily verified that

$$x_k = A^k x_0 + \left(\sum_{i=0}^{k-1} A^i\right) y.$$

Under the given condition, $\lim_{k\to\infty} A^k = 0$. Also, since

$$\begin{aligned}\sum_{i=0}^{k-1} A^i &= T^{-1}\left[\sum_{i=0}^{k-1} \Lambda^i\right] T \\ &= T^{-1}\text{diag}\left[\sum_{i=0}^{k-1} \lambda_1^i, \sum_{i=0}^{k-1} \lambda_2^i, \ldots, \sum_{i=0}^{k-1} \lambda_n^i\right] T,\end{aligned}$$

we have

$$\begin{aligned}\lim_{k\to\infty} \sum_{i=0}^{k-1} A^i &= T^{-1} \lim_{k\to\infty} \text{diag}\left[\sum_{i=0}^{k-1} \lambda_1^i, \sum_{i=0}^{k-1} \lambda_2^i, \ldots, \sum_{i=0}^{k-1} \lambda_n^i\right] T. \\ &= T^{-1}\text{diag}\left[\frac{1}{1-\lambda_1}, \frac{1}{1-\lambda_2}, \ldots, \frac{1}{1-\lambda_n}\right] T.\end{aligned}$$ □

References

1. Sankaran, S.G., Beex, A.A.: Convergence behavior of affine projection algorithms. IEEE Trans. Signal Process. **48**(4), 1086–1096 (2000)
2. Slock, D.T.M.: On the convergence behavior of the LMS and the normalized LMS algorithms. IEEE Trans. Signal Process. **41**(9), 2811–2825 (1993)
3. Rupp, M.: A family of adaptive filter algorithms with decorrelating properties. IEEE Trans. Signal Process. **46**(3), 771–775 (1998)

4. Paul, T.K., Ogunfunmi, T.: On the convergence behavior of the affine projection algorithm for adaptive filters. IEEE Trans. Circ. Syst. I **58**(8), 1813–1826 (2011)
5. Bershad, N.J., Linebarger, D., McLaughlin, S.: A stochastic analysis of the affine projection algorithm for Gaussian autoregressive inputs. In: Proceedings of the ICASSP 2001, vol. 6, pp. 3837–3840. Salt Lake City, UT, 07–11 May 2001
6. de Almeida, S.J.M., Bermudez, J.C.M., Bershad, N.J., Costa, M.H.: A statistical analysis of the affine projection algorithm for unity step size and autoregressive inputs. IEEE Trans. Circ. Syst. I **52**(7), 1394–1405 (2005)
7. Oppenheim, A.V., Schafer, R.W.: Digital Signal Processing. Prentice-Hall, Englewood Cliffs (1975)
8. Wilks, S.S.: Mathematical Statistics. Wiley, New York (1962)
9. Costa, M.H., de Almeida, S.J.M., Bermudez, J.C.M., Barcelos, R.B.: New insight into the weight behaviour of the affine projection algorithm. In: Proceedings of the EUSIPCO 2012, pp. 2610–2614. Bucharest, 27–31 Aug 2012
10. Anderson, T.W.: An Introduction to Multivariate Statistical Analysis. Wiley, New York (1958)
11. Shin, H.-C., Sayed, A.H.: Mean-square performance of a family of affine projection algorithms. IEEE Trans. Signal Process. **52**(1), 90–102 (2004)
12. Sayed, A.H.: Adaptive Filters. Wiley, Hoboken (2008)
13. Al-Naffouri, T.Y., Sayed, A.H.: Transient analysis of data-normalized adaptive filters. IEEE Trans. Signal Process. **51**(3), 639–652 (2003)
14. Satake, I.: Linear Algebra. Marcel Dekker, New York (1975)

Chapter 6
Reduction of Computational Complexity

6.1 Computational Complexity of APA

The computational complexity of the APA can be measured by the number of arithmetic operations such as additions, multiplications, multiplication–additions, and divisions necessary to execute the update of coefficient vector per sample period. We assume that each of these arithmetic operations, except for division, consumes the same amount of computational resource.

Let us recall the update equation for the R-APA:

$$\begin{aligned} e_k &= d_k - X_k w_k, \\ \Delta w_k &= X_k^t (X_k X_k^t + \delta I)^{-1} e_k, \\ w_{k+1} &= w_k + \mu \Delta w_k. \end{aligned} \tag{6.1}$$

We can estimate the amount of computation to obtain w_{k+1} by a naive implementation along (6.1) as shown in Table 6.1, where we assume the use of the Cholesky decomposition for computing $A^{-1}e_k$. Thus, approximately $p^3/6 + np^2/2 + 2np + n$ arithmetic operations are necessary in total per sample period. When n and p are large, this amount of computation becomes a heavy burden for real-time processing. In order to overcome this difficulty, many efforts have been made to reduce the computational complexity of the APA. In the following sections, two methods for complexity reduction will be reviewed.

K. Ozeki, *Theory of Affine Projection Algorithms for Adaptive Filtering*,
Mathematics for Industry 22, DOI 10.1007/978-4-431-55738-8_6

Table 6.1 Number of operations for updating the coefficient vector by a naive method in the R-APA

Expression	Number of operations
$e_k = d_k - X_k w_k$	np
$A = X_k X_k^t + \delta I$	$np(p+1)/2$
$b = A^{-1} e_k$	$p^3/6$
$\Delta w_k = X_k^t b$	np
$w_{k+1} = w_k + \mu \Delta w_k$	n
Total (approximately)	$p^3/6 + np^2/2 + 5np/2 + n$

6.2 Fast Affine Projection (FAP) Algorithm

6.2.1 Time-Shift Property of Regressor Blocks

In the update equation (6.1), regressors $x_k, x_{k-1}, \ldots, x_{k-(p-1)}$ are used at time k. At time $k-1$, regressors $x_{k-1}, x_{k-2}, \ldots, x_{k-p}$ are used. Thus, much information overlaps in the equation at time k and $k-1$. Therefore, the quantities appearing in (6.1) at time k are not independent of the corresponding quantities at time $k-1$. By exploiting this time-shift property, recurrence equations for those quantities are derived, which lead to reduction of computational complexity of the APA as described below [1, 2]. The paper [1] is concerned with the R-APA, whereas [2] is concerned with the B-APA.

6.2.2 Recursion for the Error Vector e_k

Let us first derive a recurrence equation for $e_k = (e_{k,0}, e_{k,1}, \ldots, e_{k,p-1})^t$. If $\delta = 0$, $w_{k-1} + \Delta w_{k-1}$ is the affine projection of w_{k-1} onto $\Pi_{k-1} \cap \Pi_{k-2} \cap \cdots \cap \Pi_{k-p}$. Therefore, if δ is sufficiently small, we have

$$d(k-j) - x_{k-j}^t (w_{k-1} + \Delta w_{k-1}) \approx 0 \quad (j = 1, 2, \ldots, p). \tag{6.2}$$

Thus, for $j = 1, 2, \ldots, p-1$,

$$\begin{aligned} e_{k,j} &= d(k-j) - x_{k-j}^t w_k \\ &= d(k-j) - x_{k-j}^t (w_{k-1} + \mu \Delta w_{k-1}) \\ &= \mu \left\{ d(k-j) - x_{k-j}^t (w_{k-1} + \Delta w_{k-1}) \right\} \\ &\quad + (1-\mu) \left\{ d(k-j) - x_{k-j}^t w_{k-1} \right\} \\ &\approx (1-\mu) \left\{ d((k-1)-(j-1)) - x_{(k-1)-(j-1)}^t w_{k-1} \right\} \\ &= (1-\mu) e_{k-1,j-1}. \end{aligned}$$

Therefore, we have an approximate recurrence equation

$$e_k \approx \begin{bmatrix} e_{k,0} \\ (1-\mu)(e_{k-1})_u \end{bmatrix}, \tag{6.3}$$

where $(e_{k-1})_u$ is the $(p-1)$-dimensional vector consisting of the upper $(p-1)$ components of e_{k-1}. When $\delta = 0$, (6.3) becomes an exact equation. A recursive computation method for $e_{k,0}$ will be discussed later.

6.2.3 Expression for w_{k+1} as a Linear Combination of Past Regressors

Let us denote

$$\begin{aligned} R_k &\triangleq X_k X_k^t + \delta I, \\ \epsilon_k &\triangleq R_k^{-1} e_k \\ &= (\epsilon_{k,0}, \epsilon_{k,1}, \ldots, \epsilon_{k,p-1})^t. \end{aligned}$$

Then, from (6.1),

$$\begin{aligned} w_{k+1} &= w_k + \mu X_k^t \epsilon_k \\ &= w_{k-1} + \mu X_{k-1}^t \epsilon_{k-1} + \mu X_k^t \epsilon_k \\ &\ \ \vdots \\ &= w_0 + \mu \sum_{i=0}^{k} X_{k-i}^t \epsilon_{k-i} \\ &= w_0 + \mu \sum_{i=0}^{k} \sum_{j=0}^{p-1} x_{k-i-j} \epsilon_{k-i,j}. \end{aligned} \tag{6.4}$$

In Fig. 6.1, let the rectangular area determined by the points $\{(0,0), (k,0), (k, p-1), (0, p-1)\}$ be denoted by $\square$, and the hatched triangular area determined by the points $\{(k+1,0), (k+p-1,0), (k+1, p-2)\}$ by $\triangle$. The trapezoidal area $\square \cup \triangle$ is swept by the line segments

$$L(l) \triangleq \{(i,j);\ i+j=l, 0 \le j \le \min(l, p-1)\} \quad (0 \le l \le k+p-1).$$

Also, the triangular area $\triangle$ is swept by the line segments

$$L(l) \triangleq \{(i,j);\ i+j=l, 0 \le j \le l-(k+1)\} \quad (k+1 \le l \le k+p-1).$$

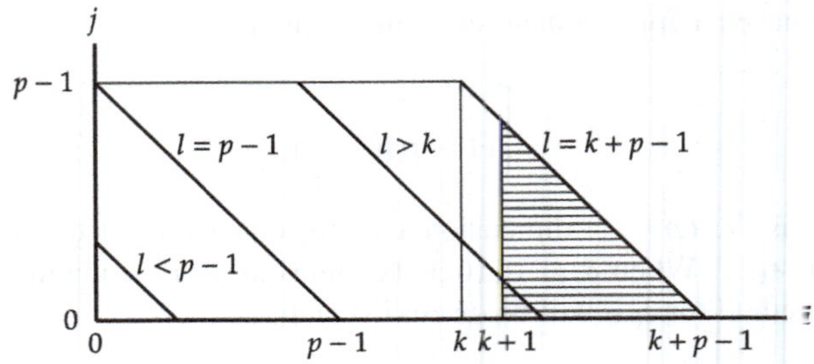

Fig. 6.1 Computation of $s_{k,l}$. The quantity $x_{k-i-j}\epsilon_{k-i,j}$ is first summed up over the trapezoidal area. Then, the sum over the hatched area is subtracted. The result gives the sum of $x_{k-i-j}\epsilon_{k-i,j}$ over the rectangular area

Therefore,

$$\begin{aligned}\sum_{i=0}^{k}\sum_{j=0}^{p-1} x_{k-i-j}\epsilon_{k-i,j} &= \sum_{(i,j)\in\square} x_{k-i-j}\epsilon_{k-i,j}\\ &= \sum_{(i,j)\in\square\cup\triangle} x_{k-i-j}\epsilon_{k-i,j} - \sum_{(i,j)\in\triangle} x_{k-i-j}\epsilon_{k-i,j}\\ &= \sum_{l=0}^{k+p-1}\sum_{j=0}^{\min(l,p-1)} x_{k-l}\epsilon_{k-(l-j),j} - \sum_{l=k+1}^{k+p-1}\sum_{j=0}^{l-(k+1)} x_{k-l}\epsilon_{k-(l-j),j}\\ &= \sum_{l=0}^{k+p-1}\left\{\sum_{j=0}^{\min(l,p-1)} \epsilon_{k-(l-j),j}\right\} x_{k-l} - \sum_{l=k+1}^{k+p-1}\left\{\sum_{j=0}^{l-(k+1)} \epsilon_{k-(l-j),j}\right\} x_{k-l}.\end{aligned} \tag{6.5}$$

From (6.4) and (6.5), we have

$$w_{k+1} = w_0 + \sum_{l=0}^{k+p-1} s_{k,l}x_{k-l}, \tag{6.6}$$

where

$$s_{k,l} \triangleq \mu \begin{cases} \displaystyle\sum_{j=0}^{\min(l,p-1)} \epsilon_{k-(l-j),j}, & \text{if } 0 \le l \le k; \\ \displaystyle\sum_{j=0}^{\min(l,p-1)} \epsilon_{k-(l-j),j} - \sum_{j=0}^{l-(k+1)} \epsilon_{k-(l-j),j}, & \text{if } k+1 \le l \le k+p-1. \end{cases} \tag{6.7}$$

From (6.7), the following recurrence equation holds for $s_{k,l}$:

For $l = 0$, it is obvious that

$$s_{k,0} = \mu\epsilon_{k,0}. \tag{6.8}$$

For $1 \le l \le p-1$, we have

$$s_{k,l} = \mu\epsilon_{k,l} + s_{k-1,l-1}. \tag{6.9}$$

In fact, if $1 \le l \le k$,

$$\begin{aligned} s_{k,l} &= \mu \sum_{j=0}^{l} \epsilon_{k-(l-j),j} \\ &= \mu\epsilon_{k,l} + \mu \sum_{j=0}^{l-1} \epsilon_{(k-1)-((l-1)-j),j} \\ &= \mu\epsilon_{k,l} + s_{k-1,l-1}. \end{aligned}$$

If $k+1 \le l \le p-1$,

$$\begin{aligned} s_{k,l} &= \mu \sum_{j=0}^{l} \epsilon_{k-(l-j),j} - \mu \sum_{j=0}^{l-(k+1)} \epsilon_{k-(l-j),j} \\ &= \mu\epsilon_{k,l} + \mu \sum_{j=0}^{l-1} \epsilon_{(k-1)-((l-1)-j),j} - \mu \sum_{j=0}^{(l-1)-((k-1)+1)} \epsilon_{(k-1)-((l-1)-j),j} \\ &= \mu\epsilon_{k,l} + s_{k-1,l-1}. \end{aligned}$$

For $p \le l \le k+p-1$, we have

$$s_{k,l} = s_{k-1,l-1}. \tag{6.10}$$

In fact, if $p \le l \le k$,

$$\begin{aligned} s_{k,l} &= \mu \sum_{j=0}^{\min(l-1,p-1)} \epsilon_{(k-1)-((l-1)-j),j} \\ &= s_{k-1,l-1}. \end{aligned}$$

If $k+1 \le l \le k+p-1$,

$$\begin{aligned} s_{k,l} &= \mu \sum_{j=0}^{\min(l-1,p-1)} \epsilon_{(k-1)-((l-1)-j),j} - \mu \sum_{j=0}^{(l-1)-((k-1)+1)} \epsilon_{(k-1)-((l-1)-j),j} \\ &= s_{k-1,l-1}. \end{aligned}$$

Equation (6.10) shows that, for $p \le l \le (k+p-1)$, no computation is needed for updating $s_{k,l}$.

6.2.4 Recursion for $\epsilon_{k,l}$

In order to implement (6.8) and (6.9) efficiently, we need a smart method for computing $\epsilon_{k,l}$ ($l = 0, 1, \ldots, p-1$).

Now, let $R = R_k$, $E_f = (E_k)_f$, $E_b = (E_k)_b$, $f = f_k$, and $b = b_k$ in Theorem 3.16 (Chap. 3, Appendix 5: "Linear Prediction"). Then, since $(R_k)_l = (R_{k-1})_u$, we have

$$R_k^{-1} = \begin{bmatrix} 0 & \mathbf{0}^t \\ \mathbf{0} & (R_{k-1})_u^{-1} \end{bmatrix} + \frac{f_k f_k^t}{(E_k)_f}, \tag{6.11}$$

$$R_k^{-1} = \begin{bmatrix} (R_k)_u^{-1} & \mathbf{0} \\ \mathbf{0}^t & 0 \end{bmatrix} + \frac{b_k b_k^t}{(E_k)_b}. \tag{6.12}$$

From (6.11),

$$\begin{aligned} \epsilon_k &= R_k^{-1} e_k \\ &= \left(\begin{bmatrix} 0 & \mathbf{0}^t \\ \mathbf{0} & (R_{k-1})_u^{-1} \end{bmatrix} + \frac{1}{(E_k)_f} f_k f_k^t \right) e_k \\ &= \begin{bmatrix} 0 \\ \tilde{\epsilon}_{k-1} \end{bmatrix} + \frac{1}{(E_k)_f} f_k f_k^t e_k, \end{aligned} \tag{6.13}$$

where

$$\tilde{\epsilon}_{k-1} \triangleq (R_{k-1})_u^{-1} (e_k)_l.$$

Using (6.3), we have

$$\begin{aligned} \tilde{\epsilon}_{k-1} &\approx (R_{k-1})_u^{-1} (1-\mu)(e_{k-1})_u \\ &= (1-\mu)\hat{\epsilon}_{k-1}. \end{aligned}$$

Substitution of this equation into (6.13) yields

$$\epsilon_k = (1-\mu) \begin{bmatrix} 0 \\ \hat{\epsilon}_{k-1} \end{bmatrix} + \frac{1}{(E_k)_f} f_k f_k^t e_k, \tag{6.14}$$

Similarly, from (6.12),

$$\begin{aligned}\epsilon_k &= R_k^{-1} e_k \\ &= \left(\begin{bmatrix} (R_k)_u^{-1} & \mathbf{0} \\ \mathbf{0}^t & 0 \end{bmatrix} + \frac{1}{(E_k)_b} b_k b_k^t \right) e_k \\ &= \begin{bmatrix} \hat{\epsilon}_k \\ 0 \end{bmatrix} + \frac{1}{(E_k)_b} b_k b_k^t e_k, \end{aligned} \tag{6.15}$$

where

$$\hat{\epsilon}_k \overset{\triangle}{=} (R_k)_u^{-1} (e_k)_u.$$

Combination of (6.14) and (6.15) enables recursive computation of ϵ_k. In fact, if ϵ_{k-1} is given, $\hat{\epsilon}_{k-1}$ can be calculated by (6.15). Then, by (6.14), ϵ_k is obtained. For these computations, the values of $f_k, (E_k)_f, b_{k-1}, (E_{k-1})_b$ are necessary. We can compute these values efficiently using the sliding window version of the fast transversal filter (FTF) algorithm (Appendix 1: "Fast Transversal Filter (FTF) Algorithm").

6.2.5 Recursion for Computing $e_{k,0}$

In (6.6), let u_{k+1} be a partial sum defined by

$$\begin{aligned} u_0 &\overset{\triangle}{=} w_0, \\ u_{k+1} &\overset{\triangle}{=} w_0 + \sum_{l=p-1}^{k+p-1} s_{k,l} x_{k-l} \quad (k \geq 0). \end{aligned}$$

Then,

$$w_{k+1} = u_{k+1} + \sum_{l=0}^{p-2} s_{k,l} x_{k-l}. \tag{6.16}$$

We can derive a recurrence equation for u_k as follows:

$$\begin{aligned} u_{k+1} &= s_{k,p-1} x_{k-(p-1)} + w_0 + \sum_{l=p}^{k+p-1} s_{k,l} x_{k-l} \\ &= s_{k,p-1} x_{k-(p-1)} + w_0 + \sum_{l=p}^{k+p-1} s_{k-1,l-1} x_{k-l} \\ &= s_{k,p-1} x_{k-(p-1)} + w_0 + \sum_{l=p-1}^{k-1+p-1} s_{k-1,l} x_{k-1-l} \\ &= s_{k,p-1} x_{k-(p-1)} + u_k. \end{aligned} \tag{6.17}$$

From (6.16),

$$w_k = u_k + \sum_{l=0}^{p-2} s_{k-1,l} x_{k-1-l}. \tag{6.18}$$

Therefore,

$$\begin{aligned} e_{k,0} &= d(k) - z(k) \\ &= d(k) - x_k^t u_k - \sum_{l=0}^{p-2} s_{k-1,l} x_k^t x_{k-1-l} \\ &= d(k) - x_k^t u_k - \sum_{l=0}^{p-2} s_{k-1,l} r_{k,l+1}, \end{aligned} \tag{6.19}$$

where $r_{k,l}$ is the correlation defined by

$$r_{k,l} \stackrel{\triangle}{=} x_k^t x_{k-l}. \tag{6.20}$$

We can calculate $r_{k,l+1}$ recursively as

$$\begin{aligned} r_{k,l+1} &= \sum_{i=0}^{n-1} x(k-i)x(k-1-l-i) \\ &= \sum_{i=0}^{n-1} x(k-1-i)x(k-2-l-i) \\ &\quad + x(k)x(k-1-l) - x(k-n)x(k-1-n-l) \\ &= r_{k-1,l+1} + x(k)x(k-1-l) - x(k-n)x(k-1-n-l). \end{aligned} \tag{6.21}$$

6.2.6 Algorithm and Computational Complexity

The above results are combined to compose Algorithm 6.1, which is called the *fast affine projection (FAP) algorithm*. It should be noted that "fast" means "computationally efficient," not "fast converging." In Algorithm 6.1, the initialization part for $k = 0$ is omitted. Steps 1–9 are repeated for $k \geq 1$. The second column shows the reference number of the equation, and the third column the computational complexity measured by the number of multiplication–additions and the number of divisions indicated in the parentheses. In Step 6, Algorithms 6.3 and 6.4 (Appendix 1: "Fast Transversal Filter (FTF) Algorithm") are used. Since Steps 13–15 in these algorithms are unnecessary for the present purpose, the complexity is about $10p$ (7). However, the forward residual filter and the backward residual filter must be denormalized which requires additional $4p$ operations. Thus, the total number of operations per sample for the FAP algorithm is approximately equal to $2n + 24p$, which is much smaller than that in the naive implementation of the R-APA.

A problem about the FAP algorithm is that errors caused by approximate equations and finite precision arithmetic are accumulated at each step of the iteration. This makes the algorithm unstable. To avoid this problem, a periodic restart strategy is recommended [1].

Algorithm 6.1 (*FAP*)

Computation	Reference	Complexity
1. for $l = 0, 1, \ldots, p-2$, $r_{k,l+1} := r_{k-1,l+1} + x(k)x(k-l-1) - x(k-n)x(k-1-n-l)$;	(6.21)	$2(p-1)$
2. $u_k := s_{k-1,p-1}x_{k-p} + u_{k-1}$;	(6.17)	n
3. $z(k) := x_k^t u_k + \sum_{l=0}^{p-2} s_{k-1,l} r_{k,l+1}$;	(6.19)	$n + p - 1$
4. $e_{k,0} := d(k) - z(k)$;	(6.19)	1
5. $e_k := \begin{bmatrix} e_{k,0} \\ (1-\mu)(e_{k-1})_u \end{bmatrix}$;	(6.3)	p
6. compute $f_k, (E_k)_f, b_{k-1}, (E_{k-1})_b$;	FTF (Appendix 1: "Fast Transversal Filter (FTF) Algorithm")	$14p$ (7)
7. $\begin{bmatrix} \hat{\epsilon}_{k-1} \\ 0 \end{bmatrix} := \epsilon_{k-1} - \frac{b_{k-1}^t e_{k-1}}{(E_{k-1})_b} b_{k-1}$;	(6.15)	$2p$ (1)
8. $\epsilon_k := (1-\mu) \begin{bmatrix} 0 \\ \hat{\epsilon}_{k-1} \end{bmatrix} + \frac{f_k^t e_k}{(E_k)_f} f_k$;	(6.14)	$3p$ (1)
9. $s_{k,0} := \mu \epsilon_{k,0}$;	(6.8)	1
$s_{k,l} := \mu \epsilon_{k,l} + s_{k-1,l-1}$ $(l = 1, 2, \ldots, p-1)$;	(6.9)	$p - 1$
	Total (approx.)	$2n + 24p$ (9)

6.3 Block Exact Fast Affine Projection (BEFAP) Algorithm

6.3.1 Block Exact Technique

The regressor blocks X_k's in the partial-rank algorithm (PRA) are concatenated to make a larger block. From (4.1),

$$\begin{aligned}\tilde{w}_{k+1} &= \tilde{w}_{k+1-p} + \mu X_k^t \tilde{\epsilon}_k \\ &= \tilde{w}_{k+1-2p} + \mu X_{k-p}^t \tilde{\epsilon}_{k-p} + \mu X_k^t \tilde{\epsilon}_k \\ &\vdots \\ &= \tilde{w}_{k+1-mp} + \mu \sum_{i=0}^{m-1} X_{k-ip}^t \tilde{\epsilon}_{k-ip} \\ &= \tilde{w}_{k-N+1} + \mu \left[X_k^t \; X_{k-p}^t \cdots X_{k-N+p}^t\right] \begin{bmatrix} \tilde{\epsilon}_k \\ \tilde{\epsilon}_{k-p} \\ \vdots \\ \tilde{\epsilon}_{k-N+p} \end{bmatrix},\end{aligned} \tag{6.22}$$

where $N \triangleq mp$ is the block length, and

$$\tilde{\epsilon}_k \triangleq (X_k X_k^t + \delta I)^{-1} \tilde{e}_k.$$

Thus, the coefficient vector is updated only once every N samples, and fixed for a period of one block as $\tilde{w}_{k-N+1} = \tilde{w}_{k-N+2} = \cdots = \tilde{w}_k$. Before updating the coefficient vector $\tilde{w}_{k-N+1}$ as above, the filter output is computed as

$$\begin{aligned}&\left[\tilde{z}(k) \; \tilde{z}(k-1) \cdots \tilde{z}(k-N+1)\right]^t \\ &= \left[X_k^t \; X_{k-p}^t \cdots X_{k-N+p}^t\right]^t \tilde{w}_{k-N+1}.\end{aligned} \tag{6.23}$$

We see that in both of (6.22) and (6.23), a large regressor block

$$\left[X_k^t \; X_{k-p}^t \cdots X_{k-N+p}^t\right]$$

is used. This algorithm is called the *straightforward block projection algorithm* [3]. Block processing has an advantage that it enables us to use the fast matrix–vector multiplication technique as we will see later. However, the convergence slows down compared with the R-APA if a simple skipping strategy is employed for updating the coefficient vector as in the straightforward block projection algorithm. Fortunately, there is a block version of the FAP algorithm that has exactly the same convergence rate as the R-APA and is computationally more efficient than the FAP algorithm [3].

Similarly to (6.4), we have

$$\begin{aligned}w_{k+i} &= w_{k+i-1} + \mu X_{k+i-1}^t \epsilon_{k+i-1} \\ &= w_{k+i-2} + \mu X_{k+i-2}^t \epsilon_{k+i-2} + \mu X_{k+i-1}^t \epsilon_{k+i-1} \\ &\vdots \\ &= w_k + \mu \sum_{j=0}^{i-1} X_{k+i-j-1}^t \epsilon_{k+i-j-1}\end{aligned}$$

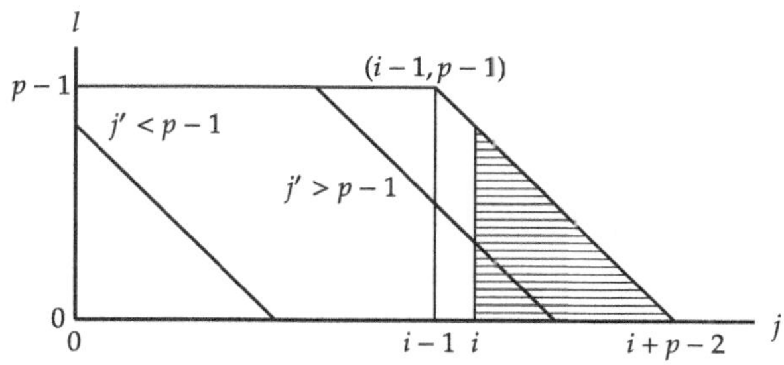

Fig. 6.2 Computation of the double summation in the right-hand side of (6.24)

$$= w_k + \mu \sum_{j=0}^{i-1} \sum_{l=0}^{p-1} x_{k+i-j-l-1} \epsilon_{k+i-j-1,l}$$

$$= w_k + \mu \sum_{j=0}^{i-1} \sum_{l=0}^{p-1} A_{k,i,j,l}, \qquad (6.24)$$

where

$$A_{k,i,j,l} \triangleq x_{k+i-j-l-1} \epsilon_{k+i-j-1,l},$$

$$\epsilon_{k+i-j-1} \triangleq R_{k+i-j-1}^{-1} e_{k+i-j-1},$$

$$R_{k+i-j-1} \triangleq X_{k+i-j-1} X_{k+i-j-1}^{t} + \delta I.$$

To calculate the sum of $A_{k,i,j,l}$, we use a technique similar to the one we used to derive (6.6). In Fig. 6.2, let the rectangular area determined by the points $\{(0, 0), (i-1, 0), (i-1, p-1), (0, p-1)\}$ be denoted by $\square$, and the hatched triangular area determined by the points $\{(i, 0), (i+p-2, 0), (i, p-2)\}$ by $\triangle$. Then,

$$\sum_{j=0}^{i-1} \sum_{l=0}^{p-1} A_{k,i,j,l} = \sum_{(j,l)\in\square} A_{k,i,j,l}$$

$$= \sum_{(j,l)\in\square\cup\triangle} A_{k,i,j,k} - \sum_{(j,l)\in\triangle} A_{k,i,j,l}. \qquad (6.25)$$

Noting that the area $\square \cup \triangle$ is swept by the line segments

$$L(j') \triangleq \{(j, l);\ j + l = j', 0 \le l \le \min(j', p-1)\} \quad (0 \le j' \le i + p - 2),$$

we have

$$
\begin{aligned}
\sum_{(j,l)\in\square\cup\triangle} A_{k,i,j,l} &= \sum_{j'=0}^{i+p-2} \sum_{(j,l)\in L(j')} A_{k,i,j,l} \\
&= \sum_{j'=0}^{i+p-2} \sum_{l=0}^{\min(j',p-1)} x_{k+i-j'-1}\epsilon_{k+i-j'+l-1,l} \\
&= \sum_{j'=0}^{i+p-2} \left\{ \sum_{l=0}^{\min(j',p-1)} \epsilon_{k+i-j'+l-1,l} \right\} x_{k+i-j'-1} \\
&= \sum_{j=0}^{i+p-2} \left\{ \sum_{l=0}^{\min(j,p-1)} \epsilon_{k-i-j+l-1,l} \right\} x_{k+i-j-1}.
\end{aligned}
\tag{6.26}
$$

Also, since the area $\triangle$ is swept by the line segments

$$
L(j') \triangleq \{(j,l);\ j+l=j',\ 0\le l\le j'-i\} \quad (i\le j'\le i+p-2),
$$

we have

$$
\begin{aligned}
\sum_{(j,l)\in\triangle} A_{k,i,j,l} &= \sum_{j'=i}^{i+p-2} \sum_{l=0}^{j'-i} x_{k+i-j'-1}\epsilon_{k+i-j'+l-1,l} \\
&= \sum_{j'=i}^{i+p-2} \left\{ \sum_{l=0}^{j'-i} \epsilon_{k+i-j'+l-1,l} \right\} x_{k+i-j'-1} \\
&= \sum_{j=0}^{p-2} \left\{ \sum_{l=0}^{j} \epsilon_{k-j+l-1,l} \right\} x_{k-j-1} \\
&= \sum_{j=0}^{p-2} \left\{ \sum_{l=0}^{\min(j,p-1)} \epsilon_{k-j+l-1,l} \right\} x_{k-j-1}.
\end{aligned}
\tag{6.27}
$$

In the last line of the above equation, we used $j=\min(j,p-1)$ because $0\le j\le p-2$. From (6.25)–(6.27), we have

$$
w_{k+i} = w_k + \sum_{j=0}^{i+p-2} s_{k+i-1,j}x_{k+i-j-1} - \sum_{j=0}^{p-2} s_{k-1,j}x_{k-j-1},
\tag{6.28}
$$

where

$$s_{k,j} \triangleq \mu \sum_{l=0}^{\min(j,p-1)} \epsilon_{k-j+l,l}. \tag{6.29}$$

We can derive a recurrence relation for $s_{k+i-1,j}$ as follows:
For $j = 0$,

$$\begin{aligned} s_{k+i-1,0} &= \mu \sum_{l=0}^{\min(0,p-1)} \epsilon_{k+i-1+l,l} \\ &= \mu\epsilon_{k+i-1,0}. \end{aligned} \tag{6.30}$$

For $j \geq 1$,

$$s_{k+i-1,j} = \mu \sum_{l=0}^{\min(j,p-1)} \epsilon_{k+i-1-j+l,l}, \tag{6.31}$$

and

$$\begin{aligned} s_{k+i-2,j-1} &= \mu \sum_{l=0}^{\min(j-1,p-1)} \epsilon_{k+i-2-(j-1)+l,l} \\ &= \mu \sum_{l=0}^{\min(j-1,p-1)} \epsilon_{k+i-1-j+l,l}. \end{aligned} \tag{6.32}$$

If $j \geq p$, then $\min(j-1, p-1) = p-1 = \min(j, p-1)$. If $j \leq p-1$, then $\min(j-1, p-1) = j-1$ and $\min(j, p-1) = j$. Therefore, from (6.31) and (6.32), if $j \geq p$

$$s_{k+i-1,j} = s_{k+i-2,j-1},$$

and if $j \leq p-1$

$$\begin{aligned} s_{k+i-1,j} &= \mu \sum_{l=0}^{j} \epsilon_{k+i-1-j+l,l} \\ &= \mu \sum_{l=0}^{j-1} \epsilon_{k+i-1-j+l,l} + \mu\epsilon_{k+i-1,j} \\ &= s_{k+i-2,j-1} + \mu\epsilon_{k+i-1,j}. \end{aligned}$$

Summarizing the above results in a concise form, we have

$$s_{k+i-1} = (0, s_{k+i-2}^t)^t + \mu(\epsilon_{k+i-1}^t, \underbrace{0, \ldots, 0}_{i-1})^t, \tag{6.33}$$

where

$$
\begin{aligned}
s_{k+i-1} &= (s_{k+i-1,0}, s_{k+i-1,1}, \ldots, s_{k+i-1,i+p-2})^t, \\
s_{k+i-2} &= (s_{k+i-2,0}, s_{k+i-2,1}, \ldots, s_{k+i-2,i+p-3})^t, \\
\epsilon_{k+i-1} &= (\epsilon_{k+i-1,0}, \epsilon_{k+i-1,1}, \ldots, \epsilon_{k+i-1,p-1})^t.
\end{aligned}
$$

From (6.28), the output $z(k+i)$ $(i \geq 0)$ is given by

$$
\begin{aligned}
z(k+i) &= x_{k+i}^t w_{k+i} \\
&= x_{k+i}^t w_k + \sum_{j=0}^{i+p-2} s_{k+i-1,j} x_{k+i}^t x_{k+i-j-1} - \sum_{j=0}^{p-2} s_{k-1,j} x_{k+i}^t x_{k-j-1} \\
&= x_{k+i}^t w_k + \sum_{j=0}^{i+p-2} s_{k+i-1,j} r_{k+i,j+1} - \sum_{j=0}^{p-2} s_{k-1,j} r_{k+i,i+j+1}, \qquad (6.34)
\end{aligned}
$$

where $r_{k,j} = x_k^t x_{k-j}$ is the correlation as in (6.20). Note that the first term in the right-hand side of (6.34) is the output of the filter when the coefficient vector w_{k+i} is fixed at w_k for $i = 1, 2, \ldots$. Thus, the second and the third terms compensate the error caused by fixing the coefficient vector to give exactly the same output as given by the R-APA.

The correlation $r_{k+i,j}$ can be calculated recursively as

$$
\begin{aligned}
r_{k+i,j} &= x_{k+i}^t x_{k+i-j} \\
&= \sum_{l=0}^{n-1} x(k+i-l)x(k+i-j-l) \\
&= \sum_{l=1}^{n} x(k+i-l)x(k+i-j-l) \\
&\quad + x(k+i)x(k+i-j) - x(k+i-n)x(k+i-j-n) \\
&= \sum_{l=0}^{n-1} x(k+i-(l+1))x(k+i-j-(l+1)) \\
&\quad + x(k+i)x(k+i-j) - x(k+i-n)x(k+i-j-n) \\
&= r_{k+i-1,j} + x(k+i)x(k+i-j) \\
&\quad - x(k+i-n)x(k+i-j-n) \quad (j = 1, 2, \ldots, i+p-1). \qquad (6.35)
\end{aligned}
$$

Now let us derive an equation for determining the blockwise output of the filter. Similarly to (6.18), let

$$
u_k \triangleq w_k - \sum_{j=0}^{p-2} s_{k-1,j} x_{k-j-1} \qquad (6.36)
$$

be the modified coefficient vector. Then, from (6.28),

$$w_{k+i} = u_k + \sum_{j=0}^{i+p-2} s_{k+i-1,j} x_{k+i-j-1}. \tag{6.37}$$

Therefore,

$$\begin{aligned} z(k+i) &= x_{k+i}^t w_{k+i} \\ &= x_{k+i}^t u_k + \sum_{j=0}^{i+p-2} s_{k+i-1,j} x_{k+i}^t x_{k+i-j-1} \\ &= x_{k+i}^t u_k + \sum_{j=0}^{i+p-2} s_{k+i-1,j} r_{k+i,j+1}. \end{aligned} \tag{6.38}$$

Applying (6.38) for $i = N-1, N-2, \ldots, 0$, where N is the block length, we have

$$\begin{bmatrix} z(k+N-1) \\ z(k+N-2) \\ \vdots \\ z(k) \end{bmatrix} = \begin{bmatrix} x_{k+N-1}^t \\ x_{k+N-2}^t \\ \vdots \\ x_k^t \end{bmatrix} u_k + \begin{bmatrix} \sum_{j=0}^{N+p-3} s_{k+N-2,j} r_{k+N-1,j+1} \\ \sum_{j=0}^{N+p-4} s_{k+N-3,j} r_{k+N-2,j+1} \\ \vdots \\ \sum_{j=0}^{p-2} s_{k-1,j} r_{k,j+1} \end{bmatrix}. \tag{6.39}$$

More concisely,

$$z_{k+N-1} = H_{k+N-1} u_k + \left(s_{k+N-2}^t r_{k+N-1}, s_{k+N-3}^t r_{k+N-2}, \ldots, s_{k-1}^t r_k \right)^t, \tag{6.40}$$

where

$$z_{k+N-1} \triangleq (z(k+N-1), z(k+N-2), \ldots, z(k))^t,$$

$$\begin{aligned} H_{k+N-1} &\triangleq \begin{bmatrix} x_{k+N-1}^t \\ x_{k+N-2}^t \\ \vdots \\ x_k^t \end{bmatrix} \\ &= \begin{bmatrix} x(k+N-1) & x(k+N-2) & \cdots & x(k+N-n) \\ x(k+N-2) & x(k+N-3) & \cdots & x(k+N-n-1) \\ \vdots & \vdots & \ddots & \vdots \\ x(k) & x(k-1) & \cdots & x(k-n+1) \end{bmatrix}, \end{aligned}$$

$$s_{k+i-1} \triangleq (s_{k+i-1,0}, s_{k+i-1,1}, \ldots, s_{k+i-1,i+p-2})^t,$$

and

$$r_{k+i} \triangleq (r_{k+i,1}, r_{k+i,2}, \ldots, r_{k+i,i+p-1})^t.$$

Equation (6.40) determines the blockwise output of the adaptive filter.

Next, let us derive an update equation for u_k. From (6.36) and (6.37),

$$\begin{aligned} u_{k+N} &= w_{k+N} - \sum_{j=0}^{p-2} s_{k+N-1,j} x_{k+N-j-1} \\ &= u_k + \sum_{j=0}^{N+p-2} s_{k+N-1,j} x_{k+N-j-1} - \sum_{j=0}^{p-2} s_{k-1,j} x_{k-j-1} \\ &= u_k + \sum_{j=p-1}^{N+p-2} s_{k+N-1,j} x_{k+N-j-1}. \end{aligned} \tag{6.41}$$

That is,

$$u_{k+N} = u_k + H_{k+N-p}^t s_{k+N-1}^{p-1,N+p-2}, \tag{6.42}$$

where

$$H_{k+N-p} \triangleq \begin{bmatrix} x(k+N-p) & x(k+N-p-1) & \cdots & x(k+N-p-(n-1)) \\ x(k+N-p-1) & x(k+N-p-2) & \cdots & x(k+N-p-1-(n-1)) \\ \vdots & \vdots & \ddots & \vdots \\ x(k+1-p) & x(k+1-p-1) & \cdots & x(k+1-p-(n-1)) \end{bmatrix},$$

and

$$s_{k+N-1}^{p-1,N+p-2} \triangleq (s_{k+N-1,p-1}, s_{k+N-1,p}, \ldots, s_{k+N-1,N+p-2})^t.$$

Although the modified coefficient vector is updated by (6.41) only once every N samples, the output sequence from the filter given by (6.39) is exactly the same as that given by the R-APA.

In order to evaluate (6.40) and (6.42), the vectors $s_{k+N-1}, s_{k+N-2}, \ldots, s_k, s_{k-1}$ are necessary. The vector s_{k-1} is propagated from the preceding block. Other vectors are calculated recursively by (6.33). In doing so, ϵ_{k+i} $(i = 0, 1, \ldots, N-1)$ are needed. These vectors can be calculated efficiently by the technique stated in Sect. 6.2.4.

6.3.2 Fast Matrix–Vector Multiplication

In (6.40), we need to compute the matrix–vector product of the form $H_k u$, where

$$H_k \triangleq \begin{bmatrix} x(k) & x(k-1) & \cdots & x(k-(n-1)) \\ x(k-1) & x(k-2) & \cdots & x(k-n) \\ \vdots & \vdots & \ddots & \vdots \\ x(k-(N-1)) & h(k-N) & \cdots & h(k-(N+n-2)) \end{bmatrix},$$

and u is an n-dimensional vector. Let us suppose that the block length N is a divisor of the filter length n: $n = N \cdot M$. Then,

$$\begin{aligned} H_k u &= \left[H_k^0 \; H_k^1 \; \cdots \; H_k^{M-1}\right]((u^0)^t, (u^1)^t, \ldots, (u^{M-1})^t)^t \\ &= H_k^0 u^0 + H_k^1 u^1 + \cdots + H_k^{M-1} u^{M-1}, \end{aligned} \tag{6.43}$$

where H_k^j is a submatrix of H_k defined by

$$H_k^j \triangleq \begin{bmatrix} x(k-jN) & x(k-1-jN) & \cdots & x(k-N+1-jN) \\ x(k-1-jN) & x(k-2-jN) & \cdots & x(k-N-jN) \\ \vdots & \vdots & \ddots & \vdots \\ x(k-N+1-jN) & x(k-N-jN) & \cdots & x(k-2N+2-jN) \end{bmatrix},$$

and u^j is a subvector of $u = (u_0, u_1, \ldots, u_{n-1})^t$ defined by

$$u^j \triangleq (u_{jN}, u_{jN+1}, \ldots, u_{jN+N-1})^t.$$

Note that each H_k^j is an $N \times N$ square matrix of the form

$$\begin{bmatrix} h_0 & h_1 & \cdots & h_{N-1} \\ h_1 & h_2 & \cdots & h_N \\ \vdots & \vdots & \ddots & \vdots \\ h_{N-1} & h_N & \cdots & h_{2N-2} \end{bmatrix}, \tag{6.44}$$

which is called a *Hankel matrix*. Thus, the problem of computing the matrix–vector product $H_k u$ reduces to the problem of computing the product of a Hankel matrix and a vector.

In (6.42), the matrix–vector product of the form $H_k^t u$ is required, where u is an N-dimensional vector in this case. Noting that each H_k^j is a symmetric matrix, we have

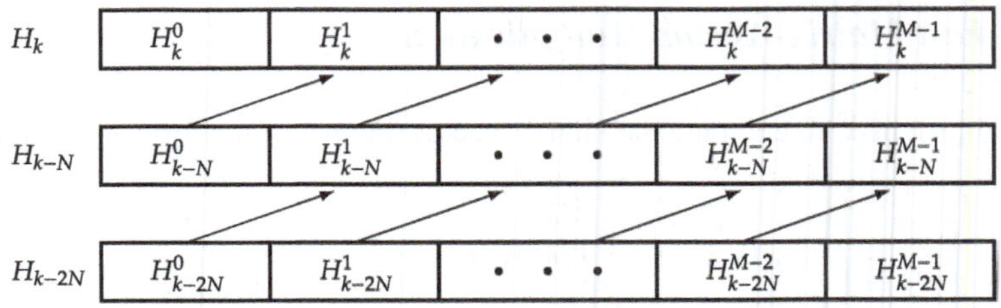

Fig. 6.3 $H_k^j = H_{k-N}^{j-1}$ in successive blocks except for H_k^0

$$H_k^t u = \begin{bmatrix} H_k^0 u \\ H_k^1 u \\ \vdots \\ H_k^{M-1} u. \end{bmatrix} \tag{6.45}$$

Thus, the problem of computing the matrix–vector product $H_k^t u$ also reduces to the problem of computing the product of a Hankel matrix and a vector.

The matrix H_k has a property that

$$H_k^j = H_{k-N}^{j-1} \quad (j = 1, 2, \ldots, M-1) \tag{6.46}$$

as shown in Fig. 6.3, which is exploited in successive block processing.

There are a couple of efficient algorithms to compute the product of a Hankel matrix and a vector [3, 4]. In the following, we assume that N is a power of 2: $N = 2^\nu$.

6.3.2.1 Circular Convolution Method

Let H be a Hankel matrix of the form (6.44), and $u = (u_0, u_1, \ldots, u_{N-1})^t$ an arbitrary N-dimensional vector. Also, let

$$f \triangleq \underbrace{(0, h_0, h_1, \ldots, h_{2N-2})}_{2N},$$

$$g \triangleq (\underbrace{0, 0, \ldots, 0}_{N}, \underbrace{u_{N-1}, u_{N-2}, \ldots, u_0}_{N}).$$

Then, the matrix–vector product $(v_0, v_1, \ldots, v_{N-1})^t \triangleq Hu$ is given by the vector consisting of the first N components $e_0, e_1, \ldots, e_{N-1}$ of the circular convolution e of f and g (Appendix 2: "Circular Convolution"). This is easily verified if we note

$$e_0 = \underbrace{\begin{pmatrix} 0 \\ \times \\ 0 \end{pmatrix} + \begin{pmatrix} h_{2N-2} \\ \times \\ 0 \end{pmatrix} + \cdots + \begin{pmatrix} h_N \\ \times \\ 0 \end{pmatrix}}_{=0} + \underbrace{\begin{pmatrix} h_{N-1} \\ \times \\ u_{N-1} \end{pmatrix} + \begin{pmatrix} h_{N-2} \\ \times \\ u_{N-2} \end{pmatrix} + \cdots + \begin{pmatrix} h_0 \\ \times \\ u_0 \end{pmatrix}}_{=v_0} = v_0,$$

$$e_1 = \underbrace{\begin{pmatrix} h_0 \\ \times \\ 0 \end{pmatrix} + \begin{pmatrix} 0 \\ \times \\ 0 \end{pmatrix} + \cdots + \begin{pmatrix} h_{N+1} \\ \times \\ 0 \end{pmatrix}}_{=0} + \underbrace{\begin{pmatrix} h_N \\ \times \\ u_{N-1} \end{pmatrix} + \begin{pmatrix} h_{N-1} \\ \times \\ u_{N-2} \end{pmatrix} + \cdots + \begin{pmatrix} h_1 \\ \times \\ u_0 \end{pmatrix}}_{=v_1} = v_1,$$

$$\vdots$$

$$e_{N-1} = \underbrace{\begin{pmatrix} h_{N-2} \\ \times \\ 0 \end{pmatrix} + \begin{pmatrix} h_{N-3} \\ \times \\ 0 \end{pmatrix} + \cdots + \begin{pmatrix} 0 \\ \times \\ 0 \end{pmatrix}}_{=0} + \underbrace{\begin{pmatrix} h_{2N-2} \\ \times \\ u_{N-1} \end{pmatrix} + \begin{pmatrix} h_{2N-1} \\ \times \\ u_{N-2} \end{pmatrix} - \cdots + \begin{pmatrix} h_{N-1} \\ \times \\ u_0 \end{pmatrix}}_{=v_{N-1}} = v_{N-1}.$$

The circular convolution can be calculated by using the FFT and the IFFT technique (Appendix 2: "Circular Convolution"). Since one complex multiplication–addition needs four real multiplication–additions, the number of real arithmetic operations is counted as follows:

I. $8N \log_2(2N)$ real multiplication–additions to compute the FFT F of f.
II. $8N \log_2(2N)$ real multiplication–additions to compute the FFT G of g.
III. $8N$ real multiplications to compute $F \cdot G$.
IV. $8N \log_2(2N)$ real multiplication–additions and $2N$ divisions to compute the IFFT of $F \cdot G$.

Because of the property (6.46), the FFT of f needs to be calculated only for the first subblock of H_k. If we put aside this portion, the complexity of computing a subblock Hu in the case of (6.43) is the sum of Part II–IV, that is, the sum of $16N \log_2(2N)$ multiplication–additions, $8N$ multiplication–additions, and $2N$ divisions. In the case of (6.45), Part II needs to be calculated only once, which is to be computed for the first subblock. If we put aside this portion, the complexity of computing a subblock Hu is the sum of Part III and IV, that is, the sum of $8N$ real multiplication–additions, $8N \log_2(2N)$ real multiplication–additions, and $2N$ divisions. For simplicity, we assume that all arithmetic operations consume the same amount of computing resources. Thus, the number of operations $f_{cc}(N)$ to compute a subblock Hu is given by

$$f_{cc}(N) = q(8N \log_2(2N)) + 10N, \tag{6.47}$$

where $q = 2$ in the case of (6.43), and $q = 1$ in the case of (6.45).

For the first subblock, all of Part I–Part IV are necessary. The complexity for this portion is given by

$$g_{cc}(N) \triangleq 24N \log_2(2N) + 10N. \tag{6.48}$$

Therefore, the complexity for computing the full-block matrix–vector product in (6.43) or (6.45) by using the circular convolution method is given by

$$(M-1) f_{cc}(N) + g_{cc}(N). \tag{6.49}$$

6.3.2.2 Divide and Conquer Method

Let H be a Hankel matrix of the form (6.44). Divide H as

$$H = \begin{bmatrix} H_{11} & H_{12} \\ H_{21} & H_{22} \end{bmatrix},$$

where

$$H_{11} \triangleq \begin{bmatrix} h_0 & h_1 & \cdots & h_{N/2-1} \\ h_1 & h_2 & \cdots & h_{N/2} \\ \vdots & \vdots & \ddots & \vdots \\ h_{N/2-1} & h_{N/2} & \cdots & h_{N-2} \end{bmatrix},$$

$$H_{12} \triangleq \begin{bmatrix} h_{N/2} & h_{N/2+1} & \cdots & h_{N-1} \\ h_{N/2+1} & h_{N/2+2} & \cdots & h_N \\ \vdots & \vdots & \ddots & \vdots \\ h_{N-1} & h_N & \cdots & h_{3N/2-2} \end{bmatrix},$$

$$H_{21} \triangleq H_{12},$$

$$H_{22} \triangleq \begin{bmatrix} h_N & h_{N+1} & \cdots & h_{3N/2-1} \\ h_{N+1} & h_{N+2} & \cdots & h_{3N/2} \\ \vdots & \vdots & \ddots & \vdots \\ h_{3N/2-1} & h_{3N/2} & \cdots & h_{2N-2} \end{bmatrix}.$$

Note that H_{11}, H_{12}, H_{21}, and H_{22} are again Hankel matrices. Also, divide an N-dimensional vector u into halves as $u = (u_a^t, u_b^t)^t$. Then,

$$\begin{aligned} Hu &= \begin{bmatrix} H_{11} & H_{12} \\ H_{21} & H_{22} \end{bmatrix} \begin{bmatrix} u_a \\ u_b \end{bmatrix} \\ &= \begin{bmatrix} H_{11}u_a + H_{12}u_b \\ H_{21}u_a + H_{22}u_b \end{bmatrix} \\ &= \begin{bmatrix} H_{11}u_a + H_{12}u_b \\ H_{12}u_a + H_{22}u_b \end{bmatrix} \\ &= \begin{bmatrix} (H_{11} - H_{12})u_a + H_{12}(u_a + u_b) \\ H_{12}(u_a + u_b) + (H_{22} - H_{12})u_b \end{bmatrix}. \end{aligned} \tag{6.50}$$

The matrix–vector products $(H_{11} - H_{12})u_a$, $H_{12}(u_a + u_b)$, and $(H_{22} - H_{12})u_b$ can be further decomposed into smaller problems. Thus, in order to compute Hu by using (6.50), the following operations are necessary for $N/2 \times N/2$ matrices and $N/2$-dimensional vectors:

I. 2 matrix–matrix subtractions,
II. q vector–vector additions ($q = 2$ or 3),
III. 3 matrix–vector multiplications.

Among these, the part I is necessary only for the first subblock H_k^0 of each H_k because of the property (6.46), so we can put aside this part. In computing $H_k u$, a vector–vector addition $u_a + u_b$ is necessary for every one of $H_k^j u^j$ $(j = 0, 1, \ldots, M)$. Therefore, $q = 3$ in this case. In computing $H_k^t u$, however, u is common for all of the $H_k^j u$ $(j = 0, 1, \ldots, M)$. If we put aside the amount of computation for $u_a + u_b$ in this case, then $q = 2$. Thus, we have the following recursion from Part II and III:

$$f_{\mathrm{dc}}(N) = 3 f_{dc}(N/2) + q(N/2), \tag{6.51}$$

where $f_{\mathrm{dc}}(N)$ denotes the complexity of computing Hu by the divide and conquer method for an $N \times N$ Hankel matrix H and an N-dimensional vector u. From (6.51), we have

$$\begin{aligned} f_{\mathrm{dc}}(N) &= 3 f_{dc}(N/2) + q(N/2) \\ &= 3((3(f_{\mathrm{dc}}(N/2^2) + q(N/2^2)) + q(N/2) \\ &= 3^2 f_{\mathrm{dc}}(N/2^2) + (qN/2)(3/2 + 1) \\ &= 3^3 f_{\mathrm{dc}}(N/2^3) + (qN/2)((3/2)^2 + (3/2) + 1) \\ &\;\;\vdots \\ &= 3^{\nu-\nu_0} f_{\mathrm{dc}}(N/2^{\nu-\nu_0}) + (qN/2) \sum_{i=0}^{\nu-\nu_0-1} (3/2)^i \\ &= 3^{\nu-\nu_0} f_{\mathrm{dc}}(2^{\nu_0}) + q(3^\nu (2/3)^{\nu_0} - 2^\nu), \end{aligned} \tag{6.52}$$

where ν_0 is an arbitrary integer within the range $0 \le \nu_0 < \nu$. If we stop dividing the matrix at the size 2^{ν_0}, and calculate the matrix–vector product in an obvious way there, then we can replace $f_{dc}(2^{\nu_0})$ in (6.52) with $(2^{\nu_0})^2$, so that

$$\begin{aligned} f_{\mathrm{dc}}(N) &= 3^\nu (4/3)^{\nu_0} + q(3^\nu (2/3)^{\nu_0} - 2^\nu) \\ &= 3^{\log_2 N} (4/3)^{\nu_0} + q(3^{\log_2 N} (2/3)^{\nu_0} - N). \end{aligned} \tag{6.53}$$

By numerical investigation, we see that $\nu_0 = 2$ is the best choice giving the minimum complexity.

For the first subblock, we need to compute Hu without receiving information from other subblocks. In the right-hand side of (6.50), it is necessary to compute two matrix–matrix subtractions, three matrix–vector multiplications, and three vector–vector additions, all of half size. Since H_{ij}'s are Hankel matrices, $2(N/2) - 1$ subtractions are necessary to compute each of $H_{11} - H_{12}$ and $H_{22} - H_{12}$. Therefore, denoting the complexity to compute Hu based on (6.50) by $g_{dc}(N)$, we have a recursion:

$$g_{\rm dc}(N) = 3g_{\rm dc}(N/2) + 3(N/2) + 2(N-1).$$

In the same way as we derived (6.53), we obtain

$$\begin{aligned} g_{\rm dc}(N) &= 3^{\log_2 N}(4/3)^{\nu_0} + 3(3^{\log_2 N}(2/3)^{\nu_0} - N) \\ &\quad + 3^{\log_2 N}\{4(2/3)^{\nu_0} - (1/3)^{\nu_0}\} - 4N + 1. \end{aligned} \tag{6.54}$$

Note that the sum of the first two terms in the right-hand side of (6.54) coincides with (6.53) for $q = 3$.

Thus, the complexity for computing the full-block matrix–vector product in (6.43) or (6.45) by using the divide and conquer method is given by

$$(M-1)f_{\rm dc}(N) + g_{\rm dc}(N). \tag{6.55}$$

Table 6.2 shows the computational complexity of the circular convolution method, the divide and conquer method, and the obvious method. In the obvious method, the computational complexity is given by $f_{\rm ob}(N) = N^2$. For $N \leq 512$, the divide and conquer method shows the best performance. For $1024 \leq N$, the circular convolution method outperforms the divide and conquer method.

6.3.3 Algorithm and Computational Complexity

All the results above are summarized as the *block exact fast affine projection (BEFAP) algorithm* [3] in Algorithm 6.2. The initialization part is omitted there. With a suitable initial value of k that ensures the existence of data, Steps 1–4 are repeated. The last column shows the computational complexity measured by the number of multiplication–additions and the number of divisions indicated in the parentheses. The quantities $C_1(N)$ and $C_2(N)$ are defined, respectively, by

Table 6.2 Comparison of computational complexity for the three methods

N	256	512	1024	2048
$f_{\rm cc}(N)$ $(q = 1.5)$	30208	66560	145408	315392
$f_{\rm dc}(N)$ $(q = 2.5)$	18314	55582	168026	506638
$f_{\rm ob}(N)$	65536	262144	1048576	4194304

$f_{\rm cc}(N)$ The circular convolution method, $f_{\rm dc}(N)$ The divide and conquer method ($\nu_0 = 2$), and $f_{\rm ob}(N)$ The obvious method

$$C_1(N) \triangleq \begin{cases} (M-1)f_{\mathrm{cc}}(N)\ (q=2)+g_{\mathrm{cc}}(N), & \text{circular convolution;} \\ (M-1)f_{\mathrm{dc}}(N)\ (q=3)+g_{dc}(N), & \text{divide and conquer,} \end{cases}$$

and

$$C_2(N) \triangleq \begin{cases} (M-1)f_{\mathrm{cc}}(N)\ (q=1)+g_{\mathrm{cc}}(N), & \text{circular convolution;} \\ (M-1)f_{\mathrm{dc}}(N)\ (q=2)+g_{\mathrm{dc}}(N), & \text{divide and conquer.} \end{cases}$$

Algorithm 6.2 (*BEFAP*)

Computation	Reference	Complexity
1. $z'_{k+N-1} := H_{k+N-1}u_k$;	first term of (6.40)	$C_1(N)$
2. for $i = 0, 1, \ldots, N-1$,		
2.1 compute r_{k+i} from r_{k+i-1};	(6.35)	$2(i+p-1)$
2.2 $z(k+i) := z'(k+i) + s^t_{k+i-1}r_{k+i}$;	(6.40)	$i+p-1$
2.3 $e(k+i) := d(k+i) - z(k+i)$;		1
2.4 $e_{k+i} := \begin{bmatrix} e(k+i) \\ (1-\mu)(e_{k+i-1})_u \end{bmatrix}$;	(6.3)	p
2.5 compute $f_{k+i}, (E_{k+i})_f, b_{k+i-1}, (E_{k+i-1})_b$;	FTF (Appendix 1: "Fast Transversal Filter (FTF) Algorithm")	$14p$ (7)
2.6 $\begin{bmatrix} \hat{\epsilon}_{k+i-1} \\ 0 \end{bmatrix} := \epsilon_{k+i-1} - \frac{b^t_{k+i-1}e_{k+i-1}}{(E_{k+i-1})_b}b_{k+i-1}$;	(6.15)	$2p$ (1)
2.7 $\epsilon_{k+i} := (1-\mu)\begin{bmatrix} 0 \\ \hat{\epsilon}_{k+i-1} \end{bmatrix} + \frac{f^t_{k+i}e_{k+i}}{(E_{k+i})_f}f_{k+i}$;	(6.14)	$3p$ (1)
2.8 $s_{k+i} := (0, s^t_{k+i-1})^t + \mu(\epsilon^t_{k+i}, \underbrace{0, \ldots, 0}_{i})^t$;	(6.33)	p
3. $u_{k+N} := u_k + H^t_{k+N-p}s^{p-1,N+p-2}_{k+N-1}$;	(6.42)	$C_2(N)$
4. $k := k+N$;		1

Let C_{befap} denote the number of multiplication–additions and divisions per sample. Then, C_{befap} is given by

$$C_{\mathrm{befap}} = \frac{1}{N}(C_1(N)+C_2(N)) + \frac{3}{2}N + 24p + 4.$$

Table 6.3 shows the computational complexity of the BEFAP algorithm (C_{befap}) and that of the FAP algorithm (C_{fap}) for $p = 8$. The divide and conquer method is employed for matrix–vector multiplication with $q = 3$ for $C_1(N)$, and $q = 2$ for

Table 6.3 Computational complexity of the BEFAP algorithm (C_{befap}) and that of the FAP algorithm (C_{fap}) for $p = 8$ and $\nu_0 = 2$

n	128	256	512	1024	2048
N	16	32	64	128	256
C_{befap}	426	577	815	1195	1814
C_{fap}	455	711	1223	2247	4295

The integers n and N are the filter length and the block length, respectively

$C_2(N)$. We set $\nu_0 = 2$. As shown in Algorithm 6.1, $C_{\text{fap}} = 2n + 24p$. It is seen that the superiority of the BEFAP algorithm to the FAP algorithm becomes more evident for large n and N.

Although the block length for computing the filter output and that for updating the coefficient vector were set equal in the above discussion, these do not have to be the same. By adjusting the block lengths independently, one can maximize the filter length given the computational complexity and the delay time [3].

Appendix 1: Fast Transversal Filter (FTF) Algorithm

Recursive-Least-Squares (RLS) Algorithm

In this appendix, we review the theory of *fast transversal filter (FTF) algorithm* [5, 6], which is a fast version of the recursive-least-squares (RLS) algorithm. We use the same notation as appears in the above papers. In particular, the coefficient vector of an adaptive filter is expressed as a row vector. This is different from the notation used in other parts of the book, but it makes easy to see the correspondence of equations appearing in this appendix and in the original papers. We treat the case where signals are of single channel and real.

The coefficient vector of an adaptive filter as in Fig. 2.1 can be estimated by the least-squares method. We will treat the sliding window case [6]: the front edge of the window is at T, and the rear edge of the window is at $T - L + 1$. As the time T progresses, the front edge and the rear edge move forward, but the window length L is kept constant as shown in Fig. 6.4. Let $\{x(k)\}$ be the input signal, and $\{d(k)\}$ the desired signal. Also, let the coefficient vector be denoted by $W_{N,L,T}$, where N is the filter length, L is the window length, and T is the time index at the front edge of the window. Then, $W_{N,L,T}$ is estimated by minimizing the quantity

$$\xi_{N,L}(T) \triangleq \sum_{k=T-L+1}^{T} \{d(k) + W_{N,L,T} X_N(k)\}^2, \tag{6.56}$$

where

$$X_N(k) \triangleq [x(k)\; x(k-1)\; \cdots\; x(k-N+1)]^t.$$

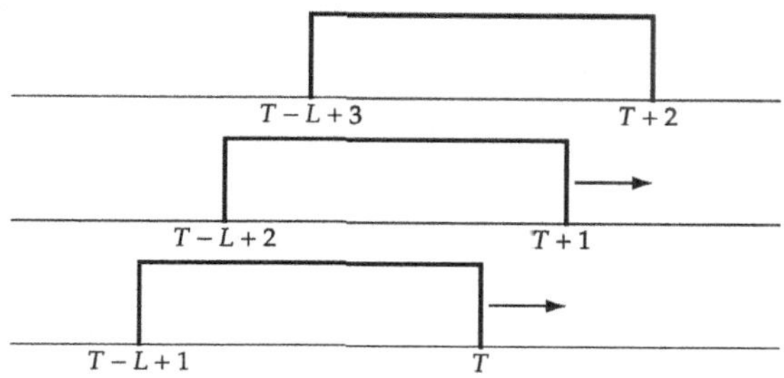

Fig. 6.4 Sliding window of length L

By introducing the vector notation as

$$\epsilon_{N,L,T} \stackrel{\triangle}{=} d_{L,T} + X_{N,L,T} W^t_{N,L,T},$$

where

$$X_{N,L,T} \stackrel{\triangle}{=} [x_{L,T}\; x_{L,T-1}\; \cdots\; x_{L,T-N+1}],$$
$$x_{L,k} \stackrel{\triangle}{=} [x(k)\; x(k-1)\; \cdots\; x(k-L+1)]^t,$$
$$d_{L,T} \stackrel{\triangle}{=} [d(T)\; d(T-1)\; \cdots\; d(T-L+1)]^t,$$

Eq. (6.56) is rewritten as

$$\xi_{N,L}(T) = \|\epsilon_{N,L,T}\|^2. \tag{6.57}$$

By Theorem 3.8, the minimum norm $W_{N,L,T}$ that minimizes (6.57) is given by

$$W^t_{N,L,T} = -X^+_{N,L,T} d_{L,T}, \tag{6.58}$$

where "+" indicates the Moore–Penrose pseudoinverse of a matrix (Chap. 3, Appendix 1: "Affine Projection"). By Corollary 3.2, we have

$$(X^+_{N,L,T} = (X^t_{N,L,T} X_{N,L,T})^+ X^t_{N,L,T}.$$

Since

$$(X^t_{N,L,T} X_{N,L,T})^+)^t = (X^t_{N,L,T} X_{N,L,T})^+$$

by Theorem 3.6(2), (6.58) is rewritten as

$$W_{N,L,T} = -(d^t_{L,T} X_{N,L,T})(X^t_{N,L,T} X_{N,L,T})^+. \tag{6.59}$$

We exploit this expression in the subsequent arguments.

If $X^t_{N,L,T}X_{N,L,T}$ is nonsingular, then by Lemma 3.1 (Chap. 3, Appendix 1 "Affine Projection"),

$$(X^t_{N,L,T}X_{N,L,T})^+ = (X^t_{N,L,T}X_{N,L,T})^{-1}.$$

Since $X^t_{N,L,T}X_{N,L,T}$ is an $N \times N$ matrix, the complexity of computing its inverse is $O(N^3)$ per sample if a standard algorithm like the Cholesky decomposition is used. The recursive-least-squares (RLS) algorithm reduces it to $O(N^2)$ per sample by using the matrix inversion lemma [5] (Chap. 3, Sect. 3.6). The fast transversal filter (FTF) algorithm further reduces the complexity to $O(N)$ per sample.

Update of Orthogonal Projection

Let X and Y be matrices having the same number of rows. By Theorem 3.6(4) (Chap. 3, Appendix 1: "Affine Projection"),

$$X(X^tX)^+X^t = P_{\mathcal{R}(X)}, \tag{6.60}$$

which is the orthogonal projection onto $\mathcal{R}(X)$. For simplicity, we denote

$$P_X \triangleq P_{\mathcal{R}(X)},$$

and

$$P_{X,Y} \triangleq P_{[X\ Y]},$$

where $[X\ Y]$ is the matrix formed by juxtaposing X and Y. The next theorem shows that $P_{X,Y}$ can be decomposed into the sum of two orthogonal projections.

We denote

$$\begin{aligned} P_X^{\perp} &\triangleq I - P_X \\ &= I - P_{\mathcal{R}(X)} \\ &= P_{\mathcal{R}(X)^{\perp}}. \end{aligned}$$

Theorem 6.1 *Let $Z \triangleq P_X^{\perp}Y$. Then,*

$$\begin{aligned} P_{X,Y} &= P_X + P_Z \\ &= P_X + P_X^{\perp}Y(Y^tP_X^{\perp}Y)^+Y^tP_X^{\perp}. \end{aligned} \tag{6.61}$$

Proof Because P_X and P_Z are symmetric and idempotent, $P_X + P_Z$ is also symmetric and idempotent. In fact,

$$\begin{aligned} (P_X + P_Z)^t &= P_X^t + P_Z^t \\ &= P_X + P_Z. \end{aligned}$$

Since $X \perp Z$, $P_X P_Z = P_Z P_X = 0$. Therefore,

$$\begin{aligned}(P_X + P_Z)^2 &= P_X^2 + P_X P_Z + P_Z P_X + P_Z^2 \\ &= P_X + P_Z.\end{aligned}$$

Thus, $P_X + P_Z$ is an orthogonal projection. Furthermore, it is a mapping onto $\mathcal{R}([X\ Z])$. Because each column of $P_X Y$ is a linear combination of the columns of X, each column of $Z = Y - P_X Y$ is a linear combination of the columns of $[X\ Y]$. Therefore, $\mathcal{R}([X\ Z]) \subseteq \mathcal{R}([X\ Y])$. Conversely, each column of $Y = Z + P_X Y$ is a linear combination of the columns of $[X\ Z]$. Therefore, $\mathcal{R}([X\ Z]) \supseteq \mathcal{R}([X\ Y])$. Thus, $\mathcal{R}([X\ Z]) = \mathcal{R}([X\ Y])$. Hence, $P_X + P_Z$ is a mapping onto $\mathcal{R}([X\ Y])$, so it is the orthogonal projection onto $\mathcal{R}([X\ Y])$:

$$P_{X,Y} = P_X + P_Z.$$

The second equality is obtained by replacing X in (6.60) with $Z = P_X^{\perp} Y$. □

By Theorem 3.3 (Chap. 3, Appendix 1: "Affine Projection"),

$$\begin{aligned}P_{X,Y}^{\perp} &= I - P_{X,Y} \\ &= P_X^{\perp} - P_X^{\perp} Y (Y^t P_X^{\perp} Y)^+ Y^t P_X^{\perp}.\end{aligned}$$

Therefore, we immediately obtain

$$U^t P_{X,Y}^{\perp} V = U^t P_X^{\perp} V - (U^t P_X^{\perp} Y)(Y^t P_X^{\perp} Y)^+ (Y^t P_X^{\perp} V), \tag{6.62}$$

where U and V are arbitrary matrices having the same number of rows as that of X and Y.

Let $K_X \triangleq X(X^t X)^+$. For a row vector u,

$$u P_X = (u K_X) X^t.$$

Thus, $u K_X$ gives the vector of combination coefficients when $u P_X$ is expressed as a linear combination of the rows of X^t. This leads to

$$X^t K_X = I, \tag{6.63}$$

if the columns of X are linearly independent. Similarly,

$$\begin{bmatrix} X^t \\ Y^t \end{bmatrix} K_{X,Y} = \begin{bmatrix} I & 0 \\ 0 & I \end{bmatrix}, \tag{6.64}$$

from which we have

$$X^t K_{X,Y} = [I\ \ 0], \tag{6.65}$$

and

$$Y^t K_{X,Y} = [0 \;\; I], \tag{6.66}$$

if the columns of $[X \;\; Y]$ are linearly independent. From (6.65),

$$\begin{aligned} P_X K_{X,Y} &= X(X^t X)^+ X^t K_{X,Y} \\ &= X(X^t X)^+ [I \;\; 0] \\ &= [K_X \;\; 0]. \end{aligned} \tag{6.67}$$

By Theorem 3.6(3) (Chap. 3, Appendix 1: "Affine Projection"),

$$\begin{aligned} P_X K_X &= (X(X^t X)^+ X^t)(X(X^t X)^+) \\ &= (X(X^t X)^+ X^t X)(X^t X)^+ \\ &= X(X^t X)^+ \\ &= K_X. \end{aligned} \tag{6.68}$$

Similarly,

$$P_{X,Y} K_{X,Y} = K_{X,Y}. \tag{6.69}$$

Multiplying $K_{X,Y}$ to both sides of (6.61) from the right, and using (6.69), we have

$$K_{X,Y} = P_X K_{X,Y} + P_X^{\perp} Y (Y^t P_X^{\perp} Y)^+ Y^t (I - P_X) K_{X,Y}. \tag{6.70}$$

Substitution of (6.66) and (6.67) into (6.70) yields

$$K_{X,Y} = [K_X \;\; 0] + P_X^{\perp} Y (Y^t P_X^{\perp} Y)^+ [-Y^t K_X \;\; I]. \tag{6.71}$$

In a similar manner, we can prove

$$K_{Y,X} = [0 \;\; K_X] + P_X^{\perp} Y (Y^t P_X^{\perp} Y)^+ [I - Y^t K_X]. \tag{6.72}$$

From (6.71) and (6.72), we have

$$U^t K_{X,Y} = [U^t K_X \;\; 0] + U^t P_X^{\perp} Y (Y^T P_X^{\perp} Y)^+ [-Y^t K_X \;\; I], \tag{6.73}$$

and

$$U^t K_{Y,X} = [0 \;\; U^t K_X] + U^t P_X^{\perp} Y (Y^t P_X^{\perp} Y)^+ [I - Y^t K_X]. \tag{6.74}$$

Now, let us define

$$P_{N,L,T} \triangleq X_{N,L,T} (X_{N,L,T}^t X_{N,L,T})^+ X_{N,L,T}^t,$$

which is the orthogonal projection onto $\mathcal{R}(X_{N,L,T})$. By Theorem 3.3 (Chap. 3, Appendix 1: "Affine Projection"), the orthogonal projection onto $\mathcal{R}(X_{N,L,T})^{\perp}$ is given by

$$P^{\perp}_{N,L,T} \triangleq I - P_{N,L,T}.$$

Let us also define

$$K_{N,L,T} \triangleq X_{N,L,T}(X^t_{N,L,T}X_{N,L,T})^+.$$

For time updating $P_{N,L,T}$ and $K_{N,L,T}$, the *pinning vectors* defined by

$$\sigma_L \triangleq [1\,0\,\cdots\,0]^t$$

and

$$\pi_L \triangleq [0\,\cdots\,0\,1]^t$$

are useful. These are column vectors of length L. We denote

$$P_{\sigma_L,N,L,T} \triangleq P_{[\sigma_L\ \ X_{N,L,T}]}, \quad K_{\sigma_L,N,L,T} \triangleq K_{[\sigma_L\ \ X_{N,L,T}]},$$
$$P_{\pi_L,N,L,T} \triangleq P_{[\pi_L\ \ X_{N,L,T}]}, \quad K_{\pi_L,N,L,T} \triangleq K_{[\pi_L\ \ X_{N,L,T}]}.$$

Note that $P_{\sigma_L,N,L,T} = P_{N,L,T,\sigma_L}$, but $K_{\sigma_L,N,L,T} \neq K_{N,L,T,\sigma_L}$.

Let

$$\Omega \triangleq P^{\perp}_{\sigma_L}X_{N,L,T} = \begin{bmatrix} 0 & 0 \\ 0 & I_{L-1} \end{bmatrix} X_{N,L,T} = \begin{bmatrix} 0 \\ X_{N,L-1,T-1} \end{bmatrix}.$$

Then,

$$\begin{aligned} P_\Omega &= \Omega(\Omega^t\Omega)^+\Omega^t \\ &= \begin{bmatrix} 0 \\ X_{N,L-1,T-1} \end{bmatrix} (X^t_{N,L-1,T-1}X_{N,L-1,T-1})^+ \left[0\ X^t_{N,L-1,T-1}\right] \\ &= \begin{bmatrix} 0 & 0 \\ 0 & X_{N,L-1,T-1}(X^t_{N,L-1,T-1}X_{N,L-1,T-1})^+X^t_{N,L-1,T-1} \end{bmatrix} \\ &= \begin{bmatrix} 0 & 0 \\ 0 & P_{N,L-1,T-1} \end{bmatrix}. \end{aligned}$$

Therefore, by Theorem 6.1,

$$\begin{aligned} P_{\sigma_L,N,L,T} &= P_{\sigma_L} + P_\Omega \\ &= \begin{bmatrix} 1 & 0 \\ 0 & 0 \end{bmatrix} + \begin{bmatrix} 0 & 0 \\ 0 & P_{N,L-1,T-1} \end{bmatrix} \\ &= \begin{bmatrix} 1 & 0 \\ 0 & P_{N,L-1,T-1} \end{bmatrix}. \end{aligned}$$

This equation shows the role of the pinning vector σ_L very clearly.

Theorem 6.2 $K_{N,L,T}$ *is updated as*

$$\begin{bmatrix} 0 \\ K_{N,L-1,T-1} \end{bmatrix} = K_{N,L,T} - P^{\perp}_{N,L,T}\sigma_L(\sigma_L^t P^{\perp}_{N,L,T}\sigma_L)^{+}(\sigma_L^t K_{N,L,T}), \tag{6.75}$$

and

$$\begin{bmatrix} K_{N,L-1,T} \\ 0 \end{bmatrix} = K_{N,L,T} - P^{\perp}_{N,L,T}\pi_L(\pi_L^t P^{\perp}_{N,L,T}\pi_L)^{+}(\pi_L^t K_{N,L,T}). \tag{6.76}$$

Proof We first show

$$K_{\sigma_L,N,L,T} = \begin{bmatrix} 1 & 0 \\ * & K_{N,L-1,T-1} \end{bmatrix}, \tag{6.77}$$

where "$*$" denotes components that are irrelevant to the proof of the theorem. Let

$$K_{\sigma_L,N,L,T} = \begin{bmatrix} a & b \\ c & S \end{bmatrix}, \tag{6.78}$$

where a is 1×1, b is $1 \times (L-1)$, c is $(L-1) \times 1$, and S is $(L-1) \times (L-1)$. Because

$$\sigma_L^t K_{\sigma_L,N,L,T} = [1\ 0\ \cdots\ 0],$$

and

$$\sigma_L^t \begin{bmatrix} a & b \\ c & S \end{bmatrix} = [a\ \ b],$$

we have

$$[a\ \ b] = [1\ 0\ \cdots\ 0].$$

Thus, $K_{\sigma_L,N,L,T}$ must be of the form

$$K_{\sigma_L,N,L,T} = \begin{bmatrix} 1 & 0 \\ c & S \end{bmatrix}. \tag{6.79}$$

Let

$$y_{L,T} = [y(T)\ y(T-1)\ \cdots\ y(T-L+1)]^t$$

be an arbitrary column vector of length L. Then, $y_{L,T}^t K_{\sigma_L,N,L,T}$ gives the vector of coefficients

$$[\alpha_{\sigma_L}\ \alpha_T\ \cdots\ \alpha_{T-N+1}]$$

that minimizes the quantity

$$\xi^2 \triangleq \|y_{L,T} + (\alpha_{\sigma_L}\sigma_L + \alpha_T x_{L,T} + \cdots + \alpha_{T-N+1}x_{L,T-N-1})\|^2.$$

Since $\sigma_L = [1\,0\,\cdots\,0]^t$, the vector $y_{L-1,T-1}$ does not contribute to determining α_{σ_L}. Also, in view of (6.79), $y(T)$ does not contribute to determining $\alpha_T, \ldots, \alpha_{T-N+1}$. Therefore, ξ^2 can be expressed as a sum of two terms:

$$\xi^2 = (y(T)+\alpha_{\sigma_L})^2 + \|y_{L-1,T-1} + (\alpha_T x_{L-1,T-1} + \cdots + \alpha_{T-N+1}x_{L-1,T-1-N+1})\|^2.$$

Since the two terms can be minimized independently, $[\alpha_T\ \cdots\ \alpha_{T-N+1}]$ gives the vector of coefficients that minimizes the second term. In addition, by (6.79),

$$y^t_{L-1,T-1}S = [\alpha_T\ \cdots\ \alpha_{T-N+1}].$$

Therefore, we have $S = K_{N,L-1,T-1}$.

In (6.72), replace $X = X_{N,L,T}$ and $Y = \sigma_L$. Then we have

$$K_{\sigma_L,N,L,T} = [0\ K_{N,L,T}] + P^{\perp}_{N,L,T}\sigma_L(\sigma^t_L P^{\perp}_{N,L,T}\sigma_L)^{+}[1\ \ -\sigma^t_L K_{N,L,T}],$$

which, together with (6.77), yields

$$\begin{bmatrix} 1 & 0 \\ * & K_{N,L-1,T-1} \end{bmatrix} = [0\ K_{N,L,T}] + P^{\perp}_{N,L,T}\sigma_L(\sigma^t_L P^{\perp}_{N,L,T}\sigma_L)^{+}[1\ \ -\sigma^t_L K_{N,L,T}].$$

Neglecting the left-most column, we finally obtain (6.75).

By using π_L instead of σ_L, and

$$K_{\pi_L,N,L,T} = \begin{bmatrix} * & K_{N,L-1,T} \\ 1 & 0 \end{bmatrix}$$

instead of (6.77), Eq. (6.76) can be proved in a similar manner. □

Equations for Updating Transversal Filter Coefficients

Definitions of RLS Variables

The adaptive filter coefficient $W_{N,L,T}$ in (6.59) is rewritten, by using $K_{N,L,T}$, as

$$W_{N,L,T} = -d^t_{L,T}K_{N,L,T}.$$

The final goal is to derive a set of update equations for $W_{N,L,T}$, in which update equations for other "transversal filters" defined as

$$A_{N,L,T} \triangleq [1 \quad -x_{L,T}^t K_{N,L,T-1}],$$
$$B_{N,L,T} \triangleq [-x_{L,T-N}^t K_{N,L,T} \quad 1],$$
$$C_{N,L,T} \triangleq -\sigma_L^t K_{N,L,T},$$
$$D_{N,L,T} \triangleq -\pi_L^t K_{N,L,T}.$$

are exploited. Since $x_{L,T}^t K_{N,L,T-1}$ gives the forward predictor, $A_{N,L,T}$ is interpreted as the forward residual filter. In the same way, $B_{N,L,T}$ is interpreted as the backward residual filter. To describe the update equations for the above transversal filters, the following variables are used. These variables are also updated recursively.

Residuals

$$e_{N,L}(T) \triangleq A_{N,L,T} X_{N+1}(T) = x_{L,T}^t P_{N,L,T-1}^{\perp} \sigma_L,$$
$$r_{N,L}(T) \triangleq B_{N,L,T} X_{N+1}(T) = x_{L,T-N}^t P_{N,L,T}^{\perp} \sigma_L,$$
$$\gamma_{N,L}(T) \triangleq 1 + C_{N,L,T} X_N(T) = \sigma_L^t P_{N,L,T}^{\perp} \sigma_L,$$
$$\delta_{N,L}(T) \triangleq 1 + D_{N,L,T} X_N(T-L+1) = \pi_L^t P_{N,L,T}^{\perp} \pi_L,$$
$$\epsilon_{N,L}(T) \triangleq d(T) + W_{N,L,T} X_N(T) = d_{L,T}^t P_{N,L,T}^{\perp} \sigma_L,$$
$$\nu_{N,L}(T) \triangleq d(T-L+1) + W_{N,L,T} X_N(T-L+1) = d_{L,T}^t P_{N,L,T}^{\perp} \pi_L,$$
$$a_{N,L}(T) \triangleq A_{N,L,T} X_{N+1}(T-L+1) = x_{L,T}^t P_{N,L,T-1}^{\perp} \pi_L,$$
$$b_{N,L}(T) \triangleq B_{N,L,T} X_{N+1}(T-L+1) = x_{L,T-N}^t P_{N,L,T}^{\perp} \pi_L.$$

Prediction Errors

$$e_{N,L-1}^p(T) \triangleq A_{N,L-1,T-1} X_{N+1}(T),$$
$$r_{N,L-1}^p(T) \triangleq B_{N,L-1,T-1} X_{N+1}(T),$$
$$\epsilon_{N,L-1}^p(T) \triangleq d(T) + W_{N,L-1,T-1} X_N(T).$$

Smoothed Errors

$$a_{N,L-1}^s(T) \triangleq A_{N,L-1,T} X_{N+1}(T-L+1),$$
$$b_{N,L-1}^s(T) \triangleq B_{N,L-1,T} X_{N+1}(T-L+1),$$
$$\nu_{N,L-1}^s(T) \triangleq d(T-L+1) + W_{N,L-1,T} X_N(T-L-1)$$

Residual Powers

$$\alpha_{N,L}(T) \triangleq x^t_{L,T} P^{\perp}_{N,L,T-1} x_{L,T},$$

$$\beta_{N,L}(T) \triangleq x^t_{L,T-N} P^{\perp}_{N,L,T} x_{L,T-N}.$$

The second equalities for Residuals are verified by direct calculation.

Although the original paper [6] is concerned with three types of algorithms (unnormalized, normalized, and overnormalized ones), we confine ourselves to the overnormalized one. In this algorithm, the unnormalized filters $A_{N,L,T}$, $B_{N,L,T}$, and $C_{N,L,T}$ are divided by the associated powers as

$$\tilde{A}_{N,L,T} \triangleq \alpha^{-1}_{N,L}(T)\, A_{N,L,T},$$

$$\tilde{B}_{N,L,T} \triangleq \beta^{-1}_{N,L}(T)\, B_{N,L,T},$$

$$\tilde{C}_{N,L,T} \triangleq \gamma^{-1}_{N,L}(T)\, C_{N,L,T}.$$

The filters $W_{N,L,T}$ and $D_{N,L,T}$ are used without normalization.

Sliding the window is performed in two phases as illustrated in Fig. 6.5. In the first phase, the rear edge of the window is moved forward by one sample keeping the front edge untouched: $(N, L, T) \to (N, L-1, T)$. Thus, the window length is shortened temporarily. Then, in the second phase, the front edge is moved forward by one sample keeping the rear edge untouched: $(N, L-1, T) \to (N, L, T+1)$.

Equations for Updating the Front Edge of the Window

In this section, a set of equations for updating $W_{N,L-1,T-1}$ to $W_{N,L,T}$ are presented. When the algorithm is actually used, T is replaced with $T+1$. The derivations of those equations are straightforward (but lengthy) from the results of the preceding section, so we omit the details.

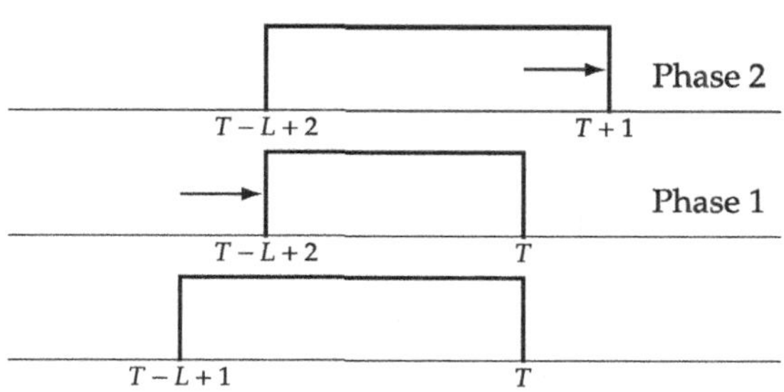

Fig. 6.5 Sliding the window in two phases

From (6.75) in Theorem 6.2,

$$\begin{bmatrix} 0 \\ K_{N,L-1,T-2} \end{bmatrix} = K_{N,L,T-1} - P^{\perp}_{N,L,T-1}\sigma_L(\sigma^t_L P^{\perp}_{N,L,T-1}\sigma_L)^{+}(\sigma^t_L K_{N,L,T-1})$$

Multiplying $x^t_{L,T}$ to both sides of the above equation from the left, we can derive

$$A_{N,L-1,T-1} = A_{N,L,T} - e_{N,L}(T)\gamma^{-1}_{N,L}(T-1)[0 \;\; C_{N,L,T-1}]. \tag{6.80}$$

Also, multiplying $x^t_{L,T-N}$ to both sides of (6.75) from the left, we obtain

$$B_{N,L-1,T-1} = B_{N,L,T} - r_{N,L}(T)\gamma^{-1}_{N,L}(T)[C_{N,L,T} \;\; 0]. \tag{6.81}$$

In (6.74), replace $U = \sigma_L$, $X = X_{N,L,T-1}$, and $Y = x_{L,T}$. Then, it is verified that

$$C_{N+1,L,T} = [0 \;\; C_{N,L,T-1}] - e_{N,L}(T)\alpha^{-1}_{N,L}(T)A_{N,L,T}. \tag{6.82}$$

In a similar way, replace $U = \sigma_L$, $X = X_{N,L,T}$, and $Y = x_{L,T-N}$ in (6.73). Then, we get

$$C_{N+1,L,T} = [C_{N,L,T} \;\; 0] - r_{N,L}(T)\beta^{-1}_{N,L}(T)B_{N,L,T}. \tag{6.83}$$

Since $r_{N,L}(T)\beta^{-1}_{N,L}(T)$ is a scalar, and the last component of $B_{N,L,T}$ is unity, we have

$$r_{N,L}(T)\beta^{-1}_{N,L}(T) = -C^N_{N+1,L,T}, \tag{6.84}$$

where $C^N_{N+1,L,T}$ is the last component of $C_{N+1,L,T}$. Therefore, (6.83) is rewritten as

$$C_{N+1,L,T} = [C_{N,L,T} \;\; 0] + C^N_{N+1,L,T}B_{N,L,T}. \tag{6.85}$$

Replace $U = V = x_{L,T}$, $X = X_{N,L,T-1}$, and $Y = \sigma_L$ in (6.62). Then, we obtain

$$\alpha_{N,L-1}(T-1) = \alpha_{N,L}(T) - e^2_{N,L}(T)\gamma^{-1}_{N,L}(T-1). \tag{6.86}$$

Similarly, replace $U = V = x_{L,T-N}$, $X = X_{N,L,T}$, and $Y = \sigma_L$ in (6.62). Then, we get

$$\beta_{N,L-1}(T-1) = \beta_{N,L}(T) - r^2_{N,L}(T)\gamma^{-1}_{N,L}(T). \tag{6.87}$$

Multiplication of $X_{N+1}(T)$ to both sides of (6.80) from the right yields

$$e^p_{N,L-1}(T) = e_{N,L}(T)\gamma^{-1}_{N,L}(T-1). \tag{6.88}$$

Similarly, multiplication of $X_{N+1}(T)$ to both sides of (6.81) from the right yields

$$r^p_{N,L-1}(T) = r_{N,L}(T)\gamma^{-1}_{N,L}(T). \tag{6.89}$$

Replace $U = V = \sigma_L$, $X = X_{N,L,T}$, and $Y = x_{L,T+1}$ in (6.62). Then, we see that

$$\gamma_{N+1,L}(T) = \gamma_{N,L}(T-1) - e^2_{N,L}(T)\alpha^{-1}_{N,L}(T). \tag{6.90}$$

By (6.86), it is shown that

$$e^2_{N,L}(T)\alpha^{-1}_{N,L}(T) = \gamma_{N,L}(T-1)\{1 - \alpha^{-1}_{N,L}(T)\alpha_{N,L-1}(T-1)\},$$

substitution of which into (6.90) yields

$$\gamma_{N+1,L}(T) = \gamma_{N,L}(T-1)\alpha^{-1}_{N,L}(T)\alpha_{N,L-1}(T-1). \tag{6.91}$$

In a similar way, replace $U = V = \sigma_L$, $X = X_{N,L,T}$, and $Y = x_{L,T-N}$ in (6.62). Then, we have

$$\gamma_{N+1,L}(T) = \gamma_{N,L}(T) - r^2_{N,L}(T)\beta^{-1}_{N,L}(T). \tag{6.92}$$

By (6.87), it is shown that

$$r^2_{N,L}(T)\beta^{-1}_{N,L}(T) = \gamma_{N,L}(T)\{1 - \beta^{-1}_{N,L}(T)\beta_{N,L-1}(T-1)\},$$

substitution of which into (6.92) yields

$$\gamma_{N+1,L}(T) = \gamma_{N,L}(T)\beta^{-1}_{N,L}(T)\beta_{N,L-1}(T-1). \tag{6.93}$$

Substitution of (6.80) and (6.91) into (6.82) yields

$$\tilde{C}_{N+1,L,T} = [0\ \ \tilde{C}_{N,L,T-1}] - e^p_{N,L-1}(T)\tilde{A}_{N,L-1,T-1}. \tag{6.94}$$

Similarly, substitution of (6.81) and (6.93) into (6.83) yields

$$\tilde{C}_{N+1,L,T} = [\tilde{C}_{N,L,T}\ \ 0] - r^p_{N,L-1}(T)\tilde{B}_{N,L-1,T-1}. \tag{6.95}$$

Substituting (6.81) into (6.85), we obtain

$$C_{N+1,L,T} = [C_{N,L,T}\ \ 0](1 + r^p_{N,L-1}(T)C^N_{N+1,L,T}) + C^N_{N+1,L,T}B_{N,L-1,T-1}. \tag{6.96}$$

Multiplying $X_{N+1}(T)$ to both sides of the above equation from the right, we find

$$1 + r^p_{N,L-1}(T)C^N_{N+1,L,T} = \gamma_{N+1,L}(T)\gamma_{N,L}(T)^{-1} = \beta_{N,L-1}(T-1)\beta_{N,L}(T)^{-1}. \tag{6.97}$$

From (6.97), we can derive

$$r^p_{N,L-1}(T) = -\beta_{N,L-1}(T-1)\tilde{C}^N_{N+1,L,T}. \tag{6.98}$$

Equations (6.80) and (6.82), together with (6.86) and (6.91), yield

$$\tilde{A}_{N,L,T} = \tilde{A}_{N,L-1,T-1} + \alpha_{N,L}^{-1}(T)e_{N,L}(T)\tilde{C}_{N+1,L,T}. \tag{6.99}$$

Similarly, Eqs. (6.81) and (6.83), together with (6.84) and (6.87), yield

$$\tilde{B}_{N,L,T} = \tilde{B}_{N,L-1,T-1} + \tilde{C}_{N+1,L,T}^{N}\gamma_{N+1,L}(T)\tilde{C}_{N+1,L,T}. \tag{6.100}$$

Finally, multiply $d_{L,T}^{t}$ to both sides of (6.75) from the left. Then, we can derive

$$W_{N,L,T} = W_{N,L-1,T-1} + \epsilon_{N,L}(T)\tilde{C}_{N,L,T}. \tag{6.101}$$

Multiplying $X_N(T)$ to both sides of (6.101) from the right, we find

$$\epsilon_{N,L-1}^{p}(T) = \epsilon_{N,L}(T)\gamma_{N,L}^{-1}(T). \tag{6.102}$$

Equations for Updating the Rear Edge of the Window

Derivations of update equations in this case are performed in parallel with those in the previous section.

From (6.76) in Theorem 6.2,

$$\begin{bmatrix} K_{N,L-1,T-1} \\ 0 \end{bmatrix} = K_{N,L,T-1} - P_{N,L,T-1}^{\perp}\pi_L(\pi_L^t P_{N,L,T-1}^{\perp}\pi_L)^{+}(\pi_L^t K_{N,L,T-1}).$$

Multiplying $x_{L,T}^{t}$ to both sides of the above equation from the left, we can derive

$$A_{N,L-1,T} = A_{N,L,T} - a_{N,L}(T)\delta_{N,L}^{-1}(T-1)[0 \;\; D_{N,L,T-1}]. \tag{6.103}$$

Also, multiplying $x_{L,T-N}^{t}$ to both sides of (6.76) from the left, we obtain

$$B_{N,L-1,T} = B_{N,L,T} - b_{N,L}(T)\delta_{N,L}^{-1}(T)[D_{N,L,T} \;\; 0] \tag{6.104}$$

In (6.74), replace $U = \pi_L$, $X = X_{N,L,T-1}$, and $Y = x_{L,T}$. Then, it is verified that

$$D_{N+1,L,T} = [0 \;\; D_{N,L,T-1}] - a_{N,L}(T)\tilde{A}_{N,L,T}. \tag{6.105}$$

In a similar way, replace $U = \pi_L$, $X = X_{N,L,T}$, and $Y = x_{L,T-N}$ in (6.73). Then, we get

$$D_{N+1,L,T} = [D_{N,L,T} \;\; 0] - b_{N,L}(T)\tilde{B}_{N,L,T}. \tag{6.106}$$

Since $b_{N,L}(T)\beta_{N,L}^{-1}(T)$ is a scalar, and the last component of $B_{N,L,T}$ is unity, we have

$$b_{N,L}(T)\beta_{N,L}^{-1}(T) = -D_{N+1,L,T}^{N}, \tag{6.107}$$

where $D_{N+1,L,T}^{N}$ is the last component of $D_{N+1,L,T}$. Therefore, (6.106) is rewritten as

$$D_{N+1,L,T} = [D_{N,L,T} \;\; 0] + D_{N+1,L,T}^{N} B_{N,L,T}. \tag{6.108}$$

Replace $U = V = x_{L,T}$, $X = X_{N,L,T-1}$, and $Y = \pi_L$ in (6.62). Then, we obtain

$$\alpha_{N,L-1}(T) = \alpha_{N,L}(T) - a_{N,L}^{2}(T)\delta_{N,L}^{-1}(T-1). \tag{6.109}$$

Similarly, replace $U = V = x_{L,T-N}$, $X = X_{N,L,T}$, and $Y = \pi_L$ in (6.62). Then, we get

$$\beta_{N,L-1}(T) = \beta_{N,L}(T) - b_{N,L}^{2}(T)\delta_{N,L}^{-1}(T). \tag{6.110}$$

Multiplication of $X_{N+1}(T-L+1)$ to both sides of (6.103) from the right yields

$$a_{N,L-1}^{s}(T) = a_{N,L}(T)\delta_{N,L}^{-1}(T-1). \tag{6.111}$$

From (6.111) and (6.109), we obtain

$$\alpha_{N,L-1}(T) = \alpha_{N,L}(T) - a_{N,L-1}^{s}(T)a_{N,L}(T). \tag{6.112}$$

Similarly, multiplication of $X_{N+1}(T-L+1)$ to both sides of (6.104) from the right yields

$$b_{N,L-1}^{s}(T) = b_{N,L}(T)\delta_{N,L}^{-1}(T). \tag{6.113}$$

From (6.113) and (6.110), we get

$$\beta_{N,L-1}(T) = \beta_{N,L}(T) - b_{N,L-1}^{s}(T)b_{N,L}(T). \tag{6.114}$$

Replace $U = V = \pi_L$, $X = X_{N,L,T}$, and $Y = x_{L,T+1}$ in (6.62). Then, we see

$$\delta_{N+1,L}(T+1) = \delta_{N,L}(T) - a_{N,L}^{2}(T+1)\alpha_{N,L}^{-1}(T+1),$$

that is

$$\delta_{N+1,L}(T) = \delta_{N,L}(T-1) - a_{N,L}^{2}(T)\alpha_{N,L}^{-1}(T). \tag{6.115}$$

By (6.109), it is shown that

$$a_{N,L}^{2}(T)\alpha_{N,L}^{-1}(T) = \delta_{N,L}(T-1)\{1 - \alpha_{N,L}^{-1}(T)\alpha_{N,L-1}(T)\},$$

substitution of which into (6.115) yields

$$\delta_{N+1,L}(T) = \delta_{N,L}(T-1)\alpha_{N,L}^{-1}(T)\alpha_{N,L-1}(T). \tag{6.116}$$

In a similar way, replace $U = V = \pi_L$, $X = X_{N,L,T}$, and $Y = x_{L,T-N}$ in (6.62). Then, we have

$$\delta_{N+1,L}(T) = \delta_{N,L}(T) - b_{N,L}^2(T)\beta_{N,L}^{-1}(T). \tag{6.117}$$

By (6.110), it is shown that

$$b_{N,L}^2(T)\beta_{N,L}^{-1}(T) = \delta_{N,L}(T)\{1 - \beta_{N,L}^{-1}(T)\beta_{N,L-1}(T)\},$$

substitution of which into (6.117) yields

$$\delta_{N+1,L}(T) = \delta_{N,L}(T)\beta_{N,L}^{-1}(T)\beta_{N,L-1}(T). \tag{6.118}$$

From (6.107) and (6.117), we obtain

$$\delta_{N,L}(T) = \delta_{N+1,L}(T) - D_{N-1,L,T}^N b_{N,L}(T). \tag{6.119}$$

Equations (6.103) and (6.105), together with (6.109) and (6.116), yield

$$\tilde{A}_{N,L-1,T} = \tilde{A}_{N,L,T} - \alpha_{N,L}^{-1}(T)a_{N,L}(T)\delta_{N+1,L}^{-1}(T)D_{N+1,L,T}. \tag{6.120}$$

Similarly, Eqs. (6.104) and (6.106), together with (6.110) and (6.113), yield

$$\tilde{B}_{N,L-1,T} = \tilde{B}_{N,L,T} - b_{N,L-1}^s(T)\beta_{N,L-1}^{-1}(T)D_{N+1,L,T}. \tag{6.121}$$

Finally, multiply $d_{L,T}^t$ to both sides of (6.76) from the left. Then, we can derive

$$W_{N,L-1,T} = W_{N,L,T} - \nu_{N,L}(T)\delta_{N,L}^{-1}(T)D_{N,L,T}. \tag{6.122}$$

Multiplying $X_N(T-L+1)$ to both sides of (6.122) from the right, we find

$$\nu_{N,L-1}^s(T) = \nu_{N,L}(T)\delta_{N,L}^{-1}(T). \tag{6.123}$$

From (6.122) and (6.123), we have

$$W_{N,L-1,T} = W_{N,L,T} - \nu_{N,L-1}^s(T)D_{N,L,T}. \tag{6.124}$$

Given value	Phase 1	Phase 2
$\alpha_{N,L}(T)$	$\alpha_{N,L-1}(T)$	$\alpha_{N,L}(T+1)$
$\beta_{N,L}(T)$	$\beta_{N,L-1}(T)$	$\beta_{N,L}(T+1)$
$\gamma_{N,L}(T)$	unchanged	$\gamma_{N,L}(T+1)$
$\delta_{N,L}(T-1)$	$\delta_{N,L}(T)$	unchanged
$\tilde{C}_{N,L,T}$	unchanged	$\tilde{C}_{N,L,T+1}$
$D_{N,L,T-1}$	$D_{N,L,T}$	unchanged
$\tilde{A}_{N,L,T}$	$\tilde{A}_{N,L-1,T}$	$\tilde{A}_{N,L,T+1}$
$\tilde{B}_{N,L,T}$	$\tilde{B}_{N,L-1,T}$	$\tilde{B}_{N,L,T+1}$
$W_{N,L,T}$	$W_{N,L-1,T}$	$W_{N,L,T+1}$

Fig. 6.6 Update of variables in two phases

Fast Transversal Filter (FTF) Algorithm

The results of the previous section are summarized in the following two algorithms: one for updating the front edge of the window, and the other for updating the rear edge of the window. The two algorithms make up the fast transversal filter (FTF) algorithm in the sliding window case. The second column in each algorithm shows the reference to equation numbers in the previous section, and the third column the complexity of each step. The complexity is measured by the number of operations, where each of a multiplication, a multiplication–addition, and a division are equally counted as one operation.

In Phase 1 of the update, Algorithm 6.4 is executed. Then, in Phase 2, Algorithm 6.3 is applied, T being replaced with $T+1$. Given $\alpha_{N,L}(T)$, $\beta_{N,L}(T)$, $\gamma_{N,L}(T)$, $\delta_{N,L}(T-1)$, $\tilde{C}_{N,L,T}$, $D_{N,L,T-1}$, $\tilde{A}_{N,L,T}$, $\tilde{B}_{N,L,T}$, $W_{N,L,T}$, the FTF algorithm updates them at $T+1$ as shown in Fig. 6.6.

The FTF algorithm computes, as by-products, prediction residual filters $\tilde{A}_{N,L,T}$ and $\tilde{B}_{N,L,T}$, as well as corresponding residual powers $\alpha_{N,L}(T)$ and $\beta_{N,L}(T)$.

Algorithm 6.3 (*Update of the Front Edge of the Window*)

Computation	Reference	Complexity
1. $e^p_{N,L-1}(T) := \alpha_{N,L-1}(T-1)\tilde{A}_{N,L-1,T-1}X_{N+1}(T)$;	definition	$N+2$
2. $e_{N,L}(T) := e^p_{N,L-1}(T)\gamma_{N,L}(T-1)$;	(6.88)	1
3. $\alpha_{N,L}(T) := \alpha_{N,L-1}(T-1) + e^p_{N,L-1}(T)e_{N,L}(T)$;	(6.86), (6.88)	1
4. $\gamma_{N+1,L}(T) := \gamma_{N,L}(T-1) - e^2_{N,L}(T)\alpha^{-1}_{N,L}(T)$;	(6.90)	1 (1)
5. $\tilde{C}_{N+1,L,T} := [0\ \tilde{C}_{N,L,T-1}] - e^p_{N,L-1}(T)\tilde{A}_{N,L-1,T-1}$;	(6.94)	$N+1$
6. $\tilde{A}_{N,L,T} := \tilde{A}_{N,L-1,T-1} + \alpha^{-1}_{N,L}(T)e_{N,L}(T)\tilde{C}_{N+1,L,T}$;	(6.99)	$N+1$
7. $r^p_{N,L-1}(T) := -\beta_{N,L-1}(T-1)\tilde{C}^N_{N+1,L,T}$;	(6.98)	1
8. $\gamma_{N,L}(T) := (1 + \gamma_{N+1,L}(T)\tilde{C}^N_{N+1,L,T}r^p_{N,L-1}(T))^{-1}\gamma_{N+1,L}(T)$;	(6.97)	2 (1)

9. $r_{N,L}(T) := r^{p}_{N,L-1}(T)\gamma_{N,L}(T)$;	(6.89)	1
10. $\beta_{N,L}(T) := \beta_{N,L-1}(T-1) + r^{p}_{N,L-1}(T)r_{N,L}(T)$;	(6.87), (6.89)	1
11. $[\tilde{C}_{N,L,T}\ \ 0] := \tilde{C}_{N+1,L,T} + r^{p}_{N,L-1}(T)\tilde{B}_{N,L-1,T-1}$;	(6.95)	N
12. $\tilde{B}_{N,L,T} := \tilde{B}_{N,L-1,T-1} + \tilde{C}^{N}_{N+1,L,T}\gamma_{N+1,L}(T)\tilde{C}_{N+1,L,T}$;	(6.100)	$N+2$
13. $\epsilon^{p}_{N,L-1}(T) := d(T) + W_{N,L-1,T-1}X_N(T)$;	definition	N
14. $\epsilon_{N,L}(T) := \epsilon^{p}_{N,L-1}(T)\gamma_{N,L}(T)$;	(6.102)	1
15. $W_{N,L,T} := W_{N,L-1,T-1} + \epsilon_{N,L}(T)\tilde{C}_{N,L,T}$;	(6.101)	N
	Total	$7N - 15\ (2)$

Algorithm 6.4 (*Update of the Rear Edge of the Window*)

Computation	Reference	Complexity
1. $a_{N,L}(T) := \alpha_{N,L}(T)\tilde{A}_{N,L,T}X_{N+1}(T-L+1)$;	definition	$N+2$
2. $a^{s}_{N,L-1}(T) := a_{N,L}(T)\delta^{-1}_{N,L}(T-1)$;	(6.111)	(1)
3. $\alpha_{N,L-1}(T) := \alpha_{N,L}(T) - a^{s}_{N,L-1}(T)a_{N,L}(T)$;	(6.112)	1
4. $\delta_{N+1,L}(T) := \delta_{N,L}(T-1) - a^{2}_{N,L}(T)\alpha^{-1}_{N,L}(T)$;	(6.115)	1 (1)
5. $D_{N+1,L,T} := [0\ \ D_{N,L,T-1}] - a_{N,L}(T)\tilde{A}_{N,L,T}$;	(6.105)	$N+1$
6. $\tilde{A}_{N,L-1,T} := \tilde{A}_{N,L,T} - \alpha^{-1}_{N,L}(T)a_{N,L}(T)\delta^{-1}_{N+1,L}(T)D_{N+1,L,T}$;	(6.120)	$N+1$ (1)
7. $b_{N,L}(T) := -\beta_{N,L}(T)D^{N}_{N+1,L,T}$;	(6.107)	1
8. $\delta_{N,L}(T) := \delta_{N+1,L}(T) - D^{N}_{N+1,L,T}b_{N,L}(T)$;	(6.119)	1
9. $b^{s}_{N,L-1}(T) := b_{N,L}(T)\delta^{-1}_{N,L}(T)$;	(6.113)	(1)
10. $\beta_{N,L-1}(T) := \beta_{N,L}(T) - b^{s}_{N,L-1}(T)b_{N,L}(T)$;	(6.114)	1
11. $[D_{N,L,T}\ \ 0] := D_{N+1,L,T} + b_{N,L}(T)\tilde{B}_{N,L,T}$;	(6.106)	N
12. $\tilde{B}_{N,L-1,T} := \tilde{B}_{N,L,T} - \beta^{-1}_{N,L-1}(T)b^{s}_{N,L-1}(T)D_{N+1,L,T}$;	(6.121)	$N+1$ (1)
13. $\nu_{N,L}(T) := d(T-L+1) + W_{N,L,T}X_N(T-L+1)$;	definition	N
14. $\nu^{s}_{N,L-1}(T) := \nu_{N,L}(T)\delta^{-1}_{N,L}(T)$;	(6.123)	(1)
15. $W_{N,L-1,T} := W_{N,L,T} - \nu^{s}_{N,L-1}(T)D_{N,L,T}$;	(6.124)	N
	Total	$7N + 10\ (6)$

Appendix 2: Circular Convolution

Let $f = (f_0, f_1, \ldots, f_{n-1})^t$ and $g = (g_0, g_1, \ldots, g_{n-1})^t$ be sequences of real numbers of length n. The circular convolution $e = (e_0, e_1, \ldots, e_{n-1})^t$ of f and g is defined by

$$e_i \triangleq \sum_{k=0}^{n-1} f_{((i-k))} g_k,$$

where $((i-k)) \triangleq i-k \pmod{n}$. It may help us to understand what is being done if we write e_i as

$$e_i = \begin{pmatrix} f_i \\ \times \\ g_0 \end{pmatrix} + \begin{pmatrix} f_{i-1} \\ \times \\ g_1 \end{pmatrix} + \cdots + \begin{pmatrix} f_0 \\ \times \\ g_i \end{pmatrix} + \begin{pmatrix} f_{n-1} \\ \times \\ g_{i+1} \end{pmatrix} + \begin{pmatrix} f_{n-2} \\ \times \\ g_{i+2} \end{pmatrix} + \cdots + \begin{pmatrix} f_{i+1} \\ \times \\ g_{n-1} \end{pmatrix}. \tag{6.125}$$

By changing the order of terms in (6.125), we also have

$$\begin{aligned} e_i &= \begin{pmatrix} f_0 \\ \times \\ g_i \end{pmatrix} + \begin{pmatrix} f_1 \\ \times \\ g_{i-1} \end{pmatrix} + \cdots + \begin{pmatrix} f_i \\ \times \\ g_0 \end{pmatrix} + \begin{pmatrix} f_{i+1} \\ \times \\ g_{n-1} \end{pmatrix} + \begin{pmatrix} f_{i+2} \\ \times \\ g_{n-2} \end{pmatrix} + \cdots + \begin{pmatrix} f_{n-1} \\ \times \\ g_{i+1} \end{pmatrix} \\ &= \sum_{k=0}^{n-1} f_k g_{((i-k))} \\ &= \sum_{k=0}^{n-1} g_{((i-k))} f_k. \end{aligned}$$

That is, the circular convolution of f and g coincides with the circular convolution of g and f.

The circular convolution can be computed by using the discrete Fourier transform as in the following theorem [7, p. 17]:

Theorem 6.3 (Convolution Theorem) *Let* $F = (F_0, F_1, \ldots, F_{n-1})^t$ *and* $G=(G_0, G_1, \ldots, G_{n-1})^t$ *be the discrete Fourier transforms of* $f = (f_0, f_1, \ldots, f_{n-1})^t$ *and* $g = (g_0, g_1, \ldots, g_{n-1})^t$*, respectively:*

$$F_k \triangleq \sum_{i=0}^{n-1} \omega^{ik} f_i, \quad G_k \triangleq \sum_{i=0}^{n-1} \omega^{ik} g_i \quad (k = 0, 1, \ldots, n-1), \tag{6.126}$$

where $\omega \triangleq e^{-j2\pi/n}$ *and* $j = \sqrt{-1}$*. Then, the circular convolution* e *of* f *and* g *is given by the inverse discrete Fourier transform of*

$$FG = (F_0G_0, F_1G_1, \ldots, F_{n-1}G_{n-1})^t,$$

that is,

$$e_i = \frac{1}{n} \sum_{k=0}^{n-1} \omega^{-ik} F_k G_k \quad (i = 0, 1, \ldots, n-1). \tag{6.127}$$

Proof Noting

$$g_l = \frac{1}{n}\sum_{k=0}^{n-1}\omega^{-kl}G_k \quad (l = 0, 1, \ldots, n-1)$$

and $\omega^{(i-l)} = \omega^{((i-l))}$, we have

$$\begin{aligned} e_i &= \sum_{l=0}^{n-1} f_{((i-l))}\left(\frac{1}{n}\sum_{k=0}^{n-1}\omega^{-lk}G_k\right) \\ &= \frac{1}{n}\sum_{k=0}^{n-1}\omega^{-ik}G_k\left(\sum_{l=0}^{n-1}\omega^{(i-l)k}f_{((i-l))}\right) \\ &= \frac{1}{n}\sum_{k=0}^{n-1}\omega^{-ik}G_k\left(\sum_{l=0}^{n-1}\omega^{((i-l))k}f_{((i-l))}\right) \\ &= \frac{1}{n}\sum_{k=0}^{n-1}\omega^{-ik}G_k\left(\sum_{l=0}^{n-1}\omega^{lk}f_l\right) \\ &= \frac{1}{n}\sum_{k=0}^{n-1}\omega^{-ik}G_kF_k. \end{aligned}$$

□

The discrete Fourier transform in (6.126) and the inverse discrete Fourier transform in (6.127) are computed efficiently by using the fast Fourier transform (FFT) and the inverse fast Fourier transform (IFFT), respectively [8, Chap. 6]. When n is a power of 2, the n-point FFT requires $n \log_2 n$ complex multiplication–additions [8, p. 294]. Since one complex multiplication–addition requires 4 real multiplication–additions, the FFT requires $4n \log_2 n$ real multiplication–additions. For the IFFT, additional n real divisions are needed. Thus, to perform a circular convolution of real sequences of length n by the FFT technique, $12n \log_2 n$ real multiplication–additions, $4n$ real multiplications, and n real divisions are needed.

References

1. Gay, S.L., Tavathia, S.: The fast affine projection algorithm. In: Proceedinds of ICASSP 95, vol. 5, pp. 3023–3026. Detroit, MI, 9–12 May 1995
2. Tanaka, M., Kaneda, Y., Makino, S., Kojima, J.: A fast projection algorithm for adaptive filtering. IEICE Trans. Fundam **E78-A**(10), 1355–1361 (1995)
3. Tanaka, M., Makino, S., Kojima, J.: A block exact fast affine projection algorithm. IEEE Trans. Speech Audio Process. **7**(1), 79–86 (1999)
4. Mou, Z.J., Duhamel, P.: Fast FIR filtering: algorithms and implementations. Signal Process. **13**, 377–384 (1987)
5. Cioffi J.M., Kailath, T.: Fast, recursive-least-squares transversal filters for adaptive filtering. IEEE Trans. Acoust. Speech Signal Process. **ASSP-32**(2), 304–337 (1984)

6. Cioffi J.M., Kailath, T.: Windowed fast transversal filters adaptive algorithms with normalization. IEEE Trans. Acoust. Speech Signal Process. **ASSP-33**(3), 607–625 (1985)
7. Blahut, R.E.: Fast Algorithms for Digital Signal Processing. Addison-Wesley, Reading (1987)
8. Oppenheim, A.V., Schafer, R.W.: Digital Signal Processing. Prentice-Hall, Englewood Cliffs (1975)

Chapter 7
Kernel Affine Projection Algorithm

7.1 Kernel Trick

We have so far assumed that the unknown system is linear. Based on this assumption, we modeled the unknown system by a transversal filter. In reality, however, there are cases where a linear filter is inadequate. In spite of this problem, considering a very general nonlinear model is not a good idea. A general nonlinear model is often intractable especially when we want to devise adaptation algorithms. Moreover, an adaptive model that has too many free parameters is not desirable from a machine learning theoretic point of view, because such a model exhibits poor generalization. Fortunately, there is a method called the *kernel trick* that can extend a linear model to a nonlinear model preserving the simplicity of the original linear model. The kernel trick is reported to be applicable to a wide range of problems such as the support vector machine, the kernel Fisher discriminant function, the kernel principal component analysis [1], and the kernel least-mean-squares algorithm [2].

Let us start with the *kernel perceptron* as a simple example to show how the kernel trick is used to extend a linear model to a nonlinear model. A linear discriminant function is a linear function

$$f(x) = \langle w, x \rangle + b \ \ (w \in \mathbb{R}^n, b \in \mathbb{R}),$$

which is used to classify patterns into two classes. The vector $x \in \mathbb{R}^n$ is the feature of a pattern, w is the weight vector, and b is the bias. If $f(x) > 0$, then a pattern with the feature x is classified into the class C_{+1}. Otherwise, it is classified into the other class C_{-1}. Suppose that we are given a set of labeled training patterns

$$\{(x_1, y_1), (x_2, y_2), \ldots, (x_M, y_M)\}, \tag{7.1}$$

where $x_m \in \mathbb{R}^n$ is the feature vector of the pattern (x_m, y_m), and $y_m = \pm 1$ is the label to indicate the class to which the pattern belongs: $y_m = +1, -1$ indicates that the pattern belongs to C_{+1}, C_{-1}, respectively. The perceptron is a learning machine that

K. Ozeki, *Theory of Affine Projection Algorithms for Adaptive Filtering*,
Mathematics for Industry 22, DOI 10.1007/978-4-431-55738-8_7

automatically acquires a linear discriminant function by the following algorithm. We assume $b = 0$ without loss of generality.

Algorithm 7.1 (*perceptron*)

1. $w := 0$;
2. for $m = 1, 2, \ldots, M$, if $y_m f(x_m) = y_m \langle w, x_m \rangle \leq 0$, then $w := w + y_m x_m$
3. if no correction is made on w in Step 2, then stop; otherwise return to Step 2

The convergence theorem of perceptron says that if the pattern set (7.1) is *linearly separable*, i.e., if there exists a linear function f_0 that satisfies

$$y_m f_0(x_m) > 0 \quad (m = 1, 2, \ldots, M), \tag{7.2}$$

then the above algorithm terminates after finite iteration giving a linear discriminant function f that classifies the training patterns (7.1) perfectly [3].

In Step 2, we see that w is always the sum of $y_i x_i$'s. If we denote by α_i the number of times $y_i x_i$ is added, w is represented as $w = \sum_{i=1}^{M} \alpha_i y_i x_i$. Therefore,

$$\begin{aligned} f(x_m) &= \langle w, x_m \rangle \\ &= \sum_{i=1}^{M} \alpha_i y_i \langle x_i, x_m \rangle. \end{aligned}$$

Thus we have the *dual form* algorithm corresponding to Algorithm 7.1 [3].

Algorithm 7.2 (*dual form perceptron*)

1. for $m = 1, 2, \ldots, M$, $\alpha_m := 0$;
2. for $m = 1, 2, \ldots, M$, if $y_m \sum_{i=1}^{M} \alpha_i y_i \langle x_i, x_m \rangle \leq 0$, then $\alpha_m := \alpha_m + 1$;
3. if no correction is made on any of α_m's in Step 2, then stop; otherwise return to Step 2;

An important fact about Algorithm 7.2 is that feature vectors appear only in the form of inner product $\langle x_i, x_m \rangle$.

In reality, the linear separability assumption on the training patterns does not always hold. This means we need a nonlinear discriminant function to attain a perfect classification of the training patterns. However, adaptation of a nonlinear model often gives rise to complications. We want to maintain the simplicity of a linear model. Therefore, we transform the training patterns, and try to classify the transformed patterns using a linear discriminant function. Let φ be a mapping for the transformation:

$$\varphi : \mathbb{R}^n \to \mathbb{G},$$

where $\mathbb{G}$ is a vector space with an inner product $\langle\cdot,\cdot\rangle_{\mathbb{G}}$. If the mapping φ is appropriately chosen, the transformed pattern set

$$\{(\varphi(x_1), y_1), (\varphi(x_2), y_2), \ldots, (\varphi(x_M), y_M)\}$$

may become linearly separable. Algorithm 7.2 on this transformed pattern set in $\mathbb{G}$ is written as follows:

Algorithm 7.3 (*transformed perceptron*)

1. for $m = 1, 2, \ldots, M$, $\alpha_m := 0$;
2. for $m = 1, 2, \ldots, M$, if $y_m \sum_{i=1}^{M} \alpha_i y_i \langle\varphi(x_i), \varphi(x_m)\rangle_{\mathbb{G}} \leq 0$, then $\alpha_m := \alpha_m + 1$;
3. if no correction is made on any of α_m's in Step 2, then stop, otherwise return to Step 2;

Focusing on the inner product $\langle\varphi(x_i), \varphi(x_m)\rangle_{\mathbb{G}}$ in Algorithm 7.3, we proceed one more step forward. Let $\mathbb{S}$ be a closed, bounded domain in $\mathbb{R}^n$ that includes all of the feature vectors appearing in the training pattern set (7.1), and K a function on $\mathbb{S}$. If K is a kernel on $\mathbb{S}$ (Appendix 1: "Mercer Kernel"), there exists a vector space $\mathbb{F}$ with an inner product $\langle\cdot,\cdot\rangle_{\mathbb{F}}$ and a mapping $\Phi : \mathbb{S} \to \mathbb{F}$ such that for any $x, x' \in \mathbb{S}$

$$K(x, x') = \langle\Phi(x), \Phi(x')\rangle_{\mathbb{F}}.$$

The mapping Φ is called the feature mapping, and $\mathbb{F}$ the feature space.

Using Φ for φ and $\mathbb{F}$ for $\mathbb{G}$ in Step 2 of Algorithm 7.3, we have the following *kernel perceptron* algorithm.

Algorithm 7.4 (*kernel perceptron*)

1. for $m = 1, 2, \ldots, M$, $\alpha_m := 0$;
2. for $m = 1, 2, \ldots, M$, if $y_m \sum_{i=1}^{M} \alpha_i y_i K(x_i, x_m) \leq 0$, then $\alpha_m := \alpha_m + 1$;
3. if no correction is made on any of α_m's in Step 2, then stop; otherwise return to Step 2;

At any step of the above algorithm, the weight vector is given by

$$\sum_{i=1}^{M} \alpha_i y_i \Phi(x_i).$$

Therefore, the discriminant function f is expressed as

$$\begin{aligned} f(x) &= \left\langle \textstyle\sum_{i=1}^{M} \alpha_i y_i \Phi(x_i), \Phi(x) \right\rangle_{\mathbb{F}} \\ &= \sum_{i=1}^{M} \alpha_i y_i K(x_i, x). \end{aligned}$$

This shows that f is a linear combination of $K(x_i, \cdot)$'s. If K is chosen to be a radial basis function such as a Gaussian kernel, then x_i is the center of the function (Appendix 1: "Mercer Kernel"). In this sense, a feature vector is often called a *center* in the context of kernel trick.

We can thus extend the perceptron algorithm into the kernel perceptron algorithm that acquires a nonlinear discriminant function. Everything is represented by the kernel without explicitly using the feature mapping Φ.

7.2 Kernel Affine Projection Algorithm (KAPA)

The kernel trick can also be applied to the affine projection algorithm to derive the *kernel affine projection algorithm (KAPA)* [4, 5]. Let us assume that all the regressors are included in a closed, bounded domain $\mathbb{S}$ in $\mathbb{R}^n$. If we choose a kernel K on $\mathbb{S}$, and map all the regressors using the feature mapping Φ, then the coefficient update equation for the R-APA (3.12) is translated into the following update equation in the feature space $\mathbb{F}$:

$$\hat{w}_{k+1} = \hat{w}_k + \mu \hat{X}_k^t (\hat{G}_k + \delta I)^{-1} \hat{e}_k, \tag{7.3}$$

where $\hat{w}_k$ is the coefficient vector in $\mathbb{F}$, and

$$\begin{aligned}
\hat{X}_k &\stackrel{\triangle}{=} [\Phi(x_k) \;\; \Phi(x_{k-1}) \;\; \cdots \;\; \Phi(x_{k-(p-1)})]^t, \\
f_k(x) &\stackrel{\triangle}{=} \langle \Phi(x), \hat{w}_k \rangle_{\mathbb{F}}, \\
\hat{e}_k &\stackrel{\triangle}{=} \begin{bmatrix} d(k) - f_k(x_k) \\ d(k-1) - f_k(x_{k-1}) \\ \vdots \\ d(k-(p-1)) - f_k(x_{k-(p-1)}) \end{bmatrix}, \\
\hat{G}_k &\stackrel{\triangle}{=} \hat{X}_k \hat{X}_k^t \\
&= \begin{bmatrix} \langle \Phi(x_k), \Phi(x_k) \rangle_{\mathbb{F}} & \cdots & \langle \Phi(x_k), \Phi(x_{k-(p-1)}) \rangle_{\mathbb{F}} \\ \vdots & \ddots & \vdots \\ \langle \Phi(x_{k-(p-1)}), \Phi(x_k) \rangle_{\mathbb{F}} & \cdots & \langle \Phi(x_{k-(p-1)}), \Phi(x_{k-(p-1)}) \rangle_{\mathbb{F}} \end{bmatrix} \\
&= \begin{bmatrix} K(x_k, x_k) & K(x_k, x_{k-1}) & \cdots & K(x_k, x_{k-(p-1)}) \\ K(x_{k-1}, x_k) & K(x_{k-1}, x_{k-1}) & \cdots & K(x_{k-1}, x_{k-(p-1)}) \\ \vdots & \vdots & \ddots & \vdots \\ K(x_{k-(p-1)}, x_k) & K(x_{k-(p-1)}, x_{k-1}) & \cdots & K(x_{k-(p-1)}, x_{k-(p-1)}) \end{bmatrix}.
\end{aligned}$$

Here, we face a problem that the right-hand side of (7.3) cannot be calculated directly, because it includes the inaccessible feature mapping Φ. Moreover, the coefficient vector $\hat{w}_k$ is of infinite dimension in general, which makes it impossible to treat by a digital processor. To avoid these difficulties we make use of the following lemma:

Lemma 7.1 *Suppose $\hat{w}_0 = 0$. Then, for $k \geq 1$, the coefficient vector $\hat{w}_k$ is expressed as a linear combination of $\Phi(x_{-(p-1)})$, $\Phi(x_{-(p-1)+1})$, ..., $\Phi(x_{k-1})$:*

$$\hat{w}_k = \sum_{j=-(p-1)}^{k-1} a_{k,j}\Phi(x_j),$$

where the coefficients $a_{k,j}$'s are given recursively as in the proof.

Proof Since

$$f_0(x) = \langle \Phi(x), \hat{w}_0 \rangle = \langle \Phi(x), 0 \rangle = 0,$$

we have

$$\begin{aligned} \hat{e}_0 &= \begin{bmatrix} d(0) - f_0(x_0) \\ d(-1) - f_0(x_{-1}) \\ \vdots \\ d(-(p-1)) - f_0(x_{-(p-1)}) \end{bmatrix} \\ &= (d(0), d(-1), \ldots, d(-(p-1)))^t. \end{aligned}$$

Therefore,

$$\begin{aligned} \hat{w}_1 &= \hat{w}_0 + \mu \hat{X}_0^t (\hat{G}_0 + \delta I)^{-1} \hat{e}_0 \\ &= \sum_{j=-(p-1)}^{0} a_{1,j}\Phi(x_j), \end{aligned}$$

where

$$(a_{1,0}, a_{1,-1}, \ldots, a_{1,-(p-1)})^t \stackrel{\triangle}{=} \mu(\hat{G}_0 + \delta I)^{-1}(d(0), d(-1), \ldots, d(-(p-1)))^t.$$

This shows that $\hat{w}_1$ is a linear combination of $\Phi(x_{-(p-1)}), \Phi(x_{-(p-1)+1}), \ldots, \Phi(x_0)$.

Next, assume that $\hat{w}_k$ is expressed as

$$\hat{w}_k = \sum_{j=-(p-1)}^{k-1} a_{k,j}\Phi(x_j).$$

Then,

$$\begin{aligned} f_k(x) &= \langle \Phi(x), \hat{w}_k \rangle_{\mathbb{F}} \\ &= \langle \Phi(x), \textstyle\sum_{j=-(p-1)}^{k-1} a_{k,j} \Phi(x_j) \rangle_{\mathbb{F}} \\ &= \sum_{j=-(p-1)}^{k-1} a_{k,j} \langle \Phi(x), \Phi(x_j) \rangle_{\mathbb{F}} \\ &= \sum_{j=-(p-1)}^{k-1} a_{k,j} K(x, x_j). \end{aligned} \tag{7.4}$$

This enables us to calculate $\hat{e}_k$ as

$$\hat{e}_k = \begin{bmatrix} d(k) - f_k(x_k) \\ d(k-1) - f_k(x_{k-1}) \\ \vdots \\ d(k-(p-1)) - f_k(x_{k-(p-1)}) \end{bmatrix}. \tag{7.5}$$

If we define

$$(b_{k,k}, b_{k,k-1}, \ldots, b_{k,k-(p-1)})^t \triangleq \mu(\hat{G}_k + \delta I)^{-1} \hat{e}_k, \tag{7.6}$$

then,

$$\begin{aligned} \hat{w}_{k+1} &= \hat{w}_k + \mu \hat{X}_k^t (\hat{G}_k + \delta I)^{-1} \hat{e}_k \\ &= \sum_{j=-(p-1)}^{k-1} a_{k,j} \Phi(x_j) + \hat{X}_k^t (b_{k,k}, b_{k,k-1}, \ldots, b_{k,k-(p-1)})^t \\ &= \sum_{j=-(p-1)}^{k-1} a_{k,j} \Phi(x_j) + \sum_{j=k-(p-1)}^{k} b_{k,j} \Phi(x_j) \\ &= \sum_{j=-(p-1)}^{k} a_{k+1,j} \Phi(x_j), \end{aligned}$$

where

$$a_{k+1,j} \triangleq \begin{cases} a_{k,j}, & \text{if } -(p-1) \le j \le k-p; \\ a_{k,j} + b_{k,j}, & \text{if } k-(p-1) \le j \le k-1; \\ b_{k,k}, & \text{if } j = k. \end{cases} \tag{7.7}$$

□

The following diagram illustrates how the coefficients $a_{k,-(p-1)}, a_{k,1-(p-1)}, \ldots, a_{k,k-1}$ are updated in (7.7).

$$\begin{array}{ccccccc}
\Phi(x_{-(p-1)}) & \cdots & \Phi(x_{k-p}) & \Phi(x_{k-(p-1)}) & \cdots & \Phi(x_{k-1}) & \Phi(x_k) \\
 & & & & & & \\
a_{k,-(p-1)} & \cdots & a_{k,k-p} & a_{k,k-(p-1)} & \cdots & a_{k,k-1} & \\
\downarrow & \cdots & \downarrow & + & \cdots & + & \\
\downarrow & \cdots & \downarrow & b_{k,k-(p-1)} & \cdots & b_{k,k-1} & b_{k,k} \\
\downarrow & \cdots & \downarrow & \downarrow & \cdots & \downarrow & \downarrow \\
a_{k+1,-(p-1)} & \cdots & a_{k+1,k-p} & a_{k+1,k-(p-1)} & \cdots & a_{k+1,k-1} & a_{k+1,k}
\end{array}$$

In Lemma 7.1, it is necessary to compute $(\hat{G}_k + \delta I)^{-1}$. The computational complexity to invert a $p \times p$ matrix is $O(p^3)$ in general. However, the upper-left $(p-1) \times (p-1)$ sub-matrix of $\hat{G}_{k-1} + \delta I$ equals the lower-right $(p-1) \times (p-1)$ sub-matrix of $\hat{G}_k + \delta I$. Noting this fact, and using the following lemma, we can invert $\hat{G}_k + \delta I$ in $O(p^2)$ if successive computation is performed.

Lemma 7.2 *Let A and B be $n \times n$ symmetric, nonsingular matrices of the form*

$$A = \begin{bmatrix} a & b^t \\ b & C \end{bmatrix}, \quad B = \begin{bmatrix} C & h \\ h^t & g \end{bmatrix},$$

where a and g are scalars, b and h are $(n-1)$-dimensional vectors, and C is an $(n-1) \times (n-1)$ matrix. Note that C is common in A and B. If

$$B^{-1} = \begin{bmatrix} H & f \\ f^t & e \end{bmatrix},$$

where e is a scalar, f is an $(n-1)$-dimensional vector, and H is an $(n-1) \times (n-1)$ matrix, then,

$$C^{-1} = H - ff^t/e,$$

and

$$A^{-1} = \begin{bmatrix} s^{-1} & -(C^{-1}b)^t s^{-1} \\ -(C^{-1}b)s^{-1} & C^{-1} + (C^{-1}b)(C^{-1}b)^t s^{-1} \end{bmatrix},$$

where $s = a - b^t C^{-1} b$ [6].

Proof By (A.3) (Appendix A),

$$\begin{aligned}
\begin{bmatrix} C & h \\ h^t & g \end{bmatrix} &= \begin{bmatrix} H & f \\ f^t & e \end{bmatrix}^{-1} \\
&= \begin{bmatrix} (H - fe^{-1}f^t)^{-1} & * \\ * & * \end{bmatrix}.
\end{aligned}$$

Therefore,

$$C^{-1} = H - ff^t/e.$$

By (A.3) again,

$$\begin{aligned}
A^{-1} &= \begin{bmatrix} a & b^t \\ b & C \end{bmatrix}^{-1} \\
&= \begin{bmatrix} (a - b^t C^{-1} b)^{-1} & -(a - b^t C^{-1} b)^{-1} b^t C^{-1} \\ -C^{-1} b (a - b^t C^{-1} b)^{-1} & C^{-1} + C^{-1} b (a - b^t C^{-1} b)^{-1} b^t C^{-1} \end{bmatrix} \\
&= \begin{bmatrix} s^{-1} & -(C^{-1} b)^t s^{-1} \\ -(C^{-1} b) s^{-1} & C^{-1} + (C^{-1} b)(C^{-1} b)^t s^{-1} \end{bmatrix}.
\end{aligned} \tag{7.8}$$

□

Now we are ready to present the kernel affine projection algorithm (KAPA).

Algorithm 7.5 (*KAPA*)

initialization
1. compute $(\hat{G}_0 + \delta I)^{-1}$;
2. $(a_{1,0}, a_{1,-1}, \ldots, a_{1,-(p-1)})^t := \mu(\hat{G}_0 + \delta I)^{-1}(d(0), d(-1), \ldots, d(-(p-1)))^t$;
3. $k := 1$;

while {regressors are available} repeat 1–6 below
1. compute $f_k(x_k), f_k(x_{k-1}), \ldots, f_k(x_{k-(p-1)})$, by (7.4);
2. compute $\hat{e}_k$ by (7.5);
3. compute $(\hat{G}_k + \delta I)^{-1}$ from $(\hat{G}_{k-1} + \delta I)^{-1}$ by Lemma 7.2;
4. compute $b_{k,k-(p-1)}, b_{k,k-(p-1)+1}, \ldots, b_{k,k}$ by (7.6);
5. update $a_{k,-(p-1)}, a_{k,1-(p-1)}, \ldots, a_{k,k-1}$ by (7.7);
6. $k := k + 1$;

Figure 7.1 illustrates the structure of the output function $f_k(x)$ constructed by Algorithm 7.5. This is similar to the *resource-allocating network (RAN)* [7]. When

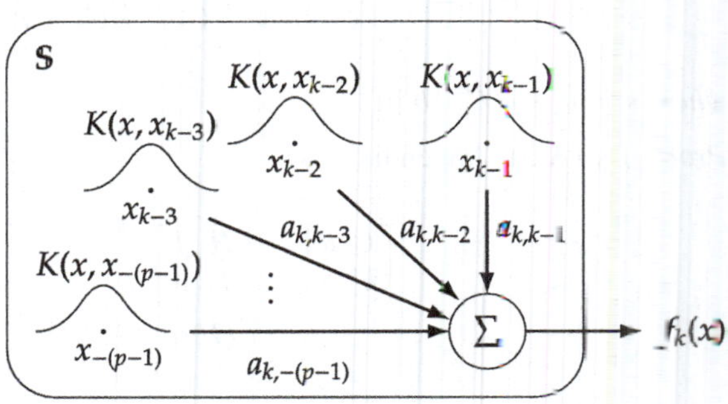

Fig. 7.1 Construction of the output function

a new regressor (center) x_k arrives, a unit $K(x, x_k)$ is added and the coefficients are updated.

7.3 Sparsification of Regressors

There is a large difference in the number of regressors between the kernel perceptron and the kernel affine projection algorithm. While the number of regressors that have to be stored is fixed in the kernel perceptron, it grows unboundedly in the KAPA. Since it is not desirable to store a huge set of regressors and perform computation on it, we want to reduce the number of regressors and keep it bounded. For that purpose, we use a method presented in [7] that tests if a newly arrived regressor is informative enough for updating the output function in the framework of resource-allocation network. In the method, the *novelty criterion (NC)* consisting of two stages is used. In the first stage, the distance $d(\mathcal{C}, x_k)$ is calculated between a set $\mathcal{C}$ called the *dictionary* and a newly arrived regressor x_k. The dictionary $\mathcal{C}$ is the set of regressors that have already passed the test, and

$$d(\mathcal{C}, x_k) \triangleq \min\{\|x - x_k\|;\ x \in \mathcal{C}\}.$$

If $d(\mathcal{C}, x_k) \leq \delta_1$ for a prescribed threshold δ_1, then x_k is discarded. Otherwise, $\hat{e}(k)$ is calculated based on regressors in $\mathcal{C}$. If $|\hat{e}(k)| \leq \delta_2$ for a second threshold δ_2, then x_k is discarded. Otherwise, x_k is added to $\mathcal{C}$ as a new member. Thus, the regressors in $\mathcal{C}$ are sparsified and renumbered version of the original regressors. The KAPA with the novelty criterion is described as follows:

Algorithm 7.6 (*KAPA with NC*)

initialization

1. $\mathcal{C} := \{x_0, x_{-1}, \ldots, x_{-(p-1)}\}$;
2. compute $(\hat{G}_0 + \delta I)^{-1}$;
3. $(a_{1,0}, a_{1,-1}, \ldots, a_{1,-(p-1)})^t := \mu(\hat{G}_0 + \delta I)^{-1}(d(0), d(-1), \ldots, d(-(p-1)))^t$;
4. $k := 1$;
5. $m := 1$;

while {regressors are available} repeat Steps 1–12 below:

1. if $d(\mathcal{C}, x_k) \leq \delta_1$, then goto Step 12;
2. compute the filter output $f_m(x_k)$ by (7.4);
3. if $|d(k) - f_m(x_k)| \leq \delta_2$, then goto Step 12;
4. $x_m := x_k$ (renumber);
5. $\mathcal{C} := \mathcal{C} \cup \{x_m\}$;
6. compute $f_m(x_{m-1}), f_m(x_{m-2}), \ldots, f_m(x_{m-(p-1)})$ by (7.4);

7. $\hat{e}_m := \begin{bmatrix} d(m) - f_m(x_m) \\ d(m-1) - f_m(x_{m-1}) \\ \vdots \\ d(m-(p-1)) - f_m(x_{m-(p-1)}) \end{bmatrix}$;
8. compute $(\hat{G}_m + \delta I)^{-1}$ using $(\hat{G}_{m-1} + \delta I)^{-1}$ by Lemma 7.2;
9. $(b_{m,m}, b_{m,m-1}, \ldots, b_{m,m-(p-1)})^t := \mu(\hat{G}_m + \delta I)^{-1}\hat{e}_m$;
10. $m := m + 1$;
11. compute $a_{m,-(p-1)}, a_{m,1-(p-1)}, \ldots, a_{m,m-1}$ by (7.7);
12. $k := k + 1$;

Let us make a realistic assumption that all the regressors are included in a bounded subset of $\mathbb{R}^n$. Since $\|x_i - x_j\| > \delta_1$ for every pair (x_i, x_j) in $\mathcal{C}$ except for the first p regressors, the size of the dictionary $\mathcal{C}$ is upper-bounded. Although this fact is intuitively apparent, we give a formal proof in the following lemma:

Lemma 7.3 *Let $\mathbb{D}$ be a bounded subset of $\mathbb{R}^n$, and $P \triangleq \{p_1, p_2, \ldots, p_m\}$ a set of distinct points in $\mathbb{D}$. If $\|p_i - p_j\| > \delta$ for every pair (p_i, p_j) in P, then the number of points in P is upper-bounded.*

Proof Let $B_r(c)$ be a ball in $\mathbb{R}^n$ centered at c with the radius r:

$$B_r(c) \triangleq \{x \in \mathbb{R}^n;\ \|x - c\| \le r\}.$$

Since $\mathbb{D}$ is a bounded subset, there exists a radius r_0 such that $\mathbb{D} \subseteq B_{r_0}(0)$. Therefore, $B_{\delta/2}(p_i) \subseteq B_{r_1}(0)$ $(1 \le i \le m)$ for $r_1 \triangleq r_0 + \delta$. Thus, as illustrated in Fig. 7.2,

$$\bigcup_{i=1}^{m} B_{\delta/2}(p_i) \subseteq B_{r_1}(0). \tag{7.9}$$

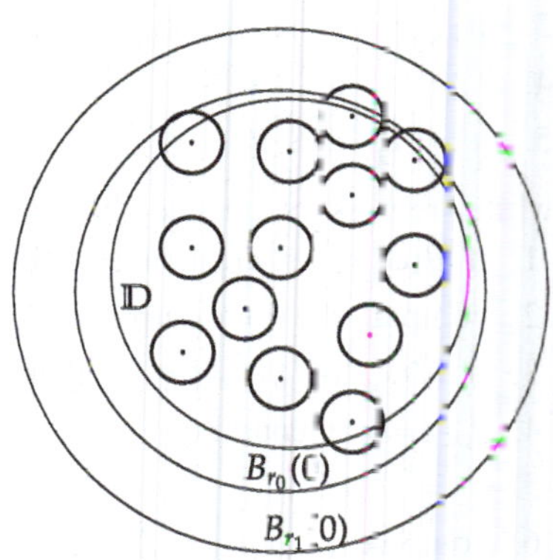

Fig. 7.2 The small balls show $B_{\delta/2}(p_i)$'s. They are included in the large ball $B_{r_1}(0)$, where $r_1 = r_0 + \delta$. Their centers are all in $\mathbb{D}$, and they are mutually disjoint

Because $B_{\delta/2}(p_i)$'s are mutually disjoint, we have from (7.9)

$$mV(\delta/2) \leq V(r_1),$$

where $V(r)$ is the volume of a ball with the radius r. This shows that m is upper-bounded as

$$m \leq V(r_1)/V(\delta/2).$$

□

Appendix 1: Mercer Kernel

Let $\mathbb{S}$ be a bounded, closed domain in $\mathbb{R}^n$. A real-valued, symmetric, continuous function K on $\mathbb{S} \times \mathbb{S}$ is called a *Mercer kernel* or simply a *kernel* iff there exists a vector space $\mathbb{F}$ with an inner product $\langle \cdot, \cdot \rangle_{\mathbb{F}}$, and a mapping $\Phi : \mathbb{S} \to \mathbb{F}$ such that

$$K(s, t) = \langle \Phi(s), \Phi(t) \rangle_{\mathbb{F}}.$$

The vector space $\mathbb{F}$ is referred to as the *feature space*, and the mapping Φ the *feature mapping*. The space $\mathbb{F}$ may be of infinite dimensions.

A real-valued symmetric function K on $\mathbb{S} \times \mathbb{S}$ is said to be of *nonnegative type* iff for any finite number of points $s_1, s_2, \ldots, s_m \in \mathbb{S}$, the $m \times m$ matrix

$$\tilde{K} \triangleq [k_{ij}], \quad k_{ij} \triangleq K(s_i, s_j)$$

is nonnegative definite.

Theorem 7.1 *Let $\mathbb{S}$ be a bounded, closed domain in $\mathbb{R}^n$. A real-valued, symmetric, continuous function K on $\mathbb{S} \times \mathbb{S}$ is a kernel iff it is of nonnegative type.*

The proof of Theorem 7.1 requires a knowledge of functional analysis. Therefore, let us start with the following restricted theorem, which will help us to see the essential point of the problem.

Theorem 7.2 *Let $\mathbb{S}$ be a finite set in $\mathbb{R}^n$. A real-valued symmetric function K on $\mathbb{S} \times \mathbb{S}$ is a kernel iff it is of nonnegative type.*

Proof Suppose K is a kernel. Then there exists a vector space $\mathbb{F}$ with an inner product $\langle \cdot, \cdot \rangle_{\mathbb{F}}$ and a mapping $\Phi : \mathbb{S} \to \mathbb{F}$ such that

$$K(s, t) = \langle \Phi(s), \Phi(t) \rangle_{\mathbb{F}}.$$

Therefore, for any $s_1, s_2, \ldots, s_m \in \mathbb{S}$, and for any real numbers $\alpha_1, \alpha_2, \ldots, \alpha_m$,

$$\begin{aligned}\sum_{i=1}^{m}\sum_{j=1}^{m}\alpha_i\alpha_j k_{ij} &= \sum_{i=1}^{m}\sum_{j=1}^{m}\alpha_i\alpha_j K(s_i, s_j)\\ &= \sum_{i=1}^{m}\sum_{j=1}^{m}\alpha_i\alpha_j \langle\Phi(s_i), \Phi(s_j)\rangle_{\mathbb{F}}\\ &= \left\langle \textstyle\sum_{i=1}^{m}\alpha_i\Phi(s_i), \sum_{j=1}^{m}\alpha_j\Phi(s_j)\right\rangle_{\mathbb{F}}\\ &= \left\|\textstyle\sum_{i=1}^{m}\alpha_i\Phi(s_i)\right\|_{\mathbb{F}}^2\\ &\geq 0,\end{aligned}$$

which shows $\tilde{K} = [k_{ij}]$ is nonnegative definite. That is, K is of nonnegative type.

Conversely, suppose K is of nonnegative type, Then, for any sequence $s_1, s_2, \ldots, s_m \in \mathbb{S}$, the $m \times m$ matrix $\tilde{K} = [k_{ij}]$ is nonnegative definite. In particular, for all the elements $s_1, s_2, \ldots, s_M \in \mathbb{S}$, the $M \times M$ matrix $\tilde{K} = [k_{ij}] = [K(s_i, s_j)]$ is nonnegative definite. Therefore, all the eigenvalues of K are nonnegative by Theorem A.6 (Appendix A). Now, let

$$t_i \triangleq (t_{i1}, t_{i2}, \ldots, t_{iM})^t$$

be the normalized eigenvector corresponding to the ith eigenvalue λ_i, and let

$$T \triangleq \left[t_1\ t_2 \cdots t_M\right].$$

Note that T is an $M \times M$ orthogonal matrix. Then, by Theorem A.5 (Appendix A), $\tilde{K}$ is diagonalized as

$$\begin{aligned}\tilde{K} &= T\Lambda T^t,\\ \Lambda &= \mathrm{diag}\,[\lambda_1, \lambda_2, \ldots, \lambda_M].\end{aligned} \tag{7.10}$$

We can rewrite $T\Lambda T^t$ as

$$T\Lambda T^t = \begin{bmatrix}\lambda_1 t_{11} & \lambda_2 t_{21} & \cdots & \lambda_M t_{M1}\\ \lambda_1 t_{12} & \lambda_2 t_{22} & \cdots & \lambda_M t_{M2}\\ \vdots & \vdots & \ddots & \vdots\\ \lambda_1 t_{1M} & \lambda_2 t_{2M} & \cdots & \lambda_M t_{MM}\end{bmatrix}\begin{bmatrix}t_{11} & t_{12} & \cdots & t_{1M}\\ t_{21} & t_{22} & \cdots & t_{2M}\\ \vdots & \vdots & \ddots & \vdots\\ t_{M1} & t_{M2} & \cdots & t_{MM}\end{bmatrix},$$

from which, it is seen that $k_{ij} = \sum_{k=1}^{M}\lambda_k t_{ki}t_{kj}$. Thus, if we define a mapping $\Phi : \mathbb{S} \to \mathbb{R}^M$ by

$$\Phi(s_i) \triangleq \left(\sqrt{\lambda_1}\,t_{1i}, \sqrt{\lambda_2}\,t_{2i}, \ldots, \sqrt{\lambda_M}\,t_{Mi}\right)^t,$$

then,

$$\begin{aligned}K(s_i,\ s_j) &= k_{ij}\\ &= \sum_{k=1}^{M} \lambda_k t_{ki} t_{kj}\\ &= \langle \Phi(s_i),\ \Phi(s_j)\rangle.\end{aligned}$$

This shows that K is a kernel. □

Let us make preparations for the proof of Theorem 7.1. Let $\mathbb{S}$ be a bounded, closed domain in $\mathbb{R}^n$. The set of complex-valued square-integrable functions on $\mathbb{S}$ is denoted by $L^2(\mathbb{S})$:

$$L^2(\mathbb{S}) \triangleq \left\{\varphi\ ;\ \varphi : \mathbb{S} \to \mathbb{C},\ \int_{\mathbb{S}} |\varphi(s)|^2\, \mathrm{d}s < \infty\right\}.$$

Likewise, the set of complex-valued square-integrable functions on $\mathbb{S} \times \mathbb{S}$ is denoted by $L^2(\mathbb{S} \times \mathbb{S})$. $L^2(\mathbb{S})$ and $L^2(\mathbb{S} \times \mathbb{S})$ are vector spaces over $\mathbb{C}$. A linear mapping from $L^2(\mathbb{S})$ to $L^2(\mathbb{S})$ is called a *linear operator*. If, for a linear operator T, the equation

$$T\psi = \lambda\psi$$

has a solution $\psi \in L^2(\mathbb{S}),\ \ \psi \neq 0$, then $\lambda \in \mathbb{C}$ is called an *eigenvalue* of T, and ψ the *eigenfunction* corresponding to λ.

For a function $K \in L^2(\mathbb{S} \times \mathbb{S})$, a linear operator T_K is defined by

$$(T_K\varphi)(s) \triangleq \int_{\mathbb{S}} K(s,t)\varphi(t)\,\mathrm{d}t \quad (\varphi \in L^2(\mathbb{S})).$$

T_K is called the linear operator with the *integral kernel* K.

A function $K \in L^2(\mathbb{S} \times \mathbb{S})$ is said to be *Hermitian* iff it satisfies $K(s,t) = \overline{K(t,s)}$, where the overline denotes complex conjugate.

Lemma 7.4 *If K is Hermitian, then the eigenvalues of T_K are real numbers.*

Proof Let λ be an eigenvalue of T_K, and ψ the corresponding eigenfunction:

$$\int_{\mathbb{S}} K(s,t)\psi(t)\,\mathrm{d}t = \lambda\psi(s). \tag{7.11}$$

Multiplying $\overline{\psi(s)}$ to both sides of the above equation, and integrating with respect to s, we have

$$\int_{\mathbb{S}\times\mathbb{S}} K(s,t)\overline{\psi(s)}\psi(t)\,\mathrm{d}s\mathrm{d}t = \lambda \int_{\mathbb{S}} \overline{\psi(s)}\psi(s)\,\mathrm{d}s. \tag{7.12}$$

On the other hand, taking the complex conjugate of (7.11), multiplying $\psi(s)$ to both sides, and integrating with respect to s, we obtain

$$\int_{\mathbb{S}\times\mathbb{S}} \overline{K(s,t)\psi(t)}\psi(s)\,\mathrm{d}s\mathrm{d}t = \overline{\lambda}\int_{\mathbb{S}} \overline{\psi(s)}\psi(s)\,\mathrm{d}s.$$

Since K is Hermitian, the left-hand side of this equation can be rewritten as

$$\begin{aligned}\int_{\mathbb{S}\times\mathbb{S}} \overline{K(s,t)\psi(t)}\psi(s)\,\mathrm{d}s\mathrm{d}t &= \int_{\mathbb{S}\times\mathbb{S}} K(t,s)\overline{\psi(t)}\psi(s)\,\mathrm{d}s\mathrm{d}t\\ &= \int_{\mathbb{S}\times\mathbb{S}} K(s,t)\overline{\psi(s)}\psi(t)\,\mathrm{d}s\mathrm{d}t.\end{aligned}$$

Therefore,

$$\int_{\mathbb{S}\times\mathbb{S}} K(s,t)\overline{\psi(s)}\psi(t)\,\mathrm{d}s\mathrm{d}t = \overline{\lambda}\int_{\mathbb{S}} \overline{\psi(s)}\psi(s)\,\mathrm{d}s. \tag{7.13}$$

From (7.12) and (7.13),

$$\lambda\int_{\mathbb{S}} |\psi(s)|^2\,\mathrm{d}s = \overline{\lambda}\int_{\mathbb{S}} |\psi(s)|^2\,\mathrm{d}s.$$

Since $\psi \neq 0$, $\int_{\mathbb{S}} |\psi(s)|^2\,\mathrm{d}s \neq 0$. Thus, $\lambda = \overline{\lambda}$, showing that λ is a real number. □

A Hermitian function K is said to be *nonnegative definite* iff for any function $\varphi \in L^2(\mathbb{S})$,

$$\iint_{\mathbb{S}\times\mathbb{S}} K(s,t)\overline{\varphi(s)}\varphi(t)\,\mathrm{d}s\mathrm{d}t \geq 0. \tag{7.14}$$

A linear operator T_K is said to be nonnegative definite iff K is nonnegative definite.

Lemma 7.5 *An eigenvalue of a nonnegative definite operator is nonnegative.*

Proof Let T_K be a nonnegative definite operator, λ its eigenvalue, and ψ the corresponding eigenfunction. Then, in (7.12), the left-hand side is nonnegative. Also, since $\psi \neq 0$, $\int_{\mathbb{S}} \overline{\psi(s)}\psi(s)\,\mathrm{d}s = \int_{\mathbb{S}} |\psi(s)|^2\,\mathrm{d}s$ is positive. Therefore, λ is nonnegative. □

For our purpose in this book, we only have to consider a real-valued function for the integral kernel. In that case, a symmetric function is Hermitian. Therefore, if $K \in L^2(\mathbb{S}\times\mathbb{S})$ is a real-valued symmetric function, the eigenvalues of T_K are all real numbers. Also, just as in the eigenvalue problem of real symmetric matrices, we can take a real-valued function for an eigenfunction.

A real-valued symmetric function K is said to be nonnegative definite iff for any real-valued function $\varphi \in L^2(\mathbb{S})$,

$$\iint_{\mathbb{S}\times\mathbb{S}} K(s,t)\varphi(s)\varphi(t)\,\mathrm{d}s\mathrm{d}t \geq 0. \tag{7.15}$$

Lemma 7.6 *A real-valued, symmetric, continuous function $K(s,t)$ on $\mathbb{S}\times\mathbb{S}$ is nonnegative definite iff it is of nonnegative type.*

Proof Suppose K is nonnegative definite. Let $s_1, s_2, \ldots, s_m$ be arbitrary points in $\mathbb{S}$, and let φ_{jk} be a real-valued continuous function on $\mathbb{S}$ satisfying

$$\varphi_{jk}(s)\begin{cases} > 0, & \text{if } \|s - s_j\| < 1/k; \\ = 0, & \text{otherwise,} \end{cases}$$

$$\int_{\mathbb{S}} \varphi_{jk}(s)\,\mathrm{d}s = 1,$$

where k is a positive integer. Then, for any real numbers $\alpha_1, \alpha_2, \ldots, \alpha_m$,

$$\iint_{\mathbb{S}\times\mathbb{S}} K(s,t)\left(\sum_{i=1}^m \alpha_i\varphi_{ik}(s)\right)\left(\sum_{j=1}^m \alpha_j\varphi_{jk}(t)\right)\mathrm{d}s\mathrm{d}t \geq 0. \tag{7.16}$$

The left-hand side of the above inequality can be rewritten as

$$\iint_{\mathbb{S}\times\mathbb{S}} K(s,t)\left(\sum_{i=1}^m \alpha_i\varphi_{ik}(s)\right)\left(\sum_{j=1}^m \alpha_j\varphi_{jk}(t)\right)\mathrm{d}s\mathrm{d}t$$

$$= \sum_{i,j=1}^m \alpha_i\alpha_j \iint_{\mathbb{S}\times\mathbb{S}} K(s,t)\varphi_{ik}(s)\varphi_{jk}(t)\,\mathrm{d}s\mathrm{d}t. \tag{7.17}$$

Moreover,

$$\lim_{k\to\infty} \iint_{\mathbb{S}\times\mathbb{S}} K(s,t)\varphi_{ik}(s)\varphi_{jk}(t)\,\mathrm{d}s\mathrm{d}t = K(s_i, s_j). \tag{7.18}$$

Combining (7.16)–(7.18), we have

$$\sum_{i,j=1}^m \alpha_i\alpha_j K(s_i, s_j) \geq 0,$$

which shows that $\tilde{K}$ is nonnegative definite. That is, K is of nonnegative type.

Conversely, suppose that K is of nonnegative type. We first show (7.15) when φ is a continuous function. Partition $\mathbb{S}$ into a finite number of subdomains as $\mathbb{S} = \mathbb{S}_1 \cup \mathbb{S}_2 \cup \cdots \cup \mathbb{S}_{k(\delta)}$ so that $\max\{\|s-t\|; s, t \in \mathbb{S}_i\} \leq \delta$, where δ is a positive real number. Then, denoting the volume of $\mathbb{S}_i$ by $\mu(\mathbb{S}_i)$, we have

$$\sum_{i,j=1}^{k(\delta)} K(s_i, s_j)\varphi(s_i)\varphi(s_j)\mu(\mathbb{S}_i)\mu(\mathbb{S}_j) = \sum_{i,j=1}^{k(\delta)} K(s_i, s_j)(\varphi(s_i)\mu(\mathbb{S}_i))(\varphi(s_j)\mu(\mathbb{S}_j))$$
$$\geq 0, \tag{7.19}$$

where s_i is a point in $\mathbb{S}_i$. By definition of the Riemann integral,

$$\iint_{\mathbb{S}\times\mathbb{S}} K(s,t)\varphi(s)\varphi(t)\,\mathrm{d}s\mathrm{d}t = \lim_{\delta\to 0}\sum_{i,j=1}^{k(\delta)} K(s_i, s_j)\varphi(s_i)\varphi(s_j)\mu(\mathbb{S}_i)\mu(\mathbb{S}_j). \tag{7.20}$$

Combining (7.19) and (7.20), we see that (7.15) holds when φ is a continuous function. Because the set of continuous functions on $\mathbb{S}$ is dense [8, p. 40] in $L^2(\mathbb{S})$, (7.15) holds for any function $\varphi \in L^2(\mathbb{S})$. □

The next theorem, *Mercer's theorem* [8, p. 344], plays an essential role in discussing kernels. Unfortunately, however, its proof is out of scope of this book. We only point out that it corresponds to diagonalization of a real symmetric matrix as in (7.10).

Theorem 7.3 (Mercer) *A real-valued, symmetric, continuous, nonnegative definite function $K \in L^2(\mathbb{S}\times\mathbb{S})$ can be expanded as*

$$K(s,t) = \sum_{i=1}^{\infty} \lambda_i \psi_i(s)\psi_i(t),$$

where λ_i's are the eigenvalues of T_K, and ψ_i's corresponding eigenfunctions. The series on the right-hand side converges absolutely and uniformly. □

Let us denote the set of square-summable real sequences by

$$l^2 \triangleq \left\{a\,;\ a = (a_1, a_2, \ldots), a_i \in \mathbb{R}, \textstyle\sum_{i=1}^{\infty} a_i^2 < \infty\right\}.$$

Then, l^2 is a vector space over $\mathbb{R}$ with an inner product $\langle a, b\rangle = \sum_{i=1}^{\infty} a_i b_i$ for $a = (a_1, a_2, \ldots)$, $b = (b_1, b_2, \ldots)$.

Now we are ready to give the proof of Theorem 7.1.

Proof of Theorem 7.1 Since the proof of "only if" part is the same as that of Theorem 7.2, we prove the "if" part. Suppose that K is of nonnegative type. Then, by Lemma 7.6, K is nonnegative definite. Thus, by Theorem 7.3, $K(s,t)$ can be expanded as

$$K(s,t) = \sum_{i=1}^{\infty} \lambda_i \psi_i(s)\psi_i(t) \tag{7.21}$$

with the set of eigenvalues λ_i's and the set of eigenfunctions ψ_i's of T_K. Since $\lambda_i \geq 0$, $\sqrt{\lambda_i}$ is a real number. The sequence $\left(\sqrt{\lambda_1}\psi_1(s), \sqrt{\lambda_2}\psi_2(s), \ldots\right)$ belongs to l^2, because

$$\sum_{i=1}^{\infty}\left(\sqrt{\lambda_i}\psi_i(s)\right)^2 = \sum_{i=1}^{\infty}\lambda_i\psi_i^2(s) = K(s,s) < \infty.$$

Therefore, if we define

$$\Phi(s) \triangleq \left(\sqrt{\lambda_1}\psi_1(s), \sqrt{\lambda_2}\psi_2(s), \ldots\right),$$

then Φ is a mapping from $\mathbb{S}$ to l^2, and by (7.21),

$$K(s,t) = \langle\Phi(s), \Phi(t)\rangle.$$

This shows that K is a kernel. □

From the above argument, it is seen that l^2 can be taken for the feature space $\mathbb{F}$.

Theorem 7.1 is very useful to verify whether a function is a kernel as shown in the next theorem [3].

Theorem 7.4 *Let K_1 and K_2 be kernels. Then, the following K's are kernels, where α, β are positive real numbers, p is a polynomial with positive coefficients, f is a real-valued function, and σ is a nonzero real number.*

(1) $K(s,t) \triangleq \alpha K_1(s,t) + \beta K_2(s,t)$.

(2) $K(s,t) \triangleq K_1(s,t) \cdot K_2(s,t)$.

(3) $K(s,t) \triangleq p(K_1(s,t))$.

(4) $K(s,t) \triangleq f(s) \cdot f(t)$.

(5) $K(s,t) \triangleq \exp(K_1(s,t))$.

(6) $K(s,t) \triangleq \exp(-\|s-t\|^2/\sigma^2)$.

Proof Obviously, K's are all real-valued, symmetric, continuous functions. Therefore, it suffices to show that K's are of nonnegative type.

Let $s_1, s_2, \ldots, s_m$ be arbitrary points in $\mathbb{S}$, and for simplicity, let us denote

$$k_{ij} \triangleq K(s_i, s_j),\ k_{ij}^{(1)} \triangleq K_1(s_i, s_j),\ k_{ij}^{(2)} \triangleq K_2(s_i, s_j).$$

(1) Since K_1 and K_2 are kernels, they are of nonnegative type by Theorem 7.1. Therefore, for arbitrary real numbers $c_1, c_2, \ldots, c_m$,

$$\begin{aligned}\sum_{i,j=1}^{m} c_i c_j k_{ij} &= \alpha\sum_{i,j=1}^{m} c_i c_j k_{ij}^{(1)} + \beta\sum_{i,j=1}^{m} c_i c_j k_{ij}^{(2)} \\ &\geq 0.\end{aligned}$$

Thus, the matrix $[k_{ij}]$ is nonnegative definite, that is, K is of nonnegative type. Therefore, again by Theorem 7.1, K is a kernel.
(2) We have $k_{ij} = k_{ij}^{(1)} \cdot k_{ij}^{(2)}$, i.e., $[k_{ij}] = [k_{ij}^{(1)}] \boxtimes [k_{ij}^{(2)}]$, where $\boxtimes$ denotes the Schur product of two matrices (Appendix A). By Corollary A.5, the Schur product of two nonnegative definite matrices is also nonnegative definite. Therefore, $[k_{ij}]$ is nonnegative definite, resulting in that K is a kernel.
(3) By (2) above, the power of a kernel is again a kernel. Moreover, by (1) above, a linear combination of kernels with positive coefficients is a kernel. Therefore, $p(K_1)$ is a kernel.
(4) The set of real numbers $\mathbb{R}$ is a vector space over $\mathbb{R}$. Also, $a \cdot b$ $(a, b \in \mathbb{R})$ is an inner product in $\mathbb{R}$. Therefore, by definition, K is a kernel.
(5) Using the power series expansion of exp,

$$K(s,t) = \lim_{j\to\infty} \sum_{k=1}^{j} \frac{1}{k!}(K_1(s,t))^k.$$

Since

$$\sum_{k=1}^{j} \frac{1}{k!}(K_1(s,t))^k \tag{7.22}$$

is a polynomial of $K_1(s,t)$ with positive coefficients, it is of nonnegative type by (3). Since K is the limit of (7.22), it is of nonnegative type. Hence K is a kernel.
(6) The right-hand side can be decomposed as

$$\exp(-\|s-t\|^2/\sigma^2) = \{\exp(-\|s\|^2/\sigma^2)\cdot\exp(-\|t\|^2/\sigma^2)\}\cdot\exp(2\langle s,t\rangle/\sigma^2).$$

By (4), $\exp(-\|s\|^2/\sigma^2)\cdot\exp(-\|t\|^2/\sigma^2)$ is a kernel. Also, since $2\langle s,t\rangle/\sigma^2$ is a kernel, $\exp(2\langle s,t\rangle/\sigma^2)$ is a kernel by (5). Therefore, by (4) again, K is a kernel. □

The last kernel in the above theorem, $\exp(-\|s-t\|^2/\sigma^2)$, is called a *Gaussian kernel*, which is very often used. A function of the form $f(x) = G(\|c-x\|)$ is called a *radial basis function (RBF)*, where G is a function of one variable, and c is a fixed vector called the *center*. If s is fixed, the above Gaussian kernel is a radial basis function of t with the center s.

Appendix 2: Reproducing Kernel Hilbert Space (RKHS)

This section outlines the relationship between a Mercer kernel and a reproducing kernel Hilbert space (RKHS).

Let K be a kernel, and denote

$$\tilde{\mathbb{H}}_0 \triangleq \{K(s,\cdot)\,;\ s\in\mathbb{S}\} \text{ and } \mathbb{H}_0 \triangleq \text{span}\,(\tilde{\mathbb{H}}_0),$$

where span (Ω) is the vector space spanned by the set Ω:

$$\text{span}\,(\Omega) \triangleq \left\{\sum_{i=1}^{k} a_i\omega_i\,;\; a_i \in \mathbb{R}, \omega_i \in \Omega, k \in \mathbb{N}\right\}.$$

In a similar way, denote

$$\tilde{\mathbb{G}}_0 \triangleq \{\Phi(s)\,;\; s \in \mathbb{S}\} \text{ and } \mathbb{G}_0 \triangleq \text{span}\,(\tilde{\mathbb{G}}_0).$$

Let $\Psi : \tilde{\mathbb{G}}_0 \to \tilde{\mathbb{H}}_0$ be a mapping that maps

$$\Phi(s) = \left(\sqrt{\lambda_1}\psi_1(s), \sqrt{\lambda_2}\psi_2(s), \ldots\right)$$

to

$$K(s,\cdot) = \sum_{i=1}^{\infty} \lambda_i\psi_i(s)\psi_i(\cdot),$$

where λ_i's are the eigenvalues of K, and ψ_i's corresponding eigenfunctions. That is,

$$\Psi(\Phi(s)) \triangleq \Phi(s)\,\text{diag}\left[\sqrt{\lambda_1}, \sqrt{\lambda_2}, \ldots\right](\psi_1(\cdot), \psi_2(\cdot), \ldots)^t.$$

The mapping Ψ is one-to-one and onto. In fact, since $\{\psi_i(\cdot)\}$ is a basis, $\Psi(\Phi(s)) = \Psi(\Phi(t))$ implies $\lambda_k\psi_k(s) = \lambda_k\psi_k(t)$ $(k = 1, 2, \ldots)$. Therefore, if $\lambda_k \neq 0$, then

$$\sqrt{\lambda_k}\psi_k(s) = \frac{1}{\sqrt{\lambda_k}}\lambda_k\psi_k(s) = \frac{1}{\sqrt{\lambda_k}}\lambda_k\psi_k(t) = \sqrt{\lambda_k}\psi_k(t).$$

If $\lambda_k = 0$, then

$$\sqrt{\lambda_k}\psi_k(s) = 0 = \sqrt{\lambda_k}\psi_k(t).$$

Thus, $\Phi(s) = \Phi(t)$ in either case, showing that Ψ is a one-to-one mapping. Furthermore, for any $K(s,\cdot) = \sum_{i=1}^{\infty} \lambda_i\psi_i(s)\psi_i(\cdot)$, Ψ maps $(\sqrt{\lambda_1}\psi_1(s), \sqrt{\lambda_2}\psi_2(s), \ldots)$ to $K(s,\cdot)$. Hence, Ψ is a mapping onto $\tilde{\mathbb{H}}_0$.

The mapping Ψ is uniquely extended to a linear mapping from the vector space $\mathbb{G}_0$ to the vector space $\mathbb{H}_0$, which is denoted by the same symbol Ψ.

Next, we define a function $\langle\cdot,\cdot\rangle_{\mathbb{H}_0}$ on $\mathbb{H}_0 \times \mathbb{H}_0$ by

$$\langle h_1, h_2\rangle_{\mathbb{H}_0} \triangleq \sum_{i=1}^{k}\sum_{j=1}^{l} a_i b_j K(s_i, t_j)$$

for

$$h_1 \triangleq \sum_{i=1}^{k} a_i K(s_i,\cdot), \quad h_2 \triangleq \sum_{j=1}^{l} b_i K(t_j,\cdot).$$

We show that $\langle\cdot,\cdot\rangle_{\mathbb{H}_0}$ is an inner product. Obviously, this function is symmetric and bilinear. Let $h \triangleq \sum_{i=1}^{k} a_i K(s_i,\cdot)$ be an arbitrary element in $\mathbb{H}_0$. Then, for $g \triangleq \sum_{i=1}^{k} a_i \Phi(s_i)$,

$$\begin{aligned}\langle h, h\rangle_{\mathbb{H}_0} &= \sum_{i=1}^{k}\sum_{j=1}^{k} a_i a_j K(s_i, s_j)\\ &= \sum_{i=1}^{k}\sum_{j=1}^{k} a_i a_j \langle \Phi(s_i), \Phi(s_j)\rangle_{\mathbb{F}}\\ &= \left\langle \textstyle\sum_{i=1}^{k} a_i \Phi(s_i), \sum_{j=1}^{k} a_j \Phi(s_j)\right\rangle_{\mathbb{F}}\\ &= \langle g, g\rangle_{\mathbb{F}}\\ &\geq 0.\end{aligned}$$

Thus, $\langle h, h\rangle_{\mathbb{H}_0}$ is nonnegative. Furthermore, if $\langle h, h\rangle_{\mathbb{H}_0} = 0$, then $\langle g, g\rangle_{\mathbb{F}} = 0$. Hence it follows that $g = 0$, and $h = \Psi(g) = 0$. These facts show that $\langle\cdot,\cdot\rangle_{\mathbb{H}_0}$ is an inner product.

For an arbitrary element $h \triangleq \sum_{i=1}^{k} a_i K(s_i,\cdot) \in \mathbb{H}_0$,

$$\begin{aligned}\langle h, K(t,\cdot)\rangle_{\mathbb{H}_0} &= \sum_{i=1}^{k} \langle a_i K(s_i,\cdot), K(t,\cdot)\rangle_{\mathbb{H}_0}\\ &= \sum_{i=1}^{k} a_i K(s_i, t)\\ &= h(t).\end{aligned}$$

Because of this property, K is called a *reproducing kernel* in $\mathbb{H}_0$.

As shown above, if $h = 0$ then $g = 0$. Therefore, the (extended) mapping Ψ is one-to-one. Obviously, it is a mapping onto $\mathbb{H}_0$. Thus, Ψ is an *isomorphism* [8, p. 118] from $\mathbb{G}_0$ to $\mathbb{H}_0$.

Finally, let $\mathbb{H} \triangleq \mathrm{cl}(\mathbb{H}_0)$ and $\mathbb{G} \triangleq \mathrm{cl}(\mathbb{G}_0)$, where $\mathrm{cl}(\Omega)$ is the *closure* [8, p. 37] of the set Ω. It is verified that the spaces $\mathbb{H}$ and $\mathbb{G}$ satisfy the definition of *Hilbert space* [8, p. 119]. The isomorphism Ψ from $\mathbb{G}_0$ to $\mathbb{H}_0$ is extended to an isomorphism from $\mathbb{G}$ to $\mathbb{H}$. Also, inner product $\langle\cdot,\cdot\rangle_{\mathbb{H}_0}$ and reproducing property $\langle h, K(t,\cdot)\rangle_{\mathbb{H}_0} = h(t)$ are extended into $\langle\cdot,\cdot\rangle_{\mathbb{H}}$ and reproducing property $\langle h, K(t,\cdot)\rangle_{\mathbb{H}} = h(t)$, respectively. The kernel K becomes a reproducing kernel in $\mathbb{H}$. The vector space $\mathbb{H}$, together with the inner product $\langle\cdot,\cdot\rangle_{\mathbb{H}}$, is called the *reproducing kernel Hilbert space (RKHS)* associated with a reproducing kernel K. The mapping Ψ preserves not only the linear structure of $\mathbb{H}$ and $\mathbb{G}$ but also their inner products: $\langle \Psi(g), \Psi(g')\rangle_{\mathbb{H}} = \langle g, g'\rangle_{\mathbb{G}}$, where

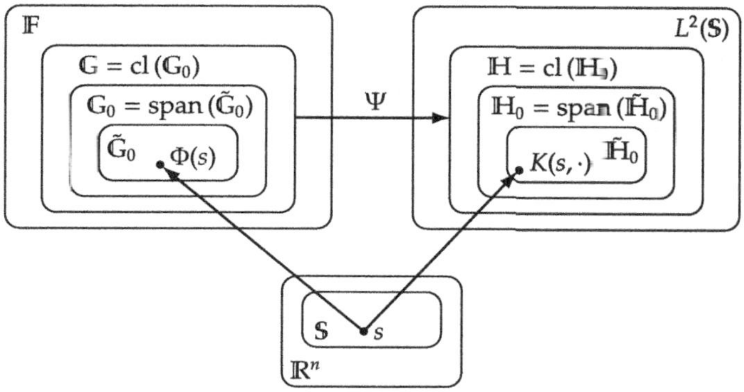

Fig. 7.3 Structure of the reproducing kernel Hilbert space $\mathbb{H}$

$\langle g, g'\rangle_{\mathbb{G}}$ is the restriction of $\langle g, g'\rangle_{\mathbb{F}}$ on $\mathbb{G} \times \mathbb{G}$. Thus, we can identify $\mathbb{H}$ with $\mathbb{G}$ as Hilbert spaces. Figure 7.3 illustrates the relationship among spaces that appear in the above arguments.

References

1. Müller, K.-R., Mika, S., Rätsch, G., Tsuda, K., Schölkopf, B.: An introduction to kernel-based learning algorithms. IEEE Trans. Neural Netw. **12**(2), 181–201 (2001)
2. Liu, W., Pokharel, P.P., Principe, J.C.: The kernel least-mean-square algorithm. IEEE Trans. Sig. Process. **56**(2), 543–554 (2008)
3. Christianini, N., Shawe-Taylor, J.: An Introduction to Support Vector Machines and Other Kernel-based Learning Methods. Cambridge University Press, Cambridge (2000)
4. Liu, W., Príncipe, J.C.: Kernel affine projection algorithms. EURASIP J. Adv. Sig. Process. **2008**, 1–12 (2008)
5. Liu, W., Príncipe, J.C., Haykin. S.: Kernel Adaptive Filtering—A Comprehensive Introduction. Wiley, Hoboken (2010)
6. Van Vaerenbergh, S., Via, J., Santamaría, I.: A sliding-window kernel RLS algorithm and its application to nonlinear channel identification. In: Proceedings of ICASSP 2006, vol. 5, pp. 789–792. Toulouse, 14–19 May (2006)
7. Platt, J.: A resource-allocating network for function interpolation. Neural Comput. **3**(2), 213–225 (1991)
8. Dieudonné, J.: Foundations of Modern Analysis. Academic Press, New York (1969)

Chapter 8
Variable Parameter APAs

8.1 Variable Step-Size APA (VSS-APA)

8.1.1 VSS-APA Based on Optimization of Step-Size

From the update equation (3.10) for the B-APA, we have

$$\tilde{\boldsymbol{w}}_{k+1} = \tilde{\boldsymbol{w}}_k - \mu \boldsymbol{X}_k^t (\boldsymbol{X}_k \boldsymbol{X}_k^t)^{-1} \boldsymbol{e}_k, \tag{8.1}$$

where $\tilde{\boldsymbol{w}}_k = w^o - \boldsymbol{w}_k$ as defined before. This is an equation in the stochastic framework: the variables except for μ are random variables. Calculating the squared norm of both sides of (8.1), and taking expectations, we find

$$\begin{aligned} \mathrm{E}\left\{\|\tilde{\boldsymbol{w}}_{k+1}\|^2\right\} &= \mathrm{E}\left\{\|\tilde{\boldsymbol{w}}_k\|^2\right\} - 2\mu \mathrm{E}\left\{\boldsymbol{e}_k^t (\boldsymbol{X}_k \boldsymbol{X}_k^t)^{-1} \boldsymbol{X}_k \tilde{\boldsymbol{w}}_k\right\} + \mu^2 \mathrm{E}\left\{\boldsymbol{e}_k^t (\boldsymbol{X}_k \boldsymbol{X}_k^t)^{-1} \boldsymbol{e}_k\right\} \\ &= \mathrm{E}\left\{\|\tilde{\boldsymbol{w}}_k\|^2\right\} - \Delta(\mu), \end{aligned} \tag{8.2}$$

where

$$\Delta(\mu) \triangleq 2\mu \mathrm{E}\left\{\boldsymbol{e}_k^t (\boldsymbol{X}_k \boldsymbol{X}_k^t)^{-1} \boldsymbol{X}_k \tilde{\boldsymbol{w}}_k\right\} - \mu^2 \mathrm{E}\left\{\boldsymbol{e}_k^t (\boldsymbol{X}_k \boldsymbol{X}_k^t)^{-1} \boldsymbol{e}_k\right\}.$$

Thus, by maximizing $\Delta(\mu)$ with respect to μ, the step-size $\mu^o(k)$ that gives the maximum decrease in $\mathrm{E}\left\{\|\tilde{\boldsymbol{w}}_k\|^2\right\}$ will be obtained. Based on this idea, a *variable step-size APA (VSS-APA)* was developed in [1] as reviewed below.

Maximization of $\Delta(\mu)$ is performed by differentiating it with respect to μ, and equating it with 0. This leads to

$$\mu^o(k) = \frac{\mathrm{E}\left\{\boldsymbol{e}_k^t (\boldsymbol{X}_k \boldsymbol{X}_k^t)^{-1} \boldsymbol{X}_k \tilde{\boldsymbol{w}}_k\right\}}{\mathrm{E}\left\{\boldsymbol{e}_k^t (\boldsymbol{X}_k \boldsymbol{X}_k^t)^{-1} \boldsymbol{e}_k\right\}}. \tag{8.3}$$

K. Ozeki, *Theory of Affine Projection Algorithms for Adaptive Filtering*, Mathematics for Industry 22, DOI 10.1007/978-4-431-55738-8_8

Let us assume that the observation noise $\{v(k)\}$ is of zero mean, identically and independently distributed (i.i.d.), and statistically independent of the regressor blocks X_k's. In addition, we assume that the statistical dependence of $\tilde{w}_k$ on the past noise is negligible. Since $e_k = X_k \tilde{w}_k + v_k$, where $v_k = (v(k), v(k-1), \ldots, v(k-p+1))^t$, we have

$$\begin{aligned} \mathrm{E}\left\{e_k^t (X_k X_k^t)^{-1} X_k \tilde{w}_k\right\} &= \mathrm{E}\left\{(X_k \tilde{w}_k + v_k)^t (X_k X_k^t)^{-1} X_k \tilde{w}_k\right\} \\ &= \mathrm{E}\left\{\tilde{w}_k^t X_k^t (X_k X_k^t)^{-1} X_k \tilde{w}_k\right\} + \mathrm{E}\left\{v_k^t (X_k X_k^t)^{-1} X_k \tilde{w}_k\right\} \\ &= \mathrm{E}\left\{\tilde{w}_k^t X_k^t (X_k X_k^t)^{-1} X_k \tilde{w}_k\right\}. \end{aligned} \tag{8.4}$$

Let $p_k \triangleq X_k^t (X_k X_k^t)^{-1} X_k \tilde{w}_k$. Then $\tilde{w}_k^t X_k^t (X_k X_k^t)^{-1} X_k \tilde{w}_k = \|p_k\|^2$. Therefore, from (8.4),

$$\mathrm{E}\left\{e_k^t (X_k X_k^t)^{-1} X_k \tilde{w}_k\right\} = \mathrm{E}\left\{\|p_k\|^2\right\}. \tag{8.5}$$

Similarly,

$$\begin{aligned} \mathrm{E}\left\{e_k^t (X_k X_k^t)^{-1} e_k\right\} &= \mathrm{E}\left\{(X_k \tilde{w}_k + v_k)^t (X_k X_k^t)^{-1} (X_k \tilde{w}_k + v_k)\right\} \\ &= \mathrm{E}\left\{\tilde{w}_k^t X_k^t (X_k X_k^t)^{-1} X_k \tilde{w}_k\right\} + \mathrm{E}\left\{v_k^t (X_k X_k^t)^{-1} v_k\right\} \\ &\quad + 2\mathrm{E}\left\{\tilde{w}_k^t X_k^t (X_k X_k^t)^{-1} v_k\right\} \\ &= \mathrm{E}\left\{\tilde{w}_k^t X_k^t (X_k X_k^t)^{-1} X_k \tilde{w}_k\right\} + \mathrm{E}\left\{v_k^t (X_k X_k^t)^{-1} v_k\right\} \\ &= \mathrm{E}\left\{\|p_k\|^2\right\} + \mathrm{E}\left\{\mathrm{tr}\left[v_k v_k^t (X_k X_k^t)^{-1}\right]\right\} \\ &= \mathrm{E}\left\{\|p_k\|^2\right\} + \sigma_v^2 \,\mathrm{tr}\left[\mathrm{E}\left\{(X_k X_k^t)^{-1}\right\}\right]. \end{aligned} \tag{8.6}$$

By using (8.5) and (8.6), Eq. (8.3) is rewritten as

$$\mu^o(k) = \frac{\mathrm{E}\left\{\|p_k\|^2\right\}}{\mathrm{E}\left\{\|p_k\|^2\right\} + \sigma_v^2 \,\mathrm{tr}\left[\mathrm{E}\left\{(X_k X_k^t)^{-1}\right\}\right]}. \tag{8.7}$$

The problem here is how to compute $\mathrm{E}\left\{\|p_k\|^2\right\}$. We note

$$\begin{aligned} \mathrm{E}\left\{p_k\right\} &= \mathrm{E}\left\{X_k^t (X_k X_k^t)^{-1} X_k \tilde{w}_k\right\} \\ &= \mathrm{E}\left\{X_k^t (X_k X_k^t)^{-1} (e_k - v_k)\right\} \\ &= \mathrm{E}\left\{X_k^t (X_k X_k^t)^{-1} e_k\right\} - \mathrm{E}\left\{X_k^t (X_k X_k^t)^{-1} v_k\right\} \\ &= \mathrm{E}\left\{X_k^t (X_k X_k^t)^{-1} e_k\right\}. \end{aligned}$$

This equation motivates us to approximate $\boldsymbol{p}_k$ with $\hat{p}_k$ obtained by time-averaging $X_k^t(X_kX_k^t)^{-1}e_k$ as

$$\hat{p}_k = \alpha\hat{p}_{k-1} + (1-\alpha)X_k^t(X_kX_k^t)^{-1}e_k,$$

where α is a smoothing factor ($0 < \alpha < 1$). Replacing $\mathrm{E}\left\{\|\boldsymbol{p}_k\|^2\right\}$ in (8.7) with $\|\hat{p}_k\|^2$, we finally obtain a VSS-APA

$$\begin{aligned} w_{k+1} &= w_k + \mu(k)X_k^t(X_kX_k^t)^{-1}e_k, \\ \mu(k) &= \mu_{\max}\frac{\|\hat{p}_k\|^2}{\|\hat{p}_k\|^2 + C}, \end{aligned} \tag{8.8}$$

where $C = \sigma_v^2\,\mathrm{tr}\left[\mathrm{E}\left\{(\boldsymbol{X}_k\boldsymbol{X}_k^t)^{-1}\right\}\right]$. If an approximation

$$\begin{aligned} \mathrm{E}\left\{(\boldsymbol{X}_k\boldsymbol{X}_k^t)^{-1}\right\} &\approx \mathrm{diag}\left[1/\mathrm{E}\left\{\|x_k\|^2\right\}, 1/\mathrm{E}\left\{\|x_{k-1}\|^2\right\}, \ldots, 1/\mathrm{E}\left\{\|x_{k-p+1}\|^2\right\}\right] \\ &\approx \mathrm{diag}\left[1/\mathrm{E}\left\{\|x_k\|^2\right\}, 1/\mathrm{E}\left\{\|x_k\|^2\right\}, \ldots, 1/\mathrm{E}\left\{\|x_k\|^2\right\}\right], \end{aligned}$$

is employed, then

$$C \approx p \cdot \frac{\sigma_v^2}{\mathrm{E}\left\{\|x_k\|^2\right\}} \approx \frac{p}{n}\frac{\sigma_v^2}{\sigma_x^2},$$

where $\sigma_x^2 \triangleq \mathrm{E}\left\{x^2(k)\right\}$. Thus, C is proportional to the ratio of the projection order p to the filter length n, and inversely proportional to SNR.

8.1.2 VSS-APA for Acoustic Echo Cancellation

Acoustic echo cancellation (AEC) is one of the important application areas of adaptive filters. Figure 8.1 depicts a situation where acoustic echo cancellation is needed. The loudspeaker, the near-end speaker, and the microphone are in the same acoustic space. The near-end speaker listens to the far-end signal coming from the loudspeaker. He/she also talks to the microphone. The far-end signal emitted from the loudspeaker as a sound wave, and the near-end sounds comprised of the near-end speaker's voice and the ambient noise are mixed in the acoustic space, and picked up by the microphone. Thus, if nothing is done, the far-end signal, modified by the acoustic space, is echoed back to the far-end. The function of an echo canceller is to cancel the far-end signal, and to send back only the near-end signal to the far-end.

Figure 8.2 illustrates a basic scheme of an echo canceller using an adaptive filter. When the far-end speaker and the near-end speaker speak simultaneously (double-talk), the signal from the near-end speaker disturbs the adaptation. To cope with this problem, a *double-talk detector (DTD)* is employed to control the adaptation, e.g., to

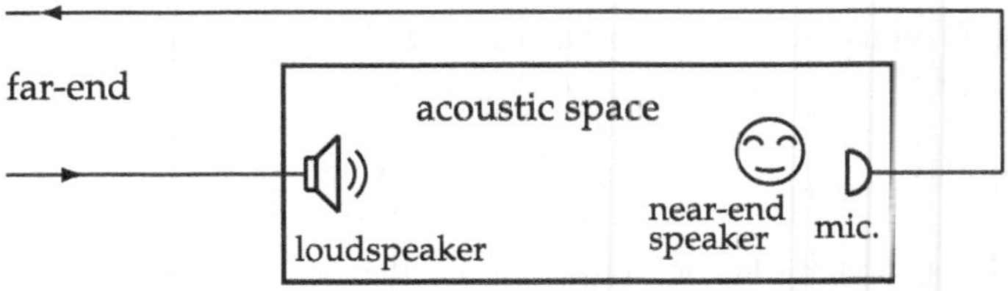

Fig. 8.1 Situation where acoustic echo cancellation is needed. The loudspeaker, the microphone, and the near-end speaker are in the same acoustic space

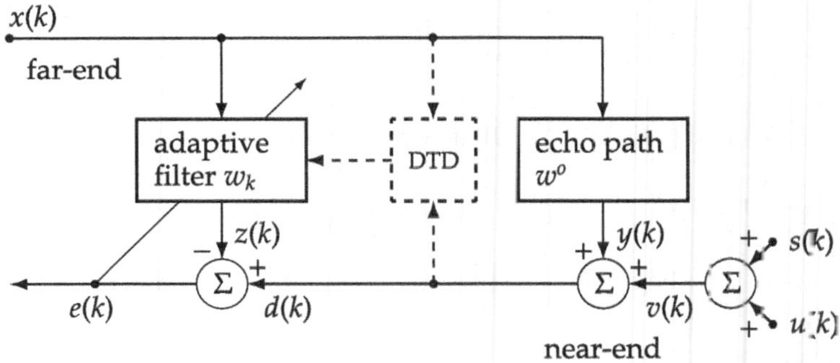

Fig. 8.2 Schematic diagram of an acoustic echo canceller. The signal $s(k)$ is the speech sound from the near-end speaker, and $u(k)$ the ambient noise, both picked up by the microphone. DTD denotes the double-talk detector

suspend the adaptation when a double-talk is detected. We will not discuss the issue of DTD here.

AEC presents a lot of challenges. The echo path can be very long depending on the size, shape, etc., of the acoustic space. This means we need a long transversal filter to simulate the impulse response of the echo path. Also, the far-end signal is often a speech signal, which is highly correlated. These factors affect the convergence rate of the adaptive filter. In addition, the echo path often exhibits a rapid change, e.g., a slight movement of the near-end speaker can give a large influence on the impulse response of the echo path. Therefore, fast convergence is needed for the adaptive filter to keep track of the change. Under these conflicting conditions, the APA is considered to be an attractive choice because of its fast convergence even for correlated input signals.

What should be done in AEC is basically system identification. We want to make a replica of the impulse response of the echo path from the loudspeaker to the microphone by an adaptive filter. However, the purpose of AEC is to send the near-end signal to the far-end without mixing the far-end signal. That is, it is desirable that the error signal $e(k)$ is equal to the near-end signal $v(k)$, whereas in the usual formulation of system identification, the objective is to make the magnitude of the error signal

$e(k)$ as small as possible. By taking notice of this subtle difference, a VSS-APA for AEC was developed [2] as described below.

Let us start with the update equation (3.11) for the B-APA

$$w_{k+1} = w_k + \mu X_k^t (X_k X_k^t)^{-1} e_k. \tag{8.9}$$

We modify this equation into

$$w_{k+1} = w_k + X_k^t (X_k X_k^t)^{-1} M_k e_k, \tag{8.10}$$

where

$$M_k = \mathrm{diag}\,[\mu_0(k), \mu_1(k), \ldots, \mu_{p-1}(k)]$$

is a $p \times p$ diagonal matrix. When $\mu_0(k) = \mu_1(k) = \cdots = \mu_{p-1}(k) = \mu$, Eq. (8.10) reduces to (8.9). Thus, (8.10) generalizes (8.9) so that each component of e_k is given a step-size individually.

Let us define the a posteriori error vector[1]

$$\epsilon_k \stackrel{\triangle}{=} d_k - X_k w_{k+1}. \tag{8.11}$$

From (8.10) and (8.11), we have

$$\begin{aligned}\epsilon_k &= d_k - X_k \left(w_k + X_k^t (X_k X_k^t)^{-1} M_k e_k \right)\\ &= d_k - X_k w_k - X_k X_k^t (X_k X_k^t)^{-1} M_k e_k\\ &= e_k - M_k e_k\\ &= (I_p - M_k) e_k. \end{aligned}\tag{8.12}$$

If we require $\epsilon_k = 0$ just as we did to derive the B-APA by the method of Lagrange multipliers in Chap. 3, Sect. 3.4, we have $M_k = I_p$ unless each component of e_k is zero. Thus, $\mu_i(k) = 1$ $(i = 0, 1, \ldots, p-1)$. This corresponds to the case $\mu = 1$ in (8.9). However, we want the error signal to coincide with the near-end signal for the present purpose. Thus, it will be more appropriate to require $\epsilon_k = v_k$, where $v_k = (v(k), v(k-1), \ldots, v(k-p+1))^t$. Therefore, denoting $\epsilon_k = (\epsilon_{k,0}, \epsilon_{k,1}, \ldots, \epsilon_{k,p-1})^t$, $e_k = (e_{k,0}, e_{k,1}, \ldots, e_{k,p-1})^t$, and using (8.12), we obtain

$$\epsilon_{k,l} = (1 - \mu_l(k)) e_{k,l} = v(k-l) \quad (l = 0, 1, \ldots, p-1). \tag{8.13}$$

---

[1] This is different from the a posteriori error vector $e_{p,k}$ defined in Chap. 5. The a posteriori error vectors ϵ_k and $e_{p,k}$ are related by $\epsilon_k = e_{p,k} + v_k$.

However, instead of exact equality, we postulate this equation to hold in the mean square, i.e.,

$$(1-\mu_l(k))^2\mathrm{E}\left\{\boldsymbol{e}_{k,l}^2\right\}=\mathrm{E}\left\{\boldsymbol{v}^2(l-l)\right\}, \tag{8.14}$$

where $\boldsymbol{e}_{k,l}$ and $\boldsymbol{v}(k-l)$ are random variables corresponding to $e_{k,l}$ and $v(k-l)$, respectively. The quadratic equation (8.14) has two solutions

$$\mu_l(k)=1\pm\sqrt{\frac{\mathrm{E}\left\{\boldsymbol{v}^2(k-l)\right\}}{\mathrm{E}\left\{\boldsymbol{e}_{k,l}^2\right\}}}.$$

As we have seen in Chap. 5, for a real number a between 0 and 1, $\mu=1-a$ and $\mu=1+a$ give the same convergence rate. However, $\mu=1-a$ gives less MSE at steady state. Therefore, we adopt

$$\mu_l(k)=1-\sqrt{\frac{\mathrm{E}\left\{\boldsymbol{v}^2(k-l)\right\}}{\mathrm{E}\left\{\boldsymbol{e}_{k,l}^2\right\}}}. \tag{8.15}$$

By denoting $\hat{\sigma}_v^2(k-l)\triangleq\mathrm{E}\left\{\boldsymbol{v}^2(k-l)\right\}$ and $\hat{\sigma}_{e_l}^2(k)\triangleq\mathrm{E}\left\{\boldsymbol{e}_{k,l}^2\right\}$, (8.15) is expressed as

$$\mu_l(k)=1-\frac{\hat{\sigma}_v(k-l)}{\hat{\sigma}_{e_l}(k)}. \tag{8.16}$$

The quantity $\hat{\sigma}_{e_l}(k)$ in the denominator of (8.16) can be estimated recursively as

$$\hat{\sigma}_{e_l}^2(k)=\lambda\hat{\sigma}_{e_l}^2(k-1)+(1-\lambda)e_{k,l}^2, \tag{8.17}$$

where λ is the smoothing factor between 0 and 1.

The estimation of $\hat{\sigma}_v^2(k-l)$ is not simple, because we can observe only $d(k)$, which is the mixture of $v(k)$ and the echo signal $y(k)$:

$$d(k)=y(k)+v(k).$$

Assuming that $\boldsymbol{y}(k)$ and $\boldsymbol{v}(k)$ are uncorrelated, we have

$$\mathrm{E}\left\{\boldsymbol{v}^2(k)\right\}=\mathrm{E}\left\{\boldsymbol{d}^2(k)\right\}-\mathrm{E}\left\{\boldsymbol{y}^2(k)\right\}. \tag{8.18}$$

Here, we make an assumption that

$$\mathrm{E}\left\{\boldsymbol{y}^2(k)\right\}\approx\mathrm{E}\left\{\boldsymbol{z}^2(k)\right\}$$

at steady state, where $z(k)$ is the output of the adaptive filter as shown in Fig. 8.2. Thus, (8.18) is rewritten as

$$\begin{aligned}\hat{\sigma}_v^2(k) &= \mathrm{E}\left\{\boldsymbol{v}^2(k)\right\} \\ &\approx \mathrm{E}\left\{\boldsymbol{d}^2(k)\right\} - \mathrm{E}\left\{z^2(k)\right\}.\end{aligned}$$

Similarly to (8.17), we can estimate $\hat{\sigma}_d^2(k) \triangleq \mathrm{E}\{\boldsymbol{d}^2(k)\}$ and $\hat{\sigma}_z^2(k) \triangleq \mathrm{E}\{z^2(k)\}$ recursively as

$$\hat{\sigma}_d^2(k) = \lambda\hat{\sigma}_d^2(k-1) + (1-\lambda)d^2(k),$$

and

$$\hat{\sigma}_z^2(k) = \lambda\hat{\sigma}_z^2(k-1) + (1-\lambda)z^2(k).$$

Substituting $\hat{\sigma}_v^2(k) \approx \hat{\sigma}_d^2(k) - \hat{\sigma}_z^2(k)$ into (8.16), we obtain

$$\mu_l(k) = 1 - \frac{\sqrt{\hat{\sigma}_d^2(k-l) - \hat{\sigma}_z^2(k-l)}}{\hat{\sigma}_{e_l}(k)}. \tag{8.19}$$

It is verified by a simple calculation that this equation also covers the case where the length of the adaptive filter is insufficient to model the impulse response of the echo path (under-modeling case). In actual computation, it is possible that the quantity $\hat{\sigma}_d^2(k-l) - \hat{\sigma}_z^2(k-l)$ in (8.19) becomes negative. Also, $\mu_l(k)$ may take a negative value. In order to avoid such problems, (8.19) is modified into

$$\mu_l(k) = \left|1 - \frac{\sqrt{\left|\hat{\sigma}_d^2(k-l) - \hat{\sigma}_z^2(k-l)\right|}}{\hat{\sigma}_{e_l}(k) + \xi}\right|, \tag{8.20}$$

where ξ is a small positive real number for preventing division by a very small number.

By using (8.20), the step-size matrix

$$M_k = \mathrm{diag}\left[\mu_0(k), \mu_1(k), \ldots, \mu_{p-1}(k)\right]$$

is defined, which is then applied to the update equation for the R-APA to make a VSS-APA as

$$w_{k+1} = w_k + X_k^t(X_kX_k^t + \delta I)^{-1}M_k e_k. \tag{8.21}$$

8.2 Proportionate APA (PAPA)

8.2.1 Proportionate NLMS (PNLMS)

The proportionate adaptation algorithm was first proposed as a variant of the NLMS algorithm for network echo cancellation. It is known that an echo path has a sparse impulse response, that is, its nonzero part is usually concentrated on a very short region, whereas its duration is long representing the delay in the echo path. This observation motivates us to distribute the magnitude of the step-size unevenly over the coefficients of the adaptive filter, so that an active coefficient receives a larger change.

Let us rewrite the update equation (2.5) for the NLMS algorithm into a component-wise form as

$$
\begin{aligned}
z(k) &= x_k^t w_k = \sum_{i=0}^{n-1} w_{k,i} x(k-i), \\
e(k) &= d(k) - z(k), \\
\hat{\sigma}_x^2(k) &= \frac{1}{n} \sum_{i=0}^{n-1} x^2(k-i), \\
w_{k+1,i} &= w_{k,i} + \mu \frac{e(k)x(k-i)}{n\,\hat{\sigma}_x^2(k)} \quad (i = 0, 1, \ldots, n-1),
\end{aligned}
\tag{8.22}
$$

where $w_{k,i}$ is the ith component of the coefficient vector w_k. In the *proportionate NLMS (PNLMS) algorithm* [3], the set of Eqs. (8.22) are modified as

$$
\begin{aligned}
z(k) &= x_k^t w_k = \sum_{i=0}^{n-1} w_{k,i} x(k-i), \\
e(k) &= d(k) - z(k), \\
l_\infty(k) &= \max\{|w_{k,0}|, |w_{k,1}|, \ldots, |w_{k,n-1}|\}, \\
l'_\infty(k) &= \max\{\delta_p, l_\infty(k)\}, \\
g_i(k) &= \max\{\rho l'_\infty(k), |w_{k,i}|\} \quad (i = 0, 1, \ldots, n-1), \\
g_{k,i} &= \frac{g_i(k)}{\frac{1}{n}\sum_{i=0}^{n-1} g_i(k)}, \\
\hat{\sigma}_x^2(k) &= \frac{1}{n} \sum_{i=0}^{n-1} x^2(k-i), \\
w_{k+1,i} &= w_{k,i} + \mu\, g_{k,i} \frac{e(k)x(k-i)}{n\,\hat{\sigma}_x^2(k)} \quad (i = 0, 1, \ldots, n-1),
\end{aligned}
\tag{8.23}
$$

where δ_p and ρ are regularization parameters. Thus, each coefficient of the adaptive filter receives a change that is approximately proportional to its own magnitude.

Using the matrix notation, and replacing the norm in the denominator with the weighted norm, we can rewrite (8.23) as

$$w_{k+1} = w_k + \mu \frac{G_k x_k e(k)}{x_k^t G_k x_k + \delta}, \qquad G_k = \text{diag}\left[g_{k,0}, g_{k,1}, \ldots, g_{k,n-1}\right], \tag{8.24}$$

where δ is a regularization parameter [4].

Later, the weight $g_{k,i}$ was improved as

$$g_{k,i} \triangleq \frac{1-\alpha}{2n} + (1+\alpha)\frac{|w_{k,i}|}{2\sum_{j=0}^{n-1}|w_{k,j}| + \xi}, \tag{8.25}$$

where α $(-1 \le \alpha < 1)$ is a parameter to adjust the degree of proportionality [5]. When α approaches -1, the distribution of $g_i(k)$'s becomes flatter. When the improved weight (8.25) is used, the PNLMS algorithm is called the *improved PNLMS (IPNLMS) algorithm*.

8.2.2 Proportionate APA (PAPA)

The idea of proportionate adaptation can also be applied to the APA. A *proportionate APA (PAPA)* [6] is obtained by generalizing (8.24) as

$$w_{k+1} = w_k + \mu G_k X_k^t (X_k G_k X_k^t + \delta I)^{-1} e_k. \tag{8.26}$$

When the improved weight in (8.25) is used for defining G_k, the PAPA obtained is called the *improved proportionate APA (IPAPA)*.

Let $P_k \triangleq G_k X_k^t$ in the IPAPA. By denoting $g_k \triangleq (g_{k,0}, g_{k,1}, \ldots, g_{k,n-1})^t$, the matrix P_k is rewritten as

$$P_k = \left[g_k \boxtimes x_k \;\; g_k \boxtimes x_{k-1} \;\cdots\; g_k \boxtimes x_{k-p+1}\right], \tag{8.27}$$

where the symbol $\boxtimes$ denotes the Schur product (Appendix A). In the matrix P_k, the history of regressors $x_k, x_{k-1}, \ldots, x_{k-p+1}$ are included. However, the time index of the weight g_k is kept at k. By taking into account the *proportionate memory*, the modification of P_k into

$$P_k' = \left[g_k \boxtimes x_k \;\; g_{k-1} \boxtimes x_{k-1} \;\cdots\; g_{k-p+1} \boxtimes x_{k-p-1}\right] \tag{8.28}$$

was proposed [6]. The adaptation algorithm using the update equation

$$w_{k+1} = w_k + \mu P'_k (X_k P'_k + \delta I)^{-1} e_k \tag{8.29}$$

is called the *improved proportionate APA with proportionate memory* (*MIPAPA*). It has been shown experimentally that the MIPAPA outperforms the IPAPA both in the convergence rate and in the misadjustment. The modification of P_k into P'_k also contributes to the reduction of computational complexity. In fact, P'_k has the time-shift property: the last $p-1$ columns of P'_k are the first $p-1$ columns of P'_{k-1}. Therefore, only the first column of P'_k needs to be computed in the successive sample-by-sample update computation.

The MIPAPA can be combined with the variable step-size technique to compose the VSS-MIPAPA [7]. Low complexity versions of the MIPAPA have also been proposed [8, 9].

8.3 Evolving Order APA (E-APA)

In Chap. 5, we have seen that the projection order p gives a definite influence on the convergence behavior of the APA. If we use a larger p, then the convergence rate is higher, but the MSE is larger at steady state. This fact motivates us to use a large p in the initial phase, where fast convergence is important, but to use a small p in the steady-state phase, where a low MSE is desirable. This idea leads to the *evolving order APA (E-APA)* [10]. In the E-APA, the projection order $p(k)$ at time k is a function of the squared error $e^2(k)$ and the preceding projection order $p(k-1)$ as defined by

$$p(k) = \begin{cases} \min\{p(k-1)+1, p_{\max}\}, & \text{if } \eta(k) < e^2(k); \\ p(k-1), & \text{if } \theta(k) < e^2(k) \le \eta(k); \\ \min\{p(k-1)-1, 1\}, & \text{if } e^2(k) \le \theta(k), \end{cases} \tag{8.30}$$

where $\eta(k)$ and $\theta(k)$ are the upper and the lower threshold at time k, respectively. In order to determine the thresholds, we utilize

$$\text{MSE} = \frac{\mu\sigma_v^2}{(2-\mu)} \mathrm{E}\left\{\frac{p}{\|\boldsymbol{x}_k\|^2}\right\} \mathrm{tr}[R_x] + \sigma_v^2,$$

which is derived from (5.76) and (5.77) in Chap. 5. Employing an approximation

$$\mathrm{E}\left\{\frac{p}{\|x_k\|^2}\right\} \approx \frac{p}{\mathrm{E}\{\|x_k\|^2\}},$$

and noting $\text{tr}[R_x] = \text{E}\left\{\|x_k\|^2\right\}$, we can approximate the MSE as

$$\text{MSE} \approx \frac{\mu\sigma_v^2 p}{2-\mu} + \sigma_v^2 = \sigma_v^2 \frac{\mu(p-1)+2}{2-\mu} \triangleq \epsilon(p).$$

Thus, we can expect that the MSE of the APA with the projection order $p(k-1)$ will converge to $\epsilon(p(k-1))$. If $e^2(k) < \epsilon(p(k-1))$, then a lower MSE might be attained. So, we decrease the projection order by one at k. On the other hand, if $e^2(k) > \epsilon(p(k-1)+1)$, then further convergence is needed. So, we increase the projection order by one. Therefore, we determine the thresholds $\eta(k)$ and $\theta(k)$ in (8.30) as

$$\eta(k) \triangleq \epsilon(p(k-1)+1) = \sigma_v^2 \frac{\mu p(k-1)+2}{2-\mu},$$

and

$$\theta(k) \triangleq \epsilon(p(k-1)) = \sigma_v^2 \frac{\mu(p(k-1)-1)+2}{2-\mu}.$$

Note that $p(k)$ given by (8.30) is always between 1 and $p_{\max}$.

The effectiveness of the E-APA thus composed has been experimentally demonstrated [10]. Equation (8.30) is also adopted in [11]. A similar method is employed in [12].

8.4 Variable Regularized APA (VR-APA)

8.4.1 Use of a Posteriori Error

Besides the step-size μ and the projection order p, the regularization parameter δ gives a heavy influence on the behavior of the R-APA. It has been experimentally observed that a larger regularization parameter results in a lower convergence rate with a smaller MSE. Thus, optimizing δ and controlling it adaptively can enhance the performance of the R-APA. Various methods for such *variable regularized APA (VR-APA)* have been proposed so far.

The method discussed in [13] takes a similar approach used in Sect. 8.1.2. In the system identification scheme in Fig. 2.1, let $\mu = 1$ and $\delta = \delta(k)$. Then, the update equation is rewritten as

$$w_{k+1} = w_k + X_k^t P_k e_k, \tag{8.31}$$

where

$$P_k \triangleq (X_k X_k^t + \delta(k))^{-1}.$$

Multiplication of X_k to both sides of (8.31) from the left leads to

$$\epsilon_k = (I - X_k X_k^t P_k) e_k, \tag{8.32}$$

where ϵ_k is the a posteriori error vector defined in (8.11).

If $w_{k+1} = w^o$ is attained, then

$$\begin{aligned}\epsilon_k &= d_k - X_k w_{k+1} \\ &= d_k - X_k w^o \\ &= v_k,\end{aligned} \tag{8.33}$$

where $v_k = (v(k), v(k-1), \ldots, v(k-p+1))^t$. However, it is impossible to postulate (8.33), because the waveform of v_k is not observable. Therefore, we postulate instead,

$$\mathrm{E}\left\{\|\boldsymbol{\epsilon}_k\|^2\right\} = \mathrm{E}\left\{\|\boldsymbol{v}_k\|^2\right\}. \tag{8.34}$$

Substituting (8.32) into (8.34), we obtain

$$\begin{aligned}\mathrm{E}\left\{\|\boldsymbol{v}_k\|^2\right\} &= \mathrm{E}\left\{\left\|\left(I - X_k X_k^t P_k\right) e_k\right\|^2\right\} \\ &= \mathrm{E}\left\{e_k^t \left(I - X_k X_k^t P_k\right)^t \left(I - X_k X_k^t P_k\right) e_k\right\}\end{aligned} \tag{8.35}$$

By Theorem A.5 (Appendix A), the Gramian matrix $X_k X_k^t$ is diagonalized as

$$X_k X_k^t = T_k \Lambda_k T_k^t, \tag{8.36}$$

where T_k is an orthogonal matrix, $\lambda_0(k), \lambda_1(k), \ldots, \lambda_{p-1}(k)$ are eigenvalues of $X_k X_k^t$, and $\Lambda_k = \mathrm{diag}\,[\lambda_0(k), \lambda_1(k), \ldots, \lambda_{p-1}(k)]$. Using this fact, we can rewrite P_k as

$$P_k = T_k(\Lambda_k + d(k)I)^{-1} T_k^t. \tag{8.37}$$

Substitution of (8.36) and (8.37) into (8.35) yields

$$\begin{aligned}\mathrm{E}\left\{\|\boldsymbol{v}_k\|^2\right\} &= \mathrm{E}\left\{e_k^t \left(I - X_k X_k^t P_k\right)^t \left(I - X_k X_k^t P_k\right) e_k\right\} \\ &= \mathrm{E}\left\{e_k^t T_k Q_k T_k^t e_k\right\},\end{aligned} \tag{8.38}$$

where $Q_k = (I - \Lambda_k(\Lambda_k + \delta(k)I)^{-1})^2$. The ith diagonal component of Q_k equals

$$\left(\frac{\delta(k)}{\lambda_i(k) + \delta(k)}\right)^2.$$

To obtain a closed-form expression for $\delta(k)$, we observe

$$\begin{aligned}\mathrm{E}\left\{\mathrm{tr}\left[\boldsymbol{X}_k\boldsymbol{X}_k^t\right]\right\} &= \mathrm{E}\left\{\sum_{i=0}^{p-1}\boldsymbol{x}_{k-i}^t\boldsymbol{x}_{k-i}\right\} \\ &\approx np\sigma_x^2(k),\end{aligned}$$

where $\sigma_x^2(k) \triangleq \mathrm{E}\left\{\boldsymbol{x}^2(k)\right\}$. On the other hand,

$$\mathrm{E}\left\{\mathrm{tr}\left[\boldsymbol{X}_k\boldsymbol{X}_k^t\right]\right\} = \mathrm{E}\left\{\sum_{i=0}^{p-1}\boldsymbol{\lambda}_i(\boldsymbol{k})\right\} = \sum_{i=0}^{p-1}\mathrm{E}\{\boldsymbol{\lambda}_i(k)\}.$$

Therefore,

$$\sum_{i=0}^{p-1}\mathrm{E}\{\boldsymbol{\lambda}_i(k)\} \approx np\sigma_x^2(k). \tag{8.39}$$

With this fact in mind, we make an assumption

$$\lambda_i(k) \approx n\sigma_x^2(k) \quad (0 \le i \le p-1). \tag{8.40}$$

Using (8.40), we can express Q_k as

$$Q_k = \left(\frac{\delta(k)}{n\sigma_x^2(k) + \delta(k)}\right)^2 I. \tag{8.41}$$

Substituting (8.41) into (8.38), and solving the equation for $\delta(k)$, we finally obtain

$$\delta(k) = \frac{n\sigma_x^2(k)\sqrt{\mathrm{E}\left\{\|\boldsymbol{v}_k\|^2\right\}}}{\sqrt{\mathrm{E}\left\{\|\boldsymbol{e}_k\|^2\right\} - \mathrm{E}\left\{\|\boldsymbol{v}_k\|^2\right\}}}. \tag{8.42}$$

The quantity $\mathrm{E}\left\{\|\boldsymbol{e}_k\|^2\right\}$ can be estimated as

$$\begin{aligned}\mathrm{E}\left\{\|\boldsymbol{e}_k\|^2\right\} &\approx p\hat{\sigma}_e^2(k), \\ \hat{\sigma}_e^2(k) &= \alpha\hat{\sigma}_e^2(k-1) + (1-\alpha)e^2(k),\end{aligned}$$

where α $(0 < \alpha < 1)$ is the smoothing factor. To estimate the variance $\sigma_x^2(k)$ of the input signal, we use an instantaneous approximation

$$\sigma_x^2(k) \approx \frac{1}{np}\mathrm{tr}\left[X_k X_k^t\right].$$

We assume that we know the noise variance σ_v^2 beforehand, or can estimate it, so that $\mathrm{E}\left\{\|v_k\|^2\right\} = p\sigma_v^2$ is available.

In [14, 15], the use of an error vector, which is different from ϵ_k, is discussed as described below. In this work, δ is not time-varying. Let us rewrite the update equation (3.12) for the R-APA as

$$w_{k+1} = S_k w_k + \mu \hat{w}_{k+1},$$

where

$$S_k \triangleq I_n - \mu X_k^t (X_k X_k^t + \delta I_p)^{-1} X_k,$$

and

$$\hat{w}_{k+1} \triangleq X_k^t (X_k X_k^t + \delta I_p)^{-1} d_k. \tag{8.43}$$

The vector $\hat{w}_{k+1}$ is the solution of the minimization problem

$$\min_{\hat{w}_{k+1}} \left\{ \|d_k - X_k \hat{w}_{k+1}\|^2 + \delta \|\hat{w}_{k+1}\|^2 \right\}.$$

In fact, differentiating the quantity $\|d_k - X_k \hat{w}_{k+1}\|^2 + \delta \|\hat{w}_{k+1}\|^2$ with respect to $\hat{w}_{k+1}$, and equating it with 0, we get an equation

$$(X_k^t X_k + \delta I_n)\hat{w}_{k+1} = X_t^t d_k.$$

The solution of this equation is given by

$$\begin{aligned}\hat{w}_{k+1} &= (X_k^t X_k + \delta I_n)^{-1} X_k^t d_k \\ &= X_k^t (X_k X_k^t + \delta I_p)^{-1} d_k,\end{aligned} \tag{8.44}$$

where we used (3.16) to derive the last expression. Thus, $\hat{w}_{k+1}$ is a regularized least-mean-squares solution of the linear equation $d_k = X_k \hat{w}_{k+1}$. Using $\hat{w}_{k+1}$, we define a new error vector

$$\hat{\epsilon}_k \triangleq d_k - X_k \hat{w}_{k+1}, \tag{8.45}$$

and postulate

$$\mathrm{E}\left\{\|\hat{\epsilon}_k\|^2\right\} = \mathrm{E}\left\{\|v_k\|^2\right\} \tag{8.46}$$

instead of (8.34). From (8.43) and (8.45), we have

$$\hat{\epsilon}_k = \left(I_p - X_k X_k^t (X_k X_k^t + \delta I)^{-1}\right) d_k. \tag{8.47}$$

Applying the eigenvalue decomposition (8.36) to $X_k X_k^t$ in (8.47), we find

$$\|\hat{\epsilon}_k\|^2 = d_k^t T_k \left(I_p - \Lambda_k(\Lambda_k + \delta I_p)^{-1}\right)^2 T_k^t d_k.$$

To make the calculation tractable, we adopt the assumption (8.40), obtaining

$$\|\hat{\epsilon}_k\|^2 = \|d_k\|^2 \left(\frac{\delta}{\delta + n\sigma_x^2}\right)^2. \tag{8.48}$$

Taking the expectation of both sides of (8.48), and using the postulate (8.46), we have

$$\mathrm{E}\left\{\|\boldsymbol{d}_k\|^2\right\} \left(\frac{\delta}{\delta + n\sigma_x^2}\right)^2 = \mathrm{E}\left\{\|\boldsymbol{v}_k\|^2\right\}. \tag{8.49}$$

Since $\boldsymbol{y}(k)$ and $\boldsymbol{v}(k)$ are uncorrelated, $\mathrm{E}\left\{\|\boldsymbol{d}_k\|^2\right\}$ is expressed as

$$\mathrm{E}\left\{\|\boldsymbol{d}_k\|^2\right\} = \mathrm{E}\left\{\|\boldsymbol{y}_k\|^2\right\} + \mathrm{E}\left\{\|\boldsymbol{v}_k\|^2\right\},$$

where $\boldsymbol{y}_k = (\boldsymbol{y}(k), \boldsymbol{y}(k-1), \ldots, \boldsymbol{y}(k-p+1))^t$. Consequently, (8.49) is rewritten as

$$\left(\mathrm{E}\left\{\|\boldsymbol{y}_k\|^2\right\} + \mathrm{E}\left\{\|\boldsymbol{v}_k\|^2\right\}\right) \left(\frac{\delta}{\delta + n\sigma_x^2}\right)^2 = \mathrm{E}\left\{\|\boldsymbol{v}_k\|^2\right\}. \tag{8.50}$$

Replacing $\mathrm{E}\left\{\|\boldsymbol{y}_k\|^2\right\} = p\sigma_y^2$, and $\mathrm{E}\left\{\|\boldsymbol{v}_k\|^2\right\} = p\sigma_v^2$ in (8.50), we get

$$\left(\frac{\delta}{\delta + n\sigma_x^2}\right)^2 = \frac{1}{1 + \mathrm{ENR}}, \tag{8.51}$$

where ENR is the *echo-to-noise ratio* defined as $\mathrm{ENR} \triangleq \sigma_y^2/\sigma_v^2$. The solution of (8.51) is given by

$$\begin{aligned} \delta &= \frac{n\left(1 + \sqrt{1 + \mathrm{ENR}}\right)}{\mathrm{ENR}} \sigma_x^2 \\ &= \beta_{\mathrm{APA}} \sigma_x^2, \end{aligned} \tag{8.52}$$

where

$$\beta_{\mathrm{APA}} = \frac{n\left(1 + \sqrt{1 + \mathrm{ENR}}\right)}{\mathrm{ENR}}$$

is the *normalized regularization parameter* of the R-APA.

The technique to derive (8.52) can also be used in combination with the IPAPA, yielding the normalized regularization parameter of the IPAPA β_{IPAPA} [16].

8.4.2 Adaptive Regularization Matrix

This subsection gives a review of a work on the adaptive regularization matrix [17]. In the R-APA, the regularization parameter δ is added evenly to all the eigenvalues of $X_k X_k^t$. However, it is possible to give different regularization parameters to different eigenvalues. This is realized by an update equation

$$w_{k+1} = w_k + \mu X_k^t (X_k X_k^t + \Delta_k)^{-1} e_k, \tag{8.53}$$

where Δ_k is a diagonal matrix defined by

$$\Delta_k \triangleq \mathrm{diag}\,[\delta_0(k), \delta_1(k), \ldots, \delta_{p-1}(k)]. \tag{8.54}$$

Let us adjust $\delta_i(k)$ so that $J(k) \triangleq (1/2)e^2(k)$ is minimized. For this purpose, we use the steepest-descent method (Chap. 2, Appendix 1: "Steepest-Descent Method"), that is, the regularization parameter $\delta_i(k)$ is updated as

$$\delta_i(k) = \delta_i(k-1) - \rho \frac{\partial J(k)}{\partial \delta_i(k-1)} \quad (i = 0, 1, \ldots, p-1), \tag{8.55}$$

where ρ is a small positive real number. For simplicity, we denote $\nabla_i J(k) \triangleq \partial J(k)/\partial \delta_i(k-1)$. By the chain rule of differentiation, we see

$$\begin{aligned} \nabla_i J(k) &= \frac{\partial J(k)}{\partial e(k)} \left(\frac{\partial e(k)}{\partial w_k} \right)^t \frac{\partial w_k}{\partial \delta_i(k-1)} \\ &= e(k)(-x_k^t) \frac{\partial w_k}{\partial \delta_i(k-1)}. \end{aligned} \tag{8.56}$$

Since

$$w_k = w_{k-1} + \mu X_{k-1}^t (X_{k-1} X_{k-1}^t + \Delta_{k-1})^{-1} e_{k-1},$$

the last factor in (8.56) is calculated by using (A.15) (Appendix A) as

$$\begin{aligned} \frac{\partial w_k}{\partial \delta_i(k-1)} &= \mu X_{k-1}^t \frac{\partial (X_{k-1} X_{k-1}^t + \Delta_{k-1})^{-1}}{\partial \delta_i(k-1)} e_{k-1} \\ &= -\mu X_{k-1}^t (X_{k-1} X_{k-1}^t + \Delta_{k-1})^{-1} \frac{\partial \Delta_{k-1}}{\partial \delta_i(k-1)} (X_{k-1} X_{k-1}^t + \Delta_{k-1})^{-1} e_{k-1}. \end{aligned} \tag{8.57}$$

From (8.56) and (8.57), we have

$$\nabla_i J(k) = \mu e(k) x_k^t X_{k-1}^t \Gamma_i e_{k-1}, \tag{8.58}$$

where

$$\Gamma_i \triangleq (X_{k-1}X_{k-1}^t + \Delta_{k-1})^{-1} \frac{\partial \Delta_{k-1}}{\partial \delta_i(k-1)} (X_{k-1}X_{k-1}^t + \Delta_{k-1})^{-1}.$$

Let $\Delta_i(k) \triangleq \delta_i(k) - \delta_i(k-1)$. From (8.55) and (8.58), we see $|\Delta_i(k)| = \mu\rho|e(k)| \cdot \left|x_k^t X_{k-1}^t \Gamma_i e_{k-1}\right|$. This implies that when $|e(k)|$ becomes small after the initial convergence phase, $\delta_i(k-1)$ receives only a small change in (8.55). This is not desirable because the regularization parameter should increase as the time goes to make the MSE smaller. Thus, we modify (8.55) into

$$\begin{aligned} \delta_i(k) &= \begin{cases} \delta_i(k-1) - \rho \frac{\nabla_i J(k)}{|\nabla_i J(k)|}, & \text{if } \nabla_i J(k) \neq 0; \\ \delta_i(k-1), & \text{if } \nabla_i J(k) = 0 \end{cases} \\ &= \delta_i(k-1) - \rho \,\mathrm{sgn}\,(\nabla_i J(k)), \end{aligned} \tag{8.59}$$

where sgn is the signum function defined as

$$\mathrm{sgn}\,(x) \triangleq \begin{cases} 1, & \text{if } x > 0; \\ 0, & \text{if } x = 0; \\ -1, & \text{if } x < 0. \end{cases}$$

To protect $\delta_i(k)$ from becoming too small, Eq. (8.59) is further modified as

$$\begin{aligned} \delta_i'(k) &\triangleq \delta_i(k-1) - \rho \,\mathrm{sgn}\,(\nabla_i J(k)), \\ \delta_i(k) &= \max\{\delta_{\min},\ \delta_i'(k)\}, \end{aligned} \tag{8.60}$$

where $\delta_{\min}$ is a prescribed lower bound for $\delta_i(k)$. The diagonal matrix Δ_k in (8.54) is defined by using $\delta_i(k)$, which is then used in the update equation (8.53).

References

1. Shin, H.-C., Sayed, A.H., Song, W.-J.: Variable step-size NLMS and affine projection algorithms. IEEE Signal Process. Lett. **11**(2), 132–135 (2004)
2. Paleologu, C., Benesty, J., Ciochină, S.: A variable step-size affine projection algorithm designed for acoustic echo cancellation. IEEE Trans. Audio Speech Lang. Process **16**(8), 1466–1478 (2008)
3. Duttweiler, D.L.: Proportionate normalized least-mean-squares adaptation in echo cancelers. IEEE Trans. Speech Audio Process. **8**(5), 508–518 (2000)
4. Benesty, J., Gansler, T., Morgan, D.R., Sondhi, M.M., Gay, S.L.: Advances in Network and Acoustic Echo Cancellation. Springer, Berlin (2001)
5. Benesty, J., Gay, S.L.: An improved PNLMS algorithm. In: Proceedings of the ICASSP 2002, vol. 2, pp. 1881–1884. Orlando, 13–17 May (2002)

6. Paleologu, C., Ciochină, S., Benesty, J.: An efficient proportionate affine projection algorithm for echo cancellation. IEEE Signal Process. Lett. **17**(2), 165–168 (2010)
7. Paleologu, C., Benesty, J., Albu, F., Ciochină, S.: An efficient variable step-size proportionate affine projection algorithm. In: Proceedings of the ICASSP, pp. 77–80. Prague, 22–27 May (2011)
8. Albu, F., Paleologu, C., Benesty, J., Ciochina, S.: A low complexity proportionate affine projection algorithm for echo cancellation. In: Proceedings of the EUSIPCO, pp. 6–10. Aalborg, 23–27 Aug 2010
9. Albu, F.: Simplified proportionate affine projection algorithms. In: Proceedings of the IWSSIP, pp. 382–385. Vienna, 11–13 April (2012)
10. Kim, S.-E., Kong, S.-J., Song, W.-J.: An affine projection algorithm with evolving order. IEEE Signal Process. Lett. **16**(11), 937–940 (2009)
11. Ji, C.-P., Ji, H.-H., Guo, W.-P., Wang, J.: APA with evolving order and variable regularization for echo cancellation. J. Netw. **9**(1), 210–215 (2014)
12. Yoo, J.W., Shin, J.W., Choi, H.-T., Park, P.G.: An affine projection algorithm with evolving order using variable step-size. Int. J. Comput. Electr. Eng. **5**(1), 5–8 (2013)
13. Yin, W., Mehr, A.S.: A variable regularization method for affine projection algorithm. IEEE Trans. Circuits Syst. II **57**(6), 476–480 (2010)
14. Paleologu, C., Benesty, J., Albu, F., Ciochină, S.: Practical regularization of the affine projection algorithm. In: Proceedings of the EUSIPCO, pp. 704–708. Barcelona, 29 Aug.–2 Sep. 2011
15. Paleologu, C., Benesty, J., Albu, F., Ciochină, S.: Regularization of the affine projection algorithm. IEEE Trans. Circuits Syst. II **58**(6), 366–370 (2011)
16. Paleologu, C., Benesty, J., Albu, F.: Regularization of the improved proportionate affine projection algorithm. In: Proceedings of the ICASSP, pp. 169–172. Kyoto, 25–30 March 2012
17. Choi, Y.-S., Shin, H.-C., Song, W.-J.: Adaptive regularization matrix for affine projection algorithm. IEEE Trans. Circuits Syst. II **54**(12), 1087–1091 (2007)

Appendix A
Matrices

A.1 Formulas

Formula A.1 *Let P_{ij} be an $n_i \times m_j$ matrix, and Q_{ij} an $m_i \times l_j$ matrix. Then*

$$\begin{bmatrix} P_{11} & P_{12} \\ P_{21} & P_{22} \end{bmatrix} \begin{bmatrix} Q_{11} & Q_{12} \\ Q_{21} & Q_{22} \end{bmatrix} = \begin{bmatrix} P_{11}Q_{11} + P_{12}Q_{21} & P_{11}Q_{12} + P_{12}Q_{22} \\ P_{21}Q_{11} + P_{22}Q_{21} & P_{21}Q_{12} + P_{22}Q_{22} \end{bmatrix}. \tag{A.1}$$

The proof is immediate from the definition of matrix multiplication.

Formula A.2 *Let A, B, C, D be $n \times n$, $n \times m$, $m \times n$, and $m \times m$ matrices, respectively. If A and $D - CA^{-1}B$ are nonsingular, then we have the following equality:*

$$\begin{bmatrix} A & B \\ C & D \end{bmatrix}^{-1} = \begin{bmatrix} A^{-1} + A^{-1}B(D - CA^{-1}B)^{-1}CA^{-1} & -A^{-1}B(D - CA^{-1}B)^{-1} \\ -(D - CA^{-1}B)^{-1}CA^{-1} & (D - CA^{-1}B)^{-1} \end{bmatrix}. \tag{A.2}$$

Similarly, if D and $A - BD^{-1}C$ are nonsingular, then

$$\begin{bmatrix} A & B \\ C & D \end{bmatrix}^{-1} = \begin{bmatrix} (A - BD^{-1}C)^{-1} & -(A - BD^{-1}C)^{-1}BD^{-1} \\ -D^{-1}C(A - BD^{-1}C)^{-1} & D^{-1} + D^{-1}C(A - BD^{-1}C)^{-1}BD^{-1} \end{bmatrix}. \tag{A.3}$$

Proof By using (A.1), it is easily verified that

$$\begin{bmatrix} A & B \\ C & D \end{bmatrix} \begin{bmatrix} A^{-1} + A^{-1}B(D - CA^{-1}B)^{-1}CA^{-1} & -A^{-1}B(D - CA^{-1}B)^{-1} \\ -(D - CA^{-1}B)^{-1}CA^{-1} & (D - CA^{-1}B)^{-1} \end{bmatrix} = I,$$

and

$$\begin{bmatrix} (A - BD^{-1}C)^{-1} & -(A - BD^{-1}C)^{-1}BD^{-1} \\ -D^{-1}C(A - BD^{-1}C)^{-1} & D^{-1} + D^{-1}C(A - BD^{-1}C)^{-1}BD^{-1} \end{bmatrix} \begin{bmatrix} A & B \\ C & D \end{bmatrix} = I,$$

K. Ozeki, *Theory of Affine Projection Algorithms for Adaptive Filtering*,
Mathematics for Industry 22, DOI 10.1007/978-4-431-55738-8

where I is the $(n+m) \times (n+m)$ identity matrix. □

Equating the right-hand sides of (A.2) and (A.3) leads to the following set of formulas, which is referred to as the *matrix inversion lemma*.

Formula A.3

$$(A - BD^{-1}C)^{-1} = A^{-1} + A^{-1}B(D - CA^{-1}B)^{-1}CA^{-1}, \tag{A.4}$$

$$(A - BD^{-1}C)^{-1}BD^{-1} = A^{-1}B(D - CA^{-1}B)^{-1}, \tag{A.5}$$

$$D^{-1}C(A - BD^{-1}C)^{-1} = (D - CA^{-1}B)^{-1}CA^{-1}, \tag{A.6}$$

$$D^{-1} + D^{-1}C(A - BD^{-1}C)^{-1}BD^{-1} = (D - CA^{-1}B)^{-1}. \tag{A.7}$$

Formula A.4 *Let A be a square matrix of the form*

$$A = \begin{bmatrix} A_{11} & A_{12} \\ 0 & A_{22} \end{bmatrix},$$

where A_{11} and A_{22} are square matrices. Then, $\det A$ *is given by*

$$\det A = \det A_{11} \cdot \det A_{22}. \tag{A.8}$$

Note that for a matrix of the form

$$A = \begin{bmatrix} A_{11} & 0 \\ A_{21} & A_{22} \end{bmatrix},$$

we also have

$$\det A = \det A_{11} \cdot \det A_{22}. \tag{A.9}$$

Formula A.5 *Let a square matrix A be partitioned as*

$$A = \begin{bmatrix} A_{11} & A_{12} \\ A_{21} & A_{22} \end{bmatrix},$$

where A_{11} and A_{22} are $p \times p$ and $q \times q$ matrices, respectively. If A_{22} is nonsingular, then $\det A$ *is given by*

$$\det A = \det(A_{11} - A_{12}A_{22}^{-1}A_{21}) \cdot \det A_{22}. \tag{A.10}$$

Proof Let us define

$$B \triangleq \begin{bmatrix} I_p & -A_{12}A_{22}^{-1} \\ 0 & I_q \end{bmatrix}.$$

Then,

$$BA = \begin{bmatrix} A_{11} - A_{12}A_{22}^{-1}A_{21} & 0 \\ A_{21} & A_{22} \end{bmatrix}.$$

Taking determinants of both sides of this equation, and using (A.9), we have

$$\det B \cdot \det A = \det(A_{11} - A_{12}A_{22}^{-1}A_{21}) \cdot \det A_{22}.$$

Since $\det B = 1$ by (A.8), we immediately obtain (A.10) from this equation. □

Let A be an $n \times n$ matrix, and C_{ij} the submatrix of A obtained by removing the ith row and the jth column. The quantity $\Delta_{ij} \triangleq (-1)^{i+j} \det C_{ij}$ is called the (i, j)*-cofactor* of A.

Formula A.6 *The cofactor matrix*

$$\Delta \triangleq \begin{bmatrix} \Delta_{11} & \Delta_{12} & \cdots & \Delta_{1n} \\ \Delta_{21} & \Delta_{22} & \cdots & \Delta_{2n} \\ \vdots & \vdots & \ddots & \vdots \\ \Delta_{n1} & \Delta_{n2} & \cdots & \Delta_{nn} \end{bmatrix}$$

and A are related by the following equation. For the proof, see "Satake, I.: Linear Algebra, p. 70. Marcel Dekker, New York (1975)" for example.

$$\Delta^t A = A\Delta^t = (\det A)I. \tag{A.11}$$

Formula A.7 *The* trace *of a square matrix A,* $\mathrm{tr}[A]$, *is define by the sum of its diagonal elements. For two square matrices* $A = [a_{ij}]$ *and* $B = [b_{ij}]$ *of the same size,*

$$\mathrm{tr}[AB] = \sum_{i,j} a_{ij}b_{ji} = \sum_{j,i} b_{ji}a_{ij} = \mathrm{tr}[BA]. \tag{A.12}$$

Formula A.8 *Let A and T be square matrices of the same size. If T is nonsingular, then from* (A.12),

$$\mathrm{tr}[T^{-1}AT] = \mathrm{tr}[TT^{-1}A] = \mathrm{tr}[A]. \tag{A.13}$$

Formula A.9 *Let* $A(x) = [a_{ij}(x)]$ *be a square matrix each component of which is a function of x. The derivative of* $A(x)$ *with respect to x is defined by*

$$\frac{dA(x)}{\mathrm{d}x} \triangleq \left[\frac{da_{ij}(x)}{\mathrm{d}x}\right].$$

By definition of the product of two matrices, we have

$$\frac{d(A(x)B(x))}{\mathrm{d}x} = \frac{dA(x)}{\mathrm{d}x}B(x) + A(x)\frac{dB(x)}{\mathrm{d}x}. \tag{A.14}$$

If $A(x)$ is nonsingular, then, by differentiating both sides of $A^{-1}(x)A(x) = I$, we immediately have

$$\frac{dA^{-1}(x)}{\mathrm{d}x} = -A^{-1}(x)\frac{dA(x)}{\mathrm{d}x}A^{-1}(x). \tag{A.15}$$

A.2 Gradient and Hessian Matrix of a Quadratic Form

For a real-valued differentiable function f defined on $\mathbb{R}^n$,

$$\nabla_x f(x) \triangleq \left(\frac{\partial f(x)}{\partial x_1}, \frac{\partial f(x)}{\partial x_2}, \ldots, \frac{\partial f(x)}{\partial x_n}\right)^t,$$
$$x = (x_1, x_2, \ldots, x_n)^t \in \mathbb{R}^n$$

is called the *gradient* of f at x.

For an $n \times n$ real symmetric matrix $A = [a_{ij}]$ and an n-dimensional vector $x = (x_1, x_2, \ldots, x_n)^t \in \mathbb{R}^n$, a function of the form

$$x^t Ax = \sum_{i,j=1}^{n} a_{ij}x_i x_j \tag{A.16}$$

is called a *quadratic form*. The gradient of a quadratic form $x^t Ax$ is given by

$$\nabla_x(x^t Ax) = 2Ax. \tag{A.17}$$

In fact,

$$\begin{aligned}
\frac{\partial x^t Ax}{\partial x_k} &= \frac{\partial}{\partial x_k}\sum_{i,j=1}^{n} a_{ij}x_i x_j \\
&= \sum_{i,j=1}^{n} a_{ij}\frac{\partial(x_i x_j)}{\partial x_k} \\
&= \sum_{i,j=1}^{n} a_{ij}\frac{dx_i}{dx_k}x_j + \sum_{i,j=1}^{n} a_{ij}x_i\frac{dx_j}{dx_k}
\end{aligned}$$

$$= \sum_{j=1}^{n} a_{kj}x_j + \sum_{i=1}^{n} a_{ik}x_i$$
$$= 2\sum_{j=1}^{n} a_{kj}x_j. \tag{A.18}$$

Since $\sum_{j=1}^{n} a_{kj}x_j$ is the kth component of Ax, we have

$$\nabla_x(x^t Ax) = 2Ax.$$

The Hessian matrix $\nabla_x^2 f(x)$ of f at x is defined by

$$\nabla_x^2 f(x) \triangleq \begin{bmatrix} \frac{\partial^2 f(x)}{\partial x_1 \partial x_1} & \frac{\partial^2 f(x)}{\partial x_1 \partial x_2} & \cdots & \frac{\partial^2 f(x)}{\partial x_1 \partial x_n} \\ \frac{\partial^2 f(x)}{\partial x_2 \partial x_1} & \frac{\partial^2 f(x)}{\partial x_2 \partial x_2} & \cdots & \frac{\partial^2 f(x)}{\partial x_2 \partial x_n} \\ \vdots & \vdots & \ddots & \vdots \\ \frac{\partial^2 f(x)}{\partial x_n \partial x_1} & \frac{\partial^2 f(x)}{\partial x_n \partial x_2} & \cdots & \frac{\partial^2 f(x)}{\partial x_n \partial x_n} \end{bmatrix}.$$

The Hessian matrix of the quadratic form (A.16) is given by

$$\nabla_x^2 x^t Ax = 2A. \tag{A.19}$$

In fact, by (A.18),

$$\frac{\partial x^t Ax}{\partial x_k} = \sum_{j=1}^{n} a_{kj}x_j + \sum_{i=1}^{n} a_{ik}x_i.$$

Therefore,

$$\frac{\partial^2 x^t Ax}{\partial x_l \partial x_k} = \frac{\partial}{\partial x_l}\sum_{j=1}^{n} a_{kj}x_j + \frac{\partial}{\partial x_l}\sum_{i=1}^{n} a_{ik}x_i$$
$$= a_{kl} + a_{lk}$$
$$= 2a_{lk}.$$

A.3 Eigenvalues and Eigenvectors

Let A be an $n \times n$ matrix. If, for a complex number λ, there exists a vector $x \neq 0$ in $\mathbb{C}^n$ such that

$$Ax = \lambda x,$$

then λ is called an *eigenvalue* of A and x the *eigenvector* corresponding, or belonging to λ. A complex number λ is an eigenvalue of A iff it is a root of the *characteristic polynomial*

$$f_A(\lambda) \triangleq \det(\lambda I_n - A).$$

The next theorem is known as the *Caley-Hamilton theorem.*

Theorem A.1 *Let A be an $n \times n$ matrix, and $f_A(\lambda) = \sum_{i=0}^{n} a_i \lambda^{n-i}$ its characteristic polynomial. Then,*

$$f_A(A) = \sum_{i=0}^{n} a_i A^{n-i} = 0,$$

where $A^0 = I_n$.

Proof Let Δ_{ij} be the (i, j)-cofactor of $\lambda I_n - A$. From the definition of the cofactor, we see that Δ_{ij} is a polynomial of λ whose degree is at most $n-1$. Let it be denoted by

$$\Delta_{ij} = b_{ij,0}\lambda^{n-1} + b_{ij,1}\lambda^{n-2} + \cdots + b_{ij,n-1}.$$

If we let $B_k \triangleq [b_{ij,k}]^t$, we have from Formula A.6

$$\begin{aligned} \det(\lambda I_n - A)\, I &= (\lambda I_n - A)(B_0\lambda^{n-1} + B_1\lambda^{n-2} + \cdots + B_{n-1}) \\ &= (B_0\lambda^{n-1} + B_1\lambda^{n-2} + \cdots + B_{n-1})(\lambda I_n - A). \end{aligned} \quad \text{(A.20)}$$

Thus, A commutes with any of $B_0, B_1, \ldots, B_{n-1}$. Therefore, we can substitute A for λ in (A.20), obtaining

$$f_A(A) = (A - A)(B_0 A^{n-1} + B_1 A^{n-2} + \cdots + B_{n-1}) = 0.$$ □

Let $X^* \triangleq \overline{X^t}$, where X is either a matrix or a vector, and the overline denotes complex conjugate. A matrix A that satisfies $A = A^*$ is called a *Hermitian matrix*. It is easily verified that the eigenvalues of a Hermitian matrix are real numbers.

Theorem A.2 *For a Hermitian matrix A, let x_1 and x_2 be eigenvectors corresponding to eigenvalues λ_1 and λ_2, respectively. If $\lambda_1 \neq \lambda_2$, then x_1 and x_2 are orthogonal: $x_1^* x_2 = 0$.*

Proof From $Ax_1 = \lambda_1 x_1$, we have $x_2^* A x_1 = \lambda_1 x_2^* x_1$. Similarly, $x_1^* A x_2 = \lambda_2 x_1^* x_2$. Since $(x_2^* A x_1)^* = x_1^* A^* x_2 = x_1^* A x_2$ and $(x_2^* x_1)^* = x_1^* x_2$, we have $\lambda_1 x_1^* x_2 = \lambda_2 x_1^* x_2$. Because $\lambda_1 \neq \lambda_2$, this shows $x_1^* x_2 = 0$. □

If there exist multiple eigenvectors corresponding to a single eigenvalue, they are not determined uniquely. However, we can choose eigenvectors $x_1, x_2, \ldots, x_n$ so that they constitute an *orthonormal basis* of $\mathbb{C}^n$, i.e., $x_j^* x_k = \delta_{jk}$, where δ_{jk} is the Kronecker delta defined by

$$\delta_{jk} \triangleq \begin{cases} 1, & \text{if } j = k; \\ 0, & \text{otherwise.} \end{cases}$$

The linear subspace spanned by the eigenvectors corresponding to an eigenvalue λ is called the *eigenspace* corresponding to λ.

An $n \times n$ matrix U is called a *unitary matrix* iff $UU^* = U^*U = I$. A unitary matrix U has the property $U^{-1} = U^*$.

A Hermitian matrix can be diagonalized by using a unitary matrix as in the following theorem.

Theorem A.3 *For an $n \times n$ Hermitian matrix A, there exists an $n \times n$ unitary matrix U such that*

$$U^{-1}AU = U^*AU = \operatorname{diag}[\lambda_1, \lambda_2, \ldots, \lambda_n], \tag{A.21}$$

where $\lambda_1, \lambda_2, \ldots, \lambda_n$ are the eigenvalues of A.

Proof Let u_i be the normalized eigenvector of A corresponding to λ_i. If there are multiple eigenvectors corresponding to a single eigenvalue, we choose $u_1, u_2, \ldots, u_n$ so that they constitute an orthonormal basis of $\mathbb{C}^n$. Then, it is immediately verified that

$$U \triangleq \begin{bmatrix} u_1 \; u_2 \cdots u_n \end{bmatrix}$$

is an unitary matrix which satisfies (A.21). □

Noting that $\det(U^*U) = \det I = 1$, and that the determinant of a diagonal matrix is the product of the diagonal components, we have

$$\begin{aligned} \det A &= \det(U^*U) \det A = \det U^* \det U \det A = \det U^* \det A \det U \\ &= \det(U^*AU) = \lambda_1\lambda_2 \cdots \lambda_n. \end{aligned} \tag{A.22}$$

Because a matrix is nonsingular iff its determinant is nonzero, a Hermitian matrix is nonsingular iff its eigenvalues are all nonzero.

By Formula A.8, we also have

$$\operatorname{tr}[A] = \operatorname{tr}[U^{-1}AU] = \lambda_1 + \lambda_2 + \cdots + \lambda_n. \tag{A.23}$$

An $n \times n$ Hermitian matrix A is said to be *nonnegative definite* iff for any $x \in \mathbb{C}^n$, $x^*Ax \geq 0$. A nonnegative definite Hermitian matrix A is said to be *positive definite* iff $x^*Ax = 0$ implies $x = 0$. We denote $A \geq 0$ to mean that A is nonnegative definite, and $A > 0$ to mean that A is positive definite. Also, we denote $A > B$ to mean $A - B > 0$. Obviously, $A \geq B$ means $A - B \geq 0$.

From Theorem A.3, we can easily prove the following theorem.

Theorem A.4 *A Hermitian matrix is nonnegative definite iff its eigenvalues are all nonnegative. It is positive definite iff its eigenvalues are all positive.*

Corollary A.1 *A nonnegative definite Hermitian matrix is nonsingular iff it is positive definite.*

A matrix is called a *real matrix* iff all of its components are real numbers. A real matrix A is said to be *symmetric* iff $A = A^t$. Because a real symmetric matrix is Hermitian, its eigenvalues are all real numbers. For a real symmetric matrix, the imaginary part of an eigenvector becomes redundant. Therefore, we only have to consider a real vector for an eigenvector.

A square matrix T is called an *orthogonal matrix* iff $TT^t = T^tT = I$. An orthogonal matrix T has the property $T^{-1} = T^t$.

Theorem A.3 is modified as follows for a real symmetric matrix.

Theorem A.5 *For an $n \times n$ real symmetric matrix A, there exists an orthogonal matrix T such that*

$$T^{-1}AT = T^tAT = \text{diag}\,[\lambda_1, \lambda_2, \ldots, \lambda_n]$$

where $\lambda_1, \lambda_2, \ldots, \lambda_n$ are the eigenvalues of A.

Note that the matrix T in the above theorem can be chosen as

$$T \triangleq \left[t_1\ t_2 \cdots t_n\right],$$

where t_i is the normalized eigenvector of A corresponding to λ_i.

An $n \times n$ real symmetric matrix A is said to be *nonnegative definite* iff for any $x \in \mathbb{R}^n$, $x^tAx \geq 0$. A nonnegative definite real symmetric matrix A is said to be *positive definite* iff $x^tAx = 0$ implies $x = 0$.

Corresponding to Theorem A.4 and Corollary A.1, we have the following theorem and corollary.

Theorem A.6 *A real symmetric matrix is nonnegative definite iff all the eigenvalues are nonnegative. It is positive definite iff all the eigenvalues are positive.*

Corollary A.2 *A real symmetric, nonnegative definite matrix is nonsingular iff it is positive definite.*

Lemma A.1 *For a real symmetric, positive definite matrix A, there exists a nonsingular matrix S such that*

$$S^tAS = I.$$

Proof By Theorem A.5, there exists an orthogonal matrix T such that

$$T^tAT = \Lambda$$

where $\Lambda = \text{diag}\,[\lambda_1, \lambda_2, \ldots, \lambda_n]$, λ_k's being the eigenvalues of A. Since λ_k's are positive by Theorem A.6, we can define a matrix

$$\Lambda^{-1/2} \triangleq \text{diag}\left[\lambda_1^{-1/2}, \lambda_2^{-1/2}, \ldots, \lambda_n^{-1/2}\right].$$

Let $S \triangleq T\Lambda^{-1/2}$. Then, S is nonsingular, and

$$\begin{aligned} S^t AS &= (T\Lambda^{-1/2})^t A(T\Lambda^{-1/2}) \\ &= \Lambda^{-1/2}T^t AT\Lambda^{-1/2} \\ &= \Lambda^{-1/2}\Lambda\Lambda^{-1/2} \\ &= I. \end{aligned}$$

□

Theorem A.7 *For $n \times n$ real symmetric matrices $A > 0$ and $B \geq 0$, there exists a nonsingular matrix F such that*

$$F^t AF = I,$$

and

$$F^t BF = \text{diag}\,[\lambda_1, \lambda_2, \ldots, \lambda_n],$$

where λ_k's are roots of the equation $\det(\lambda A - B) = 0$.

Proof By Lemma A.1, there exists a nonsingular matrix E such that

$$E^t AE = I.$$

Using E, define

$$B' \triangleq E^t BE.$$

Since B' is symmetric, Theorem A.5 guarantees the existence of an orthogonal matrix C such that

$$C^t B'C = \text{diag}\,[\lambda_1, \lambda_2, \ldots, \lambda_n],$$

where λ_k's are the eigenvalues of B'. Let us define $F \triangleq EC$. Then, F is nonsingular, and

$$\begin{aligned} F^t AF &= (EC)^t A(EC) \\ &= C^t(E^t AE)C \\ &= C^t IC \\ &= I, \\ F^t BF &= (EC)^t B(EC) \\ &= C^t(E^t BE)C \\ &= C^t B'C \\ &= \text{diag}\,[\lambda_1, \lambda_2, \ldots, \lambda_n]. \end{aligned}$$

Also, because

$$\begin{aligned}\det(\lambda I - B') &= \det(E^t(\lambda A - B)E) \\ &= (\det E)^2 \cdot \det(\lambda A - B),\end{aligned}$$

and $\det E \neq 0$, λ_k's are the roots of the equation $\det(\lambda A - B) = 0$. □

A.4 Gramian

Theorem A.8 *Vectors $x_1, x_2, \ldots, x_m \in \mathbb{R}^n$ are linearly dependent iff*

$$\det \begin{bmatrix} \langle x_1, x_1\rangle & \langle x_1, x_2\rangle & \cdots & \langle x_1, x_m\rangle \\ \langle x_2, x_1\rangle & \langle x_2, x_2\rangle & \cdots & \langle x_2, x_m\rangle \\ \vdots & \vdots & \ddots & \vdots \\ \langle x_m, x_1\rangle & \langle x_m, x_2\rangle & \cdots & \langle x_m, x_m\rangle \end{bmatrix} = 0. \tag{A.24}$$

Proof Let G be the matrix in (A.24). Suppose the vectors $x_1, x_2, \ldots, x_m$ are linearly dependent. Then there exist scalars $\alpha_1, \alpha_2, \ldots, \alpha_m \in \mathbb{R}$, at least one of which is not equal to 0, such that

$$\alpha_1 x_1 + \alpha_2 x_2 + \cdots + \alpha_m x_m = 0. \tag{A.25}$$

This leads to

$$\begin{aligned}&\alpha_1 \begin{bmatrix} \langle x_1, x_1\rangle \\ \langle x_2, x_1\rangle \\ \vdots \\ \langle x_m, x_1\rangle \end{bmatrix} + \alpha_2 \begin{bmatrix} \langle x_1, x_2\rangle \\ \langle x_2, x_2\rangle \\ \vdots \\ \langle x_m, x_2\rangle \end{bmatrix} + \cdots + \alpha_m \begin{bmatrix} \langle x_1, x_m\rangle \\ \langle x_2, x_m\rangle \\ \vdots \\ \langle x_m, x_m\rangle \end{bmatrix} \\ &= \begin{bmatrix} \langle x_1, \alpha_1 x_1 + \alpha_2 x_2 + \cdots + \alpha_m x_m\rangle \\ \langle x_2, \alpha_1 x_1 + \alpha_2 x_2 + \cdots + \alpha_m x_m\rangle \\ \vdots \\ \langle x_m, \alpha_1 x_1 + \alpha_2 x_2 + \cdots + \alpha_m x_m\rangle \end{bmatrix} \\ &= \begin{bmatrix} \langle x_1, 0\rangle \\ \langle x_2, 0\rangle \\ \vdots \\ \langle x_m, 0\rangle \end{bmatrix} \\ &= 0,\end{aligned}$$

which shows that the column vectors of G are linearly dependent. Hence $\det G = 0$.

Conversely, suppose that $\det G = 0$. Then the column vectors of G must be linearly dependent. Therefore, there exist scalars $\alpha_1, \alpha_2, \ldots, \alpha_m \in \mathbb{R}$, at least one of which is not equal to 0, such that

$$\alpha_1 \begin{bmatrix} \langle x_1, x_1 \rangle \\ \langle x_2, x_1 \rangle \\ \vdots \\ \langle x_m, x_1 \rangle \end{bmatrix} + \alpha_2 \begin{bmatrix} \langle x_1, x_2 \rangle \\ \langle x_2, x_2 \rangle \\ \vdots \\ \langle x_m, x_2 \rangle \end{bmatrix} + \cdots + \alpha_m \begin{bmatrix} \langle x_1, x_m \rangle \\ \langle x_2, x_m \rangle \\ \vdots \\ \langle x_m, x_m \rangle \end{bmatrix} = 0,$$

which is equivalent to

$$\begin{aligned} \langle x_1, \alpha_1 x_1 + \alpha_2 x_2 + \cdots + \alpha_m x_m \rangle &= 0, \\ \langle x_2, \alpha_1 x_1 + \alpha_2 x_2 + \cdots + \alpha_m x_m \rangle &= 0, \\ &\vdots \\ \langle x_m, \alpha_1 x_1 + \alpha_2 x_2 + \cdots + \alpha_m x_m \rangle &= 0. \end{aligned}$$

Multiplying α_k to the kth equation above and adding them together, we obtain

$$\langle \alpha_1 x_1 + \alpha_2 x_2 + \cdots + \alpha_m x_m, \alpha_1 x_1 + \alpha_2 x_2 + \cdots + \alpha_m x_m \rangle = 0,$$

from which

$$\alpha_1 x_1 + \alpha_2 x_2 + \cdots + \alpha_m x_m = 0.$$

This shows that $x_1, x_2, \ldots, x_m$ are linearly dependent. □

The matrix G is called the *Gramian matrix*, and $\det G$ the *Gramian*.

A.5 Schur Product and Tensor Product

Let $A = [a_{ij}]$ and $B = [b_{ij}]$ be real $m \times n$ matrices. The $m \times n$ matrix $C = [c_{ij}]$ defined by component-wise multiplication as $c_{ij} \triangleq a_{ij} \cdot b_{ij}$ is called the *Schur product*, or *Hadamard product* of A and B, and denoted by $C = A \boxtimes B$.

For an $n \times n$ matrix $A = [a_{ij}]$, and for integers $1 \leq k_1 < k_2 < \cdots < k_m \leq n$, a matrix of the form

$$\begin{bmatrix} a_{k_1k_1} & a_{k_1k_2} & \cdots & a_{k_1k_m} \\ a_{k_2k_1} & a_{k_2k_2} & \cdots & a_{k_2k_m} \\ \vdots & \vdots & \ddots & \vdots \\ a_{k_mk_1} & a_{k_mk_2} & \cdots & a_{k_mk_m} \end{bmatrix} \tag{A.26}$$

is called a *principal submatrix* of A. The determinant of a principal submatrix is called a *principal minor* of A.

Lemma A.2 *If a real symmetric matrix A is nonnegative definite, then the principal submatrices of A are all nonnegative definite.*

Proof For a principal submatrix (A.26), let $x_1, x_2, \ldots, x_m$ be arbitrary real numbers, and for $s = 1, 2, \ldots, n$, let

$$y_s \triangleq \begin{cases} x_i, & \text{if } s = k_i; \\ 0, & \text{otherwise.} \end{cases}$$

Then, since A is nonnegative definite,

$$\begin{aligned} \sum_{i,j=1}^{m} x_i x_j a_{k_i k_j} &= \sum_{s,t=1}^{n} y_s y_t a_{st} \\ &\geq 0, \end{aligned}$$

which shows that the principal submatrix (A.26) is nonnegative definite. □

Let A be an $m \times n$ matrix:

$$A = \begin{bmatrix} a_{11} & a_{12} & \cdots & a_{1n} \\ a_{21} & a_{22} & \cdots & a_{2n} \\ \vdots & \vdots & \ddots & \vdots \\ a_{m1} & a_{m2} & \cdots & a_{mn} \end{bmatrix},$$

and B an $r \times s$ matrix:

$$B = \begin{bmatrix} b_{11} & b_{12} & \cdots & b_{1s} \\ b_{21} & b_{22} & \cdots & b_{2s} \\ \vdots & \vdots & \ddots & \vdots \\ b_{r1} & b_{r2} & \cdots & b_{rs} \end{bmatrix}.$$

The *tensor product*, or *Kronecker product* of A and B, denoted by $A \otimes B$, is defined as

$$A \otimes B \triangleq \begin{bmatrix} a_{11}B & a_{12}B & \cdots & a_{1n}B \\ a_{21}B & a_{22}B & \cdots & a_{2n}B \\ \vdots & \vdots & \ddots & \vdots \\ a_{m1}B & a_{m2}B & \cdots & a_{mn}B \end{bmatrix},$$

where $a_{ij}B$ is the matrix defined by

$$a_{ij}B \triangleq \begin{bmatrix} a_{ij}b_{11} & a_{ij}b_{12} & \cdots & a_{ij}b_{1s} \\ a_{ij}b_{21} & a_{ij}b_{22} & \cdots & a_{ij}b_{2s} \\ \vdots & \vdots & \ddots & \vdots \\ a_{ij}b_{r1} & a_{ij}b_{r2} & \cdots & a_{ij}b_{rs} \end{bmatrix}.$$

Note that $A \otimes B$ is an $(mr) \times (ns)$ matrix.

Theorem A.9 *Let $A = [a_{ij}]$ be an $m \times n$ matrix, and B an $r \times s$ matrix. Then, for arbitrary vectors $u \in \mathbb{R}^n$ and $v \in \mathbb{R}^s$,*

$$(A \otimes B)(u \otimes v) = (Au) \otimes (Bv).$$

Proof Let us express $u = (u_1, u_2, \ldots, u_n)^t$. Then,

$$\begin{aligned}
(A \otimes B)(u \otimes v) &= \begin{bmatrix} a_{11}B & a_{12}B & \cdots & a_{1n}B \\ a_{21}B & a_{22}B & \cdots & a_{2n}B \\ \vdots & \vdots & \ddots & \vdots \\ a_{m1}B & a_{m2}B & \cdots & a_{mn}B \end{bmatrix} \begin{bmatrix} u_1 v \\ u_2 v \\ \vdots \\ u_n v \end{bmatrix} \\
&= \begin{bmatrix} a_{11}u_1 Bv + a_{12}u_2 Bv + \cdots + a_{1n}u_n Bv \\ a_{21}u_1 Bv + a_{22}u_2 Bv + \cdots + a_{2n}u_n Bv \\ \vdots \\ a_{m1}u_1 Bv + a_{m2}u_2 Bv + \cdots + a_{mn}u_n Bv \end{bmatrix} \\
&= \begin{bmatrix} (a_{11}u_1 + a_{12}u_2 + \cdots + a_{1n}u_n)Bv \\ (a_{21}u_1 + a_{22}u_2 + \cdots + a_{2n}u_n)Bv \\ \vdots \\ (a_{m1}u_1 + a_{m2}u_2 + \cdots + a_{mn}u_n)Bv \end{bmatrix} \\
&= (Au) \otimes (Bv).
\end{aligned}$$

□

Corollary A.3 *Let $\mu_1, \mu_2, \ldots, \mu_n$ be the eigenvalues of an $n \times n$ matrix A, and $\lambda_1, \lambda_2, \ldots, \lambda_m$ the eigenvalues of an $m \times m$ matrix B. Then, $\mu_i \lambda_j$ ($i = 1, 2, \ldots, n$; $j = 1, 2, \ldots, m$) are the eigenvalues of $A \otimes B$.*

Proof Let u_i be the eigenvector corresponding to μ_i, and v_j the eigenvector corresponding to λ_j. Then, by Theorem A.9,

$$(A \otimes B)(u_i \otimes v_j) = (Au_i) \otimes (Bv_j) = (\mu_i u_i) \otimes (\lambda_j v_j) = \mu_i \lambda_j u_i \otimes v_j.$$

This shows that $\mu_i \lambda_j$ is an eigenvalue of $A \otimes B$. Since $A \otimes B$ is an $mn \times mn$ matrix, it has at most mn eigenvalues. Therefore, $\mu_i \lambda_j$'s ($i = 1, 2, \ldots, m$; $j = 1, 2, \ldots, n$) exhaust the eigenvalues of $A \otimes B$. □

Corollary A.4 *Let A and B be real symmetric matrices of sizes $m \times m$ and $n \times n$, respectively. If A and B are nonnegative definite, then $A \otimes B$ is also nonnegative definite. If A and B are positive definite, then $A \otimes B$ is also positive definite.*

Proof If A and B are nonnegative definite, then the eigenvalues μ_i ($i = 1, 2, \ldots, m$) of A and the eigenvalues λ_j ($j = 1, 2, \ldots, n$) of B are all nonnegative by Theorem A.6. Therefore, the eigenvalues $\mu_i \lambda_j$ ($i = 1, 2, \ldots, m$; $j = 1, 2, \ldots, n$) of

$A \otimes B$ are all nonnegative. Thus, by Theorem A.6 again, $A \otimes B$ is nonnegative definite. If A and B are positive definite, μ_i's and λ_j's are all positive. Therefore, $\mu_i \lambda_j$'s are all positive, that is, $A \otimes B$ is positive definite. □

Corollary A.5 *Let A, B be real symmetric matrices of the same size. If A, B are nonnegative definite, then the Schur product $A \boxtimes B$ is also nonnegative definite.*

Proof Because $A \boxtimes B$ is a principal submatrix of $A \otimes B$, this corollary is a direct consequence of Lemma A.2 and Corollary A.4. □

The inner product of $x \otimes y$ and $u \otimes v$ is defined by

$$\langle x \otimes y, u \otimes v \rangle \triangleq \langle x, u \rangle \cdot \langle y, v \rangle.$$

For an $n \times m$ matrix $A = [a_{ij}]$, we define a vector vec $\{A\}$ by concatenating the columns of A as

$$\text{vec}\,\{A\} \triangleq (a_{11}, a_{21}, \ldots, a_{n1}, a_{12}, a_{22}, \ldots, a_{n2}, \ldots, a_{1m}, a_{2m}, \ldots, a_{nm})^t.$$

We denote the inverse operation by vec^{-1}: $\text{vec}^{-1}\{\text{vec}\,\{A\}\} = A$.

Theorem A.10 *Let $P = [p_{ij}]$, $R = [r_{ij}]$, and $Q = [q_{ij}]$ be $n \times m$, $m \times k$, and $k \times l$ matrices, respectively. Then,*

$$\text{vec}\,\{PRQ\} = (Q^t \otimes P)\text{vec}\,\{R\}. \tag{A.27}$$

Proof Since the (i, j) th component of PRQ is given by

$$(PRQ)_{ij} = \sum_{s,t} p_{is} r_{st} q_{tj},$$

we have

$$\text{vec}\,\{PRQ\} = \begin{bmatrix} \sum_{s,t} p_{1s} r_{st} q_{t1} \\ \vdots \\ \sum_{s,t} p_{ns} r_{st} q_{t1} \\ \sum_{s,t} p_{1s} r_{st} q_{t2} \\ \vdots \\ \sum_{s,t} p_{ns} r_{st} q_{t2} \\ \vdots \\ \sum_{s,t} p_{1s} r_{st} q_{tl} \\ \vdots \\ \sum_{s,t} p_{ns} r_{st} q_{tl} \end{bmatrix}. \tag{A.28}$$

On the other hand,

$$Q^t \otimes P = \begin{bmatrix} q_{11} & \cdots & q_{k1} \\ \vdots & \ddots & \vdots \\ q_{1l} & \cdots & q_{kl} \end{bmatrix} \otimes \begin{bmatrix} p_{11} & \cdots & p_{1m} \\ \vdots & \ddots & \vdots \\ p_{n1} & \cdots & p_{nm} \end{bmatrix}$$

$$= \begin{bmatrix} q_{11} \begin{bmatrix} p_{11} & \cdots & p_{1m} \\ \vdots & \ddots & \vdots \\ p_{n1} & \cdots & p_{nm} \end{bmatrix} & \cdots & q_{k1} \begin{bmatrix} p_{11} & \cdots & p_{1m} \\ \vdots & \ddots & \vdots \\ p_{n1} & \cdots & p_{nm} \end{bmatrix} \\ \vdots & \ddots & \vdots \\ q_{1l} \begin{bmatrix} p_{11} & \cdots & p_{1m} \\ \vdots & \ddots & \vdots \\ p_{n1} & \cdots & p_{nm} \end{bmatrix} & \cdots & q_{kl} \begin{bmatrix} p_{11} & \cdots & p_{1m} \\ \vdots & \ddots & \vdots \\ p_{n1} & \cdots & p_{nm} \end{bmatrix} \end{bmatrix}.$$

Since

$$\text{vec}\,\{R\} = (r_{11}, r_{21}, \ldots, r_{m1}, r_{12}, r_{22}, \ldots, r_{m2}, \ldots, r_{1k}, r_{2k}, \ldots, r_{mk})^t,$$

the components of $(Q^t \otimes P)\text{vec}\,\{R\}$ are given as follows:

$$((Q^t \otimes P)\text{vec}\,\{R\})_1 = \sum_{s,t} q_{t1} p_{1s} r_{st} = \sum_{s,t} p_{1s} r_{st} q_{t1}$$

$$((Q^t \otimes P)\text{vec}\,\{R\})_2 = \sum_{s,t} q_{t1} p_{2s} r_{st} = \sum_{s,t} p_{2s} r_{st} q_{t1}$$

$$\vdots$$

$$((Q^t \otimes P)\text{vec}\,\{R\})_n = \sum_{s,t} q_{t1} p_{ns} r_{st} = \sum_{s,t} p_{ns} r_{st} q_{t1}$$

$$((Q^t \otimes P)\text{vec}\,\{R\})_{n+1} = \sum_{s,t} q_{t2} p_{1s} r_{st} = \sum_{s,t} p_{1s} r_{st} q_{t2}$$

$$\vdots$$

$$((Q^t \otimes P)\text{vec}\,\{R\})_{l\times n} = \sum_{s,t} q_{tl} p_{ns} r_{st} = \sum_{s,t} p_{ns} r_{st} q_{tl}.$$

By comparing this result with (A.28), Eq. (A.27) is verified. □

Index

K. Ozeki, *Theory of Affine Projection Algorithms for Adaptive Filtering*, Mathematics for Industry 22, DOI 10.1007/978-4-431-55738-8

The manufacturer's authorised representative in the EU is Springer Nature Customer Service Centre GmbH, Europaplatz 3, 69115 Heidelberg, Germany. If you have any concerns regarding our products, please contact ProductSafety@springernature.com

Printed and bound by CPI Group (UK) Ltd, Croydon, CR0 4YY

29/07/2026

02180621-0004